The UK Pesticide Guide 2018

Editor: M.A. Lainsbury, BSc(Hons)

 BCPC

www.cabi.org

CABI is one of the world's foremost publishers of databases, books and compendia in agriculture and applied life sciences. It has a worldwide reputation for producing high quality, value-added information, drawing on its links with the scientific community. CABI produces CAB Abstracts – the leading agricultural A&I database; together with the Crop Protection Compendium which provides solutions for identifying, preventing and solving crop health problems. CABI is a not-for-profit organisation. For further information, please contact CABI, Nosworthy Way, Wallingford, Oxon OX10 8DE, UK.

Telephone: (01491) 832111
Fax: (01491) 833508
e-mail: enquiries@cabi.org
Web: www.cabi.org

BCPC (British Crop Production Council) promotes the use of good science and technology in understanding and applying effective and sustainable crop production. Its key objectives are to identify developing issues in the science and practice of crop protection and production and provide informed, independent analysis; to publish definitive information for growers, advisors and other stakeholders; to organise conferences and symposia; and to stimulate interest and learning. BCPC is a registered Charity and a Company limited by Guarantee. Further details available from BCPC, Garden Studio, 4 Hillside, Aldershot, Surrey, UK. GU11 3NB

Telephone: (01252) 285223
e-mail: md@bcpc.org
Web: www.bcpc.org

ISBN: 978-1-9998966-0-7

Typeset and Printed in the United Kingdom by
Hobbs the Printers Ltd, Totton, Hampshire, SO40 3WX
Website: www.hobbs.uk.com

Contents

Editor's Note

This is the 31st edition of *The UK Pesticide Guide* and 2017 has seen the introduction of several new pesticide molecules that will play an important part in future crop protection. New molecules bring new modes of action – vital in combating resistance. Besides new fungicides we also have a new herbicide and while blackgrass has dominated the discussion on weed control in recent years, it has a limited persistence in the soil (about 3 years) and so is amenable to both chemical and cultural control. Poppy resistant to ALS herbicides is beginning to appear around the UK and with a longevity in the soil of perhaps 80 years, it is not amenable to cultural control. With the re-registration of pendimethalin and the introduction of halauxifen-methyl products, we still have effective chemical control methods and it is vital that we guard the efficacy of these actives by using combinations of residual and contact products applied at optimal timing for efficacy and by maintaining dose rates. Let's not make the same mistakes with poppy that we made with blackgrass because regaining control will be even more difficult.

For 2018 the book contains 15 new active ingredient profiles which include two new actives as full entries (amisulbrom for blight control in potatoes and benzovindiflupyr mixtures for disease control in cereals).

The full list of gains and losses is presented here with new actives shown in **bold**:

Herbicide gains are:
- florasulam + halauxifen-me
- flufenacet + picolinafen
- fluroxypyr + metsulfuron-me + thifensulfuron-me
- mesosulfuron-me + propoxycarbazone-sodium
- clomazone + metazachlor
- clopyralid + 2,4-D + MCPA
- diflufenican + flupyrsulfuron-me
- flupyrsulfuron-me
- flupyrsulfuron-me + pyroxsulam
- linuron

Fungicide gains are:
- **amisulbrom**
- **Bacillus amyloliquifaciens D747**
- **benzovindiflupyr**
- benzovindiflupyr + prothioconazole
- fludioxonil + tebuconazole
- fludioxonil + **sedaxane**
- fluopyram + trifloxystrobin

Insecticide & PGR full entry gains are:
- **Bacillus thuringiensis israeliensis AM65-52**
- **docecadienol + tetradienyl acetate**
- fatty acids
- 1-napthylacetic acid

Herbicides demoted to PAR or lost are:
- clomazone + linuron

Fungicides demoted to PAR or lost are:
- carbendazim
- chlorothalonil + pyraclostrobin
- cymoxanil + famoxadone
- cyproconazole + picoxystrobin
- cyproconazole + propiconazole
- fenbuconazole
- penthiopyrad + picoxystrobin
- picoxystrobin
- prochloraz + thiram
- pyrimethanil
- quinoxyfen
- spiroxamine
- tolclofos-me

Insecticides & PGRs demoted to PAR or lost are:
- chlormequat + ethephon + mepiquat chloride

Another new feature of the book this year is the inclusion of arthropod buffer zone restrictions and information on which products require the use of low drift spraying equipment for the first 30m from the top of the bank of any surface water bodies.

Many older products have been deleted from Section 3 – Products also Registered (PAR) as their approvals expire. As always, this book is a snapshot of the products with a valid approval on 9 October 2017 when the text went for initial typesetting. If you cannot see a product in the main section of the book, please check the index because it may be listed in the Products also Registered.

For those who must have the most up-to-date information I can recommend the on-line version – *ukpesticideguide.co.uk*. This is a searchable database with much more information than we can squeeze into the book.

We express our gratitude to all those who have provided the information for the compilation of this guide. Criticisms and suggestions for improvement are always welcome and if you notice any errors or omissions, please let me know via the email address below.

Martin Lainsbury, Editor
email: ukpg@bcpc.org

Disclaimer

Every effort has been made to ensure that the information in this book is correct at the time of going to press, but the Editor and the publishers do not accept liability for any error or omission in the content, or for any loss, damage or other accident arising from the use of the products listed herein. Omission of a product does not necessarily mean that it is not approved or that it is unavailable.

It is essential to follow the instructions on the approved label when handling, storing or using any crop protection product. Approved 'off-label' extensions of use now termed Extension of Authorisation for Minor Use (EAMUs) are undertaken entirely at the risk of the user.

The information in this publication has been selected from official sources, from the suppliers product labels and from the product manuals of the pesticides approved for use in the UK under EU Regulation 1107/2009.

Changes Since 2017 Edition

Pesticides and adjuvants added or deleted since the 2017 edition are listed below. Every effort has been made to ensure the accuracy of this list, but late notifications to the Editor are not included.

Products Added

Products new to this edition are shown here. In addition, products that were listed in the previous edition whose PSD/HSE registration number and/or supplier have changed are included.

Product	Reg. No.	Supplier	Product	Reg. No.	Supplier
AC 650	17966	UPL Europe	Devrinol	17853	UPL Europe
Allstar	18138	BASF	Difcon 250	17851	Harvest
Ambarac	17971	Life Scientific	Dualitas	18000	ProKlass
Amec	18035	Harvest	Dualitas	18070	ProKlass
Amenity Whippet	17931	Harvest	Dursban WG	17932	Dow
Amistar	18039	Syngenta	Elatus Era	17889	Syngenta
Amistar Opti	18156	Syngenta	Electis 75WG	17871	Gowan
Amistar Top	18050	Syngenta	Emblem	18066	Pure Amenity
Amylo X WG	17978	Certis	Envy	17901	Dow
Atlantis OD	18100	Bayer CropScience	Epee	17601	Harvest
			Equity	17933	Dow
Aubrac Xtra	18083	Harvest	Ethefol	18224	UPL Europe
Banner Maxx II	18038	Syngenta	Exteris Stressgard	17825	Bayer CropScience
Basilico	18028	Life Scientific			
BEC CCC 750	17930	Becesane	Fade	17982	Life Scientific
Belimar	18090	AgChem Access	Finalsan Plus	17928	Certis
			Flame	17842	Albaugh UK
Blaster Pro	18074	Headland Amenity	Flexidor	18042	Landseer
			Frozen	17691	Harvest
Buzz Ultra DF	17924	Clayton	Gando	18001	Harvest
Cintac	18222	Life Scientific	Garrison	18174	Pan Agriculture
Clayton Bonsai	17983	Clayton			
Clayton Mohawk	18086	Clayton	Gyo	17875	Belchim
Clayton Mohawk	18133	Clayton	Harvest Proteb	17692	Harvest
Clayton NFP	A0834	Clayton	Intellis Plus	17908	DuPont
Clayton Repel	18037	Clayton	Intracrop Zodiac	A0755	Intracrop
Clayton Roulette	18017	Clayton	Isonet T	17929	Fargro
CloverMaster	17927	Nufarm UK	Kabuki	18187	Belchim
Colzamid	17967	UPL Europe	Katoun Gold	17879	Belchim
Contans WG	17985	Bayer CropScience	Kideka	18033	Nufarm UK
			Killgerm ULV 1500	H6961	Killgerm
Curator	18157	Syngenta	Kipota	17993	Life Scientific
Cymozeb	18085	Belchim	Knight	18218	Harvest
Cyren	17934	Headland	Krug Flow	17964	Belchim
Daneva	18029	Rotam	Lanista	18118	Life Scientific
Danso Flow	17963	Belchim	Laya	17984	Life Scientific
Defiant	18216	UPL Europe	Leash	17969	Life Scientific

Product	Reg. No.	Supplier	Product	Reg. No.	Supplier
Lenazar Flow 500 SC	18124	Belcrop	Sakarat Brodikill Whole Wheat	UK17-1051	Killgerm
Leystar	17921	Dow	Sakarat Warfarin Whole Wheat	UK17-1059	Killgerm
Macho	18122	Albaugh UK	Savannah	18058	Rotam
Mandoprid	18182	Generica	Sirtaki CS	18032	Sipcam
Master Wett	A0826	Global Adjuvants	Sixte	18142	Headland
Monsanto Amenity Glyphosate XL	17997	Monsanto	Socom	17155	Clayton
			Spartan	17920	Harvest
Moose 800 EC	17968	Belchim	Spur	A0387	Amega
Naceto	18063	Globachem	Starpro	18059	Rotam
Nimbus Gold	18192	BASF	Stature	18064	Pure Amenity
Odin	18016	Rotam	Sudo	17979	Life Scientific
Olympus	18158	Syngenta	Sumir	17974	Life Scientific
Ormet Plus	17816	SFP Europe	Tachigaren 70 WP	17977	Sumi Agro
Ovarb	17972	Life Scientific	Tailout	17854	ProKlass
Palermo	17900	SFP Europe	Tandus	18071	Nufarm UK
Pamino	18069	Harvest	Tebusha 25 EW	18034	Sharda
Pan Penco	18180	Pan Amenity	Telsee	16325	Clayton
PastureMaster	17994	Nufarm UK	Thyram Plus	17890	Agrichem
Presidium	18119	Gowan	Tronix 700 SC	18137	Belchim
Pure Azoxy	18065	Pure Amenity	Trumpet	18183	Harvest
Recital	17909	Bayer CropScience	Unikat 75WG	17868	Gowan
			Venzar 500 SC	17743	DuPont
Revel	17973	Life Scientific	Vibrance Duo	17838	Syngenta
Rosh	17673	Harvest	Vivando	18026	BASF
Roundup Vista Plus	18002	Monsanto	Zypar	17938	Dow
Roxam 75WG	17869	Gowan			
Safari Lite WSB	17954	DuPont			

Products Deleted

The appearance of a product name in the following list may not necessarily mean that it is no longer available. In cases of doubt, refer to Section 3 or the supplier.

Product	Reg. No.	Supplier	Product	Reg. No.	Supplier
Agrovista Radni	14404	Agrovista	Menara	14398	Syngenta
Bromag Fresh Bait	UK12-0657	Killgerm	Pastor Pro	16465	Dow
			Perseo	17230	Sipcam
Centium Plus	16611	Belchim	Renovator Pro	15815	Everris Ltd
Contans WG	17085	Bayer CropScience	Ringer	13289	Barclay
			Sangue	16042	Agroquimicos
Cyclade	16350	BASF	Solar	A0225	Intracrop
Flurox 200	13938	Sipcam	Speed	A0710	Headland Amenity
Governer	16849	Dow			
Intracrop Brigand	A0718	Intracrop	Teridox	15876	Syngenta
Intracrop Vantage	A0719	Intracrop	Torch	11258	Bayer CropScience
Madex Top	18227	Certis			
Mascot Systemic	14488	Rigby Taylor	Zamide 400	16002	Becesane

Introduction

Purpose

The primary aim of this book is to provide a practical handbook of pesticides, plant growth regulators and adjuvants that the farmer or grower can realistically and legally obtain in the UK and to identify the purposes for which they can be used. It is designed to help in the identification of products appropriate to a particular problem. In addition to uses recommended on product labels, details are provided of those uses which do not appear on product labels but which have been granted Extension of Authorisation for Minor Use (EAMUs). As well as identifying the products available, the book provides guidance on how to use them safely and effectively but without giving details of doses, volumes, spray schedules or approved tank mixtures. Sections 5 and 6 provide essential background information on a wide range of pesticide-related issues including: legislation, biodiversity, codes of practice, poisons and treatment of poisoning, products for use in special situations and weed and crop growth stage keys.

While we have tried to cover all other important factors, this book does not provide a full statement of product recommendations. Before using any pesticide product it is essential that the user should read the label carefully and comply strictly with the instructions it contains. **This Guide does not substitute for the product label.**

Scope

The *Guide* is confined to pesticides registered in the UK for professional use in arable agriculture, horticulture (including amenity use), forestry and areas in or near water. Within these fields of use there are about 2837 products in the UK with current approval. The *Guide* is a reference for virtually all of them given in two sections. Section 2 gives full details of the products notified to the editor as available on the market. Products are included in this section only if requested by the supplier and supported by evidence of approval. Section 3 gives brief details of all other registered products with extant approval.

Section 2 lists some 421 individual active ingredients and mixtures. For each entry a list is shown of the available approved products together with the name of the supplier from whom each may be obtained. All types of pesticide covered by Control of Pesticides Regulations or Plant Protection Products Regulations are included. This embraces acaricides, algicides, herbicides, fungicides, insecticides, lumbricides, molluscicides, nematicides and rodenticides together with plant growth regulators used as straw shorteners, sprout inhibitors and for various horticultural purposes. The total number of products included in section 2 (i.e. available in the market in 2018) is about 1283.

The *Guide* also gives information (in section 4) on around 163 authorised adjuvants which, although not active pesticides themselves, may be added to pesticide products to improve their effectiveness. The *Guide* does not include products solely approved for amateur, home and garden, domestic, food storage, public or animal health uses.

Sources of Information

The information in this edition has been drawn from these authoritative sources:

- approved labels and product manuals received from suppliers of pesticides up to October 2017.
- websites of the Chemicals Regulation Division (CRD) *www.hse.gov.uk/crd/index.htm* and the Health and Safety Executive (HSE), *www.hse.gov.uk*.

Criteria for Inclusion

To be included, a pesticide must meet the following conditions:

- it must have extant approval under UK pesticides legislation

- information on the approved uses must have been provided by the supplier
- it must be expected to be on the UK market during the currency of this edition.

When a company changes its name, whether by merger, takeover or joint venture, it is obliged to re-register its products in the name of the new company, and new MAPP numbers are normally assigned. Where stocks of the previously registered product remain legally available, both old and new identities are included in the *Guide* and remain until approval of the former lapses or stocks are notified as exhausted.

Products that have been withdrawn from the market and whose approval will finally lapse during 2018 are identified in each profile. After the indicated date, sale, storage or use of the product bearing that approval number becomes illegal. Where there is a direct replacement product, this is indicated.

The Voluntary Initiative

The Voluntary Initiative (VI) is a programme of measures, agreed by the crop protection industry and related organisations with Government, to minimise the environmental impacts of crop protection products. The programme has provided a framework of practices and principles to help Government achieve its objective of protection of water quality and enhancement of farmland biodiversity. Many of the environmental protection schemes launched under the VI represent current best practice.

The first five-year phase of the programme formally concluded in March 2006, but it is continuing with a rolling two-year review of proposals and targets; current proposals include a new online crop protection management plan and greater involvement in catchment-sensitive farming (see below).

A key element of the VI has been the provision of environmental information on crop protection products. Members of the Crop Protection Association (CPA) have committed to do this by producing Environmental Information Sheets (EISs) for all their marketed professional products.

EISs reinforce and supplement information on a product label by giving specific environmental impact information in a standardised format. They highlight any situations where risk management is essential to ensure environmental protection. Their purpose is to provide user-friendly information to advisers (including those in the amenity sector), farmers and growers on the environmental impact of crop protection products, to allow planning with a better understanding of the practical implications.

Links are provided to give users rapid access to EISs on the VI website: www.voluntaryinitiative.org.uk.

Catchment-sensitive Farming

Catchment-sensitive Farming (CSF) is the government's response to climate change and the new European Water Directive. Rainfall events are likely to become heavier, with a greater risk of soil, nutrients and pesticides ending up in the waterways. The CSF programme is investigating the effects of specific targeted advice on 20 catchment areas perceived as being at risk, and new entries in this *Guide* can help identify the pesticides that pose the greatest risk of water contamination (see Environmental Safety).

Notes on Contents and Layout

The book consists of six main sections:

1. Crop/Pest Guide
2. Pesticide Profiles
3. Products Also Registered
4. Adjuvants
5. Useful Information
6. Appendices

1 Crop/Pest Guide

This section enables the user to identify which active ingredients are approved for a particular crop/pest combination. The crops are grouped as shown in the Crop Index. For convenience, some crops and pests have been grouped into generic units (for example, 'cereals' are covered as a group, not individually). Indications from the Crop/Pest Guide must always be checked against the specific entry in the pesticide profiles section because a product may not be approved on all crops in the group. Chemicals indicated as having uses in cereals, for example, may be approved for use only on winter wheat and winter barley, not for other cereals. Because of the difference in wording of product labels, it may sometimes be necessary to refer to broader categories of organism in addition to specific organisms (e.g. annual grasses and annual weeds in addition to annual meadow-grass).

2 Pesticide Profiles

Each active ingredient and mixture of ingredients has a separate numbered profile entry. The entries are arranged in alphabetical order using the common names approved by the British Standards Institution and used in *The Pesticide Manual* (17th Edition; BCPC, 2015) now available as an online subscription resource visit: *www.bcpc.org/shop* Where an active ingredient is available only in mixtures, this is stated. The ingredients of the mixtures are themselves ordered alphabetically, and the entries appear at the correct point for the first named active ingredient.

Below the profile title, a brief phrase describes the main use of the active ingredient(s). This is followed where appropriate by the mode of action code(s) as published by the Fungicide, Herbicide and Insecticide Resistance Action Committees.

Within each profile entry, a table lists the products approved and available on the market in the following style:

Product name	Main supplier	Active ingredient contents	Formulation type	Registration number
1 Amistar Opti	Syngenta	100:500g/l	SC	18156
2 Olympus	Syngenta	80:400g/l	SC	18158

Many of the **product names** are registered trademarks, but no special indication of this is given. Individual products are indexed under their entry number in the Index of Proprietary Names at the back of the book. The **main supplier** indicates the marketing outlet through which the product may be purchased. Full addresses and contact details of suppliers are listed in Appendix 1. For mixtures, the **active ingredient contents** are given in the same order as they appear in the profile heading. The **formulation types** are listed in full in the Key to Abbreviations and Acronyms (Appendix 5). The **registration number** normally refers to the registration with the Chemicals Regulation Division (CRD). In cases where the products registered with the Health and Safety Executive are included, HSE numbers are quoted, e.g. H0462 and approved biocides are numbered UK12-XXXX.

Below the product table, a **Uses** section lists all approved uses notified by the suppliers to the editor by October 2017, giving both the principal target organisms and the recommended crops or situations (identified in ***bold italics***). Where there is an important condition of use, or where the approval is off-label, this is shown in parentheses as (*off-label*). Numbers in square brackets refer to the numbered products in the table above. Thus, in the example shown above, a use approved for Olympus (product 2) but not for Amistar Opti (products 1) appears as:

- Rust in ***asparagus*** (moderate control only) [2]

Any Extensions of Authorisation for Minor Uses (EAMUs), for products in the profile are detailed below the list of approved uses. Each EAMU has a separate entry and shows the crops to which it applies, the Notice of Approval number, the expiry date (if due to fall during the current edition), and the product reference number in square brackets. EAMUs do not generally appear on the product label and are undertaken entirely at the risk of the user.

Below the EAMU paragraph, **Notes** are listed under the following headings. Unless otherwise stated, any reference to dose is made in terms of product rather than active ingredient. Where a note refers to a particular product rather than to the entry generally, this is indicated by numbers in square brackets as described above.

Approval information	Information of a general nature about the approval status of the active ingredient or products in the profile is given here. Notes on approval for aerial or ULV application are given.
	Inclusion of the active ingredient in Annex I under EC Directive 1107/2009 is shown, as well as any acceptance from the British Beer and Pub Association (BBPA) for its use on barley or hops for brewing or malting.
	Where a product approval will finally expire in 2018, the expiry date is shown here.
Efficacy	This entry identifies factors important in making the most effective use of the treatment. Certain factors, such as the need to apply chemicals uniformly, the need to use appropriate volume rates, and the need to prevent settling out of the active ingredient in the spray tank, are assumed to be important for all treatments and are not emphasised in individual profiles.
Restrictions	Notes are included in this section where products are subject to the Poisons Law, and where the label warns about organophosphorus and/or anticholinesterase compounds. Factors important in optimising performance and minimising the risk of crop damage, including statutory conditions relating to the maximum permitted number of applications, are listed. Any restrictions on crop varieties that may not be sprayed are mentioned here.
Environmental safety	Where the label specifies an environmental hazard classification, it is noted together with the associated risk phrases. Any other special operator precautions and any conditions concerning withholding livestock from treated areas are specified. Where any of the products in the profile are subject to Category A, Category B or broadcast air-assisted buffer zone restrictions under the LERAP scheme, the relevant classification is shown. Where a LERAP requires a buffer zone greater than 5m, detail of the larger buffer zone is given here. Arthropod buffer zones and statutory drift reduction limitations are also listed here.
	Other environmental hazards are also noted here, including potential dangers to livestock, game, wildlife, bees and fish. The need to avoid drift onto neighbouring crops and the need to wash out equipment thoroughly after use are important with all pesticide treatments, but may receive special mention here if of particular significance.
Crop-specific information	Instructions about the timing of application or cultivations that are specific to a particular crop, rather than generally applicable, are mentioned here. The latest permitted use and harvest intervals (if any) are shown for individual crops.

Following crops guidance
Any specific label instructions about what may be grown after a treated crop, whether harvested normally or after failure, are shown here.

Hazard classification and safety precautions
The label hazard classifications and precautions are shown using a series of letter and number codes which are explained in Appendix 4. Hazard warnings are now given in full, with the other precautions listed under the following subheadings:

- Risk phrases
- Operator protection
- Environmental protection
- Consumer protection
- Storage and disposal
- Treated seed
- Vertebrate/rodent control products
- Medical advice

This section is given for information only and should not be used for the purpose of making a COSHH assessment without reference to the actual label of the product to be used.

3 Products also Registered

Products with an extant approval for all or part of the year of the edition in which they appear are listed in this section if they have not been requested by their supplier or manufacturer for inclusion in Section 2. Details shown are the same (apart from formulation) as those in the product tables of Section 2 and the approval expiry date is shown. Not all products listed here will be available in the market, but this list, together with the products in Section 2, comprises a comprehensive listing of all approved products in the UK for uses within the scope of the *Guide at the time of going to print*.

4 Adjuvants

Adjuvants are listed in a table ordered alphabetically by product name. For each product, details are shown of the main supplier, the authorisation number and the mode of action (e.g. wetter, sticker, etc.) as shown on the label. A brief statement of the uses of the adjuvant is given. Protective clothing requirements and label precautions are listed using the codes from Appendix 4.

5 Useful Information

This section summarises legislation covering approval, storage, sale and use of pesticides in the UK. There are brief articles on the broader issues concerning the use of crop protection chemicals, including resistance management. Lists are provided of products approved for use in or near water, in forestry, as seed treatments and for aerial application. Chemicals subject to the Poisons Laws are listed, and there is a summary of first aid measures if pesticide poisoning should be suspected. Finally, this section provides guidance on environmental protection issues and covers the protection of bees and the use of pesticides in or near water.

6 Appendices

Appendix 1
Gives contact details of all companies listed as main suppliers in the pesticide profiles.

Where a supplier is no longer trading under that name (usually following a merger or takeover), but products under that name are still listed in the Guide because they are still available in the supply chain, a cross-reference indicates the new 'parent' company from which technical or commercial information can be obtained.

Appendix 2
Gives contact details of useful contacts, including the National Poisons Information Service.

Appendix 3	Gives details of the keys to crop and weed growth stages, including the publication reference for each. The numeric codes are used in the descriptive sections of the pesticide profiles (Section 2).
Appendix 4	Shows the full text for code letters and numbers used in the pesticide profiles (Section 2) to indicate personal protective equipment requirements and label precautions.
Appendix 5	Shows the full text for the formulation abbreviations used in the pesticide profiles (Section 2). Other abbreviations and acronyms used in the Guide are also explained here.
Appendix 6	Provides full definitions of officially agreed descriptive phrases for crops or situations used in the pesticide profiles (Section 2) where misunderstandings can occur.
Appendix 7	Shows a list of useful reference publications, which amplify the summarised information provided in this section.

SECTION 1
CROP/PEST GUIDE

Crop/Pest Guide Index

Important note: The Crop/Pest Guide Index refers to pages on which the subject can be located.

Crop/Pest Guide

Important note: For convenience, some crops and pests or targets have been brought together into genetic groups in this guide, e.g. 'cereals', 'annual grasses'. It is essential to check the profile entry in Section 2 *and* the label to ensure that a product is approved for a specific crop/pest combination, e.g. winter wheat/blackgrass.

Arable and vegetable crops

Agricultural herbage - Grass

Crop control	Desiccation	glyphosate
Diseases	Crown rust	propiconazole
	Drechslera leaf spot	propiconazole
	Powdery mildew	propiconazole
	Rhynchosporium	propiconazole
Pests	Aphids	esfenvalerate
	Birds/mammals	aluminium ammonium sulphate
	Flies	esfenvalerate
	Slugs/snails	ferric phosphate
Plant growth regulation	Growth control	gibberellins
Weeds	Broad-leaved weeds	2,4-D, 2,4-D + dicamba, 2,4-D + MCPA, 2,4-DB, 2,4-DB + MCPA, amidosulfuron, aminopyralid + triclopyr, clopyralid, clopyralid + florasulam + fluroxypyr, clopyralid + triclopyr, florasulam + fluroxypyr, fluroxypyr, fluroxypyr (*off-label*), fluroxypyr + triclopyr, glyphosate, glyphosate (*wiper application*), MCPA, thifensulfuron-methyl, tribenuron-methyl, tribenuron-methyl (*off-label*)
	Crops as weeds	florasulam + fluroxypyr, fluroxypyr (*off-label*)
	Grass weeds	glyphosate
	Weeds, miscellaneous	aminopyralid + triclopyr, clopyralid + florasulam + fluroxypyr, clopyralid + triclopyr, fluroxypyr (*off-label*), glyphosate, MCPA
	Woody weeds/scrub	clopyralid + triclopyr

Agricultural herbage - Herbage legumes

Crop control	Desiccation	diquat
Plant growth regulation	Growth control	trinexapac-ethyl (*off-label*)
Weeds	Broad-leaved weeds	2,4-DB, diquat (*off-label*), propyzamide
	Grass weeds	diquat (*seed crop*), fluazifop-P-butyl (*off-label*), propyzamide
	Weeds, miscellaneous	diquat (*seed crop*)

Agricultural herbage - Herbage seed

Crop control	Desiccation	diquat
Diseases	Crown rust	bixafen + prothioconazole (*off-label*), boscalid + epoxiconazole (*off-label*), propiconazole

	Damping off	thiram (*seed treatment*)
	Disease control	epoxiconazole (*off-label*), epoxiconazole + fenpropimorph + kresoxim-methyl (*off-label*), epoxiconazole + pyraclostrobin (*off-label*), fenpropimorph + pyraclostrobin (*off-label*), tebuconazole (*off-label*)
	Drechslera leaf spot	bixafen + prothioconazole (*off-label*), fenpropimorph + pyraclostrobin, propiconazole
	Foliar diseases	azoxystrobin + cyproconazole (*off-label*), epoxiconazole + fenpropimorph (*off-label*), epoxiconazole + fenpropimorph + kresoxim-methyl (*off-label*), fenpropimorph (*off-label*), pyraclostrobin (*off-label*), trifloxystrobin (*off-label*)
	Leaf spot	prothioconazole + tebuconazole (*off-label*)
	Net blotch	boscalid + epoxiconazole (*off-label*)
	Powdery mildew	azoxystrobin (*off-label*), bixafen + prothioconazole (*off-label*), cyflufenamid (*off-label*), epoxiconazole (*off-label*), propiconazole, prothioconazole + tebuconazole (*off-label*), sulphur (*off-label*), tebuconazole (*off-label*)
	Rhynchosporium	bixafen + prothioconazole (*off-label*), boscalid + epoxiconazole (*off-label*), propiconazole, prothioconazole + tebuconazole (*off-label*)
	Rust	azoxystrobin (*off-label*), bixafen + prothioconazole (*off-label*), boscalid + epoxiconazole (*off-label*), epoxiconazole (*off-label*), fenpropimorph + pyraclostrobin, fenpropimorph + pyraclostrobin (*off-label*), prothioconazole + tebuconazole (*off-label*), tebuconazole (*off-label*)
	Septoria	fenpropimorph + pyraclostrobin, fenpropimorph + pyraclostrobin (*off-label*)
Pests	Aphids	deltamethrin (*off-label*), esfenvalerate (*off-label*), tau-fluvalinate (*off-label*), zeta-cypermethrin (*off-label*)
	Flies	esfenvalerate
	Pests, miscellaneous	deltamethrin (*off-label*), lambda-cyhalothrin (*off-label*), zeta-cypermethrin (*off-label*)
Plant growth regulation	Growth control	trinexapac-ethyl, trinexapac-ethyl (*off-label*)
Weeds	Broad-leaved weeds	bifenox (*off-label*), bromoxynil + diflufenican (*off-label*), clopyralid + florasulam + fluroxypyr, dicamba + MCPA + mecoprop-P, diflufenican (*off-label*), diflufenican + flufenacet (*off-label*), ethofumesate, florasulam (*off-label*), florasulam + fluroxypyr, florasulam + fluroxypyr (*off-label*), fluroxypyr, MCPA, mecoprop-P, pendimethalin (*off-label*), propyzamide
	Crops as weeds	bromoxynil + diflufenican (*off-label*), clopyralid (*off-label*), diflufenican (*off-label*), florasulam (*off-label*), florasulam + fluroxypyr
	Grass weeds	clodinafop-propargyl (*off-label*), diflufenican (*off-label*), diflufenican + flufenacet (*off-label*), ethofumesate, fenoxaprop-P-ethyl (*off-label*), pendimethalin (*off-label*), propyzamide
	Weeds, miscellaneous	clopyralid + florasulam + fluroxypyr, diflufenican (*off-label*), florasulam (*off-label*), MCPA

Brassicas - Brassica seed crops

Diseases	Alternaria	iprodione, iprodione (*pre-storage only*)
	Botrytis	iprodione
Weeds	Broad-leaved weeds	propyzamide
	Grass weeds	propyzamide

Brassicas - Brassicas, general

Diseases	Alternaria	azoxystrobin + difenoconazole (*off-label*), cyprodinil + fludioxonil (*off-label*)
	Botrytis	Bacillus subtilis (*off-label*), boscalid + pyraclostrobin (*off-label*), cyprodinil + fludioxonil (*off-label*)
	Downy mildew	dimethomorph (*off-label*)
	Pythium	metalaxyl-M (*off-label*)
	Rhizoctonia	boscalid + pyraclostrobin (*off-label*), cyprodinil + fludioxonil (*off-label*)
	Ring spot	azoxystrobin + difenoconazole (*off-label*)
	Sclerotinia	boscalid + pyraclostrobin (*off-label*)
	Stem canker	cyprodinil + fludioxonil (*off-label*)
Pests	Aphids	pymetrozine (*off-label*), spirotetramat (*off-label*), thiacloprid (*off-label*)
	Caterpillars	Bacillus thuringiensis (*off-label*), deltamethrin (*off-label - caterpillars*), indoxacarb (*off-label*)
	Cutworms	Bacillus thuringiensis (*off-label*)
	Flies	Metarhizium anisopliae (*off-label*)
	Whiteflies	spirotetramat (*off-label*)
Weeds	Broad-leaved weeds	clopyralid (*off-label*), pyridate (*off-label*)
	Crops as weeds	clopyralid (*off-label*)
	Grass weeds	fluazifop-P-butyl (*off-label*)

Brassicas - Fodder brassicas

Diseases	Alternaria	azoxystrobin, cyprodinil + fludioxonil (*off-label*), difenoconazole
	Botrytis	cyprodinil + fludioxonil (*off-label*), Gliocladium catenulatum (*off-label*)
	Damping off	Bacillus subtilis (*off-label*), thiram (*seed treatment*)
	Downy mildew	fluopicolide + propamocarb hydrochloride (*off-label*), metalaxyl-M (*off-label*)
	Fusarium	Gliocladium catenulatum (*off-label*)
	Phytophthora	Gliocladium catenulatum (*off-label*)
	Powdery mildew	azoxystrobin + difenoconazole
	Pythium	Bacillus subtilis (*off-label*)
	Rhizoctonia	cyprodinil + fludioxonil (*off-label*), Gliocladium catenulatum (*off-label*)
	Ring spot	azoxystrobin, difenoconazole
	Stem canker	cyprodinil + fludioxonil (*off-label*)
	White blister	azoxystrobin, azoxystrobin + difenoconazole, boscalid + pyraclostrobin (*off-label*)
Pests	Aphids	esfenvalerate, lambda-cyhalothrin, pymetrozine (*off-label*), thiacloprid (*off-label*), thiamethoxam (*off-label*)
	Beetles	alpha-cypermethrin, thiamethoxam (*off-label*)

	Caterpillars	alpha-cypermethrin, Bacillus thuringiensis, Bacillus thuringiensis (*off-label*), deltamethrin (*off-label - caterpillars*), esfenvalerate, indoxacarb (*off-label*), spinosad (*off-label*)
	Flies	spinosad (*off-label*), thiamethoxam (*off-label*)
	Whiteflies	lambda-cyhalothrin
Weeds	Broad-leaved weeds	clomazone (*off-label*), clopyralid (*off-label*), dimethenamid-p + metazachlor (*off-label*), metazachlor (*off-label*), napropamide, pendimethalin (*off-label*), pyridate (*off-label*)
	Crops as weeds	clopyralid (*off-label*), fluazifop-P-butyl (*stockfeed only*)
	Grass weeds	dimethenamid-p + metazachlor (*off-label*), fluazifop-P-butyl (*stockfeed only*), napropamide, pendimethalin (*off-label*), S-metolachlor (*off-label*)

Brassicas - Leaf and flowerhead brassicas

Diseases	Alternaria	azoxystrobin, boscalid + pyraclostrobin, chlorothalonil + metalaxyl-M (*moderate control*), cyprodinil + fludioxonil (*off-label*), difenoconazole, iprodione, prothioconazole, tebuconazole
	Black rot	azoxystrobin (*off-label*)
	Botrytis	cyprodinil + fludioxonil (*off-label*), Gliocladium catenulatum (*off-label*), iprodione, iprodione (*off-label*)
	Damping off	Bacillus subtilis (*off-label*), thiram (*seed treatment*)
	Downy mildew	chlorothalonil + metalaxyl-M, fluopicolide + propamocarb hydrochloride (*off-label*), metalaxyl-M (*off-label*)
	Fusarium	Gliocladium catenulatum (*off-label*)
	Light leaf spot	prothioconazole, tebuconazole
	Phytophthora	Gliocladium catenulatum (*off-label*)
	Powdery mildew	azoxystrobin + difenoconazole, prothioconazole, tebuconazole
	Pythium	Bacillus subtilis (*off-label*), metalaxyl-M
	Rhizoctonia	cyprodinil + fludioxonil (*off-label*), Gliocladium catenulatum (*off-label*)
	Ring spot	azoxystrobin, boscalid + pyraclostrobin, chlorothalonil + metalaxyl-M (*reduction*), difenoconazole, prothioconazole, tebuconazole, tebuconazole (*off-label*)
	Stem canker	cyprodinil + fludioxonil (*off-label*), prothioconazole
	Storage rots	1-methylcyclopropene (*off-label*), metalaxyl-M (*off-label*)
	White blister	azoxystrobin, azoxystrobin (*off-label*), azoxystrobin + difenoconazole, boscalid + pyraclostrobin, boscalid + pyraclostrobin (*off-label*), boscalid + pyraclostrobin (*qualified minor use*), chlorothalonil + metalaxyl-M, mancozeb + metalaxyl-M (*off-label*)
Pests	Aphids	acetamiprid (*off-label*), beta-cyfluthrin, cypermethrin, deltamethrin, esfenvalerate, lambda-cyhalothrin, pymetrozine, pymetrozine (*off-label*), pymetrozine (*useful levels of control*), pyrethrins, spirotetramat, thiacloprid, thiacloprid (*off-label*)
	Beetles	alpha-cypermethrin, alpha-cypermethrin (*off-label*), deltamethrin, pyrethrins

	Birds/mammals	aluminium ammonium sulphate
	Caterpillars	alpha-cypermethrin, alpha-cypermethrin (*off-label*), Bacillus thuringiensis, Bacillus thuringiensis (*off-label*), beta-cyfluthrin, cypermethrin, deltamethrin, deltamethrin (*off-label - caterpillars*), diflubenzuron, esfenvalerate, indoxacarb, indoxacarb (*off-label*), lambda-cyhalothrin, pyrethrins, spinosad, spinosad (*off-label*)
	Flies	cyantraniliprole, spinosad (*off-label*)
	Slugs/snails	metaldehyde
	Whiteflies	acetamiprid (*off-label*), cypermethrin, deltamethrin, lambda-cyhalothrin, pyrethrins, spirotetramat
Weeds	Broad-leaved weeds	clomazone (*off-label*), clopyralid, clopyralid (*off-label*), dimethenamid-p + metazachlor (*off-label*), dimethenamid-p + pendimethalin (*off-label*), metazachlor, metazachlor (*off-label*), napropamide, napropamide (*off-label*), pendimethalin, pendimethalin (*off-label*), propyzamide (*off-label*), pyridate, pyridate (*off-label*)
	Crops as weeds	clopyralid (*off-label*), cycloxydim, cycloxydim (*off-label*), pendimethalin
	Grass weeds	clethodim (*off-label*), cycloxydim, cycloxydim (*off-label*), dimethenamid-p + metazachlor (*off-label*), dimethenamid-p + pendimethalin (*off-label*), fluazifop-P-butyl (*off-label*), metazachlor, napropamide, napropamide (*off-label*), pendimethalin, pendimethalin (*off-label*), propyzamide (*off-label*), S-metolachlor (*off-label*)

Brassicas - Mustard

Crop control	Desiccation	diquat (*off-label*), glyphosate
Diseases	Alternaria	azoxystrobin (*off-label*), Bacillus amyloliquefaciens D747, cyprodinil + fludioxonil (*off-label*), difenoconazole (*off-label*)
	Botrytis	cyprodinil + fludioxonil (*off-label*)
	Disease control	boscalid (*off-label*), mandipropamid, prothioconazole (*off-label*), tebuconazole (*off-label*)
	Downy mildew	fenamidone + fosetyl-aluminium (*off-label*), fluopicolide + propamocarb hydrochloride (*off-label*), mandipropamid
	Fusarium	Bacillus amyloliquefaciens D747
	Powdery mildew	Bacillus amyloliquefaciens D747, difenoconazole (*off-label*), tebuconazole (*off-label*)
	Pythium	Bacillus amyloliquefaciens D747
	Rhizoctonia	Bacillus amyloliquefaciens D747, cyprodinil + fludioxonil (*off-label*)
	Rust	tebuconazole (*off-label*)
	Sclerotinia	azoxystrobin (*off-label*), difenoconazole (*off-label*)
	Stem canker	cyprodinil + fludioxonil (*off-label*)
Pests	Aphids	cypermethrin, spirotetramat (*off-label*), tau-fluvalinate (*off-label*)
	Beetles	deltamethrin, indoxacarb (*off-label*), thiacloprid
	Caterpillars	deltamethrin, spinosad (*off-label*)
	Flies	tefluthrin (*off-label - seed treatment*)

	Pests, miscellaneous	cypermethrin, lambda-cyhalothrin (*off-label*)
	Thrips	spinosad (*off-label*)
	Weevils	deltamethrin
	Whiteflies	spirotetramat (*off-label*)
Weeds	Broad-leaved weeds	bifenox (*off-label*), clopyralid (*off-label*), clopyralid + picloram (*off-label*), glyphosate, metazachlor (*off-label*), napropamide (*off-label*)
	Crops as weeds	glyphosate
	Grass weeds	fluazifop-P-butyl (*off-label*), glyphosate, metazachlor (*off-label*), napropamide (*off-label*)
	Weeds, miscellaneous	glyphosate, napropamide (*off-label*)

Brassicas - Root brassicas

Diseases	Alternaria	azoxystrobin (*off-label*), fenpropimorph (*off-label*), prothioconazole, tebuconazole
	Botrytis	boscalid + pyraclostrobin (*off-label*), Gliocladium catenulatum (*off-label*)
	Crown rot	fenpropimorph (*off-label*)
	Damping off	thiram (*off-label*), thiram (*seed treatment*)
	Disease control	boscalid + pyraclostrobin (*off-label*)
	Downy mildew	azoxystrobin (*off-label*), dimethomorph (*off-label*), fluopicolide + propamocarb hydrochloride (*off-label*), metalaxyl-M (*off-label*), propamocarb hydrochloride (*off-label*)
	Fungus diseases	azoxystrobin + difenoconazole (*off-label*)
	Fusarium	Gliocladium catenulatum (*off-label*)
	Phytophthora	Gliocladium catenulatum (*off-label*)
	Powdery mildew	azoxystrobin (*off-label*), fenpropimorph (*off-label*), isopyrazam (*off-label*), prothioconazole, sulphur, tebuconazole, tebuconazole (*off-label*)
	Pythium	metalaxyl-M, metalaxyl-M (*off-label*)
	Rhizoctonia	azoxystrobin (*off-label*), Gliocladium catenulatum (*off-label*)
	Rust	azoxystrobin (*off-label*), boscalid + pyraclostrobin (*off-label*)
	Sclerotinia	boscalid + pyraclostrobin (*off-label*), cyprodinil + fludioxonil (*off-label*)
	Stem canker	azoxystrobin + difenoconazole (*off-label*), thiram (*off-label*)
	White blister	azoxystrobin (*off-label*), metalaxyl-M (*off-label*)
Pests	Aphids	cypermethrin, esfenvalerate, pirimicarb (*off-label*), pymetrozine (*off-label*), spirotetramat (*off-label*), thiacloprid (*off-label*), thiamethoxam (*off-label*)
	Beetles	alpha-cypermethrin (*off-label*), deltamethrin, thiamethoxam (*off-label*)
	Caterpillars	alpha-cypermethrin (*off-label*), Bacillus thuringiensis, Bacillus thuringiensis (*off-label*), chlorantraniliprole (*off-label*), deltamethrin, esfenvalerate, indoxacarb (*off-label*), lambda-cyhalothrin (*off-label*), spinosad (*off-label*)
	Cutworms	Bacillus thuringiensis (*off-label*), cypermethrin, deltamethrin (*off-label*), lambda-cyhalothrin (*off-label*)

	Flies	chlorantraniliprole (*off-label*), cypermethrin, deltamethrin (*off-label*), lambda-cyhalothrin (*off-label*), spinosad (*off-label*), thiamethoxam (*off-label*)
	Pests, miscellaneous	cypermethrin, deltamethrin (*off-label*), lambda-cyhalothrin (*off-label*), spinosad (*off-label*)
	Slugs/snails	ferric phosphate
	Thrips	spinosad (*off-label*)
	Weevils	lambda-cyhalothrin (*off-label*)
Plant growth regulation	Quality/yield control	sulphur
Weeds	Broad-leaved weeds	clomazone (*off-label*), clopyralid, dimethenamid-p + metazachlor (*off-label*), glyphosate, metamitron (*off-label*), metazachlor (*off-label*), napropamide (*off-label*), pendimethalin (*off-label*), prosulfocarb (*off-label*), S-metolachlor (*off-label*)
	Crops as weeds	cycloxydim, fluazifop-P-butyl (*stockfeed only*), glyphosate
	Grass weeds	cycloxydim, cycloxydim (*off-label*), dimethenamid-p + metazachlor (*off-label*), fluazifop-P-butyl (*off-label*), fluazifop-P-butyl (*stockfeed only*), glyphosate, metamitron (*off-label*), napropamide (*off-label*), propaquizafop, prosulfocarb (*off-label*), S-metolachlor (*off-label*)
	Weeds, miscellaneous	glyphosate, glyphosate (*off-label*)

Brassicas - Salad greens

Diseases	Alternaria	azoxystrobin + difenoconazole (*off-label*), cyprodinil + fludioxonil (*off-label*), difenoconazole (*off-label*)
	Botrytis	cyprodinil + fludioxonil (*off-label*), fenhexamid (*off-label - baby leaf production*), Gliocladium catenulatum (*off-label*)
	Damping off	Bacillus subtilis (*off-label*)
	Disease control	dimethomorph + mancozeb (*off-label*), mandipropamid
	Downy mildew	dimethomorph (*off-label*), dimethomorph + mancozeb (*off-label*), fenamidone + fosetyl-aluminium (*off-label*), mancozeb (*off-label*), mancozeb + metalaxyl-M (*off-label*), mandipropamid, metalaxyl-M (*off-label*)
	Fusarium	Gliocladium catenulatum (*off-label*), Trichoderma asperellum (Strain T34) (*off-label*)
	Leaf spot	difenoconazole (*off-label*)
	Phytophthora	Gliocladium catenulatum (*off-label*)
	Powdery mildew	tebuconazole (*off-label*)
	Pythium	metalaxyl-M, Trichoderma asperellum (Strain T34) (*off-label*)
	Rhizoctonia	cyprodinil + fludioxonil (*off-label*), Gliocladium catenulatum (*off-label*)
	Ring spot	azoxystrobin + difenoconazole (*off-label*), difenoconazole (*off-label*), tebuconazole (*off-label*)
	Seed-borne diseases	thiram (*seed soak*)
	Stem canker	cyprodinil + fludioxonil (*off-label*)
	White blister	boscalid + pyraclostrobin (*off-label*)

Pests	Aphids	acetamiprid (*off-label*), alpha-cypermethrin (*off-label*), esfenvalerate, pymetrozine (*off-label*), pymetrozine (*off-label - for baby leaf production*), spirotetramat (*off-label*), thiacloprid (*off-label*), thiacloprid (*off-label - baby leaf production*)
	Beetles	alpha-cypermethrin (*off-label*)
	Birds/mammals	aluminium ammonium sulphate
	Caterpillars	alpha-cypermethrin (*off-label*), Bacillus thuringiensis, Bacillus thuringiensis (*off-label*), diflubenzuron (*off-label - for baby leaf production*), esfenvalerate, indoxacarb (*off-label*), spinosad, spinosad (*off-label*)
	Flies	tefluthrin (*off-label - seed treatment*)
	Pests, miscellaneous	lambda-cyhalothrin (*off-label*)
	Spider mites	Lecanicillium muscarium (*off-label*)
	Thrips	Lecanicillium muscarium (*off-label*), spinosad (*off-label*)
	Whiteflies	Lecanicillium muscarium (*off-label*), spirotetramat (*off-label*)
Weeds	Broad-leaved weeds	chlorpropham, chlorpropham (*off-label*), clomazone (*off-label*), clopyralid (*off-label*), dimethenamid-p + metazachlor (*off-label*), dimethenamid-p + pendimethalin (*off-label*), metazachlor (*off-label*), napropamide (*off-label*), pendimethalin (*off-label*), pendimethalin (*off-label - for baby leaf production*), propyzamide (*off-label*), S-metolachlor (*off-label*)
	Crops as weeds	clopyralid (*off-label*), cycloxydim (*off-label*)
	Grass weeds	chlorpropham, chlorpropham (*off-label*), cycloxydim (*off-label*), dimethenamid-p + metazachlor (*off-label*), dimethenamid-p + pendimethalin (*off-label*), fluazifop-P-butyl (*off-label*), napropamide (*off-label*), propyzamide (*off-label*), S-metolachlor (*off-label*)

Cereals - Barley

Crop control	Desiccation	diquat (*off-label*), glyphosate
Diseases	Covered smut	clothianidin + prothioconazole (*seed treatment*), fludioxonil (*seed treatment*), fludioxonil + tefluthrin, fluopyram + prothioconazole + tebuconazole, prothioconazole + tebuconazole
	Disease control	chlorothalonil + penthiopyrad, epoxiconazole + folpet, fluxapyroxad + metconazole, isopyrazam
	Ear blight	prothioconazole + tebuconazole
	Eyespot	azoxystrobin + cyproconazole (*reduction*), bixafen + prothioconazole (*reduction of incidence and severity*), boscalid + epoxiconazole (*moderate control*), boscalid + epoxiconazole + pyraclostrobin (*moderate control*), chlorothalonil + cyproconazole, cyprodinil, epoxiconazole (*reduction*), epoxiconazole + fenpropimorph (*reduction*), epoxiconazole + fenpropimorph + kresoxim-methyl (*reduction*), epoxiconazole + metconazole (*reduction*), fluoxastrobin + prothioconazole (*reduction*), fluoxastrobin + prothioconazole + trifloxystrobin (*reduction*), prochloraz + propiconazole, prochloraz + proquinazid + tebuconazole, prochloraz + tebuconazole, prothioconazole, prothioconazole + spiroxamine, prothioconazole + spiroxamine +

	tebuconazole (*reduction*), prothioconazole + tebuconazole, prothioconazole + tebuconazole (*reduction*), prothioconazole + trifloxystrobin (*reduction*), prothioconazole + trifloxystrobin (*reduction in severity*)
Foot rot	clothianidin + prothioconazole (*seed treatment*), fludioxonil (*seed treatment*), fludioxonil + tefluthrin
Fusarium	fluopyram + prothioconazole + tebuconazole, prothioconazole + tebuconazole
Late ear diseases	bixafen + prothioconazole, fluoxastrobin + prothioconazole, fluoxastrobin + prothioconazole + trifloxystrobin, prothioconazole, prothioconazole + spiroxamine + tebuconazole, prothioconazole + tebuconazole
Leaf stripe	clothianidin + prothioconazole (*seed treatment*), fludioxonil (*seed treatment - reduction*), fludioxonil + tefluthrin (*partial control*), fluopyram + prothioconazole + tebuconazole, prothioconazole + tebuconazole
Loose smut	clothianidin + prothioconazole (*seed treatment*), fluopyram + prothioconazole + tebuconazole, prothioconazole + tebuconazole
Net blotch	azoxystrobin, azoxystrobin + chlorothalonil, azoxystrobin + cyproconazole, benzovindiflupyr , benzovindiflupyr + prothioconazole, bixafen + prothioconazole, boscalid + epoxiconazole, boscalid + epoxiconazole + pyraclostrobin, chlorothalonil + cyproconazole, chlorothalonil + cyproconazole + propiconazole (*reduction*), chlorothalonil + penthiopyrad, cyprodinil, cyprodinil + isopyrazam, epoxiconazole, epoxiconazole (*moderate control*), epoxiconazole + fenpropimorph, epoxiconazole + fenpropimorph + kresoxim-methyl, epoxiconazole + fenpropimorph + metrafenone, epoxiconazole + fluxapyroxad, epoxiconazole + fluxapyroxad + pyraclostrobin, epoxiconazole + isopyrazam, epoxiconazole + metconazole, epoxiconazole + metrafenone, epoxiconazole + pyraclostrobin, fenpropimorph + pyraclostrobin, fluopyram + prothioconazole + tebuconazole (*seed borne*), fluoxastrobin + prothioconazole, fluoxastrobin + prothioconazole + trifloxystrobin, fluxapyroxad, fluxapyroxad + metconazole, fluxapyroxad + pyraclostrobin, metconazole (*reduction*), penthiopyrad, prochloraz + proquinazid + tebuconazole (*moderate control*), prochloraz + tebuconazole, prothioconazole, prothioconazole + spiroxamine, prothioconazole + spiroxamine + tebuconazole, prothioconazole + tebuconazole, prothioconazole + trifloxystrobin, pyraclostrobin, tebuconazole, trifloxystrobin
Powdery mildew	azoxystrobin, azoxystrobin (*moderate control*), azoxystrobin + cyproconazole, bixafen + prothioconazole, boscalid + epoxiconazole (*moderate control*), boscalid + epoxiconazole + pyraclostrobin (*moderate control*), chlorothalonil + cyproconazole, chlorothalonil + cyproconazole + propiconazole, chlorothalonil + proquinazid, cyflufenamid, cyprodinil, cyprodinil + isopyrazam, epoxiconazole,

epoxiconazole + fenpropimorph, epoxiconazole + fenpropimorph + kresoxim-methyl, epoxiconazole + fenpropimorph + metrafenone, epoxiconazole + fluxapyroxad (*moderate control*), epoxiconazole + fluxapyroxad + pyraclostrobin (*moderate control*), epoxiconazole + metconazole (*moderate control*), epoxiconazole + metrafenone, epoxiconazole + pyraclostrobin, fenpropimorph, fenpropimorph + pyraclostrobin, fluoxastrobin + prothioconazole, fluoxastrobin + prothioconazole + trifloxystrobin, flutriafol, fluxapyroxad (*reduction*), fluxapyroxad + metconazole (*moderate control*), fluxapyroxad + pyraclostrobin (*moderate control*), metconazole, metconazole (*moderate control*), metrafenone, prochloraz + proquinazid + tebuconazole, prochloraz + tebuconazole, propiconazole, proquinazid, prothioconazole, prothioconazole + spiroxamine, prothioconazole + spiroxamine + tebuconazole, prothioconazole + tebuconazole, prothioconazole + trifloxystrobin, sulphur, tebuconazole, tebuconazole (*moderate control*)

Ramularia leaf spots — benzovindiflupyr, benzovindiflupyr + prothioconazole, bixafen + prothioconazole, boscalid + epoxiconazole (*moderate control*), boscalid + epoxiconazole + pyraclostrobin (*moderate control*), chlorothalonil + penthiopyrad, cyprodinil + isopyrazam, epoxiconazole (*reduction*), epoxiconazole + fluxapyroxad, epoxiconazole + fluxapyroxad + pyraclostrobin, epoxiconazole + isopyrazam, epoxiconazole + metconazole (*moderate control*), epoxiconazole + pyraclostrobin (*moderate control*), epoxiconazole + pyraclostrobin (*reduction*), fluxapyroxad + metconazole, fluxapyroxad + pyraclostrobin, penthiopyrad (*moderate control*)

Rhynchosporium — azoxystrobin, azoxystrobin (*reduction*), azoxystrobin + chlorothalonil, azoxystrobin + chlorothalonil (*moderate control*), azoxystrobin + cyproconazole (*moderate control*), benzovindiflupyr (*moderate control*), benzovindiflupyr + prothioconazole (*moderate control*), bixafen + prothioconazole, boscalid + epoxiconazole (*moderate control*), boscalid + epoxiconazole + pyraclostrobin, chlorothalonil (*moderate control*), chlorothalonil + cyproconazole, chlorothalonil + cyproconazole + propiconazole (*reduction*), chlorothalonil + penthiopyrad, chlorothalonil + propiconazole, chlorothalonil + proquinazid (*suppression only*), chlorothalonil + tebuconazole (*useful reduction*), cyprodinil, cyprodinil + isopyrazam, epoxiconazole, epoxiconazole + fenpropimorph, epoxiconazole + fenpropimorph + kresoxim-methyl, epoxiconazole + fenpropimorph + metrafenone, epoxiconazole + fluxapyroxad, epoxiconazole + fluxapyroxad + pyraclostrobin, epoxiconazole + folpet, epoxiconazole + isopyrazam, epoxiconazole + metconazole, epoxiconazole + metrafenone, epoxiconazole + pyraclostrobin, fenpropimorph, fenpropimorph + pyraclostrobin, fluoxastrobin + prothioconazole, fluoxastrobin + prothioconazole + trifloxystrobin, flutriafol, fluxapyroxad, fluxapyroxad + metconazole, fluxapyroxad + pyraclostrobin, folpet (*reduction*),

		metconazole, metconazole (*reduction*), penthiopyrad, prochloraz + propiconazole, prochloraz + proquinazid + tebuconazole (*moderate control*), prochloraz + tebuconazole, propiconazole, prothioconazole, prothioconazole + spiroxamine, prothioconazole + spiroxamine + tebuconazole, prothioconazole + tebuconazole, prothioconazole + trifloxystrobin, pyraclostrobin (*moderate control*), tebuconazole, tebuconazole (*moderate control*), trifloxystrobin
	Rust	azoxystrobin, azoxystrobin + chlorothalonil, azoxystrobin + cyproconazole, benzovindiflupyr , benzovindiflupyr + prothioconazole, bixafen + prothioconazole, boscalid + epoxiconazole, boscalid + epoxiconazole + pyraclostrobin, chlorothalonil + cyproconazole, chlorothalonil + cyproconazole + propiconazole, chlorothalonil + propiconazole, cyprodinil + isopyrazam, epoxiconazole, epoxiconazole + fenpropimorph, epoxiconazole + fenpropimorph + kresoxim-methyl, epoxiconazole + fenpropimorph + metrafenone, epoxiconazole + fluxapyroxad, epoxiconazole + fluxapyroxad + pyraclostrobin, epoxiconazole + isopyrazam, epoxiconazole + metconazole, epoxiconazole + metrafenone, epoxiconazole + pyraclostrobin, fenpropimorph, fenpropimorph + pyraclostrobin, fluoxastrobin + prothioconazole, fluoxastrobin + prothioconazole + trifloxystrobin, flutriafol, fluxapyroxad, fluxapyroxad (*moderate control*), fluxapyroxad + metconazole, fluxapyroxad + pyraclostrobin, metconazole, penthiopyrad, penthiopyrad (*moderate control*), prochloraz + proquinazid + tebuconazole, prochloraz + tebuconazole, propiconazole, prothioconazole, prothioconazole + spiroxamine, prothioconazole + spiroxamine + tebuconazole, prothioconazole + tebuconazole, prothioconazole + trifloxystrobin, pyraclostrobin, tebuconazole, trifloxystrobin
	Seed-borne diseases	fludioxonil + tebuconazole, fluopyram + prothioconazole + tebuconazole, ipconazole
	Snow mould	fludioxonil (*seed treatment*), fludioxonil + tefluthrin
	Soil-borne diseases	ipconazole
	Sooty moulds	prothioconazole + tebuconazole
	Take-all	azoxystrobin (*reduction*), azoxystrobin + chlorothalonil (*reduction*), azoxystrobin + chlorothalonil (*reduction only*), azoxystrobin + cyproconazole (*reduction*), fluoxastrobin + prothioconazole (*reduction*)
Pests	Aphids	alpha-cypermethrin, beta-cyfluthrin, clothianidin (*seed treatment*), clothianidin + prothioconazole (*seed treatment*), cypermethrin, cypermethrin (*autumn sown*), deltamethrin, esfenvalerate, lambda-cyhalothrin, tau-fluvalinate, zeta-cypermethrin
	Beetles	lambda-cyhalothrin
	Birds/mammals	aluminium ammonium sulphate
	Caterpillars	lambda-cyhalothrin
	Flies	alpha-cypermethrin, cypermethrin, cypermethrin (*autumn sown*), fludioxonil + tefluthrin

	Leafhoppers	clothianidin (*seed treatment*), clothianidin + prothioconazole (*seed treatment*)
	Slugs/snails	clothianidin (*seed treatment*), clothianidin + prothioconazole (*seed treatment*)
	Weevils	lambda-cyhalothrin
	Wireworms	clothianidin (*seed treatment*), clothianidin + prothioconazole (*seed treatment*), fludioxonil + tefluthrin
Plant growth regulation	Growth control	chlormequat, chlormequat + 2-chloroethylphosphonic acid, ethephon, ethephon + mepiquat chloride, mepiquat chloride + prohexadione-calcium, prohexadione-calcium + trinexapac-ethyl, trinexapac-ethyl
	Quality/yield control	ethephon + mepiquat chloride (*low lodging situations*), mepiquat chloride + prohexadione-calcium, sulphur
Weeds	Broad-leaved weeds	2,4-D, 2,4-D + MCPA, 2,4-DB, 2,4-DB + MCPA, amidosulfuron, amidosulfuron + iodosulfuron-methyl-sodium, bifenox, bromoxynil, bromoxynil + diflufenican, bromoxynil + diflufenican (*at 1.9 l/ha only*), carfentrazone-ethyl, carfentrazone-ethyl + mecoprop-P, chlorotoluron + diflufenican + pendimethalin, clopyralid, clopyralid + florasulam, clopyralid + florasulam + fluroxypyr, dicamba + MCPA + mecoprop-P, dicamba + mecoprop-P, dichlorprop-P + MCPA + mecoprop-P, diflufenican, diflufenican + florasulam, diflufenican + flufenacet, diflufenican + flufenacet (*off-label*), diflufenican + flufenacet + flurtamone, diflufenican + flurtamone, diflufenican + metsulfuron-methyl, diflufenican + pendimethalin, florasulam, florasulam + fluroxypyr, florasulam + halauxifen-methyl, flufenacet, flufenacet + pendimethalin, flufenacet + pendimethalin (*off-label*), flufenacet + picolinafen, fluroxypyr, fluroxypyr + halauxifen-methyl, fluroxypyr + metsulfuron-methyl + thifensulfuron-methyl, glyphosate, imazosulfuron, isoxaben, MCPA, mecoprop-P, metsulfuron-methyl, metsulfuron-methyl + thifensulfuron-methyl, metsulfuron-methyl + tribenuron-methyl, pendimethalin, pendimethalin + picolinafen, pendimethalin + picolinafen (*off-label*), prosulfocarb, prosulfocarb (*off-label*), thifensulfuron-methyl, thifensulfuron-methyl + tribenuron-methyl, tribenuron-methyl
	Crops as weeds	amidosulfuron + iodosulfuron-methyl-sodium, bromoxynil, bromoxynil + diflufenican, chlorotoluron + diflufenican + pendimethalin, clopyralid + florasulam, diflufenican, diflufenican + florasulam, diflufenican + flufenacet, diflufenican + flurtamone, florasulam, florasulam + fluroxypyr, florasulam + halauxifen-methyl, flufenacet + picolinafen, fluroxypyr, glyphosate, imazosulfuron, metsulfuron-methyl, metsulfuron-methyl + tribenuron-methyl, pendimethalin
	Grass weeds	chlorotoluron + diflufenican + pendimethalin, diflufenican, diflufenican + flufenacet, diflufenican + flufenacet (*off-label*), diflufenican + flufenacet + flurtamone, diflufenican + flurtamone, diflufenican + pendimethalin, fenoxaprop-P-ethyl, flufenacet,

flufenacet + pendimethalin, flufenacet +
pendimethalin (*off-label*), flufenacet + picolinafen,
glyphosate, pendimethalin, pendimethalin +
picolinafen, pendimethalin + picolinafen (*off-label*),
pinoxaden, prosulfocarb, prosulfocarb (*off-label*), tri-
allate

	Weeds, miscellaneous	chlorotoluron + diflufenican + pendimethalin, clopyralid + florasulam, diflufenican, diflufenican + metsulfuron-methyl, florasulam, glyphosate, metsulfuron-methyl, metsulfuron-methyl + thifensulfuron-methyl

Cereals - Cereals, general

Crop control	Desiccation	diquat, glyphosate, glyphosate (*off-label*)
Diseases	Alternaria	bromuconazole + tebuconazole, difenoconazole (*off-label*)
	Blue mould	prothioconazole + tebuconazole
	Bunt	prothioconazole + tebuconazole
	Cladosporium	bromuconazole + tebuconazole, difenoconazole (*off-label*)
	Crown rust	boscalid + epoxiconazole, boscalid + epoxiconazole + pyraclostrobin, epoxiconazole, epoxiconazole + fluxapyroxad, epoxiconazole + fluxapyroxad + pyraclostrobin
	Disease control	benzovindiflupyr + prothioconazole, chlorothalonil (*off-label*), fluxapyroxad + metconazole
	Ear blight	metconazole (*reduction*)
	Eyespot	boscalid + epoxiconazole (*moderate control*), boscalid + epoxiconazole + pyraclostrobin (*moderate control*), epoxiconazole + fluxapyroxad (*moderate control*), epoxiconazole + fluxapyroxad + pyraclostrobin (*moderate control*), epoxiconazole + metconazole (*reduction*), fluoxastrobin + prothioconazole + trifloxystrobin (*reduction*), fluxapyroxad + metconazole (*good reduction*), prothioconazole + trifloxystrobin
	Foot rot	fludioxonil (*seed treatment*)
	Fungus diseases	prothioconazole + tebuconazole (*qualified minor use*)
	Fusarium	bromuconazole + tebuconazole, prothioconazole + tebuconazole
	Late ear diseases	fluoxastrobin + prothioconazole + trifloxystrobin, prothioconazole + trifloxystrobin
	Loose smut	prothioconazole + tebuconazole
	Powdery mildew	azoxystrobin, azoxystrobin (*moderate control*), azoxystrobin + chlorothalonil, boscalid + epoxiconazole (*moderate control*), boscalid + epoxiconazole + pyraclostrobin (*moderate control*), bromuconazole + tebuconazole (*moderate control*), cyflufenamid (*off-label*), epoxiconazole, epoxiconazole + fluxapyroxad (*moderate control*), epoxiconazole + fluxapyroxad + pyraclostrobin (*moderate control*), epoxiconazole + metconazole (*moderate control*), epoxiconazole + pyraclostrobin, fluoxastrobin + prothioconazole + trifloxystrobin, fluxapyroxad (*reduction*), fluxapyroxad + metconazole (*moderate control*), fluxapyroxad + pyraclostrobin

		(*moderate control*), metconazole (*moderate control*), prothioconazole + trifloxystrobin, tebuconazole
	Ramularia leaf spots	epoxiconazole + pyraclostrobin (*moderate control*)
	Rhynchosporium	azoxystrobin, azoxystrobin (*reduction*), azoxystrobin + chlorothalonil, benzovindiflupyr (*moderate control*), benzovindiflupyr + prothioconazole (*moderate control*), boscalid + epoxiconazole (*moderate control*), boscalid + epoxiconazole + pyraclostrobin, epoxiconazole, epoxiconazole + fluxapyroxad + pyraclostrobin, epoxiconazole + metconazole (*moderate control*), epoxiconazole + pyraclostrobin, fluxapyroxad, fluxapyroxad + metconazole, fluxapyroxad + pyraclostrobin, metconazole (*moderate control*), penthiopyrad, penthiopyrad (*moderate control*), tebuconazole
	Rust	azoxystrobin, azoxystrobin + chlorothalonil, benzovindiflupyr , benzovindiflupyr + prothioconazole, boscalid + epoxiconazole, boscalid + epoxiconazole + pyraclostrobin, bromuconazole + tebuconazole, difenoconazole (*off-label*), epoxiconazole, epoxiconazole + fluxapyroxad, epoxiconazole + fluxapyroxad + pyraclostrobin, epoxiconazole + metconazole, epoxiconazole + pyraclostrobin, fluoxastrobin + prothioconazole + trifloxystrobin, fluxapyroxad, fluxapyroxad (*moderate control*), fluxapyroxad + metconazole, fluxapyroxad + pyraclostrobin, metconazole, penthiopyrad, penthiopyrad (*moderate control*), prothioconazole + trifloxystrobin, tebuconazole
	Seed-borne diseases	fludioxonil (*seed treatment*)
	Septoria	azoxystrobin + chlorothalonil, bromuconazole + tebuconazole (*moderate control*), difenoconazole (*off-label*), fluoxastrobin + prothioconazole + trifloxystrobin, metconazole, prothioconazole + trifloxystrobin
	Snow mould	fludioxonil + sedaxane
	Stripe smut	fludioxonil, fludioxonil + sedaxane
	Take-all	azoxystrobin (*reduction*), azoxystrobin + chlorothalonil (*reduction*)
	Tan spot	fluoxastrobin + prothioconazole + trifloxystrobin
Pests	Aphids	beta-cyfluthrin, cypermethrin, deltamethrin (*off-label*), dimethoate, lambda-cyhalothrin, tau-fluvalinate (*off-label*), zeta-cypermethrin (*off-label*)
	Beetles	deltamethrin (*off-label*), deltamethrin (*off-label - cereal*), lambda-cyhalothrin
	Caterpillars	lambda-cyhalothrin
	Flies	deltamethrin (*off-label*)
	Midges	beta-cyfluthrin, deltamethrin (*off-label*)
	Pests, miscellaneous	cypermethrin, deltamethrin (*off-label*), lambda-cyhalothrin (*off-label*), zeta-cypermethrin (*off-label*)
	Weevils	lambda-cyhalothrin
Plant growth regulation	Growth control	chlormequat, chlormequat + imazaquin (*off-label*), mepiquat chloride + prohexadione-calcium, trinexapac-ethyl, trinexapac-ethyl (*off-label*)
	Quality/yield control	mepiquat chloride + prohexadione-calcium

Weeds	Broad-leaved weeds	2,4-DB + MCPA, bromoxynil, bromoxynil (*off-label*), carfentrazone-ethyl (*off-label*), clopyralid + florasulam, dicamba, dicamba + mecoprop-P (*off-label*), dichlorprop-P + MCPA + mecoprop-P, diflufenican, diflufenican + florasulam, diflufenican + flufenacet (*off-label*), florasulam + fluroxypyr, florasulam + halauxifen-methyl, flufenacet + picolinafen, fluroxypyr, fluroxypyr + halauxifen-methyl, mecoprop-P (*off-label*), mesosulfuron-methyl + propoxycarbazone-sodium, metsulfuron-methyl + tribenuron-methyl (*off-label*), pendimethalin + picolinafen, prosulfocarb (*off-label*)
	Crops as weeds	carfentrazone-ethyl (*off-label*), clopyralid + florasulam, diflufenican + florasulam, florasulam + fluroxypyr, florasulam + halauxifen-methyl, flufenacet + picolinafen, mesosulfuron-methyl + propoxycarbazone-sodium
	Grass weeds	diflufenican + flufenacet (*off-label*), flufenacet (*off-label*), flufenacet + picolinafen, glyphosate, iodosulfuron-methyl-sodium + mesosulfuron-methyl (*off-label*), mesosulfuron-methyl + propoxycarbazone-sodium, pendimethalin + picolinafen, prosulfocarb (*off-label*)
	Weeds, miscellaneous	bromoxynil (*off-label*), clopyralid + florasulam, dicamba, glyphosate (*off-label*)

Cereals - Maize/sweetcorn

Diseases	Damping off	fludioxonil + metalaxyl-M, thiram (*seed treatment*)
	Disease control	azoxystrobin + propiconazole, azoxystrobin + propiconazole (*useful reduction*), epoxiconazole + pyraclostrobin (*off-label*), pyraclostrobin
	Eyespot	azoxystrobin + propiconazole (*useful reduction*), epoxiconazole + pyraclostrobin, epoxiconazole + pyraclostrobin (*off-label*), pyraclostrobin, pyraclostrobin (*off-label*)
	Foliar diseases	azoxystrobin + propiconazole (*off-label*), epoxiconazole + pyraclostrobin, pyraclostrobin, pyraclostrobin (*moderate control*)
	Fusarium	fludioxonil + metalaxyl-M
	Pythium	fludioxonil + metalaxyl-M
	Rust	epoxiconazole + pyraclostrobin (*off-label*)
Pests	Aphids	cypermethrin, pymetrozine (*off-label*)
	Caterpillars	Bacillus thuringiensis (*off-label*), indoxacarb (*off-label*)
	Cutworms	Bacillus thuringiensis (*off-label*)
	Flies	lambda-cyhalothrin (*off-label*), methiocarb
	Pests, miscellaneous	cypermethrin, lambda-cyhalothrin (*off-label*)
	Wireworms	thiacloprid
Weeds	Bindweeds	dicamba + prosulfuron
	Broad-leaved weeds	bromoxynil, clopyralid, clopyralid (*off-label*), clopyralid + florasulam + fluroxypyr, dicamba, dicamba + prosulfuron, dicamba + prosulfuron (*seedlings only*), dimethenamid-p + pendimethalin, dimethenamid-p + pendimethalin (*off-label*), flufenacet + isoxaflutole, fluroxypyr, fluroxypyr (*off-label*), foramsulfuron + iodosulfuron-methyl-sodium, mesotrione, mesotrione (*off-label*), mesotrione + nicosulfuron, mesotrione +

		terbuthylazine, mesotrione + terbuthylazine (*off-label*), nicosulfuron, nicosulfuron (*off-label*), pendimethalin, pendimethalin (*off-label*), pendimethalin (*off-label - under covers*), prosulfuron, pyridate, rimsulfuron, S-metolachlor, S-metolachlor (*moderately susceptible*)
	Crops as weeds	clopyralid (*off-label*), fluroxypyr (*off-label*), mesotrione, mesotrione + nicosulfuron, nicosulfuron, nicosulfuron (*off-label*), pendimethalin, rimsulfuron
	Grass weeds	dimethenamid-p + pendimethalin, dimethenamid-p + pendimethalin (*off-label*), flufenacet + isoxaflutole, foramsulfuron + iodosulfuron-methyl-sodium, mesotrione, mesotrione (*off-label*), mesotrione + nicosulfuron, mesotrione + terbuthylazine, mesotrione + terbuthylazine (*off-label*), nicosulfuron, nicosulfuron (*off-label*), pendimethalin, S-metolachlor
	Weed grasses	foramsulfuron + iodosulfuron-methyl-sodium, mesotrione, mesotrione + nicosulfuron, nicosulfuron
	Weeds, miscellaneous	bromoxynil, dicamba, mesotrione + nicosulfuron, nicosulfuron (*off-label*)

Cereals - Oats

Crop control	Desiccation	diquat (*off-label*), diquat (*stockfeed only*), glyphosate
Diseases	Blue mould	prothioconazole + tebuconazole
	Bunt	prothioconazole + tebuconazole
	Covered smut	prothioconazole + tebuconazole
	Crown rust	azoxystrobin, azoxystrobin + cyproconazole, bixafen + prothioconazole, boscalid + epoxiconazole, epoxiconazole + fenpropimorph + metrafenone, epoxiconazole + metrafenone, epoxiconazole + pyraclostrobin, fenpropimorph + pyraclostrobin, fluoxastrobin + prothioconazole, fluxapyroxad, fluxapyroxad + pyraclostrobin, prochloraz + proquinazid + tebuconazole (*qualified minor use recommendation*), propiconazole, prothioconazole, prothioconazole + spiroxamine, prothioconazole + spiroxamine + tebuconazole, prothioconazole + tebuconazole, pyraclostrobin, tebuconazole, tebuconazole (*reduction*)
	Eyespot	bixafen + prothioconazole (*reduction of incidence and severity*), epoxiconazole + fenpropimorph + kresoxim-methyl (*reduction*), fluoxastrobin + prothioconazole, fluoxastrobin + prothioconazole (*reduction of incidence and severity*), prothioconazole, prothioconazole + spiroxamine, prothioconazole + spiroxamine + tebuconazole, prothioconazole + tebuconazole
	Foot rot	clothianidin + prothioconazole (*seed treatment*), difenoconazole + fludioxonil, fludioxonil (*seed treatment*), fludioxonil + tefluthrin
	Fusarium	prothioconazole + tebuconazole
	Leaf spot	difenoconazole + fludioxonil, fludioxonil (*seed treatment*), fludioxonil + tefluthrin
	Loose smut	clothianidin + prothioconazole (*seed treatment*), fludioxonil + sedaxane, prothioconazole + tebuconazole

	Net blotch	tebuconazole
	Powdery mildew	azoxystrobin, azoxystrobin (*moderate control*), azoxystrobin + cyproconazole, bixafen + prothioconazole, cyflufenamid, epoxiconazole, epoxiconazole + fenpropimorph, epoxiconazole + fenpropimorph + kresoxim-methyl, epoxiconazole + fenpropimorph + metrafenone, epoxiconazole + metrafenone, epoxiconazole + pyraclostrobin, fenpropimorph, fenpropimorph + pyraclostrobin, fluoxastrobin + prothioconazole, fluxapyroxad (*reduction*), fluxapyroxad + pyraclostrobin (*moderate control*), metrafenone (*evidence of mildew control on oats is limited*), prochloraz + proquinazid + tebuconazole, propiconazole, proquinazid, prothioconazole, prothioconazole + spiroxamine, prothioconazole + spiroxamine + tebuconazole, prothioconazole + tebuconazole, sulphur, tebuconazole
	Rhynchosporium	tebuconazole
	Rust	tebuconazole
	Seed-borne diseases	difenoconazole + fludioxonil, fludioxonil + tebuconazole
	Snow mould	fludioxonil (*seed treatment*)
	Take-all	azoxystrobin (*reduction*)
Pests	Aphids	clothianidin (*seed treatment*), clothianidin + prothioconazole (*seed treatment*), cypermethrin, deltamethrin, lambda-cyhalothrin, zeta-cypermethrin
	Beetles	lambda-cyhalothrin
	Birds/mammals	aluminium ammonium sulphate
	Caterpillars	lambda-cyhalothrin
	Flies	cypermethrin
	Leafhoppers	clothianidin (*seed treatment*), clothianidin + prothioconazole (*seed treatment*)
	Slugs/snails	clothianidin (*seed treatment*), clothianidin + prothioconazole (*seed treatment*)
	Weevils	lambda-cyhalothrin
	Wireworms	clothianidin (*seed treatment*), clothianidin + prothioconazole (*seed treatment*), fludioxonil + tefluthrin
Plant growth regulation	Growth control	chlormequat, prohexadione-calcium + trinexapac-ethyl, trinexapac-ethyl
	Quality/yield control	sulphur
Weeds	Broad-leaved weeds	2,4-D, 2,4-D + MCPA, 2,4-DB, amidosulfuron, bromoxynil, carfentrazone-ethyl, carfentrazone-ethyl + mecoprop-P, clopyralid, clopyralid + florasulam, clopyralid + florasulam + fluroxypyr, dicamba + MCPA + mecoprop-P, dicamba + mecoprop-P, dichlorprop-P + MCPA + mecoprop-P, diflufenican (*off-label*), diflufenican + flufenacet (*off-label*), florasulam, florasulam + fluroxypyr, flumioxazin (*off-label*), fluroxypyr, glyphosate, isoxaben, MCPA, mecoprop-P, metsulfuron-methyl, metsulfuron-methyl + thifensulfuron-methyl, metsulfuron-methyl + tribenuron-methyl, thifensulfuron-methyl + tribenuron-methyl, tribenuron-methyl

	Crops as weeds	bromoxynil, clopyralid + florasulam, diflufenican (off-label), florasulam, flumioxazin (off-label), fluroxypyr, glyphosate, metsulfuron-methyl, metsulfuron-methyl + tribenuron-methyl
	Grass weeds	diflufenican (off-label), diflufenican + flufenacet (off-label), diquat (animal feed), flumioxazin (off-label), glyphosate
	Weeds, miscellaneous	clopyralid + florasulam, diflufenican (off-label), diquat (animal feed only), florasulam, glyphosate, metsulfuron-methyl, metsulfuron-methyl + thifensulfuron-methyl

Cereals - Rye/triticale

Diseases	Alternaria	bromuconazole + tebuconazole, difenoconazole (off-label)
	Blue mould	prothioconazole + tebuconazole
	Bunt	clothianidin + prothioconazole (seed treatment), prothioconazole + tebuconazole
	Cladosporium	bromuconazole + tebuconazole, difenoconazole (off-label)
	Crown rust	tebuconazole
	Disease control	chlorothalonil (off-label), epoxiconazole + folpet, fenpropimorph + pyraclostrobin (off-label), fluxapyroxad + metconazole, tebuconazole (off-label)
	Drechslera leaf spot	fenpropimorph + pyraclostrobin
	Ear blight	epoxiconazole + metconazole (good reduction), epoxiconazole + metconazole (moderate control), epoxiconazole + pyraclostrobin (good reduction), metconazole (reduction), tebuconazole, thiophanate-methyl (reduction)
	Eyespot	bixafen + prothioconazole (reduction of incidence and severity), bixafen + prothioconazole + spiroxamine, bixafen + prothioconazole + tebuconazole (reduction of incidence and severity), boscalid + epoxiconazole (moderate control), boscalid + epoxiconazole + pyraclostrobin (moderate control), epoxiconazole + fenpropimorph + kresoxim-methyl (reduction), epoxiconazole + fluxapyroxad (moderate control), epoxiconazole + fluxapyroxad + pyraclostrobin (moderate control), epoxiconazole + metconazole (reduction), fluoxastrobin + prothioconazole (reduction), fluoxastrobin + prothioconazole + trifloxystrobin (reduction), fluxapyroxad + metconazole (good reduction), prochloraz + tebuconazole, prothioconazole, prothioconazole + spiroxamine, prothioconazole + spiroxamine + tebuconazole (reduction), prothioconazole + tebuconazole, prothioconazole + tebuconazole (reduction), prothioconazole + trifloxystrobin
	Foliar diseases	azoxystrobin + cyproconazole (off-label), pyraclostrobin (off-label), trifloxystrobin (off-label)
	Foot rot	clothianidin + prothioconazole (seed treatment), fludioxonil (seed treatment)
	Fusarium	bromuconazole + tebuconazole, prothioconazole + tebuconazole, thiophanate-methyl, thiophanate-methyl (reduction)

Late ear diseases	bixafen + fluoxastrobin + prothioconazole, bixafen + prothioconazole, bixafen + prothioconazole + spiroxamine, bixafen + prothioconazole + tebuconazole, fluoxastrobin + prothioconazole + trifloxystrobin, prothioconazole + spiroxamine, prothioconazole + trifloxystrobin
Loose smut	prothioconazole + tebuconazole
Mycosphaerella	benzovindiflupyr
Net blotch	azoxystrobin + chlorothalonil, tebuconazole
Powdery mildew	azoxystrobin, azoxystrobin (*moderate control*), azoxystrobin + chlorothalonil, azoxystrobin + cyproconazole, bixafen + fluoxastrobin + prothioconazole, bixafen + prothioconazole, bixafen + prothioconazole + spiroxamine, bixafen + prothioconazole + tebuconazole, boscalid + epoxiconazole + pyraclostrobin (*moderate control*), bromuconazole + tebuconazole (*moderate control*), cyflufenamid, epoxiconazole, epoxiconazole + fenpropimorph, epoxiconazole + fenpropimorph + kresoxim-methyl, epoxiconazole + fenpropimorph + metrafenone, epoxiconazole + fluxapyroxad (*moderate control*), epoxiconazole + fluxapyroxad + pyraclostrobin (*moderate control*), epoxiconazole + metconazole (*moderate control*), epoxiconazole + metrafenone, fenpropimorph, fluoxastrobin + prothioconazole, fluoxastrobin + prothioconazole + trifloxystrobin, fluxapyroxad (*reduction*), fluxapyroxad + metconazole (*moderate control*), fluxapyroxad + pyraclostrobin (*moderate control*), metconazole (*moderate control*), prochloraz + tebuconazole, propiconazole, proquinazid, prothioconazole, prothioconazole + spiroxamine, prothioconazole + spiroxamine + tebuconazole, prothioconazole + tebuconazole, prothioconazole + trifloxystrobin, tebuconazole, tebuconazole (*off-label*)
Ramularia leaf spots	epoxiconazole + pyraclostrobin (*moderate control*)
Rhynchosporium	azoxystrobin, azoxystrobin (*reduction*), azoxystrobin + chlorothalonil, azoxystrobin + chlorothalonil (*moderate control*), azoxystrobin + cyproconazole (*moderate control*), benzovindiflupyr + prothioconazole (*moderate control*), bixafen + fluoxastrobin + prothioconazole, bixafen + prothioconazole, bixafen + prothioconazole + spiroxamine, bixafen + prothioconazole + tebuconazole, epoxiconazole, epoxiconazole + fenpropimorph, epoxiconazole + fenpropimorph + kresoxim-methyl, epoxiconazole + fenpropimorph + metrafenone, epoxiconazole + folpet, epoxiconazole + isopyrazam, epoxiconazole + metrafenone, fluoxastrobin + prothioconazole, penthiopyrad (*moderate control*), prochloraz + tebuconazole, propiconazole, prothioconazole, prothioconazole + spiroxamine, prothioconazole + spiroxamine + tebuconazole, prothioconazole + tebuconazole, tebuconazole, tebuconazole (*moderate control*)
Rust	azoxystrobin, azoxystrobin + chlorothalonil, azoxystrobin + cyproconazole, benzovindiflupyr , benzovindiflupyr + prothioconazole, bixafen + fluoxastrobin + prothioconazole, bixafen +

prothioconazole, bixafen + prothioconazole + spiroxamine, bixafen + prothioconazole + tebuconazole, boscalid + epoxiconazole, boscalid + epoxiconazole + pyraclostrobin, bromuconazole + tebuconazole, difenoconazole (*off-label*), epoxiconazole, epoxiconazole + fenpropimorph, epoxiconazole + fenpropimorph + kresoxim-methyl, epoxiconazole + fenpropimorph + metrafenone, epoxiconazole + fluxapyroxad, epoxiconazole + fluxapyroxad + pyraclostrobin, epoxiconazole + folpet, epoxiconazole + isopyrazam, epoxiconazole + metconazole, epoxiconazole + metrafenone, epoxiconazole + pyraclostrobin, fenpropimorph, fenpropimorph + pyraclostrobin, fenpropimorph + pyraclostrobin (*off-label*), fluoxastrobin + prothioconazole, fluoxastrobin + prothioconazole + trifloxystrobin, fluxapyroxad, fluxapyroxad (*moderate control*), fluxapyroxad + metconazole, fluxapyroxad + pyraclostrobin, metconazole, penthiopyrad, penthiopyrad (*moderate control*), prochloraz + tebuconazole, propiconazole, prothioconazole, prothioconazole + spiroxamine, prothioconazole + spiroxamine + tebuconazole, prothioconazole + tebuconazole, prothioconazole + trifloxystrobin, tebuconazole, tebuconazole (*off-label*)

Seed-borne diseases fludioxonil, fludioxonil + tebuconazole, fludioxonil + tefluthrin (*off-label*)

Septoria azoxystrobin + chlorothalonil, benzovindiflupyr , benzovindiflupyr + prothioconazole, bixafen + fluoxastrobin + prothioconazole, bixafen + prothioconazole, bixafen + prothioconazole + spiroxamine, bixafen + prothioconazole + tebuconazole, boscalid + epoxiconazole, boscalid + epoxiconazole + pyraclostrobin, bromuconazole + tebuconazole (*moderate control*), difenoconazole (*off-label*), epoxiconazole, epoxiconazole + fenpropimorph, epoxiconazole + fenpropimorph + kresoxim-methyl, epoxiconazole + fenpropimorph + metrafenone, epoxiconazole + fluxapyroxad, epoxiconazole + fluxapyroxad + pyraclostrobin, epoxiconazole + folpet, epoxiconazole + isopyrazam, epoxiconazole + metconazole, epoxiconazole + metrafenone, epoxiconazole + pyraclostrobin, epoxiconazole + pyraclostrobin (*moderate control*), fenpropimorph + pyraclostrobin, fenpropimorph + pyraclostrobin (*off-label*), fluoxastrobin + prothioconazole + trifloxystrobin, fluxapyroxad, fluxapyroxad + metconazole, fluxapyroxad + pyraclostrobin, metconazole, propiconazole, prothioconazole + trifloxystrobin, tebuconazole

Snow mould fludioxonil + sedaxane

Sooty moulds tebuconazole

Stripe smut difenoconazole + fludioxonil

Take-all azoxystrobin (*reduction*), azoxystrobin (*reduction in severity*), azoxystrobin + chlorothalonil (*reduction*)

Tan spot bixafen + fluoxastrobin + prothioconazole, bixafen + prothioconazole, bixafen + prothioconazole + spiroxamine, bixafen + prothioconazole + tebuconazole, epoxiconazole + pyraclostrobin

		(*moderate control*), fluoxastrobin + prothioconazole + trifloxystrobin
Pests	Aphids	beta-cyfluthrin, clothianidin (*seed treatment*), clothianidin + prothioconazole (*seed treatment*), cypermethrin, deltamethrin (*off-label*), dimethoate, lambda-cyhalothrin, tau-fluvalinate (*off-label*), zeta-cypermethrin (*off-label*)
	Beetles	deltamethrin (*off-label*), deltamethrin (*off-label - cereal*), lambda-cyhalothrin
	Caterpillars	lambda-cyhalothrin
	Flies	cypermethrin, deltamethrin (*off-label*)
	Leafhoppers	clothianidin (*seed treatment*), clothianidin + prothioconazole (*seed treatment*)
	Midges	beta-cyfluthrin, deltamethrin (*off-label*)
	Pests, miscellaneous	deltamethrin (*off-label*), lambda-cyhalothrin (*off-label*), zeta-cypermethrin (*off-label*)
	Slugs/snails	clothianidin (*seed treatment*), clothianidin + prothioconazole (*seed treatment*)
	Weevils	lambda-cyhalothrin
	Wireworms	clothianidin (*seed treatment*), clothianidin + prothioconazole (*seed treatment*)
Plant growth regulation	Growth control	chlormequat, chlormequat + imazaquin (*off-label*), ethephon, ethephon + mepiquat chloride, mepiquat chloride + prohexadione-calcium, prohexadione-calcium + trinexapac-ethyl, trinexapac-ethyl
	Quality/yield control	mepiquat chloride + prohexadione-calcium
Weeds	Broad-leaved weeds	2,4-D, amidosulfuron, amidosulfuron + iodosulfuron-methyl-sodium, bifenox, bromoxynil, carfentrazone-ethyl, chlorotoluron + diflufenican + pendimethalin, clopyralid (*off-label*), clopyralid + florasulam, dicamba + mecoprop-P (*off-label*), dichlorprop-P + MCPA + mecoprop-P, diflufenican, diflufenican + florasulam, diflufenican + flufenacet, diflufenican + flufenacet (*off-label*), diflufenican + pendimethalin, florasulam, florasulam + fluroxypyr, florasulam + fluroxypyr (*off-label*), florasulam + halauxifen-methyl, florasulam + pyroxsulam, flufenacet + picolinafen, fluroxypyr, fluroxypyr + halauxifen-methyl, imazosulfuron, isoxaben, MCPA, mecoprop-P (*off-label*), mesosulfuron-methyl + propoxycarbazone-sodium, metsulfuron-methyl, metsulfuron-methyl + thifensulfuron-methyl, metsulfuron-methyl + tribenuron-methyl, metsulfuron-methyl + tribenuron-methyl (*off-label*), pendimethalin, pendimethalin + picolinafen, prosulfocarb (*off-label*), thifensulfuron-methyl + tribenuron-methyl, tribenuron-methyl
	Crops as weeds	amidosulfuron + iodosulfuron-methyl-sodium, chlorotoluron + diflufenican + pendimethalin, clopyralid + florasulam, diflufenican, diflufenican + florasulam, diflufenican + flufenacet, florasulam, florasulam + fluroxypyr, florasulam + halauxifen-methyl, florasulam + pyroxsulam, flufenacet + picolinafen, imazosulfuron, mesosulfuron-methyl + propoxycarbazone-sodium, metsulfuron-methyl, metsulfuron-methyl + tribenuron-methyl, pendimethalin

	Grass weeds	chlorotoluron + diflufenican + pendimethalin, clodinafop-propargyl, diflufenican, diflufenican + flufenacet, diflufenican + flufenacet (*off-label*), diflufenican + pendimethalin, florasulam + pyroxsulam, flufenacet (*off-label*), flufenacet + picolinafen, iodosulfuron-methyl-sodium + mesosulfuron-methyl (*off-label*), mesosulfuron-methyl + propoxycarbazone-sodium, pendimethalin, pendimethalin + picolinafen, prosulfocarb (*off-label*), tri-allate (*off-label*)
	Weeds, miscellaneous	chlorotoluron + diflufenican + pendimethalin, clopyralid + florasulam, diflufenican, florasulam, metsulfuron-methyl, metsulfuron-methyl + thifensulfuron-methyl

Cereals - Undersown cereals

Weeds	Broad-leaved weeds	2,4-D, 2,4-DB, 2,4-DB + MCPA, MCPA, MCPA (*red clover or grass*), tribenuron-methyl

Cereals - Wheat

Crop control	Desiccation	glyphosate
Diseases	Alternaria	bromuconazole + tebuconazole, difenoconazole (*off-label*), tebuconazole
	Blue mould	prothioconazole + tebuconazole
	Bunt	clothianidin + prothioconazole (*seed treatment*), difenoconazole, difenoconazole + fludioxonil, fludioxonil (*seed treatment*), fludioxonil + sedaxane, fludioxonil + tefluthrin, prothioconazole + tebuconazole
	Cladosporium	bromuconazole + tebuconazole, difenoconazole (*off-label*), fluxapyroxad + metconazole (*moderate control*), tebuconazole
	Disease control	chlorothalonil, chlorothalonil (*off-label*), chlorothalonil + penthiopyrad, epoxiconazole (*off-label*), fluxapyroxad + metconazole
	Drechslera leaf spot	fenpropimorph + pyraclostrobin
	Ear blight	boscalid + epoxiconazole (*good reduction*), boscalid + epoxiconazole + pyraclostrobin (*good reduction*), chlorothalonil + tebuconazole, dimoxystrobin + epoxiconazole, epoxiconazole, epoxiconazole (*good reduction*), epoxiconazole + fenpropimorph, epoxiconazole + fenpropimorph + kresoxim-methyl (*reduction*), epoxiconazole + fenpropimorph + metrafenone (*reduction*), epoxiconazole + fluxapyroxad (*good reduction*), epoxiconazole + fluxapyroxad + pyraclostrobin (*good reduction*), epoxiconazole + metconazole (*good reduction*), epoxiconazole + metconazole (*moderate control*), epoxiconazole + metrafenone (*reduction*), epoxiconazole + pyraclostrobin (*good reduction*), fludioxonil + sedaxane (*moderate control*), fluxapyroxad + metconazole (*good reduction*), metconazole (*reduction*), prothioconazole + tebuconazole, tebuconazole, thiophanate-methyl (*reduction*)
	Eyespot	azoxystrobin + cyproconazole (*reduction*), bixafen + fluopyram + prothioconazole (*reduction of incidence*

and severity), bixafen + prothioconazole (*reduction of incidence and severity*), bixafen + prothioconazole + spiroxamine, bixafen + prothioconazole + tebuconazole (*reduction of incidence and severity*), boscalid + epoxiconazole (*moderate control*), boscalid + epoxiconazole + pyraclostrobin (*moderate control*), chlorothalonil + cyproconazole, epoxiconazole (*reduction*), epoxiconazole + fenpropimorph (*reduction*), epoxiconazole + fenpropimorph + kresoxim-methyl (*reduction*), epoxiconazole + fluxapyroxad (*moderate control*), epoxiconazole + fluxapyroxad + pyraclostrobin (*moderate control*), epoxiconazole + metconazole (*reduction*), fenpropidin + prochloraz + tebuconazole, fluoxastrobin + prothioconazole (*reduction*), fluoxastrobin + prothioconazole + trifloxystrobin (*reduction*), fluxapyroxad + metconazole (*good reduction*), metrafenone (*reduction*), prochloraz + propiconazole, prochloraz + proquinazid + tebuconazole, prochloraz + tebuconazole, prothioconazole, prothioconazole + spiroxamine, prothioconazole + spiroxamine + tebuconazole (*reduction*), prothioconazole + tebuconazole, prothioconazole + tebuconazole (*reduction*), prothioconazole + trifloxystrobin

Foliar diseases	azoxystrobin + cyproconazole (*off-label*), epoxiconazole + fenpropimorph (*off-label*), epoxiconazole + fenpropimorph + kresoxim-methyl (*off-label*), fenpropimorph (*off-label*), pyraclostrobin (*off-label*), trifloxystrobin (*off-label*)
Foot rot	clothianidin + prothioconazole (*seed treatment*), difenoconazole + fludioxonil, fludioxonil (*seed treatment*), fludioxonil + tefluthrin, fluoxastrobin + prothioconazole (*reduction*)
Fusarium	benzovindiflupyr + prothioconazole (*moderate control*), bromuconazole + tebuconazole, fludioxonil + sedaxane, prothioconazole + tebuconazole, tebuconazole, thiophanate-methyl, thiophanate-methyl (*reduction*)
Late ear diseases	azoxystrobin, bixafen + fluopyram + prothioconazole (*reduction of incidence and severity*), bixafen + fluoxastrobin + prothioconazole, bixafen + prothioconazole, bixafen + prothioconazole + spiroxamine, bixafen + prothioconazole + tebuconazole, fenpropidin + prochloraz + tebuconazole, fluoxastrobin + prothioconazole, fluoxastrobin + prothioconazole + trifloxystrobin, prochloraz + proquinazid + tebuconazole (*reduction only*), prochloraz + tebuconazole, prothioconazole, prothioconazole + spiroxamine, prothioconazole + spiroxamine + tebuconazole, prothioconazole + tebuconazole, prothioconazole + trifloxystrobin, tebuconazole
Loose smut	clothianidin + prothioconazole (*seed treatment*), fludioxonil + sedaxane, prothioconazole + tebuconazole
Mycosphaerella	benzovindiflupyr , benzovindiflupyr + prothioconazole
Net blotch	metconazole (*reduction*), tebuconazole
Powdery mildew	azoxystrobin (*off-label*), azoxystrobin + cyproconazole, bixafen + fluopyram +

prothioconazole, bixafen + fluoxastrobin + prothioconazole, bixafen + prothioconazole, bixafen + prothioconazole + spiroxamine, bixafen + prothioconazole + tebuconazole, bromuconazole + tebuconazole (*moderate control*), chlorothalonil + cyproconazole, chlorothalonil + cyproconazole + propiconazole (*moderate control*), chlorothalonil + proquinazid, cyflufenamid, epoxiconazole, epoxiconazole + fenpropimorph, epoxiconazole + fenpropimorph + metrafenone, epoxiconazole + fluxapyroxad (*moderate control*), epoxiconazole + fluxapyroxad + pyraclostrobin (*moderate control*), epoxiconazole + metconazole (*moderate control*), epoxiconazole + metrafenone, fenpropidin + prochloraz + tebuconazole, fenpropimorph, fenpropimorph (*off-label*), fluoxastrobin + prothioconazole, fluoxastrobin + prothioconazole + trifloxystrobin, flutriafol, fluxapyroxad (*reduction*), fluxapyroxad + metconazole (*moderate control*), fluxapyroxad + pyraclostrobin (*moderate control*), metconazole (*moderate control*), metrafenone, prochloraz + proquinazid + tebuconazole, prochloraz + tebuconazole, propiconazole, proquinazid, proquinazid (*off-label*), prothioconazole, prothioconazole + spiroxamine, prothioconazole + spiroxamine + tebuconazole, prothioconazole + tebuconazole, prothioconazole + trifloxystrobin, pyriofenone, sulphur, tebuconazole, tebuconazole (*moderate control*)

Rhynchosporium	tebuconazole
Rust	azoxystrobin, azoxystrobin (*off-label*), azoxystrobin + chlorothalonil, azoxystrobin + cyproconazole, benzovindiflupyr, benzovindiflupyr + prothioconazole, bixafen + fluopyram + prothioconazole, bixafen + fluoxastrobin + prothioconazole, bixafen + prothioconazole, bixafen + prothioconazole + spiroxamine, bixafen + prothioconazole + tebuconazole, boscalid + epoxiconazole, boscalid + epoxiconazole + pyraclostrobin, bromuconazole + tebuconazole, chlorothalonil + cyproconazole, chlorothalonil + cyproconazole + propiconazole, chlorothalonil + penthiopyrad, chlorothalonil + propiconazole, chlorothalonil + tebuconazole, difenoconazole, difenoconazole (*off-label*), dimoxystrobin + epoxiconazole, epoxiconazole, epoxiconazole + fenpropimorph, epoxiconazole + fenpropimorph + kresoxim-methyl, epoxiconazole + fenpropimorph + metrafenone, epoxiconazole + fluxapyroxad, epoxiconazole + fluxapyroxad + pyraclostrobin, epoxiconazole + folpet, epoxiconazole + isopyrazam, epoxiconazole + metconazole, epoxiconazole + metrafenone, epoxiconazole + pyraclostrobin, fenpropidin + prochloraz + tebuconazole, fenpropimorph, fenpropimorph + pyraclostrobin, fluoxastrobin + prothioconazole, fluoxastrobin + prothioconazole + trifloxystrobin, flutriafol, fluxapyroxad, fluxapyroxad (*moderate control*), fluxapyroxad + metconazole, fluxapyroxad + pyraclostrobin, mancozeb, mancozeb (*useful control*), metconazole, penthiopyrad, penthiopyrad (*moderate control*), prochloraz + proquinazid + tebuconazole,

prochloraz + tebuconazole, propiconazole, prothioconazole, prothioconazole + spiroxamine, prothioconazole + spiroxamine + tebuconazole, prothioconazole + tebuconazole, prothioconazole + trifloxystrobin, pyraclostrobin, tebuconazole, trifloxystrobin

Seed-borne diseases	difenoconazole + fludioxonil, fludioxonil + tebuconazole, ipconazole
Septoria	azoxystrobin, azoxystrobin + chlorothalonil, azoxystrobin + cyproconazole, benzovindiflupyr , benzovindiflupyr (*moderate control*), benzovindiflupyr + prothioconazole, benzovindiflupyr + prothioconazole (*moderate control*), bixafen + fluopyram + prothioconazole, bixafen + fluoxastrobin + prothioconazole, bixafen + prothioconazole, bixafen + prothioconazole + spiroxamine, bixafen + prothioconazole + tebuconazole, boscalid + epoxiconazole, boscalid + epoxiconazole + pyraclostrobin, bromuconazole + tebuconazole (*moderate control*), chlorothalonil, chlorothalonil + cyproconazole, chlorothalonil + cyproconazole + propiconazole, chlorothalonil + mancozeb, chlorothalonil + penthiopyrad, chlorothalonil + propiconazole, chlorothalonil + proquinazid (*suppression only*), chlorothalonil + tebuconazole, cyproconazole + penthiopyrad, difenoconazole, difenoconazole (*off-label*), difenoconazole + fludioxonil, dimoxystrobin + epoxiconazole, epoxiconazole, epoxiconazole + fenpropimorph, epoxiconazole + fenpropimorph + kresoxim-methyl, epoxiconazole + fenpropimorph + metrafenone, epoxiconazole + fluxapyroxad, epoxiconazole + fluxapyroxad + pyraclostrobin, epoxiconazole + folpet, epoxiconazole + isopyrazam, epoxiconazole + metconazole, epoxiconazole + metrafenone, epoxiconazole + pyraclostrobin, epoxiconazole + pyraclostrobin (*moderate control*), fenpropidin + prochloraz + tebuconazole, fenpropimorph + pyraclostrobin, fludioxonil (*seed treatment*), fludioxonil + sedaxane, fludioxonil + tefluthrin, fluoxastrobin + prothioconazole, fluoxastrobin + prothioconazole + trifloxystrobin, flutriafol, fluxapyroxad, fluxapyroxad + metconazole, fluxapyroxad + pyraclostrobin, fluxapyroxad + pyraclostrobin (*moderate control*), folpet (*reduction*), mancozeb, mancozeb (*reduction*), metconazole, penthiopyrad, prochloraz + propiconazole, prochloraz + proquinazid + tebuconazole, prochloraz + proquinazid + tebuconazole (*moderate control*), prochloraz + tebuconazole, propiconazole, prothioconazole, prothioconazole + spiroxamine, prothioconazole + spiroxamine + tebuconazole, prothioconazole + tebuconazole, prothioconazole + trifloxystrobin, pyraclostrobin, tebuconazole, trifloxystrobin
Sharp eyespot	fluoxastrobin + prothioconazole, fluoxastrobin + prothioconazole (*reduction*)
Snow mould	difenoconazole + fludioxonil, fludioxonil (*seed treatment*), fludioxonil + sedaxane, fludioxonil + tefluthrin

	Soil-borne diseases	ipconazole
	Sooty moulds	boscalid + epoxiconazole, boscalid + epoxiconazole + pyraclostrobin (*good reduction*), chlorothalonil + tebuconazole, epoxiconazole (*reduction*), epoxiconazole + fenpropimorph (*reduction*), epoxiconazole + fenpropimorph + kresoxim-methyl (*reduction*), epoxiconazole + fenpropimorph + metrafenone (*reduction*), epoxiconazole + fluxapyroxad (*reduction*), epoxiconazole + fluxapyroxad + pyraclostrobin (*reduction*), epoxiconazole + metconazole (*good reduction*), epoxiconazole + metconazole (*moderate control*), epoxiconazole + metrafenone (*reduction*), epoxiconazole + pyraclostrobin (*reduction*), fluoxastrobin + prothioconazole (*reduction*), mancozeb, propiconazole, prothioconazole + tebuconazole, tebuconazole
	Take-all	azoxystrobin (*reduction*), azoxystrobin + chlorothalonil (*reduction*), azoxystrobin + cyproconazole (*reduction*), fluoxastrobin + prothioconazole (*reduction*)
	Tan spot	bixafen + fluopyram + prothioconazole, bixafen + fluoxastrobin + prothioconazole, bixafen + prothioconazole, bixafen + prothioconazole + spiroxamine, bixafen + prothioconazole + tebuconazole, boscalid + epoxiconazole (*reduction*), boscalid + epoxiconazole + pyraclostrobin (*moderate control*), dimoxystrobin + epoxiconazole, epoxiconazole + fluxapyroxad (*moderate control*), epoxiconazole + fluxapyroxad + pyraclostrobin (*moderate control*), epoxiconazole + metconazole (*moderate control*), epoxiconazole + pyraclostrobin, epoxiconazole + pyraclostrobin (*moderate control*), fluoxastrobin + prothioconazole, fluoxastrobin + prothioconazole + trifloxystrobin, fluxapyroxad (*reduction*), fluxapyroxad + metconazole (*good reduction*), fluxapyroxad + pyraclostrobin, penthiopyrad (*reduction*), prothioconazole, prothioconazole + spiroxamine, prothioconazole + tebuconazole
Pests	Aphids	alpha-cypermethrin, beta-cyfluthrin, clothianidin (*seed treatment*), clothianidin + prothioconazole (*seed treatment*), cypermethrin, cypermethrin (*autumn sown*), deltamethrin, deltamethrin (*off-label*), dimethoate, esfenvalerate, flonicamid, lambda-cyhalothrin, tau-fluvalinate, tau-fluvalinate (*off-label*), thiacloprid, zeta-cypermethrin, zeta-cypermethrin (*off-label*)
	Beetles	deltamethrin (*off-label*), deltamethrin (*off-label - cereal*), lambda-cyhalothrin
	Birds/mammals	aluminium ammonium sulphate
	Caterpillars	lambda-cyhalothrin
	Flies	alpha-cypermethrin, cypermethrin, cypermethrin (*autumn sown*), deltamethrin (*off-label*), fludioxonil + tefluthrin, lambda-cyhalothrin
	Leafhoppers	clothianidin (*seed treatment*), clothianidin + prothioconazole (*seed treatment*)

	Midges	beta-cyfluthrin, deltamethrin (*off-label*), lambda-cyhalothrin, thiacloprid
	Pests, miscellaneous	deltamethrin (*off-label*), zeta-cypermethrin (*off-label*)
	Slugs/snails	clothianidin (*seed treatment*), clothianidin + prothioconazole (*seed treatment*)
	Suckers	lambda-cyhalothrin
	Weevils	lambda-cyhalothrin
	Wireworms	clothianidin (*seed treatment*), clothianidin + prothioconazole (*seed treatment*), fludioxonil + tefluthrin
Plant growth regulation	Growth control	chlormequat, chlormequat + 2-chloroethylphosphonic acid, chlormequat + imazaquin, ethephon, ethephon (*off-label*), ethephon + mepiquat chloride, mepiquat chloride + prohexadione-calcium, prohexadione-calcium + trinexapac-ethyl, trinexapac-ethyl, trinexapac-ethyl (*off-label*)
	Quality/yield control	chlormequat + imazaquin, ethephon + mepiquat chloride (*low lodging situations*), mepiquat chloride + prohexadione-calcium, sulphur
Weeds	Broad-leaved weeds	2,4-D, 2,4-D + MCPA, 2,4-DB, 2,4-DB + MCPA, amidosulfuron, amidosulfuron + iodosulfuron-methyl-sodium, amidosulfuron + iodosulfuron-methyl-sodium + mesosulfuron-methyl, bifenox, bifenox (*off-label*), bromoxynil, bromoxynil + diflufenican, carfentrazone-ethyl, carfentrazone-ethyl (*off-label*), carfentrazone-ethyl + mecoprop-P, chlorotoluron + diflufenican + pendimethalin, clodinafop-propargyl + prosulfocarb, clopyralid, clopyralid (*off-label*), clopyralid + florasulam, clopyralid + florasulam + fluroxypyr, dicamba + MCPA + mecoprop-P, dicamba + mecoprop-P, dicamba + mecoprop-P (*off-label*), dichlorprop-P + MCPA + mecoprop-P, diflufenican, diflufenican + florasulam, diflufenican + flufenacet, diflufenican + flufenacet (*off-label*), diflufenican + flufenacet + flurtamone, diflufenican + flurtamone, diflufenican + iodosulfuron-methyl-sodium + mesosulfuron-methyl, diflufenican + metsulfuron-methyl, diflufenican + pendimethalin, ethofumesate, florasulam, florasulam + fluroxypyr, florasulam + fluroxypyr (*off-label*), florasulam + halauxifen-methyl, florasulam + pyroxsulam, flufenacet, flufenacet + pendimethalin, flufenacet + picolinafen, flumioxazin, fluroxypyr, fluroxypyr + halauxifen-methyl, fluroxypyr + metsulfuron-methyl + thifensulfuron-methyl, glyphosate, imazosulfuron, iodosulfuron-methyl-sodium + mesosulfuron-methyl, isoxaben, MCPA, mecoprop-P, mecoprop-P (*off-label*), mesosulfuron-methyl + propoxycarbazone-sodium, metsulfuron-methyl, metsulfuron-methyl + thifensulfuron-methyl, metsulfuron-methyl + thifensulfuron-methyl (*off-label*), metsulfuron-methyl + tribenuron-methyl, metsulfuron-methyl + tribenuron-methyl (*off-label*), pendimethalin, pendimethalin (*off-label*), pendimethalin + picolinafen, pendimethalin + picolinafen (*off-label*), prosulfocarb, prosulfocarb (*off-label*), sulfosulfuron, thifensulfuron-methyl, thifensulfuron-methyl + tribenuron-methyl, tribenuron-methyl

	Crops as weeds	amidosulfuron + iodosulfuron-methyl-sodium, bromoxynil, bromoxynil + diflufenican, carfentrazone-ethyl (*off-label*), chlorotoluron + diflufenican + pendimethalin, clopyralid + florasulam, diflufenican, diflufenican + florasulam, diflufenican + flufenacet, diflufenican + flurtamone, diflufenican + iodosulfuron-methyl-sodium + mesosulfuron-methyl, florasulam, florasulam + fluroxypyr, florasulam + halauxifen-methyl, florasulam + pyroxsulam, flufenacet + picolinafen, flumioxazin, fluroxypyr, glyphosate, imazosulfuron, mesosulfuron-methyl + propoxycarbazone-sodium, metsulfuron-methyl, metsulfuron-methyl + tribenuron-methyl, pendimethalin
	Grass weeds	amidosulfuron + iodosulfuron-methyl-sodium + mesosulfuron-methyl, chlorotoluron + diflufenican + pendimethalin, clodinafop-propargyl, clodinafop-propargyl + cloquintocet-mexyl, clodinafop-propargyl + pinoxaden, clodinafop-propargyl + pinoxaden (*from seed*), clodinafop-propargyl + prosulfocarb, diflufenican, diflufenican + flufenacet, diflufenican + flufenacet (*off-label*), diflufenican + flufenacet + flurtamone, diflufenican + flufenacet + flurtamone (*moderately susceptible*), diflufenican + flurtamone, diflufenican + iodosulfuron-methyl-sodium + mesosulfuron-methyl, diflufenican + pendimethalin, ethofumesate, fenoxaprop-P-ethyl, florasulam + pyroxsulam, flufenacet, flufenacet + pendimethalin, flufenacet + picolinafen, flumioxazin, glyphosate, iodosulfuron-methyl-sodium + mesosulfuron-methyl, iodosulfuron-methyl-sodium + mesosulfuron-methyl (*off-label*), mesosulfuron-methyl + propoxycarbazone-sodium, pendimethalin, pendimethalin (*off-label*), pendimethalin + picolinafen, pendimethalin + picolinafen (*off-label*), pinoxaden, propoxycarbazone-sodium, prosulfocarb, prosulfocarb (*off-label*), sulfosulfuron, sulfosulfuron (*moderate control of barren brome*), sulfosulfuron (*moderate control only*), tri-allate
	Weeds, miscellaneous	chlorotoluron + diflufenican + pendimethalin, clopyralid + florasulam, diflufenican, diflufenican + metsulfuron-methyl, florasulam, glyphosate, metsulfuron-methyl, metsulfuron-methyl + thifensulfuron-methyl

Edible fungi - Mushrooms

Diseases	Alternaria	Bacillus amyloliquefaciens D747
	Bacterial blotch	sodium hypochlorite (commodity substance)
	Cobweb	metrafenone
	Fusarium	Bacillus amyloliquefaciens D747
	Pythium	Bacillus amyloliquefaciens D747
	Rhizoctonia	Bacillus amyloliquefaciens D747
Pests	Flies	diflubenzuron (*off-label - other than mushrooms*), Metarhizium anisopliae (*off-label*)

Fruiting vegetables - Aubergines

Diseases	Alternaria	Bacillus amyloliquefaciens D747
	Botrytis	azoxystrobin (*off-label*)

	Foot rot	Trichoderma asperellum (Strain T34)
	Fusarium	Bacillus amyloliquefaciens D747
	Phytophthora	azoxystrobin (*off-label*)
	Powdery mildew	azoxystrobin (*off-label*), Bacillus amyloliquefaciens D747
	Pythium	Bacillus amyloliquefaciens D747, Bacillus subtilis (*off-label*)
	Rhizoctonia	Bacillus amyloliquefaciens D747
Pests	Caterpillars	methoxyfenozide (*off-label*)
	Pests, miscellaneous	methoxyfenozide (*off-label*)

Fruiting vegetables - Cucurbits

Diseases	Alternaria	Bacillus amyloliquefaciens D747
	Botrytis	azoxystrobin (*off-label*), cyprodinil + fludioxonil (*off-label*), mepanipyrim (*off-label*)
	Cladosporium	boscalid + pyraclostrobin (*off-label*)
	Disease control	cyflufenamid, mancozeb
	Downy mildew	azoxystrobin (*off-label*)
	Fusarium	Bacillus amyloliquefaciens D747
	Leaf spot	boscalid + pyraclostrobin (*off-label*)
	Phytophthora	Bacillus subtilis (*off-label*)
	Powdery mildew	azoxystrobin (*off-label*), Bacillus amyloliquefaciens D747, boscalid + pyraclostrobin (*off-label*), bupirimate (*off-label*), bupirimate (*outdoor only*), cyflufenamid, cyflufenamid (*off-label*), penconazole (*off-label*), potassium bicarbonate (commodity substance) (*off-label*), proquinazid (*off-label*)
	Pythium	Bacillus amyloliquefaciens D747, Bacillus subtilis (*off-label*)
	Rhizoctonia	Bacillus amyloliquefaciens D747
Pests	Aphids	flonicamid (*off-label*)
	Caterpillars	Bacillus thuringiensis
	Leaf miners	deltamethrin (*off-label*)
	Pests, miscellaneous	spinosad (*off-label*), thiacloprid (*off-label*)
	Thrips	deltamethrin (*off-label*), spinosad (*off-label*)
	Whiteflies	flonicamid (*off-label*)
Weeds	Broad-leaved weeds	clomazone (*off-label*), propyzamide (*off-label*)
	Grass weeds	propyzamide (*off-label*)

Fruiting vegetables - Peppers

Diseases	Alternaria	Bacillus amyloliquefaciens D747
	Foot rot	Trichoderma asperellum (Strain T34)
	Fusarium	Bacillus amyloliquefaciens D747
	Powdery mildew	azoxystrobin (*off-label*), Bacillus amyloliquefaciens D747
	Pythium	Bacillus amyloliquefaciens D747, Bacillus subtilis (*off-label*)
	Rhizoctonia	Bacillus amyloliquefaciens D747
Pests	Caterpillars	methoxyfenozide (*off-label*)
	Pests, miscellaneous	methoxyfenozide (*off-label*)

Fruiting vegetables - Tomatoes

Diseases	Alternaria	Bacillus amyloliquefaciens D747
	Botrytis	azoxystrobin (*off-label*), fenhexamid (*off-label*)
	Foot rot	Trichoderma asperellum (Strain T34)
	Fusarium	Bacillus amyloliquefaciens D747
	Phytophthora	azoxystrobin (*off-label*), mancozeb
	Powdery mildew	azoxystrobin (*off-label*), Bacillus amyloliquefaciens D747
	Pythium	Bacillus amyloliquefaciens D747, Bacillus subtilis (*off-label*)
	Rhizoctonia	Bacillus amyloliquefaciens D747
Pests	Aphids	pyrethrins
	Beetles	pyrethrins
	Caterpillars	methoxyfenozide (*off-label*), pyrethrins
	Pests, miscellaneous	methoxyfenozide (*off-label*)
	Whiteflies	pyrethrins

Herb crops - Herbs

Crop control	Desiccation	diquat (*off-label*)
Diseases	Alternaria	Bacillus amyloliquefaciens D747, boscalid + dimoxystrobin (*off-label*), cyprodinil + fludioxonil (*off-label*)
	Botrytis	Bacillus subtilis (*off-label*), cyprodinil + fludioxonil (*off-label*), fenhexamid (*off-label*), Gliocladium catenulatum (*off-label*)
	Damping off	Bacillus subtilis (*off-label*), fosetyl-aluminium + propamocarb hydrochloride (*off-label*), thiram (*off-label*)
	Disease control	boscalid (*off-label*), dimethomorph + mancozeb (*off-label*), mandipropamid
	Downy mildew	dimethomorph (*off-label*), dimethomorph + mancozeb (*off-label*), fenamidone + fosetyl-aluminium (*off-label*), fluopicolide + propamocarb hydrochloride (*off-label*), fosetyl-aluminium + propamocarb hydrochloride (*off-label*), mancozeb + metalaxyl-M (*off-label*), mandipropamid, metalaxyl-M (*off-label*)
	Fusarium	Bacillus amyloliquefaciens D747, Gliocladium catenulatum (*off-label*)
	Phytophthora	Gliocladium catenulatum (*off-label*)
	Powdery mildew	azoxystrobin + difenoconazole, Bacillus amyloliquefaciens D747, prothioconazole (*off-label*), tebuconazole (*off-label*)
	Pythium	Bacillus amyloliquefaciens D747, metalaxyl-M (*off-label*)
	Rhizoctonia	Bacillus amyloliquefaciens D747, cyprodinil + fludioxonil (*off-label*), Gliocladium catenulatum (*off-label*)
	Rust	tebuconazole (*off-label*)
	Sclerotinia	boscalid + dimoxystrobin (*off-label*), cyprodinil + fludioxonil (*off-label*), prothioconazole (*off-label*)
	Stem canker	cyprodinil + fludioxonil (*off-label*), prothioconazole (*off-label*)
	White blister	boscalid + pyraclostrobin (*off-label*)

Pests	Aphids	acetamiprid (*off-label*), alpha-cypermethrin (*off-label*), deltamethrin (*off-label*), Metarhizium anisopliae (*off-label*), pirimicarb (*off-label*), spirotetramat (*off-label*)
	Beetles	alpha-cypermethrin (*off-label*), deltamethrin (*off-label*), lambda-cyhalothrin (*off-label*)
	Capsid bugs	deltamethrin (*off-label*)
	Caterpillars	alpha-cypermethrin (*off-label*), Bacillus thuringiensis, Bacillus thuringiensis (*off-label*), deltamethrin (*off-label*), diflubenzuron (*off-label*), spinosad (*off-label*)
	Cutworms	deltamethrin (*off-label*), lambda-cyhalothrin (*off-label*)
	Flies	lambda-cyhalothrin (*off-label*), Metarhizium anisopliae (*off-label*), spinosad (*off-label*), tefluthrin (*off-label - seed treatment*)
	Leaf miners	spinosad (*off-label*)
	Midges	Metarhizium anisopliae (*off-label*)
	Pests, miscellaneous	deltamethrin (*off-label*), lambda-cyhalothrin (*off-label*)
	Thrips	Metarhizium anisopliae (*off-label*), spinosad (*off-label*)
	Whiteflies	spirotetramat (*off-label*)
Weeds	Broad-leaved weeds	amidosulfuron (*off-label*), bentazone (*off-label*), bromoxynil (*off-label*), carbetamide (*off-label*), chloridazon (*off-label*), chlorpropham (*off-label*), clomazone (*off-label*), clopyralid (*off-label*), clopyralid + picloram (*off-label*), dimethenamid-p + metazachlor (*off-label*), dimethenamid-p + metazachlor + quinmerac (*off-label*), dimethenamid-p + pendimethalin (*off-label*), metazachlor + quinmerac (*off-label*), napropamide (*off-label*), pendimethalin, pendimethalin (*off-label*), phenmedipham (*off-label*), propyzamide (*off-label*), prosulfocarb (*off-label*), pyridate (*off-label*), S-metolachlor (*off-label*)
	Crops as weeds	carbetamide (*off-label*), clethodim (*off-label*), cycloxydim (*off-label*), pendimethalin
	Grass weeds	carbetamide (*off-label*), chloridazon (*off-label*), chlorpropham (*off-label*), clethodim (*off-label*), cycloxydim (*off-label*), dimethenamid-p + metazachlor (*off-label*), dimethenamid-p + metazachlor + quinmerac (*off-label*), dimethenamid-p + pendimethalin (*off-label*), fluazifop-P-butyl (*off-label*), metazachlor + quinmerac (*off-label*), napropamide (*off-label*), pendimethalin, pendimethalin (*off-label*), propyzamide (*off-label*), prosulfocarb (*off-label*), S-metolachlor (*off-label*)
	Weeds, miscellaneous	diquat (*off-label*)

Leafy vegetables - Endives

Diseases	Alternaria	Bacillus amyloliquefaciens D747, cyprodinil + fludioxonil (*off-label*)
	Botrytis	cyprodinil + fludioxonil (*off-label*), fenhexamid (*off-label*), Gliocladium catenulatum (*off-label*)
	Damping off	thiram (*off-label*)
	Disease control	dimethomorph + mancozeb (*off-label*), mandipropamid

	Downy mildew	azoxystrobin, boscalid + pyraclostrobin (*off-label*), dimethomorph + mancozeb (*off-label*), fenamidone + fosetyl-aluminium (*off-label*), mandipropamid
	Fusarium	Bacillus amyloliquefaciens D747, Gliocladium catenulatum (*off-label*)
	Phytophthora	Gliocladium catenulatum (*off-label*)
	Powdery mildew	Bacillus amyloliquefaciens D747
	Pythium	Bacillus amyloliquefaciens D747
	Rhizoctonia	Bacillus amyloliquefaciens D747, cyprodinil + fludioxonil (*off-label*), Gliocladium catenulatum (*off-label*)
	Stem canker	cyprodinil + fludioxonil (*off-label*)
Pests	Aphids	acetamiprid (*off-label*), alpha-cypermethrin (*off-label*), deltamethrin (*off-label*), pymetrozine (*off-label*), spirotetramat (*off-label*)
	Beetles	alpha-cypermethrin (*off-label*), deltamethrin (*off-label*)
	Caterpillars	alpha-cypermethrin (*off-label*), Bacillus thuringiensis, Bacillus thuringiensis (*off-label*), deltamethrin (*off-label*), diflubenzuron (*off-label*), indoxacarb (*off-label*), spinosad (*off-label*)
	Pests, miscellaneous	deltamethrin (*off-label*), lambda-cyhalothrin (*off-label*)
	Thrips	spinosad (*off-label*)
	Whiteflies	spirotetramat (*off-label*)
Weeds	Broad-leaved weeds	chlorpropham, pendimethalin (*off-label*), propyzamide, propyzamide (*off-label*), S-metolachlor (*off-label*)
	Crops as weeds	cycloxydim (*off-label*)
	Grass weeds	chlorpropham, cycloxydim (*off-label*), diquat, propyzamide, propyzamide (*off-label*), S-metolachlor (*off-label*)
	Weeds, miscellaneous	diquat

Leafy vegetables - Lettuce

Diseases	Alternaria	Bacillus amyloliquefaciens D747, cyprodinil + fludioxonil (*off-label*), iprodione
	Botrytis	boscalid + pyraclostrobin, boscalid + pyraclostrobin (*off-label*), cyprodinil + fludioxonil (*off-label*), fenhexamid (*off-label*), fluopyram + trifloxystrobin (*off-label*), Gliocladium catenulatum (*off-label*), iprodione
	Bottom rot	boscalid + pyraclostrobin
	Damping off	Bacillus subtilis (*off-label*), thiram (*off-label*), thiram (*seed treatment*)
	Disease control	boscalid + pyraclostrobin (*off-label*), dimethomorph + mancozeb (*off-label*), mandipropamid
	Downy mildew	azoxystrobin, dimethomorph (*off-label*), dimethomorph + mancozeb (*off-label*), fenamidone + fosetyl-aluminium (*off-label*), fluopicolide + propamocarb hydrochloride (*off-label*), fosetyl-aluminium + propamocarb hydrochloride, mancozeb (*off-label*), mancozeb + metalaxyl-M (*off-label*), mandipropamid, potassium bicarbonate (commodity substance) (*off-label*)

	Fungus diseases	Bacillus subtilis (*off-label*)
	Fusarium	Bacillus amyloliquefaciens D747, Gliocladium catenulatum (*off-label*)
	Phytophthora	Gliocladium catenulatum (*off-label*)
	Powdery mildew	Bacillus amyloliquefaciens D747, fluopyram + trifloxystrobin (*off-label*)
	Pythium	Bacillus amyloliquefaciens D747, fosetyl-aluminium + propamocarb hydrochloride
	Rhizoctonia	Bacillus amyloliquefaciens D747, Bacillus subtilis (*off-label*), cyprodinil + fludioxonil (*off-label*), Gliocladium catenulatum (*off-label*)
	Root rot	fenamidone + fosetyl-aluminium (*off-label*)
	Sclerotinia	fluopyram + trifloxystrobin (*off-label*)
	Soft rot	boscalid + pyraclostrobin
	Stem canker	cyprodinil + fludioxonil (*off-label*)
Pests	Aphids	acetamiprid (*off-label*), alpha-cypermethrin (*off-label*), deltamethrin, deltamethrin (*off-label*), lambda-cyhalothrin, pymetrozine (*off-label*), pyrethrins, spirotetramat, spirotetramat (*off-label*)
	Beetles	alpha-cypermethrin (*off-label*), deltamethrin, deltamethrin (*off-label*)
	Caterpillars	alpha-cypermethrin (*off-label*), Bacillus thuringiensis, Bacillus thuringiensis (*off-label*), deltamethrin, deltamethrin (*off-label*), diflubenzuron (*off-label*), indoxacarb (*off-label*), pyrethrins, spinosad (*off-label*)
	Cutworms	deltamethrin, lambda-cyhalothrin
	Leafhoppers	deltamethrin
	Pests, miscellaneous	deltamethrin, deltamethrin (*off-label*), lambda-cyhalothrin (*off-label*)
	Thrips	deltamethrin, spinosad (*off-label*)
	Whiteflies	spirotetramat (*off-label*)
Plant growth regulation	Growth control	trinexapac-ethyl (*off-label*)
Weeds	Broad-leaved weeds	chlorpropham, dimethenamid-p + pendimethalin (*off-label*), napropamide (*off-label*), pendimethalin (*off-label*), propyzamide, propyzamide (*off-label*), S-metolachlor (*off-label*)
	Crops as weeds	cycloxydim (*off-label*)
	Grass weeds	chlorpropham, cycloxydim (*off-label*), dimethenamid-p + pendimethalin (*off-label*), napropamide (*off-label*), pendimethalin (*off-label*), propyzamide, propyzamide (*off-label*), S-metolachlor (*off-label*)

Leafy vegetables - Spinach

Diseases	Alternaria	Bacillus amyloliquefaciens D747, cyprodinil + fludioxonil (*off-label*)
	Botrytis	cyprodinil + fludioxonil (*off-label*), Gliocladium catenulatum (*off-label*)
	Damping off	Bacillus subtilis (*off-label*)
	Disease control	mandipropamid
	Downy mildew	boscalid + pyraclostrobin (*off-label*), dimethomorph (*off-label*), fenamidone + fosetyl-aluminium (*off-label*), mandipropamid, metalaxyl-M (*off-label*)

	Fusarium	Bacillus amyloliquefaciens D747, Gliocladium catenulatum (*off-label*)
	Phytophthora	Gliocladium catenulatum (*off-label*)
	Powdery mildew	Bacillus amyloliquefaciens D747
	Pythium	Bacillus amyloliquefaciens D747, metalaxyl-M (*off-label*)
	Rhizoctonia	Bacillus amyloliquefaciens D747, cyprodinil + fludioxonil (*off-label*), Gliocladium catenulatum (*off-label*)
	Stem canker	cyprodinil + fludioxonil (*off-label*)
	White tip	spirotetramat (*off-label*)
Pests	Aphids	acetamiprid (*off-label*), alpha-cypermethrin (*off-label*), pymetrozine (*off-label*), spirotetramat (*off-label*)
	Beetles	alpha-cypermethrin (*off-label*)
	Caterpillars	alpha-cypermethrin (*off-label*), Bacillus thuringiensis, Bacillus thuringiensis (*off-label*), diflubenzuron (*off-label*), spinosad (*off-label*)
	Flies	tefluthrin (*off-label - seed treatment*)
	Thrips	spinosad (*off-label*)
	Whiteflies	spirotetramat (*off-label*)
Weeds	Broad-leaved weeds	chlorpropham (*off-label*), clopyralid (*off-label*), phenmedipham (*off-label*)
	Crops as weeds	cycloxydim (*off-label*)
	Grass weeds	chlorpropham (*off-label*), cycloxydim (*off-label*)

Leafy vegetables - Watercress

Diseases	Alternaria	Bacillus amyloliquefaciens D747
	Damping off	metalaxyl-M (*off-label*)
	Downy mildew	metalaxyl-M (*off-label*), propamocarb hydrochloride (*off-label*)
	Fusarium	Bacillus amyloliquefaciens D747, Trichoderma asperellum (Strain T34) (*off-label*)
	Phytophthora	propamocarb hydrochloride (*off-label*)
	Powdery mildew	Bacillus amyloliquefaciens D747
	Pythium	Bacillus amyloliquefaciens D747, propamocarb hydrochloride (*off-label*), Trichoderma asperellum (Strain T34) (*off-label*)
	Rhizoctonia	Bacillus amyloliquefaciens D747
Pests	Aphids	alpha-cypermethrin (*off-label*), pyrethrins (*off-label*)
	Beetles	alpha-cypermethrin (*off-label*), pyrethrins (*off-label*)
	Caterpillars	alpha-cypermethrin (*off-label*), Bacillus thuringiensis (*off-label*)
	Cutworms	Bacillus thuringiensis (*off-label*)
Weeds	Crops as weeds	cycloxydim (*off-label*)
	Grass weeds	cycloxydim (*off-label*)

Legumes - Beans (Phaseolus)

Crop control	Desiccation	diquat (*off-label*)
Diseases	Alternaria	iprodione (*off-label*)
	Ascochyta	azoxystrobin (*useful control*)

	Botrytis	azoxystrobin (*some control*), boscalid + pyraclostrobin (*off-label*), cyprodinil + fludioxonil (*moderate control*), Gliocladium catenulatum (*off-label*), iprodione (*off-label*)
	Damping off	thiram (*seed treatment*)
	Downy mildew	azoxystrobin (*reduction*)
	Fusarium	Gliocladium catenulatum (*off-label*)
	Mycosphaerella	azoxystrobin (*some control*)
	Phytophthora	Gliocladium catenulatum (*off-label*)
	Powdery mildew	tebuconazole (*off-label*)
	Rhizoctonia	Gliocladium catenulatum (*off-label*)
	Rust	tebuconazole (*off-label*)
	Sclerotinia	cyprodinil + fludioxonil
Pests	Aphids	cypermethrin, pirimicarb (*off-label*)
	Caterpillars	Bacillus thuringiensis, Bacillus thuringiensis (*off-label*), lambda-cyhalothrin (*off-label*), spinosad (*off-label*)
	Cutworms	Bacillus thuringiensis (*off-label*)
	Pests, miscellaneous	cypermethrin, lambda-cyhalothrin (*off-label*)
Weeds	Broad-leaved weeds	bentazone, clomazone (*off-label*), pendimethalin, pendimethalin (*off-label*)
	Crops as weeds	cycloxydim, cycloxydim (*off-label*)
	Grass weeds	clethodim (*off-label*), cycloxydim, cycloxydim (*off-label*), fluazifop-P-butyl (*off-label*), pendimethalin (*off-label*), propaquizafop, S-metolachlor (*off-label*)

Legumes - Beans (Vicia)

Crop control	Desiccation	diquat, glyphosate
Diseases	Botrytis	boscalid + pyraclostrobin (*off-label*), cyprodinil + fludioxonil (*moderate control*), Gliocladium catenulatum (*off-label*)
	Chocolate spot	azoxystrobin + tebuconazole, azoxystrobin + tebuconazole (*moderate control*), boscalid + pyraclostrobin (*moderate control*), chlorothalonil + cyproconazole, tebuconazole
	Damping off	thiram (*seed treatment*)
	Disease control	azoxystrobin + tebuconazole, azoxystrobin + tebuconazole (*moderate control*), chlorothalonil, metconazole (*off-label*), tebuconazole
	Downy mildew	chlorothalonil + metalaxyl-M, metalaxyl-M (*off-label*)
	Fusarium	Gliocladium catenulatum (*off-label*)
	Phytophthora	Gliocladium catenulatum (*off-label*)
	Powdery mildew	tebuconazole (*off-label*)
	Rhizoctonia	Gliocladium catenulatum (*off-label*)
	Rust	azoxystrobin, boscalid + pyraclostrobin, chlorothalonil + cyproconazole, metconazole, tebuconazole, tebuconazole (*off-label*)
	Sclerotinia	cyprodinil + fludioxonil
Pests	Aphids	esfenvalerate, lambda-cyhalothrin, pirimicarb, zeta-cypermethrin (*off-label*)
	Beetles	lambda-cyhalothrin, lambda-cyhalothrin (*off-label*), thiacloprid

	Birds/mammals	aluminium ammonium sulphate
	Caterpillars	Bacillus thuringiensis (*off-label*), lambda-cyhalothrin
	Cutworms	Bacillus thuringiensis (*off-label*)
	Pests, miscellaneous	lambda-cyhalothrin (*off-label*), zeta-cypermethrin (*off-label*)
	Thrips	lambda-cyhalothrin (*off-label*)
	Weevils	alpha-cypermethrin, cypermethrin, deltamethrin, esfenvalerate, lambda-cyhalothrin, zeta-cypermethrin
Weeds	Broad-leaved weeds	bentazone, carbetamide, clomazone, clomazone (*off-label*), clomazone + pendimethalin, glyphosate, imazamox + pendimethalin, imazamox + pendimethalin (*off-label*), pendimethalin (*off-label*), propyzamide, prosulfocarb (*off-label*), S-metolachlor (*off-label*)
	Crops as weeds	carbetamide, cycloxydim, cycloxydim (*off-label*), fluazifop-P-butyl, glyphosate, pendimethalin (*off-label*), propyzamide, quizalofop-P-ethyl, quizalofop-P-tefuryl
	Grass weeds	carbetamide, clethodim (*off-label*), cycloxydim, cycloxydim (*off-label*), diquat, fluazifop-P-butyl, fluazifop-P-butyl (*off-label*), glyphosate, pendimethalin (*off-label*), propaquizafop, propyzamide, prosulfocarb (*off-label*), quizalofop-P-ethyl, quizalofop-P-tefuryl, S-metolachlor (*off-label*)
	Weeds, miscellaneous	diquat, glyphosate, pendimethalin (*off-label*)

Legumes - Forage legumes, general

Weeds	Broad-leaved weeds	pendimethalin (*off-label*)
	Grass weeds	pendimethalin (*off-label*)

Legumes - Lupins

Crop control	Desiccation	diquat (*off-label*), glyphosate (*off-label*)
Diseases	Anthracnose	thiram (*off-label*)
	Ascochyta	metconazole (*qualified minor use*)
	Botrytis	cyprodinil + fludioxonil (*off-label*), metconazole (*qualified minor use*)
	Damping off	thiram (*off-label*)
	Powdery mildew	azoxystrobin (*off-label*)
	Rust	azoxystrobin, azoxystrobin (*off-label*), metconazole (*qualified minor use*)
Pests	Aphids	cypermethrin, lambda-cyhalothrin (*off-label*), zeta-cypermethrin (*off-label*)
	Pests, miscellaneous	cypermethrin, lambda-cyhalothrin (*off-label*), zeta-cypermethrin (*off-label*)
Weeds	Broad-leaved weeds	clomazone (*off-label*), pendimethalin (*off-label*), pyridate (*off-label*)
	Grass weeds	fluazifop-P-butyl (*off-label*), pendimethalin (*off-label*), propaquizafop (*off-label*)
	Weeds, miscellaneous	glyphosate (*off-label*)

Legumes - Peas

Crop control	Desiccation	diquat, glyphosate
Diseases	Alternaria	iprodione (*off-label*)
	Ascochyta	azoxystrobin, azoxystrobin (*useful control*), boscalid + pyraclostrobin (*moderate control only*), cyprodinil + fludioxonil, metconazole (*reduction*)
	Botrytis	azoxystrobin (*some control*), azoxystrobin + chlorothalonil, chlorothalonil + cyproconazole, cyprodinil + fludioxonil (*moderate control*), Gliocladium catenulatum (*off-label*), iprodione (*off-label*), metconazole (*reduction*)
	Damping off	thiram (*seed treatment*)
	Downy mildew	azoxystrobin (*reduction*), azoxystrobin + chlorothalonil, cymoxanil + fludioxonil + metalaxyl-M
	Fusarium	Gliocladium catenulatum (*off-label*)
	Mycosphaerella	azoxystrobin (*some control*), cyprodinil + fludioxonil, metconazole (*reduction*)
	Phytophthora	Gliocladium catenulatum (*off-label*)
	Powdery mildew	azoxystrobin + chlorothalonil, sulphur (*off-label*)
	Rhizoctonia	Gliocladium catenulatum (*off-label*)
	Rust	metconazole
	Sclerotinia	cyprodinil + fludioxonil
Pests	Aphids	alpha-cypermethrin (*reduction*), cypermethrin, esfenvalerate, lambda-cyhalothrin, pirimicarb, pirimicarb (*off-label*), thiacloprid, zeta-cypermethrin
	Beetles	lambda-cyhalothrin
	Birds/mammals	aluminium ammonium sulphate
	Caterpillars	alpha-cypermethrin, Bacillus thuringiensis, cypermethrin, deltamethrin, lambda-cyhalothrin, zeta-cypermethrin
	Midges	deltamethrin, lambda-cyhalothrin, thiacloprid
	Pests, miscellaneous	aluminium phosphide, cypermethrin
	Weevils	alpha-cypermethrin, cypermethrin, deltamethrin, esfenvalerate, lambda-cyhalothrin, zeta-cypermethrin
Weeds	Broad-leaved weeds	bentazone, clomazone, clomazone + pendimethalin, flumioxazin (*off-label*), glyphosate, imazamox + pendimethalin, MCPB, pendimethalin, pendimethalin (*off-label*)
	Crops as weeds	cycloxydim, cycloxydim (*off-label*), fluazifop-P-butyl, flumioxazin (*off-label*), glyphosate, pendimethalin, quizalofop-P-ethyl, quizalofop-P-tefuryl
	Grass weeds	clethodim (*off-label*), cycloxydim, cycloxydim (*off-label*), diquat (*dry harvested*), fluazifop-P-butyl, flumioxazin (*off-label*), glyphosate, pendimethalin, pendimethalin (*off-label*), propaquizafop, quizalofop-P-ethyl, quizalofop-P-tefuryl, S-metolachlor (*off-label*)
	Weeds, miscellaneous	diquat (*dry harvested only*), glyphosate, glyphosate (*off-label*)

Miscellaneous arable - Industrial crops

Crop control	Desiccation	glyphosate (*off-label*)
Pests	Aphids	lambda-cyhalothrin (*off-label*)
	Pests, miscellaneous	lambda-cyhalothrin (*off-label*)

Weeds	Broad-leaved weeds	2,4-D (*off-label*), bromoxynil (*off-label*), clopyralid (*off-label*), fluroxypyr (*off-label*), MCPA (*off-label*), mecoprop-P (*off-label*), metsulfuron-methyl (*off-label*), metsulfuron-methyl + thifensulfuron-methyl (*off-label*), pendimethalin (*off-label*), prosulfocarb (*off-label*), tribenuron-methyl (*off-label*)
	Crops as weeds	fluroxypyr (*off-label*)
	Grass weeds	pendimethalin (*off-label*), propoxycarbazone-sodium (*off-label*), prosulfocarb (*off-label*), tri-allate (*off-label*)
	Weeds, miscellaneous	bromoxynil (*off-label*), glyphosate (*off-label*)

Miscellaneous arable - Miscellaneous arable crops

Crop control	Desiccation	glyphosate (*off-label*)
	Miscellaneous non-selective situations	benzoic acid
Diseases	Alternaria	Bacillus amyloliquefaciens D747
	Botrytis	azoxystrobin (*off-label*), Bacillus subtilis (*off-label*)
	Didymella stem rot	Gliocladium catenulatum (*moderate control*)
	Disease control	boscalid (*off-label*), boscalid + pyraclostrobin (*off-label*), chlorothalonil (*off-label*), potassium bicarbonate (commodity substance)
	Downy mildew	azoxystrobin (*off-label*), boscalid + pyraclostrobin (*off-label*)
	Fusarium	Bacillus amyloliquefaciens D747, Gliocladium catenulatum (*moderate control*)
	Phytophthora	Gliocladium catenulatum (*moderate control*)
	Powdery mildew	azoxystrobin (*off-label*), Bacillus amyloliquefaciens D747, cyflufenamid (*off-label*)
	Pythium	Bacillus amyloliquefaciens D747, Gliocladium catenulatum (*moderate control*)
	Rhizoctonia	Bacillus amyloliquefaciens D747, Gliocladium catenulatum (*moderate control*)
	Rust	azoxystrobin (*off-label*)
	Sclerotinia	Coniothyrium minitans
Pests	Aphids	flonicamid (*off-label*), lambda-cyhalothrin (*off-label*), maltodextrin, pyrethrins, thiacloprid (*off-label*)
	Beetles	deltamethrin (*off-label*)
	Birds/mammals	aluminium ammonium sulphate
	Caterpillars	Bacillus thuringiensis, Bacillus thuringiensis (*off-label*), indoxacarb (*off-label*), pyrethrins
	Pests, miscellaneous	indoxacarb (*off-label*), lambda-cyhalothrin (*off-label*)
	Slugs/snails	metaldehyde
	Spider mites	maltodextrin, pyrethrins
	Thrips	pyrethrins
	Whiteflies	maltodextrin
Plant growth regulation	Growth control	trinexapac-ethyl (*off-label*)
Weeds	Broad-leaved weeds	2,4-D (*off-label*), amidosulfuron (*off-label*), bentazone (*off-label*), bromoxynil (*off-label*), carfentrazone-ethyl (*before planting*), clomazone (*off-label*), clopyralid (*off-label*), clopyralid + picloram (*off-label*), clopyralid + triclopyr (*off-label*), dicamba + MCPA + mecoprop-P

		(*off-label*), diflufenican (*off-label*), dimethenamid-p + metazachlor (*off-label*), diquat, diquat (*around*), florasulam + fluroxypyr (*off-label*), flufenacet + pendimethalin (*off-label*), fluroxypyr (*off-label*), MCPA (*off-label*), MCPB (*off-label*), mecoprop-P (*off-label*), mesotrione (*off-label*), metsulfuron-methyl (*off-label*), napropamide (*off-label*), nicosulfuron (*off-label*), pendimethalin (*off-label*), pendimethalin + picolinafen (*off-label*), propyzamide (*off-label*), prosulfocarb (*off-label*), prosulfuron (*off-label*), pyridate (*off-label*), thifensulfuron-methyl + tribenuron-methyl (*off-label*)
	Crops as weeds	bromoxynil (*off-label*), carfentrazone-ethyl (*before planting*), diflufenican (*off-label*), fluroxypyr (*off-label*), nicosulfuron (*off-label*), pendimethalin (*off-label*)
	Grass weeds	diflufenican (*off-label*), dimethenamid-p + metazachlor (*off-label*), diquat, diquat (*around only*), flufenacet + pendimethalin (*off-label*), mesotrione (*off-label*), napropamide (*off-label*), nicosulfuron (*off-label*), pendimethalin (*off-label*), pendimethalin + picolinafen (*off-label*), pinoxaden (*off-label*), propaquizafop (*off-label*), propyzamide (*off-label*), prosulfocarb (*off-label*), tri-allate (*off-label*)
	Weeds, miscellaneous	2,4-D + glyphosate, bromoxynil (*off-label*), carfentrazone-ethyl, diflufenican (*off-label*), diquat (*around only*), glyphosate, glyphosate (*before planting*), glyphosate (*off-label*), glyphosate (*pre-sowing/planting*), napropamide (*off-label*)

Miscellaneous arable - Miscellaneous arable situations

Weeds	Broad-leaved weeds	citronella oil
	Crops as weeds	fluazifop-P-butyl, glyphosate
	Grass weeds	fluazifop-P-butyl, glyphosate
	Weeds, miscellaneous	cycloxydim, fluazifop-P-butyl, glyphosate, metsulfuron-methyl, thifensulfuron-methyl

Miscellaneous field vegetables - All vegetables

Weeds	Broad-leaved weeds	clomazone (*off-label*)

Oilseed crops - Linseed/flax

Crop control	Desiccation	diquat, glyphosate
Diseases	Alternaria	difenoconazole (*off-label*)
	Botrytis	tebuconazole, tebuconazole (*reduction*)
	Disease control	boscalid (*off-label*), metconazole (*off-label*), prothioconazole (*off-label*), tebuconazole
	Foliar diseases	difenoconazole + paclobutrazol (*off-label*), prothioconazole (*off-label*)
	Powdery mildew	difenoconazole (*off-label*), difenoconazole + paclobutrazol (*off-label*), tebuconazole
	Rust	tebuconazole
	Sclerotinia	difenoconazole (*off-label*)
	Septoria	difenoconazole + paclobutrazol (*off-label*), prothioconazole (*off-label*)
Pests	Aphids	cypermethrin, tau-fluvalinate (*off-label*)
	Beetles	zeta-cypermethrin
	Pests, miscellaneous	cypermethrin, lambda-cyhalothrin (*off-label*)

Weeds	Broad-leaved weeds	amidosulfuron, bentazone, bifenox (*off-label*), bromoxynil, carbetamide (*off-label*), clopyralid, glyphosate, mesotrione (*off-label*), metazachlor (*off-label*), metsulfuron-methyl, napropamide (*off-label*), prosulfocarb (*off-label*)
	Crops as weeds	carbetamide (*off-label*), clethodim (*off-label*), cycloxydim, fluazifop-P-butyl, glyphosate, mesotrione (*off-label*), metsulfuron-methyl, quizalofop-P-ethyl, quizalofop-P-tefuryl
	Grass weeds	carbetamide (*off-label*), clethodim (*off-label*), cycloxydim, diquat, fluazifop-P-butyl, fluazifop-P-butyl (*off-label*), glyphosate, mesotrione (*off-label*), metazachlor (*off-label*), napropamide (*off-label*), propaquizafop, prosulfocarb (*off-label*), quizalofop-P-ethyl, quizalofop-P-tefuryl, tri-allate (*off-label*)
	Weeds, miscellaneous	bromoxynil, diquat, glyphosate, metsulfuron-methyl, napropamide (*off-label*)

Oilseed crops - Miscellaneous oilseeds

Crop control	Desiccation	diquat (*off-label*), glyphosate (*off-label*)
Diseases	Alternaria	azoxystrobin (*off-label*), boscalid + pyraclostrobin (*off-label*), difenoconazole (*off-label*)
	Botrytis	boscalid (*off-label*), boscalid + pyraclostrobin (*off-label*)
	Disease control	boscalid (*off-label*), chlorothalonil + metalaxyl-M (*off-label*), dimethomorph + mancozeb (*off-label*), propiconazole (*off-label*), tebuconazole (*off-label*)
	Downy mildew	chlorothalonil (*off-label*), dimethomorph + mancozeb (*off-label*), mancozeb + metalaxyl-M (*off-label*)
	Fire	boscalid + pyraclostrobin (*off-label*)
	Powdery mildew	azoxystrobin (*off-label*), difenoconazole (*off-label*), sulphur (*off-label*), tebuconazole (*off-label*)
	Rust	tebuconazole (*off-label*)
	Sclerotinia	azoxystrobin (*off-label*), boscalid (*off-label*), difenoconazole (*off-label*)
Pests	Aphids	deltamethrin (*off-label*), lambda-cyhalothrin (*off-label*), tau-fluvalinate (*off-label*)
	Beetles	lambda-cyhalothrin (*off-label*)
	Caterpillars	deltamethrin (*off-label*)
	Pests, miscellaneous	deltamethrin (*off-label*), lambda-cyhalothrin (*off-label*)
Weeds	Broad-leaved weeds	bifenox (*off-label*), clomazone (*off-label*), clopyralid (*off-label*), clopyralid + picloram (*off-label*), fluroxypyr (*off-label*), mesotrione (*off-label*), metazachlor (*off-label*), napropamide (*off-label*), pendimethalin (*off-label*), prosulfocarb (*off-label*)
	Crops as weeds	clethodim (*off-label*), fluroxypyr (*off-label*), propaquizafop (*off-label*)
	Grass weeds	clethodim (*off-label*), fluazifop-P-butyl (*off-label*), metazachlor (*off-label*), napropamide (*off-label*), propaquizafop (*off-label*), prosulfocarb (*off-label*)
	Weeds, miscellaneous	diquat (*off-label*), glyphosate (*off-label*), napropamide (*off-label*)

Oilseed crops - Oilseed rape

Crop control	Desiccation	diquat, glyphosate
Diseases	Alternaria	azoxystrobin, azoxystrobin + cyproconazole, boscalid, boscalid + dimoxystrobin (*moderate control*), difenoconazole, iprodione, iprodione + thiophanate-methyl, metconazole, tebuconazole
	Black scurf and stem canker	bixafen + prothioconazole + tebuconazole, boscalid + dimoxystrobin (*moderate control*), difenoconazole, fluopyram + prothioconazole, iprodione + thiophanate-methyl, prothioconazole, prothioconazole + tebuconazole, tebuconazole
	Botrytis	iprodione, iprodione + thiophanate-methyl
	Damping off	thiram (*seed treatment*)
	Disease control	boscalid + dimoxystrobin , difenoconazole + paclobutrazol, fluopyram + prothioconazole
	Ear blight	thiophanate-methyl (*reduction*)
	Fusarium	thiophanate-methyl
	Light leaf spot	bixafen + prothioconazole + tebuconazole, boscalid + dimoxystrobin (*reduction*), difenoconazole, difenoconazole + paclobutrazol, fluopyram + prothioconazole, iprodione + thiophanate-methyl, metconazole, metconazole (*reduction*), prochloraz + propiconazole, propiconazole (*reduction*), prothioconazole, prothioconazole + tebuconazole, prothioconazole + tebuconazole (*moderate control only*), tebuconazole
	Phoma leaf spot	prothioconazole + tebuconazole, tebuconazole
	Powdery mildew	fluopyram + prothioconazole
	Ring spot	tebuconazole, tebuconazole (*reduction*)
	Sclerotinia	azoxystrobin, azoxystrobin + cyproconazole, azoxystrobin + isopyrazam, azoxystrobin + tebuconazole (*moderate control*), bixafen + prothioconazole + tebuconazole, boscalid, boscalid + dimoxystrobin , boscalid + metconazole, fluopyram + prothioconazole, iprodione (*moderate control*), iprodione + thiophanate-methyl, prochloraz + tebuconazole, prothioconazole, prothioconazole + tebuconazole, tebuconazole, tebuconazole (*reduction*), thiophanate-methyl
	Stem canker	bixafen + prothioconazole + tebuconazole, boscalid + dimoxystrobin (*moderate control*), difenoconazole + paclobutrazol, fluopyram + prothioconazole, metconazole (*reduction*), prochloraz + propiconazole, prothioconazole, prothioconazole + tebuconazole, tebuconazole, tebuconazole (*good reduction*)
Pests	Aphids	acetamiprid, deltamethrin, lambda-cyhalothrin, tau-fluvalinate
	Beetles	alpha-cypermethrin, beta-cyfluthrin, cypermethrin, deltamethrin, etofenprox (*moderate control*), indoxacarb, lambda-cyhalothrin, pymetrozine, tau-fluvalinate, thiacloprid, zeta-cypermethrin
	Birds/mammals	aluminium ammonium sulphate
	Caterpillars	lambda-cyhalothrin
	Midges	alpha-cypermethrin, beta-cyfluthrin, cypermethrin, lambda-cyhalothrin, zeta-cypermethrin

	Weevils	alpha-cypermethrin, beta-cyfluthrin, cypermethrin, deltamethrin, lambda-cyhalothrin, zeta-cypermethrin
Plant growth regulation	Growth control	mepiquat chloride + metconazole, metconazole, tebuconazole
	Quality/yield control	sulphur
Weeds	Broad-leaved weeds	aminopyralid + metazachlor + picloram, aminopyralid + propyzamide, bifenox (*off-label*), carbetamide, clomazone, clomazone + dimethenamid-p + metazachlor, clomazone + napropamide, clopyralid, clopyralid + picloram, clopyralid + picloram (*off-label*), dimethachlor, dimethenamid-p + metazachlor, dimethenamid-p + metazachlor + quinmerac, dimethenamid-p + quinmerac, glyphosate, imazamox + metazachlor, imazamox + quinmerac, imazamox + quinmerac (*cut-leaved*), metazachlor, metazachlor + quinmerac, napropamide, phenmedipham, phenmedipham (*moderately susceptible*), propyzamide, pyridate (*off-label*)
	Crops as weeds	carbetamide, clethodim, cycloxydim, fluazifop-P-butyl, glyphosate, imazamox + quinmerac, propyzamide, quizalofop-P-ethyl, quizalofop-P-tefuryl
	Grass weeds	aminopyralid + propyzamide, carbetamide, clethodim, clomazone + dimethenamid-p + metazachlor, clomazone + napropamide, cycloxydim, dimethachlor, dimethenamid-p + metazachlor + quinmerac, diquat, fluazifop-P-butyl, glyphosate, imazamox + metazachlor, metazachlor, metazachlor + quinmerac, napropamide, propaquizafop, propyzamide, quizalofop-P-ethyl, quizalofop-P-tefuryl
	Weeds, miscellaneous	diquat, glyphosate

Oilseed crops - Soya

Diseases	Botrytis	Gliocladium catenulatum (*off-label*)
	Damping off	thiram (*seed treatment - qualified minor use*)
	Fusarium	Gliocladium catenulatum (*off-label*)
	Phytophthora	Gliocladium catenulatum (*off-label*)
	Rhizoctonia	Gliocladium catenulatum (*off-label*)
Weeds	Broad-leaved weeds	bentazone (*off-label*), clomazone (*off-label*), flufenacet + metribuzin (*off-label*), imazamox + pendimethalin (*off-label*), thifensulfuron-methyl (*off-label*)
	Crops as weeds	cycloxydim (*off-label*)
	Grass weeds	cycloxydim (*off-label*), flufenacet + metribuzin (*off-label*)
	Weeds, miscellaneous	imazamox + pendimethalin (*off-label*)

Oilseed crops - Sunflowers

Crop control	Desiccation	diquat (*off-label*)
Weeds	Broad-leaved weeds	pendimethalin
	Crops as weeds	pendimethalin
	Grass weeds	pendimethalin

Root and tuber crops - Beet crops

Diseases	Alternaria	Bacillus amyloliquefaciens D747, cyprodinil + fludioxonil (*off-label*), isopyrazam (*off-label*)
	Black leg	hymexazol (*seed treatment*)
	Botrytis	cyprodinil + fludioxonil (*off-label*), Gliocladium catenulatum (*off-label*), iprodione (*off-label*)
	Cercospora leaf spot	azoxystrobin + cyproconazole, cyproconazole + trifloxystrobin, cyproconazole + trifloxystrobin (*off-label*), difenoconazole + propiconazole, epoxiconazole + pyraclostrobin, isopyrazam (*off-label*)
	Damping off	Bacillus subtilis (*off-label*)
	Disease control	boscalid + pyraclostrobin (*off-label*), mandipropamid
	Downy mildew	boscalid + pyraclostrobin (*off-label*), dimethomorph (*off-label*), fenamidone + fosetyl-aluminium (*off-label*), fluopicolide + propamocarb hydrochloride (*off-label*), fosetyl-aluminium + propamocarb hydrochloride (*off-label*), mandipropamid, metalaxyl-M (*off-label*)
	Fusarium	Bacillus amyloliquefaciens D747, Gliocladium catenulatum (*off-label*)
	Leaf spot	epoxiconazole (*off-label*)
	Phytophthora	Gliocladium catenulatum (*off-label*)
	Powdery mildew	azoxystrobin + cyproconazole, Bacillus amyloliquefaciens D747, cyproconazole + trifloxystrobin, cyproconazole + trifloxystrobin (*off-label*), difenoconazole + propiconazole, epoxiconazole, epoxiconazole + pyraclostrobin, epoxiconazole + pyraclostrobin (*off-label*), fenpropimorph (*off-label*), isopyrazam (*off-label*), sulphur, sulphur (*off-label*)
	Pythium	Bacillus amyloliquefaciens D747, metalaxyl-M, metalaxyl-M (*off-label*)
	Ramularia leaf spots	azoxystrobin + cyproconazole, cyproconazole + trifloxystrobin, cyproconazole + trifloxystrobin (*off-label*), difenoconazole + propiconazole (*moderate control*), epoxiconazole + pyraclostrobin, isopyrazam (*off-label*), propiconazole (*reduction*)
	Rhizoctonia	Bacillus amyloliquefaciens D747, cyprodinil + fludioxonil (*off-label*), Gliocladium catenulatum (*off-label*)
	Root malformation disorder	azoxystrobin (*off-label*), metalaxyl-M (*off-label*)
	Rust	azoxystrobin + cyproconazole, cyproconazole + trifloxystrobin, cyproconazole + trifloxystrobin (*off-label*), difenoconazole + propiconazole, epoxiconazole, epoxiconazole + pyraclostrobin, epoxiconazole + pyraclostrobin (*off-label*), fenpropimorph (*off-label*), propiconazole
	Sclerotinia	cyprodinil + fludioxonil (*off-label*)
	Seed-borne diseases	thiram (*seed soak*)
	Stem canker	cyprodinil + fludioxonil (*off-label*)
Pests	Aphids	acetamiprid (*off-label*), alpha-cypermethrin (*off-label*), beta-cyfluthrin, beta-cyfluthrin + clothianidin (*seed treatment*), lambda-cyhalothrin, lambda-cyhalothrin (*off-label*), pirimicarb (*off-label*), pymetrozine (*off-*

		label), spirotetramat (*off-label*), thiacloprid (*off-label*), thiamethoxam, zeta-cypermethrin (*off-label*)
	Beetles	alpha-cypermethrin (*off-label*), deltamethrin, lambda-cyhalothrin, lambda-cyhalothrin (*off-label*), tefluthrin (*seed treatment*), thiamethoxam (*seed treatment*)
	Birds/mammals	aluminium ammonium sulphate
	Caterpillars	alpha-cypermethrin (*off-label*), Bacillus thuringiensis, Bacillus thuringiensis (*off-label*), chlorantraniliprole (*off-label*), lambda-cyhalothrin, lambda-cyhalothrin (*off-label*), spinosad (*off-label*)
	Cutworms	Bacillus thuringiensis (*off-label*), cypermethrin, lambda-cyhalothrin, lambda-cyhalothrin (*off-label*), zeta-cypermethrin
	Flies	chlorantraniliprole (*off-label*), lambda-cyhalothrin
	Free-living nematodes	garlic extract (*off-label*), oxamyl
	Leaf miners	lambda-cyhalothrin, lambda-cyhalothrin (*off-label*), thiamethoxam (*seed treatment*)
	Millipedes	tefluthrin (*seed treatment*), thiamethoxam (*seed treatment*)
	Pests, miscellaneous	lambda-cyhalothrin (*off-label*), zeta-cypermethrin (*off-label*)
	Slugs/snails	ferric phosphate
	Springtails	tefluthrin (*seed treatment*), thiamethoxam (*seed treatment*)
	Symphylids	tefluthrin (*seed treatment*), thiamethoxam (*seed treatment*)
	Thrips	spinosad (*off-label*)
	Weevils	lambda-cyhalothrin
	Whiteflies	spirotetramat (*off-label*)
	Wireworms	thiamethoxam (*seed treatment - reduction*)
Plant growth regulation	Quality/yield control	sulphur
Weeds	Broad-leaved weeds	chloridazon, chloridazon (*off-label*), chloridazon + metamitron, chloridazon + quinmerac, chlorpropham (*off-label*), clomazone (*off-label*), clopyralid, clopyralid (*off-label*), desmedipham + ethofumesate + lenacil + phenmedipham, desmedipham + ethofumesate + lenacil + phenmedipham (*off-label*), desmedipham + ethofumesate + phenmedipham, desmedipham + phenmedipham, diquat, ethofumesate, ethofumesate + metamitron, ethofumesate + metamitron + phenmedipham, ethofumesate + phenmedipham, glyphosate, lenacil, lenacil + triflusulfuron-methyl, metamitron, phenmedipham, phenmedipham (*off-label*), propyzamide, S-metolachlor (*off-label*), triflusulfuron-methyl, triflusulfuron-methyl (*off-label*)
	Crops as weeds	clethodim, cycloxydim, cycloxydim (*off-label*), fluazifop-P-butyl, glyphosate, glyphosate (*wiper application*), lenacil + triflusulfuron-methyl, propaquizafop (*off-label*), quizalofop-P-ethyl, quizalofop-P-tefuryl
	Grass weeds	chloridazon, chloridazon (*off-label*), chloridazon + metamitron, chloridazon + quinmerac, chlorpropham (*off-label*), clethodim, cycloxydim, cycloxydim (*off-*

label), desmedipham + ethofumesate + lenacil + phenmedipham, desmedipham + ethofumesate + lenacil + phenmedipham (*off-label*), desmedipham + ethofumesate + phenmedipham, desmedipham + ethofumesate + phenmedipham (*moderately susceptible*), diquat, ethofumesate, ethofumesate + metamitron, ethofumesate + metamitron + phenmedipham, ethofumesate + phenmedipham, fluazifop-P-butyl, fluazifop-P-butyl (*off-label*), glyphosate, lenacil, metamitron, propaquizafop, propaquizafop (*off-label*), propyzamide, quizalofop-P-ethyl, quizalofop-P-tefuryl, S-metolachlor (*off-label*)

	Weeds, miscellaneous	chloridazon + quinmerac, desmedipham + ethofumesate + lenacil + phenmedipham (*off-label*), desmedipham + ethofumesate + phenmedipham, diquat, glyphosate, glyphosate (*off-label*)

Root and tuber crops - Carrots/parsnips

Diseases	Alternaria	azoxystrobin, azoxystrobin (*off-label*), azoxystrobin + difenoconazole, boscalid + pyraclostrobin (*moderate control*), cyprodinil + fludioxonil (*moderate control*), fenpropimorph (*off-label*), isopyrazam, mancozeb, prothioconazole, tebuconazole
	Botrytis	cyprodinil + fludioxonil (*moderate control*), Gliocladium catenulatum (*off-label*)
	Cavity spot	Bacillus subtilis (*off-label*), metalaxyl-M, metalaxyl-M (*off-label*), metalaxyl-M (*off-label - reduction*), metalaxyl-M (*reduction only*)
	Crown rot	fenpropimorph (*off-label*)
	Damping off	Bacillus subtilis (*off-label*), cymoxanil + fludioxonil + metalaxyl-M (*off-label*)
	Disease control	mancozeb
	Fungus diseases	azoxystrobin + difenoconazole (*off-label*)
	Fusarium	Gliocladium catenulatum (*off-label*)
	Phytophthora	Gliocladium catenulatum (*off-label*)
	Powdery mildew	azoxystrobin, azoxystrobin (*off-label*), azoxystrobin + difenoconazole, boscalid + pyraclostrobin, fenpropimorph (*off-label*), isopyrazam, prothioconazole, sulphur (*off-label*), tebuconazole
	Pythium	Bacillus subtilis (*off-label*), cymoxanil + fludioxonil + metalaxyl-M (*off-label*)
	Rhizoctonia	Gliocladium catenulatum (*off-label*)
	Rust	azoxystrobin (*off-label*)
	Sclerotinia	boscalid + pyraclostrobin (*moderate control*), cyprodinil + fludioxonil (*moderate control*), cyprodinil + fludioxonil (*off-label*), prothioconazole, tebuconazole
	Seed-borne diseases	thiram (*seed soak*)
	Stem canker	azoxystrobin + difenoconazole (*off-label*), isopyrazam (*off-label*)
Pests	Aphids	cypermethrin, spirotetramat (*off-label*), thiacloprid
	Birds/mammals	aluminium ammonium sulphate
	Cutworms	Bacillus thuringiensis (*off-label*), deltamethrin (*off-label*), lambda-cyhalothrin

	Flies	chlorantraniliprole (*off-label*), cypermethrin, deltamethrin (*off-label*), lambda-cyhalothrin (*off-label*), tefluthrin (*off-label - seed treatment*)
	Free-living nematodes	garlic extract
	Pests, miscellaneous	cypermethrin, deltamethrin (*off-label*), lambda-cyhalothrin (*off-label*)
	Slugs/snails	ferric phosphate
	Stem nematodes	oxamyl
Plant growth regulation	Growth control	maleic hydrazide (*off-label*)
Weeds	Broad-leaved weeds	clomazone, flumioxazin (*off-label*), metamitron (*off-label*), metribuzin (*off-label*), pendimethalin, pendimethalin (*off-label*), prosulfocarb (*off-label*)
	Crops as weeds	cycloxydim, fluazifop-P-butyl, flumioxazin (*off-label*), metribuzin (*off-label*), pendimethalin
	Grass weeds	clethodim (*off-label*), cycloxydim, fluazifop-P-butyl, fluazifop-P-butyl (*off-label*), flumioxazin (*off-label*), metamitron (*off-label*), metribuzin (*off-label*), pendimethalin, propaquizafop, prosulfocarb (*off-label*)
	Weeds, miscellaneous	glyphosate (*off-label*)

Root and tuber crops - Miscellaneous root crops

Crop control	Desiccation	diquat (*off-label*)
Diseases	Alternaria	azoxystrobin (*off-label*), fenpropimorph (*off-label*)
	Botrytis	azoxystrobin + difenoconazole (*off-label*), chlorothalonil (*off-label*), cyprodinil + fludioxonil (*moderate control*), Gliocladium catenulatum (*off-label*)
	Crown rot	fenpropimorph (*off-label*)
	Damping off	Bacillus subtilis (*off-label*)
	Disease control	tebuconazole (*off-label*)
	Fungus diseases	azoxystrobin + difenoconazole (*off-label*)
	Fusarium	Gliocladium catenulatum (*off-label*)
	Leaf spot	boscalid + pyraclostrobin (*off-label*), chlorothalonil (*off-label*)
	Phytophthora	azoxystrobin (*off-label*), Gliocladium catenulatum (*off-label*)
	Powdery mildew	azoxystrobin (*off-label*), boscalid + pyraclostrobin (*off-label*), fenpropimorph (*off-label*), tebuconazole (*off-label*)
	Rhizoctonia	Gliocladium catenulatum (*off-label*)
	Rust	azoxystrobin (*off-label*), boscalid + pyraclostrobin (*off-label*), tebuconazole (*off-label*)
	Sclerotinia	azoxystrobin (*off-label*), azoxystrobin + difenoconazole (*off-label*), boscalid + pyraclostrobin (*off-label*), cyprodinil + fludioxonil (*off-label*), isopyrazam (*off-label*)
	Septoria	isopyrazam (*off-label*)
	Stem canker	azoxystrobin + difenoconazole (*off-label*)
Pests	Aphids	cypermethrin, pirimicarb (*off-label*), pymetrozine (*off-label*), thiacloprid (*off-label*)
	Cutworms	Bacillus thuringiensis (*off-label*), lambda-cyhalothrin (*off-label*)

	Flies	chlorantraniliprole (*off-label*), cypermethrin, lambda-cyhalothrin (*off-label*)
	Pests, miscellaneous	cypermethrin, lambda-cyhalothrin (*off-label*), spinosad (*off-label*)
	Slugs/snails	ferric phosphate
	Thrips	spinosad (*off-label*)
	Weevils	lambda-cyhalothrin (*off-label*)
	Wireworms	tefluthrin (*off-label*)
Weeds	Broad-leaved weeds	clomazone (*off-label*), metribuzin (*off-label*), napropamide (*off-label*), pendimethalin (*off-label*), propyzamide (*off-label*), prosulfocarb (*off-label*), S-metolachlor (*off-label*)
	Crops as weeds	metribuzin (*off-label*)
	Grass weeds	cycloxydim (*off-label*), diquat, fluazifop-P-butyl (*off-label*), metribuzin (*off-label*), napropamide (*off-label*), propyzamide (*off-label*), prosulfocarb (*off-label*), S-metolachlor (*off-label*)
	Weeds, miscellaneous	diquat, glyphosate (*off-label*), metribuzin (*off-label*)

Root and tuber crops - Potatoes

Crop control	Desiccation	carfentrazone-ethyl, diquat, glufosinate-ammonium, pyraflufen-ethyl
Diseases	Alternaria	difenoconazole + mandipropamid, fenamidone + propamocarb hydrochloride, mancozeb
	Bacterial blight	Bacillus subtilis (*off-label*)
	Black dot	azoxystrobin, fludioxonil (*some reduction*)
	Black scurf and stem canker	azoxystrobin, fludioxonil, flutolanil (*tuber treatment*), pencycuron (*tuber treatment*)
	Blight	ametoctradin + dimethomorph (*reduction*), amisulbrom
	Disease control	azoxystrobin + fluazinam, difenoconazole, fluxapyroxad
	Dry rot	imazalil, thiabendazole
	Gangrene	imazalil, thiabendazole
	Helminthosporium seedling rot	Bacillus subtilis (*off-label*)
	Phytophthora	ametoctradin + dimethomorph, amisulbrom, azoxystrobin + fluazinam, benthiavalicarb-isopropyl + mancozeb, boscalid + pyraclostrobin (*off-label*), chlorothalonil + cymoxanil, cyazofamid, cymoxanil, cymoxanil + fluazinam, cymoxanil + mancozeb, cymoxanil + mandipropamid, cymoxanil + zoxamide, difenoconazole + mandipropamid, dimethomorph + fluazinam, dimethomorph + mancozeb, dimethomorph + zoxamide, fenamidone + propamocarb hydrochloride, fluazinam, fluopicolide + propamocarb hydrochloride, mancozeb, mancozeb + metalaxyl-M, mancozeb + metalaxyl-M (*reduction*), mancozeb + zoxamide
	Powdery scab	fluazinam (*off-label*)
	Rhizoctonia	Bacillus subtilis (*off-label*)
	Scab	fludioxonil (*off-label*)
	Silver scurf	fludioxonil (*reduction*), imazalil, thiabendazole
	Skin spot	imazalil, thiabendazole

Pests	Aphids	acetamiprid, esfenvalerate, flonicamid, lambda-cyhalothrin, pymetrozine, thiacloprid, thiamethoxam
	Beetles	lambda-cyhalothrin
	Caterpillars	lambda-cyhalothrin
	Cutworms	cypermethrin, zeta-cypermethrin
	Cyst nematodes	fosthiazate, oxamyl
	Free-living nematodes	fosthiazate (*reduction*), oxamyl
	Slugs/snails	metaldehyde
	Weevils	lambda-cyhalothrin
	Wireworms	fosthiazate (*reduction*)
Plant growth regulation	Growth control	chlorpropham, chlorpropham (*thermal fog*), ethylene, maleic hydrazide, spearmint oil
Weeds	Broad-leaved weeds	bentazone, carfentrazone-ethyl, clomazone, clomazone + metribuzin, diquat, flufenacet + metribuzin, glyphosate, metobromuron, metribuzin, pendimethalin, prosulfocarb, pyraflufen-ethyl, rimsulfuron
	Crops as weeds	carfentrazone-ethyl, cycloxydim, glyphosate, metribuzin, pendimethalin, quizalofop-P-ethyl, quizalofop-P-tefuryl, rimsulfuron
	Grass weeds	clomazone + metribuzin, cycloxydim, diquat, diquat (*weed control and desiccation*), flufenacet + metribuzin, glyphosate, metobromuron, metribuzin, pendimethalin, propaquizafop, prosulfocarb, quizalofop-P-ethyl, quizalofop-P-tefuryl
	Weeds, miscellaneous	carfentrazone-ethyl, diquat, diquat (*weed control or desiccation*), glyphosate

Stem and bulb vegetables - Asparagus

Diseases	Ascochyta	azoxystrobin + difenoconazole (*off-label*)
	Botrytis	azoxystrobin + chlorothalonil (*qualified minor use recommendation*), boscalid + pyraclostrobin (*off-label*), cyprodinil + fludioxonil (*off-label*), Gliocladium catenulatum (*off-label*)
	Downy mildew	metalaxyl-M (*off-label*)
	Fusarium	Gliocladium catenulatum (*off-label*)
	Phytophthora	Gliocladium catenulatum (*off-label*)
	Rhizoctonia	Gliocladium catenulatum (*off-label*)
	Rust	azoxystrobin, azoxystrobin + chlorothalonil (*moderate control only*), azoxystrobin + difenoconazole (*off-label*), boscalid + pyraclostrobin (*off-label*), difenoconazole (*off-label*)
	Stemphylium	azoxystrobin, azoxystrobin + chlorothalonil (*qualified minor use recommendation*)
Pests	Aphids	cypermethrin
	Beetles	spinosad (*off-label*)
	Pests, miscellaneous	cypermethrin
	Thrips	spinosad (*off-label*)
Weeds	Broad-leaved weeds	bromoxynil (*off-label*), carfentrazone-ethyl (*off-label*), clomazone (*off-label*), clopyralid (*off-label*), isoxaben, mesotrione (*off-label*), metribuzin (*off-label*), metribuzin (*off-label - from seed*), pendimethalin (*off-label*), pyridate (*off-label*)

	Crops as weeds	metribuzin (*off-label*)
	Grass weeds	fluazifop-P-butyl (*off-label*), metribuzin (*off-label*)
	Weeds, miscellaneous	clomazone (*off-label - spring-germinating*), glyphosate, glyphosate (*off-label*), metribuzin (*off-label*)

Stem and bulb vegetables - Celery/chicory

Diseases	Alternaria	Bacillus amyloliquefaciens D747, cyprodinil + fludioxonil (*off-label*)
	Botrytis	azoxystrobin (*off-label*), azoxystrobin + difenoconazole (*off-label*), cyprodinil + fludioxonil (*off-label*), Gliocladium catenulatum (*off-label*)
	Damping off	Bacillus subtilis (*off-label*)
	Disease control	dimethomorph + mancozeb (*off-label*), mandipropamid
	Downy mildew	dimethomorph + mancozeb (*off-label*), fenamidone + fosetyl-aluminium (*off-label*), mandipropamid
	Fusarium	Bacillus amyloliquefaciens D747, Gliocladium catenulatum (*off-label*)
	Leaf spot	azoxystrobin (*off-label*), cyprodinil + fludioxonil (*off-label*)
	Phytophthora	fenamidone + fosetyl-aluminium (*off-label*), Gliocladium catenulatum (*off-label*)
	Powdery mildew	Bacillus amyloliquefaciens D747
	Pythium	Bacillus amyloliquefaciens D747
	Rhizoctonia	azoxystrobin (*off-label*), Bacillus amyloliquefaciens D747, Gliocladium catenulatum (*off-label*)
	Sclerotinia	azoxystrobin (*off-label*), azoxystrobin + difenoconazole (*off-label*), cyprodinil + fludioxonil (*off-label*)
	Seed-borne diseases	thiram (*seed soak*)
Pests	Aphids	acetamiprid (*off-label*), alpha-cypermethrin (*off-label*), deltamethrin (*off-label*), pirimicarb (*off-label*), pymetrozine (*off-label*)
	Beetles	alpha-cypermethrin (*off-label*), deltamethrin (*off-label*)
	Capsid bugs	deltamethrin (*off-label*)
	Caterpillars	alpha-cypermethrin (*off-label*), Bacillus thuringiensis, Bacillus thuringiensis (*off-label*), deltamethrin (*off-label*), diflubenzuron (*off-label*), lambda-cyhalothrin (*off-label*), spinosad (*off-label*)
	Cutworms	deltamethrin (*off-label*), lambda-cyhalothrin (*off-label*)
	Flies	lambda-cyhalothrin (*off-label*), spinosad (*off-label*)
	Leaf miners	spinosad (*off-label*)
	Pests, miscellaneous	lambda-cyhalothrin (*off-label*), spinosad (*off-label*)
	Thrips	spinosad (*off-label*)
	Wireworms	tefluthrin (*off-label*)
Weeds	Broad-leaved weeds	clomazone (*off-label*), pendimethalin (*off-label*), propyzamide, propyzamide (*off-label*), prosulfocarb (*off-label*), S-metolachlor (*off-label*), triflusulfuron-methyl (*off-label*)
	Crops as weeds	cycloxydim (*off-label*)

| | Grass weeds | cycloxydim (*off-label*), fluazifop-P-butyl (*off-label*), pendimethalin (*off-label*), propyzamide, propyzamide (*off-label*), prosulfocarb (*off-label*), S-metolachlor (*off-label*) |

Stem and bulb vegetables - Onions/leeks/garlic

Diseases	Alternaria	iprodione, iprodione (*off-label*)
	Blight	fluoxastrobin + prothioconazole (*useful reduction*)
	Botrytis	azoxystrobin + chlorothalonil, cyprodinil + fludioxonil (*off-label*), fluoxastrobin + prothioconazole (*useful reduction*), Gliocladium catenulatum (*off-label*), iprodione, iprodione (*off-label*)
	Cladosporium	propiconazole (*off-label*)
	Collar rot	iprodione, iprodione (*off-label*)
	Damping off	Bacillus subtilis (*off-label*), thiram (*seed treatment*)
	Disease control	dimethomorph + mancozeb, dimethomorph + mancozeb (*off-label*), mancozeb, propiconazole (*off-label*)
	Downy mildew	azoxystrobin, azoxystrobin (*moderate control*), azoxystrobin (*off-label*), azoxystrobin (*reduction*), azoxystrobin + chlorothalonil, benthiavalicarb-isopropyl + mancozeb (*off-label*), dimethomorph + mancozeb, dimethomorph + mancozeb (*off-label*), dimethomorph + pyraclostrobin, fluopicolide + propamocarb hydrochloride (*off-label*), fluoxastrobin + prothioconazole (*control*), mancozeb, mancozeb + metalaxyl-M (*off-label*), mancozeb + metalaxyl-M (*useful control*), metalaxyl-M (*off-label*), propiconazole (*off-label*)
	Fusarium	boscalid + pyraclostrobin (*off-label*), Gliocladium catenulatum (*off-label*)
	Phytophthora	Gliocladium catenulatum (*off-label*)
	Powdery mildew	azoxystrobin (*off-label*), tebuconazole (*off-label*)
	Purple blotch	azoxystrobin, azoxystrobin + difenoconazole (*moderate control*), prothioconazole
	Pythium	metalaxyl-M, metalaxyl-M (*off-label*)
	Rhizoctonia	Gliocladium catenulatum (*off-label*)
	Rhynchosporium	prothioconazole (*useful reduction*)
	Rust	azoxystrobin, azoxystrobin (*off-label*), azoxystrobin + difenoconazole, fenpropimorph, mancozeb, propiconazole (*off-label*), prothioconazole, tebuconazole
	Stemphylium	prothioconazole (*useful reduction*)
	White rot	Bacillus subtilis (*off-label*), boscalid + pyraclostrobin (*off-label*), tebuconazole (*off-label*)
	White tip	azoxystrobin, azoxystrobin + difenoconazole (*qualified minor use*), boscalid + pyraclostrobin (*off-label*), dimethomorph + mancozeb (*reduction*)
Pests	Caterpillars	Bacillus thuringiensis, Bacillus thuringiensis (*off-label*)
	Cutworms	Bacillus thuringiensis (*off-label*), deltamethrin (*off-label*)
	Flies	tefluthrin (*off-label*)
	Free-living nematodes	garlic extract (*off-label*)

	Pests, miscellaneous	deltamethrin (*off-label*), lambda-cyhalothrin (*off-label*), Metarhizium anisopliae
	Root-knot nematodes	garlic extract (*off-label*)
	Stem nematodes	garlic extract (*off-label*), oxamyl (*off-label*)
	Thrips	abamectin (*off-label*), lambda-cyhalothrin (*off-label*), spinosad
Plant growth regulation	Growth control	maleic hydrazide, maleic hydrazide (*off-label*)
Weeds	Broad-leaved weeds	bentazone (*off-label*), bromoxynil, bromoxynil (*off-label*), chloridazon (*off-label*), chlorpropham, chlorpropham (*off-label*), clopyralid, clopyralid (*off-label*), dimethenamid-p + metazachlor (*off-label*), dimethenamid-p + pendimethalin (*off-label*), flumioxazin (*off-label*), fluroxypyr (*off-label*), glyphosate, pendimethalin, pendimethalin (*off-label*), pendimethalin (*pre + post emergence treatment*), prosulfocarb (*off-label*), pyridate, pyridate (*off-label*), S-metolachlor (*off-label*)
	Crops as weeds	bromoxynil, bromoxynil (*off-label*), clopyralid (*off-label*), cycloxydim, fluazifop-P-butyl, flumioxazin (*off-label*), fluroxypyr (*off-label*), glyphosate, pendimethalin
	Grass weeds	chloridazon (*off-label*), chlorpropham, clethodim (*off-label*), cycloxydim, dimethenamid-p + metazachlor (*off-label*), dimethenamid-p + pendimethalin (*off-label*), diquat, fluazifop-P-butyl, fluazifop-P-butyl (*off-label*), flumioxazin (*off-label*), glyphosate, pendimethalin, pendimethalin (*off-label*), propaquizafop, propaquizafop (*off-label*), prosulfocarb (*off-label*), S-metolachlor (*off-label*)
	Weeds, miscellaneous	bromoxynil (*off-label*), diquat, fluroxypyr (*off-label*), glyphosate, glyphosate (*off-label*)

Stem and bulb vegetables - Vegetables

Diseases	Botrytis	Bacillus subtilis (*off-label*), Gliocladium catenulatum (*off-label*)
	Downy mildew	azoxystrobin (*off-label*)
	Fusarium	Gliocladium catenulatum (*off-label*)
	Phytophthora	Gliocladium catenulatum (*off-label*)
	Powdery mildew	azoxystrobin (*off-label*), penconazole (*off-label*)
	Rhizoctonia	Gliocladium catenulatum (*off-label*)
Pests	Aphids	deltamethrin (*off-label*), Metarhizium anisopliae (*off-label*)
	Beetles	deltamethrin (*off-label*)
	Caterpillars	Bacillus thuringiensis, deltamethrin (*off-label*)
	Flies	Metarhizium anisopliae (*off-label*)
	Midges	Metarhizium anisopliae (*off-label*)
	Pests, miscellaneous	deltamethrin (*off-label*)
	Thrips	Metarhizium anisopliae (*off-label*)
Weeds	Grass weeds	cycloxydim (*off-label*), fluazifop-P-butyl (*off-label*)

Flowers and ornamentals

Flowers - Bedding plants, general

Diseases	Powdery mildew	bupirimate
Pests	Aphids	thiacloprid
	Beetles	thiacloprid
	Weevils	thiacloprid
	Whiteflies	thiacloprid
Plant growth regulation	Flowering control	paclobutrazol
	Growth control	paclobutrazol

Flowers - Bulbs/corms

Crop control	Desiccation	carfentrazone-ethyl (*off-label*)
Diseases	Basal stem rot	chlorothalonil (*off-label*), cyprodinil + fludioxonil (*off-label*), thiabendazole (*off-label*)
	Botrytis	Bacillus subtilis (*off-label*), boscalid + epoxiconazole (*off-label*), chlorothalonil (*off-label*), Gliocladium catenulatum (*off-label*), mancozeb, tebuconazole (*off-label*), thiabendazole (*off-label*)
	Crown rot	metalaxyl-M (*off-label*)
	Downy mildew	metalaxyl-M (*off-label*)
	Fusarium	cyprodinil + fludioxonil (*off-label*), Gliocladium catenulatum (*off-label*), thiabendazole
	Phytophthora	Gliocladium catenulatum (*off-label*)
	Powdery mildew	bupirimate, tebuconazole (*off-label*)
	Rhizoctonia	Gliocladium catenulatum (*off-label*)
	White mould	boscalid + epoxiconazole (*off-label*)
Weeds	Broad-leaved weeds	bentazone, bromoxynil (*off-label*)
	Grass weeds	S-metolachlor (*off-label*)

Flowers - Miscellaneous flowers

Diseases	Botrytis	Gliocladium catenulatum (*off-label*)
	Crown rot	fenpropimorph (*off-label*)
	Disease control	tebuconazole (*off-label*)
	Downy mildew	fenamidone + fosetyl-aluminium (*off-label*)
	Fusarium	Gliocladium catenulatum (*off-label*)
	Phytophthora	Gliocladium catenulatum (*off-label*)
	Powdery mildew	bupirimate, fenpropimorph (*off-label*), tebuconazole (*off-label*)
	Rhizoctonia	Gliocladium catenulatum (*off-label*)
	Rust	tebuconazole (*off-label*)
Pests	Aphids	alpha-cypermethrin (*off-label*)
	Beetles	alpha-cypermethrin (*off-label*)
	Caterpillars	alpha-cypermethrin (*off-label*)
	Cutworms	deltamethrin (*off-label*)
	Flies	deltamethrin (*off-label*), lambda-cyhalothrin (*off-label*)
	Pests, miscellaneous	deltamethrin (*off-label*), lambda-cyhalothrin (*off-label*)

Plant growth regulation	Flowering control	paclobutrazol
	Growth control	paclobutrazol
Weeds	Broad-leaved weeds	metribuzin (*off-label*), pendimethalin (*off-label*)
	Crops as weeds	metribuzin (*off-label*)
	Grass weeds	fluazifop-P-butyl (*off-label*), metribuzin (*off-label*)

Flowers - Pot plants

Diseases	Rust	azoxystrobin (*off-label*)
Pests	Aphids	thiacloprid
	Beetles	thiacloprid
	Weevils	thiacloprid
	Whiteflies	thiacloprid
Plant growth regulation	Flowering control	paclobutrazol
	Growth control	benzyladenine, paclobutrazol

Flowers - Protected flowers

Diseases	Crown rot	metalaxyl-M (*off-label*)
	Downy mildew	metalaxyl-M (*off-label*)
	Rust	azoxystrobin (*off-label*)
Pests	Spider mites	tebufenpyrad
Plant growth regulation	Flowering control	benzyladenine

Miscellaneous flowers and ornamentals - Miscellaneous uses

Diseases	Botrytis	boscalid + pyraclostrobin (*off-label*), isopyrazam (*off-label*)
	Disease control	boscalid + pyraclostrobin (*off-label*), potassium bicarbonate (commodity substance)
	Leaf spot	boscalid + pyraclostrobin (*off-label*)
	Powdery mildew	boscalid + pyraclostrobin (*off-label*), isopyrazam (*off-label*), physical pest control
	Rust	boscalid + pyraclostrobin (*off-label*)
	Sclerotinia	Coniothyrium minitans, isopyrazam (*off-label*)
Pests	Aphids	maltodextrin, physical pest control, pyrethrins
	Birds/mammals	aluminium ammonium sulphate
	Caterpillars	Bacillus thuringiensis (*off-label*), cypermethrin, diflubenzuron, pyrethrins
	Flies	cypermethrin
	Mealybugs	physical pest control
	Scale insects	physical pest control
	Slugs/snails	ferric phosphate, metaldehyde
	Spider mites	maltodextrin, physical pest control, pyrethrins
	Suckers	physical pest control
	Thrips	pyrethrins
	Wasps	cypermethrin
	Whiteflies	maltodextrin, physical pest control
Weeds	Broad-leaved weeds	carfentrazone-ethyl (*before planting*), diquat, diquat (*around*), propyzamide

Crops as weeds	carfentrazone-ethyl (*before planting*)	
Grass weeds	diquat, diquat (*around only*), propyzamide	
Weeds, miscellaneous	carfentrazone-ethyl, diquat (*around only*), glyphosate, glyphosate (*before planting*), glyphosate (*pre-sowing/ planting*)	

Ornamentals - Nursery stock

Diseases	Alternaria	iprodione
	Basal stem rot	azoxystrobin (*off-label*)
	Black root rot	azoxystrobin (*off-label*)
	Botrytis	azoxystrobin (*off-label*), Bacillus subtilis (*off-label*), boscalid + pyraclostrobin (*off-label*), captan (*off-label*), cyprodinil + fludioxonil, Gliocladium catenulatum (*off-label*), iprodione, iprodione (*off-label*)
	Disease control	chlorothalonil (*off-label*), chlorothalonil + metalaxyl-M (*off-label*), dimethomorph + mancozeb (*off-label*), iprodione (*off-label*), mancozeb, mancozeb (*off-label*), mancozeb + metalaxyl-M (*off-label*), mancozeb + zoxamide (*off-label*), metrafenone (*off-label*), propiconazole (*off-label*)
	Downy mildew	ametoctradin + dimethomorph (*off-label*), azoxystrobin (*off-label*), benthiavalicarb-isopropyl + mancozeb (*off-label*), captan (*off-label*), dimethomorph (*off-label*), dimethomorph + mancozeb (*off-label*), fenamidone + fosetyl-aluminium (*off-label*), fluopicolide + propamocarb hydrochloride (*off-label*), fosetyl-aluminium + propamocarb hydrochloride (*off-label*), mancozeb, mandipropamid (*off-label*)
	Foliar diseases	pyraclostrobin (*off-label*), trifloxystrobin (*off-label*)
	Foot rot	Trichoderma asperellum (Strain T34)
	Fungus diseases	captan (*off-label*), propamocarb hydrochloride (*off-label*)
	Fusarium	Gliocladium catenulatum (*off-label*), thiophanate-methyl (*off-label*), Trichoderma asperellum (Strain T34) (*off-label*)
	Leaf spot	azoxystrobin (*off-label*)
	Phytophthora	fenamidone + fosetyl-aluminium, Gliocladium catenulatum (*off-label*), metalaxyl-M, propamocarb hydrochloride (*off-label*)
	Powdery mildew	Ampelomyces quisqualis (Strain AQ10) (*off-label*), azoxystrobin (*off-label*), bupirimate, cyflufenamid (*off-label*), isopyrazam (*off-label*), kresoxim-methyl, metrafenone, physical pest control
	Pythium	fenamidone + fosetyl-aluminium, fosetyl-aluminium + propamocarb hydrochloride (*off-label*), metalaxyl-M, metalaxyl-M (*off-label*), propamocarb hydrochloride (*off-label*), Trichoderma asperellum (Strain T34) (*off-label*)
	Rhizoctonia	Gliocladium catenulatum (*off-label*)
	Root diseases	thiophanate-methyl (*off-label*)
	Root rot	fenamidone + fosetyl-aluminium (*off-label*)
	Rust	azoxystrobin (*off-label*), propiconazole (*off-label*)
	Scab	azoxystrobin (*off-label*), captan (*off-label*)

	Silver leaf	oxamyl (*off-label*)
	White blister	azoxystrobin (*off-label*)
Pests	Aphids	acetamiprid, Beauveria bassiana, cypermethrin, deltamethrin, dimethoate, esfenvalerate, imidacloprid, lambda-cyhalothrin (*off-label*), physical pest control, pymetrozine, pymetrozine (*off-label*), pyrethrins, spirotetramat (*off-label*), thiacloprid, thiacloprid (*off-label*)
	Beetles	Beauveria bassiana, thiacloprid, thiacloprid (*off-label*)
	Capsid bugs	deltamethrin
	Caterpillars	Bacillus thuringiensis, deltamethrin, diflubenzuron, indoxacarb, indoxacarb (*off-label*), pyrethrins, spinosad (*off-label*)
	Flies	imidacloprid, Metarhizium anisopliae (*off-label*), thiacloprid (*off-label*)
	Free-living nematodes	fosthiazate (*off-label*)
	Gall, rust and leaf & bud mites	clofentezine (*off-label*)
	Leaf miners	deltamethrin (*off-label*), dimethoate, oxamyl (*off-label*), thiacloprid (*off-label*)
	Mealybugs	deltamethrin, physical pest control
	Midges	Metarhizium anisopliae (*off-label*)
	Mites	clofentezine (*off-label*)
	Pests, miscellaneous	cypermethrin, deltamethrin (*off-label*), esfenvalerate, indoxacarb (*off-label*), lambda-cyhalothrin (*off-label*), Metarhizium anisopliae, methoxyfenozide (*off-label*), spinosad (*off-label*), spirotetramat (*off-label*), thiacloprid (*off-label*)
	Scale insects	deltamethrin, physical pest control
	Spider mites	abamectin, clofentezine (*off-label*), dimethoate, etoxazole (*off-label*), physical pest control, spirodiclofen (*off-label*)
	Stem nematodes	oxamyl (*off-label*)
	Suckers	physical pest control
	Thrips	abamectin, deltamethrin, deltamethrin (*off-label*), Metarhizium anisopliae (*off-label*), spinosad, thiacloprid (*off-label*), thiamethoxam (*off-label*)
	Weevils	imidacloprid, Metarhizium anisopliae, thiacloprid, thiacloprid (*off-label*)
	Whiteflies	acetamiprid, Beauveria bassiana, deltamethrin, flonicamid (*off-label*), imidacloprid, Lecanicillium muscarium, oxamyl (*off-label*), physical pest control, pymetrozine (*off-label*), thiacloprid, thiacloprid (*off-label*)
Plant growth regulation	Flowering control	paclobutrazol, paclobutrazol (*off-label*)
	Growth control	4-indol-3-ylbutyric acid, benzyladenine, benzyladenine + gibberellin, chlormequat (*off-label*), daminozide, ethephon (*off-label*), maleic hydrazide (*off-label*), paclobutrazol, paclobutrazol (*off-label*), prohexadione-calcium (*off-label*), trinexapac-ethyl (*off-label*)
Weeds	Broad-leaved weeds	2,4-D (*off-label*), amidosulfuron (*off-label*), bentazone, bentazone (*off-label*), bromoxynil (*off-label*), chlorpropham, clomazone (*off-label*), clopyralid,

		clopyralid + picloram (*off-label*), diflufenican (*off-label*), dimethenamid-p + metazachlor (*off-label*), dimethenamid-p + pendimethalin (*off-label*), diquat, florasulam (*off-label*), florasulam + fluroxypyr (*off-label*), flufenacet (*off-label*), flumioxazin (*off-label*), fluroxypyr (*off-label*), imazamox + pendimethalin (*off-label*), isoxaben, metamitron (*off-label*), metazachlor, metribuzin (*off-label*), metsulfuron-methyl (*off-label*), nicosulfuron (*off-label*), pendimethalin (*off-label*), phenmedipham (*off-label*), propyzamide, propyzamide (*off-label*), prosulfocarb (*off-label*), rimsulfuron (*off-label*), S-metolachlor (*off-label*)
	Crops as weeds	diflufenican (*off-label*), florasulam (*off-label*), flumioxazin (*off-label*), fluroxypyr (*off-label*), metribuzin (*off-label*), nicosulfuron (*off-label*)
	Grass weeds	chlorpropham, cycloxydim, cycloxydim (*off-label*), diflufenican (*off-label*), dimethenamid-p + metazachlor (*off-label*), dimethenamid-p + pendimethalin (*off-label*), diquat, fluazifop-P-butyl (*off-label*), flufenacet (*off-label*), flumioxazin (*off-label*), metamitron (*off-label*), metazachlor, metribuzin (*off-label*), nicosulfuron (*off-label*), pendimethalin (*off-label*), propyzamide, propyzamide (*off-label*), prosulfocarb (*off-label*), S-metolachlor (*off-label*)
	Mosses	maleic hydrazide + pelargonic acid, quinoclamine (*off-label*)
	Weeds, miscellaneous	bromoxynil (*off-label*), carfentrazone-ethyl (*off-label*), diflufenican (*off-label*), diquat, glyphosate, glyphosate + pyraflufen-ethyl, maleic hydrazide + pelargonic acid, metribuzin (*off-label*)

Ornamentals - Trees and shrubs

Crop control	Desiccation	glyphosate
Diseases	Botrytis	boscalid + pyraclostrobin (*off-label*)
	Disease control	mancozeb
	Downy mildew	benthiavalicarb-isopropyl + mancozeb (*off-label*)
	Leaf spot	boscalid + pyraclostrobin (*off-label*)
	Phytophthora	Bacillus subtilis (*off-label*), metalaxyl-M (*off-label*)
	Powdery mildew	boscalid + pyraclostrobin (*off-label*)
	Rust	boscalid + pyraclostrobin (*off-label*)
Pests	Aphids	deltamethrin, lambda-cyhalothrin (*off-label*)
	Birds/mammals	aluminium ammonium sulphate
	Capsid bugs	deltamethrin
	Caterpillars	Bacillus thuringiensis, deltamethrin, diflubenzuron
	Flies	Metarhizium anisopliae (*off-label*)
	Mealybugs	deltamethrin
	Midges	Metarhizium anisopliae (*off-label*)
	Pests, miscellaneous	lambda-cyhalothrin (*off-label*)
	Scale insects	deltamethrin
	Slugs/snails	ferric phosphate
	Thrips	deltamethrin, Metarhizium anisopliae (*off-label*)
	Weevils	Metarhizium anisopliae (*off-label*)
	Whiteflies	deltamethrin

Plant growth regulation	Growth control	glyphosate, maleic hydrazide (*off-label*)
Weeds	Broad-leaved weeds	clopyralid (*off-label*), isoxaben, pelargonic acid, propyzamide
	Grass weeds	propyzamide
	Mosses	maleic hydrazide + pelargonic acid
	Weeds, miscellaneous	2,4-D + glyphosate, flazasulfuron, glyphosate, glyphosate + pyraflufen-ethyl, glyphosate + sulfosulfuron, maleic hydrazide + pelargonic acid, propyzamide
	Woody weeds/scrub	glyphosate

Forestry

Forest nurseries, general - Forest nurseries

Crop control	Chemical stripping/ thinning	glyphosate
Diseases	Basal stem rot	azoxystrobin (*off-label*)
	Black root rot	azoxystrobin (*off-label*)
	Blight	copper oxychloride (*off-label*)
	Botrytis	azoxystrobin (*off-label*), boscalid + pyraclostrobin (*off-label*), cyprodinil + fludioxonil, fenhexamid (*off-label*), iprodione (*off-label*), mepanipyrim (*off-label*)
	Damping off	fosetyl-aluminium + propamocarb hydrochloride (*off-label*)
	Disease control	chlorothalonil (*off-label*), chlorothalonil + metalaxyl-M (*off-label*), cyprodinil (*off-label*), dimethomorph + mancozeb (*off-label*), epoxiconazole (*off-label*), mancozeb (*off-label*), mancozeb + metalaxyl-M (*off-label*), propiconazole (*off-label*)
	Downy mildew	azoxystrobin (*off-label*), benthiavalicarb-isopropyl + mancozeb (*off-label*), dimethomorph + mancozeb (*off-label*), fosetyl-aluminium + propamocarb hydrochloride (*off-label*), mancozeb (*off-label*)
	Fungus diseases	propamocarb hydrochloride (*off-label*)
	Fusarium	Trichoderma asperellum (Strain T34) (*off-label*)
	Leaf spot	azoxystrobin (*off-label*), difenoconazole (*off-label*)
	Phytophthora	Bacillus subtilis (*off-label*), difenoconazole (*off-label*), metalaxyl-M (*off-label*), propamocarb hydrochloride (*off-label*)
	Powdery mildew	Ampelomyces quisqualis (Strain AQ10) (*off-label*), azoxystrobin (*off-label*), cyflufenamid (*off-label*), proquinazid (*off-label*)
	Pythium	propamocarb hydrochloride (*off-label*), Trichoderma asperellum (Strain T34) (*off-label*)
	Root rot	Phlebiopsis gigantea
	Rust	azoxystrobin (*off-label*), difenoconazole (*off-label*)
	Scab	azoxystrobin (*off-label*)
	White blister	azoxystrobin (*off-label*)

Pests	Aphids	acetamiprid (*off-label*), Beauveria bassiana (*off-label*), deltamethrin (*off-label*), lambda-cyhalothrin (*off-label*), pymetrozine (*off-label*), spirotetramat (*off-label*)
	Beetles	acetamiprid (*off-label*), imidacloprid (*reduction in damage*)
	Birds/mammals	aluminium ammonium sulphate
	Caterpillars	Bacillus thuringiensis (*off-label*), chlorantraniliprole (*off-label*), indoxacarb (*off-label*)
	Cutworms	Bacillus thuringiensis (*off-label*)
	Flies	acetamiprid (*off-label*)
	Gall, rust and leaf & bud mites	clofentezine (*off-label*)
	Mites	bifenazate (*off-label*), clofentezine (*off-label*)
	Pests, miscellaneous	deltamethrin (*off-label*), dimethoate (*off-label*), indoxacarb (*off-label*), lambda-cyhalothrin (*off-label*), spirotetramat (*off-label*), thiacloprid (*off-label*)
	Spider mites	clofentezine (*off-label*)
	Wasps	acetamiprid (*off-label*)
	Weevils	acetamiprid (*off-label*), chlorantraniliprole (*off-label*), imidacloprid
	Whiteflies	Beauveria bassiana (*off-label*), pymetrozine (*off-label*)
Plant growth regulation	Growth control	ethephon + mepiquat chloride (*off-label*), trinexapac-ethyl (*off-label*)
Weeds	Broad-leaved weeds	2,4-D (*off-label*), amidosulfuron (*off-label*), amidosulfuron + iodosulfuron-methyl-sodium (*off-label*), chlorpropham (*off-label*), clomazone (*off-label*), clopyralid (*off-label*), clopyralid + picloram (*off-label*), desmedipham + ethofumesate + lenacil + phenmedipham (*off-label*), diflufenican (*off-label*), dimethenamid-p + metazachlor (*off-label*), flufenacet (*off-label*), flumioxazin (*off-label*), fluroxypyr (*off-label*), isoxaben, metsulfuron-methyl (*off-label*), nicosulfuron (*off-label*), pendimethalin (*off-label*), pendimethalin + picolinafen (*off-label*), propyzamide, propyzamide (*off-label*), prosulfocarb (*off-label*), rimsulfuron (*off-label*), S-metolachlor (*off-label*), tribenuron-methyl (*off-label*)
	Crops as weeds	amidosulfuron + iodosulfuron-methyl-sodium (*off-label*), clopyralid (*off-label*), diflufenican (*off-label*), flumioxazin (*off-label*), fluroxypyr (*off-label*), nicosulfuron (*off-label*)
	Grass weeds	chlorpropham (*off-label*), cycloxydim, desmedipham + ethofumesate + lenacil + phenmedipham (*off-label*), diflufenican (*off-label*), dimethenamid-p + metazachlor (*off-label*), flufenacet (*off-label*), nicosulfuron (*off-label*), pendimethalin (*off-label*), pendimethalin + picolinafen (*off-label*), propaquizafop (*off-label*), propyzamide, propyzamide (*off-label*), prosulfocarb (*off-label*), quizalofop-P-ethyl (*off-label*), S-metolachlor (*off-label*)
	Weeds, miscellaneous	diflufenican (*off-label*), diquat (*off-label*), glyphosate
	Woody weeds/scrub	glyphosate

Forestry plantations, general - Forestry plantations

Crop control	Chemical stripping/ thinning	glyphosate, glyphosate (*stump treatment*)
Diseases	Root rot	Phlebiopsis gigantea
Pests	Birds/mammals	aluminium ammonium sulphate
	Caterpillars	Bacillus thuringiensis (*off-label*), diflubenzuron
	Weevils	acetamiprid (*off-label*), cypermethrin
Plant growth regulation	Growth control	glyphosate
Weeds	Broad-leaved weeds	clopyralid (*off-label*), isoxaben, pendimethalin (*off-label*), propyzamide
	Grass weeds	cycloxydim, glyphosate, propaquizafop (*off-label*), propyzamide
	Weeds, miscellaneous	carfentrazone-ethyl (*off-label*), glyphosate
	Woody weeds/scrub	glyphosate

Miscellaneous forestry situations - Cut logs/timber

Pests	Beetles	cypermethrin

Woodland on farms - Woodland

Crop control	Chemical stripping/ thinning	glyphosate
Diseases	Root rot	Phlebiopsis gigantea
Pests	Aphids	lambda-cyhalothrin (*off-label*)
	Beetles	lambda-cyhalothrin (*off-label*)
	Pests, miscellaneous	lambda-cyhalothrin (*off-label*)
	Sawflies	lambda-cyhalothrin (*off-label*)
Plant growth regulation	Growth control	gibberellins
Weeds	Broad-leaved weeds	2,4-D (*off-label*), amidosulfuron (*off-label*), clopyralid (*off-label*), flumioxazin (*off-label*), fluroxypyr (*off-label*), lenacil (*off-label*), MCPA (*off-label*), metsulfuron-methyl (*off-label*), pendimethalin (*off-label*), propyzamide, tribenuron-methyl (*off-label*)
	Crops as weeds	fluazifop-P-butyl, flumioxazin (*off-label*), fluroxypyr (*off-label*)
	Grass weeds	fluazifop-P-butyl, lenacil (*off-label*), propyzamide
	Weeds, miscellaneous	glyphosate, glyphosate (*off-label*)

Fruit and hops

All bush fruit - All currants

Diseases	Alternaria	Bacillus amyloliquefaciens D747
	Botrytis	Bacillus subtilis (*off-label*), boscalid + pyraclostrobin, boscalid + pyraclostrobin (*off-label*), cyprodinil + fludioxonil (*qualified minor use recommendation*), fenhexamid, Gliocladium catenulatum (*off-label*)

	Disease control	mancozeb + metalaxyl-M (*off-label*), sulphur (*off-label*)
	Fusarium	Bacillus amyloliquefaciens D747, Gliocladium catenulatum (*off-label*)
	Leaf spot	boscalid + pyraclostrobin (*moderate control only*), boscalid + pyraclostrobin (*off-label*)
	Phytophthora	Bacillus subtilis (*off-label*), Gliocladium catenulatum (*off-label*)
	Powdery mildew	azoxystrobin (*off-label*), Bacillus amyloliquefaciens D747, boscalid + pyraclostrobin (*moderate control only*), boscalid + pyraclostrobin (*off-label*), bupirimate, bupirimate (*off-label*), fenpropimorph (*off-label*), kresoxim-methyl, kresoxim-methyl (*off-label*), penconazole, sulphur
	Pythium	Bacillus amyloliquefaciens D747
	Rhizoctonia	Bacillus amyloliquefaciens D747, Gliocladium catenulatum (*off-label*)
	Rust	azoxystrobin (*off-label*)
Pests	Aphids	pymetrozine (*off-label*), spirotetramat (*off-label*)
	Caterpillars	Bacillus thuringiensis (*off-label*), chlorantraniliprole (*off-label*), diflubenzuron, diflubenzuron (*off-label*), spinosad (*off-label*)
	Cutworms	Bacillus thuringiensis (*off-label*)
	Flies	lambda-cyhalothrin (*off-label*), Metarhizium anisopliae (*off-label*)
	Gall, rust and leaf & bud mites	sulphur, tebufenpyrad (*off-label*)
	Leatherjackets	Metarhizium anisopliae (*off-label*)
	Midges	lambda-cyhalothrin (*off-label*), Metarhizium anisopliae (*off-label*)
	Pests, miscellaneous	Bacillus thuringiensis (*off-label*), lambda-cyhalothrin (*off-label*), spinosad (*off-label*), thiacloprid (*off-label*)
	Sawflies	lambda-cyhalothrin (*off-label*), spinosad (*off-label*)
	Spider mites	spirodiclofen (*off-label*)
	Thrips	lambda-cyhalothrin (*off-label*), Metarhizium anisopliae (*off-label*)
	Weevils	Metarhizium anisopliae
Plant growth regulation	Growth control	maleic hydrazide (*off-label*)
Weeds	Broad-leaved weeds	carfentrazone-ethyl (*off-label*), clopyralid (*off-label*), flufenacet + metribuzin (*off-label*), isoxaben, pendimethalin, pendimethalin (*off-label*), propyzamide
	Crops as weeds	fluazifop-P-butyl, pendimethalin
	Grass weeds	fluazifop-P-butyl, fluazifop-P-butyl (*off-label*), flufenacet + metribuzin (*off-label*), pendimethalin, propyzamide
	Weeds, miscellaneous	glyphosate (*off-label*)

All bush fruit - All protected bush fruit

Diseases	Anthracnose	boscalid + pyraclostrobin (*off-label*), cyprodinil + fludioxonil (*off-label*)

	Botrytis	boscalid + pyraclostrobin (*off-label*), cyprodinil + fludioxonil (*off-label*), fenhexamid (*off-label*)
	Disease control	mancozeb + metalaxyl-M (*off-label*)
	Fusarium	Trichoderma asperellum (Strain T34) (*off-label*)
	Leaf spot	boscalid + pyraclostrobin (*off-label*)
	Powdery mildew	Ampelomyces quisqualis (Strain AQ10) (*off-label*)
	Pythium	Trichoderma asperellum (Strain T34) (*off-label*)
Pests	Aphids	pymetrozine (*off-label*), pyrethrins
	Caterpillars	Bacillus thuringiensis (*off-label*), pyrethrins, spinosad (*off-label*), thiacloprid (*off-label*)
	Flies	spinosad (*off-label*)
	Midges	thiacloprid (*off-label*)
	Spider mites	Lecanicillium muscarium (*off-label*)
	Thrips	Lecanicillium muscarium (*off-label*)
	Whiteflies	Lecanicillium muscarium (*off-label*), pymetrozine (*off-label*)

All bush fruit - All vines

Diseases	Black rot	kresoxim-methyl (*off-label*)
	Botrytis	Bacillus subtilis (*off-label*), fenhexamid, fenpyrazamine, Gliocladium catenulatum (*off-label*), potassium bicarbonate (commodity substance) (*off-label*), prohexadione-calcium (*off-label*)
	Disease control	mancozeb, mancozeb + zoxamide
	Downy mildew	ametoctradin + dimethomorph (*off-label*), benthiavalicarb-isopropyl + mancozeb (*off-label*), copper oxychloride, cymoxanil (*off-label*), fenamidone + fosetyl-aluminium (*off-label*), potassium bicarbonate (commodity substance) (*off-label*)
	Fusarium	Gliocladium catenulatum (*off-label*)
	Phytophthora	Gliocladium catenulatum (*off-label*)
	Powdery mildew	Ampelomyces quisqualis (Strain AQ10) (*off-label*), cyflufenamid (*off-label*), kresoxim-methyl (*off-label*), meptyldinocap, penconazole, potassium bicarbonate (commodity substance) (*off-label*), sulphur
	Rhizoctonia	Gliocladium catenulatum (*off-label*)
Pests	Aphids	spirotetramat (*off-label*)
	Capsid bugs	lambda-cyhalothrin (*off-label*)
	Caterpillars	Bacillus thuringiensis, Bacillus thuringiensis (*off-label*), lambda-cyhalothrin (*off-label*), methoxyfenozide (*off-label*)
	Flies	Metarhizium anisopliae (*off-label*)
	Leatherjackets	Metarhizium anisopliae (*off-label*)
	Midges	Metarhizium anisopliae (*off-label*)
	Spider mites	Lecanicillium muscarium (*off-label*)
	Thrips	Lecanicillium muscarium (*off-label*), Metarhizium anisopliae (*off-label*)
	Wasps	lambda-cyhalothrin (*off-label*)
	Whiteflies	Lecanicillium muscarium (*off-label*)
Weeds	Broad-leaved weeds	carfentrazone-ethyl (*off-label*), propyzamide (*off-label*)
	Grass weeds	propyzamide (*off-label*)

All bush fruit - Bilberries/blueberries/cranberries

Diseases	Alternaria	Bacillus amyloliquefaciens D747
	Anthracnose	boscalid + pyraclostrobin (*off-label*)
	Botrytis	Bacillus subtilis (*off-label*), boscalid + pyraclostrobin (*off-label*), cyprodinil + fludioxonil (*qualified minor use recommendation*), fenhexamid (*off-label*), Gliocladium catenulatum (*off-label*)
	Disease control	mancozeb + metalaxyl-M (*off-label*), sulphur (*off-label*)
	Fusarium	Bacillus amyloliquefaciens D747, Gliocladium catenulatum (*off-label*)
	Leaf spot	boscalid + pyraclostrobin (*off-label*)
	Phytophthora	Bacillus subtilis (*off-label*), Gliocladium catenulatum (*off-label*)
	Powdery mildew	azoxystrobin (*off-label*), Bacillus amyloliquefaciens D747, fenpropimorph (*off-label*), kresoxim-methyl (*off-label*)
	Pythium	Bacillus amyloliquefaciens D747
	Rhizoctonia	Bacillus amyloliquefaciens D747, Gliocladium catenulatum (*off-label*)
	Rust	azoxystrobin (*off-label*)
Pests	Aphids	pymetrozine (*off-label*), spirotetramat (*off-label*)
	Caterpillars	Bacillus thuringiensis (*off-label*), chlorantraniliprole (*off-label*), diflubenzuron (*off-label*), indoxacarb (*off-label*), spinosad (*off-label*)
	Cutworms	Bacillus thuringiensis (*off-label*)
	Flies	lambda-cyhalothrin (*off-label*), Metarhizium anisopliae (*off-label*), spinosad (*off-label*)
	Gall, rust and leaf & bud mites	tebufenpyrad (*off-label*)
	Leatherjackets	Metarhizium anisopliae (*off-label*)
	Midges	Metarhizium anisopliae (*off-label*)
	Pests, miscellaneous	Bacillus thuringiensis (*off-label*), indoxacarb (*off-label*), thiacloprid (*off-label*)
	Thrips	lambda-cyhalothrin (*off-label*), Metarhizium anisopliae (*off-label*)
	Weevils	Metarhizium anisopliae
Plant growth regulation	Growth control	maleic hydrazide (*off-label*)
Weeds	Broad-leaved weeds	carfentrazone-ethyl (*off-label*), clopyralid (*off-label*), flufenacet + metribuzin (*off-label*), pendimethalin (*off-label*), propyzamide (*off-label*)
	Grass weeds	fluazifop-P-butyl (*off-label*), flufenacet + metribuzin (*off-label*), propyzamide (*off-label*)
	Weeds, miscellaneous	glyphosate (*off-label*)

All bush fruit - Bush fruit, general

Pests	Aphids	pyrethrins
	Birds/mammals	aluminium ammonium sulphate
	Caterpillars	pyrethrins
Weeds	Broad-leaved weeds	clopyralid (*off-label*)

All bush fruit - Gooseberries

Diseases	Alternaria	Bacillus amyloliquefaciens D747
	Anthracnose	boscalid + pyraclostrobin (*off-label*)
	Botrytis	Bacillus subtilis (*off-label*), boscalid + pyraclostrobin (*off-label*), chlorothalonil + metalaxyl-M (*off-label*), cyprodinil + fludioxonil (*qualified minor use recommendation*), fenhexamid, Gliocladium catenulatum (*off-label*)
	Disease control	mancozeb + metalaxyl-M (*off-label*)
	Foliar diseases	chlorothalonil + metalaxyl-M (*off-label*)
	Fusarium	Bacillus amyloliquefaciens D747, Gliocladium catenulatum (*off-label*)
	Leaf spot	boscalid + pyraclostrobin (*off-label*), chlorothalonil + metalaxyl-M (*off-label*)
	Phytophthora	Bacillus subtilis (*off-label*), Gliocladium catenulatum (*off-label*)
	Powdery mildew	azoxystrobin (*off-label*), Bacillus amyloliquefaciens D747, bupirimate, fenpropimorph (*off-label*), kresoxim-methyl (*off-label*), sulphur
	Pythium	Bacillus amyloliquefaciens D747
	Rhizoctonia	Bacillus amyloliquefaciens D747, Gliocladium catenulatum (*off-label*)
	Rust	azoxystrobin (*off-label*)
Pests	Aphids	pymetrozine (*off-label*), spirotetramat (*off-label*)
	Caterpillars	Bacillus thuringiensis (*off-label*), chlorantraniliprole (*off-label*), diflubenzuron (*off-label*), spinosad (*off-label*)
	Cutworms	Bacillus thuringiensis (*off-label*)
	Flies	spinosad (*off-label*)
	Gall, rust and leaf & bud mites	tebufenpyrad (*off-label*)
	Midges	lambda-cyhalothrin (*off-label*)
	Pests, miscellaneous	Bacillus thuringiensis (*off-label*), lambda-cyhalothrin (*off-label*), thiacloprid (*off-label*)
	Sawflies	lambda-cyhalothrin (*off-label*)
	Spider mites	spirodiclofen (*off-label*)
	Weevils	Metarhizium anisopliae
Plant growth regulation	Growth control	maleic hydrazide (*off-label*)
Weeds	Broad-leaved weeds	carfentrazone-ethyl (*off-label*), clopyralid (*off-label*), flufenacet + metribuzin (*off-label*), isoxaben, pendimethalin, propyzamide
	Crops as weeds	fluazifop-P-butyl, pendimethalin
	Grass weeds	fluazifop-P-butyl, flufenacet + metribuzin (*off-label*), pendimethalin, propyzamide
	Weeds, miscellaneous	glyphosate (*off-label*)

All bush fruit - Miscellaneous bush fruit

Diseases	Alternaria	iprodione (*off-label*)
	Botrytis	cyprodinil + fludioxonil (*off-label*), iprodione (*off-label*)

	Downy mildew	metalaxyl-M (*off-label*)
	Powdery mildew	boscalid (*off-label*), proquinazid (*off-label*), sulphur
Pests	Capsid bugs	lambda-cyhalothrin (*off-label*)
	Caterpillars	indoxacarb (*off-label*)
	Pests, miscellaneous	lambda-cyhalothrin (*off-label*), spinosad (*off-label*)
	Wasps	lambda-cyhalothrin (*off-label*)
Weeds	Grass weeds	fluazifop-P-butyl (*off-label*)
	Weeds, miscellaneous	glyphosate (*off-label*)

Cane fruit - All outdoor cane fruit

Crop control	Desiccation	carfentrazone-ethyl (*off-label*)
Diseases	Alternaria	Bacillus amyloliquefaciens D747, iprodione
	Botrytis	Bacillus subtilis (*off-label*), fenhexamid, Gliocladium catenulatum (*off-label*), iprodione
	Botrytis fruit rot	cyprodinil + fludioxonil, tebuconazole (*off-label*)
	Cane blight	boscalid + pyraclostrobin (*off-label*), tebuconazole (*off-label*)
	Disease control	mancozeb + metalaxyl-M (*off-label*)
	Downy mildew	metalaxyl-M (*off-label*)
	Fusarium	Bacillus amyloliquefaciens D747, Gliocladium catenulatum (*off-label*)
	Phytophthora	Bacillus subtilis (*off-label*), Gliocladium catenulatum (*off-label*), tebuconazole (*off-label*)
	Powdery mildew	azoxystrobin (*off-label*), Bacillus amyloliquefaciens D747, bupirimate (*outdoor only*), fenpropimorph (*off-label*), tebuconazole (*off-label*)
	Purple blotch	boscalid + pyraclostrobin (*off-label*)
	Pythium	Bacillus amyloliquefaciens D747
	Rhizoctonia	Bacillus amyloliquefaciens D747, Gliocladium catenulatum (*off-label*)
	Root rot	dimethomorph, dimethomorph (*moderate control*)
	Rust	azoxystrobin (*off-label*)
Pests	Aphids	pymetrozine (*off-label*), pyrethrins
	Beetles	deltamethrin, deltamethrin (*off-label*)
	Birds/mammals	aluminium ammonium sulphate
	Capsid bugs	lambda-cyhalothrin (*off-label*)
	Caterpillars	Bacillus thuringiensis, Bacillus thuringiensis (*off-label*), deltamethrin (*off-label*), diflubenzuron (*off-label*), indoxacarb (*off-label*), pyrethrins
	Cutworms	Bacillus thuringiensis (*off-label*)
	Flies	Metarhizium anisopliae (*off-label*), spinosad (*off-label*)
	Gall, rust and leaf & bud mites	tebufenpyrad (*off-label*)
	Leatherjackets	Metarhizium anisopliae (*off-label*)
	Midges	Metarhizium anisopliae (*off-label*)
	Pests, miscellaneous	Bacillus thuringiensis (*off-label*), deltamethrin (*off-label*), indoxacarb (*off-label*), lambda-cyhalothrin (*off-label*), spinosad (*off-label*), thiacloprid (*off-label*)
	Slugs/snails	metaldehyde

	Spider mites	tebufenpyrad (*off-label*)
	Thrips	Metarhizium anisopliae (*off-label*), spinosad (*off-label*)
	Weevils	lambda-cyhalothrin (*off-label*), Metarhizium anisopliae
Plant growth regulation	Growth control	maleic hydrazide (*off-label*)
Weeds	Broad-leaved weeds	isoxaben, pendimethalin, propyzamide, propyzamide (*England only*)
	Crops as weeds	fluazifop-P-butyl, pendimethalin
	Grass weeds	fluazifop-P-butyl, fluazifop-P-butyl (*off-label*), pendimethalin, propyzamide, propyzamide (*England only*)
	Weeds, miscellaneous	glyphosate (*off-label*)

Cane fruit - All protected cane fruit

Crop control	Desiccation	carfentrazone-ethyl (*off-label*)
Diseases	Cane blight	boscalid + pyraclostrobin (*off-label*)
	Disease control	mancozeb + metalaxyl-M (*off-label*)
	Fusarium	Trichoderma asperellum (Strain T34) (*off-label*)
	Powdery mildew	azoxystrobin (*off-label*)
	Purple blotch	boscalid + pyraclostrobin (*off-label*)
	Pythium	Trichoderma asperellum (Strain T34) (*off-label*)
	Root rot	dimethomorph (*moderate control*)
	Rust	boscalid + pyraclostrobin (*off-label*)
Pests	Aphids	pymetrozine (*off-label*), pyrethrins
	Caterpillars	indoxacarb (*off-label*), pyrethrins
	Gall, rust and leaf & bud mites	abamectin (*off-label*), tebufenpyrad (*off-label*)
	Mites	abamectin (*off-label*)
	Pests, miscellaneous	indoxacarb (*off-label*), spinosad (*off-label*), thiacloprid (*off-label*)
	Spider mites	abamectin (*off-label*), Lecanicillium muscarium (*off-label*), tebufenpyrad (*off-label*)
	Thrips	Lecanicillium muscarium (*off-label*), spinosad (*off-label*)
	Whiteflies	Lecanicillium muscarium (*off-label*), pymetrozine (*off-label*)

Hops, general - Hops

Crop control	Chemical stripping/ thinning	diquat
	Desiccation	diquat, diquat (*off-label*), pyraflufen-ethyl (*off-label*)
Diseases	Botrytis	Bacillus subtilis (*off-label*), fenhexamid (*off-label*), Gliocladium catenulatum (*off-label*), iprodione (*off-label*)
	Disease control	boscalid + pyraclostrobin (*off-label*), captan (*off-label*), cymoxanil + mancozeb (*off-label*), mancozeb (*off-label*), mancozeb + metalaxyl-M (*off-label*), propiconazole (*off-label*), thiophanate-methyl (*off-label - do not harvest for human or animal consumption (including idling) within 12 months of treatment.*)

	Downy mildew	ametoctradin + dimethomorph (*off-label*), benthiavalicarb-isopropyl + mancozeb (*off-label*), cymoxanil (*off-label*), mandipropamid (*off-label*), metalaxyl-M (*off-label*)
	Fusarium	Gliocladium catenulatum (*off-label*)
	Phytophthora	Gliocladium catenulatum (*off-label*)
	Powdery mildew	bupirimate, cyflufenamid (*off-label*), fenpropimorph (*off-label*), penconazole (*off-label*), sulphur
	Rhizoctonia	Gliocladium catenulatum (*off-label*)
Pests	Aphids	acetamiprid (*off-label*), flonicamid (*off-label*), lambda-cyhalothrin (*off-label*), pymetrozine (*off-label*), spirotetramat (*off-label*), tebufenpyrad, thiamethoxam (*off-label*)
	Beetles	lambda-cyhalothrin (*off-label*)
	Capsid bugs	lambda-cyhalothrin (*off-label*)
	Caterpillars	Bacillus thuringiensis, indoxacarb (*off-label*), lambda-cyhalothrin (*off-label*), spinosad (*off-label*)
	Free-living nematodes	fosthiazate (*off-label*)
	Pests, miscellaneous	indoxacarb (*off-label*), lambda-cyhalothrin (*off-label*), methoxyfenozide (*off-label*), spinosad (*off-label*), thiacloprid (*off-label*)
	Spider mites	abamectin (*off-label*), spirodiclofen (*off-label*), tebufenpyrad
	Thrips	thiamethoxam (*off-label*)
Plant growth regulation	Growth control	diquat, maleic hydrazide (*off-label*), prohexadione-calcium (*off-label*)
Weeds	Broad-leaved weeds	bentazone (*off-label*), clomazone (*off-label*), clopyralid (*off-label*), flumioxazin (*off-label*), imazamox + pendimethalin (*off-label*), isoxaben, pendimethalin, pendimethalin (*off-label*), propyzamide (*off-label*), pyraflufen-ethyl (*off-label*)
	Crops as weeds	fluazifop-P-butyl, flumioxazin (*off-label*), pendimethalin
	Grass weeds	diquat, fluazifop-P-butyl, flumioxazin (*off-label*), pendimethalin, propyzamide (*off-label*)
	Weeds, miscellaneous	diquat, glyphosate (*off-label*)

Miscellaneous fruit situations - Fruit crops, general

Diseases	Botrytis	Bacillus subtilis (*off-label*), fenhexamid (*off-label*), iprodione (*off-label*)
	Damping off	fosetyl-aluminium + propamocarb hydrochloride (*off-label*)
	Disease control	boscalid + pyraclostrobin (*off-label*), mancozeb (*off-label*), propiconazole (*off-label*), thiophanate-methyl (*off-label*)
	Downy mildew	benthiavalicarb-isopropyl + mancozeb (*off-label*), fosetyl-aluminium + propamocarb hydrochloride (*off-label*), metalaxyl-M (*off-label*)
	Powdery mildew	cyflufenamid (*off-label*), kresoxim-methyl (*reduction*), proquinazid (*off-label*)
	Scab	kresoxim-methyl
Pests	Aphids	acetamiprid (*off-label*), lambda-cyhalothrin (*off-label*), pymetrozine (*off-label*)

	Caterpillars	Bacillus thuringiensis (*off-label*), indoxacarb (*off-label*), spinosad (*off-label*)
	Flies	Metarhizium anisopliae (*off-label*)
	Free-living nematodes	fosthiazate (*off-label*)
	Midges	Metarhizium anisopliae (*off-label*)
	Pests, miscellaneous	aluminium phosphide, lambda-cyhalothrin (*off-label*), methoxyfenozide (*off-label*), spinosad (*off-label*), thiacloprid (*off-label*)
	Thrips	Metarhizium anisopliae (*off-label*)
	Weevils	Metarhizium anisopliae (*off-label*)
Weeds	Broad-leaved weeds	flumioxazin (*off-label*), imazamox + pendimethalin (*off-label*), nicosulfuron (*off-label*), pendimethalin (*off-label*), prosulfocarb (*off-label*)
	Crops as weeds	flumioxazin (*off-label*), nicosulfuron (*off-label*)
	Grass weeds	fluazifop-P-butyl (*off-label*), flumioxazin (*off-label*), nicosulfuron (*off-label*), pendimethalin (*off-label*), prosulfocarb (*off-label*)
	Weeds, miscellaneous	glyphosate (*off-label*)

Miscellaneous fruit situations - Fruit nursery stock

Diseases	Disease control	captan (*off-label*)
	Fungus diseases	potassium bicarbonate (commodity substance)
	Powdery mildew	penconazole (*off-label*)
	Sooty moulds	potassium bicarbonate (commodity substance)
Pests	Aphids	thiamethoxam (*off-label*)
	Spider mites	spirodiclofen (*off-label*)
	Thrips	thiamethoxam (*off-label*)
Plant growth regulation	Growth control	prohexadione-calcium (*off-label*)
Weeds	Broad-leaved weeds	clopyralid (*off-label*), metazachlor
	Grass weeds	metazachlor

Other fruit - All outdoor strawberries

Diseases	Alternaria	Bacillus amyloliquefaciens D747, iprodione
	Anthracnose	azoxystrobin
	Black spot	boscalid + pyraclostrobin, cyprodinil + fludioxonil (*qualified minor use*)
	Botrytis	Bacillus subtilis, boscalid + pyraclostrobin, chlorothalonil (*off-label*), cyprodinil + fludioxonil, fenhexamid, fenpyrazamine, iprodione, mepanipyrim
	Botrytis fruit rot	Bacillus subtilis
	Collar rot	fenamidone + fosetyl-aluminium (*moderate control from foliar spray*)
	Crown rot	dimethomorph (*moderate control*), fenamidone + fosetyl-aluminium (*moderate control from foliar spray*)
	Didymella stem rot	Gliocladium catenulatum (*moderate control*)
	Disease control	azoxystrobin + difenoconazole, mancozeb + metalaxyl-M (*off-label*)
	Fusarium	Bacillus amyloliquefaciens D747, Gliocladium catenulatum (*moderate control*)

	Leaf spot	chlorothalonil (*off-label*)
	Phytophthora	fenamidone + fosetyl-aluminium, Gliocladium catenulatum (*moderate control*)
	Powdery mildew	azoxystrobin, azoxystrobin (*off-label*), Bacillus amyloliquefaciens D747, boscalid + pyraclostrobin, bupirimate, cyflufenamid (*off-label*), fenpropimorph (*off-label*), kresoxim-methyl, penconazole, potassium bicarbonate (commodity substance) (*off-label*), sulphur
	Pythium	Bacillus amyloliquefaciens D747, fenamidone + fosetyl-aluminium, Gliocladium catenulatum (*moderate control*)
	Red core	fenamidone + fosetyl-aluminium
	Rhizoctonia	Bacillus amyloliquefaciens D747, Gliocladium catenulatum (*moderate control*)
	Rust	azoxystrobin (*off-label*)
Pests	Aphids	pymetrozine (*off-label*)
	Birds/mammals	aluminium ammonium sulphate
	Capsid bugs	indoxacarb (*off-label*), thiacloprid (*off-label*)
	Caterpillars	Bacillus thuringiensis, indoxacarb (*off-label*)
	Pests, miscellaneous	lambda-cyhalothrin (*off-label*)
	Spider mites	abamectin (*off-label*), spirodiclofen (*off-label*), tebufenpyrad
	Tarsonemid mites	abamectin (*off-label*), abamectin (*off-label - in propagation*)
	Thrips	Beauveria bassiana (*off-label*), deltamethrin (*off-label*)
	Weevils	Metarhizium anisopliae
	Whiteflies	Beauveria bassiana (*off-label*)
Plant growth regulation	Growth control	maleic hydrazide (*off-label*)
Weeds	Broad-leaved weeds	carfentrazone-ethyl (*off-label*), clopyralid (*off-label*), dimethenamid-p + pendimethalin (*off-label*), isoxaben, metamitron (*off-label*), pendimethalin, propyzamide
	Crops as weeds	cycloxydim, fluazifop-P-butyl, pendimethalin
	Grass weeds	cycloxydim, dimethenamid-p + pendimethalin (*off-label*), diquat (*around only*), fluazifop-P-butyl, metamitron (*off-label*), pendimethalin, propyzamide, S-metolachlor (*off-label*)
	Weeds, miscellaneous	diquat (*around only*), glyphosate (*off-label*)

Other fruit - Protected miscellaneous fruit

Diseases	Alternaria	iprodione
	Anthracnose	azoxystrobin
	Black spot	boscalid + pyraclostrobin, cyprodinil + fludioxonil (*qualified minor use*)
	Botrytis	Bacillus subtilis (*reduction of damage to fruit*), boscalid + pyraclostrobin, cyprodinil + fludioxonil, fenpyrazamine, fluopyram + trifloxystrobin (*moderate control*), iprodione, iprodione (*off-label*), mepanipyrim
	Botrytis fruit rot	Bacillus subtilis (*reduction of damage to fruit*)
	Crown rot	dimethomorph (*moderate control*)

	Damping off	fosetyl-aluminium + propamocarb hydrochloride (*off-label*)
	Disease control	azoxystrobin + difenoconazole, boscalid + pyraclostrobin (*off-label*), mancozeb + metalaxyl-M (*off-label*)
	Downy mildew	fosetyl-aluminium + propamocarb hydrochloride (*off-label*)
	Fusarium	Trichoderma asperellum (Strain T34) (*off-label*)
	Powdery mildew	Ampelomyces quisqualis (Strain AQ10), azoxystrobin, boscalid + pyraclostrobin, bupirimate, fluopyram + trifloxystrobin, kresoxim-methyl, meptyldinocap, penconazole, penconazole (*off-label*), proquinazid (*off-label*)
	Pythium	Trichoderma asperellum (Strain T34) (*off-label*)
Pests	Aphids	acetamiprid (*off-label*), Beauveria bassiana, fatty acids C7-C20, pymetrozine (*off-label*)
	Beetles	Beauveria bassiana
	Caterpillars	Bacillus thuringiensis (*off-label*), deltamethrin (*off-label*), indoxacarb (*off-label*)
	Mites	bifenazate (*off-label*)
	Pests, miscellaneous	indoxacarb (*off-label*), lambda-cyhalothrin (*off-label*), Metarhizium anisopliae, spinosad, thiacloprid (*off-label*)
	Spider mites	abamectin (*off-label*), bifenazate, etoxazole (*off-label*), fatty acids C7-C20, Lecanicillium muscarium (*off-label*), spirodiclofen (*off-label*)
	Tarsonemid mites	abamectin (*off-label*)
	Thrips	Beauveria bassiana (*off-label*), Lecanicillium muscarium (*off-label*), spinosad
	Whiteflies	Beauveria bassiana, Beauveria bassiana (*off-label*), fatty acids C7-C20, Lecanicillium muscarium, Lecanicillium muscarium (*off-label*), pymetrozine (*off-label*)
Plant growth regulation	Growth control	gibberellins (*off-label*)

Other fruit - Rhubarb

Diseases	Botrytis	Gliocladium catenulatum (*off-label*)
	Downy mildew	mancozeb + metalaxyl-M (*off-label*)
	Fusarium	Gliocladium catenulatum (*off-label*)
	Phytophthora	Gliocladium catenulatum (*off-label*)
	Rhizoctonia	Gliocladium catenulatum (*off-label*)
Pests	Aphids	deltamethrin (*off-label*), pirimicarb (*off-label*)
	Beetles	deltamethrin (*off-label*)
	Capsid bugs	deltamethrin (*off-label*)
	Caterpillars	Bacillus thuringiensis (*off-label*), deltamethrin (*off-label*)
	Cutworms	Bacillus thuringiensis (*off-label*), deltamethrin (*off-label*)
Weeds	Broad-leaved weeds	clomazone (*off-label*), mesotrione (*off-label*), pendimethalin (*off-label*), propyzamide
	Grass weeds	pendimethalin (*off-label*), propyzamide

| | Weeds, miscellaneous | glyphosate (*off-label*), mesotrione (*off-label*), metribuzin (*off-label*) |

Tree fruit - All nuts

Diseases	Botrytis	Gliocladium catenulatum (*off-label*)
	Fusarium	Gliocladium catenulatum (*off-label*)
	Phytophthora	Gliocladium catenulatum (*off-label*)
	Powdery mildew	tebuconazole (*off-label*)
	Rhizoctonia	Gliocladium catenulatum (*off-label*)
	Stem canker	tebuconazole (*off-label*)
Pests	Aphids	lambda-cyhalothrin (*off-label*), pymetrozine (*off-label*)
	Caterpillars	Bacillus thuringiensis (*off-label*), diflubenzuron (*off-label*)
	Mites	tebufenpyrad (*off-label*), thiacloprid (*off-label*)
	Pests, miscellaneous	aluminium phosphide, lambda-cyhalothrin (*off-label*)
	Whiteflies	pymetrozine (*off-label*)
Weeds	Broad-leaved weeds	fluroxypyr (*off-label*), pendimethalin (*off-label*), propyzamide (*off-label*), prosulfocarb (*off-label*)
	Grass weeds	fluazifop-P-butyl (*off-label*), propyzamide (*off-label*), prosulfocarb (*off-label*)
	Weeds, miscellaneous	glyphosate (*off-label*)

Tree fruit - All pome fruit

Crop control	Sucker/shoot control	glyphosate
Diseases	Alternaria	cyprodinil + fludioxonil, fludioxonil (*reduction*)
	Botrytis	fludioxonil (*reduction*), Gliocladium catenulatum (*off-label*), iprodione (*off-label*)
	Botrytis fruit rot	cyprodinil + fludioxonil
	Disease control	captan (*off-label*), fludioxonil, mancozeb
	Fungus diseases	fludioxonil (*reduction*), mancozeb + metalaxyl-M (*off-label*), potassium bicarbonate (commodity substance)
	Fusarium	cyprodinil + fludioxonil, Gliocladium catenulatum (*off-label*)
	Penicillium rot	cyprodinil + fludioxonil, fludioxonil (*reduction*)
	Phytophthora	Gliocladium catenulatum (*off-label*)
	Powdery mildew	boscalid + pyraclostrobin, bupirimate, cyflufenamid, dithianon + pyraclostrobin, fluxapyroxad, kresoxim-methyl (*reduction*), meptyldinocap (*off-label*), penconazole, penthiopyrad (*suppression*), proquinazid (*off-label*), sulphur, tebuconazole (*off-label*)
	Rhizoctonia	Gliocladium catenulatum (*off-label*)
	Scab	boscalid + pyraclostrobin, captan, difenoconazole, dithianon, dithianon + potassium phosphonates, dithianon + pyraclostrobin, fluxapyroxad, kresoxim-methyl, mancozeb, penthiopyrad, sulphur
	Sooty moulds	potassium bicarbonate (commodity substance), potassium bicarbonate (commodity substance) (*reduction*)
	Stem canker	tebuconazole (*off-label*)
	Storage rots	captan, cyprodinil + fludioxonil

Pests	Aphids	acetamiprid, deltamethrin, flonicamid, lambda-cyhalothrin, thiacloprid
	Capsid bugs	deltamethrin
	Caterpillars	Bacillus thuringiensis, Bacillus thuringiensis (*off-label*), chlorantraniliprole, Cydia pomonella GV, deltamethrin, diflubenzuron, diflubenzuron (*off-label*), dodeca-8,10-dienyl acetate, dodecadienol, tetradecenyl acetate and tetradecylacetate, fenoxycarb, indoxacarb, lambda-cyhalothrin, methoxyfenozide, spinosad
	Gall, rust and leaf & bud mites	diflubenzuron, spirodiclofen
	Midges	thiacloprid (*off-label*)
	Pests, miscellaneous	potassium bicarbonate (commodity substance), thiacloprid (*off-label*)
	Sawflies	deltamethrin
	Scale insects	spirodiclofen
	Spider mites	clofentezine, spirodiclofen, tebufenpyrad
	Suckers	deltamethrin, diflubenzuron, lambda-cyhalothrin, potassium bicarbonate (commodity substance) (*reduction*), spirodiclofen
	Whiteflies	acetamiprid
Plant growth regulation	Fruiting control	1-methylcyclopropene (*off-label - post harvest only*), 1-methylcyclopropene (*post-harvest use*), 1-naphthylacetic acid, benzyladenine, gibberellins, metamitron
	Growth control	benzyladenine, ethephon (*off-label*), gibberellins, gibberellins (*off-label*), metamitron, prohexadione-calcium, prohexadione-calcium (*off-label*)
	Quality/yield control	gibberellins, gibberellins (*off-label*)
Weeds	Broad-leaved weeds	2,4-D, clopyralid (*off-label*), fluroxypyr (*off-label*), isoxaben, pendimethalin, pendimethalin (*off-label*), propyzamide, propyzamide (*off-label*), prosulfocarb (*off-label*)
	Crops as weeds	pendimethalin
	Grass weeds	fluazifop-P-butyl (*off-label*), pendimethalin, propyzamide, propyzamide (*off-label*), prosulfocarb (*off-label*)
	Weeds, miscellaneous	2,4-D + glyphosate, glyphosate, glyphosate (*off-label*)

Tree fruit - All stone fruit

Crop control	Sucker/shoot control	glyphosate
Diseases	Blossom wilt	boscalid + pyraclostrobin (*off-label*), cyprodinil + fludioxonil (*off-label*)
	Botrytis	cyprodinil + fludioxonil (*off-label*), fenhexamid (*off-label*), Gliocladium catenulatum (*off-label*)
	Fusarium	Gliocladium catenulatum (*off-label*)
	Phytophthora	Gliocladium catenulatum (*off-label*)
	Rhizoctonia	Gliocladium catenulatum (*off-label*)
	Sclerotinia	boscalid + pyraclostrobin (*off-label*)
Pests	Aphids	acetamiprid, pymetrozine (*off-label*), tebufenpyrad (*off-label*), thiacloprid (*off-label*)

	Capsid bugs	indoxacarb (*off-label*)
	Caterpillars	Bacillus thuringiensis (*off-label*), diflubenzuron, dodecadienol, tetradecenyl acetate and tetradecylacetate, indoxacarb (*off-label*), methoxyfenozide (*off-label*)
	Flies	deltamethrin (*off-label*)
	Gall, rust and leaf & bud mites	diflubenzuron, spirodiclofen (*off-label*)
	Pests, miscellaneous	lambda-cyhalothrin (*off-label*), thiacloprid (*off-label - under temporary protective rain covers*)
	Spider mites	spirodiclofen (*off-label*)
	Whiteflies	acetamiprid, pymetrozine (*off-label*)
Plant growth regulation	Growth control	prohexadione-calcium (*off-label*)
Weeds	Broad-leaved weeds	isoxaben, pendimethalin, propyzamide, propyzamide (*off-label*), prosulfocarb (*off-label*)
	Crops as weeds	pendimethalin
	Grass weeds	fluazifop-P-butyl (*off-label*), pendimethalin, propyzamide, propyzamide (*off-label*), prosulfocarb (*off-label*)
	Weeds, miscellaneous	glyphosate, glyphosate (*off-label*)

Tree fruit - Miscellaneous top fruit

Diseases	Botrytis	Gliocladium catenulatum (*off-label*)
	Fusarium	Gliocladium catenulatum (*off-label*)
	Phytophthora	Gliocladium catenulatum (*off-label*)
	Rhizoctonia	Gliocladium catenulatum (*off-label*)
Pests	Birds/mammals	aluminium ammonium sulphate
Weeds	Broad-leaved weeds	prosulfocarb (*off-label*)
	Grass weeds	prosulfocarb (*off-label*)
	Weeds, miscellaneous	glyphosate (*off-label*)

Tree fruit - Protected fruit

Diseases	Powdery mildew	Ampelomyces quisqualis (Strain AQ10) (*off-label*), penconazole (*off-label*)
Pests	Aphids	Beauveria bassiana, pymetrozine (*off-label*)
	Beetles	Beauveria bassiana
	Caterpillars	Bacillus thuringiensis (*off-label*)
	Pests, miscellaneous	Metarhizium anisopliae
	Spider mites	Lecanicillium muscarium (*off-label*)
	Thrips	Lecanicillium muscarium (*off-label*)
	Whiteflies	Beauveria bassiana, Lecanicillium muscarium (*off-label*), pymetrozine (*off-label*)

Grain/crop store uses

Stored produce - Food/produce storage

Pests	Pests, miscellaneous	aluminium phosphide, magnesium phosphide

Stored seed - Stored grain/rape/linseed

Pests	Birds/mammals	aluminium ammonium sulphate
	Food/grain storage pests	chlorpyrifos-methyl, d-phenothrin + tetramethrin, pirimiphos-methyl
	Mites	chlorpyrifos-methyl, physical pest control, pirimiphos-methyl
	Pests, miscellaneous	aluminium phosphide, chlorpyrifos-methyl, deltamethrin, magnesium phosphide, pirimiphos-methyl

Grass

Turf/amenity grass - Amenity grassland

Diseases	Anthracnose	azoxystrobin, azoxystrobin + propiconazole (*moderate control*), chlorothalonil + fludioxonil + propiconazole (*reduction*), fludioxonil (*reduction*), propiconazole (*qualified minor use*), tebuconazole + trifloxystrobin
	Brown patch	azoxystrobin, chlorothalonil + fludioxonil + propiconazole, iprodione, propiconazole (*qualified minor use*)
	Crown rust	azoxystrobin
	Dollar spot	azoxystrobin + propiconazole, chlorothalonil + fludioxonil + propiconazole, iprodione, propiconazole, tebuconazole + trifloxystrobin
	Drechslera leaf spot	fludioxonil (*useful levels of control*)
	Fairy rings	azoxystrobin
	Fusarium	azoxystrobin, azoxystrobin + propiconazole, chlorothalonil + fludioxonil + propiconazole, fludioxonil, iprodione, propiconazole, tebuconazole + trifloxystrobin, trifloxystrobin
	Melting out	azoxystrobin, iprodione, tebuconazole + trifloxystrobin
	Red thread	iprodione, tebuconazole + trifloxystrobin, trifloxystrobin
	Rust	tebuconazole + trifloxystrobin
	Seed-borne diseases	azoxystrobin + propiconazole, fludioxonil
	Snow mould	iprodione
	Take-all patch	azoxystrobin
Pests	Birds/mammals	aluminium ammonium sulphate, aluminium ammonium sulphate (*anti-fouling*)
	Slugs/snails	metaldehyde
Plant growth regulation	Growth control	maleic hydrazide (*off-label*), trinexapac-ethyl
Weeds	Broad-leaved weeds	2,4-D, 2,4-D + dicamba, 2,4-D + dicamba + fluroxypyr, 2,4-D + florasulam, aminopyralid + fluroxypyr, aminopyralid + triclopyr, citronella oil, clopyralid + florasulam + fluroxypyr, clopyralid + triclopyr, dicamba + MCPA + mecoprop-P, dicamba + MCPA + mecoprop-P (*moderately susceptible*), dicamba + mecoprop-P, ethofumesate, florasulam + fluroxypyr, fluroxypyr + triclopyr, mecoprop-P

	Crops as weeds	2,4-D + dicamba + fluroxypyr, 2,4-D + florasulam, dicamba + MCPA + mecoprop-P (*moderately susceptible*), florasulam + fluroxypyr, pinoxaden (*reduction only*)
	Grass weeds	ethofumesate, pinoxaden (*off-label*)
	Mosses	carfentrazone-ethyl + mecoprop-P, ferrous sulphate
	Weeds, miscellaneous	2,4-D + glyphosate, aminopyralid + triclopyr, carfentrazone-ethyl + mecoprop-P, clopyralid + florasulam + fluroxypyr, dicamba + MCPA + mecoprop-P, florasulam + fluroxypyr
	Woody weeds/scrub	aminopyralid + triclopyr, clopyralid + triclopyr

Turf/amenity grass - Managed amenity turf

Diseases	Anthracnose	azoxystrobin, azoxystrobin + propiconazole (*moderate control*), chlorothalonil + fludioxonil + propiconazole (*reduction*), fludioxonil (*reduction*), iprodione, iprodione + trifloxystrobin (*moderate control*), propiconazole (*qualified minor use*), tebuconazole + trifloxystrobin
	Brown patch	azoxystrobin, chlorothalonil + fludioxonil + propiconazole, iprodione, propiconazole (*qualified minor use*)
	Crown rust	azoxystrobin
	Disease control	iprodione, iprodione + trifloxystrobin
	Dollar spot	azoxystrobin + propiconazole, chlorothalonil + fludioxonil + propiconazole, fluopyram + trifloxystrobin, iprodione, iprodione + trifloxystrobin, propiconazole, pyraclostrobin (*useful reduction*), tebuconazole + trifloxystrobin
	Drechslera leaf spot	fludioxonil (*useful levels of control*)
	Fairy rings	azoxystrobin
	Fusarium	azoxystrobin, azoxystrobin + propiconazole, chlorothalonil + fludioxonil + propiconazole, fludioxonil, iprodione, iprodione + trifloxystrobin, propiconazole, pyraclostrobin (*moderate control*), pyraclostrobin (*moderate control only*), tebuconazole + trifloxystrobin, trifloxystrobin
	Leaf spot	iprodione + trifloxystrobin
	Melting out	azoxystrobin, iprodione, tebuconazole + trifloxystrobin
	Red thread	iprodione, iprodione + trifloxystrobin, pyraclostrobin, tebuconazole + trifloxystrobin, trifloxystrobin
	Rust	iprodione, iprodione + trifloxystrobin, tebuconazole + trifloxystrobin
	Seed-borne diseases	azoxystrobin + propiconazole, fludioxonil, fluopyram + trifloxystrobin
	Snow mould	iprodione
	Take-all patch	azoxystrobin
Pests	Aphids	esfenvalerate
	Birds/mammals	aluminium ammonium sulphate, aluminium ammonium sulphate (*anti-fouling*)
	Flies	esfenvalerate
	Slugs/snails	ferric phosphate, metaldehyde

Plant growth regulation	Growth control	trinexapac-ethyl
Weeds	Broad-leaved weeds	2,4-D, 2,4-D + dicamba + fluroxypyr, 2,4-D + dicamba + MCPA + mecoprop-P, 2,4-D + florasulam, clopyralid + florasulam + fluroxypyr, clopyralid + fluroxypyr + MCPA, dicamba + MCPA + mecoprop-P, dicamba + MCPA + mecoprop-P (*moderately susceptible*), dicamba + mecoprop-P, dichlorprop-P + ferrous sulphate + MCPA, ferrous sulphate + MCPA + mecoprop-P, florasulam + fluroxypyr, MCPA + mecoprop-P, mecoprop-P
	Crops as weeds	2,4-D + dicamba + fluroxypyr, 2,4-D + dicamba + MCPA + mecoprop-P, 2,4-D + florasulam, dicamba + MCPA + mecoprop-P (*moderately susceptible*), florasulam + fluroxypyr, pinoxaden (*reduction only*)
	Grass weeds	cycloxydim (*off-label*), pinoxaden (*off-label*)
	Mosses	carfentrazone-ethyl + mecoprop-P, dichlorprop-P + ferrous sulphate + MCPA, ferrous sulphate, ferrous sulphate + MCPA + mecoprop-P, quinoclamine (*moderate control*)
	Weeds, miscellaneous	carfentrazone-ethyl + mecoprop-P, clopyralid + florasulam + fluroxypyr, dicamba + MCPA + mecoprop-P, florasulam + fluroxypyr, glufosinate-ammonium, glyphosate (*pre-establishment*), glyphosate (*pre-establishment only*)

Non-crop pest control

Farm buildings/yards - Farm buildings

Pests	Ants	cypermethrin + tetramethrin
	Bedbugs	cypermethrin + tetramethrin
	Birds/mammals	aluminium ammonium sulphate, bromadiolone, difenacoum, flocoumafen, warfarin
	Caterpillars	cypermethrin, cypermethrin + tetramethrin
	Cockroaches	cypermethrin + tetramethrin
	Flies	cypermethrin, cypermethrin + tetramethrin, diflubenzuron, d-phenothrin + tetramethrin
	Pests, miscellaneous	physical pest control, pyrethrins
	Silverfish	cypermethrin + tetramethrin
	Wasps	cypermethrin, cypermethrin + tetramethrin, d-phenothrin + tetramethrin

Farm buildings/yards - Farmyards

Pests	Birds/mammals	bromadiolone, difenacoum

Farmland pest control - Farmland situations

Crop control	Desiccation	diquat (*off-label*)
Pests	Birds/mammals	aluminium phosphide

Miscellaneous non-crop pest control - Manure/rubbish

Pests	Ants	cypermethrin + tetramethrin
	Bedbugs	cypermethrin + tetramethrin

Caterpillars	cypermethrin + tetramethrin
Cockroaches	cypermethrin + tetramethrin
Flies	cypermethrin + tetramethrin, diflubenzuron
Silverfish	cypermethrin + tetramethrin
Wasps	cypermethrin + tetramethrin

Miscellaneous non-crop pest control - Miscellaneous pest control situations

Pests	Birds/mammals	carbon dioxide (commodity substance)

Protected salad and vegetable crops

Protected brassicas - Protected brassica vegetables

Diseases	Damping off	fosetyl-aluminium + propamocarb hydrochloride, propamocarb hydrochloride
	Downy mildew	fluopicolide + propamocarb hydrochloride (*off-label*), fosetyl-aluminium + propamocarb hydrochloride, propamocarb hydrochloride
	Fungus diseases	propamocarb hydrochloride
	Fusarium	Trichoderma asperellum (Strain T34) (*off-label*)
	Phytophthora	propamocarb hydrochloride
	Pythium	propamocarb hydrochloride, Trichoderma asperellum (Strain T34) (*off-label*)
	Root rot	propamocarb hydrochloride
	White blister	boscalid + pyraclostrobin (*off-label*)
Pests	Aphids	acetamiprid (*off-label*), pymetrozine (*off-label*), pyrethrins, spirotetramat (*off-label*)
	Beetles	alpha-cypermethrin (*off-label*), pyrethrins
	Caterpillars	alpha-cypermethrin (*off-label*), Bacillus thuringiensis (*off-label*), indoxacarb (*off-label*), pyrethrins
	Flies	chlorpyrifos
	Mites	abamectin (*off-label*)
	Pests, miscellaneous	chlorpyrifos
	Whiteflies	pyrethrins, spirotetramat (*off-label*)

Protected brassicas - Protected salad brassicas

Diseases	Damping off	fosetyl-aluminium + propamocarb hydrochloride
	Downy mildew	fosetyl-aluminium + propamocarb hydrochloride
	Fusarium	Trichoderma asperellum (Strain T34) (*off-label*)
	Pythium	Trichoderma asperellum (Strain T34) (*off-label*)
Pests	Aphids	lambda-cyhalothrin (*off-label*), pymetrozine (*off-label*)
	Beetles	alpha-cypermethrin (*off-label*), lambda-cyhalothrin (*off-label*)
	Caterpillars	alpha-cypermethrin (*off-label*), Bacillus thuringiensis (*off-label*), lambda-cyhalothrin (*off-label*)
	Pests, miscellaneous	lambda-cyhalothrin (*off-label*)

Protected crops, general - All protected crops

Pests	Aphids	Beauveria bassiana, pyrethrins
	Beetles	pyrethrins
	Caterpillars	Bacillus thuringiensis (*off-label*), indoxacarb (*off-label*), pyrethrins
	Mealybugs	pyrethrins
	Pests, miscellaneous	lambda-cyhalothrin (*off-label*)
	Scale insects	pyrethrins
	Slugs/snails	metaldehyde
	Spider mites	pyrethrins
	Thrips	pyrethrins
	Whiteflies	Beauveria bassiana, pyrethrins

Protected crops, general - Protected nuts

Pests	Aphids	pymetrozine (*off-label*)
	Whiteflies	pymetrozine (*off-label*)

Protected crops, general - Protected vegetables, general

Diseases	Powdery mildew	Ampelomyces quisqualis (Strain AQ10) (*off-label*)
Pests	Aphids	pymetrozine (*off-label*)
	Spider mites	Lecanicillium muscarium (*off-label*)
	Thrips	Lecanicillium muscarium (*off-label*)
	Whiteflies	Lecanicillium muscarium (*off-label*), pymetrozine (*off-label*)

Protected crops, general - Soils and compost

Diseases	Root rot	metam-sodium
	Sclerotinia	metam-sodium
	Soil-borne diseases	dazomet
Pests	Cyst nematodes	metam-sodium
	Millipedes	metam-sodium
	Root-knot nematodes	metam-sodium
	Soil pests	dazomet
	Symphylids	metam-sodium
	Wireworms	metam-sodium
Weeds	Weeds, miscellaneous	metam-sodium

Protected fruiting vegetables - Protected aubergines

Diseases	Botrytis	boscalid + pyraclostrobin (*off-label*), cyprodinil + fludioxonil (*off-label*), fenhexamid (*off-label*), fenpyrazamine, iprodione (*off-label*), potassium bicarbonate (commodity substance) (*off-label*)
	Damping off	fosetyl-aluminium + propamocarb hydrochloride (*off-label*)
	Powdery mildew	Ampelomyces quisqualis (Strain AQ10), isopyrazam, potassium bicarbonate (commodity substance) (*off-label*), proquinazid (*off-label*), sulphur (*off-label*)
Pests	Aphids	acetamiprid, Beauveria bassiana, fatty acids C7-C20 (*off-label*), lambda-cyhalothrin (*off-label*), pymetrozine (*off-label*)

	Beetles	Beauveria bassiana
	Caterpillars	Bacillus thuringiensis, indoxacarb
	Leaf miners	deltamethrin (*off-label*), tetradecadienyl acetate, thiacloprid (*off-label*)
	Mealybugs	flonicamid (*off-label*)
	Mites	etoxazole
	Pests, miscellaneous	lambda-cyhalothrin (*off-label*), Metarhizium anisopliae
	Spider mites	abamectin (*off-label*), fatty acids C7-C20 (*off-label*)
	Thrips	abamectin (*off-label*), deltamethrin (*off-label*), spinosad, thiacloprid (*off-label*)
	Whiteflies	acetamiprid, Beauveria bassiana, fatty acids C7-C20 (*off-label*), flonicamid (*off-label*), pymetrozine (*off-label*), thiacloprid (*off-label*)

Protected fruiting vegetables - Protected cucurbits

Diseases	Botrytis	boscalid + pyraclostrobin (*off-label*), cyprodinil + fludioxonil (*off-label*), fenhexamid (*off-label*), fenpyrazamine, potassium bicarbonate (commodity substance) (*off-label*)
	Cladosporium	boscalid + pyraclostrobin (*off-label*)
	Damping off	fosetyl-aluminium + propamocarb hydrochloride, fosetyl-aluminium + propamocarb hydrochloride (*off-label*)
	Disease control	isopyrazam
	Downy mildew	azoxystrobin (*off-label*), fosetyl-aluminium + propamocarb hydrochloride, metalaxyl-M (*off-label*)
	Fungus diseases	boscalid + pyraclostrobin (*off-label*)
	Fusarium	cyprodinil + fludioxonil (*off-label*), Trichoderma asperellum (Strain T34) (*off-label*)
	Powdery mildew	Ampelomyces quisqualis (Strain AQ10), Ampelomyces quisqualis (Strain AQ10) (*off-label*), azoxystrobin (*off-label*), boscalid + pyraclostrobin (*off-label*), bupirimate, isopyrazam, penconazole, penconazole (*off-label*), potassium bicarbonate (commodity substance) (*off-label*), proquinazid (*off-label*), sulphur (*off-label*)
	Pythium	fosetyl-aluminium + propamocarb hydrochloride, Trichoderma asperellum (Strain T34) (*off-label*)
Pests	Aphids	acetamiprid, Beauveria bassiana, deltamethrin, fatty acids C7-C20, fatty acids C7-C20 (*off-label*), flonicamid (*off-label*), lambda-cyhalothrin (*off-label*), pirimicarb (*off-label*), pymetrozine, pymetrozine (*off-label*), pyrethrins (*off-label*)
	Beetles	Beauveria bassiana
	Caterpillars	Bacillus thuringiensis, Bacillus thuringiensis (*off-label*), deltamethrin, indoxacarb
	Cutworms	Bacillus thuringiensis (*off-label*)
	Leaf miners	deltamethrin (*off-label*), tetradecadienyl acetate, thiacloprid (*off-label*)
	Mealybugs	deltamethrin
	Pests, miscellaneous	lambda-cyhalothrin (*off-label*), Metarhizium anisopliae, spinosad (*off-label*)
	Scale insects	deltamethrin

	Spider mites	abamectin, abamectin (*off-label*), fatty acids C7-C20, fatty acids C7-C20 (*off-label*), Lecanicillium muscarium (*off-label*), spirodiclofen (*off-label*)
	Thrips	abamectin, deltamethrin, deltamethrin (*off-label*), Lecanicillium muscarium (*off-label*), spinosad, spinosad (*off-label*), thiacloprid (*off-label*)
	Whiteflies	acetamiprid, Beauveria bassiana, deltamethrin, fatty acids C7-C20, fatty acids C7-C20 (*off-label*), flonicamid (*off-label*), Lecanicillium muscarium, Lecanicillium muscarium (*off-label*), pymetrozine (*off-label*), thiacloprid (*off-label*)

Protected fruiting vegetables - Protected tomatoes

Diseases	Alternaria	iprodione
	Botrytis	boscalid + pyraclostrobin (*off-label*), cyprodinil + fludioxonil (*off-label*), fenhexamid (*off-label*), fenpyrazamine, iprodione, potassium bicarbonate (commodity substance) (*off-label*)
	Damping off	propamocarb hydrochloride (*off-label*)
	Phytophthora	propamocarb hydrochloride (*off-label*)
	Powdery mildew	Ampelomyces quisqualis (Strain AQ10), bupirimate (*off-label*), cyflufenamid (*off-label*), isopyrazam, penconazole (*off-label*), potassium bicarbonate (commodity substance) (*off-label*), proquinazid (*off-label*), sulphur (*off-label*)
	Pythium	fosetyl-aluminium + propamocarb hydrochloride
	Root rot	propamocarb hydrochloride (*off-label*)
	Verticillium wilt	thiophanate-methyl (*off-label*)
Pests	Aphids	acetamiprid, Beauveria bassiana, deltamethrin, fatty acids C7-C20, lambda-cyhalothrin (*off-label*), pymetrozine (*off-label*), pyrethrins
	Beetles	Beauveria bassiana, pyrethrins
	Caterpillars	Bacillus thuringiensis, chlorantraniliprole (*off-label*), deltamethrin, indoxacarb, pyrethrins
	Leaf miners	abamectin, abamectin (*off-label*), spinosad (*off-label*), tetradecadienyl acetate, thiacloprid (*off-label*)
	Mealybugs	deltamethrin, flonicamid (*off-label*), pyrethrins, pyrethrins (*off-label*)
	Mirid bugs	pyrethrins, pyrethrins (*off-label*)
	Mites	etoxazole
	Pests, miscellaneous	lambda-cyhalothrin (*off-label*), Metarhizium anisopliae, spinosad (*off-label*)
	Scale insects	deltamethrin
	Spider mites	abamectin, fatty acids C7-C20, spirodiclofen (*off-label*)
	Thrips	deltamethrin, spinosad, spinosad (*off-label*), thiacloprid (*off-label*)
	Whiteflies	acetamiprid, Beauveria bassiana, deltamethrin, fatty acids C7-C20, flonicamid (*off-label*), Lecanicillium muscarium, pymetrozine (*off-label*), pyrethrins, thiacloprid (*off-label*)
Plant growth regulation	Growth control	ethephon (*off-label*)

Protected herb crops - Protected herbs

Diseases	Alternaria	cyprodinil + fludioxonil (*off-label*), iprodione (*off-label*)
	Botrytis	cyprodinil + fludioxonil (*off-label*), fenhexamid (*off-label*), iprodione (*off-label*), potassium bicarbonate (commodity substance) (*off-label*)
	Damping off	fosetyl-aluminium + propamocarb hydrochloride (*off-label*)
	Disease control	mandipropamid
	Downy mildew	dimethomorph (*off-label*), mancozeb + metalaxyl-M (*off-label*), mandipropamid, metalaxyl-M (*off-label*)
	Fusarium	Trichoderma asperellum (Strain T34) (*off-label*)
	Powdery mildew	Ampelomyces quisqualis (Strain AQ10), Ampelomyces quisqualis (Strain AQ10) (*off-label*), isopyrazam, penconazole (*off-label*), potassium bicarbonate (commodity substance) (*off-label*), sulphur (*off-label*)
	Pythium	Trichoderma asperellum (Strain T34) (*off-label*)
	Rhizoctonia	cyprodinil + fludioxonil (*off-label*)
	Stem canker	cyprodinil + fludioxonil (*off-label*)
	White blister	boscalid + pyraclostrobin (*off-label*)
Pests	Aphids	acetamiprid (*off-label*), alpha-cypermethrin (*off-label*), Beauveria bassiana, deltamethrin, fatty acids C7-C20 (*off-label*), pirimicarb (*off-label*), pymetrozine (*off-label*), pyrethrins (*off-label*), spirotetramat (*off-label*), thiacloprid (*off-label*)
	Beetles	alpha-cypermethrin (*off-label*), Beauveria bassiana
	Caterpillars	alpha-cypermethrin (*off-label*), Bacillus thuringiensis, Bacillus thuringiensis (*off-label*), deltamethrin, indoxacarb (*off-label*), spinosad (*off-label*)
	Cutworms	Bacillus thuringiensis (*off-label*)
	Leaf miners	tetradecadienyl acetate
	Mealybugs	deltamethrin
	Mites	abamectin (*off-label*)
	Pests, miscellaneous	abamectin (*off-label*), Metarhizium anisopliae
	Scale insects	deltamethrin
	Spider mites	fatty acids C7-C20 (*off-label*), Lecanicillium muscarium (*off-label*), spirodiclofen (*off-label*)
	Thrips	deltamethrin, Lecanicillium muscarium (*off-label*), spinosad (*off-label*)
	Whiteflies	Beauveria bassiana, deltamethrin, fatty acids C7-C20 (*off-label*), Lecanicillium muscarium (*off-label*), pymetrozine (*off-label*), spirotetramat (*off-label*)
Weeds	Broad-leaved weeds	napropamide (*off-label*), propyzamide (*off-label*)
	Grass weeds	napropamide (*off-label*), propyzamide (*off-label*)

Protected leafy vegetables - Mustard and cress

Diseases	Alternaria	Bacillus amyloliquefaciens D747, cyprodinil + fludioxonil (*off-label*)
	Botrytis	cyprodinil + fludioxonil (*off-label*), Gliocladium catenulatum (*off-label*)
	Disease control	dimethomorph + mancozeb (*off-label*), mandipropamid

	Downy mildew	dimethomorph + mancozeb (*off-label*), fenamidone + fosetyl-aluminium (*off-label*), mandipropamid
	Fusarium	Bacillus amyloliquefaciens D747, Gliocladium catenulatum (*off-label*)
	Phytophthora	Gliocladium catenulatum (*off-label*)
	Powdery mildew	Bacillus amyloliquefaciens D747
	Pythium	Bacillus amyloliquefaciens D747
	Rhizoctonia	Bacillus amyloliquefaciens D747, cyprodinil + fludioxonil (*off-label*), Gliocladium catenulatum (*off-label*)
	Stem canker	cyprodinil + fludioxonil (*off-label*)
Pests	Aphids	acetamiprid (*off-label*), alpha-cypermethrin (*off-label*), deltamethrin (*off-label*), spirotetramat (*off-label*)
	Beetles	alpha-cypermethrin (*off-label*), deltamethrin (*off-label*)
	Caterpillars	alpha-cypermethrin (*off-label*), deltamethrin (*off-label*), spinosad (*off-label*)
	Pests, miscellaneous	deltamethrin (*off-label*)
	Thrips	spinosad (*off-label*)
	Whiteflies	spirotetramat (*off-label*)
Weeds	Broad-leaved weeds	pendimethalin (*off-label*), propyzamide (*off-label*)
	Crops as weeds	cycloxydim (*off-label*)
	Grass weeds	cycloxydim (*off-label*), propyzamide (*off-label*)

Protected leafy vegetables - Protected leafy vegetables

Diseases	Alternaria	cyprodinil + fludioxonil (*off-label*), iprodione
	Botrytis	boscalid + pyraclostrobin, boscalid + pyraclostrobin (*off-label*), cyprodinil + fludioxonil (*off-label*), fluopyram + trifloxystrobin (*off-label*), iprodione
	Bottom rot	boscalid + pyraclostrobin
	Disease control	mandipropamid
	Downy mildew	azoxystrobin, boscalid + pyraclostrobin (*off-label*), dimethomorph (*off-label*), fosetyl-aluminium + propamocarb hydrochloride, mandipropamid, potassium bicarbonate (commodity substance) (*off-label*)
	Fusarium	Trichoderma asperellum (Strain T34) (*off-label*)
	Powdery mildew	fluopyram + trifloxystrobin (*off-label*)
	Pythium	fosetyl-aluminium + propamocarb hydrochloride, Trichoderma asperellum (Strain T34) (*off-label*)
	Rhizoctonia	boscalid + pyraclostrobin (*off-label*), cyprodinil + fludioxonil (*off-label*)
	Sclerotinia	boscalid + pyraclostrobin (*off-label*), fluopyram + trifloxystrobin (*off-label*)
	Soft rot	boscalid + pyraclostrobin
	Stem canker	cyprodinil + fludioxonil (*off-label*)
Pests	Aphids	acetamiprid (*off-label*), alpha-cypermethrin (*off-label*), lambda-cyhalothrin (*off-label*), pymetrozine (*off-label*), pyrethrins, spirotetramat (*off-label*), thiacloprid (*off-label*)
	Beetles	alpha-cypermethrin (*off-label*), lambda-cyhalothrin (*off-label*)

	Caterpillars	alpha-cypermethrin (*off-label*), Bacillus thuringiensis (*off-label*), lambda-cyhalothrin (*off-label*), pyrethrins
	Cutworms	Bacillus thuringiensis (*off-label*)
	Mites	abamectin (*off-label*)
	Pests, miscellaneous	abamectin (*off-label*), lambda-cyhalothrin (*off-label*)
	Whiteflies	spirotetramat (*off-label*)
Weeds	Broad-leaved weeds	napropamide (*off-label*), propyzamide (*off-label*)
	Grass weeds	napropamide (*off-label*), propyzamide (*off-label*)

Protected leafy vegetables - Protected spinach

Diseases	Alternaria	cyprodinil + fludioxonil (*off-label*)
	Botrytis	boscalid + pyraclostrobin (*off-label*), cyprodinil + fludioxonil (*off-label*)
	Disease control	mandipropamid
	Downy mildew	mandipropamid, metalaxyl-M (*off-label*)
	Rhizoctonia	boscalid + pyraclostrobin (*off-label*), cyprodinil + fludioxonil (*off-label*)
	Sclerotinia	boscalid + pyraclostrobin (*off-label*)
	Stem canker	cyprodinil + fludioxonil (*off-label*)
Pests	Aphids	acetamiprid (*off-label*), spirotetramat (*off-label*)
	Caterpillars	Bacillus thuringiensis (*off-label*)
	Whiteflies	spirotetramat (*off-label*)

Protected legumes - Protected peas and beans

Diseases	Damping off	fosetyl-aluminium + propamocarb hydrochloride (*off-label*)
Pests	Caterpillars	Bacillus thuringiensis, Bacillus thuringiensis (*off-label*)
	Cutworms	Bacillus thuringiensis (*off-label*)

Protected root and tuber vegetables - Protected root brassicas

Diseases	Alternaria	cyprodinil + fludioxonil (*off-label*)
	Damping off	fosetyl-aluminium + propamocarb hydrochloride
	Disease control	boscalid + pyraclostrobin (*off-label*)
	Downy mildew	fosetyl-aluminium + propamocarb hydrochloride
	Leaf spot	cyprodinil + fludioxonil (*off-label*)
	Sclerotinia	cyprodinil + fludioxonil (*off-label*)

Protected stem and bulb vegetables - Protected asparagus

Diseases	Botrytis	cyprodinil + fludioxonil (*off-label*)

Protected stem and bulb vegetables - Protected celery/chicory

Diseases	Alternaria	azoxystrobin (*off-label*)
	Botrytis	azoxystrobin (*off-label*)
	Leaf spot	azoxystrobin (*off-label*)
	Phytophthora	azoxystrobin (*off-label*)
	Powdery mildew	azoxystrobin (*off-label*)
	Rhizoctonia	azoxystrobin (*off-label*)
	Rust	azoxystrobin (*off-label*)
	Sclerotinia	azoxystrobin (*off-label*)
Pests	Aphids	pymetrozine (*off-label*)

	Caterpillars	Bacillus thuringiensis (*off-label*), deltamethrin (*off-label*)
	Cutworms	Bacillus thuringiensis (*off-label*)
	Mites	abamectin (*off-label*)
	Pests, miscellaneous	abamectin (*off-label*)

Protected stem and bulb vegetables - Protected onions/leeks/garlic

Pests	Leaf miners	deltamethrin (*off-label*)
	Pests, miscellaneous	deltamethrin (*off-label*)

Total vegetation control

Aquatic situations, general - Aquatic situations

Pests	Flies	Bacillus thuringiensis israelensis, physical pest control
Weeds	Aquatic weeds	glyphosate
	Grass weeds	glyphosate
	Weeds, miscellaneous	glyphosate

Non-crop areas, general - Miscellaneous non-crop situations

Pests	Birds/mammals	difenacoum, flocoumafen
Weeds	Weeds, miscellaneous	2,4-D + glyphosate, flazasulfuron

Non-crop areas, general - Non-crop farm areas

Weeds	Broad-leaved weeds	metsulfuron-methyl, tribenuron-methyl
	Crops as weeds	metsulfuron-methyl
	Weeds, miscellaneous	2,4-D + glyphosate, glyphosate, glyphosate + pyraflufen-ethyl, metsulfuron-methyl

Non-crop areas, general - Paths/roads etc

Pests	Birds/mammals	aluminium ammonium sulphate, aluminium ammonium sulphate (*anti-fouling*)
	Slugs/snails	metaldehyde
Weeds	Grass weeds	glyphosate
	Mosses	maleic hydrazide + pelargonic acid, starch, protein, oil and water
	Weeds, miscellaneous	acetic acid, diflufenican + glyphosate, flazasulfuron, glufosinate-ammonium, glyphosate, glyphosate + pyraflufen-ethyl, glyphosate + sulfosulfuron, maleic hydrazide + pelargonic acid, starch, protein, oil and water

SECTION 2
PESTICIDE PROFILES

1   abamectin

A selective acaricide and insecticide for use in ornamentals and other protected crops
IRAC mode of action code: 6

Products

1 Amec	Harvest	18 g/l	EC	18035
2 Clayton Abba	Clayton	18 g/l	EC	13808
3 Dynamec	Syngenta	18 g/l	EC	13331
4 Killermite	Pan Agriculture	18 g/l	EC	16453
5 Smite	Pan Agriculture	18 g/l	EC	16255

Uses

- Insect control in **protected chives** *(off-label)*, **protected cress** *(off-label)*, **protected frise** *(off-label)*, **protected herbs (see appendix 6)** *(off-label)*, **protected lamb's lettuce** *(off-label)*, **protected lettuce** *(off-label)*, **protected parsley** *(off-label)*, **protected radicchio** *(off-label)*, **protected scarole** *(off-label)* [3]
- Leaf and bud mite in **protected blackberries** *(off-label)*, **protected raspberries** *(off-label)* [3]
- Leaf miner in **protected cherry tomatoes** *(off-label)* [3]; **protected tomatoes** [1-5]
- Mites in **protected blackberries** *(off-label)*, **protected cress** *(off-label)*, **protected frise** *(off-label)*, **protected herbs (see appendix 6)** *(off-label)*, **protected lamb's lettuce** *(off-label)*, **protected leaf brassicas** *(off-label)*, **protected lettuce** *(off-label)*, **protected radicchio** *(off-label)*, **protected raspberries** *(off-label)*, **protected scarole** *(off-label)* [2]
- Red spider mites in **hops** *(off-label)*, **protected blackberries** *(off-label)*, **protected cayenne peppers** *(off-label)*, **protected peppers** *(off-label)*, **protected raspberries** *(off-label)* [3]; **protected strawberries** *(off-label)* [2, 3]; **strawberries** *(off-label)* [2]
- Tarsonemid mites in **protected strawberries** *(off-label)*, **strawberries** *(off-label)* [2]; **strawberries** *(off-label - in propagation)* [3]
- Thrips in **leeks** *(off-label)*, **spring onions** *(off-label)* [3]
- Two-spotted spider mite in **ornamental plant production, protected cucumbers, protected tomatoes** [1-5]; **protected aubergines** *(off-label)* [3]; **protected ornamentals** [1-3]
- Western flower thrips in **ornamental plant production, protected cucumbers** [1-5]; **protected aubergines** *(off-label)* [3]; **protected ornamentals** [1-3]

Extension of Authorisation for Minor Use (EAMUs)

- **hops** *20170278* [3]
- **leeks** *20160259* [3]
- **protected aubergines** *20070421* [3]
- **protected blackberries** *20101940* [2], *20072290* [3]
- **protected cayenne peppers** *20070422* [3]
- **protected cherry tomatoes** *20070422* [3]
- **protected chives** *20070430* [3]
- **protected cress** *20101939* [2], *20070430* [3]
- **protected frise** *20101939* [2], *20070430* [3]
- **protected herbs (see appendix 6)** *20101939* [2], *20070430* [3]
- **protected lamb's lettuce** *20101939* [2], *20070430* [3]
- **protected leaf brassicas** *20101939* [2]
- **protected lettuce** *20101939* [2], *20070430* [3]
- **protected parsley** *20070430* [3]
- **protected peppers** *20070422* [3]
- **protected radicchio** *20101939* [2], *20070430* [3]
- **protected raspberries** *20101940* [2], *20072290* [3]
- **protected scarole** *20101939* [2], *20070430* [3]
- **protected strawberries** *20090342* [2], *20070423* [3]
- **spring onions** *20160259* [3]
- **strawberries** *20090342* [2], *(in propagation) 20070423* [3]

Approval information

- Abamectin included in Annex I under EC Regulation 1107/2009

SEE SECTION 3 FOR PRODUCTS ALSO REGISTERED

Efficacy guidance

- Abamectin controls adults and immature stages of the two-spotted spider mite, the larval stages of leaf miners and the nymphs of Western Flower Thrips, plus a useful reduction in adults
- Treat at first sign of infestation. Repeat sprays may be required
- For effective control total cover of all plant surfaces is essential, but avoid run-off
- Target pests quickly become immobilised but 3-5 d may be required for maximum mortality
- Indoor applications should be made through a hydraulic nozzle applicator or a knapsack applicator. Outdoors suitable high volume hydraulic nozzle applicators should be used
- Limited data shows abamectin only slightly harmful to Anthocorid bugs and so compatible with biological control systems in which Anthocorid bugs are important

Restrictions

- Number of treatments 6 on protected tomatoes and cucumbers (only 4 of which can be made when flowers or fruit present); not restricted on flowers but rotation with other products advised
- Maximum concentration must not exceed 50 ml per 100 l water
- Do not mix with wetters, stickers or other adjuvants
- Do not use on ferns (*Adiantum* spp) or Shasta daisies
- Do not treat protected crops which are in flower or have set fruit between 1 Nov and 28 Feb
- Do not treat cherry tomatoes (but protected cherry tomatoes may be treated off-label)
- Consult manufacturer for list of plant varieties tested for safety
- There is insufficient evidence to support product compatibility with integrated and biological pest control programmes
- Unprotected persons must be kept out of treated areas until the spray has dried

Crop-specific information

- HI 3 d for protected edible crops
- On tomato or cucumber crops that are in flower, or have started to set fruit, treat only between 1 Mar and 31 Oct. Seedling tomatoes or cucumbers that have not started to flower or set fruit may be treated at any time
- Some spotting or staining may occur on carnation, kalanchoe and begonia foliage

Environmental safety

- Dangerous for the environment
- Very toxic to aquatic organisms
- High risk to bees. Do not apply to crops in flower or to those in which bees are actively foraging. Do not apply when flowering weeds are present
- Keep in original container, tightly closed, in a safe place, under lock and key
- Where bumble bees are used in tomatoes as pollinators, keep them out for 24 h after treatment

Hazard classification and safety precautions

Hazard Harmful, Dangerous for the environment [1-5]; Harmful if swallowed [4, 5]; Very toxic to aquatic organisms [2]
Transport code 9
Packaging group III
UN Number 3082
Risk phrases R22a [1-3]; R50, R53a [1, 3]
Operator protection A, C, D, H, K, M; U02a, U05a, U20a
Environmental protection E02a [1-5] (until spray has dried); E12a, E12e, E15b, E34, E38 [1-5]; H410 [2, 4, 5]
Storage and disposal D01, D02, D05, D09b, D10c, D12a
Medical advice M04a

2   acetamiprid

A neonicotinoid insecticide for use in top fruit and horticulture
IRAC mode of action code: 4A

Products

1	Gazelle SG	Certis	20% w/w	SG	13725
2	Insyst	Certis	20% w/w	SP	13414

FOR FULL CONDITIONS OF USE ALWAYS READ THE PRODUCT LABEL

Products – continued

| 3 | Sangue | Agroquimicos | 20% w/w | SG | 16042 |
| 4 | Vulcan | Pan Agriculture | 20% w/w | SG | 16689 |

Uses

- Aphids in **apples, cherries, ornamental plant production, pears, plums, protected aubergines, protected ornamentals, protected peppers, protected tomatoes** [1, 3, 4]; **brussels sprouts** *(off-label)*, **seed potatoes, spring oilseed rape, ware potatoes, winter oilseed rape** [2]; **chard** *(off-label)*, **cress** *(off-label)*, **forest nurseries** *(off-label)*, **frise** *(off-label)*, **herbs (see appendix 6)** *(off-label)*, **hops** *(off-label)*, **lamb's lettuce** *(off-label)*, **leaf brassicas** *(off-label)*, **lettuce** *(off-label)*, **parsley** *(off-label)*, **protected cress** *(off-label)*, **protected forest nurseries** *(off-label)*, **protected herbs (see appendix 6)** *(off-label)*, **protected hops** *(off-label)*, **protected lamb's lettuce** *(off-label)*, **protected leaf brassicas** *(off-label)*, **protected lettuce** *(off-label)*, **protected parsley** *(off-label)*, **protected rocket** *(off-label)*, **protected soft fruit** *(off-label)*, **protected spinach** *(off-label)*, **radicchio** *(off-label)*, **rocket** *(off-label)*, **scarole** *(off-label)*, **soft fruit** *(off-label)*, **spinach** *(off-label)*, **spinach beet** *(off-label)*, **top fruit** *(off-label)* [1]
- Common oak thelaxid in **forest nurseries** *(off-label)*, **protected forest nurseries** *(off-label)* [1]
- Gall wasp in **forest nurseries** *(off-label)*, **protected forest nurseries** *(off-label)* [1]
- Great spruce bark beetle in **forest nurseries** *(off-label)*, **protected forest nurseries** *(off-label)* [1]
- Green aphid in **forest nurseries** *(off-label)*, **protected forest nurseries** *(off-label)* [1]
- Large pine weevil in **forest** *(off-label)* [1]
- Pine weevil in **forest nurseries** *(off-label)* [1]
- Whitefly in **apples, cherries, ornamental plant production, pears, plums, protected aubergines, protected ornamentals, protected peppers, protected tomatoes** [1, 3, 4]; **brussels sprouts** *(off-label)* [2]
- Woolly aphid in **forest nurseries** *(off-label)*, **protected forest nurseries** *(off-label)* [1]

Extension of Authorisation for Minor Use (EAMUs)

- **brussels sprouts** *20072866* [2]
- **chard** *20113144* [1]
- **cress** *20101101* [1]
- **forest** *20121068* [1]
- **forest nurseries** *20111313* [1], *20160084* [1], *20162653* [1]
- **frise** *20111994* [1]
- **herbs (see appendix 6)** *20101101* [1]
- **hops** *20082857* [1]
- **lamb's lettuce** *20111994* [1]
- **leaf brassicas** *20111994* [1]
- **lettuce** *20111994* [1]
- **parsley** *20111994* [1]
- **protected cress** *20101101* [1]
- **protected forest nurseries** *20162653* [1]
- **protected herbs (see appendix 6)** *20101101* [1]
- **protected hops** *20082857* [1]
- **protected lamb's lettuce** *20111994* [1]
- **protected leaf brassicas** *20111994* [1]
- **protected lettuce** *20111994* [1]
- **protected parsley** *20111994* [1]
- **protected rocket** *20111994* [1]
- **protected soft fruit** *20082857* [1]
- **protected spinach** *20101101* [1]
- **radicchio** *20111994* [1]
- **rocket** *20111994* [1]
- **scarole** *20111994* [1]
- **soft fruit** *20082857* [1]
- **spinach** *20101101* [1]
- **spinach beet** *20113144* [1]
- **top fruit** *20082857* [1]

SEE SECTION 3 FOR PRODUCTS ALSO REGISTERED

SECTION 2

Approval information
- Acetamiprid included in Annex I under EC Regulation 1107/2009

Efficacy guidance
- Best results obtained from application at the first sign of pest attack or when appropriate thresholds are reached
- Thorough coverage of foliage is essential to ensure best control. Acetamiprid has contact, systemic and translaminar activity

Restrictions
- Maximum number of treatments 1 per yr for cherries, ware potatoes; 2 per yr or crop for apples, pears, plums, ornamental plant production, protected crops, seed potatoes
- Do not use more than two applications of any neonicotinoid insecticide (e.g. acetamiprid, clothianidin, imidacloprid, thiacloprid) on any crop. Previous soil or seed treatment with a neonicotinoid counts as one such treatment
- 2 treatments are permitted on seed potatoes but must not be used consecutively [2]

Crop-specific information
- HI 3 d for protected crops; 14 d for apples, cherries, pears, plums, potatoes

Environmental safety
- Harmful to aquatic organisms
- Acetamiprid is slightly toxic to predatory mites and generally slightly toxic to other beneficials
- Broadcast air-assisted LERAP [1, 3, 4] (18 m); LERAP Category B [1-4]

Hazard classification and safety precautions
Hazard Dangerous for the environment [1-4]; Harmful if swallowed [1, 4]
UN Number N/C
Risk phrases R22a, R52, R53a [3]
Operator protection A [1-4]; H [1, 3, 4]; U02a, U19a [1-4]; U20a [2]; U20b [1, 3, 4]
Environmental protection E15b, E16a, E16b [1-4]; E17b [1, 3, 4] (18 m); E38 [2]; H412 [1, 2, 4]
Storage and disposal D05 [1, 3, 4]; D09a, D10b [1-4]; D12a [2]

3   acetic acid

A non-selective herbicide for non-crop situations

Products

New Way Weed Spray	Headland Amenity	240 g/l	SL	15319

Uses
- General weed control in **hard surfaces**, **natural surfaces not intended to bear vegetation**, **permeable surfaces overlying soil**

Approval information
- Acetic acid is included in Annex 1 under EC Regulation 1107/2009

Efficacy guidance
- Best results obtained from treatment of young tender weeds less than 10 cm high
- Treat in spring and repeat as necessary throughout the growing season
- Treat survivors as soon as fresh growth is seen
- Ensure complete coverage of foliage to the point of run-off
- Rainfall after treatment may reduce efficacy

Restrictions
- No restriction on number of treatments

Following crops guidance
- There is no residual activity in the soil and sowing or planting may take place as soon as treated weeds have died

Environmental safety
- Harmful to aquatic organisms

FOR FULL CONDITIONS OF USE ALWAYS READ THE PRODUCT LABEL

- High risk to bees
- Do not apply to crops in flower or to those in which bees are actively foraging. Do not apply when flowering weeds are present
- Keep people and animals off treated dense weed patches until spray has dried. This is not necessary for areas with occasional low growing prostrate weeds such as on pathways

Hazard classification and safety precautions
Hazard Irritant
UN Number N/C
Risk phrases R36, R37, R38, R52
Operator protection A, C, D, H; U05a, U09a, U19a, U20b
Environmental protection E12a, E12e, E15a
Storage and disposal D01, D02, D09a, D12a

4   alpha-cypermethrin

A contact and ingested pyrethroid insecticide for use in arable crops and agricultural buildings
IRAC mode of action code: 3

Products

1	Al-cyper Ec	Harvest	100 g/l	EC	17801
2	Eribea	Belchim	100 g/l	EC	17270

Uses

- Aphids in **baby leaf crops** (off-label), **celery leaves** (off-label), **chicory** (off-label), **chives** (off-label), **coriander** (off-label), **cress** (off-label), **edible flowers** (off-label), **endives** (off-label), **fennel leaves** (off-label), **herbs (see appendix 6)** (off-label), **hyssop** (off-label), **lamb's lettuce** (off-label), **land cress** (off-label), **lettuce** (off-label), **lovage** (off-label), **protected baby leaf crops** (off-label), **protected endives** (off-label), **protected herbs (see appendix 6)** (off-label), **protected lettuce** (off-label), **rocket** (off-label), **rosemary** (off-label), **sage** (off-label), **salad burnet** (off-label), **savory** (off-label), **spinach** (off-label), **spinach beet** (off-label), **sweet ciceley** (off-label), **tarragon** (off-label), **thyme** (off-label), **watercress** (off-label) [2]
- Brassica pod midge in **winter oilseed rape** [2]
- Cabbage seed weevil in **spring oilseed rape**, **winter oilseed rape** [1, 2]
- Cabbage stem flea beetle in **winter oilseed rape** [2]
- Caterpillars in **baby leaf crops** (off-label), **celery leaves** (off-label), **chicory** (off-label), **chives** (off-label), **choi sum** (off-label), **collards** (off-label), **coriander** (off-label), **cress** (off-label), **edible flowers** (off-label), **endives** (off-label), **fennel leaves** (off-label), **herbs (see appendix 6)** (off-label), **hyssop** (off-label), **kohlrabi** (off-label), **lamb's lettuce** (off-label), **land cress** (off-label), **lettuce** (off-label), **lovage** (off-label), **oriental cabbage** (off-label), **protected baby leaf crops** (off-label), **protected choi sum** (off-label), **protected collards** (off-label), **protected endives** (off-label), **protected herbs (see appendix 6)** (off-label), **protected kohlrabi** (off-label), **protected lettuce** (off-label), **protected oriental cabbage** (off-label), **rocket** (off-label), **rosemary** (off-label), **sage** (off-label), **salad burnet** (off-label), **savory** (off-label), **spinach** (off-label), **spinach beet** (off-label), **sweet ciceley** (off-label), **tarragon** (off-label), **thyme** (off-label), **watercress** (off-label) [2]; **broccoli, brussels sprouts, cabbages, calabrese, cauliflowers, kale** [1, 2]
- Cereal aphid in **spring barley, spring wheat, winter barley, winter wheat** [1, 2]
- Flea beetle in **baby leaf crops** (off-label), **celery leaves** (off-label), **chicory** (off-label), **chives** (off-label), **choi sum** (off-label), **collards** (off-label), **coriander** (off-label), **cress** (off-label), **edible flowers** (off-label), **endives** (off-label), **fennel leaves** (off-label), **herbs (see appendix 6)** (off-label), **hyssop** (off-label), **kohlrabi** (off-label), **lamb's lettuce** (off-label), **land cress** (off-label), **lettuce** (off-label), **lovage** (off-label), **oriental cabbage** (off-label), **protected baby leaf crops** (off-label), **protected choi sum** (off-label), **protected collards** (off-label), **protected endives** (off-label), **protected herbs (see appendix 6)** (off-label), **protected kohlrabi** (off-label), **protected lettuce** (off-label), **protected oriental cabbage** (off-label), **rocket** (off-label), **rosemary** (off-label), **salad burnet** (off-label), **savory** (off-label), **spinach** (off-label), **spinach beet** (off-label), **sweet ciceley** (off-label), **tarragon** (off-label), **thyme** (off-label), **watercress** (off-label) [2]; **broccoli, brussels sprouts, cabbages, calabrese, cauliflowers, kale** [1, 2]

SEE SECTION 3 FOR PRODUCTS ALSO REGISTERED

- Pea and bean weevil in **broad beans**, **combining peas**, **spring field beans**, **vining peas**, **winter field beans** [1, 2]
- Pea aphid in **combining peas** *(reduction)*, **vining peas** *(reduction)* [1, 2]
- Pea moth in **combining peas**, **vining peas** [1, 2]
- Pollen beetle in **spring oilseed rape**, **winter oilseed rape** [1, 2]
- Rape winter stem weevil in **winter oilseed rape** [2]
- Silver Y moth in **baby leaf crops** *(off-label)*, **celery leaves** *(off-label)*, **chicory** *(off-label)*, **chives** *(off-label)*, **coriander** *(off-label)*, **cress** *(off-label)*, **edible flowers** *(off-label)*, **endives** *(off-label)*, **fennel leaves** *(off-label)*, **herbs (see appendix 6)** *(off-label)*, **hyssop** *(off-label)*, **lamb's lettuce** *(off-label)*, **land cress** *(off-label)*, **lettuce** *(off-label)*, **lovage** *(off-label)*, **protected baby leaf crops** *(off-label)*, **protected endives** *(off-label)*, **protected herbs (see appendix 6)** *(off-label)*, **protected lettuce** *(off-label)*, **rocket** *(off-label)*, **rosemary** *(off-label)*, **sage** *(off-label)*, **salad burnet** *(off-label)*, **savory** *(off-label)*, **spinach** *(off-label)*, **spinach beet** *(off-label)*, **sweet ciceley** *(off-label)*, **tarragon** *(off-label)*, **thyme** *(off-label)*, **watercress** *(off-label)* [2]
- Yellow cereal fly in **winter barley**, **winter wheat** [2]

Extension of Authorisation for Minor Use (EAMUs)
- **baby leaf crops** *20170518* [2]
- **celery leaves** *20170518* [2]
- **chicory** *20170518* [2]
- **chives** *20170518* [2]
- **choi sum** *20170515* [2]
- **collards** *20170515* [2]
- **coriander** *20170518* [2]
- **cress** *20170518* [2]
- **edible flowers** *20170518* [2]
- **endives** *20170518* [2]
- **fennel leaves** *20170518* [2]
- **herbs (see appendix 6)** *20170518* [2]
- **hyssop** *20170518* [2]
- **kohlrabi** *20170515* [2]
- **lamb's lettuce** *20170518* [2]
- **land cress** *20170518* [2]
- **lettuce** *20170518* [2]
- **lovage** *20170518* [2]
- **oriental cabbage** *20170515* [2]
- **protected baby leaf crops** *20170518* [2]
- **protected choi sum** *20170515* [2]
- **protected collards** *20170515* [2]
- **protected endives** *20170518* [2]
- **protected herbs (see appendix 6)** *20170518* [2]
- **protected kohlrabi** *20170515* [2]
- **protected lettuce** *20170518* [2]
- **protected oriental cabbage** *20170515* [2]
- **rocket** *20170518* [2]
- **rosemary** *20170518* [2]
- **sage** *20170518* [2]
- **salad burnet** *20170518* [2]
- **savory** *20170518* [2]
- **spinach** *20170518* [2]
- **spinach beet** *20170518* [2]
- **sweet ciceley** *20170518* [2]
- **tarragon** *20170518* [2]
- **thyme** *20170518* [2]
- **watercress** *20170518* [2]

Approval information
- Alpha-cypermethrin included in Annex I under EC Regulation 1107/2009
- Accepted by BBPA for use on malting barley

Efficacy guidance
- For cabbage stem flea beetle control spray oilseed rape when adult or larval damage first seen and about 1 mth later
- For flowering pests on oilseed rape apply at any time during flowering, against pollen beetle best results achieved at green to yellow bud stage (GS 3.3-3.7), against seed weevil between 20 pods set and 80% petal fall (GS 4.7-5.8)
- Spray cereals in autumn for control of cereal aphids, in spring/summer for grain aphids. (See label for details)
- For flea beetle, caterpillar and cabbage aphid control on brassicas apply when the pest or damage first seen or as a preventive spray. Repeat if necessary
- For pea and bean weevil control in peas and beans apply when pest attack first seen and repeat as necessary

Restrictions
- Apply up to 2 sprays on cereals in autumn and spring, 1 in summer between 1 Apr and 31 Aug. See label for details of rates and maximum total dose
- Only 1 aphicide treatment may be applied in cereals between 1 Apr and 31 Aug and spray volume must not be reduced in this period
- Do not apply to a cereal crop if any product containing a pyrethroid or dimethoate has been applied after the start of ear emergence (GS 51)
- Use low drift spraying equipment up to 30m from the top of the bank of any surface water bodies [1, 2]

Crop-specific information
- Latest use: before the end of flowering for oilseed rape; before 31 Aug for cereals
- HI vining peas 1 d; brassicas, combining peas, broad beans, field beans 7 d
- For summer cereal application do not spray within 6 m from edge of crop and do not reduce volume when used after 31 Mar

Environmental safety
- Very toxic to aquatic organisms
- Extremely dangerous to fish or other aquatic life. Do not contaminate surface waters or ditches with chemical or used container
- Dangerous to bees. Where possible spray oilseed rape crops in the late evening or early morning or in dull weather. Give local beekeepers warning when using in flowering crops
- Do not spray within 6 m of the edge of a cereal crop after 31 Mar in yr of harvest
- Buffer zone requirement 12m [1, 2]
- To protect non target insects/arthropods respect an unsprayed buffer zone of 5 m to non-crop land [1, 2]
- Horizontal boom sprayers must be fitted with three star drift reduction technology for all uses
- LERAP Category A [1]; LERAP Category B [2]

Hazard classification and safety precautions
Hazard Harmful, Dangerous for the environment, Flammable liquid and vapour, Toxic if swallowed, Harmful if inhaled, Very toxic to aquatic organisms [1, 2]; Flammable [1]
Transport code 6.1
Packaging group III
UN Number 2903
Risk phrases H304, H315, H317, H318, H335, H336, H373 [1, 2]; R43 [2]
Operator protection A [1, 2]; H [1]; U05a, U19c [1, 2]; U10, U20b [1]; U14 [2]
Environmental protection E12c, E38, H410 [1, 2]; E15a, E16c, E16d [1]; E16a [2]
Storage and disposal D01, D02, D05, D12a [1, 2]; D09a, D10b [1]
Medical advice M04a [2]; M05a [1]; M05b [1, 2]

SEE SECTION 3 FOR PRODUCTS ALSO REGISTERED

5   aluminium ammonium sulphate

An inorganic bird and animal repellent

Products

1	Curb Crop Spray Powder	Sphere	88% w/w	WP	02480
2	Liquid Curb Crop Spray	Sphere	83 g/l	SC	03164
3	Rezist	Sphere	88% w/w	WP	08576
4	Rezist Liquid	Sphere	83 g/l	SC	14643
5	Sphere ASBO	Sphere	83 g/l	SC	15064

Uses

- Animal repellent in *agricultural premises, all top fruit, broad beans, bush fruit, cane fruit, carrots, flowerhead brassicas, forest nursery beds, forestry plantations, grain stores, leaf brassicas, peas, permanent grassland, spring barley, spring field beans, spring oats, spring oilseed rape, spring wheat, strawberries, sugar beet, winter barley, winter field beans, winter oats, winter oilseed rape, winter wheat* [1, 2]; *all edible crops (outdoor), all non-edible crops (outdoor), forest* [5]; *amenity grassland, hard surfaces, managed amenity turf* [4, 5]; *amenity vegetation* [1]
- Bird repellent in *agricultural premises, all top fruit, broad beans, bush fruit, cane fruit, carrots, flowerhead brassicas, forest nursery beds, forestry plantations, grain stores, leaf brassicas, peas, permanent grassland, spring barley, spring field beans, spring oats, spring oilseed rape, spring wheat, strawberries, sugar beet, winter barley, winter field beans, winter oats, winter oilseed rape, winter wheat* [1, 2]; *all edible crops (outdoor), all non-edible crops (outdoor), forest* [5]; *amenity grassland, hard surfaces, managed amenity turf* [4, 5]; *amenity vegetation* [1]
- Dogs in *amenity grassland* (anti-fouling), *hard surfaces* (anti-fouling), *managed amenity turf* (anti-fouling) [3]
- Moles in *amenity grassland, hard surfaces, managed amenity turf* [3]

Approval information

- aluminium ammonium sulphate is included in Annex 1 under EC Regulation 1107/2009

Efficacy guidance

- Apply as overall spray to growing crops before damage starts or mix powder with seed depending on type of protection required
- Spray deposit protects growth present at spraying but gives little protection to new growth
- Product must be sprayed onto dry foliage to be effective and must dry completely before dew or frost forms. In winter this may require some wind

Crop-specific information

- Latest use: no restriction

Hazard classification and safety precautions

UN Number N/C
Operator protection U05a [1, 2, 5]; U15 [4]; U20a [1-5]
Environmental protection E15a [1-5]; E19b [1, 2, 4, 5]
Storage and disposal D01, D02, D05 [1, 2, 4, 5]; D09a [3, 4]; D10b [4]; D11a [1-5]; D12a [1, 2, 5]
Medical advice M03 [1, 2, 4, 5]

6   aluminium phosphide

A phosphine generating compound used against vertebrates and grain store pests. Under new regulations any person wishing to purchase or use Aluminium Phosphide must be suitably qualified and licensed by January 2015. A new licensing body has been set up.
IRAC mode of action code: 24A

Products

1	Degesch Fumigation Tablets	Rentokil	56% w/w	GE	17035
2	Detia Gas-Ex-B	Rentokil	57% w/w	GE	17097

FOR FULL CONDITIONS OF USE ALWAYS READ THE PRODUCT LABEL

Products – continued

3	Phostoxin	Rentokil	56% w/w	GE	17000
4	Talunex	Killgerm	56% w/w	GE	17001

Uses

- Insect pests in **carob**, **cocoa**, **coffee**, **herbal infusions**, **stored dried spices**, **stored grain**, **stored linseed**, **stored nuts**, **stored oilseed rape**, **stored pulses**, **tea**, **tobacco** [1]; **coconuts**, **dried fruit**, **grain stores**, **nuts**, **pulses** [2]; **crop handling & storage structures**, **processed consumable products** [1, 2]
- Moles in **all situations** [3, 4]; **farmland** [3]
- Rabbits in **all situations** [3, 4]; **farmland** [3]
- Rats in **all situations** [3, 4]; **farmland** [3]

Approval information

- Aluminium phosphide is included in Annex 1 under EC Regulation 1107/2009
- Accepted by BBPA for use in stores for malting barley

Efficacy guidance

- Product releases poisonous hydrogen phosphide gas in contact with moisture
- Place fumigation tablets in grain stores as directed [1, 2]
- Place pellets in burrows or runs and seal hole by heeling in or covering with turf. Do not cover pellets with soil. Inspect daily and treat any new or re-opened holes [3, 4]

Restrictions

- Aluminium phosphide is subject to the Poisons Rules 1982 and the Poisons Act 1972. See Section 5 for more information
- Only to be used by operators instructed or trained in the use of aluminium phosphide and familiar with the precautionary measures to be taken. See label and HSE Guidance Notes for full precautions
- Only open container outdoors [3, 4] and for immediate use. Keep away from liquid or water as this causes immediate release of gas. Do not use in wet weather
- Do not use within 3 m of human or animal habitation. Before application ensure that no humans or domestic animals are in adjacent buildings or structures. Allow a minimum airing-off period of 4 h before re-admission

Environmental safety

- Product liberates very toxic, highly flammable gas
- Dangerous to fish or other aquatic life. Do not contaminate surface waters or ditches with chemical or used container
- Prevent access by livestock, pets and other non-target mammals and birds to buildings under fumigation and ventilation
- Pellets must never be placed or allowed to remain on ground surface
- Do not use adjacent to watercourses
- Take particular care to avoid gassing non-target animals, especially those protected under the Wildlife and Countryside Act (e.g. badgers, polecat, reptiles, natterjack toads, most birds). Do not use in burrows where there is evidence of badger or fox activity, or when burrows might be occupied by birds
- Dust remaining after decomposition is harmless and of no environmental hazard
- Keep in original container, tightly closed, in a safe place, under lock and key
- Dispose of empty containers as directed on label

Hazard classification and safety precautions

Hazard Very toxic, In contact with water releases flammable gases which may ignite spontaneously, Fatal if swallowed, Toxic in contact with skin, Fatal if inhaled, Very toxic to aquatic organisms [1-4]; Harmful [2-4]; Highly flammable [2, 3]; Dangerous for the environment [1-3]

Transport code 4.1 6.1 [4]; 4.3 [1-3]

Packaging group I

UN Number 1397

Risk phrases H315, H318 [1, 3, 4]

Operator protection A, D, H; U01, U07, U13, U19a, U20a [1-4]; U05a [2-4]; U05b, U18 [1]

Environmental protection E02a [1] (4 h min); E02a [2-4] (4 h); E02b [1]; E13b, E34 [1-4]

SEE SECTION 3 FOR PRODUCTS ALSO REGISTERED

Storage and disposal D01, D02, D07, D09b [1-4]; D11b [2-4]; D11c [1]
Vertebrate/rodent control products V04a [2-4]
Medical advice M04a

7 ametoctradin

A potato blight fungicide available only in mixtures
FRAC mode of action code: 45

See also ametoctradin + dimethomorph
ametoctradin + mancozeb

8 ametoctradin + dimethomorph

A systemic and protectant fungicide for potato blight control
FRAC mode of action code: 40 + 45

Products

1	Percos	BASF	300:225 g/l	SC	15248
2	Zampro DM	BASF	300:225 g/l	SC	15013

Uses
- Blight in **potatoes** [1, 2]
- Downy mildew in **hops** *(off-label)*, **ornamental plant production** *(off-label)*, **protected ornamentals** *(off-label)*, **table grapes** *(off-label)*, **wine grapes** *(off-label)* [1]
- Tuber blight in **potatoes** *(reduction)* [1, 2]

Extension of Authorisation for Minor Use (EAMUs)
- **hops** *20170556* [1]
- **ornamental plant production** *20130819* [1]
- **protected ornamentals** *20130819* [1]
- **table grapes** *20150254* [1]
- **wine grapes** *20150254* [1]

Approval information
- Ametoctradin and dimethomorph are included in Annex 1 under EC Regulation 1107/2009.

Efficacy guidance
- Application to dry foliage is rainfast within 1 hour of drying on the leaf.
- Will give effective control of phenylamide-resistant strains of blight

Crop-specific information
- HI 7 days for potatoes

Hazard classification and safety precautions
Hazard Harmful, Dangerous for the environment, Harmful if swallowed
UN Number N/C
Operator protection U05a
Environmental protection E15b, E34, E38, H412
Storage and disposal D01, D02, D09a, D10c
Medical advice M03, M05a

9 amidosulfuron

A post-emergence sulfonylurea herbicide for cleavers and other broad-leaved weed control in cereals
HRAC mode of action code: B

Products

1	Eagle	Interfarm	75% w/w	WG	16490
2	Squire Ultra	Interfarm	75% w/w	WG	16491

FOR FULL CONDITIONS OF USE ALWAYS READ THE PRODUCT LABEL

Uses

- Annual dicotyledons in **durum wheat, farm forestry** *(off-label)*, **forest nurseries** *(off-label)*, **game cover** *(off-label)*, **linseed, spring barley, spring oats, spring rye, spring wheat, triticale, winter barley, winter oats, winter rye, winter wheat** [1]; **grassland** [2]; **ornamental plant production** *(off-label)* [1, 2]
- Charlock in **corn gromwell** *(off-label)*, **durum wheat, farm forestry** *(off-label)*, **forest nurseries** *(off-label)*, **game cover** *(off-label)*, **linseed, spring barley, spring oats, spring rye, spring wheat, triticale, winter barley, winter oats, winter rye, winter wheat** [1]; **ornamental plant production** *(off-label)* [1, 2]
- Cleavers in **corn gromwell** *(off-label)*, **durum wheat, farm forestry** *(off-label)*, **forest nurseries** *(off-label)*, **game cover** *(off-label)*, **linseed, spring barley, spring oats, spring rye, spring wheat, triticale, winter barley, winter oats, winter rye, winter wheat** [1]; **ornamental plant production** *(off-label)* [1, 2]
- Docks in **grassland** [2]
- Forget-me-not in **corn gromwell** *(off-label)*, **durum wheat, farm forestry** *(off-label)*, **forest nurseries** *(off-label)*, **game cover** *(off-label)*, **linseed, spring barley, spring oats, spring rye, spring wheat, triticale, winter barley, winter oats, winter rye, winter wheat** [1]; **ornamental plant production** *(off-label)* [1, 2]
- Shepherd's purse in **corn gromwell** *(off-label)*, **durum wheat, farm forestry** *(off-label)*, **forest nurseries** *(off-label)*, **game cover** *(off-label)*, **linseed, spring barley, spring oats, spring rye, spring wheat, triticale, winter barley, winter oats, winter rye, winter wheat** [1]; **ornamental plant production** *(off-label)* [1, 2]

Extension of Authorisation for Minor Use (EAMUs)

- **corn gromwell** *20150903* [1]
- **farm forestry** *20142508* [1]
- **forest nurseries** *20142508* [1]
- **game cover** *20142508* [1]
- **ornamental plant production** *20142508* [1], *20142511* [2]

Approval information

- Accepted by BBPA for use on malting barley
- Amidosulfuron included in Annex I under EC Regulation 1107/2009

Efficacy guidance

- For best results apply in spring (from 1 Feb) in warm weather when soil moist and weeds growing actively. When used in grassland following cutting or grazing, docks should be allowed to regrow before treatment
- Weed kill is slow, especially under cool, dry conditions. Weeds may sometimes only be stunted but will have little or no competitive effect on crop
- May be used on all soil types unless certain sequences are used on linseed. See label
- Spray is rainfast after 1 h
- Cleavers controlled from emergence to flower bud stage. If present at application charlock (up to flower bud), shepherds purse (up to flower bud) and field forget-me-not (up to 6 leaves) will also be controlled
- Amidosulfuron is a member of the ALS-inhibitor group of herbicides and products should be used in a planned Resistance Management strategy. See Section 5 for more information

Restrictions

- Maximum number of treatments 1 per crop [1]
- Use after 1 Feb and do not apply to rotational grass after 30 Jun, or to permanent grassland after 15 Oct [2]
- Do not apply to crops undersown or due to be undersown with clover or alfalfa [1]
- Do not spray crops under stress, suffering drought, waterlogged, grazed, lacking nutrients or if soil compacted
- Do not spray if frost expected
- Do not roll or harrow within 1 wk of spraying
- Specific restrictions apply to use in sequence or tank mixture with other sulfonylurea or ALS-inhibiting herbicides. See label for details. There are no recommendations for mixtures with metsulfuron-methyl products on linseed
- Certain mixtures with fungicides are expressly forbidden. See label for details

SEE SECTION 3 FOR PRODUCTS ALSO REGISTERED

SECTION 2

Crop-specific information
- Latest use: before first spikelets just visible (GS 51) for cereals; before flower buds visible for linseed; 15 Oct for grassland
- Broadcast cereal crops should be sprayed post-emergence after plants have a well established root system

Following crops guidance
- If a treated crop fails cereals may be sown after 15 d and thorough cultivation
- After normal harvest of a treated crop only cereals, winter oilseed rape, mustard, turnips, winter field beans or vetches may be sown in the same year as treatment and these must be preceded by ploughing or thorough cultivation
- Only cereals may be sown within 12 mth of application to grassland [2]
- Cereals or potatoes must be sown as the following crop after use of permitted mixtures or sequences with other sulfonylurea herbicides in cereals. Only cereals may be sown after the use of such sequences in linseed

Environmental safety
- Take care to wash out sprayers thoroughly. See label for details
- Avoid drift onto neighbouring broad-leaved plants or onto surface waters or ditches

Hazard classification and safety precautions
Hazard Dangerous for the environment
Transport code 9
Packaging group III
UN Number 3077
Operator protection U20a [1]; U20b [2]
Environmental protection E07b [2] (1 week); E15a, E38, H410 [1, 2]; E41 [2]
Storage and disposal D10a, D12a

10   amidosulfuron + iodosulfuron-methyl-sodium

A post-emergence sulfonylurea herbicide mixture for cereals
HRAC mode of action code: B + B

See also iodosulfuron-methyl-sodium

Products

1	Chekker	Bayer CropScience	12.5:1.25% w/w	WG	16495
2	Sekator OD	Interfarm	100:25 g/l	OD	16494

Uses
- Annual dicotyledons in *forest nurseries* (off-label) [1]; *spring barley, spring rye, spring wheat, triticale, winter barley, winter rye, winter wheat* [1, 2]
- Chickweed in *forest nurseries* (off-label) [1]; *spring barley, spring rye, spring wheat, triticale, winter barley, winter rye, winter wheat* [1, 2]
- Cleavers in *forest nurseries* (off-label) [1]; *spring barley, spring rye, spring wheat, triticale, winter barley, winter rye, winter wheat* [1, 2]
- Mayweeds in *forest nurseries* (off-label) [1]; *spring barley, spring rye, spring wheat, triticale, winter barley, winter rye, winter wheat* [1, 2]
- Volunteer oilseed rape in *forest nurseries* (off-label) [1]; *spring barley, spring rye, spring wheat, triticale, winter barley, winter rye, winter wheat* [1, 2]

Extension of Authorisation for Minor Use (EAMUs)
- *forest nurseries* 20142722 [1]

Approval information
- Amidosulfuron and iodosulfuron-methyl-sodium included in Annex I under EC Regulation 1107/2009
- Accepted by BBPA for use on malting barley

Efficacy guidance
- Best results obtained from treatment in warm weather when soil is moist and the weeds are growing actively
- Weeds must be present at application to be controlled
- Dry conditions resulting in moisture stress may reduce effectiveness
- Weed control is slow especially under cool dry conditions
- Occasionally weeds may only be stunted but they will normally have little or no competitive effect on the crop
- Amidosulfuron and iodosulfuron are members of the ALS-inhibitor group of herbicides and products should be used in a planned Resistance Management strategy. See Section 5 for more information

Restrictions
- Maximum number of treatments 1 per crop
- Must only be applied between 1 Feb in yr of harvest and specified latest time of application
- Do not apply to crops undersown or to be undersown with grass, clover or alfalfa
- Do not roll or harrow within 1 wk of spraying
- Do not spray crops under stress from any cause or if the soil is compacted
- Do not spray if rain or frost expected
- Do not apply in mixture or in sequence with any other ALS inhibitor

Crop-specific information
- Latest use: before first spikelet of inflorescence just visible (GS 51)
- Treat drilled crops after the 2-leaf stage; treat broadcast crops after the plants have a well-established root system
- Applications to spring barley may cause transient crop yellowing

Following crops guidance
- Cereals, winter oilseed rape and winter field beans may be sown in the same yr as treatment provided they are preceded by ploughing or thorough cultivation. Any crop may be sown in the spring of the yr following treatment
- A minimum of 3 mth must elapse between treatment and sowing winter oilseed rape

Environmental safety
- Dangerous for the environment
- Toxic to aquatic organisms
- Take extreme care to avoid damage by drift onto broad-leaved plants outside the target area or onto ponds, waterways and ditches
- Observe carefully label instructions for sprayer cleaning
- LERAP Category B

Hazard classification and safety precautions
Hazard Irritant, Dangerous for the environment [1, 2]; Very toxic to aquatic organisms [1]
Transport code 9
Packaging group III
UN Number 3077 [1]; 3082 [2]
Risk phrases H317 [2]; H319 [1, 2]
Operator protection A, C, H; U05a, U08, U11, U14, U15, U20b
Environmental protection E15a, E16a, E16b, E38, H410
Storage and disposal D01, D02, D10a, D12a

11 amidosulfuron + iodosulfuron-methyl-sodium + mesosulfuron-methyl

A herbicide mixture for weed control in winter wheat
HRAC mode of action code: B + B + B

See also amidosulfuron
amidosulfuron + iodosulfuron-methyl-sodium
iodosulfuron-methyl-sodium
mesosulfuron-methyl

SEE SECTION 3 FOR PRODUCTS ALSO REGISTERED

Products

Pacifica Plus	Bayer CropScience	0.5:0.1:0.3% w/w	WG	17272

Uses
- Annual dicotyledons in *winter wheat*
- Annual grasses in *winter wheat*
- Blackgrass in *winter wheat*

Approval information
- Amidosulfuron, iodosulfuron-methyl-sodium and mesosulfuron-methyl included in Annex I under EC Regulation 1107/2009
- Accepted by BBPA for use on malting barley

Efficacy guidance
- Apply as early as possible in spring and before stem extension stage of any grass weeds
- Monitor efficacy and investigate areas of poor control

Restrictions
- Do not apply in tank-mixture or sequence with any product containing any other sulfonylurea or ALS inhibiting herbicide
- Do not use on crops undersown with grasses, clover, legumes or any other broad-leaved crop
- Take care to avoid drift on to non-target plants

Following crops guidance
- Winter wheat and winter barley may be sown in the same year as harvest of a crop treated with 0.5 kg/ha. Winter oilseed rape may be sown after ploughing following a crop treated with 0.4 kg/ha. In the event of crop failure, sow only winter or spring wheat.

Environmental safety
- LERAP Category B

Hazard classification and safety precautions
   Hazard Very toxic to aquatic organisms
   Transport code 9
   Packaging group III
   UN Number 3077
   Risk phrases H317, H318
   Operator protection A, C, H; U05a, U11, U20a
   Environmental protection E15a, E16a, H410
   Storage and disposal D01, D02, D05, D09a, D10b

12   aminopyralid

A pyridine carboxylic acid herbicide available only in mixtures
HRAC mode of action code: O

13   aminopyralid + fluroxypyr

A foliar acting herbicide mixture for use in grassland
HRAC mode of action code: O + O

See also fluroxypyr

Products

Synero	Nomix Enviro	30:100 g/l	EW	14708

Uses
- Buttercups in *amenity grassland*
- Chickweed in *amenity grassland*
- Dandelions in *amenity grassland*

FOR FULL CONDITIONS OF USE ALWAYS READ THE PRODUCT LABEL

- Docks in **amenity grassland**
- Stinging nettle in **amenity grassland**
- Thistles in **amenity grassland**

Approval information
- Aminopyralid and fluroxypyr are included in Annex I under EC Regulation 1107/2009.

Efficacy guidance
- For best results and to avoid crop check, grass and weeds must be growing actively
- Allow 2-3 wk after cutting for hay or silage for sufficient regrowth to occur before spraying and leave 7 d afterwards to allow maximum translocation
- Where there is a high reservoir of weed seed or a historically high weed population a programmed approach may be needed involving a second treatment in the following yr
- Control may be reduced if rain falls within 1 h of spraying

Restrictions
- Maximum number of treatments 1 per yr.
- Do not apply to leys less than 1 year old.
- Do not apply by hand-held equipment.
- Do not use on grassland that will be used for animal feed, bedding, composting or mulching within 1 calender year of application.
- Do not use on grassland that will be grazed by animals other than cattle or sheep.
- Use of an antifoam is compulsory
- Do not use any treated plant material, or manure from animals fed on treated crops, for composting or mulching
- Do not use on crops grown for seed
- Manure from animals fed on pasture or silage treated with this herbicide should not leave the farm

Crop-specific information
- Treatment will kill clover
- Treatment may occasionally cause transient yellowing of the sward which is quickly outgrown
- Late treatments may lead to a slight transient leaning of grass that does not affect yield

Following crops guidance
- Do not drill clover or other legumes within 4 mth of treatment, or potatoes in the spring following treatment in the previous autumn
- In the event of failure of newly seeded treated grassland, grass may be re-seeded immediately, or wheat may be sown provided 4 mth have elapsed since application
- Residues in plant tissues, including manure, may affect succeeding susceptible crops of peas, beans, other legumes, carrots, other Umbelliferae, potatoes, tomatoes, lettuce and other Compositae. These crops should not be sown within 3 mth of ploughing up treated grassland

Environmental safety
- Dangerous for the environment
- Toxic to aquatic organisms
- Keep livestock out of treated areas for at least 7 d following treatment and until poisonous weeds, such as ragwort, have died down and become unpalatable
- Avoid damage by drift onto susceptible crops, non-target plants or waterways

Hazard classification and safety precautions
Hazard Irritant, Dangerous for the environment
Transport code 9
Packaging group III
UN Number 3082
Risk phrases H304, H315, H318, H336
Operator protection A, C; U05a, U08, U11, U14, U23a
Environmental protection E07a, E15b, E38, H411
Consumer protection C01
Storage and disposal D01, D02, D05, D09a, D10b, D12a

SEE SECTION 3 FOR PRODUCTS ALSO REGISTERED

14   aminopyralid + metazachlor + picloram

A herbicide mixture for weed control in oilseed rape.
HRAC mode of action code: K3 + O

See also metazachlor

Products

Ralos	Dow	5.3:500:13.3 g/l	SC	16737

Uses
- Chickweed in *winter oilseed rape*
- Dead nettle in *winter oilseed rape*
- Fat hen in *winter oilseed rape*
- Penny cress in *winter oilseed rape*
- Poppies in *winter oilseed rape*
- Scentless mayweed in *winter oilseed rape*

Approval information
- Aminopyralid, metazachlor and picloram are included in Annex 1 under 1107/2009.

Restrictions
- Applications shall be limited to a total dose of not more than 1.0 kg metazachlor/ha in a three year period on the same field.
- Do not use on sands and very light soils or on soils with more than 10% organic matter.
- Straw from treated crops must not be baled and must stay on the field.
- Do not use treated plant material for composting or mulching.
- Do not use digestate from anaerobic digesters on crops such as peas, beans and other legumes, carrots and other Umbelliferae, potatoes, lettuce or other Compositae, glasshouse or protected crops.

Following crops guidance
- Only wheat, barley, oats, maize or oilseed rape should be planted within 4 months (120 days) of application. All other crops should wait until 12 months after application.
- In the event of crop failure, only spring wheat, spring barley, spring oats, maize or ryegrass should be planted after ploughing or thorough cultivation of the soil.

Environmental safety
- Metazachlor stewardship guidelines advise a maximum dose of 750 g.a.i/ha/annum. Applications to drained fields should be complete by 15th Oct but, if drains are flowing, complete applications by 1st Oct.
- LERAP Category B

Hazard classification and safety precautions
 Hazard Dangerous for the environment
 Transport code 9
 Packaging group III
 UN Number 3082
 Risk phrases H351
 Operator protection A; U05a, U20a
 Environmental protection E06c (1 week); E15b, E16a, E34, H410
 Storage and disposal D01, D02, D09a, D10c
 Medical advice M03

15   aminopyralid + propyzamide

A herbicide mixture for weed control in winter oilseed rape
HRAC mode of action code: K1 + O

Products

1	AstroKerb	Dow	6.3:500 g/l	SC	16184

FOR FULL CONDITIONS OF USE ALWAYS READ THE PRODUCT LABEL

SECTION 2

Products – continued

2	Clayton Propel Plus	Clayton	6.3:500 g/l	SC	17227
3	Dymid	BASF	5.3:500 g/l	SC	17585
4	Pamino	Harvest	5.3:500 g/l	SC	18069

Uses

- Annual dicotyledons in *winter oilseed rape*
- Annual grasses in *winter oilseed rape*
- Mayweeds in *winter oilseed rape*
- Poppies in *winter oilseed rape*

Approval information

- Aminopyralid and propyzamide are included in Annex 1 under EC Regulation 1107/2009.

Efficacy guidance

- Allow soil temperatures to drop before application to winter oilseed rape to optimise activity of propyzamide component

Restrictions

- Users must have received adequate instruction, training and guidance on the safe and efficient use of the product.
- Must not be used on land where vegetation will be cut for animal feed, fodder, bedding nor for composting or mulching within one calendar year of treatment.
- Do not remove oilseed rape straw from the field unless it is for burning for heat or electricity production. Do not use the oilseed rape straw for animal feed, animal bedding, composting or mulching.

Crop-specific information

- Only cereals may follow application of aminopyralid + propyzamide and these should not be planted within 30 weeks of treatment.
- Aminopyralid residues in plant tissue which have not completely decayed may affect susceptible crops such as legumes (e.g. peas and beans), sugar and fodder beet, carrots and other Umbelliferae, potatoes, tomatoes, lettuce and other Compositae.

Following crops guidance

- May only be followed by spring or winter cereals. Treated land must be mouldboard ploughed to a depth of at least 15 cm before a following cereal crop is planted.

Hazard classification and safety precautions

Hazard Harmful, Dangerous for the environment
Transport code 9
Packaging group III
UN Number 3082
Risk phrases H351
Operator protection A, H; U20c
Environmental protection E07b (1); E15b, E38, E39, H411
Storage and disposal D09a, D11a, D12a

16 aminopyralid + triclopyr

A foliar acting herbicide mixture for broad-leaved weed control in grassland
HRAC mode of action code: O + O

See also triclopyr

Products

1	Forefront T	Dow	30:240g/l	EW	15568
2	Garlon Ultra	Nomix Enviro	12:120 g/l	SL	16211
3	Icade	Dow	12:120 g/l	SL	16182

Uses

- Annual dicotyledons in *amenity grassland* [2, 3]
- Brambles in *amenity grassland* [3]

SEE SECTION 3 FOR PRODUCTS ALSO REGISTERED

- Broom in **amenity grassland** [3]
- Buddleia in **amenity grassland** [3]
- Buttercups in **grassland** [1]
- Common mugwort in **amenity grassland** [3]
- Common nettle in **amenity grassland** [3]; **grassland** [1]
- Creeping thistle in **amenity grassland** [3]
- Dandelions in **grassland** [1]
- Docks in **grassland** [1]
- Gorse in **amenity grassland** [3]
- Hogweed in **amenity grassland** [3]
- Japanese knotweed in **amenity grassland** [3]
- Perennial dicotyledons in **amenity grassland** [3]
- Rosebay willowherb in **amenity grassland** [3]
- Thistles in **grassland** [1]

Approval information
- Aminopyralid and triclopyr are included in Annex I under EC Regulation 1107/2009.

Efficacy guidance
- For best results and to avoid crop check, grass and weeds must be growing actively
- Use adeqate water volume to ensure good weed coverage. Increase water volume if necessary where the weed population is high and where the grass is dense
- Allow 2-3 wk after cutting for hay or silage for sufficient regrowth to occur before spraying and leave 7 d afterwards to allow maximum translocation
- Where there is a high reservoir of weed seed or a historically high weed population a programmed approach may be needed involving a second treatment in the following yr
- Control may be reduced if rain falls within 1 h of spraying

Restrictions
- Maximum number of treatments 1 per yr.
- Do not apply to leys less than 1 year old.
- Do not apply by hand-held equipment [1]
- Do not use on grassland that will be used for animal feed, bedding, composting or mulching within 1 calender year of application.
- Do not use on grassland that will be grazed by animals other than cattle or sheep.
- Do not use any treated plant material, or manure from animals fed on treated crops, for composting or mulching
- Do not use on crops grown for seed
- Manure from animals fed on pasture or silage from treated crops should not leave the farm

Crop-specific information
- Latest use: 7 d before grazing or harvest for grassland
- Late applications may lead to transient leaning of grass which does not affect final yield

Following crops guidance
- Ensure that all plant remains of a treated crop have completely decayed before planting susceptible crops such as peas, beans and other legumes, sugar beet, carrots and other Umbelliferae, potatoes and tomatoes, lettuce and other Compositae
- Do not plant potatoes, sugar beet, vegetables, beans or other leguminous crops in the calendar yr following application

Environmental safety
- Dangerous for the environment
- Toxic to aquatic organisms
- Keep livestock out of treated areas for at least 7 d after treatment or until foliage of any poisonous weeds such as ragwort has died and become unpalatable
- To protect groundwater do not apply to grass leys less than 1 yr old
- Take extreme care to avoid drift onto susceptible crops, non-target plants or waterways. All conifers, especially pine and larch, are very sensitive and may be damaged by vapour drift in hot conditions
- LERAP Category B [1]

FOR FULL CONDITIONS OF USE ALWAYS READ THE PRODUCT LABEL

Hazard classification and safety precautions
> **Hazard** Harmful [2, 3]; Irritant [1]; Dangerous for the environment [1-3]
> **Transport code** 9 [1]
> **Packaging group** III [1]
> **UN Number** 3082 [1]; N/C [2, 3]
> **Risk phrases** H317 [1]; H319, H335 [2, 3]
> **Operator protection** A, H [1-3]; M [2, 3]; U05a, U08 [1-3]; U14, U23a [1]
> **Environmental protection** E06a [1] (7 days); E07c [2, 3] (7 days); E15b, E38 [1-3]; E15c, E16a, H411 [1]; E39 [2, 3]
> **Consumer protection** C01
> **Storage and disposal** D01, D02, D09a [1-3]; D05, D10c [2, 3]; D10b, D12a [1]
> **Medical advice** M05a [2, 3]

17   amisulbrom

A fungicide for use in potatoes
FRAC mode of action code: 21

Products

Shinkon	Gowan	200 g/l	SC	17498

Uses
- Late blight in **potatoes**
- Tuber blight in **potatoes**

Approval information
- Amisulbrom is included in Annex 1 under EC Regulation 1107/2009

Efficacy guidance
- Make no more than 3 consecutive applications before switching to a blight fungicide with a different mode of action to protect against the risk of resistance.
- Do not use if blight is already visible on 1% of leaves.

Crop-specific information
- Rainfast within 3 hours of application but apply to dry foliage.

Following crops guidance
- Plough before planting new crops in the same soil.

Environmental safety
- LERAP Category B

Hazard classification and safety precautions
> **Hazard** Dangerous for the environment, Very toxic to aquatic organisms
> **Transport code** 9
> **Packaging group** III
> **UN Number** 3082
> **Risk phrases** H351
> **Operator protection** A, C; U05a, U15, U20b
> **Environmental protection** E13b, E15a, E16a, E34, E38, H410
> **Storage and disposal** D01, D02, D05, D09a, D10b, D12a

18   Ampelomyces quisqualis (Strain AQ10)

A biological fungicide that combats powdery mildew in fruit and ornamentals
FRAC mode of action code: 44

Products

AQ 10	Fargro	58% w/w	WG	17102

SEE SECTION 3 FOR PRODUCTS ALSO REGISTERED

Uses

- Powdery mildew in **protected apple** *(off-label)*, **protected aubergines, protected blackcurrants** *(off-label)*, **protected blueberry** *(off-label)*, **protected chilli peppers, protected chilli peppers** *(off-label)*, **protected courgettes, protected courgettes** *(off-label)*, **protected crabapple** *(off-label)*, **protected cucumbers, protected forest nurseries** *(off-label)*, **protected gooseberries** *(off-label)*, **protected herbs (see appendix 6)** *(off-label)*, **protected melons, protected ornamentals** *(off-label)*, **protected other small fruit and berries** *(off-label)*, **protected pear** *(off-label)*, **protected peppers, protected peppers** *(off-label)*, **protected pumpkins, protected quince** *(off-label)*, **protected redcurrants** *(off-label)*, **protected squashes, protected strawberries, protected summer squash** *(off-label)*, **protected table grapes** *(off-label)*, **protected tomatoes, protected watermelon** *(off-label)*, **protected wine grapes** *(off-label)*, **protected winter squash**

Extension of Authorisation for Minor Use (EAMUs)

- **protected apple** *20152646*
- **protected blackcurrants** *20152646*
- **protected blueberry** *20152646*
- **protected chilli peppers** *20152646*
- **protected courgettes** *20152646*
- **protected crabapple** *20152646*
- **protected forest nurseries** *20152646*
- **protected gooseberries** *20152646*
- **protected herbs (see appendix 6)** *20152646*
- **protected ornamentals** *20152646*
- **protected other small fruit and berries** *20152646*
- **protected pear** *20152646*
- **protected peppers** *20152646*
- **protected quince** *20152646*
- **protected redcurrants** *20152646*
- **protected summer squash** *20152646*
- **protected table grapes** *20152646*
- **protected watermelon** *20152646*
- **protected wine grapes** *20152646*

Approval information

- Ampelomyces quisqualis is included in Annex 1 under EC Regulation 1107/2009

Hazard classification and safety precautions

UN Number N/C
Operator protection A, D, H; U14, U20b
Environmental protection E15b, E34
Storage and disposal D01, D02, D05, D09a, D10a, D16, D20
Medical advice M03

19   azoxystrobin

A systemic translaminar and protectant strobilurin fungicide for a wide range of crops
FRAC mode of action code: 11

Products

1	Amistar	Syngenta	250 g/l	SC	18039
2	Azaka	Headland	250 g/l	SC	16422
3	Azoxy 250	Becesane	250 g/l	SC	15980
4	Azoxystar	Life Scientific	250 g/l	SC	17407
5	Chamane	UPL Europe	250 g/l	SC	15922
6	Conclude	Globachem	250 g/l	SC	16905
7	Globaztar SC	Globachem	250 g/l	SC	15575
8	Harness	ChemSource	50% w/w	WG	14946
9	Heritage	Syngenta	50% w/w	WG	13536
10	Heritage Maxx	Syngenta	95 g/l	DC	14787

FOR FULL CONDITIONS OF USE ALWAYS READ THE PRODUCT LABEL

Products – continued

11	Pure Azoxy	Pure Amenity	50% w/w	WG	18065
12	Sinstar	Agrii	250 g/l	SC	16852
13	Tazer	Nufarm UK	250 g/l	SC	15495
14	Valiant	Headland	250 g/l	SC	16776

Uses

- Alternaria in *broccoli*, *brussels sprouts*, *cabbages*, *calabrese*, *carrots*, *cauliflowers*, *collards*, *kale* [1, 3]; *chicory grown outside for forcing* (off-label), *crambe* (off-label), *mustard* (off-label), *protected chicory* (off-label) [7]; *horseradish* (off-label), *parsnips* (off-label) [1]; *spring oilseed rape*, *winter oilseed rape* [1, 3, 4, 12, 13]
- Alternaria blight in *carrots* [13]
- Anthracnose in *amenity grassland* [8, 10]; *managed amenity turf* [8-11]; *protected strawberries*, *strawberries* [1]
- Ascochyta in *combining peas*, *vining peas* [3-5]; *combining peas* (useful control), *dwarf beans* (useful control), *edible podded peas* (useful control), *french beans* (useful control), *mange-tout peas* (useful control), *sugar snap peas* (useful control), *vining peas* (useful control) [1]
- Basal stem rot in *forest nurseries* (off-label), *ornamental plant production* (off-label), *protected ornamentals* (off-label) [1]
- Black dot in *potatoes* [1, 3]
- Black root rot in *forest nurseries* (off-label), *ornamental plant production* (off-label), *protected ornamentals* (off-label) [1]
- Black rot in *oriental cabbage* (off-label) [1]
- Black scurf and stem canker in *potatoes* [1, 3]
- Botrytis in *celery (outdoor)* (off-label), *forest nurseries* (off-label), *ornamental plant production* (off-label), *protected celery* (off-label), *protected ornamentals* (off-label) [1]
- Brown patch in *amenity grassland* [8, 10]; *managed amenity turf* [8-11]
- Brown rust in *rye* [2, 6, 7, 13, 14]; *spring barley*, *spring wheat*, *winter barley*, *winter wheat* [1-7, 12-14]; *spring rye*, *winter rye* [1, 3-5]; *triticale* [1-7, 13, 14]
- Crown rust in *amenity grassland* [8, 10]; *managed amenity turf* [8-11]; *spring oats*, *winter oats* [1-4, 6, 7, 13, 14]
- Dark leaf spot in *spring oilseed rape*, *winter oilseed rape* [2, 13, 14]
- Downy mildew in *bulb onions* [3]; *bulb onions* (moderate control) [13]; *bulb onions* (reduction), *courgettes* (off-label), *cucumbers* (off-label), *dwarf beans* (reduction), *edible podded peas* (reduction), *endives*, *forest nurseries* (off-label), *french beans* (reduction), *garlic* (reduction), *gherkins* (off-label), *lettuce*, *mange-tout peas* (reduction), *melons* (off-label), *ornamental plant production* (off-label), *protected endives*, *protected lettuce*, *protected ornamentals* (off-label), *pumpkins* (off-label), *radishes* (off-label), *salad onions* (off-label), *shallots* (reduction), *sugar snap peas* (reduction), *summer squash* (off-label), *watermelons* (off-label), *winter squash* (off-label) [1]; *combining peas* (reduction), *vining peas* (reduction) [1, 13]; *globe artichoke* (off-label), *protected courgettes* (off-label), *protected melons* (off-label), *protected pumpkins* (off-label), *protected summer squash* (off-label), *protected winter squash* (off-label) [7]
- Ear diseases in *spring wheat*, *winter wheat* [2, 6, 7, 13, 14]
- Fairy rings in *amenity grassland* [8, 10]; *managed amenity turf* [8-11]
- Fusarium patch in *amenity grassland* [8, 10]; *managed amenity turf* [8-11]
- Glume blotch in *spring wheat*, *winter wheat* [1-7, 12-14]
- Grey mould in *aubergines* (off-label), *courgettes* (off-label), *dwarf beans* (some control), *edible podded peas* (some control), *french beans* (some control), *mange-tout peas* (some control), *melons* (off-label), *pumpkins* (off-label), *sugar snap peas* (some control), *summer squash* (off-label), *tomatoes* (off-label), *watermelons* (off-label), *winter squash* (off-label) [1]; *combining peas* (some control), *vining peas* (some control) [1, 13]
- Late blight in *aubergines* (off-label), *tomatoes* (off-label) [1]
- Late ear diseases in *spring wheat*, *winter wheat* [1, 3-5, 12]
- Leaf and pod spot in *combining peas* (useful control), *vining peas* (useful control) [1, 13]; *dwarf beans* (useful control), *edible podded peas* (useful control), *french beans* (useful control), *mange-tout peas* (useful control), *sugar snap peas* (useful control) [1]

SEE SECTION 3 FOR PRODUCTS ALSO REGISTERED

- Leaf spot in *celery (outdoor)* *(off-label)*, *forest nurseries* *(off-label)*, *ornamental plant production* *(off-label)*, *protected celery* *(off-label)*, *protected ornamentals* *(off-label)* [1]
- Melting out in *amenity grassland* [8, 10]; *managed amenity turf* [8-11]
- Mycosphaerella in *combining peas* *(some control)*, *vining peas* *(some control)* [1, 13]; *dwarf beans* *(some control)*, *edible podded peas* *(some control)*, *french beans* *(some control)*, *mange-tout peas* *(some control)*, *sugar snap peas* *(some control)* [1]
- Needle blight in *forest nurseries* *(off-label)*, *ornamental plant production* *(off-label)*, *protected ornamentals* *(off-label)* [1]
- Needle casts in *forest nurseries* *(off-label)*, *ornamental plant production* *(off-label)*, *protected ornamentals* *(off-label)* [1]
- Net blotch in *spring barley*, *winter barley* [1-7, 12-14]
- Phytophthora in *chicory grown outside for forcing* *(off-label)*, *protected chicory* *(off-label)* [7]
- Powdery mildew in *all edible seed crops grown outdoors* *(off-label)*, *all non-edible seed crops grown outdoors* *(off-label)*, *bilberries* *(off-label)*, *blackcurrants* *(off-label)*, *blueberries* *(off-label)*, *chicory grown outside for forcing* *(off-label)*, *cranberries* *(off-label)*, *durum wheat* *(off-label)*, *garlic* *(off-label)*, *globe artichoke* *(off-label)*, *gooseberries* *(off-label)*, *grass seed crops* *(off-label)*, *horseradish* *(off-label)*, *lupins* *(off-label)*, *poppies for morphine production* *(off-label)*, *protected chicory* *(off-label)*, *protected courgettes* *(off-label)*, *protected forest nurseries* *(off-label)*, *protected melons* *(off-label)*, *protected pumpkins* *(off-label)*, *protected summer squash* *(off-label)*, *protected winter squash* *(off-label)*, *redcurrants* *(off-label)*, *shallots* *(off-label)*, *strawberries* *(off-label)* [7]; *aubergines* *(off-label)*, *chillies* *(off-label)*, *cucumbers* *(off-label)*, *gherkins* *(off-label)*, *melons* *(off-label)*, *peppers* *(off-label)*, *protected blackberries* *(off-label)*, *protected loganberries* *(off-label)*, *protected raspberries* *(off-label)*, *protected rubus hybrids* *(off-label)*, *protected strawberries*, *pumpkins* *(off-label)*, *strawberries*, *tomatoes* *(off-label)*, *watermelons* *(off-label)*, *winter squash* *(off-label)* [1]; *blackberries* *(off-label)*, *courgettes* *(off-label)*, *forest nurseries* *(off-label)*, *loganberries* *(off-label)*, *ornamental plant production* *(off-label)*, *parsnips* *(off-label)*, *protected ornamentals* *(off-label)*, *raspberries* *(off-label)*, *rubus hybrids* *(off-label)*, *summer squash* *(off-label)* [1, 7]; *carrots* [1, 3, 13]; *rye* [2]; *rye* *(moderate control)*, *triticale* *(moderate control)* [6, 7, 13, 14]; *spring barley*, *winter barley* [3-5, 12]; *spring barley* *(moderate control)*, *winter barley* *(moderate control)* [2, 6, 7, 13, 14]; *spring oats*, *winter oats* [3, 4]; *spring oats* *(moderate control)*, *winter oats* *(moderate control)* [2, 6, 7, 14]; *triticale* [2-5]; *winter rye* [1, 3-5]
- Purple blotch in *leeks* [1, 3, 13]
- Rhizoctonia in *celery (outdoor)* *(off-label)*, *protected celery* *(off-label)*, *radishes* *(off-label)*, *swedes* *(off-label)*, *turnips* *(off-label)* [1]
- Rhynchosporium in *rye* [2]; *rye* *(reduction)*, *triticale* *(reduction)* [6, 7, 13, 14]; *spring barley*, *winter barley* [1, 3-5, 12]; *spring barley* *(reduction)*, *winter barley* *(reduction)* [2, 6, 7, 13, 14]; *spring rye*, *winter rye* [1, 3-5]; *triticale* [1-5]
- Ring spot in *broccoli*, *brussels sprouts*, *cabbages*, *calabrese*, *cauliflowers*, *collards*, *kale* [1, 3]
- Root malformation disorder in *red beet* *(off-label)* [1]
- Rust in *all edible seed crops grown outdoors* *(off-label)*, *all non-edible seed crops grown outdoors* *(off-label)*, *bilberries* *(off-label)*, *blackberries* *(off-label)*, *blackcurrants* *(off-label)*, *blueberries* *(off-label)*, *chicory grown outside for forcing* *(off-label)*, *cranberries* *(off-label)*, *durum wheat* *(off-label)*, *garlic* *(off-label)*, *gooseberries* *(off-label)*, *grass seed crops* *(off-label)*, *horseradish* *(off-label)*, *loganberries* *(off-label)*, *lupins* *(off-label)*, *parsnips* *(off-label)*, *protected chicory* *(off-label)*, *protected forest nurseries* *(off-label)*, *raspberries* *(off-label)*, *redcurrants* *(off-label)*, *rubus hybrids* *(off-label)*, *shallots* *(off-label)*, *strawberries* *(off-label)* [7]; *asparagus* [1, 3, 5, 13]; *broad beans*, *lupins* [1]; *forest nurseries* *(off-label)*, *ornamental plant production* *(off-label)*, *protected ornamentals* *(off-label)* [1, 7]; *leeks*, *spring field beans*, *winter field beans* [1, 3, 13]
- Scab in *forest nurseries* *(off-label)*, *ornamental plant production* *(off-label)*, *protected ornamentals* *(off-label)* [1]
- Sclerotinia in *celeriac* *(off-label)*, *celery (outdoor)* *(off-label)*, *protected celery* *(off-label)* [1]; *crambe* *(off-label)*, *mustard* *(off-label)* [7]; *spring oilseed rape*, *winter oilseed rape* [13]
- Sclerotinia stem rot in *spring oilseed rape*, *winter oilseed rape* [1-4, 12, 14]
- Septoria leaf blotch in *spring wheat* [2-7, 12-14]; *winter wheat* [1-7, 12-14]
- Stemphylium in *asparagus* [1, 3, 5, 13]

FOR FULL CONDITIONS OF USE ALWAYS READ THE PRODUCT LABEL

- Take-all in **rye** *(reduction)* [2, 6, 7, 13, 14]; **spring barley** *(reduction)*, **spring wheat** *(reduction)*, **winter barley** *(reduction)*, **winter wheat** *(reduction)* [1-7, 12-14]; **spring oats** *(reduction)*, **winter oats** *(reduction)* [2]; **spring rye** *(reduction in severity)* [1]; **triticale** *(reduction)* [1, 2, 6, 7, 13, 14]
- Take-all patch in **amenity grassland** [8, 10]; **managed amenity turf** [8-11]
- White blister in **broccoli**, **brussels sprouts**, **cabbages**, **calabrese**, **cauliflowers**, **collards**, **kale** [1, 3]; **forest nurseries** *(off-label)*, **horseradish** *(off-label)*, **oriental cabbage** *(off-label)*, **ornamental plant production** *(off-label)*, **protected ornamentals** *(off-label)* [1]
- White rust in **pot chrysanthemums** *(off-label)*, **protected chrysanthemums** *(off-label)* [7]
- White tip in **leeks** [1, 13]
- Yellow rust in **spring wheat**, **winter wheat** [1-7, 12-14]

Extension of Authorisation for Minor Use (EAMUs)

- **all edible seed crops grown outdoors** *20170179 expires 30 Jun 2018* [7]
- **all non-edible seed crops grown outdoors** *20170179 expires 30 Jun 2018* [7]
- **aubergines** *20170894* [1]
- **bilberries** *20170179 expires 30 Jun 2018* [7]
- **blackberries** *20170895* [1], *20170179 expires 30 Jun 2018* [7]
- **blackcurrants** *20170179 expires 30 Jun 2018* [7]
- **blueberries** *20170179 expires 30 Jun 2018* [7]
- **celeriac** *20170892* [1]
- **celery (outdoor)** *20170891* [1]
- **chicory grown outside for forcing** *20170176 expires 30 Jun 2018* [7]
- **chillies** *20170894* [1]
- **courgettes** *20170893* [1], *20170894* [1], *20170172 expires 30 Jun 2018* [7]
- **crambe** *20170168 expires 30 Jun 2018* [7]
- **cranberries** *20170179 expires 30 Jun 2018* [7]
- **cucumbers** *20170894* [1]
- **durum wheat** *20170170 expires 30 Jun 2018* [7]
- **forest nurseries** *20170888* [1], *20170179 expires 30 Jun 2018* [7]
- **garlic** *20170177 expires 30 Jun 2018* [7]
- **gherkins** *20170894* [1]
- **globe artichoke** *20170174 expires 30 Jun 2018* [7]
- **gooseberries** *20170179 expires 30 Jun 2018* [7]
- **grass seed crops** *20170170 expires 30 Jun 2018* [7]
- **horseradish** *20170892* [1], *20170178 expires 30 Jun 2018* [7]
- **loganberries** *20170895* [1], *20170179 expires 30 Jun 2018* [7]
- **lupins** *20170169 expires 30 Jun 2018* [7]
- **melons** *20170893* [1], *20170894* [1]
- **mustard** *20170173 expires 30 Jun 2018* [7]
- **oriental cabbage** *20170889* [1]
- **ornamental plant production** *20170888* [1], *20170179 expires 30 Jun 2018* [7]
- **parsnips** *20170892* [1], *20170178 expires 30 Jun 2018* [7]
- **peppers** *20170894* [1]
- **poppies for morphine production** *20170175 expires 30 Jun 2018* [7]
- **pot chrysanthemums** *20170171 expires 30 Jun 2018* [7]
- **protected blackberries** *20170895* [1]
- **protected celery** *20170891* [1]
- **protected chicory** *20170176 expires 30 Jun 2018* [7]
- **protected chrysanthemums** *20170171 expires 30 Jun 2018* [7]
- **protected courgettes** *20170167 expires 30 Jun 2018* [7]
- **protected forest nurseries** *20170179 expires 30 Jun 2018* [7]
- **protected loganberries** *20170895* [1]
- **protected melons** *20170167 expires 30 Jun 2018* [7]
- **protected ornamentals** *20170888* [1], *20170179 expires 30 Jun 2018* [7]
- **protected pumpkins** *20170167 expires 30 Jun 2018* [7]
- **protected raspberries** *20170895* [1]
- **protected rubus hybrids** *20170895* [1]
- **protected summer squash** *20170167 expires 30 Jun 2018* [7]

SEE SECTION 3 FOR PRODUCTS ALSO REGISTERED

- **protected winter squash** *20170167 expires 30 Jun 2018* [7]
- **pumpkins** *20170893* [1], *20170894* [1]
- **radishes** *20170892* [1]
- **raspberries** *20170895* [1], *20170179 expires 30 Jun 2018* [7]
- **red beet** *20170892* [1]
- **redcurrants** *20170179 expires 30 Jun 2018* [7]
- **rubus hybrids** *20170895* [1], *20170179 expires 30 Jun 2018* [7]
- **salad onions** *20170890* [1]
- **shallots** *20170177 expires 30 Jun 2018* [7]
- **strawberries** *20170179 expires 30 Jun 2018* [7]
- **summer squash** *20170893* [1], *20170894* [1], *20170172 expires 30 Jun 2018* [7]
- **swedes** *20170892* [1]
- **tomatoes** *20170894* [1]
- **turnips** *20170892* [1]
- **watermelons** *20170893* [1], *20170894* [1]
- **winter squash** *20170893* [1], *20170894* [1]

Approval information
- Azoxystrobin included in Annex I under EC Regulation 1107/2009
- Accepted by BBPA for use on malting barley
- Approval expiry 31 Dec 2018 [3]

Efficacy guidance
- Best results obtained from use as a protectant or during early stages of disease establishment or when a predictive assessment indicates a risk of disease development
- Azoxystrobin inhibits fungal respiration and should always be used in mixture with fungicides with other modes of action
- Treatment under poor growing conditions may give less reliable results
- For good control of *Fusarium* patch in amenity turf and grass repeat treatment at minimum intervals of 2 wk
- Azoxystrobin is a member of the QoI cross resistance group. Product should be used preventatively and not relied on for its curative potential
- Use product in cereals as part of an Integrated Crop Management strategy incorporating other methods of control, including where appropriate other fungicides with a different mode of action. Do not apply more than two foliar applications of QoI containing products to any cereal crop
- There is a significant risk of widespread resistance occurring in *Septoria tritici* populations in UK. Failure to follow resistance management action may result in reduced levels of disease control
- On cereal crops product must always be used in mixture with another product, recommended for control of the same target disease, that contains a fungicide from a different cross resistance group and is applied at a dose that will give robust control
- Strains of barley powdery mildew resistant to QoIs are common in the UK

Restrictions
- Maximum number of treatments 1 per crop for potatoes; 2 per crop for brassicas, peas, cereals, oilseed rape; 4 per crop or yr for onions, carrots, leeks, amenity turf
- Maximum total dose ranges from 2-4 times the single full dose depending on crop and product. See labels for details
- On turf the maximum number of treatments is 4 per yr but they must not exceed one third of the total number of fungicide treatments applied
- Do not use where there is risk of spray drift onto neighbouring apple crops
- The same spray equipment should not be used to treat apples

Crop-specific information
- Latest use: at planting for potatoes; grain watery ripe (GS 71) for cereals; before senescence for asparagus
- HI: 10 d for carrots;14 d for broccoli, Brussels sprouts, bulb onions, cabbages, calabrese, cauliflowers, collards, kale, vining peas; 21 d for leeks, spring oilseed rape, winter oilseed rape; 36 d for combining peas, 35 d for field beans

FOR FULL CONDITIONS OF USE ALWAYS READ THE PRODUCT LABEL

- In cereals control of established infections can be improved by appropriate tank mixtures or application as part of a programme. Always use in mixture with another product from a different cross-resistance group
- In turf use product at full dose rate in a disease control programme, alternating with fungicides of different modes of action
- In potatoes when used as incorporated treatment apply overall to the entire area to be planted, incorporate to 15 cm and plant on the same day. In-furrow spray should be directed at the furrow and not the seed tubers
- Applications to brassica crops must only be made to a developed leaf canopy and not before growth stages specified on the label
- Heavy disease pressure in brassicae and oilseed rape may require a second treatment
- All crops should be treated when not under stress. Check leaf wax on peas if necessary
- Consult processor before treating any crops for processing
- Treat asparagus after the harvest season. Where a new bed is established do not treat within 3 wk of transplanting out the crowns
- Do not apply to turf when ground is frozen or during drought

Environmental safety
- Dangerous for the environment
- Very toxic to aquatic organisms
- Avoid spray drift onto surrounding areas or crops, especially apples, plums or privet
- Buffer zone requirement 5 m in winter wheat, winter barley and winter oilseed rape [12]
- Buffer zone requirement 10 m in blackberry, loganberry, raspberry and rubus hybrid [1]
- Buffer zone requirement 6 m in bulb onion, carrots and leeks [13]
- LERAP Category B [1, 2, 4, 9, 11-13]; LERAP Category B [3] (potatoes only)

Hazard classification and safety precautions
Hazard Harmful [5-7, 13]; Dangerous for the environment [1-14]; Harmful if inhaled [5-7]; Very toxic to aquatic organisms [4, 6, 9-11, 13]
Transport code 9
Packaging group III
UN Number 3077 [8, 9, 11]; 3082 [1-7, 10, 12-14]
Risk phrases R50, R53a [3, 8, 12, 14]
Operator protection A [1-7, 10, 12-14]; H [2]; U05a, U20b [1-14]; U09a, U19a [1-7, 12-14]
Environmental protection E15a [1, 3, 4, 12]; E15b [2, 5-11, 13, 14]; E16a [1, 2]; E16a, E16b [3] (potatoes only); E16a [4, 9, 11-13]; E23 [5]; E38 [1-14]; H410 [1, 2, 4-7, 9-11, 13]
Storage and disposal D01, D02, D09a, D12a [1-14]; D03, D10b [8-11]; D05 [1, 3-12]; D10c [1-7, 12-14]
Medical advice M03 [5]; M05a [8-11]

20   azoxystrobin + chlorothalonil

A preventative and systemic fungicide mixture for cereals
FRAC mode of action code: 11 + M5

See also chlorothalonil

Products

1	Amistar Opti	Syngenta	100:500 g/l	SC	18156
2	Aubrac Optimum	AgChem Access	80:400 g/l	SC	15921
3	Curator	Syngenta	80:400 g/l	SC	18157
4	Olympus	Syngenta	80:400 g/l	SC	18158
5	Perseo	Sipcam	68:233 g/l	SC	17230

Uses
- Botrytis in **asparagus** *(qualified minor use recommendation)* [2, 4]; **bulb onions, garlic, shallots** [2]; **edible podded peas, vining peas** [4]
- Brown rust in **rye, spring barley, spring wheat, triticale, winter barley, winter wheat** [1, 3, 5]
- Downy mildew in **bulb onions, garlic, shallots** [2]; **edible podded peas, vining peas** [4]
- Glume blotch in **rye** [1, 3]; **spring wheat, winter wheat** [1, 3, 5]; **triticale** [3]

SEE SECTION 3 FOR PRODUCTS ALSO REGISTERED

- Net blotch in **spring barley**, **winter barley** [1, 3, 5]; **triticale** [1]
- Powdery mildew in **edible podded peas**, **vining peas** [4]; **rye**, **triticale** [5]
- Rhynchosporium in **rye**, **triticale** [5]; **spring barley**, **winter barley** [3]; **spring barley** *(moderate control)*, **winter barley** *(moderate control)* [1, 5]; **triticale** *(moderate control)* [1]
- Rust in **asparagus** *(moderate control only)* [2, 4]
- Septoria leaf blotch in **rye** [1, 3]; **spring wheat**, **winter wheat** [1, 3, 5]; **triticale** [3]
- Stemphylium in **asparagus** *(qualified minor use recommendation)* [2, 4]
- Take-all in **rye** *(reduction)*, **triticale** *(reduction)* [1, 3]; **spring barley** *(reduction)*, **winter barley** *(reduction)* [1, 5]; **spring barley** *(reduction only)*, **winter barley** *(reduction only)* [3]; **spring wheat** *(reduction)*, **winter wheat** *(reduction)* [1, 3, 5]
- Yellow rust in **rye** [1, 3]; **spring barley**, **triticale**, **winter barley** [3]; **spring wheat**, **winter wheat** [1, 3, 5]

Approval information
- Azoxystrobin and chlorothalonil included in Annex I under EC Regulation 1107/2009
- Accepted by BBPA for use on malting barley

Efficacy guidance
- Best results obtained from applications made as a protectant treatment or in earliest stages of disease development. Further applications may be needed if disease attack is prolonged
- Reduction of barley spotting occurs when used as part of a programme with other fungicides
- Control of *Septoria* and rust diseases may be improved by mixture with a triazole fungicide
- Azoxystrobin is a member of the QoI cross resistance group. Product should be used preventatively and not relied on for its curative potential
- Use product in cereals as part of an Integrated Crop Management strategy incorporating other methods of control, including where appropriate other fungicides with a different mode of action. Do not apply more than two foliar applications of QoI containing products to any cereal crop
- There is a significant risk of widespread resistance occurring in *Septoria tritici* populations in UK. Failure to follow resistance management action may result in reduced levels of disease control

Restrictions
- Maximum number of treatments 2 per crop
- Maximum total dose on barley equivalent to one full dose treatment
- Do not use where there is risk of spray drift onto neighbouring apple crops
- The same spray equipment should not be used to treat apples

Crop-specific information
- Latest use: before beginning of heading (GS 51) for barley; before caryopsis watery ripe (GS 71) for wheat

Environmental safety
- Dangerous for the environment
- Very toxic to aquatic organisms
- Buffer zone requirement 6m [5]
- LERAP Category B

Hazard classification and safety precautions
Hazard Toxic [1, 5]; Harmful [2-4]; Dangerous for the environment [1-5]; Toxic if inhaled [1]; Harmful if inhaled [3-5]; Very toxic to aquatic organisms [1, 3, 4]
Transport code 9
Packaging group III
UN Number 3082
Risk phrases H317, H335, H351 [1, 3-5]; H318 [1, 3, 4]; R20, R37, R40, R41, R43, R50, R53a [2]
Operator protection A, C, H; U02a, U05a, U09a, U11, U14, U15, U19a [1-5]; U20a [3]; U20b [1, 2, 4, 5]
Environmental protection E15a [1, 2, 4, 5]; E15b [3]; E16a, E34, E38 [1-5]; H410 [1, 3-5]
Storage and disposal D01, D02, D09a, D10c, D12a [1-5]; D05 [1, 2, 4, 5]
Medical advice M04a [1, 2, 4, 5]

FOR FULL CONDITIONS OF USE ALWAYS READ THE PRODUCT LABEL

21   azoxystrobin + cyproconazole

A contact and systemic broad spectrum fungicide mixture for cereals, beet crops and oilseed rape

FRAC mode of action code: 11 + 3

See also cyproconazole

Products

1	Aubrac Xtra	Harvest	200:80 g/l	SC	18083
2	Priori Xtra	Syngenta	200:80 g/l	SC	11518

Uses
- Alternaria in **spring oilseed rape**, **winter oilseed rape** [1, 2]
- Brown rust in **spring barley**, **spring rye**, **spring wheat**, **winter barley**, **winter rye**, **winter wheat** [1, 2]
- Cercospora leaf spot in **fodder beet**, **sugar beet** [1, 2]
- Crown rust in **spring oats**, **winter oats** [1, 2]
- Eyespot in **spring barley** *(reduction)*, **spring wheat** *(reduction)*, **winter barley** *(reduction)*, **winter wheat** *(reduction)* [1, 2]
- Foliar disease control in **durum wheat** *(off-label)*, **grass seed crops** *(off-label)*, **triticale** *(off-label)* [2]
- Glume blotch in **spring wheat**, **winter wheat** [1, 2]
- Net blotch in **spring barley**, **winter barley** [1, 2]
- Powdery mildew in **fodder beet**, **spring barley**, **spring oats**, **spring rye**, **spring wheat**, **sugar beet**, **winter barley**, **winter oats**, **winter rye**, **winter wheat** [1, 2]
- Ramularia leaf spots in **fodder beet**, **sugar beet** [1, 2]
- Rhynchosporium in **spring barley** *(moderate control)*, **spring rye** *(moderate control)*, **winter barley** *(moderate control)*, **winter rye** *(moderate control)* [1, 2]
- Rust in **fodder beet**, **sugar beet** [1, 2]
- Sclerotinia stem rot in **spring oilseed rape**, **winter oilseed rape** [1, 2]
- Septoria leaf blotch in **spring wheat**, **winter wheat** [1, 2]
- Take-all in **spring barley** *(reduction)*, **spring wheat** *(reduction)*, **winter barley** *(reduction)*, **winter wheat** *(reduction)* [1, 2]
- Yellow rust in **spring wheat**, **winter wheat** [1, 2]

Extension of Authorisation for Minor Use (EAMUs)
- **durum wheat** *20052891* [2]
- **grass seed crops** *20052891* [2]
- **triticale** *20052891* [2]

Approval information
- Azoxystrobin and cyproconazole included in Annex I under EC Regulation 1107/2009
- Accepted by BBPA for use on malting barley

Efficacy guidance
- Best results obtained from treatment during the early stages of disease development
- A second application may be needed if disease attack is prolonged
- Azoxystrobin is a member of the QoI cross resistance group. Product should be used preventatively and not relied on for its curative potential
- Use product as part of an Integrated Crop Management strategy incorporating other methods of control, including where appropriate other fungicides with a different mode of action. Do not apply more than two foliar applications of QoI containing products to any cereal crop
- There is a significant risk of widespread resistance occurring in *Septoria tritici* populations in UK. Failure to follow resistance management action may result in reduced levels of disease control
- Strains of wheat and barley powdery mildew resistant to QoIs are common in the UK. Control of wheat powdery mildew can only be relied upon from the triazole component
- Where specific control of wheat mildew is required this should be achieved through a programme of measures including products recommended for the control of mildew that contain a fungicide from a different cross-resistance group and applied at a dose that will give robust control

SEE SECTION 3 FOR PRODUCTS ALSO REGISTERED

SECTION 2

- Cyproconazole is a DMI fungicide. Resistance to some DMI fungicides has been identified in Septoria leaf blotch which may seriously affect performance of some products. For further advice contact a specialist advisor and visit the Fungicide Resistance Action Group (FRAG)-UK website

Restrictions
- Maximum total dose equivalent to two full dose treatments
- Do not use where there is risk of spray drift onto neighbouring apple crops
- The same spray equipment should not be used to treat apples

Crop-specific information
- Latest use: up to and including anthesis complete (GS 69) for rye and wheat; up to and including emergence of ear complete (GS 59) for barley and oats; BBCH79 (nearly all pods at final size) or 30 days before harvest, whichever is sooner for oilseed rape

Environmental safety
- Dangerous for the environment
- Very toxic to aquatic organisms

Hazard classification and safety precautions
  Hazard Harmful, Dangerous for the environment, Harmful if swallowed, Harmful if inhaled, Very toxic to aquatic organisms
  Transport code 9
  Packaging group III
  UN Number 3082
  Risk phrases H361
  Operator protection A; U05a, U09a, U19a, U20b
  Environmental protection E15b, E34, E38, H410
  Storage and disposal D01, D02, D05, D09a, D10c, D12a
  Medical advice M03

22   azoxystrobin + difenoconazole

A broad spectrum fungicide mixture for field crops
FRAC mode of action code: 11 + 3

See also difenoconazole

Products

Amistar Top	Syngenta	200:125 g/l	SC	18050

Uses
- Alternaria in **choi sum** *(off-label)*, **oriental brassicas** *(off-label)*
- Alternaria blight in **carrots**
- Black canker in **horseradish** *(off-label)*, **parsley root** *(off-label)*, **parsnips** *(off-label)*, **salsify** *(off-label)*
- Botrytis in **chicory** *(off-label)*, **chicory grown outside for forcing** *(off-label)*
- Disease control in **protected strawberries**, **strawberries**
- Phoma in **horseradish** *(off-label)*, **parsley root** *(off-label)*, **parsnips** *(off-label)*, **salsify** *(off-label)*
- Powdery mildew in **broccoli**, **brussels sprouts**, **cabbages**, **calabrese**, **carrots**, **cauliflowers**, **collards**, **kale**, **rocket**
- Purple blotch in **leeks** *(moderate control)*
- Purple spot in **asparagus** *(off-label)*
- Ring spot in **choi sum** *(off-label)*, **oriental brassicas** *(off-label)*
- Rust in **asparagus** *(off-label)*, **leeks**
- Sclerotinia in **chicory** *(off-label)*, **chicory grown outside for forcing** *(off-label)*
- White blister in **broccoli**, **brussels sprouts**, **cabbages**, **calabrese**, **cauliflowers**, **collards**, **kale**
- White tip in **leeks** *(qualified minor use)*

Extension of Authorisation for Minor Use (EAMUs)
- **asparagus** 20171501

- *chicory* 20171503
- *chicory grown outside for forcing* 20171503
- *choi sum* 20171502
- *horseradish* 20171340
- *oriental brassicas* 20171502
- *parsley root* 20171340
- *parsnips* 20171340
- *salsify* 20171340

Approval information
- Azoxystrobin and difenoconazole included in Annex I under EC Regulation 1107/2009

Efficacy guidance
- Best results obtained from applications made in the earliest stages of disease development or as a protectant treatment following a disease risk assessment
- Ensure the crop is free from any stress caused by environmental or agronomic effects
- Azoxystrobin is a member of the QoI cross resistance group. Product should be used preventatively and not relied on for its curative potential
- Use as part of an Integrated Crop Management strategy incorporating other methods of control, including where appropriate other fungicides with a different mode of action. Do not apply more than two foliar applications of QoI containing products

Restrictions
- Maximum number of treatments 2 per crop
- Do not apply where there is a risk of spray drift onto neighbouring apple crops
- Consult processors before treating a crop destined for processing

Crop-specific information
- HI 14 d for carrots; 21 d for brassicas, leeks
- Minimum spray interval of 14 d must be observed on brassicas

Environmental safety
- Dangerous for the environment
- Very toxic to aquatic organisms
- LERAP Category B

Hazard classification and safety precautions
Hazard Irritant, Dangerous for the environment, Harmful if swallowed, Harmful if inhaled
Transport code 9
Packaging group III
UN Number 3082
Risk phrases H317
Operator protection A, H; U05a, U09a, U19a, U20b
Environmental protection E15b, E16a, E38, H410
Storage and disposal D01, D02, D05, D09a, D10c, D12a

23   azoxystrobin + fluazinam

A blight fungicide for use in potatoes
FRAC mode of action code: 11 + 29

Products
Vendetta	Headland	150:375 g/l	SC	17500

Uses
- Blight in *potatoes*
- Leaf blight in *potatoes*

Approval information
- Azoxystrobin and fluazinam included in Annex I under EC Regulation 1107/2009

SEE SECTION 3 FOR PRODUCTS ALSO REGISTERED

Efficacy guidance
- Applications must begin prior to blight development. The first application should be applied at the first blight warning or when local weather conditions are favourable for disease development, whichever is the sooner. In the absence of weather conducive to disease development, the first application should be made just before the crop meets in the rows

Restrictions
- Do not use more than 3 consecutive Qol-containing sprays
- Certain apple varieties are highly sensitive. As a precaution , do not apply when there is a risk of spray drift onto neighbouring apple crops. Spray equipment used for application should not be used to treat apples

Environmental safety
- Buffer zone requirement 7m [1]

Hazard classification and safety precautions
   Hazard Very toxic to aquatic organisms
   Transport code 9
   Packaging group III
   UN Number 3082
   Risk phrases H317, H361
   Operator protection A, H
   Environmental protection H410

24   azoxystrobin + isopyrazam

A fungicide mixture for use in oilseed rape
FRAC mode of action code: 11 + 7

See also azoxystrobin

Products

Symetra	Syngenta	200:125 g/l	SC	16701

Uses
- Sclerotinia stem rot in **spring oilseed rape**, **winter oilseed rape**

Approval information
- Azoxystrobin and isopyrazam are included in Appendix 1 under EC Regulation 1107/2009

Efficacy guidance
- Rainfast within 1 hour of application.

Restrictions
- Do not use application equipment used to apply Symetra on apples or damage will occur.
- Contains a member of the Qol cross resistance group and a member of the SDHI cross resistance group. Should be used preventatively and should not be relied on for its curative potential.

Environmental safety
- LERAP Category B

Hazard classification and safety precautions
   Hazard Toxic, Dangerous for the environment, Harmful if swallowed, Toxic if inhaled, Very toxic to aquatic organisms
   Transport code 9
   Packaging group III
   UN Number 3082
   Risk phrases H361
   Operator protection A, H; U05a, U19a
   Environmental protection E16a, E38, H410
   Storage and disposal D01, D02, D12a

FOR FULL CONDITIONS OF USE ALWAYS READ THE PRODUCT LABEL

25   azoxystrobin + propiconazole

A contact and systemic broad spectrum fungicide mixture for use on maize or grass
FRAC mode of action code: 11 + 3

Products

1 Headway	Syngenta	62.5:104 g/l	EC	14396
2 PureProgress	Pure Amenity	62.5:104 g/l	EC	15471
3 Quilt Xcel	Syngenta	141.4:122.4 g/l	SE	16450

Uses

- Anthracnose in **amenity grassland** *(moderate control)*, **managed amenity turf** *(moderate control)* [1, 2]
- Disease control in **forage maize**, **grain maize** [3]
- Dollar spot in **amenity grassland**, **managed amenity turf** [1, 2]
- Eyespot in **forage maize** *(useful reduction)*, **grain maize** *(useful reduction)* [3]
- Fusarium diseases in **amenity grassland**, **managed amenity turf** [1, 2]
- Leaf blight in **forage maize** *(useful reduction)*, **grain maize** *(useful reduction)* [3]
- Microdochium nivale in **amenity grassland**, **managed amenity turf** [1, 2]
- Northern corn blight in **sweetcorn** *(off-label)* [3]

Extension of Authorisation for Minor Use (EAMUs)

- **sweetcorn** *20171534* [3]

Approval information

- Azoxystrobin and propiconazole included in Annex I under EC Regulation 1107/2009

Environmental safety

- LERAP Category B [1, 2]

Hazard classification and safety precautions

Hazard Dangerous for the environment [1, 2]; Harmful if swallowed, Harmful if inhaled [3]
Transport code 9
Packaging group III
UN Number 3082
Risk phrases H319 [3]
Operator protection A [1-3]; C, H [3]; U05a, U20a [1, 2]; U08, U11, U20b [3]; U09a, U19a [1-3]
Environmental protection E15b, E34, H410 [1-3]; E16a, E16b, E38 [1, 2]
Storage and disposal D01, D05, D09a, D10c [1-3]; D02, D12a [1, 2]; D12b [3]
Medical advice M03 [3]

26   azoxystrobin + tebuconazole

A strobilurin/triazole fungicide mixture for disease control in oilseed rape
FRAC mode of action code: 11 + 3

Products

Custodia	Adama	120:200 g/l	SC	16393

Uses

- Chocolate spot in **spring field beans** *(moderate control)*, **winter field beans**
- Disease control in **spring field beans**, **winter field beans** *(moderate control)*
- Sclerotinia in **spring oilseed rape** *(moderate control)*, **winter oilseed rape** *(moderate control)*

Approval information

- Azoxystrobin and tebuconazole included in Annex I under EC Regulation 1107/2009

Efficacy guidance

- Azoxystrobin is a member of the QoI cross resistance group. Product should be used preventatively and not relied on for its curative potential

SEE SECTION 3 FOR PRODUCTS ALSO REGISTERED

SECTION 2

- Use product as part of an Integrated Crop Management strategy incorporating other methods of control, including where appropriate other fungicides with a different mode of action. Do not apply more than two foliar applications of QoI containing products to any crop

Restrictions
- Avoid drift on to neighbouring crops since damage may occur especially to broad-leaved plants
- Newer authorisations for tebuconazole products require application to cereals only after GS 30 and applications to oilseed rape and linseed after GS20 - check label

Environmental safety
- LERAP Category B

Hazard classification and safety precautions

Hazard Harmful, Dangerous for the environment, Harmful if swallowed, Very toxic to aquatic organisms
Transport code 9
Packaging group III
UN Number 3082
Risk phrases H361
Operator protection A, H; U02a, U04a, U05a, U20b
Environmental protection E15b, E16a, E34, E38, H410
Storage and disposal D01, D02, D09a, D10b, D12a
Medical advice M03, M05a

27   Bacillus amyloliquefaciens D747

A fungicide for use on a range of horticultural crops
FRAC mode of action code: 44

Products

Amylo X WG	Certis	25% w/w	PO	17978

Uses
- Alternaria in **aubergines, blackberries, blackcurrants, blueberries, chicory, chillies, courgettes, cress, cucumbers, endives, gooseberries, lamb's lettuce, land cress, lettuce, loganberries, melons, mushrooms, peppers, pumpkins, raspberries, red mustard, redcurrants, rocket, rubus hybrids, spinach, spinach beet, strawberries, summer squash, tomatoes, watercress, watermelons, winter squash**
- Fusarium in **aubergines, blackberries, blackcurrants, blueberries, chicory, chillies, courgettes, cress, cucumbers, endives, gooseberries, lamb's lettuce, land cress, lettuce, loganberries, melons, mushrooms, peppers, pumpkins, raspberries, red mustard, redcurrants, rocket, rubus hybrids, spinach, spinach beet, strawberries, summer squash, tomatoes, watercress, watermelons, winter squash**
- Powdery mildew in **aubergines, blackberries, blackcurrants, blueberries, chicory, chillies, courgettes, cress, cucumbers, endives, gooseberries, lamb's lettuce, land cress, lettuce, loganberries, melons, peppers, pumpkins, raspberries, red mustard, redcurrants, rocket, rubus hybrids, spinach, spinach beet, strawberries, summer squash, tomatoes, watercress, watermelons, winter squash**
- Pythium in **aubergines, blackberries, blackcurrants, blueberries, chicory, chillies, courgettes, cress, cucumbers, endives, gooseberries, lamb's lettuce, land cress, lettuce, loganberries, melons, mushrooms, peppers, pumpkins, raspberries, red mustard, redcurrants, rocket, rubus hybrids, spinach, spinach beet, strawberries, summer squash, tomatoes, watercress, watermelons, winter squash**
- Rhizoctonia in **aubergines, blackberries, blackcurrants, blueberries, chicory, chillies, courgettes, cress, cucumbers, endives, gooseberries, lamb's lettuce, land cress, lettuce, loganberries, melons, mushrooms, peppers, pumpkins, raspberries, red mustard, redcurrants, rocket, rubus hybrids, spinach, spinach beet, strawberries, summer squash, tomatoes, watercress, watermelons, winter squash**

Approval information
- Bacillus amyloliquefaciens included in Annex I under EC Regulation 1107/2009

Hazard classification and safety precautions
UN Number N/C
Operator protection A, D, H; U05b
Environmental protection E15b, E34
Storage and disposal D01, D02, D10c

28 Bacillus subtilis

A bacterial fungicide for the control of Botrytis cinerea
FRAC mode of action code: 44

Products

Serenade ASO	Bayer CropScience	13.96 g/l	SC	16139

Uses
- Botrytis in *bilberries* (off-label), *blackcurrants* (off-label), *blueberries* (off-label), *bulb vegetables* (off-label), *canary grass* (off-label), *cane fruit* (off-label), *figs* (off-label), *fruiting vegetables* (off-label), *gooseberries* (off-label), *herbs (see appendix 6)* (off-label), *hops* (off-label), *leafy vegetables* (off-label), *legumes* (off-label), *ornamental plant production* (off-label), *redcurrants* (off-label), *ribes hybrids* (off-label), *root & tuber crops* (off-label), *stem vegetables* (off-label), *table grapes* (off-label), *top fruit* (off-label), *vegetable brassicas* (off-label), *wine grapes* (off-label)
- Botrytis fruit rot in *protected strawberries* (reduction of damage to fruit), *strawberries*
- Butt rot in *lettuce* (off-label)
- Cavity spot in *carrots* (off-label), *parsnips* (off-label)
- Damping off in *baby leaf crops* (off-label), *broccoli* (off-label), *brussels sprouts* (off-label), *bulb onions* (off-label), *cabbages* (off-label), *calabrese* (off-label), *carrots* (off-label), *cauliflowers* (off-label), *celeriac* (off-label), *celery (outdoor)* (off-label), *chard* (off-label), *collards* (off-label), *garlic* (off-label), *herbs (see appendix 6)* (off-label), *kale* (off-label), *leeks* (off-label), *lettuce* (off-label), *parsnips* (off-label), *salad onions* (off-label), *shallots* (off-label), *spinach* (off-label)
- Grey mould in *protected strawberries* (reduction of damage to fruit), *strawberries*
- Helminthosporium in *potatoes* (off-label)
- Phytophthora in *amenity vegetation* (off-label), *blackberries* (off-label), *blackcurrants* (off-label), *blueberries* (off-label), *courgettes* (off-label), *cranberries* (off-label), *cucumbers* (off-label), *forest nurseries* (off-label), *gooseberries* (off-label), *loganberries* (off-label), *marrows* (off-label), *pumpkins* (off-label), *raspberries* (off-label), *redcurrants* (off-label), *rubus hybrids* (off-label), *squashes* (off-label)
- Pythium in *aubergines* (off-label), *broccoli* (off-label), *brussels sprouts* (off-label), *cabbages* (off-label), *calabrese* (off-label), *carrots* (off-label), *cauliflowers* (off-label), *chillies* (off-label), *collards* (off-label), *courgettes* (off-label), *cucumbers* (off-label), *kale* (off-label), *marrows* (off-label), *parsnips* (off-label), *peppers* (off-label), *pumpkins* (off-label), *squashes* (off-label), *tomatoes (outdoor)* (off-label)
- Rhizoctonia in *lettuce* (off-label), *potatoes* (off-label)
- Streptomyces in *potatoes* (off-label)
- White rot in *bulb onions* (off-label), *garlic* (off-label), *leeks* (off-label), *salad onions* (off-label), *shallots* (off-label)

Extension of Authorisation for Minor Use (EAMUs)
- *amenity vegetation* 20130704
- *aubergines* 20150306
- *baby leaf crops* 20150306
- *bilberries* 20130706
- *blackberries* 20150306
- *blackcurrants* 20130706, 20150306
- *blueberries* 20130706, 20150306
- *broccoli* 20150306
- *brussels sprouts* 20150306

SEE SECTION 3 FOR PRODUCTS ALSO REGISTERED

- **bulb onions** *20150306*
- **bulb vegetables** *20130706*
- **cabbages** *20150306*
- **calabrese** *20150306*
- **canary grass** *20130706*
- **cane fruit** *20130706*
- **carrots** *20150306*
- **cauliflowers** *20150306*
- **celeriac** *20150306*
- **celery (outdoor)** *20150306*
- **chard** *20150306*
- **chillies** *20150306*
- **collards** *20150306*
- **courgettes** *20150306*
- **cranberries** *20150306*
- **cucumbers** *20150306*
- **figs** *20130706*
- **forest nurseries** *20130704*
- **fruiting vegetables** *20130706*
- **garlic** *20150306*
- **gooseberries** *20130706, 20150306*
- **herbs (see appendix 6)** *20130706, 20150306*
- **hops** *20130706*
- **kale** *20150306*
- **leafy vegetables** *20130706*
- **leeks** *20150306*
- **legumes** *20130706*
- **lettuce** *20150306*
- **loganberries** *20150306*
- **marrows** *20150306*
- **ornamental plant production** *20130706*
- **parsnips** *20150306*
- **peppers** *20150306*
- **potatoes** *20150306*
- **pumpkins** *20150306*
- **raspberries** *20150306*
- **redcurrants** *20130706, 20150306*
- **ribes hybrids** *20130706*
- **root & tuber crops** *20130706*
- **rubus hybrids** *20150306*
- **salad onions** *20150306*
- **shallots** *20150306*
- **spinach** *20150306*
- **squashes** *20150306*
- **stem vegetables** *20130706*
- **table grapes** *20130706*
- **tomatoes (outdoor)** *20150306*
- **top fruit** *20130706*
- **vegetable brassicas** *20130706*
- **wine grapes** *20130706*

Approval information
- Bacillus subtilis included in Annex I under EC Regulation 1107/2009

Efficacy guidance
- Alternating applications with fungicides using a different mode of action is recommended for resistance management.
- Do not apply using irrigation equipment.
- For maximum effectiveness, start applications before disease development.
- Apply in a minimum water volume of 400 l/ha.

FOR FULL CONDITIONS OF USE ALWAYS READ THE PRODUCT LABEL

Restrictions
• Consult processor before using on crops grown for processing

Hazard classification and safety precautions
Hazard Irritant
UN Number N/C
Operator protection A, D, H; U05a, U11, U14, U20b
Environmental protection E15b
Storage and disposal D01, D02, D05, D10a, D16

29 Bacillus thuringiensis

A bacterial insecticide for control of caterpillars
IRAC mode of action code: 11

Products

1	Dipel DF	Interfarm	54% w/w	WG	17499
2	Lepinox Plus	Fargro	37.5% w/w	WP	16269

Uses
• Bramble shoot moth in **bilberries** *(off-label)*, **blackberries** *(off-label)*, **blackcurrants** *(off-label)*, **blueberries** *(off-label)*, **cranberries** *(off-label)*, **gooseberries** *(off-label)*, **loganberries** *(off-label)*, **redcurrants** *(off-label)*, **rubus hybrids** *(off-label)* [2]
• Cabbage moth in **baby leaf crop** *(off-label)*, **choi sum** *(off-label)*, **collards** *(off-label)*, **kohlrabi** *(off-label)*, **oriental cabbage** *(off-label)*, **protected chicory** *(off-label)*, **protected cress** *(off-label)*, **protected lamb's lettuce** *(off-label)*, **protected rocket** *(off-label)*, **protected watercress** *(off-label)* [2]
• Cabbage white butterfly in **baby leaf crop** *(off-label)*, **choi sum** *(off-label)*, **collards** *(off-label)*, **kohlrabi** *(off-label)*, **oriental cabbage** *(off-label)*, **protected chicory** *(off-label)*, **protected cress** *(off-label)*, **protected lamb's lettuce** *(off-label)*, **protected rocket** *(off-label)*, **protected watercress** *(off-label)* [2]
• Caterpillars in **all edible seed crops grown outdoors** *(off-label)*, **all non-edible crops (outdoor)** *(off-label)*, **all protected non-edible crops** *(off-label)*, **amenity vegetation**, **apples** *(off-label)*, **apricots** *(off-label)*, **baby leaf crops** *(off-label)*, **calabrese** *(off-label)*, **cauliflowers**, **celery (outdoor)** *(off-label)*, **cherries** *(off-label)*, **chestnuts** *(off-label)*, **choi sum** *(off-label)*, **cob nuts** *(off-label)*, **collards** *(off-label)*, **edible podded peas**, **endives** *(off-label)*, **fennel** *(off-label)*, **filberts** *(off-label)*, **forest** *(off-label)*, **hazel nuts** *(off-label)*, **herbs (see appendix 6)** *(off-label)*, **kale** *(off-label)*, **kiwi fruit** *(off-label)*, **kohlrabi** *(off-label)*, **leeks**, **lettuce** *(off-label)*, **nectarines** *(off-label)*, **olives** *(off-label)*, **oriental cabbage** *(off-label)*, **ornamental plant production**, **peaches** *(off-label)*, **pears** *(off-label)*, **plums** *(off-label)*, **protected apricots** *(off-label)*, **protected baby leaf crops** *(off-label)*, **protected broad beans**, **protected broccoli** *(off-label)*, **protected brussels sprouts** *(off-label)*, **protected cabbages** *(off-label)*, **protected calabrese** *(off-label)*, **protected cauliflowers** *(off-label)*, **protected celery** *(off-label)*, **protected chestnuts** *(off-label)*, **protected chilli peppers**, **protected choi sum** *(off-label)*, **protected collards** *(off-label)*, **protected cucumbers**, **protected dwarf french beans**, **protected edible crops** *(off-label)*, **protected endives** *(off-label)*, **protected hazelnuts** *(off-label)*, **protected herbs (see appendix 6)** *(off-label)*, **protected kale** *(off-label)*, **protected kiwi fruit** *(off-label)*, **protected lettuce** *(off-label)*, **protected nectarines** *(off-label)*, **protected olives** *(off-label)*, **protected oriental cabbage** *(off-label)*, **protected peaches** *(off-label)*, **protected plums** *(off-label)*, **protected quince** *(off-label)*, **protected rhubarb** *(off-label)*, **protected runner beans**, **protected spinach** *(off-label)*, **protected spinach beet** *(off-label)*, **protected tatsoi** *(off-label)*, **protected walnuts** *(off-label)*, **protected watercress** *(off-label)*, **protected wine grapes** *(off-label)*, **quinces** *(off-label)*, **raspberries**, **rhubarb** *(off-label)*, **spinach** *(off-label)*, **spinach beet** *(off-label)*, **spring cabbage** *(off-label)*, **tatsoi** *(off-label)*, **vining peas**, **walnuts** *(off-label)*, **watercress** *(off-label)*, **wine grapes** *(off-label)* [1]; **apples**, **brussels sprouts**, **celery (outdoor)**, **chicory**, **chinese cabbage**, **courgettes**, **dwarf beans**, **endives**, **french beans**, **herbs (see appendix 6)**, **hops**, **kale**, **lettuce**, **pears**, **protected melons**, **protected watermelon**, **pumpkins**, **radishes**, **spinach**, **spinach beet**, **turnips**, **wine grapes**, **winter squash** [2]; **broccoli**, **cabbages**,

SEE SECTION 3 FOR PRODUCTS ALSO REGISTERED

calabrese, combining peas, globe artichoke, protected aubergines, protected peppers, protected tomatoes, strawberries [1, 2]

- Cherry fruit moth in **apricots** *(off-label)*, **cherries** *(off-label)*, **nectarines** *(off-label)*, **peaches** *(off-label)*, **plums** *(off-label)* [2]
- Cutworms in **baby leaf crop** *(off-label)*, **bilberries** *(off-label)*, **blackberries** *(off-label)*, **blackcurrants** *(off-label)*, **blueberries** *(off-label)*, **broad beans** *(off-label)*, **cranberries** *(off-label)*, **forest nurseries** *(off-label)*, **gooseberries** *(off-label)*, **loganberries** *(off-label)*, **protected chicory** *(off-label)*, **protected chilli peppers** *(off-label)*, **protected courgettes** *(off-label)*, **protected cress** *(off-label)*, **protected dwarf french beans** *(off-label)*, **protected edible podded peas** *(off-label)*, **protected lamb's lettuce** *(off-label)*, **protected peppers** *(off-label)*, **protected rocket** *(off-label)*, **protected runner beans** *(off-label)*, **protected summer squash** *(off-label)*, **protected watercress** *(off-label)*, **redcurrants** *(off-label)*, **rhubarb** *(off-label)*, **rubus hybrids** *(off-label)*, **sweetcorn** *(off-label)* [2]; **bulb onions** *(off-label)*, **carrots** *(off-label)*, **celeriac** *(off-label)*, **garlic** *(off-label)*, **horseradish** *(off-label)*, **mooli** *(off-label)*, **parsley root** *(off-label)*, **parsnips** *(off-label)*, **radishes** *(off-label)*, **red beet** *(off-label)*, **salsify** *(off-label)*, **shallots** *(off-label)*, **swedes** *(off-label)*, **turnips** *(off-label)* [1]; **leeks** *(off-label)* [1, 2]
- Diamond-back moth in **baby leaf crop** *(off-label)*, **broccoli, brussels sprouts, cabbages, calabrese, chinese cabbage, choi sum** *(off-label)*, **collards** *(off-label)*, **kohlrabi** *(off-label)*, **oriental cabbage** *(off-label)*, **protected chicory** *(off-label)*, **protected cress** *(off-label)*, **protected lamb's lettuce** *(off-label)*, **protected rocket** *(off-label)*, **protected watercress** *(off-label)*, **turnips** [2]
- Flax tortrix moth in **broad beans** *(off-label)*, **protected dwarf french beans** *(off-label)*, **protected edible podded peas** *(off-label)*, **protected runner beans** *(off-label)* [2]
- Leek moth in **leeks** *(off-label)*, **rhubarb** *(off-label)* [2]
- Light brown apple moth in **apricots** *(off-label)*, **cherries** *(off-label)*, **nectarines** *(off-label)*, **peaches** *(off-label)*, **plums** *(off-label)* [2]
- Oak Processionary Moth in **forest** *(off-label)* [1]
- Plum fruit moth in **apricots** *(off-label)*, **cherries** *(off-label)*, **nectarines** *(off-label)*, **peaches** *(off-label)*, **plums** *(off-label)* [2]
- Plum tortrix moth in **apricots** *(off-label)*, **cherries** *(off-label)*, **nectarines** *(off-label)*, **peaches** *(off-label)*, **plums** *(off-label)* [2]
- Plutella xylostella in **broccoli, brussels sprouts, cabbages, calabrese, chinese cabbage, turnips** [2]
- Raspberry moth in **bilberries** *(off-label)*, **blackberries** *(off-label)*, **blackcurrants** *(off-label)*, **blueberries** *(off-label)*, **cranberries** *(off-label)*, **gooseberries** *(off-label)*, **loganberries** *(off-label)*, **redcurrants** *(off-label)*, **rubus hybrids** *(off-label)* [2]
- Silver Y moth in **baby leaf crop** *(off-label)*, **protected chicory** *(off-label)*, **protected chilli peppers** *(off-label)*, **protected courgettes** *(off-label)*, **protected cress** *(off-label)*, **protected dwarf french beans** *(off-label)*, **protected edible podded peas** *(off-label)*, **protected lamb's lettuce** *(off-label)*, **protected peppers** *(off-label)*, **protected rocket** *(off-label)*, **protected runner beans** *(off-label)*, **protected summer squash** *(off-label)*, **protected watercress** *(off-label)* [2]; **broad beans** *(off-label)*, **sweetcorn** *(off-label)* [1, 2]; **dwarf beans** *(off-label)*, **french beans** *(off-label)*, **runner beans** *(off-label)* [1]
- Strawberry tortrix in **bilberries** *(off-label)*, **blackberries** *(off-label)*, **blackcurrants** *(off-label)*, **blueberries** *(off-label)*, **cranberries** *(off-label)*, **gooseberries** *(off-label)*, **loganberries** *(off-label)*, **redcurrants** *(off-label)*, **rubus hybrids** *(off-label)* [2]
- Summer-fruit tortrix moth in **apricots** *(off-label)*, **cherries** *(off-label)*, **nectarines** *(off-label)*, **peaches** *(off-label)*, **plums** *(off-label)* [2]
- Tomato moth in **protected chilli peppers** *(off-label)*, **protected courgettes** *(off-label)*, **protected peppers** *(off-label)*, **protected summer squash** *(off-label)*, **sweetcorn** *(off-label)* [2]
- Tortrix moths in **apples, pears, strawberries** [2]
- Vapourer moth in **forest nurseries** *(off-label)* [2]
- White ermine moth in **forest nurseries** *(off-label)* [2]
- Winter moth in **apricots** *(off-label)*, **cherries** *(off-label)*, **nectarines** *(off-label)*, **peaches** *(off-label)*, **plums** *(off-label)* [2]; **bilberries** *(off-label)*, **blackcurrants** *(off-label)*, **blueberries** *(off-label)*, **cranberries** *(off-label)*, **prorected vaccinium spp.** *(off-label)*, **protected bilberries** *(off-label)*, **protected blackcurrants** *(off-label)*, **protected blueberry** *(off-label)*, **protected cranberries** *(off-label)*, **protected gooseberries** *(off-label)*, **protected redcurrants** *(off-label)*, **redcurrants** *(off-label)*, **vaccinium spp.** *(off-label)* [1]

FOR FULL CONDITIONS OF USE ALWAYS READ THE PRODUCT LABEL

Extension of Authorisation for Minor Use (EAMUs)

- *all edible seed crops grown outdoors* 20162634 [1]
- *all non-edible crops (outdoor)* 20162634 [1]
- *all protected non-edible crops* 20162634 [1]
- *apples* 20162635 [1]
- *apricots* 20162632 [1], 20142700 [2]
- *baby leaf crop* 20142704 [2]
- *baby leaf crops* 20162627 [1]
- *bilberries* 20162633 [1], 20142706 [2]
- *blackberries* 20142706 [2]
- *blackcurrants* 20162633 [1], 20142706 [2]
- *blueberries* 20162633 [1], 20142706 [2]
- *broad beans* 20162625 [1], 20142702 [2]
- *bulb onions* 20162623 [1]
- *calabrese* 20162626 [1]
- *carrots* 20162631 [1]
- *celeriac* 20162631 [1]
- *celery (outdoor)* 20162626 [1]
- *cherries* 20162635 [1], 20142700 [2]
- *chestnuts* 20162632 [1]
- *choi sum* 20162626 [1], 20142701 [2]
- *cob nuts* 20162632 [1]
- *collards* 20162626 [1], 20142701 [2]
- *cranberries* 20162633 [1], 20142706 [2]
- *dwarf beans* 20162625 [1]
- *endives* 20162627 [1]
- *fennel* 20162626 [1]
- *filberts* 20162632 [1]
- *forest* 20160931 [1]
- *forest nurseries* 20142705 [2]
- *french beans* 20162625 [1]
- *garlic* 20162623 [1]
- *gooseberries* 20142706 [2]
- *hazel nuts* 20162632 [1]
- *herbs (see appendix 6)* 20162627 [1]
- *horseradish* 20162631 [1]
- *kale* 20162626 [1]
- *kiwi fruit* 20162632 [1]
- *kohlrabi* 20162626 [1], 20142701 [2]
- *leeks* 20162629 [1], 20142703 [2]
- *lettuce* 20162627 [1]
- *loganberries* 20142706 [2]
- *mooli* 20162631 [1]
- *nectarines* 20162632 [1], 20142700 [2]
- *olives* 20162632 [1]
- *oriental cabbage* 20162626 [1], 20142701 [2]
- *parsley root* 20162631 [1]
- *parsnips* 20162631 [1]
- *peaches* 20162632 [1], 20142700 [2]
- *pears* 20162635 [1]
- *plums* 20162632 [1], 20142700 [2]
- *protected apricots* 20162632 [1]
- *protected baby leaf crops* 20162627 [1]
- *protected bilberries* 20162633 [1]
- *protected blackcurrants* 20162633 [1]
- *protected blueberry* 20162633 [1]
- *protected broccoli* 20162626 [1]
- *protected brussels sprouts* 20162626 [1]
- *protected cabbages* 20162626 [1]

SEE SECTION 3 FOR PRODUCTS ALSO REGISTERED

- *protected calabrese* *20162626* [1]
- *protected cauliflowers* *20162626* [1]
- *protected celery* *20162626* [1]
- *protected chestnuts* *20162632* [1]
- *protected chicory* *20142704* [2]
- *protected chilli peppers* *20142707* [2]
- *protected choi sum* *20162626* [1]
- *protected collards* *20162626* [1]
- *protected courgettes* *20142707* [2]
- *protected cranberries* *20162633* [1]
- *protected cress* *20142704* [2]
- *protected dwarf french beans* *20142702* [2]
- *protected edible crops* *20162634* [1]
- *protected edible podded peas* *20142702* [2]
- *protected endives* *20162627* [1]
- *protected gooseberries* *20162633* [1]
- *protected hazelnuts* *20162632* [1]
- *protected herbs (see appendix 6)* *20162627* [1]
- *protected kale* *20162626* [1]
- *protected kiwi fruit* *20162632* [1]
- *protected lamb's lettuce* *20142704* [2]
- *protected lettuce* *20162627* [1]
- *protected nectarines* *20162632* [1]
- *protected olives* *20162632* [1]
- *protected oriental cabbage* *20162626* [1]
- *protected peaches* *20162632* [1]
- *protected peppers* *20142707* [2]
- *protected plums* *20162632* [1]
- *protected quince* *20162632* [1]
- *protected redcurrants* *20162633* [1]
- *protected rhubarb* *20162626* [1]
- *protected rocket* *20142704* [2]
- *protected runner beans* *20142702* [2]
- *protected spinach* *20162627* [1]
- *protected spinach beet* *20162627* [1]
- *protected summer squash* *20142707* [2]
- *protected tatsoi* *20162626* [1]
- *protected vaccinium spp.* *20162633* [1]
- *protected walnuts* *20162632* [1]
- *protected watercress* *20162628* [1], *20142704* [2]
- *protected wine grapes* *20162632* [1]
- *quinces* *20162632* [1]
- *radishes* *20162631* [1]
- *red beet* *20162624* [1]
- *redcurrants* *20162633* [1], *20142706* [2]
- *rhubarb* *20162626* [1], *20142703* [2]
- *rubus hybrids* *20142706* [2]
- *runner beans* *20162625* [1]
- *salsify* *20162631* [1]
- *shallots* *20162623* [1]
- *spinach* *20162627* [1]
- *spinach beet* *20162627* [1]
- *spring cabbage* *20162626* [1]
- *swedes* *20162631* [1]
- *sweetcorn* *20162630* [1], *20142707* [2]
- *tatsoi* *20162626* [1]
- *turnips* *20162631* [1]
- *vaccinium spp.* *20162633* [1]

FOR FULL CONDITIONS OF USE ALWAYS READ THE PRODUCT LABEL

- **walnuts** *20162632* [1]
- **watercress** *20162628* [1]
- **wine grapes** *20162632* [1]

Approval information
- Bacillus thuringiensis included in Annex I under EC Regulation 1107/2009

Efficacy guidance
- Pest control achieved by ingestion by caterpillars of the treated plant vegetation. Caterpillars cease feeding and die in 1-3 d
- Apply as soon as larvae appear on crop and repeat every 7-10 d until the end of the hatching period
- Good coverage is essential, especially of undersides of leaves. Spray onto dry foliage and do not apply if rain expected within 6 h

Restrictions
- Apply spray mixture as soon as possible after preparation

Crop-specific information
- HI zero

Environmental safety
- Store out of direct sunlight

Hazard classification and safety precautions
UN Number N/C
Operator protection A, C, D, H; U05a, U15, U20c [1]; U11, U12, U14, U16b, U20b [2]; U19a [1, 2]
Environmental protection E15a [1]; E15b, E34 [2]
Storage and disposal D01, D02, D09a [1, 2]; D05, D11a [1]; D10b [2]
Medical advice M03, M05a [2]

30 Bacillus thuringiensis israelensis

A bacterial insecticide for control of larvae of chironomid midges
IRAC mode of action code: 11

Products

Vectobac 12 AS	Resource Chemicals	12 g/l	ZZ	HSE 6205

Uses
- Mosquitoes in **open waters**

Approval information
- Bacillus thuringiensis israelensis included in Annex I under EC Regulation 1107/2009

Hazard classification and safety precautions
UN Number N/C
Operator protection A, H

31 Beauveria bassiana

It is an entomopathogenic fungus causing white muscardine disease. It can be used as a biological insecticide to control a number of pests such as termites, whitefly and some beetles
IRAC mode of action code: 11

Products

1	Botanigard WP	Certis	22% w/w	WP	17054
2	Naturalis-L	Fargro	7.16% w/w	OD	17526

SEE SECTION 3 FOR PRODUCTS ALSO REGISTERED

Uses

- Aphids in **protected aubergines, protected chilli peppers, protected courgettes, protected cucumbers, protected melons, protected nursery fruit trees, protected peppers, protected strawberries, protected summer squash, protected tomatoes** [1]; **protected edible crops, protected forest nurseries** *(off-label)* [2]; **protected ornamentals** [1, 2]
- Beetles in **protected aubergines, protected chilli peppers, protected courgettes, protected cucumbers, protected melons, protected nursery fruit trees, protected ornamentals, protected peppers, protected strawberries, protected summer squash, protected tomatoes** [1]
- Thrips in **protected strawberries** *(off-label)*, **strawberries** *(off-label)* [1]
- Whitefly in **protected aubergines, protected chilli peppers, protected courgettes, protected cucumbers, protected melons, protected nursery fruit trees, protected peppers, protected strawberries, protected strawberries** *(off-label)*, **protected summer squash, protected tomatoes, strawberries** *(off-label)* [1]; **protected edible crops, protected forest nurseries** *(off-label)* [2]; **protected ornamentals** [1, 2]

Extension of Authorisation for Minor Use (EAMUs)

- **protected forest nurseries** *20162195* [2]
- **protected strawberries** *20160110 expires 30 Sep 2018* [1]
- **strawberries** *20160110 expires 30 Sep 2018* [1]

Approval information

- Beauveria bassiana included in Annex 1 under EC Regulation 1107/2009

Hazard classification and safety precautions

UN Number N/C
Risk phrases H317, H334 [1]
Operator protection A, D, H; U05a, U11, U14, U15, U16b, U19a, U20b
Environmental protection E15b, E34
Storage and disposal D01, D02, D05, D09a, D10a
Medical advice M03, M04a

32   bentazone

A post-emergence contact benzothiadiazinone herbicide
HRAC mode of action code: C3

Products

1 Basagran SG	BASF	87% w/w	SG	08360
2 Clayton Dent 480	Clayton	480 g/l	SL	15546
3 Tanaru	Agroquimicos	480 g/l	SL	15726
4 Troy 480	UPL Europe	480 g/l	SL	16954

Uses

- Annual dicotyledons in **broad beans, linseed, potatoes, runner beans, spring field beans, winter field beans** [1-4]; **bulb onions** *(off-label)*, **chives** *(off-label)*, **garlic** *(off-label)*, **herbs (see appendix 6)** *(off-label)*, **leeks** *(off-label)*, **salad onions** *(off-label)*, **shallots** *(off-label)*, **soya beans** *(off-label)* [1]; **combining peas, dwarf beans, vining peas** [3, 4]; **french beans, peas** [1, 2]; **narcissi** [1-3]; **navy beans** [1, 3]; **ornamental plant production** [4]
- Chickweed in **game cover** *(off-label)*, **hops** *(off-label)*, **ornamental plant production** *(off-label)* [1]
- Cleavers in **game cover** *(off-label)*, **hops** *(off-label)*, **ornamental plant production** *(off-label)* [1]
- Common storksbill in **leeks** *(off-label)* [1]
- Fool's parsley in **leeks** *(off-label)* [1]
- Groundsel in **game cover** *(off-label)*, **herbs (see appendix 6)** *(off-label)*, **hops** *(off-label)*, **ornamental plant production** *(off-label)* [1]
- Mayweeds in **bulb onions** *(off-label)*, **game cover** *(off-label)*, **garlic** *(off-label)*, **herbs (see appendix 6)** *(off-label)*, **hops** *(off-label)*, **ornamental plant production** *(off-label)*, **shallots** *(off-label)* [1]

Extension of Authorisation for Minor Use (EAMUs)
- ***bulb onions*** *20061631* [1]
- ***chives*** *20102994* [1]
- ***game cover*** *20082819* [1]
- ***garlic*** *20061631* [1]
- ***herbs (see appendix 6)*** *20120744* [1]
- ***hops*** *20082819* [1]
- ***leeks*** *20061630* [1]
- ***ornamental plant production*** *20082819* [1]
- ***salad onions*** *20102994* [1]
- ***shallots*** *20061631* [1]
- ***soya beans*** *20061629* [1]

Approval information
- Bentazone included in Annex I under EC Regulation 1107/2009

Efficacy guidance
- Most effective control obtained when weeds are growing actively and less than 5 cm high or across. Good spray cover is essential
- The addition of specified adjuvant oils is recommended for use on some crops to improve fat hen control. Do not use under hot or humid conditions. See label for details
- Split dose application may be made in all recommended crops except peas and generally gives better weed control. See label for details

Restrictions
- Maximum number of treatments normally 2 per crop but check label
- Crops must be treated at correct stage of growth to avoid danger of scorch. See label for details
- Not all varieties of recommended crops are fully tolerant. Use only on tolerant varieties named in label. Do not use on forage or mange-tout varieties of peas
- Do not use on crops which have been affected by drought, waterlogging, frost or other stress conditions
- Do not apply insecticides within 7 d of treatment
- Leave 14 d after using a post-emergence grass herbicide and carry out a leaf wax test where relevant or 7 d where treatment precedes the grass herbicide
- Do not spray at temperatures above 21°C. Delay spraying until evening if necessary
- Do not apply if rain or frost expected, if unseasonably cold, if foliage wet or in drought
- A minimum of 6 h (preferably 12 h) free from rain is required after application
- May be used on selected varieties of maincrop and second early potatoes (see label for details), not on seed crops or first earlies

Crop-specific information
- Latest use: before shoots exceed 15 cm high for potatoes and spring field beans (or 6-7 leaf pairs); 4 leaf pairs (6 pairs or 15 cm high with split dose) for broad beans; before flower buds visible for French, navy, runner and winter field beans, and linseed; before flower buds can be found enclosed in terminal shoot for peas
- Best results in narcissi obtained by using a suitable pre-emergence herbicide first
- Consult processor before using on crops for processing
- A satisfactory wax test must be carried out before use on peas
- Do not treat narcissi during flower bud initiation

Environmental safety
- Dangerous for the environment
- Harmful to aquatic organisms
- Some pesticides pose a greater threat of contamination of water than others and bentazone is one of these pesticides. Take special care when applying bentazone near water and do not apply if heavy rain is forecast

Hazard classification and safety precautions
    Hazard Harmful, Harmful if swallowed [1, 2]; Irritant [3, 4]
    Packaging group III [1]
    UN Number 2588 [1]; N/C [2-4]
    Risk phrases H317 [1, 2, 4]; H318 [1]; H320 [4]; R36, R43 [3]

SEE SECTION 3 FOR PRODUCTS ALSO REGISTERED

Operator protection A [1-4]; C [1, 3, 4]; H [3, 4]; U05a, U08, U19a [1-4]; U11, U20b [1, 3, 4]; U14 [1]; U20a [2]
Environmental protection E15a, H412 [2]; E15b, E34 [1, 3, 4]; E38 [4]
Storage and disposal D01, D02, D09a [1-4]; D10b [1]; D10c [2-4]
Medical advice M03 [1, 3, 4]; M05a [1]

33 benthiavalicarb-isopropyl

An amino acid amide carbamate fungicide available only in mixtures
FRAC mode of action code: 40

34 benthiavalicarb-isopropyl + mancozeb

A fungicide mixture for potatoes
FRAC mode of action code: 40 + M3

See also mancozeb

Products

Valbon	Certis	1.75:70% w/w	WG	14868

Uses
- Blight in **potatoes**
- Downy mildew in **amenity vegetation** *(off-label)*, **bulb onions** *(off-label)*, **forest nurseries** *(off-label)*, **garlic** *(off-label)*, **hops** *(off-label)*, **ornamental plant production** *(off-label)*, **shallots** *(off-label)*, **soft fruit** *(off-label)*, **table grapes** *(off-label)*, **wine grapes** *(off-label)*

Extension of Authorisation for Minor Use (EAMUs)
- **amenity vegetation** *20142325 expires 23 Feb 2018*
- **bulb onions** *20142326 expires 23 Feb 2018*
- **forest nurseries** *20142328 expires 23 Feb 2018*
- **garlic** *20142326 expires 23 Feb 2018*
- **hops** *20142328 expires 23 Feb 2018*
- **ornamental plant production** *20142325 expires 23 Feb 2018*
- **shallots** *20142326 expires 23 Feb 2018*
- **soft fruit** *20142328 expires 23 Feb 2018*
- **table grapes** *20142327 expires 23 Feb 2018*
- **wine grapes** *20142327 expires 23 Feb 2018*

Approval information
- Benthiavalicarb-isopropyl and mancozeb included in Annex I under EC Regulation 1107/2009

Efficacy guidance
- Apply as a protectant spray commencing before blight enters the crop irrespective of growth stage following a blight warning or where there is a local source of infection
- In the absence of a warning commence the spray programme before the foliage meets in the rows
- Repeat treatment every 7-10 d depending on disease pressure
- Increase water volume as necessary to ensure thorough coverage of the plant
- Use of air assisted sprayers or drop legs may help to improve coverage

Restrictions
- Maximum number of treatments 6 per crop
- Do not use more than three consecutive treatments. Include products with a different mode of action in the programme

Crop-specific information
- HI 7 d for potatoes

Environmental safety
- Very toxic to aquatic organisms

FOR FULL CONDITIONS OF USE ALWAYS READ THE PRODUCT LABEL

- Risk to non-target insects or other arthropods. Avoid spraying within 6 m of field boundary
- LERAP Category B

Hazard classification and safety precautions

Hazard Irritant
Transport code 9
Packaging group III
UN Number 3077
Risk phrases R40, R43, R50, R53a
Operator protection A, D, H; U14
Environmental protection E15a, E16a, E22c
Storage and disposal D01, D02, D05, D09a, D11a, D12b

35   benzoic acid

An organic horticultural disinfectant

Products

Menno Florades	Fargro	90 g/kg	SL	13985

Uses

- Bacteria in *all protected non-edible crops*
- Fungal spores in *all protected non-edible crops*
- Viroids in *all protected non-edible crops*
- Virus in *all protected non-edible crops*

Approval information

- Benzoic acid is included in Annex 1 under EC Regulation 1107/2009

Hazard classification and safety precautions

Hazard Corrosive
Transport code 3
Packaging group III
UN Number 1987
Risk phrases H318, H336
Operator protection A, C, H; U05a, U11, U13, U14, U15
Environmental protection E15a
Storage and disposal D01, D02, D09a, D10a
Medical advice M05a

36   benzovindiflupyr

An SDHI fungicide for disease control in cereals
FRAC mode of action code: 7

See also benzovindiflupyr + prothioconazole

Products

1	Ceravato Plus	Syngenta	100 g/l	EC	17865
2	Elatus Plus	Syngenta	100 g/l	EC	17841
3	Velogy Plus	Syngenta	100 g/l	EC	17866

Uses

- Brown rust in *rye*, *spring barley*, *spring wheat*, *triticale*, *winter barley*, *winter wheat*
- Glume blotch in *spring wheat* (moderate control), *winter wheat* (moderate control)
- Mycosphaerella in *spring wheat*, *triticale*, *winter wheat*
- Net blotch in *spring barley*, *winter barley*
- Ramularia leaf spots in *spring barley*, *winter barley*
- Rhynchosporium in *rye* (moderate control), *spring barley* (moderate control), *winter barley* (moderate control)

SEE SECTION 3 FOR PRODUCTS ALSO REGISTERED

- Septoria leaf blotch in *spring wheat, triticale, winter wheat*
- Yellow rust in *spring wheat, triticale, winter wheat*

Approval information
- Benzovindiflupyr is included in Annex 1 under EC Regulation 1107/2009

Efficacy guidance
- Benzovindiflupyr works predominately via protective action

Restrictions
- Do not apply via hand-held equipment
- No more than 2 applications of SDHI fungicides should be applied to any cereal crop

Following crops guidance
- There are no restrictions on succeeding crops in a normal rotation

Environmental safety
- Buffer zone requirement 6 m [1-3]
- LERAP Category B

Hazard classification and safety precautions
    Hazard Harmful if swallowed, Harmful if inhaled, Very toxic to aquatic organisms
    Transport code 9
    Packaging group III
    UN Number 3082
    Risk phrases H317, H318
    Operator protection A, C, H; U09a, U10, U11, U19a, U20a
    Environmental protection E15b, E16a, H410
    Storage and disposal D01, D09a, D10c, D12a, D12b, D22

37   benzovindiflupyr + prothioconazole

An SDHI and triazole fungicide mixture for disease control in cereals
FRAC mode of action code: 3 + 7

See also benzovindiflupyr

Products

Elatus Era	Syngenta	75:150 g/l	EC	17889

Uses
- Brown rust in *rye, spring barley, spring wheat, triticale, winter barley, winter wheat*
- Disease control in *rye*
- Fusarium in *spring wheat* *(moderate control)*, *winter wheat* *(moderate control)*
- Glume blotch in *spring wheat* *(moderate control)*, *winter wheat* *(moderate control)*
- Mycosphaerella in *spring wheat, winter wheat*
- Net blotch in *spring barley, winter barley*
- Ramularia leaf spots in *spring barley, winter barley*
- Rhynchosporium in *rye* *(moderate control)*, *spring barley* *(moderate control)*, *triticale* *(moderate control)*, *winter barley* *(moderate control)*
- Septoria leaf blotch in *spring wheat, triticale, winter wheat*
- Yellow rust in *spring wheat, winter wheat*

Approval information
- Benzovindiflupyr and prothioconazole are included in Annex 1 under EC Regulation 1107/2009

Efficacy guidance
- Benzovindiflupyr is predominantly protectant while prothioconazole is a triazole with systemic and protectant activity. Use as a protectant treatment or in the earliest stages of disease development.

Restrictions
- The earliest time of application is GS31

- No more than 2 applications of SDHI fungicides should be applied to any cereal crop
- Do not apply via hand-held equipment

Following crops guidance
- There are no restrictions on succeeding crops in a normal rotation

Environmental safety
- Buffer zone requirement 6m [1]
- LERAP Category B

Hazard classification and safety precautions
Hazard Harmful if swallowed, Harmful if inhaled
Transport code 9
Packaging group III
UN Number 3082
Risk phrases H315, H317, H318
Operator protection A, C, H; U09a, U10, U11, U19a, U20a
Environmental protection E15b, E16a, H410
Storage and disposal D01, D09a, D10c, D12a, D12b
Medical advice M04a

38 benzyladenine

A cytokinin plant growth regulator for use in apples and pears

Products

1 Configure	Fargro	20 g/l	SC	17523
2 Exilis	Fine	20 g/l	SL	15706
3 Globaryll 100	Globachem	100 g/l	SL	16097
4 MaxCel	Interfarm	20 g/l	SL	15708

Uses
- Fruit thinning in **apples** [2, 4]; **pears** [2]
- Growth regulation in **apples**, **pears** [3]; **kalanchoes**, **ornamental plant production** [1]
- Increasing flowering in **phalaenopsis spp.**, **schlumbergera spp.**, **sempervivium spp.** [1]

Approval information
- Benzyladenine is included in Annex 1 under EC Directive 1107/2009

Efficacy guidance
- A post-bloom thinner for apples and pears with excessive blossom [2, 3]
- Apply when the temperature will exceed 15°C on day of application but note that temperatures above 28°C may result in excessive thinning [2-4]

Restrictions
- Use in pears is a Qualified Minor Use recommendation. Evidence of efficacy and crop safety in pears is limited to the pear variety 'Conference' [2]

Environmental safety
- Broadcast air-assisted LERAP [4] (5 m)

Hazard classification and safety precautions
Hazard Harmful, Dangerous for the environment [3]
UN Number N/C
Risk phrases H317 [1]; H318, H361 [3]
Operator protection A [1, 2, 4]; C [3]; H [1]; U02a, U09a, U19a, U20b [2, 4]; U05a [2-4]; U11, U14, U15, U20c [3]
Environmental protection E15a [3]; E15b [1, 2, 4]; E17b [4] (5 m); E38, H412 [2-4]
Storage and disposal D01 [1, 3]; D02, D09a [1-4]; D03, D05, D12a [2, 4]; D10a [1]; D10b [2-4]
Medical advice M04a [3]; M05a [2, 4]

SEE SECTION 3 FOR PRODUCTS ALSO REGISTERED

39   benzyladenine + gibberellin

A plant growth regulator mixture for use on ornamental plants

See also benzyladenine
*   gibberellins*

Products

Chrysal BVB	Chrysal	19:19 g/l	SL	17780

Uses
- Growth regulation in **ornamental plant production**

Approval information
- Benzyladenine and gibberellins are included in Annex 1 under EC Directive 1107/2009

Hazard classification and safety precautions
  UN Number N/C
  Operator protection A
  Environmental protection H412

40   beta-cyfluthrin

A non-systemic pyrethroid insecticide for insect control in cereals, oilseed rape, cabbages, cauliflowers and sugar beet
IRAC mode of action code: 3

Products

Gandalf	Adama	25 g/l	EC	12865

Uses
- Aphids in **oats**, **rye**, **spring barley**, **spring wheat**, **sugar beet**, **triticale**, **winter barley**, **winter wheat**
- Brassica pod midge in **oilseed rape**
- Cabbage moth in **cabbages**, **cauliflowers**
- Cabbage seed weevil in **oilseed rape**
- Cabbage stem flea beetle in **oilseed rape**
- Cabbage stem weevil in **oilseed rape**
- Peach-potato aphid in **cabbages**, **cauliflowers**
- Wheat-blossom midge in **rye**, **spring wheat**, **triticale**, **winter wheat**

Approval information
- Beta-cyfluthrin included in Annex I under EC Regulation 1107/2009

Environmental safety
- Dangerous to bees
- To protect non target insects/arthropods respect an unsprayed buffer zone of 5 m to non-crop land

Hazard classification and safety precautions
  Hazard Harmful, Dangerous for the environment, Flammable liquid and vapour, Harmful if swallowed, Harmful if inhaled, Very toxic to aquatic organisms
  Transport code 9
  Packaging group III
  UN Number 3082
  Risk phrases H304, H317, H319, H336
  Operator protection A, C, H; U04a, U05a, U10, U12, U14, U19a, U20a
  Environmental protection E12c, E12e, E16e, E34, E38, H410
  Storage and disposal D01, D02, D09a, D10a, D12a
  Medical advice M03, M05b

FOR FULL CONDITIONS OF USE ALWAYS READ THE PRODUCT LABEL

41   beta-cyfluthrin + clothianidin

An insecticidal seed treatment mixture for beet
IRAC mode of action code: 3 + 4A

See also clothianidin

Products

Poncho Beta	Bayer CropScience	53:400 g/l	FS	12076

Uses
- Beet virus yellows vectors in **fodder beet** *(seed treatment)*, **sugar beet** *(seed treatment)*

Approval information
- Beta-cyfluthrin and clothianidin included in Annex I under EC Regulation 1107/2009

Efficacy guidance
- In addition to control of aphid virus vectors, product improves crop establishment by reducing damage caused by symphylids, springtails, millipedes, wireworms and leatherjackets
- Additional control measures should be taken where very high populations of soil pests are present
- Product reduces direct feeding damage by foliar pests such as pygmy mangold beetle, beet flea beetle, mangold fly
- Product is not active against nematodes
- Treatment does not alter physical characteristics of pelleted seed and no change to standard drill settings should be necessary

Restrictions
- Maximum number of treatments: 1 per seed batch
- Product must be co-applied with a colouring dye
- Product must only be applied as a coating to pelleted seed using special treatment machinery

Crop-specific information
- Latest use: pre-drilling

Environmental safety
- Extremely dangerous to fish or other aquatic life. Do not contaminate surface waters or ditches with chemical or used container

Hazard classification and safety precautions
Hazard Harmful, Harmful if swallowed, Very toxic to aquatic organisms
Transport code 9
Packaging group III
UN Number 3082
Operator protection A, H; U04a, U05a, U07, U13, U14, U20b, U24
Environmental protection E13a, E34, E36a, H410
Storage and disposal D01, D02, D05, D09a, D14
Treated seed S02, S03, S04a, S04b, S05, S06a, S06b, S07
Medical advice M03

42   bifenazate

A bifenazate acaricide for use on protected strawberries
IRAC mode of action code: 25

Products

Floramite 240 SC	Certis	240 g/l	SC	13686

Uses
- Mites in **protected forest nurseries** *(off-label)*, **protected hops** *(off-label)*
- Two-spotted spider mite in **protected strawberries**

SEE SECTION 3 FOR PRODUCTS ALSO REGISTERED

Extension of Authorisation for Minor Use (EAMUs)
- *protected forest nurseries* 20142220
- *protected hops* 20142220

Approval information
- Bifenazate included in Annex I under EC Regulation 1107/2009

Efficacy guidance
- Bifenazate acts by contact knockdown after approximately 4 d and a subsequent period of residual control
- All mobile stages of mites are controlled with occasionally some ovicidal activity
- Best results obtained from a programme of two treatments started as soon as first spider mites are seen
- To minimise possible development of resistance use in a planned Resistance Management strategy. Alternate at least two products with different modes of action between treatment programmes of bifenazate. See Section 5 for more information

Restrictions
- Maximum number of treatments 2 per yr for protected strawberries
- After use in protected environments avoid re-entry for at least 2 h while full ventilation is carried out allowing spray to dry on leaves
- Personnel working among treated crops should wear suitable long-sleeved garments and gloves for 14 d after treatment
- Do not apply as a low volume spray

Crop-specific information
- HI: 7 d for protected strawberries
- Carry out small scale tolerance test on the strawberry cultivar before large scale use

Environmental safety
- Dangerous for the environment
- Toxic to aquatic organisms
- Harmful to non-target predatory mites. Avoid spray drift onto field margins, hedges, ditches, surface water and neighbouring crops

Hazard classification and safety precautions
Hazard Irritant, Dangerous for the environment
Transport code 9
Packaging group III
UN Number 3082
Risk phrases H317
Operator protection A, H; U02a, U04a, U05a, U10, U14, U15, U19a, U20b
Environmental protection E15b, E38, H411
Storage and disposal D01, D02, D05, D09a, D10b, D12b
Medical advice M03

43   bifenox

A diphenyl ether herbicide for use in cereals and (off-label) in oilseed rape.
HRAC mode of action code: E

Products

1 Clayton Belstone	Clayton	480 g/l	SC	14033
2 Fox	Adama	480 g/l	SC	11981

Uses
- Annual dicotyledons in **canary flower (echium spp.)** (off-label), **durum wheat** (off-label), **grass seed crops** (off-label) [1, 2]; **evening primrose** (off-label), **honesty** (off-label), **linseed** (off-label), **mustard** (off-label), **oilseed rape** (off-label) [2]; **triticale, winter barley, winter rye, winter wheat** [1]
- Cleavers in **triticale, winter barley, winter rye, winter wheat** [2]
- Field pansy in **triticale, winter barley, winter rye, winter wheat** [2]

- Field speedwell in *triticale, winter barley, winter rye, winter wheat* [2]
- Forget-me-not in *triticale, winter barley, winter rye, winter wheat* [2]
- Ivy-leaved speedwell in *triticale, winter barley, winter rye, winter wheat* [2]
- Poppies in *triticale, winter barley, winter rye, winter wheat* [2]
- Red dead-nettle in *triticale, winter barley, winter rye, winter wheat* [2]

Extension of Authorisation for Minor Use (EAMUs)
- *canary flower (echium spp.)* 20121422 [1], 20142222 expires 31 Aug 2018 [2], 20142318 [2]
- *durum wheat* 20121421 [1], 20142221 expires 31 Aug 2018 [2], 20142319 [2]
- *evening primrose* 20142222 expires 31 Aug 2018 [2], 20142318 [2]
- *grass seed crops* 20121421 [1], 20142221 expires 31 Aug 2018 [2], 20142319 [2]
- *honesty* 20142222 expires 31 Aug 2018 [2], 20142318 [2]
- *linseed* 20142222 expires 31 Aug 2018 [2], 20142318 [2]
- *mustard* 20142222 expires 31 Aug 2018 [2], 20142318 [2]
- *oilseed rape* 20142222 expires 31 Aug 2018 [2], 20142318 [2]

Approval information
- Accepted by BBPA for use on malting barley
- Bifenox included in Annex I under EC Regulation 1107/2009

Efficacy guidance
- Bifenox is absorbed by foliage and emerging roots of susceptible species
- Best results obtained when weeds are growing actively with adequate soil moisture

Restrictions
- Maximum number of treatments: one per crop
- Do not apply to crops suffering from stress from whatever cause
- Do not apply if the crop is wet or if rain or frost is expected
- Avoid drift onto broad-leaved plants outside the target area

Crop-specific information
- Latest use: before 2nd node detectable (GS 32) for all crops

Environmental safety
- Dangerous for the environment
- Very toxic to aquatic organisms
- Do not empty into drains

Hazard classification and safety precautions
Hazard Dangerous for the environment
Transport code 9
Packaging group III
UN Number 3082
Operator protection A; U05a, U08, U20b
Environmental protection E15a, E19b, E34, E38, H410
Storage and disposal D01, D02, D09a, D12a
Medical advice M03

44   bixafen

A succinate dehydrogenase inhibitor (SDHI) fungicide available only in mixtures
FRAC mode of action code: 7

45   bixafen + fluopyram + prothioconazole

A fungicide mixture for disease control in wheat
FRAC mode of action code: 7 + 3

See also bixafen + prothioconazole

SEE SECTION 3 FOR PRODUCTS ALSO REGISTERED

Products

Ascra Xpro	Bayer CropScience	65:65:130 g/l	EC	17623

Uses

- Brown rust in **spring wheat**, **winter wheat**
- Ear diseases in **spring wheat** *(reduction of incidence and severity)*, **winter wheat** *(reduction of incidence and severity)*
- Eyespot in **spring wheat** *(reduction of incidence and severity)*, **winter wheat** *(reduction of incidence and severity)*
- Glume blotch in **spring wheat**, **winter wheat**
- Powdery mildew in **spring wheat**, **winter wheat**
- Septoria leaf blotch in **spring wheat**, **winter wheat**
- Tan spot in **spring wheat**, **winter wheat**
- Yellow rust in **spring wheat**, **winter wheat**

Approval information

- Bixafen, fluopyram and prothioconazole are included in Annex 1 under EC Regulation 1107/2009.

Efficacy guidance

- Applications to upper leaves where *S. tritici* symptoms are present are likely to be less effective

Restrictions

- Bixafen is an SDH respiration inhibitor; Do not apply more than two foliar applications of products containing an SDH inhibitor to any cereal crop
- Resistance to some DMI fungicides has been identified in Septoria leaf blotch (*Mycosphaerella graminicola*) which may seriously affect the performance of some products. For further advice on resistance management in DMIs contact your agronomist or specialist advisor, and visit the FRAG-UK website.

Environmental safety

- LERAP Category B

Hazard classification and safety precautions

Hazard Harmful if swallowed
Transport code 9
Packaging group III
UN Number 3082
Risk phrases H317, H318
Operator protection A, C, H; U05a, U09b, U11, U20a
Environmental protection E15b, E16a, E34, H410
Storage and disposal D01, D02, D05, D09a, D10b, D12b

46   bixafen + fluoxastrobin + prothioconazole

An SDHI + strobilurin + triazole mixture for disease control in cereals
FRAC mode of action code: 7 + 11 + 3

Products

Variano Xpro	Bayer CropScience	40:50:100 g/l	EC	15218

Uses

- Brown rust in **spring wheat**, **triticale**, **winter rye**, **winter wheat**
- Ear diseases in **spring wheat**, **triticale**, **winter rye**, **winter wheat**
- Glume blotch in **spring wheat**, **triticale**, **winter rye**, **winter wheat**
- Leaf blotch in **triticale**, **winter rye**
- Powdery mildew in **spring wheat**, **triticale**, **winter rye**, **winter wheat**
- Rhynchosporium in **winter rye**
- Tan spot in **spring wheat**, **triticale**, **winter rye**, **winter wheat**
- Yellow rust in **spring wheat**, **triticale**, **winter rye**, **winter wheat**

Approval information
- Bixafen, fluroxystrobin and prothioconazole are included in Annex 1 under EC Regulation 1107/2009.

Restrictions
- Bixafen is an SDH respiration inhibitor; Do not apply more than two foliar applications of products containing an SDH inhibitor to any cereal crop

Environmental safety
- LERAP Category B

Hazard classification and safety precautions
Hazard Irritant, Dangerous for the environment
Transport code 9
Packaging group III
UN Number 3082
Risk phrases H317, H319
Operator protection A, H; U05b, U08, U14, U20b
Environmental protection E15a, E16a, E16b, E34, E38, H410
Storage and disposal D01, D02, D05, D09a, D10b, D12a
Medical advice M03

47   bixafen + prothioconazole

Succinate dehydrogenase inhibitor (SDHI) + triazole fungicide mixture for disease control in cereals
FRAC mode of action code: 3 + 7

Products
1	Aviator 235 Xpro	Bayer CropScience	75:160 g/l	EC	15026
2	Siltra Xpro	Bayer CropScience	60:200 g/l	EC	15082

Uses
- Brown rust in **grass seed crops** *(off-label)*, **spring barley**, **winter barley** [2]; **spring wheat**, **triticale**, **winter rye**, **winter wheat** [1]
- Crown rust in **grass seed crops** *(off-label)*, **spring oats**, **winter oats** [2]
- Drechslera leaf spot in **grass seed crops** *(off-label)* [2]
- Ear diseases in **spring barley**, **winter barley** [2]; **spring wheat**, **triticale**, **winter wheat** [1]
- Eyespot in **spring barley** *(reduction of incidence and severity)*, **spring oats** *(reduction of incidence and severity)*, **winter barley** *(reduction of incidence and severity)*, **winter oats** *(reduction of incidence and severity)* [2]; **spring wheat** *(reduction of incidence and severity)*, **triticale** *(reduction of incidence and severity)*, **winter wheat** *(reduction of incidence and severity)* [1]
- Glume blotch in **spring wheat**, **triticale**, **winter wheat** [1]
- Mildew in **grass seed crops** *(off-label)* [2]
- Net blotch in **spring barley**, **winter barley** [2]
- Powdery mildew in **spring barley**, **spring oats**, **winter barley**, **winter oats** [2]; **spring wheat**, **triticale**, **winter rye**, **winter wheat** [1]
- Ramularia leaf spots in **spring barley**, **winter barley** [2]
- Rhynchosporium in **grass seed crops** *(off-label)*, **spring barley**, **winter barley** [2]; **winter rye** [1]
- Septoria leaf blotch in **spring wheat**, **triticale**, **winter wheat** [1]
- Tan spot in **spring wheat**, **triticale**, **winter wheat** [1]
- Yellow rust in **spring barley**, **winter barley** [2]; **spring wheat**, **triticale**, **winter wheat** [1]

Extension of Authorisation for Minor Use (EAMUs)
- **grass seed crops** *20171034* [2]

Approval information
- Bixafen and prothioconazole are included in Annex I under EC Regulation 1107/2009.
- Accepted by BBPA for use on malting barley

SEE SECTION 3 FOR PRODUCTS ALSO REGISTERED

Efficacy guidance
- Resistance to some DMI fungicides has been identified in Septoria leaf blotch (*Mycosphaerella graminicola*) which may seriously affect the performance of some products. For further advice on resistance management in DMIs contact your agronomist or specialist advisor, and visit the FRAG-UK website.

Restrictions
- Bixafen is an SDH respiration inhibitor; Do not apply more than two foliar applications of products containing an SDH inhibitor to any cereal crop
- Resistance to some DMI fungicides has been identified in Septoria leaf blotch (*Mycosphaerella graminicola*) which may seriously affect the performance of some products. For further advice on resistance management in DMIs contact your agronomist or specialist advisor, and visit the FRAG-UK website.

Environmental safety
- LERAP Category B

Hazard classification and safety precautions
Hazard Harmful, Dangerous for the environment [1, 2]; Harmful if swallowed [2]
Transport code 9
Packaging group III
UN Number 3082
Risk phrases H317 [2]; H319 [1, 2]
Operator protection U05a, U09a, U20b
Environmental protection E15b, E16a, E16b, E34, E38, H410
Storage and disposal D01, D02, D09a, D10b, D12a

48   bixafen + prothioconazole + spiroxamine

A broad-spectrum fungicide mixture for disease control in cereals
FRAC mode of action code: 3 + 5 + 7

Products

Boogie Xpro	Bayer CropScience	50:100:250 g/l	EC	15061

Uses
- Brown rust in *spring wheat, triticale, winter rye, winter wheat*
- Ear diseases in *spring wheat, triticale, winter rye, winter wheat*
- Eyespot in *spring wheat, triticale, winter rye, winter wheat*
- Glume blotch in *spring wheat, triticale, winter rye, winter wheat*
- Powdery mildew in *spring wheat, triticale, winter rye, winter wheat*
- Rhynchosporium in *triticale, winter rye*
- Septoria leaf blotch in *spring wheat, triticale, winter rye, winter wheat*
- Tan spot in *spring wheat, triticale, winter rye, winter wheat*
- Yellow rust in *spring wheat, triticale, winter rye, winter wheat*

Approval information
- Bixafen, prothioconazole and spiroxamine included in Annex I under EC Regulation 1107/2009.

Efficacy guidance
- Resistance to some DMI fungicides has been identified in Septoria leaf blotch (*Mycosphaerella graminicola*) which may seriously affect the performance of some products. For further advice on resistance management in DMIs contact your agronomist or specialist advisor, and visit the FRAG-UK website.

Restrictions
- No more than two applications of SDH inhibitors must be applied to the same cereal crop.

Environmental safety
- LERAP Category B

FOR FULL CONDITIONS OF USE ALWAYS READ THE PRODUCT LABEL

Hazard classification and safety precautions

Hazard Harmful, Dangerous for the environment, Harmful if swallowed, Harmful if inhaled
Transport code 9
Packaging group III
UN Number 3082
Risk phrases H318
Operator protection A, C, H; U05a, U09b, U11, U20a, U26
Environmental protection E15b, E16a, E16b, E34, E38, H410
Storage and disposal D01, D02, D05, D09a, D10b, D12a
Medical advice M03

49   bixafen + prothioconazole + tebuconazole

A succinate dehydrogenase inhibitor (SDHI) + triazole fungicide mixture for disease control in cereals
FRAC mode of action code: 3 + 7

Products

1	Skyway 285 Xpro	Bayer CropScience	75:110:100 g/l	EC	15028
2	Sparticus Xpro	Bayer CropScience	75:110:90 g/l	EC	15162

Uses

- Brown rust in *spring wheat, triticale, winter rye, winter wheat*
- Ear diseases in *spring wheat, triticale, winter wheat*
- Eyespot in *spring wheat* (reduction of incidence and severity), *triticale* (reduction of incidence and severity), *winter wheat* (reduction of incidence and severity)
- Glume blotch in *spring wheat, triticale, winter wheat*
- Light leaf spot in *winter oilseed rape*
- Phoma in *winter oilseed rape*
- Powdery mildew in *spring wheat, triticale, winter rye, winter wheat*
- Rhynchosporium in *winter rye*
- Sclerotinia in *winter oilseed rape*
- Septoria leaf blotch in *spring wheat, triticale, winter wheat*
- Stem canker in *winter oilseed rape*
- Tan spot in *spring wheat, triticale, winter wheat*
- Yellow rust in *spring wheat, triticale, winter wheat*

Approval information

- Bixafen, prothioconazole and tebuconazole included in Annex I under EC Regulation 1107/2009.

Efficacy guidance

- Resistance to some DMI fungicides has been identified in Septoria leaf blotch (*Mycosphaerella graminicola*) which may seriously affect the performance of some products. For further advice on resistance management in DMIs contact your agronomist or specialist advisor, and visit the FRAG-UK website.

Restrictions

- Bixafen is an SDH respiration inhibitor; Do not apply more than two foliar applications of products containing an SDH inhibitor to any cereal crop
- Newer authorisations for tebuconazole products require application to cereals only after GS 30 and applications to oilseed rape and linseed after GS20 - check label

Environmental safety

- LERAP Category B

Hazard classification and safety precautions

Hazard Harmful, Dangerous for the environment, Harmful if swallowed
Transport code 9
Packaging group III
UN Number 3082
Risk phrases H315, H317, H361

SEE SECTION 3 FOR PRODUCTS ALSO REGISTERED

Operator protection A, H; U05a, U09a, U20b
Environmental protection E15b, E16a, E16b, E34, E38, H410
Storage and disposal D01, D02, D09a, D10b, D12a

50 boscalid

A translocated and translaminar anilide fungicide
FRAC mode of action code: 7

Products

1	Filan	BASF	50% w/w	WG	11449
2	Fulmar	AgChem Access	50% w/w	WG	15767

Uses

- Alternaria in *spring oilseed rape*, *winter oilseed rape* [1, 2]
- Botrytis in *poppies for morphine production* (off-label) [1]
- Disease control in *all edible seed crops grown outdoors* (off-label), *all non-edible seed crops grown outdoors* (off-label), *borage for oilseed production* (off-label), *canary flower (echium spp.)* (off-label), *corn gromwell* (off-label), *evening primrose* (off-label), *honesty* (off-label), *linseed* (off-label), *mustard* (off-label) [1]
- Powdery mildew in *grapevines* (off-label) [1]
- Sclerotinia in *poppies for morphine production* (off-label) [1]
- Sclerotinia stem rot in *spring oilseed rape*, *winter oilseed rape* [1, 2]

Extension of Authorisation for Minor Use (EAMUs)

- *all edible seed crops grown outdoors* 20093231 expires 31 Jul 2018 [1]
- *all non-edible seed crops grown outdoors* 20093231 expires 31 Jul 2018 [1]
- *borage for oilseed production* 20093229 expires 31 Jul 2018 [1], 20131293 expires 31 Jul 2018 [1]
- *canary flower (echium spp.)* 20093229 expires 31 Jul 2018 [1], 20131293 expires 31 Jul 2018 [1]
- *corn gromwell* 20131293 expires 31 Jul 2018 [1]
- *evening primrose* 20093229 expires 31 Jul 2018 [1], 20131293 expires 31 Jul 2018 [1]
- *grapevines* 20131947 expires 31 Jul 2018 [1]
- *honesty* 20093229 expires 31 Jul 2018 [1], 20131293 expires 31 Jul 2018 [1]
- *linseed* 20093229 expires 31 Jul 2018 [1], 20131293 expires 31 Jul 2018 [1]
- *mustard* 20093229 expires 31 Jul 2018 [1], 20131293 expires 31 Jul 2018 [1]
- *poppies for morphine production* 20093222 expires 31 Jul 2018 [1]

Approval information

- Boscalid included in Annex I under EC Regulation 1107/2009
- Accepted by BBPA on malting barley before ear emergence only and for use on hops

Efficacy guidance

- For best results apply as a protectant spray before symptoms are visible
- Applications against *Sclerotinia* should be made in high disease risk situations at early to full flower
- *Sclerotinia* control may be reduced when high risk conditions occur after flowering, leading to secondary disease spread
- Applications against *Alternaria* should be made at full flowering. Later applications may result in reduced levels of control
- Ensure adequate spray penetration and good coverage

Restrictions

- Maximum total dose equivalent to two full dose treatments
- Do no treat crops to be used for seed production later than full flowering stage
- Avoid spray drift onto neighbouring crops

Crop-specific information

- Latest use: up to and including when 50% of pods have reached final size (GS 75) for oilseed rape

FOR FULL CONDITIONS OF USE ALWAYS READ THE PRODUCT LABEL

Environmental safety
- Dangerous for the environment
- Toxic to aquatic organisms
- Product represents minimal hazard to bees when used as directed. However local bee-keepers should be notified if crops are to be sprayed when in flower

Hazard classification and safety precautions
>**Hazard** Dangerous for the environment
>**Transport code** 9
>**Packaging group** III
>**UN Number** 3077
>**Operator protection** U05a, U20c
>**Environmental protection** E15a, E38, H411
>**Storage and disposal** D01, D02, D05, D09a, D10c, D12a

51   boscalid + dimoxystrobin

A fungicide mixture for disease control in oilseed rape
FRAC mode of action code: 7 + 11

See also boscalid

Products

Pictor	BASF	200:200 g/l	SC	16783

Uses
- Alternaria in **corn gromwell** *(off-label)*, **winter oilseed rape** *(moderate control)*
- Dark leaf and pod spot in **corn gromwell** *(off-label)*
- Dark leaf spot in **winter oilseed rape** *(moderate control)*
- Disease control in **winter oilseed rape**
- Light leaf spot in **winter oilseed rape** *(reduction)*
- Phoma in **winter oilseed rape** *(moderate control)*
- Sclerotinia in **corn gromwell** *(off-label)*, **winter oilseed rape**
- Stem canker in **winter oilseed rape** *(moderate control)*

Extension of Authorisation for Minor Use (EAMUs)
- *corn gromwell 20151106*

Approval information
- Boscalid and dimoxystrobin included in Annex I under EC Regulation 1107/2009

Efficacy guidance
- Treat light leaf spot and phoma/stem canker at first sign of infection in the spring. Treat sclerotinia at early - full flower in high disease risk situations. Treat altenaria from full flower to when 50% of pods have reached full size.

Environmental safety
- LERAP Category B

Hazard classification and safety precautions
>**Hazard** Harmful if swallowed, Harmful if inhaled, Very toxic to aquatic organisms
>**Transport code** 9
>**Packaging group** III
>**UN Number** 3082
>**Risk phrases** H317, H351, H361
>**Operator protection** A, H; U05a
>**Environmental protection** E15b, E16a, E34, H410
>**Storage and disposal** D01, D02, D09a, D10c, D12b
>**Medical advice** M05a

52   boscalid + epoxiconazole

A broad spectrum fungicide mixture for cereals
FRAC mode of action code: 7 + 3

See also epoxiconazole

Products

1	Chord	BASF	210:75 g/l	SC	16065
2	Enterprise	BASF	140:50 g/l	OD	15228
3	Kingdom	BASF	140:50 g/l	OD	15240
4	Tracker	BASF	233:67 g/l	SC	16048

Uses

- Botrytis in **narcissi** *(off-label)* [4]
- Brown rust in **durum wheat** [1, 4]; **grass seed crops** *(off-label)* [4]; **rye**, **spring barley**, **spring wheat**, **triticale**, **winter barley**, **winter wheat** [1-4]
- Crown rust in **grass seed crops** *(off-label)* [4]; **oats** [2, 3]; **spring oats**, **winter oats** [1, 4]
- Eyespot in **durum wheat** *(moderate control)* [1, 4]; **rye** *(moderate control)*, **spring barley** *(moderate control)*, **spring wheat** *(moderate control)*, **winter barley** *(moderate control)*, **winter wheat** *(moderate control)* [1-4]; **triticale** *(moderate control)* [1-3]
- Fusarium ear blight in **spring wheat** *(good reduction)*, **winter wheat** *(good reduction)* [2, 3]
- Glume blotch in **spring wheat**, **triticale**, **winter wheat** [2, 3]
- Net blotch in **grass seed crops** *(off-label)* [4]; **spring barley**, **winter barley** [1-4]
- Powdery mildew in **oats** *(moderate control)*, **rye** *(moderate control)*, **spring barley** *(moderate control)*, **winter barley** *(moderate control)* [2, 3]
- Ramularia leaf spots in **spring barley** *(moderate control)*, **winter barley** *(moderate control)* [1-4]
- Rhynchosporium in **grass seed crops** *(off-label)* [4]; **rye** *(moderate control)*, **spring barley** *(moderate control)*, **winter barley** *(moderate control)* [1-4]
- Septoria leaf blotch in **durum wheat** [1, 4]; **spring wheat**, **triticale**, **winter wheat** [1-4]
- Smoulder in **narcissi** *(off-label)* [4]
- Sooty moulds in **spring wheat**, **winter wheat** [2, 3]
- Tan spot in **durum wheat** *(reduction)* [1, 4]; **spring wheat** *(reduction)*, **winter wheat** *(reduction)* [1-4]
- White mould in **narcissi** *(off-label)* [4]
- Yellow rust in **rye**, **spring barley**, **spring wheat**, **triticale**, **winter barley**, **winter wheat** [2, 3]

Extension of Authorisation for Minor Use (EAMUs)

- **grass seed crops** *20150655* [4]
- **narcissi** *20132458* [4]

Approval information

- Boscalid and epoxiconazole included in Annex I under EC Regulation 1107/2009
- Accepted by BBPA for use on malting barley (before ear emergence only)

Efficacy guidance

- Apply at the start of foliar or stem based disease attack
- Optimum effect against eyespot achieved by spraying between leaf-sheath erect and second node detectable stages (GS 30-32)
- Epoxiconazole is a DMI fungicide. Resistance to some DMI fungicides has been identified in Septoria leaf blotch which may seriously affect performance of some products. For further advice contact a specialist advisor and visit the Fungicide Resistance Action Group (FRAG)-UK website

Crop-specific information

- Latest use: before cereal ear emergence

Environmental safety

- Dangerous for the environment
- Toxic to aquatic organisms
- Avoid drift onto neighbouring crops. May cause damage to broad-leaved plant species
- LERAP Category B

FOR FULL CONDITIONS OF USE ALWAYS READ THE PRODUCT LABEL

SECTION 2

Hazard classification and safety precautions
Hazard Harmful, Dangerous for the environment [1-4]; Harmful if inhaled [2-4]
Transport code 9
Packaging group III
UN Number 3082
Risk phrases H318 [2-4]; H319 [1]; H351, H360 [1-4]
Operator protection A, C [1-4]; H [1, 4]; U05a [1-4]; U14, U20a [2, 3]; U20b [1, 4]
Environmental protection E15a [1, 4]; E15b [2, 3]; E16a, E38, H410 [1-4]
Storage and disposal D01, D02, D10c, D12a [1-4]; D05, D09a [2, 3]; D08 [1, 4]
Medical advice M03 [1-4]; M05a [2, 3]

53 boscalid + epoxiconazole + pyraclostrobin

A systemic and protectant fungicide mixture for disease control in cereals
FRAC mode of action code: 3 + 7 + 11

Products

Nebula XL	BASF	140:50:60 g/l	OD	15555

Uses
- Brown rust in *rye, spring wheat, triticale, winter barley, winter wheat*
- Crown rust in *oats*
- Eyespot in *rye* (moderate control), *spring wheat* (moderate control), *triticale* (moderate control), *winter barley* (moderate control), *winter wheat* (moderate control)
- Fusarium ear blight in *spring wheat* (good reduction), *winter wheat* (good reduction)
- Glume blotch in *spring wheat, triticale, winter wheat*
- Net blotch in *winter barley*
- Powdery mildew in *oats* (moderate control), *rye* (moderate control), *triticale* (moderate control), *winter barley* (moderate control)
- Ramularia leaf spots in *winter barley* (moderate control)
- Rhynchosporium in *rye, winter barley*
- Septoria leaf blotch in *spring wheat, triticale, winter wheat*
- Sooty moulds in *spring wheat* (good reduction), *winter wheat* (good reduction)
- Tan spot in *spring wheat* (moderate control), *winter wheat* (moderate control)
- Yellow rust in *rye, spring wheat, triticale, winter barley, winter wheat*

Approval information
- Boscalid, epoxiconazole and pyraclostrobin included in Annex I under EC Regulation 1107/2009

Following crops guidance
- Onions, oats, sugar beet, oilseed rape, cabbages, carrots, sunflowers, barley, lettuce, ryegrass, dwarf french beans, peas, potatoes, clover, wheat, field beans and maize may be sown as the following crop.

Environmental safety
- LERAP Category B

Hazard classification and safety precautions
Hazard Harmful, Dangerous for the environment, Harmful if inhaled, Very toxic to aquatic organisms
Transport code 9
Packaging group III
UN Number 3082
Risk phrases H317, H319, H351, H360, R53a
Operator protection A, H; U05a, U14, U20a
Environmental protection E15b, E16a, E34, E38, H410
Storage and disposal D01, D02, D05, D09a, D10c, D12a
Medical advice M03, M05a

SEE SECTION 3 FOR PRODUCTS ALSO REGISTERED

54   boscalid + metconazole

An anilide and triazole fungicide mixture for disease control in oilseed rape
FRAC mode of action code: 3 + 7

Products

1	Highgate	BASF	133:60 g/l	SC	15251
2	Tectura	BASF	133:60 g/l	SC	15232

Uses
- Sclerotinia stem rot in *oilseed rape*

Approval information
- Boscalid and metconazole included in Annex I under EC Regulation 1107/2009

Following crops guidance
- Any crop can follow normally-harvested oilseed rape treated with [1] or [2]

Environmental safety
- LERAP Category B

Hazard classification and safety precautions
Hazard Harmful
Transport code 9
Packaging group III
UN Number 3082
Risk phrases H361
Operator protection A, H; U05a, U20a
Environmental protection E15b, E16a, E16b, E34, E38, H412
Storage and disposal D01, D02, D05, D09a, D10c, D12a
Medical advice M03, M05a

55   boscalid + pyraclostrobin

A protectant and systemic fungicide mixture
FRAC mode of action code: 7 + 11

See also pyraclostrobin

Products

1	Bellis	BASF	25.2:12.8% w/w	WG	12522
2	Signum	BASF	26.7:6.7% w/w	WG	11450

Uses
- Alternaria in *cabbages*, *calabrese*, *carrots* (moderate control), *cauliflowers*, *poppies for morphine production* (off-label) [2]
- American gooseberry mildew in *blackcurrants* (moderate control only) [2]
- Anthracnose in *blueberries* (off-label), *gooseberries* (off-label), *protected blueberry* (off-label), *protected gooseberries* (off-label) [2]
- Black spot in *protected strawberries*, *strawberries* [2]
- Blight in *potatoes* (off-label) [2]
- Blossom wilt in *cherries* (off-label), *mirabelles* (off-label), *plums* (off-label), *protected cherries* (off-label), *protected mirabelles* (off-label), *protected plums* (off-label) [2]
- Botrytis in *amenity vegetation* (off-label), *asparagus* (off-label), *beans without pods (fresh)* (off-label), *blueberries* (off-label), *broad beans* (off-label), *forest nurseries* (off-label), *gooseberries* (off-label), *interior landscapes* (off-label), *lamb's lettuce* (off-label), *ornamental plant production* (off-label), *poppies for morphine production* (off-label), *protected aubergines* (off-label), *protected blueberry* (off-label), *protected chard* (off-label), *protected forest nurseries* (off-label), *protected gooseberries* (off-label), *protected lettuce* (off-label), *protected ornamentals* (off-label), *protected peppers* (off-label), *protected spinach beet* (off-

label), **protected tomatoes** (off-label), **redcurrants** (off-label), **swedes** (off-label), **turnips** (off-label), **whitecurrants** (off-label) [2]
- Bottom rot in **lettuce**, **protected lettuce** [2]
- Brown rot in **apricots** (off-label), **nectarines** (off-label), **peaches** (off-label) [2]
- Cane blight in **blackberries** (off-label), **protected blackberries** (off-label), **raspberries** (off-label) [2]
- Chocolate spot in **spring field beans** (moderate control), **winter field beans** (moderate control) [2]
- Cladosporium in **protected cucumbers** (off-label) [2]
- Cladosporium leaf blotch in **courgettes** (off-label), **summer squash** (off-label) [2]
- Colletotrichum orbiculare in **courgettes** (off-label), **summer squash** (off-label) [2]
- Disease control in **all edible crops (outdoor and protected)** (off-label), **all non-edible crops (outdoor)** (off-label), **all protected non-edible crops** (off-label), **beetroot** (off-label), **kohlrabi** (off-label), **lamb's lettuce** (off-label), **protected hops** (off-label), **protected radishes** (off-label), **protected soft fruit** (off-label), **protected top fruit** (off-label), **radishes** (off-label), **red beet** (off-label), **soft fruit** (off-label), **top fruit** (off-label) [2]; **hops** (off-label) [1, 2]
- Downy mildew in **bulls blood** (off-label), **chard** (off-label), **endives** (off-label), **protected endives** (off-label), **spinach** (off-label), **spinach beet** (off-label) [2]
- Fusarium in **bulb onion sets** (off-label) [2]
- Grey mould in **blackcurrants**, **lettuce**, **protected lettuce**, **protected strawberries**, **strawberries** [2]
- Gummy stem blight in **protected cucumbers** (off-label) [2]
- Late blight in **potatoes** (off-label) [2]
- Leaf and pod spot in **combining peas** (moderate control only), **vining peas** (moderate control only) [2]
- Leaf spot in **amenity vegetation** (off-label), **black salsify** (off-label), **blackcurrants** (moderate control only), **blueberries** (off-label), **gooseberries** (off-label), **interior landscapes** (off-label), **parsley root** (off-label), **protected blueberry** (off-label), **protected gooseberries** (off-label), **redcurrants** (off-label), **salsify** (off-label), **whitecurrants** (off-label) [2]
- Poppy fire in **poppies for morphine production** (off-label) [2]
- Powdery mildew in **amenity vegetation** (off-label), **black salsify** (off-label), **carrots**, **interior landscapes** (off-label), **parsley root** (off-label), **protected cucumbers** (off-label), **protected strawberries**, **pumpkins** (off-label), **redcurrants** (off-label), **salsify** (off-label), **strawberries**, **whitecurrants** (off-label) [2]; **apples** [1]
- Purple blotch in **blackberries** (off-label), **protected blackberries** (off-label), **raspberries** (off-label) [2]
- Rhizoctonia in **protected chard** (off-label), **protected lettuce** (off-label), **protected spinach beet** (off-label) [2]
- Ring spot in **broccoli**, **brussels sprouts**, **cabbages**, **calabrese**, **cauliflowers** [2]
- Rust in **amenity vegetation** (off-label), **asparagus** (off-label), **black salsify** (off-label), **interior landscapes** (off-label), **kohlrabi** (off-label), **parsley root** (off-label), **protected blackberries** (off-label), **salsify** (off-label), **spring field beans**, **winter field beans** [2]
- Scab in **apples**, **pears** [1]
- Sclerotinia in **black salsify** (off-label), **carrots** (moderate control), **celeriac** (off-label), **horseradish** (off-label), **parsley root** (off-label), **protected chard** (off-label), **protected lettuce** (off-label), **protected spinach beet** (off-label), **salsify** (off-label) [2]
- Soft rot in **lettuce**, **protected lettuce** [2]
- White blister in **broccoli**, **brussels sprouts**, **cabbages** (qualified minor use), **calabrese** (qualified minor use), **chinese cabbage** (off-label), **choi sum** (off-label), **collards** (off-label), **herbs (see appendix 6)** (off-label), **kale** (off-label), **leaf brassicas** (off-label), **pak choi** (off-label), **protected herbs (see appendix 6)** (off-label), **protected leaf brassicas** (off-label), **tatsoi** (off-label) [2]
- White rot in **bulb onions** (off-label), **garlic** (off-label), **salad onions** (off-label), **shallots** (off-label) [2]
- White rust in **kohlrabi** (off-label) [2]
- White tip in **leeks** (off-label) [2]

Extension of Authorisation for Minor Use (EAMUs)
- **all edible crops (outdoor and protected)** 20102111 [2]
- **all non-edible crops (outdoor)** 20102111 [2]

SEE SECTION 3 FOR PRODUCTS ALSO REGISTERED

SECTION 2

- **all protected non-edible crops** *20102111* [2]
- **amenity vegetation** *20122317* [2]
- **apricots** *20121721* [2]
- **asparagus** *20102105* [2]
- **beans without pods (fresh)** *20121009* [2]
- **beetroot** *20121717* [2]
- **black salsify** *20121720* [2]
- **blackberries** *20102110* [2]
- **blueberries** *20121722* [2]
- **broad beans** *20121009* [2]
- **bulb onion sets** *20103122* [2]
- **bulb onions** *20102108* [2]
- **bulls blood** *20102136* [2]
- **celeriac** *20141059* [2]
- **chard** *20102136* [2]
- **cherries** *20102109* [2]
- **chinese cabbage** *20130285* [2]
- **choi sum** *20130285* [2]
- **collards** *20130285* [2]
- **courgettes** *20142855* [2]
- **endives** *20102136* [2]
- **forest nurseries** *20102119* [2]
- **garlic** *20102108* [2]
- **gooseberries** *20121722* [2]
- **herbs (see appendix 6)** *20102115* [2]
- **hops** *20112732* [1], *20102111* [2]
- **horseradish** *20093375* [2]
- **interior landscapes** *20122317* [2]
- **kale** *20130285* [2]
- **kohlrabi** *20121719* [2]
- **lamb's lettuce** *20121718* [2]
- **leaf brassicas** *20102115* [2]
- **leeks** *20102134* [2]
- **mirabelles** *20102109* [2]
- **nectarines** *20121721* [2]
- **ornamental plant production** *20122141* [2]
- **pak choi** *20130285* [2]
- **parsley root** *20121720* [2]
- **peaches** *20121721* [2]
- **plums** *20102109* [2]
- **poppies for morphine production** *20101233* [2]
- **potatoes** *20110394* [2]
- **protected aubergines** *20120427* [2]
- **protected blackberries** *20102102* [2]
- **protected blueberry** *20121722* [2]
- **protected chard** *20131807* [2]
- **protected cherries** *20102109* [2]
- **protected cucumbers** *20151178* [2]
- **protected endives** *20102136* [2]
- **protected forest nurseries** *20102119* [2]
- **protected gooseberries** *20121722* [2]
- **protected herbs (see appendix 6)** *20102115* [2]
- **protected hops** *20102111* [2]
- **protected leaf brassicas** *20102115* [2]
- **protected lettuce** *20131807* [2]
- **protected mirabelles** *20102109* [2]
- **protected ornamentals** *20122141* [2]
- **protected peppers** *20120427* [2]
- **protected plums** *20102109* [2]

FOR FULL CONDITIONS OF USE ALWAYS READ THE PRODUCT LABEL

- **protected radishes** *20121717* [2]
- **protected soft fruit** *20102111* [2]
- **protected spinach beet** *20131807* [2]
- **protected tomatoes** *20120427* [2]
- **protected top fruit** *20102111* [2]
- **pumpkins** *20152651* [2]
- **radishes** *20121717* [2]
- **raspberries** *20102110* [2]
- **red beet** *20121717* [2]
- **redcurrants** *20102114* [2]
- **salad onions** *20102107* [2]
- **salsify** *20121720* [2]
- **shallots** *20102108* [2]
- **soft fruit** *20102111* [2]
- **spinach** *20102136* [2]
- **spinach beet** *20102136* [2]
- **summer squash** *20142855* [2]
- **swedes** *20151961* [2]
- **tatsoi** *20130285* [2]
- **top fruit** *20102111* [2]
- **turnips** *20151961* [2]
- **whitecurrants** *20102114* [2]

Approval information
- Boscalid and pyraclostrobin included in Annex I under EC Regulation 1107/2009
- Accepted by BBPA on malting barley before ear emergence only and for use on hops

Efficacy guidance
- On brassicas apply as a protectant spray or at the first sign of disease and repeat at 3-4 wk intervals depending on disease pressure [2]
- Ensure adequate spray penetration and coverage by increasing water volume in dense crops [2]
- For best results on strawberries apply as a protectant spray at the white bud stage. Applications should be made in sequence with other products as part of a fungicide spray programme during flowering at 7-10 day intervals [2]
- On carrots and field beans apply as a protectant spray or at the first sign of disease with a repeat treatment if needed, as directed on the label [2]
- On lettuce apply as a protectant spray 1-2 wk after planting [2]
- Optimum results on apples and pears obtained from a protectant treatment from bud burst [1]
- Application as the final 2 sprays on apples and pears gives reduction in storage rots [1]
- Pyraclostrobin is a member of the QoI cross-resistance group of fungicides and should be used in programmes with fungicides with a different mode of action

Restrictions
- Maximum total dose equivalent to three full dose treatments on brassicas; two full dose treatments on all other field crops [2]
- Maximum number of treatments (including other QoI treatments) on apples and pears 4 per yr if total number of applications is 12 or more, or 3 per yr if the total number is fewer than 12 [1]
- Do not use more than 2 consecutive treatments on apples and pears, and these must be separated by a minimum of 2 applications of a fungicide with a different mode of action [1]
- Use a maximum of three applications per yr on brassicas and no more than two per yr on other field crops [2]
- Do not use consecutive treatments; apply in alternation with fungicides from a different cross resistance group and effective against the target diseases [2]
- Do not apply more than 6 kg/ha to the same area of land per yr [2]
- Consult processor before use on crops for processing
- Applications to lettuce and protected salad crops may only be made between 1 Apr and 31 Oct [2]

SEE SECTION 3 FOR PRODUCTS ALSO REGISTERED

Crop-specific information
- HI 21 d for field beans; 14 d for brassica crops, carrots, lettuce; 7 d for apples, pears; 3 d for strawberries

Environmental safety
- Dangerous for the environment
- Very toxic to aquatic organisms
- Broadcast air-assisted LERAP [1] (40 m); LERAP Category B [2]

Hazard classification and safety precautions
  Hazard Harmful, Dangerous for the environment, Harmful if swallowed, Very toxic to aquatic organisms
  Transport code 9
  Packaging group III
  UN Number 3077
  Operator protection A; U05a, U20c [1, 2]; U13 [1]
  Environmental protection E15a, E38, H410 [1, 2]; E16a [2]; E17b [1] (40 m); E34 [1]
  Storage and disposal D01, D02, D10c, D12a [1, 2]; D08 [1]; D09a [2]
  Medical advice M03 [1]; M05a [1, 2]

56   bromadiolone

An anti-coagulant coumarin-derivative rodenticide

Products

1	Bromag	Killgerm	0.005% w/w	RB	UK15-0932
2	Bromag Wax Blocks	Killgerm	0.005% w/w	RB	UK13-0755
3	Formula B Rat & Mouse Killer	Killgerm	0.005% w/w	RB	UK15-0934
4	Mamba	Killgerm	0.005% w/w	RB	UK15-0933
5	Sakarat Brodikill Whole Wheat	Killgerm	0.005% w/w	RB	UK17-1051
6	Sakarat Bromabait	Killgerm	0.005% w/w	RB	UK15-0931
7	Sakarat Bromakill	Killgerm	0.005% w/w	RB	UK15-0935

Uses
- Mice in *farm buildings*, *farmyards* [2, 5-7]; *farm buildings/yards* [1, 3, 4]
- Rats in *farm buildings*, *farmyards* [2, 5-7]; *farm buildings/yards* [1, 3, 4]

Approval information
- Bromadiolone included in Annex I under EC Directive 1107/2009

Efficacy guidance
- Ready-to-use baits are formulated on a mould-resistant, whole-wheat base
- Use in baiting programme. Place baits in protected situations, sufficient for continuous feeding between treatments
- Chemical is effective against warfarin- and coumatetralyl-resistant rats and mice and does not induce bait shyness
- Use bait bags where loose baiting inconvenient (eg behind ricks, silage clamps etc)
- The resistance status of the rodent population should be assessed when considering the choice of product to use. Resistance to bromadiolone in rats is now widespread in the UK.

Restrictions
- For use only by professional operators

Environmental safety
- Access to baits by children, birds and animals, particularly cats, dogs, pigs and poultry, must be prevented
- Baits must not be placed where food, feed or water could become contaminated
- Remains of bait and bait containers must be removed after treatment and burned or buried

- Rodent bodies must be searched for and burned or buried. They must not be placed in refuse bins or on rubbish tips
- Take extreme care to prevent domestic animals having access to the bait

Hazard classification and safety precautions
  Operator protection A [1]; U13, U20b
  Storage and disposal D05, D07, D09a, D11a
  Vertebrate/rodent control products V01b, V02, V03b, V04b
  Medical advice M03

57   bromoxynil

A contact acting HBN herbicide
HRAC mode of action code: C3

Products

1	Buctril	Bayer CropScience	225 g/l	EC	16597
2	Butryflow	Nufarm UK	402 g/l	SC	14056
3	Maya	Nufarm UK	402 g/l	SC	16760

Uses

- Annual dicotyledons in *asparagus* (off-label), *chives* (off-label) [1]; *bulb onion sets* (off-label), *millet* (off-label), *salad onions* (off-label), *shallots* (off-label) [2]; *bulb onions*, *forage maize*, *grain maize*, *linseed*, *ornamental bulbs* (off-label), *shallots*, *spring barley*, *spring oats*, *spring wheat*, *sweetcorn*, *winter barley*, *winter oats*, *winter wheat* [2, 3]; *leeks* (off-label) [1, 2]
- Black bindweed in *bulb onion sets* (off-label), *canary grass* (off-label), *forage maize*, *grain maize*, *leeks* (off-label), *linseed*, *salad onions* (off-label), *shallots* (off-label), *sweetcorn* [3]; *game cover* (off-label), *millet* (off-label), *miscanthus* (off-label), *ornamental bulbs* (off-label), *ornamental plant production* (off-label) [2, 3]
- Black nightshade in *bulb onion sets* (off-label), *canary grass* (off-label), *leeks* (off-label), *linseed*, *sweetcorn* [3]; *bulb onions* (off-label), *garlic* (off-label), *rye*, *spring barley*, *spring oats*, *spring wheat*, *triticale*, *winter barley*, *winter oats*, *winter wheat* [1]; *forage maize*, *grain maize*, *salad onions* (off-label), *shallots* (off-label) [1, 3]; *game cover* (off-label), *millet* (off-label), *miscanthus* (off-label), *ornamental plant production* (off-label) [2, 3]
- Charlock in *bulb onions* (off-label), *garlic* (off-label), *salad onions* (off-label), *shallots* (off-label) [1]
- Chickweed in *bulb onion sets* (off-label), *canary grass* (off-label), *leeks* (off-label) [3]; *bulb onions* (off-label), *garlic* (off-label) [1]; *game cover* (off-label), *millet* (off-label), *miscanthus* (off-label), *ornamental plant production* (off-label) [2, 3]; *salad onions* (off-label), *shallots* (off-label) [1, 3]
- Common orache in *bulb onion sets* (off-label), *forage maize*, *grain maize*, *leeks* (off-label), *linseed*, *millet* (off-label), *salad onions* (off-label), *shallots* (off-label), *sweetcorn* [3]; *game cover* (off-label), *miscanthus* (off-label), *ornamental plant production* (off-label) [2, 3]
- Docks in *bulb onion sets* (off-label), *leeks* (off-label), *salad onions* (off-label), *shallots* (off-label) [2, 3]; *millet* (off-label) [3]
- Fat hen in *bulb onion sets* (off-label), *bulb onions*, *canary grass* (off-label), *leeks* (off-label), *linseed*, *shallots*, *sweetcorn* [3]; *bulb onions* (off-label), *garlic* (off-label), *rye*, *triticale* [1]; *forage maize*, *grain maize*, *salad onions* (off-label), *shallots* (off-label), *spring barley*, *spring oats*, *spring wheat*, *winter barley*, *winter oats*, *winter wheat* [1, 3]; *game cover* (off-label), *millet* (off-label), *miscanthus* (off-label), *ornamental bulbs* (off-label), *ornamental plant production* (off-label) [2, 3]
- Field bindweed in *bulb onion sets* (off-label), *forage maize*, *grain maize*, *leeks* (off-label), *linseed*, *millet* (off-label), *salad onions* (off-label), *shallots* (off-label), *sweetcorn* [3]; *game cover* (off-label), *miscanthus* (off-label), *ornamental plant production* (off-label) [2, 3]
- Field speedwell in *bulb onions* (off-label), *forage maize*, *garlic* (off-label), *grain maize*, *rye*, *salad onions* (off-label), *shallots* (off-label), *spring barley*, *spring oats*, *spring wheat*, *triticale*, *winter barley*, *winter oats*, *winter wheat* [1]

SEE SECTION 3 FOR PRODUCTS ALSO REGISTERED

SECTION 2

- Groundsel in *bulb onion sets* (off-label), *bulb onions*, *canary grass* (off-label), *leeks* (off-label), *shallots*, *spring barley*, *spring oats*, *spring wheat*, *winter barley*, *winter oats*, *winter wheat* [3]; *bulb onions* (off-label), *garlic* (off-label) [1]; *ornamental bulbs* (off-label) [2, 3]; *salad onions* (off-label), *shallots* (off-label) [1, 3]
- Knotgrass in *bulb onion sets* (off-label), *forage maize*, *grain maize*, *leeks* (off-label), *linseed*, *millet* (off-label), *salad onions* (off-label), *shallots* (off-label), *sweetcorn* [3]; *game cover* (off-label), *miscanthus* (off-label), *ornamental plant production* (off-label) [2, 3]
- Mayweeds in *bulb onions* (off-label), *forage maize*, *garlic* (off-label), *grain maize*, *rye*, *salad onions* (off-label), *shallots* (off-label), *spring barley*, *spring oats*, *spring wheat*, *triticale*, *winter barley*, *winter oats*, *winter wheat* [1]
- Penny cress in *bulb onions* (off-label), *forage maize*, *garlic* (off-label), *grain maize*, *rye*, *salad onions* (off-label), *shallots* (off-label), *spring barley*, *spring oats*, *spring wheat*, *triticale*, *winter barley*, *winter oats*, *winter wheat* [1]
- Poppies in *bulb onions* (off-label), *forage maize*, *garlic* (off-label), *grain maize*, *rye*, *salad onions* (off-label), *shallots* (off-label), *spring barley*, *spring oats*, *spring wheat*, *triticale*, *winter barley*, *winter oats*, *winter wheat* [1]
- Redshank in *bulb onion sets* (off-label), *leeks* (off-label), *salad onions* (off-label), *shallots* (off-label) [3]; *game cover* (off-label), *millet* (off-label), *miscanthus* (off-label), *ornamental plant production* (off-label) [2, 3]
- Runch in *bulb onions* (off-label), *garlic* (off-label), *salad onions* (off-label), *shallots* (off-label) [1]
- Scentless mayweed in *bulb onion sets* (off-label), *canary grass* (off-label), *leeks* (off-label), *salad onions* (off-label), *shallots* (off-label) [3]
- Shepherd's purse in *bulb onions* (off-label), *forage maize*, *garlic* (off-label), *grain maize*, *rye*, *salad onions* (off-label), *shallots* (off-label), *spring barley*, *spring oats*, *spring wheat*, *triticale*, *winter barley*, *winter oats*, *winter wheat* [1]
- Small nettle in *ornamental bulbs* (off-label) [3]
- Volunteer oilseed rape in *bulb onion sets* (off-label), *bulb onions*, *canary grass* (off-label), *leeks* (off-label), *salad onions* (off-label), *shallots*, *shallots* (off-label), *spring barley*, *spring oats*, *spring wheat*, *winter barley*, *winter oats*, *winter wheat* [3]
- Wild mustard in *forage maize*, *grain maize*, *rye*, *spring barley*, *spring oats*, *spring wheat*, *triticale*, *winter barley*, *winter oats*, *winter wheat* [1]
- Wild radish in *forage maize*, *grain maize*, *rye*, *spring barley*, *spring oats*, *spring wheat*, *triticale*, *winter barley*, *winter oats*, *winter wheat* [1]

Extension of Authorisation for Minor Use (EAMUs)
- *asparagus* 20152332 [1]
- *bulb onion sets* 20162683 [2], 20162882 [3]
- *bulb onions* 20152400 [1]
- *canary grass* 20171860 [3]
- *chives* 20152331 [1]
- *game cover* 20131620 [2], 20171044 [3]
- *garlic* 20152400 [1]
- *leeks* 20152332 [1], 20162683 [2], 20162882 [3]
- *millet* 20120593 [2], 20162881 [3]
- *miscanthus* 20131620 [2], 20171044 [3]
- *ornamental bulbs* 20130517 [2], 20171045 [3]
- *ornamental plant production* 20140561 [2], 20171046 [3]
- *salad onions* 20152400 [1], 20162683 [2], 20162882 [3]
- *shallots* 20152400 [1], 20162683 [2], 20162882 [3]

Approval information
- Bromoxynil included in Annex I under EC Regulation 1107/2009
- Accepted by BBPA for use on malting barley

Efficacy guidance
- Spray when main weed flush has germinated and the largest are at the 4 leaf stage
- Weed control can be enhanced by using a split treatment spraying each application when the weeds are seedling to 2 true leaves.

FOR FULL CONDITIONS OF USE ALWAYS READ THE PRODUCT LABEL

Restrictions
- Maximum number of treatments 1 per crop or yr or maximum total dose equivalent to one full dose treatment
- Do not apply with oils or other adjuvants
- Do not apply to bulb onion or shallot grown from sets [2]
- Do not apply using hand-held equipment or at concentrations higher than those recommended
- Do not apply during frosty weather, drought, when soil is waterlogged, when rain expected within 4 h or to crops under any stress
- Take particular care to avoid drift onto neighbouring susceptible crops or open water surfaces
- Do not apply if air temp is above 25°C in cereals or 20°C in maize as some necrosis may occur [1]

Crop-specific information
- Latest use: before 10 fully expanded leaf stage of crop for maize, sweetcorn; before crop 20 cm tall and before flower buds visible for linseed; before 2nd node detectable (GS 32) for cereals
- Foliar scorch, which rapidly disappears without affecting growth, will occur if treatment made in hot weather or during rapid growth

Following crops guidance
- There are no restrictions on the establishment of succeeding or replacement crops

Environmental safety
- Very toxic to aquatic organisms
- Keep livestock out of treated areas for at least 6 wk after treatment
- High risk to bees. Do not apply to crops in flower or to those in which bees are actively foraging
- Do not contaminate surface waters or ditches with chemical or used container
- LERAP Category B

Hazard classification and safety precautions
Hazard Harmful, Dangerous for the environment, Harmful if swallowed, Harmful if inhaled [1-3]; Flammable, Flammable liquid and vapour, Very toxic to aquatic organisms [1]
Transport code 3 [1]; 9 [2, 3]
Packaging group III
UN Number 1993 [1]; 3082 [2, 3]
Risk phrases H304, H315, H319, H336 [1]; H317, H361 [1-3]
Operator protection A, H [1-3]; C [1]; K [2, 3]; U05a [1-3]; U12, U12b [1]; U14, U15, U23a [2, 3]
Environmental protection E15a, E16a, E34, E38, H410
Storage and disposal D01, D02, D09a, D12a [1-3]; D05, D10c [2, 3]; D10a [1]
Medical advice M03 [1-3]; M05b [1]

58   bromoxynil + diflufenican

A contact and residual herbicide mixture for weed control in cereals
HRAC mode of action code: C3 + F1

See also diflufenican

Products
Cyclops	Nufarm UK	160:26.7 g/l	EC	16676

Uses
- Chickweed in **grass seed crops** *(off-label)*, **winter barley, winter wheat**
- Fat hen in **grass seed crops** *(off-label)*, **winter barley** *(at 1.9 l/ha only)*
- Field pansy in **grass seed crops** *(off-label)*, **winter barley, winter wheat**
- Groundsel in **grass seed crops** *(off-label)*, **winter barley** *(at 1.9 l/ha only)*
- Ivy-leaved speedwell in **grass seed crops** *(off-label)*, **winter barley** *(at 1.9 l/ha only)*
- Volunteer beans in **winter barley, winter wheat**
- Volunteer oilseed rape in **grass seed crops** *(off-label)*, **winter barley, winter wheat**

Extension of Authorisation for Minor Use (EAMUs)
- **grass seed crops** *20161452*

SEE SECTION 3 FOR PRODUCTS ALSO REGISTERED

Approval information
- Bromoxynil and diflufenican included in Annex I under EC Regulation 1107/2009

Restrictions
- Do not allow spray to drift on to non-target plants.
- Do not treat stressed crops.
- Do not overlap spray swathes.

Following crops guidance
- In the event of crop failure, winter wheat can be drilled immediately after cultivation but winter barley can only be sown after ploughing. Before spring crops of wheat, barley, oilseed rape, peas, sugar beet, field beans, potatoes, carrots, edible brassicas or onions are sown, 12 weeks should elapse since application and the land must be ploughed. Do not plant any crop other than those listed here before the subsequent autumn.

Environmental safety
- Buffer zone requirement 7 m [1]
- To protect non target insects/arthropods respect an unsprayed buffer zone of 5 m to non-crop land [1]
- LERAP Category B

Hazard classification and safety precautions
Hazard Corrosive, Dangerous for the environment, Harmful if swallowed
Transport code 9
Packaging group III
UN Number 3082
Risk phrases H317, H361, R70
Operator protection A, C, H; U04b, U05a, U09a, U11, U19a, U20a
Environmental protection E07a, E15b, E16a, E34, E38, H410
Storage and disposal D01, D02, D05, D09a, D10b, D12a, D12b
Medical advice M03, M05a

59 bromuconazole

A systemic triazole fungicide for cereals only available in mixtures
FRAC mode of action code: 3

60 bromuconazole + tebuconazole

A triazole mixture for disease control in wheat, rye and triticale
FRAC mode of action code: 3

See also tebuconazole

Products

Soleil	Interfarm	167:107 g/l	EC	16869

Uses
- Alternaria in *rye, spring wheat, triticale, winter wheat*
- Brown rust in *rye, spring wheat, triticale, winter wheat*
- Cladosporium in *rye, spring wheat, triticale, winter wheat*
- Fusarium in *rye, spring wheat, triticale, winter wheat*
- Powdery mildew in *rye* (moderate control), *spring wheat* (moderate control), *triticale* (moderate control), *winter wheat* (moderate control)
- Septoria leaf blotch in *rye* (moderate control), *spring wheat* (moderate control), *triticale* (moderate control), *winter wheat* (moderate control)

Approval information
- Bromuconazole and tebuconazole included in Annex 1 under EC Directive 1107/2009

Restrictions
- Newer authorisations for tebuconazole products require application to cereals only after GS 30 and applications to oilseed rape and linseed after GS20 - check label

Following crops guidance
- Only sugar beet, cereals, oilseed rape, field beans, peas, potatoes, linseed and Italian rye-grass may be sown as the following crop. The effect on other crops has not been assessed.
- Some effects may be seen on sugar beet crops grown following a cereal crop but the sugar beet will usually recover completely.

Environmental safety
- LERAP Category B

Hazard classification and safety precautions
Hazard Harmful, Dangerous for the environment, Very toxic to aquatic organisms
Transport code 9
Packaging group III
UN Number 3082
Risk phrases H304, H318, H336, H361
Operator protection A, C, H; U11, U19a, U19f, U20b
Environmental protection E16a, E34, E38, H410
Storage and disposal D01, D02, D05, D09a, D11a, D12a, D12b
Medical advice M03

61   bupirimate

A systemic aminopyrimidinol fungicide active against powdery mildew
FRAC mode of action code: 8

Products
1	Clayton Maritime	Clayton	250 g/l	EC	16593
2	Jatapu	Agroquimicos	250 g/l	EC	15776
3	Nimrod	Adama	250 g/l	EC	13046

Uses
- Powdery mildew in **apples, blackcurrants, courgettes** (outdoor only), **gooseberries, hops, marrows** (outdoor only), **pears, protected cucumbers, protected strawberries, raspberries** (outdoor only), **strawberries** [1-3]; **begonias, chrysanthemums, roses** [2, 3]; **ornamental plant production, protected ornamentals** [1]; **protected tomatoes** (off-label), **pumpkins** (off-label), **redcurrants** (off-label), **squashes** (off-label), **whitecurrants** (off-label) [3]

Extension of Authorisation for Minor Use (EAMUs)
- **protected tomatoes** *20070997* [3], *20141225* [3]
- **pumpkins** *20070994* [3], *20141224* [3]
- **redcurrants** *20082082* [3], *20141223* [3]
- **squashes** *20070994* [3], *20141224* [3]
- **whitecurrants** *20082082* [3], *20141223* [3]

Approval information
- Bupirimate included in Annex 1 under EC Regulation 1107/2009
- Accepted by BBPA for use on hops

Efficacy guidance
- On apples during periods that favour disease development lower doses applied weekly give better results than higher rates fortnightly
- Not effective in protected crops against strains of mildew resistant to bupirimate

Restrictions
- Maximum number of treatments or maximum total dose depends on crop and dose (see label for details)

SEE SECTION 3 FOR PRODUCTS ALSO REGISTERED

Crop-specific information

- HI: 1 d for apples, pears, strawberries; 2 d for cucurbits; 7 d for blackcurrants; 8 d for raspberries; 14 d for gooseberries, hops
- Apply before or at first signs of disease and repeat at 7-14 d intervals. Timing and maximum dose vary with crop. See label for details
- With apples, hops and ornamentals cultivars may vary in sensitivity to spray. See label for details
- If necessary to spray cucurbits in winter or early spring spray a few plants 10-14 d before spraying whole crop to test for likelihood of leaf spotting problem
- On roses some leaf puckering may occur on young soft growth in early spring or under low light intensity. Avoid use of high rates or wetter on such growth
- Never spray flowering begonias (or buds showing colour) as this can scorch petals
- Do not mix with other chemicals for application to begonias, cucumbers or gerberas

Environmental safety

- Dangerous for the environment
- Toxic to aquatic organisms
- Flammable
- Product has negligible effect on *Phytoseiulus* and *Encarsia* and may be used in conjunction with biological control of red spider mite

Hazard classification and safety precautions

Hazard Harmful [1]; Irritant [2, 3]; Flammable, Dangerous for the environment [1-3]
Transport code 3
Packaging group III
UN Number 1993
Risk phrases R22b, R38, R51, R53a [1-3]; R37, R67 [1]
Operator protection A, C; U05a, U20b [1-3]; U09a, U14, U19a [1]
Environmental protection E15a [1-3]; E34, E38 [1]
Storage and disposal D01, D02, D09a, D10c [1-3]; D12a [1]
Medical advice M05b

62   captan

A protectant phthalimide fungicide with horticultural uses
FRAC mode of action code: M4

Products

1	Captan 80 WDG	Adama	80% w/w	WG	16293
2	Clayton Core	Clayton	80% w/w	WG	16934
3	Multicap	Belcrop	80% w/w	WG	17652
4	PP Captan 80 WG	Arysta	80% w/w	WG	16294

Uses

- Botrytis in **ornamental plant production** *(off-label)* [1]
- Disease control in **hops in propagation** *(off-label)*, **nursery fruit trees** *(off-label)*, **quinces** *(off-label)* [1]
- Downy mildew in **ornamental plant production** *(off-label)* [1]
- Gloeosporium rot in **apples** [1-4]; **pears** [1]
- Scab in **apples** [1-4]; **ornamental plant production** *(off-label)* [1]; **pears** [1, 3, 4]
- Shot-hole in **ornamental plant production** *(off-label)* [1]

Extension of Authorisation for Minor Use (EAMUs)

- **hops in propagation** *20142510* [1]
- **nursery fruit trees** *20142510* [1]
- **ornamental plant production** *20151919* [1], *20151920* [1]
- **quinces** *20141276* [1]

Approval information

- Captan included in Annex I under EC Regulation 1107/2009

Restrictions
- Maximum number of treatments 12 per yr on apples and pears as pre-harvest sprays
- Do not use on apple cultivars Bramley, Monarch, Winston, King Edward, Spartan, Kidd's Orange or Red Delicious or on pear cultivar D'Anjou
- Do not mix with alkaline materials or oils
- Do not use on fruit for processing
- Powered visor respirator with hood and neck cape must be used when handling concentrate

Crop-specific information
- HI apples, pears 14 d; strawberries 7 d
- For control of scab apply at bud burst and repeat at 10-14 d intervals until danger of scab infection ceased
- For suppression of fruit storage rots apply from late Jul and repeat at 2-3 wk intervals
- For black spot control in roses apply after pruning with 3 further applications at 14 d intervals or spray when spots appear and repeat at 7-10 d intervals
- For grey mould in strawberries spray at first open flower and repeat every 7-10 d
- Do not leave diluted material for more than 2 h. Agitate well before and during spraying

Environmental safety
- Dangerous for the environment
- Very toxic to aquatic organisms
- LERAP Category B [3, 4]

Hazard classification and safety precautions
Hazard Harmful, Dangerous for the environment [1-4]; Very toxic to aquatic organisms [1-3]
Transport code 9
Packaging group III
UN Number 3077
Risk phrases H317, H351 [1-4]; H318, H320 [4]; H319 [1-3]
Operator protection A, D, E, H; U05a, U09a, U11, U19a, U20c
Environmental protection E15a [1-4]; E16a [3, 4]; H410 [4]; H412 [1-3]
Storage and disposal D01, D02, D09a, D11a, D12a

63   carbetamide

A residual pre- and post-emergence carbamate herbicide for a range of field crops
HRAC mode of action code: K2

Products
Crawler	Adama	60% w/w	WG	16856

Uses
- Annual dicotyledons in **corn gromwell** *(off-label)*, **linseed** *(off-label)*, **winter field beans**, **winter oilseed rape**
- Annual grasses in **winter field beans**, **winter oilseed rape**
- Annual meadow grass in **corn gromwell** *(off-label)*, **linseed** *(off-label)*
- Blackgrass in **corn gromwell** *(off-label)*, **linseed** *(off-label)*
- Chickweed in **corn gromwell** *(off-label)*, **linseed** *(off-label)*
- Cleavers in **corn gromwell** *(off-label)*, **linseed** *(off-label)*
- Couch in **corn gromwell** *(off-label)*, **linseed** *(off-label)*
- Fat hen in **corn gromwell** *(off-label)*, **linseed** *(off-label)*
- Fumitory in **corn gromwell** *(off-label)*, **linseed** *(off-label)*
- Knotgrass in **corn gromwell** *(off-label)*, **linseed** *(off-label)*
- Ryegrass in **corn gromwell** *(off-label)*, **linseed** *(off-label)*
- Small nettle in **corn gromwell** *(off-label)*, **linseed** *(off-label)*
- Speedwells in **corn gromwell** *(off-label)*, **linseed** *(off-label)*
- Volunteer cereals in **corn gromwell** *(off-label)*, **linseed** *(off-label)*, **winter field beans**, **winter oilseed rape**
- Wild oats in **corn gromwell** *(off-label)*, **linseed** *(off-label)*
- Yorkshire fog in **corn gromwell** *(off-label)*, **linseed** *(off-label)*

SEE SECTION 3 FOR PRODUCTS ALSO REGISTERED

Extension of Authorisation for Minor Use (EAMUs)
- *corn gromwell* 20151828
- *linseed* 20152544

Approval information
- Carbetamide included in Annex I under EC Regulation 1107/2009

Efficacy guidance
- Best results pre- or early post-emergence of weeds under cool, moist conditions. Adequate soil moisture is essential
- May be applied pre or post emergence up to end of Feb in year of harvest
- Maintain an interval of at least 7 days when using split applications
- Dicotyledons controlled include chickweed, cleavers and speedwell
- Weed growth stops rapidly after treatment but full effects may take 6-8 wk to develop
- Various tank mixes are recommended to broaden the weed spectrum. See label for details
- Always follow WRAG guidelines for preventing and managing herbicide resistant weeds. See Section 5 for more information

Restrictions
- Maximum number of treatments 1 per crop for all crops
- Do not treat any crop on waterlogged soil
- Do not use on soils with more than 10% organic matter as residual activity is impaired
- Do not apply during prolonged periods of cold weather when weeds are dormant
- Do not tank-mix with prothioconazole or liquid manganese

Crop-specific information
- HI 6 wk for all crops
- Apply to brassicas from mid-Oct to end-Feb provided crop has at least 4 true leaves (spring cabbage, spring greens), 3-4 true leaves (seed crops, oilseed rape)
- Apply to established lucerne or sainfoin from Nov to end-Feb
- Apply to established red or white clover from Feb to mid-Mar

Following crops guidance
- Succeeding crops may be sown 2 wk after treatment for brassicas, field beans, 8 wk after treatment for peas, runner beans, 16 wk after treatment for cereals, maize
- Ploughing is not necessary before sowing subsequent crops

Environmental safety
- Do not graze crops for at least 6 wk after treatment
- Some pesticides pose a greater threat of contamination of water than others and carbetamide is one of these pesticides. Take special care when applying carbetamide near water and do not apply if heavy rain is forecast

Hazard classification and safety precautions
Hazard Dangerous for the environment
Transport code 9
Packaging group III
UN Number 3077
Risk phrases H319, H361
Operator protection A, C, H; U20c
Environmental protection E15a, E38, H411
Storage and disposal D01, D02, D09a, D11a

64   carbon dioxide (commodity substance)

A gas for the control of trapped rodents and other vertebrates. Approval valid until 31/8/2019

Products

carbon dioxide	various	99.9%	GA	-

Uses
- Birds in *traps*

- Mice in **traps**
- Rats in **traps**

Approval information
- Carbon dioxide included in Annex 1 under EC Regulation 1107/2009

Efficacy guidance
- Use to destroy trapped rodent pests
- Use to control birds covered by general licences issued by the Agriculture and Environment Departments under Section 16(1) of the Wildlife and Countryside Act (1981) for the control of opportunistic bird species, where birds have been trapped or stupefied with alphachloralose/seconal

Restrictions
- Operators must wear self-contained breathing apparatus when carbon dioxide levels are greater than 0.5% v/v
- Operators must be suitably trained and competent

Environmental safety
- Unprotected persons and non-target animals must be excluded from the treatment enclosures and surrounding areas unless the carbon dioxide levels are below 0.5% v/v

Hazard classification and safety precautions
  Operator protection G

65   carfentrazone-ethyl

A triazolinone contact herbicide
HRAC mode of action code: E

Products

1	Aurora 40 WG	Headland	40% w/w	WG	17278
2	Headlite	AgChem Access	60 g/l	ME	13624
3	Shark	Headland	60 g/l	ME	17256
4	Shylock	AgChem Access	60 g/l	ME	15405
5	Spotlight Plus	Headland	60 g/l	ME	17241

Uses
- Annual and perennial weeds in **all edible crops (outdoor)**, **all non-edible crops (outdoor)**, **potatoes** [5]; **forest** (off-label), **ornamental plant production** (off-label), **protected forest** (off-label), **protected ornamentals** (off-label) [3]
- Annual dicotyledons in **all edible crops (outdoor)** (before planting), **all non-edible crops (outdoor)** (before planting), **potatoes** [3, 4]; **asparagus** (off-label), **blackcurrants** (off-label), **blueberries** (off-label), **cranberries** (off-label), **gooseberries** (off-label), **ribes species** (off-label), **whitecurrants** (off-label), **wine grapes** (off-label) [3]; **durum wheat** (off-label), **rye** (off-label), **spring barley, spring oats, spring wheat, triticale, winter barley, winter oats, winter wheat** [1]
- Black bindweed in **durum wheat** (off-label), **rye** (off-label) [1]; **potatoes** [3, 4]
- Black nightshade in **strawberries** (off-label) [3]
- Cleavers in **durum wheat** (off-label), **rye** (off-label) [1]; **potatoes** [3, 4]; **strawberries** (off-label) [3]
- Desiccation in **blackberries** (off-label), **loganberries** (off-label), **protected blackberries** (off-label), **protected loganberries** (off-label), **protected raspberries** (off-label), **protected rubus hybrids** (off-label), **raspberries** (off-label), **rubus hybrids** (off-label) [3]; **narcissi** (off-label), **potatoes** [5]
- Fat hen in **durum wheat** (off-label), **rye** (off-label) [1]; **potatoes** [3, 4]
- Field pansy in **durum wheat** (off-label), **rye** (off-label) [1]
- Groundsel in **durum wheat** (off-label), **rye** (off-label) [1]
- Hairy bittercress in **strawberries** (off-label) [3]
- Haulm destruction in **seed potatoes, ware potatoes** [2]
- Ivy-leaved speedwell in **potatoes** [3, 4]

SEE SECTION 3 FOR PRODUCTS ALSO REGISTERED

- Knotgrass in **durum wheat** *(off-label)*, **rye** *(off-label)* [1]; **potatoes** [3, 4]; **strawberries** *(off-label)* [3]
- Polygonums in **blackcurrants** *(off-label)*, **blueberries** *(off-label)*, **cranberries** *(off-label)*, **gooseberries** *(off-label)*, **ribes species** *(off-label)*, **whitecurrants** *(off-label)*, **wine grapes** *(off-label)* [3]
- Redshank in **durum wheat** *(off-label)*, **rye** *(off-label)* [1]; **potatoes** [3, 4]; **strawberries** *(off-label)* [3]
- Small nettle in **durum wheat** *(off-label)*, **rye** *(off-label)* [1]; **strawberries** *(off-label)* [3]
- Speedwells in **durum wheat** *(off-label)*, **rye** *(off-label)* [1]; **strawberries** *(off-label)* [3]
- Thistles in **blackcurrants** *(off-label)*, **blueberries** *(off-label)*, **cranberries** *(off-label)*, **gooseberries** *(off-label)*, **ribes species** *(off-label)*, **whitecurrants** *(off-label)*, **wine grapes** *(off-label)* [3]
- Volunteer oilseed rape in **all edible crops (outdoor)** *(before planting)*, **all non-edible crops (outdoor)** *(before planting)*, **potatoes** [3, 4]; **durum wheat** *(off-label)*, **rye** *(off-label)* [1]
- Willowherb in **strawberries** *(off-label)* [3]

Extension of Authorisation for Minor Use (EAMUs)
- **asparagus** *20160098* [3]
- **blackberries** *20160097* [3]
- **blackcurrants** *20160100* [3]
- **blueberries** *20160100* [3]
- **cranberries** *20160100* [3]
- **durum wheat** *20152860* [1]
- **forest** *20160099* [3]
- **gooseberries** *20160100* [3]
- **loganberries** *20160097* [3]
- **narcissi** *20152644* [5]
- **ornamental plant production** *20160099* [3]
- **protected blackberries** *20160097* [3]
- **protected forest** *20160099* [3]
- **protected loganberries** *20160097* [3]
- **protected ornamentals** *20160099* [3]
- **protected raspberries** *20160097* [3]
- **protected rubus hybrids** *20160097* [3]
- **raspberries** *20160097* [3]
- **ribes species** *20160100* [3]
- **rubus hybrids** *20160097* [3]
- **rye** *20152860* [1]
- **strawberries** *20170378* [3]
- **whitecurrants** *20160100* [3]
- **wine grapes** *20160102* [3]

Approval information
- Carfentrazone-ethyl included in Annex I under EC Regulation 1107/2009
- Accepted by BBPA for use on malting barley and hops

Efficacy guidance
- Best weed control results achieved from good spray cover applied to small actively growing weeds
- Carfentrazone-ethyl acts by contact only; see label for optimum timing on specified weeds. Weeds emerging after application will not be controlled
- For weed control in cereals use as two spray programme with one application in autumn and one in spring
- Efficacy of haulm destruction will be reduced where flailed haulm covers the stems at application
- For potato crops with very dense vigorous haulm or where regrowth occurs following a single application a second application may be necessary to achieve satisfactory desiccation. A minimum interval between applications of 7 d should be observed to achieve optimum performance

FOR FULL CONDITIONS OF USE ALWAYS READ THE PRODUCT LABEL

Restrictions

- Maximum number of treatments 2 per crop for cereals (1 in Autumn and 1 in Spring); 1 per crop for weed control in potatoes; 1 per yr for pre-planting treatments
- Maximum total dose for potato haulm destruction equivalent to 1.6 full dose treatments
- Do not treat cereal crops under stress from drought, waterlogging, cold, pests, diseases, nutrient or lime deficiency or any factors reducing plant growth
- Do not treat cereals undersown with clover or other legumes
- Allow at least 2-3 wk between application and lifting potatoes to allow skins to set if potatoes are to be stored
- Follow label instructions for sprayer cleaning
- Do not apply through knapsack sprayers
- Contact processor before using a split dose on potatoes for processing
- Do not apply within 10 days of an application of iodosulfuron-methyl-sodium + mesosulfuron-methyl formulations, e.g. MAPP 12478

Crop-specific information

- Latest use: 1 mth before planting edible and non-edible crops; before 3rd node detectable (GS 33) on cereals
- HI 7 d for potatoes
- For weed control in cereals treat from 2 leaf stage
- When used as a treatment prior to planting a subsequent crop apply before weeds exceed maximum sizes indicated in the label
- If treated potato tubers are to be stored then allow at least 2-3 wk between the final application and lifting to allow skins to set

Following crops guidance

- No restrictions apply on the planting of succeeding crops 1 mth after application to potatoes for haulm destruction or as a pre-planting treatment, or 3 mth after application to cereals for weed control
- In the event of failure of a treated cereal crop, all cereals, ryegrass, maize, oilseed rape, peas, sunflowers, *Phacelia*, vetches, carrots or onions may be planted within 1 mth of treatment

Environmental safety

- Dangerous for the environment
- Very toxic to aquatic organisms
- Some non-target crops are sensitive. Avoid drift onto broad-leaved plants outside the treated area, or onto ponds waterways or ditches

Hazard classification and safety precautions

Hazard Irritant, Dangerous for the environment, Very toxic to aquatic organisms
Transport code 9
Packaging group III
UN Number 3077 [1]; 3082 [2-5]
Risk phrases H317 [2-5]
Operator protection A, H; U05a, U14 [1-5]; U08, U13 [1]; U20a [2-5]
Environmental protection E15a [1, 3, 4]; E15b [2, 5]; E34 [1]; E38, H410 [1-5]
Storage and disposal D01, D02, D09a, D12a [1-5]; D10b [1]; D10c [2-5]
Medical advice M05a [1]

66   carfentrazone-ethyl + mecoprop-P

A foliar applied herbicide for cereals
HRAC mode of action code: E + O

See also mecoprop-P

Products

1	Jewel	Everris Ltd	1.5:60% w/w	WG	16212
2	Pan Glory	Pan Amenity	1.5:60% w/w	WG	14487
3	Platform S	Headland	1.5:60% w/w	WG	17259

SEE SECTION 3 FOR PRODUCTS ALSO REGISTERED

Uses

- Annual and perennial weeds in *amenity grassland, managed amenity turf* [1, 2]
- Charlock in *spring barley, spring oats, spring wheat, winter barley, winter oats, winter wheat* [3]
- Chickweed in *spring barley, spring oats, spring wheat, winter barley, winter oats, winter wheat* [3]
- Cleavers in *spring barley, spring oats, spring wheat, winter barley, winter oats, winter wheat* [3]
- Field speedwell in *spring barley, spring oats, spring wheat, winter barley, winter oats, winter wheat* [3]
- Ivy-leaved speedwell in *spring barley, spring oats, spring wheat, winter barley, winter oats, winter wheat* [3]
- Moss in *amenity grassland, managed amenity turf* [1, 2]
- Red dead-nettle in *spring barley, spring oats, spring wheat, winter barley, winter oats, winter wheat* [3]

Approval information

- Carfentrazone-ethyl and mecoprop-P included in Annex I under EC Regulation 1107/2009
- Accepted by BBPA for use on malting barley

Efficacy guidance

- Best results obtained when weeds have germinated and growing vigorously in warm moist conditions
- Treatment of large weeds and poor spray coverage may result in reduced weed control

Restrictions

- Maximum number of treatments 2 per crop. The total amount of mecoprop-P applied in a single yr must not exceed the maximum total dose approved for any single product for the crop per situation
- Do not treat crops suffering from stress from any cause
- Do not treat crops undersown or to be undersown
- Do not apply between 1st October and 1st March

Crop-specific information

- Latest use: before 3rd node detectable (GS 33)
- Can be used on all varieties of wheat and barley in autumn or spring from the beginning of tillering

Following crops guidance

- In the event of crop failure, any cereal, maize, oilseed rape, peas, vetches or sunflowers may be sown 1 mth after a spring treatment. Any crop may be planted 3 mth after treatment

Environmental safety

- Dangerous for the environment
- Very toxic to aquatic organisms
- Keep livestock out of treated areas for at least two weeks following treatment and until poisonous weeds, such as ragwort, have died down and become unpalatable
- Some pesticides pose a greater threat of contamination of water than others and mecoprop-P is one of these pesticides. Take special care when applying mecoprop-P near water and do not apply if heavy rain is forecast

Hazard classification and safety precautions

Hazard Harmful, Dangerous for the environment [1-3]; Harmful if swallowed [2, 3]; Very toxic to aquatic organisms [3]
Transport code 9
Packaging group III
UN Number 3077
Risk phrases H317, H318 [2, 3]; R22a, R41, R43, R50, R53a [1]
Operator protection A, C, H [1-3]; M [3]; U05a, U11, U13, U14, U20b [1-3]; U08 [3]; U09a [1, 2]
Environmental protection E07a, E34, E38 [1-3]; E15a, H411 [3]; E15b [1, 2]; H410 [2]
Storage and disposal D01, D02, D09a, D10b [1-3]; D12a [3]
Medical advice M05a

FOR FULL CONDITIONS OF USE ALWAYS READ THE PRODUCT LABEL

67   chlorantraniliprole

An ingested and contact insecticide for insect pest control in apples and pears and, in Ireland only, for colorado beetle control in potatoes.
IRAC mode of action code: 28

Products

Coragen	DuPont	200 g/l	SC	14930

Uses

- Carnation tortrix moth in **bilberries** *(off-label)*, **blackcurrants** *(off-label)*, **blueberries** *(off-label)*, **cranberries** *(off-label)*, **gooseberries** *(off-label)*, **redcurrants** *(off-label)*
- Carrot fly in **carrots** *(off-label)*, **celeriac** *(off-label)*, **horseradish** *(off-label)*, **parsley root** *(off-label)*, **parsnips** *(off-label)*, **red beet** *(off-label)*, **swedes** *(off-label)*, **turnips** *(off-label)*
- Caterpillars in **forest nurseries** *(off-label)*, **horseradish** *(off-label)*, **red beet** *(off-label)*, **swedes** *(off-label)*, **turnips** *(off-label)*
- Codling moth in **apples**, **pears**
- Large pine weevil in **forest nurseries** *(off-label)*
- Light brown apple moth in **bilberries** *(off-label)*, **blackcurrants** *(off-label)*, **blueberries** *(off-label)*, **cranberries** *(off-label)*, **gooseberries** *(off-label)*, **redcurrants** *(off-label)*
- Silver Y moth in **horseradish** *(off-label)*, **red beet** *(off-label)*, **swedes** *(off-label)*, **turnips** *(off-label)*
- South American Tomato Moth in **protected tomatoes** *(off-label)*

Extension of Authorisation for Minor Use (EAMUs)

- **bilberries** *20151087 expires 28 Feb 2018*
- **blackcurrants** *20151087 expires 28 Feb 2018*
- **blueberries** *20151087 expires 28 Feb 2018*
- **carrots** *20131156 expires 28 Feb 2018*
- **celeriac** *20131156 expires 28 Feb 2018*
- **cranberries** *20151087 expires 28 Feb 2018*
- **forest nurseries** *20162003 expires 28 Feb 2018*
- **gooseberries** *20151087 expires 28 Feb 2018*
- **horseradish** *20140302 expires 28 Feb 2018*
- **parsley root** *20131156 expires 28 Feb 2018*
- **parsnips** *20131156 expires 28 Feb 2018*
- **protected tomatoes** *20162080 expires 28 Feb 2018*
- **red beet** *20140302 expires 28 Feb 2018*
- **redcurrants** *20151087 expires 28 Feb 2018*
- **swedes** *20140302 expires 28 Feb 2018*
- **turnips** *20140302 expires 28 Feb 2018*

Approval information

- Chlorantraniliprole is included in Annex 1 under EC Regulation 1107/2009

Efficacy guidance

- Can be used as part of an Integrated Pest Management programme.
- For best fruit protection in apples and pears, apply before egg hatch.
- Best applied early morning or late evening to avoid applications when bees may be present.

Restrictions

- Maximum number of applications is two per year

Crop-specific information

- Latest time of application is 14 days before harvest

Environmental safety

- To protect bees and pollinating insects, do not apply to crops when in flower. Do not apply when bees are actively foraging or when flowering plants are present.
- Broadcast air-assisted LERAP (10 m); LERAP Category B

SEE SECTION 3 FOR PRODUCTS ALSO REGISTERED

Hazard classification and safety precautions

Hazard Dangerous for the environment
Transport code 9
Packaging group III
UN Number 3082
Risk phrases R50, R53a
Environmental protection E12e, E15b, E16a, E22b, E38; E17b (10 m)
Storage and disposal D01, D02, D05, D09a, D12a

68   chloridazon

A residual pyridazinone herbicide for beet crops
HRAC mode of action code: C1

Products

1 Better DF	Sipcam	65% w/w	WG	16101
2 Pyramin DF	BASF	65% w/w	WG	16768

Uses

- Annual dicotyledons in **bulb onions** *(off-label)*, **garlic** *(off-label)*, **herbs (see appendix 6)** *(off-label)*, **red beet** *(off-label)*, **shallots** *(off-label)* [2]; **fodder beet**, **mangels**, **sugar beet** [1, 2]
- Annual meadow grass in **bulb onions** *(off-label)*, **fodder beet**, **garlic** *(off-label)*, **mangels**, **shallots** *(off-label)*, **sugar beet** [1, 2]; **herbs (see appendix 6)** *(off-label)*, **red beet** *(off-label)* [2]
- Charlock in **bulb onions** *(off-label)*, **garlic** *(off-label)*, **shallots** *(off-label)* [1]
- Fat hen in **bulb onions** *(off-label)*, **garlic** *(off-label)*, **shallots** *(off-label)* [1]
- Knotgrass in **bulb onions** *(off-label)*, **garlic** *(off-label)*, **shallots** *(off-label)* [1]
- Mayweeds in **bulb onions** *(off-label)*, **garlic** *(off-label)*, **shallots** *(off-label)* [1]
- Poppies in **bulb onions** *(off-label)*, **garlic** *(off-label)*, **shallots** *(off-label)* [1]
- Small nettle in **bulb onions** *(off-label)*, **garlic** *(off-label)*, **shallots** *(off-label)* [1]

Extension of Authorisation for Minor Use (EAMUs)

- **bulb onions** *20140395* [1], *20152349 expires 15 Dec 2018* [2]
- **garlic** *20140395* [1], *20152349 expires 15 Dec 2018* [2]
- **herbs (see appendix 6)** *20152350 expires 15 Dec 2018* [2]
- **red beet** *20151219 expires 15 Dec 2018* [2]
- **shallots** *20140395* [1], *20152349 expires 15 Dec 2018* [2]

Approval information

- Chloridazon included in Annex I under EC Regulation 1107/2009

Efficacy guidance

- Absorbed by roots of germinating weeds and best results achieved pre-emergence of weeds or crop when soil moist and adequate rain falls after application
- Application rate for some products depends on soil type. Check label for details

Restrictions

- Maximum number of treatments generally 1 per crop (pre-emergence) for fodder beet and mangels; 1 (pre-emergence) + 3 (post-emergence) per crop for sugar beet, but labels vary slightly. Check labels for maximum total dose for the crop to be treated
- Do not use on Coarse Sands, Sands or Fine Sands or where organic matter exceeds 5%
- A maximum total dose of 2.6 kg/ha chloridazon may only be applied every third year on the same field

Crop-specific information

- Latest use: pre-emergence for fodder beet and mangels; normally before leaves of crop meet in row for sugar beet, but labels vary slightly
- Where used pre-emergence spray as soon as possible after drilling in mid-Mar to mid-Apr on fine, firm, clod-free seedbed
- Where crop drilled after mid-Apr or soil dry apply pre-drilling and incorporate to 2.5 cm immediately afterwards

FOR FULL CONDITIONS OF USE ALWAYS READ THE PRODUCT LABEL

- Various tank mixes recommended on sugar beet for pre- and post-emergence use and as repeated low dose treatments. See label for details
- Crop vigour may be reduced by treatment of crops growing under unfavourable conditions including poor tilth, drilling at incorrect depth, soil capping, physical damage, pest or disease damage, excess seed dressing, trace-element deficiency or a sudden rise in temperature after a cold spell

Following crops guidance
- Winter cereals or any spring sown crop may follow a treated beet crop harvested at the normal time and after ploughing to at least 150 mm
- In the event of failure of a treated crop only a beet crop or maize may be drilled, after cultivation

Environmental safety
- Dangerous for the environment
- Very toxic to aquatic organisms

Hazard classification and safety precautions
Hazard Harmful, Dangerous for the environment, Harmful if swallowed, Harmful if inhaled
Transport code 9
Packaging group III
UN Number 3077
Risk phrases H317 [1, 2]; R20 [2]
Operator protection A; U05a, U08, U14, U20b
Environmental protection E15a, E38 [1, 2]; H410 [1]; H411 [2]
Storage and disposal D01, D02, D09a, D11a, D12a
Medical advice M05a

69   chloridazon + metamitron

A contact and residual herbicide mixture for sugar beet
HRAC mode of action code: C1 + C1

See also metamitron

Products
Volcan Combi FL	Sipcam	300:280 g/l	SC	16681

Uses
- Annual dicotyledons in **fodder beet**, **sugar beet**
- Annual meadow grass in **fodder beet**, **sugar beet**

Approval information
- Chloridazon and metamitron included in Annex I under EC Regulation 1107/2009

Efficacy guidance
- Best results achieved from a sequential programmme of sprays
- Pre-emergence use improves efficacy of post-emergence programme. Best results obtained from application to a moist seed bed
- First post-emergence treatment should be made when weeds at early cotyledon stage and subsequent applications made when new weed flushes reach this stage
- Weeds surviving an earlier treatment should be treated again after 7-10 d even if no new weeds have appeared

Restrictions
- Maximum number of treatments 1 pre-emergence followed by 3 post-emergence
- Take advice if light soils have a high proportion of stones

Crop-specific information
- Latest use: when leaves of crop meet in rows
- Tolerance of crops growing under stress from any cause may be reduced
- Crops treated pre-emergence and subsequently subjected to frost may be checked and recovery may not be complete

SEE SECTION 3 FOR PRODUCTS ALSO REGISTERED

Following crops guidance
- After the last application only sugar beet or mangels may be sown within 4 mth; cereals may be sown after 16 wk. Land should be mouldboard ploughed to 15 cm and thoroughly cultivated before any succeeding crop

Environmental safety
- Harmful to fish or other aquatic life. Do not contaminate surface waters or ditches with chemical or used container

Hazard classification and safety precautions
Hazard Irritant, Harmful if swallowed
Transport code 9
Packaging group III
UN Number 3082
Risk phrases H317
Operator protection A, C, H; U08, U14, U15, U20a
Environmental protection E13c, E34, H410
Storage and disposal D09a, D10b

70   chloridazon + quinmerac

A herbicide mixture for use in beet crops
HRAC mode of action code: C1 + O

See also quinmerac

Products
Fiesta T	BASF	360:60 g/l	SC	17153

Uses
- Annual dicotyledons in **sugar beet**
- Annual meadow grass in **sugar beet**
- Chickweed in **sugar beet**
- Cleavers in **sugar beet**
- Common field speedwell in **sugar beet**
- Poppies in **sugar beet**
- Small nettle in **sugar beet**

Approval information
- Chloridazon and quinmerac included in Annex I under EC Regulation 1107/2009

Efficacy guidance
- Best results obtained when adequate soil moisture is present at application and afterwards to form an active herbicidal layer in the soil
- A programme of pre-emergence treatment followed by post-emergence application(s) in mixture with a contact herbicide optimises weed control and is essential for some difficult weed species such as cleavers
- Treatment pre-emergence only may not provide sufficient residual activity to give season-long weed control
- Effectiveness may be reduced under conditions of low pH
- Always follow WRAG guidelines for preventing and managing herbicide resistant weeds. See Section 5 for more information

Restrictions
- Maximum total dose on sugar beet equivalent to one treatment pre-emergence plus two treatments post-emergence, all at maximum individual dose; maximum total dose on fodder beet and mangels equivalent to one full dose pre-emergence treatment
- Heavy rain shortly after treatment may check crop growth particularly when it leaves water standing in surface depressions
- Treatment of stressed crops or those growing in unfavourable conditions may depress crop vigour and possibly reduce stand

FOR FULL CONDITIONS OF USE ALWAYS READ THE PRODUCT LABEL

- Do not use on Sands or soils of high organic matter content
- Where rates of nitrogen higher than those generally recommended are considered necessary, apply at least 3 wk before drilling
- A maximum total dose of 2.6 kg/ha chloridazon may only be applied every third year on the same field

Crop-specific information
- Latest use: before plants meet between the rows for sugar beet; pre-emergence for fodder beet, mangels

Following crops guidance
- In the spring following normal harvest of a treated crop sow only cereals, beet or mangel crops, potatoes or field beans, after ploughing
- In the event of failure of a treated crop only a beet crop may be drilled, after cultivation

Environmental safety
- Dangerous for the environment
- Toxic to aquatic organisms
- To reduce movement to groundwater do not apply to dry soil or when heavy rain is forecast

Hazard classification and safety precautions
Hazard Irritant, Dangerous for the environment
UN Number N/C
Risk phrases H317
Operator protection A, H; U05a, U08, U14, U19a, U20b
Environmental protection E15a, E38
Storage and disposal D01, D02, D08, D09a, D10c, D12a
Medical advice M03, M05a

71 chlormequat

A plant-growth regulator for reducing stem growth and lodging

Products

1	3C Chlormequat 750	BASF	750 g/l	SL	16690
2	Adjust	Taminco	620 g/l	SL	17141
3	Barleyquat B	Taminco	620 g/l	SL	17204
4	BEC CCC 750	Becesane	750 g/l	SL	17930
5	Belcocel	Taminco	720 g/l	SL	16652
6	Bettequat B	Taminco	620 g/l	SL	17208
7	CCC 750	Harvest	750 g/l	SL	17753
8	CleanCrop Transist	Agrii	720 g/l	SL	16684
9	Jadex Plus	Clayton	1045 g/l	SL	17524
10	Jadex-o-720	SFP Europe	720 g/l	SL	16284
11	K2	Taminco	620 g/l	SL	17206
12	Manipulator	Taminco	620 g/l	SL	17207
13	Palermo	SFP Europe	720 g/l	SL	17900
14	Selon	Taminco	620 g/l	SL	17209
15	Stabilan 750	Nufarm UK	750 g/l	SL	09303
16	Stefes CCC 720	Stefes	720 g/l	SL	17731

Uses
- Growth regulation in *ornamental plant production* (off-label), *ornamental specimens* (off-label) [15]; *rye* [16]; *spring barley* [2, 3, 6, 11, 12, 14, 16]; *spring oats, spring wheat, winter barley, winter oats* [2, 3, 5, 6, 8, 9, 11-14, 16]; *triticale* [4, 7, 15, 16]; *winter wheat* [2, 3, 5, 6, 8-14, 16]
- Lodging control in *oats* [1]; *rye* [1, 16]; *spring barley* [1-3, 6, 11, 12, 14, 16]; *spring oats, winter oats* [2-9, 11-16]; *spring wheat, winter barley* [1-9, 11-16]; *triticale* [1, 4, 7, 15, 16]; *winter rye* [4, 7, 15]; *winter wheat* [1-16]

SEE SECTION 3 FOR PRODUCTS ALSO REGISTERED

SECTION 2

Extension of Authorisation for Minor Use (EAMUs)
- *ornamental plant production* 20170910 [15]
- *ornamental specimens* 20171416 [15]

Approval information
- Chlormequat included in Annex 1 under EC Regulation 1107/2009
- Accepted by BBPA for use on malting barley

Efficacy guidance
- Most effective results on cereals normally achieved from application from Apr onwards, on wheat and rye from leaf sheath erect to first node detectable (GS 30-31), on oats at second node detectable (GS 32), on winter barley from mid-tillering to leaf sheath erect (GS 25-30). However, recommendations vary with product. See label for details
- Influence on growth varies with crop and growth stage. Risk of lodging reduced by application at early stem extension. Root development and yield can be improved by earlier treatment
- Results on barley can be variable
- In tank mixes with other pesticides on cereals optimum timing for herbicide action may differ from that for growth reduction. See label for details of tank mix recommendations
- Most products recommended for use on oats require addition of approved non-ionic wetter. Check label
- Some products are formulated with trace elements to help compensate for increased demand during rapid growth

Restrictions
- Maximum number of treatments or maximum total dose varies with crop and product and whether split dose treatments are recommended. Check labels
- Do not use on very late sown spring wheat or oats or on crops under stress
- Mixtures with liquid nitrogen fertilizers may cause scorch and are specifically excluded on some labels
- Do not use on soils of low fertility unless such crops regularly receive adequate dressings of nitrogen
- At least 6 h, preferably 24 h, required before rain for maximum effectiveness. Do not apply to wet crops
- Check labels for tank mixtures known to be incompatible
- Must only be applied between growth stages BBCH 31 to BBCH 39

Crop-specific information
- Latest use varies with crop and product. See label for details
- May be used on cereals undersown with grass or clovers
- Ornamentals to be treated must be well established and growing vigorously. Do not treat in strong sunlight or when temperatures are likely to fall below 10°C
- Temporary yellow spotting may occur on poinsettias. It can be minimised by use of a non-ionic wetting agent - see label

Environmental safety
- Wash equipment thoroughly with water and wetting agent immediately after use and spray out. Spray out again before storing or using for another product. Traces can cause harm to susceptible crops sprayed later
- Do not use straw from treated cereals as horticultural growth medium or mulch

Hazard classification and safety precautions
Hazard Harmful [1, 4, 5, 7, 8, 15, 16]; Harmful if swallowed [1-16]
Transport code 8
Packaging group III
UN Number 1760
Risk phrases H290 [1-3, 5-7, 9, 11-14, 16]; H312 [1, 4, 5, 7-10, 13, 15, 16]
Operator protection A [1-16]; C [9, 10, 13]; H [5, 8-10, 13, 16]; U05a, U20b [1-16]; U08 [1-3, 5, 6, 8, 11, 12, 14, 16]; U09a [4, 7, 15]; U13 [9, 10, 13]; U19a [1-8, 11, 12, 14-16]
Environmental protection E15a [1-8, 11, 12, 14-16]; E15b, E40e [9, 10, 13]; E34 [1-16]; H411 [9, 10, 16]; H412 [2-4, 6, 7, 11, 12, 14, 15]

FOR FULL CONDITIONS OF USE ALWAYS READ THE PRODUCT LABEL

Storage and disposal D01, D02 [1-16]; D05 [1, 4, 7, 15]; D09a [2-16]; D10a [4, 7, 15]; D10b [1-3, 5, 6, 8-14, 16]
Medical advice M03 [1-16]; M05a [4, 5, 7, 8, 15, 16]

72   chlormequat + 2-chloroethylphosphonic acid

A plant growth regulator for use in cereals

See also ethephon

Products

1	BOGOTA UK	SFP Europe	305:155 g/l	SL	16318
2	Chlormephon UK	Clayton	305:155 g/l	SL	16215
3	Ormet Plus	SFP Europe	305:150 g/l	SL	17816
4	Socom	Clayton	305:155 g/l	SL	17155
5	Spatial Plus	Interfarm	300:150 g/l	SL	16665
6	Vivax	Interfarm	300:150 g/l	SL	16682

Uses
- Lodging control in *spring barley, winter barley, winter wheat*

Approval information
- Chlormequat and 2-chloroethylphosphonic acid (ethephon) included in Annex I under EC Regulation 1107/2009
- Accepted by BBPA for use on malting barley
- All products containing 2-chloroethylphosphonic acid carry the warning: '2-chloroethylphosphonic acid is an anticholinesterase organophosphate. Handle with care'.

Efficacy guidance
- Best results obtained when crops growing vigorously
- Recommended dose varies with growth stage. See labels for details and recommendations for use of sequential treatments

Restrictions
- 2-chloroethylphosphonic acid is an anticholinesterase organophosphorus compound. Do not use if under medical advice not to work with such compounds
- Maximum number of treatments 1 per crop; maximum total dose equivalent to one full dose treatment
- Product must always be used with specified non-ionic wetter - see labels
- Do not use on any crop in sequence with any other product containing 2-chloroethylphosphonic acid
- Do not spray when crop wet or rain imminent
- Do not spray during cold weather or periods of night frost, when soil is very dry, when crop diseased or suffering pest damage, nutrient deficiency or herbicide stress
- If used on seed crops grown for certification inform seed merchant beforehand
- Do not use on wheat variety Moulin or on any winter varieties sown in spring
- Do not use on spring barley variety Triumph
- Do not treat barley on soils with more than 10% organic matter
- Do not use in programme with any other product containing 2-chloroethylphosphonic acid
- Only crops growing under conditions of high fertility should be treated

Crop-specific information
- Latest use: before flag leaf ligule/collar just visible (GS 39) or 1st spikelet visible (GS 51) for wheat or barley at top dose; or before flag leaf sheath opening (GS 47) for winter wheat at reduced dose
- Apply before lodging has started

Environmental safety
- Harmful to fish or other aquatic life. Do not contaminate surface waters or ditches with chemical or used container
- Do not use straw from treated cereals as a horticultural growth medium or as a mulch

SEE SECTION 3 FOR PRODUCTS ALSO REGISTERED

Hazard classification and safety precautions

    Hazard Harmful, Harmful if swallowed

    Transport code 8

    Packaging group III

    UN Number 3265

    Risk phrases H335

    Operator protection A; U05a, U08, U19a, U20b [1-6]; U14 [1-4]

    Environmental protection E15a, E34, E38 [1-6]; H411 [3]; H412 [5, 6]

    Storage and disposal D01, D02, D09a, D10b [1-6]; D12a [1-4]

    Medical advice M01, M03, M05a

73   chlormequat + imazaquin

A plant growth regulator mixture for winter wheat

See also imazaquin

Products

1 Meteor	BASF	368:0.8 g/l	SL	16800
2 Upright	BASF	368:0.8 g/l	SL	16884

Uses

- Increasing yield in *winter wheat* [1, 2]
- Lodging control in *rye (off-label)*, *triticale (off-label)* [1]; *winter wheat* [1, 2]

Extension of Authorisation for Minor Use (EAMUs)

- *rye* 20142647 [1]
- *triticale* 20142647 [1]

Approval information

- Chlormequat and imazaquin included in Annex I under EC Regulation 1107/2009

Efficacy guidance

- Apply to crops during good growing conditions or to those at risk from lodging
- On soils of low fertility, best results obtained where adequate nitrogen fertilizer used

Restrictions

- Maximum number of applications 1 per crop (2 per crop at split dose)
- Do not treat durum wheat
- Do not apply to undersown crops
- Do not apply when crop wet or rain imminent

Crop-specific information

- Latest use: before second node detectable (GS 31)
- Apply as single dose from leaf sheath lengthening up to and including 1st node detectable or as split dose, the first from tillers formed to leaf sheath lengthening, the second from leaf sheath erect up to and including 1st node detectable

Environmental safety

- Dangerous for the environment
- Toxic to aquatic organisms
- Do not use straw from treated cereals as horticultural growth medium or mulch

Hazard classification and safety precautions

    Hazard Harmful, Dangerous for the environment, Toxic if swallowed, Harmful if inhaled

    Transport code 8

    Packaging group III

    UN Number 1760

    Risk phrases H290, H318

    Operator protection A, C; U05a, U08, U11, U13, U15, U19a, U20b

    Environmental protection E15a, E34, E38, H412

    Storage and disposal D01, D02, D05, D06c, D09a, D10b, D12a

    Medical advice M03, M05a

FOR FULL CONDITIONS OF USE ALWAYS READ THE PRODUCT LABEL

74　chlorothalonil

A protectant chlorophenyl fungicide for use in many crops and turf
FRAC mode of action code: M5

See also azoxystrobin + chlorothalonil
carbendazim + chlorothalonil

Products

1	Abringo	Belcrop	500 g/l	SC	15607
2	Barclay Avoca	Barclay	500 g/l	SC	15549
3	Bravo 500	Syngenta	500 g/l	SC	14548
4	Clayton Turret	Clayton	500 g/l	SC	17220
5	Damocles	AgChem Access	500 g/l	SC	15689
6	Joules	Nufarm UK	500 g/l	SC	15730
7	Juliet	ChemSource	500 g/l	SC	15454
8	Life Scientific Chlorothalonil	Life Scientific	500 g/l	SC	15369
9	Multi-Star 500	Albaugh UK	500 g/l	SC	16784
10	Ovarb	Life Scientific	500 g/l	SC	17972
11	Piper	Nufarm UK	500 g/l	SC	15786
12	Premium 500	Agroquimicos	500 g/l	SC	16137
13	Pronto	Stefes	500 g/l	SC	17688
14	Resort	Globachem	500 g/l	SC	17861
15	Rover 500	Sipcam	500 g/l	SC	15496
16	Spirodor	Belchim	500 g/l	SC	17710
17	Stefonil	Stefes	500 g/l	SC	16862
18	Supreme	Mitsui	500 g/l	SC	15395
19	Synergy Chlorothalonil	Synergy	500 g/l	SC	17215
20	UPL Chlorothalonil	UPL Europe	500 g/l	SC	15472

Uses

- Basal stem rot in **narcissi** *(off-label)* [3, 8]
- Botrytis in **celeriac** *(off-label)*, **strawberries** *(off-label)* [3]
- Disease control in **all edible seed crops grown outdoors** *(off-label)*, **all non-edible seed crops grown outdoors** *(off-label)*, **durum wheat** *(off-label)*, **forest nurseries** *(off-label)*, **ornamental plant production** *(off-label)*, **rye** *(off-label)*, **triticale** *(off-label)* [3, 8]; **spring field beans**, **winter field beans** [15]; **spring wheat**, **winter wheat** [9]
- Downy mildew in **poppies for morphine production** *(off-label)* [3]
- Glume blotch in **spring wheat**, **winter wheat** [1-8, 10-20]
- Leaf spot in **celeriac** *(off-label)*, **strawberries** *(off-label)* [3]
- Neck rot in **narcissi** *(off-label)* [3, 8]
- Rhynchosporium in **spring barley** *(moderate control)*, **winter barley** *(moderate control)* [1-20]
- Septoria leaf blotch in **spring wheat**, **winter wheat** [1-20]

Extension of Authorisation for Minor Use (EAMUs)

- **all edible seed crops grown outdoors** *20111130* [3], *20120380* [8]
- **all non-edible seed crops grown outdoors** *20111130* [3], *20120380* [8]
- **celeriac** *20110948* [3]
- **durum wheat** *20110944* [3], *20120381* [8]
- **forest nurseries** *20111130* [3], *20120380* [8]
- **narcissi** *20110943* [3], *20120379* [8]
- **ornamental plant production** *20111130* [3], *20120380* [8]
- **poppies for morphine production** *20110947* [3]
- **rye** *20110944* [3], *20120381* [8]
- **strawberries** *20110949* [3]
- **triticale** *20110944* [3], *20120381* [8]

Approval information

- Chlorothalonil included in Annex I under EC Regulation 1107/2009.

SEE SECTION 3 FOR PRODUCTS ALSO REGISTERED

- Accepted by BBPA for use on malting barley and hops
- Approval expiry 30 Apr 2018 [18]

Efficacy guidance
- For some crops products differ in diseases listed as controlled. See label for details and for application rates, timing and number of sprays
- Apply as protective spray or as soon as disease appears and repeat as directed
- On cereals activity against Septoria may be reduced where serious mildew or rust present. In such conditions mix with suitable mildew or rust fungicide
- May be used for preventive and curative treatment of turf but treatment at a late stage of disease development may not be so successful and can leave bare patches of soil requiring renovation.

Restrictions
- Maximum number of treatments and maximum total doses vary with crop and product - see labels for details
- Operators of vehicle mounted equipment must use a vehicle fitted with a cab and a forced air filtration unit with a pesticide filter complying with HSE Guidance Note PM74 or to an equivalent or higher standard when making broadcast or air-assisted applications

Crop-specific information
- Latest time of application to cereals varies with product. Consult label
- Latest use for winter oilseed rape before flowering; end Aug in yr of harvest for blackcurrants, gooseberries, redcurrants,
- HI 8 wk for field beans; 6 wk for combining peas; 28 d or before 31 Aug for post-harvest treatment for blackcurrants, gooseberries, redcurrants; 14 d for onions, strawberries; 10 d for hops; 7-14 d for broccoli, Brussels sprouts, cabbages, cauliflowers, celery, onions, potatoes; 3 d for cane fruit; 12-48 h for protected cucumbers, protected tomatoes
- For Botrytis control in strawberries important to start spraying early in flowering period and repeat at least 3 times at 10 d intervals
- On strawberries some scorching of calyx may occur with protected crops

Environmental safety
- Dangerous for the environment
- Very toxic to aquatic organisms
- Do not spray from the air within 250 m horizontal distance of surface waters or ditches
- Broadcast air-assisted LERAP [1-8, 10, 11, 13, 14, 16-19] (18 m); LERAP Category B [1-20]

Hazard classification and safety precautions
    Hazard Harmful, Dangerous for the environment [1-20]; Harmful if swallowed [6]; Fatal if inhaled [11]; Harmful if inhaled [1-4, 6, 8-10, 13-20]; Very toxic to aquatic organisms [2, 3, 9, 13, 17, 19]
    Transport code 9
    Packaging group III
    UN Number 3082
    Risk phrases H315 [9, 15]; H317 [1-4, 6, 8-11, 13-20]; H318 [6, 11]; H319 [2-4, 8-10, 13, 14, 16, 17, 19]; H320 [1, 18, 20]; H335 [1-4, 6, 8, 10, 11, 13-20]; H351 [1-4, 8-11, 13-20]; H373 [15]; R20, R37, R40, R43, R50, R53a [5, 7, 12]; R36 [5, 7]; R38, R41 [12]
    Operator protection A, H [1-20]; C [1, 3-20]; J, M [1, 3-5, 7, 8, 10, 14, 16, 17, 19, 20]; U02a, U05a, U09a [1-8, 10-20]; U11 [3, 4, 9, 10, 12-17, 19]; U13, U14, U15, U20b [12, 15]; U19a [1-20]; U20a [1-8, 10, 11, 13, 14, 16-20]
    Environmental protection E15a [1-8, 10-20]; E16a, E38 [1-20]; E16b [1-8, 10, 11, 13, 14, 16-19]; E17b [1-8, 10, 11, 13, 14, 16-19] (18 m); E18 [1-8, 10, 11, 13, 14, 16-20]; H410 [1-4, 6, 8-11, 13-20]
    Storage and disposal D01, D02, D09a [1-8, 10-20]; D05, D10c [1-8, 10, 11, 13, 14, 16-20]; D10b [12, 15]; D12a [1-20]
    Medical advice M03 [2]

FOR FULL CONDITIONS OF USE ALWAYS READ THE PRODUCT LABEL

75   chlorothalonil + cymoxanil

A protectant and systemic fungicide for potato blight
FRAC mode of action code: M5 + 27

See also cymoxanil

Products

Mixanil	Sipcam	375:50 g/l	SC	15675

Uses
• Blight in *potatoes*

Approval information
• Cymoxanil and chlorothalonil included in Annex I under EC Regulation 1107/2009

Environmental safety
• LERAP Category B

Hazard classification and safety precautions
Hazard Harmful, Dangerous for the environment, Harmful if inhaled
Transport code 9
Packaging group III
UN Number 3082
Risk phrases H317, H335, H351
Operator protection A, H; U04a, U05b, U08, U20b
Environmental protection E15b, E16a, E16b, E38, H410
Storage and disposal D01, D02, D09a, D12a, D19
Medical advice M04a, M05a

76   chlorothalonil + cyproconazole

A systemic protectant and curative fungicide mixture for cereals, combining peas and some
products can also be used in field beans
FRAC mode of action code: M5 + 3

See also cyproconazole

Products

1	Alto Elite	Syngenta	375:40 g/l	SC	08467
2	Belimar	AgChem Access	375:40 g/l	SC	18090
3	Octolan	Syngenta	375:40 g/l	SC	11675
4	SAN 703	Syngenta	375:40 g/l	SC	11676

Uses
• Brown rust in *spring barley*, *winter barley*, *winter wheat* [1-4]
• Chocolate spot in *spring field beans*, *winter field beans* [1-4]
• Eyespot in *winter barley*, *winter wheat* [1-4]
• Glume blotch in *winter wheat* [1, 2]
• Grey mould in *combining peas* [1-4]
• Net blotch in *spring barley*, *winter barley* [1-4]
• Powdery mildew in *spring barley*, *winter barley*, *winter wheat* [1-4]
• Rhynchosporium in *spring barley*, *winter barley* [1-4]
• Rust in *spring field beans*, *winter field beans* [1, 2]
• Septoria leaf blotch in *winter wheat* [1-4]
• Yellow rust in *spring barley*, *winter barley*, *winter wheat* [1-4]

Approval information
• Chlorothalonil and cyproconazole included in Annex I under EC Regulation 1107/2009
• Accepted by BBPA for use on malting barley

SEE SECTION 3 FOR PRODUCTS ALSO REGISTERED

SECTION 2

Efficacy guidance

- Apply at first signs of infection or as soon as disease becomes active
- A repeat application may be made if re-infection occurs
- For established mildew tank-mix with an approved mildewicide
- When applied prior to third node detectable (GS 33) a useful reduction of eyespot will be obtained
- Cyproconazole is a DMI fungicide. Resistance to some DMI fungicides has been identified in Septoria leaf blotch which may seriously affect performance of some products. For further advice contact a specialist advisor and visit the Fungicide Resistance Action Group (FRAG)-UK website

Restrictions

- Maximum total dose equivalent to 2 full dose treatments
- Do not apply at concentrations higher than recommended

Crop-specific information

- Latest use: before beginning of anthesis (GS 60) for barley; before caryopsis watery ripe (GS 71) for wheat
- HI peas, field beans 6 wk
- If applied to winter wheat in spring at GS 30-33 straw shortening may occur but yield is not reduced

Environmental safety

- Dangerous for the environment
- Very toxic to aquatic organisms
- LERAP Category B

Hazard classification and safety precautions

Hazard Harmful, Dangerous for the environment [1-4]; Harmful if inhaled, Very toxic to aquatic organisms [1, 3, 4]

Transport code 9

Packaging group III

UN Number 3082

Risk phrases H317, H318, H335, H351, H361 [1, 3, 4]; R20, R37, R40, R41, R43, R50, R53a [2]

Operator protection A, C, H, M; U05a, U19a [1-4]; U11, U20a [1, 2]; U12, U20b [3, 4]

Environmental protection E15a, E16b [1, 2]; E15b [3, 4]; E16a, E38 [1-4]; H410 [1, 3, 4]

Storage and disposal D01, D02, D09a, D10c, D12a [1-4]; D05 [1, 2]

Medical advice M04a

77   chlorothalonil + cyproconazole + propiconazole

A broad-spectrum fungicide mixture for cereals
FRAC mode of action code: M5 + 3 + 3

See also cyproconazole
*   propiconazole*

Products

1 Apache	Syngenta	375:50:62.5 g/l	SE	14255
2 Cherokee	Syngenta	375:50:62.5 g/l	SE	13251

Uses

- Brown rust in **spring barley, winter barley, winter wheat**
- Net blotch in **spring barley** (reduction), **winter barley** (reduction)
- Powdery mildew in **spring barley, winter barley, winter wheat** (moderate control)
- Rhynchosporium in **spring barley** (reduction), **winter barley** (reduction)
- Septoria leaf blotch in **winter wheat**
- Yellow rust in **winter wheat**

Approval information
- Chlorothalonil, propiconazole and cyproconazole included in Annex I under EC Regulation 1107/2009

Efficacy guidance
- Best results obtained from treatment at the early stages of disease development
- Cyproconazole and propiconazole are DMI fungicides. Resistance to some DMI fungicides has been identified in Septoria leaf blotch which may seriously affect performance of some products. For further advice contact a specialist advisor and visit the Fungicide Resistance Action Group (FRAG)-UK website
- Product should be used as part of an Integrated Crop Management strategy incorporating other methods of control and including, where appropriate, other fungicides with a different mode of action
- Product should be used preventatively and not relied on for its curative potential
- Always follow FRAG guidelines for resistance management. See Section 5 for more information

Restrictions
- Maximum total dose equivalent to two full dose treatments on wheat and barley
- Do not treat crops under stress
- Spray solution must be used as soon as possible after mixing

Crop-specific information
- Latest use: before caryopsis watery ripe (GS 71) for winter wheat; before first spikelet just visible (GS 51) for barley
- When applied to winter wheat in the spring straw shortening may occur which causes no loss of yield

Environmental safety
- Dangerous for the environment
- Very toxic to aquatic organisms
- LERAP Category B

Hazard classification and safety precautions
Hazard Harmful, Dangerous for the environment, Harmful if inhaled, Very toxic to aquatic organisms
Transport code 9
Packaging group III
UN Number 3082
Risk phrases H317, H319, H335, H351, H361
Operator protection A, C, H; U02a, U05a, U09a, U10, U11, U14, U15, U19a, U20b
Environmental protection E15b, E16a, E16b, E34, E38, H410
Storage and disposal D01, D02, D05, D07, D09a, D10c, D12a
Medical advice M04a

78 chlorothalonil + fludioxonil + propiconazole

A fungicide treatment for amenity grassland
FRAC mode of action code: M5 + 12 + 3

See also fludioxonil
propiconazole

Products

Instrata	Syngenta	362:14.5:56.9 g/l	SE	16458

Uses
- Anthracnose in **amenity grassland** *(reduction)*, **managed amenity turf** *(reduction)*
- Brown patch in **amenity grassland**, **managed amenity turf**
- Dollar spot in **amenity grassland**, **managed amenity turf**
- Fusarium patch in **amenity grassland**, **managed amenity turf**

SEE SECTION 3 FOR PRODUCTS ALSO REGISTERED

SECTION 2

Approval information
- Chlorothalonil, fludioxonil and propiconazole included in Annex I under EC Regulation 1107/2009

Environmental safety
- Buffer zone requirement 12m [1]
- LERAP Category B

Hazard classification and safety precautions
  Hazard Toxic, Dangerous for the environment, Very toxic to aquatic organisms
  Transport code 9
  Packaging group III
  UN Number 3082
  Risk phrases R23a, R36, R37, R40, R43
  Operator protection A, C, H; U02a, U05a, U09a, U10, U11, U19a, U19c, U20b
  Environmental protection E15b, E16a, E34, E38, H410
  Storage and disposal D01, D02, D09a, D12a
  Medical advice M04a

79   chlorothalonil + mancozeb

A multi-site protectant fungicide mixture for disease control in winter wheat
FRAC mode of action code: M5 + M3

See also mancozeb

Products

Guru	Interfarm	286:194 g/l	SC	15268

Uses
- Glume blotch in **winter wheat**
- Septoria leaf blotch in **winter wheat**

Approval information
- Chlorothalonil and mancozeb included in Annex I under EC Regulation 1107/2009

Efficacy guidance
- Best results on winter wheat achieved from protective applications to the flag leaf. If disease is already present on the lower leaves, treat as soon as the flag leaf is just visible (GS 37).
- Activity on Septoria may be reduced in the presence of severe mildew infection.

Restrictions
- Maximum total dose on winter wheat equivalent to 1.5 x full dose treatment.
- Do not treat crops under stress from frost, drought, water-logging, trace element deficiency or pest attack.
- Broadcast air assisted applications must only be made by equipment fitted with a cab with a forced air filtration unit plus a pesticide filter complying with HSE Guidance Note PM 74 or an equivalent or higher standard

Crop-specific information
- Latest use: before grain watery ripe stage (GS 71)

Environmental safety
- Dangerous for the environment
- Very toxic to aquatic organisms
- LERAP Category B

Hazard classification and safety precautions
  Hazard Irritant, Dangerous for the environment, Fatal if inhaled, Very toxic to aquatic organisms
  Transport code 9
  Packaging group III
  UN Number 3082
  Risk phrases H317, H318, H335, H351, H361

FOR FULL CONDITIONS OF USE ALWAYS READ THE PRODUCT LABEL

Operator protection A, C, H, M; U02a, U05a, U08, U10, U11, U13, U14, U19a, U20b
Environmental protection E16a, E16b, E38, H410
Storage and disposal D01, D02, D05, D09a, D10b, D12a

80   chlorothalonil + metalaxyl-M

A systemic and protectant fungicide for various crops
FRAC mode of action code: M5 + 4

See also metalaxyl-M

Products

1	Broadsheet	AgChem Access	500:37.5 g/l	SC	15906
2	Folio Gold	Syngenta	500:37.5 g/l	SC	14368

Uses
- Alternaria in **brussels sprouts** *(moderate control)*, **cauliflowers** *(moderate control)* [1, 2]
- Botrytis in **gooseberries** *(off-label)* [2]
- Disease control in **forest nurseries** *(off-label)*, **ornamental plant production** *(off-label)*, **poppies for morphine production** *(off-label)*, **protected ornamentals** *(off-label)* [2]
- Downy mildew in **broad beans**, **brussels sprouts**, **cauliflowers** [1, 2]
- Drepanopeziza ribis in **gooseberries** *(off-label)* [2]
- Leaf spot in **gooseberries** *(off-label)* [2]
- Ring spot in **brussels sprouts** *(reduction)*, **cauliflowers** *(reduction)* [1, 2]
- White blister in **brussels sprouts**, **cauliflowers** [1, 2]

Extension of Authorisation for Minor Use (EAMUs)
- *forest nurseries 20111128* [2]
- *gooseberries 20161257* [2]
- *ornamental plant production 20111128* [2]
- *poppies for morphine production 20090251* [2]
- *protected ornamentals 20120032* [2]

Approval information
- Chlorothalonil and metalaxyl-M included in Annex I under EC Regulation 1107/2009

Efficacy guidance
- Apply at first signs of disease or when weather conditions favourable for disease pressure
- Best results obtained when used in a full and well-timed programme. Repeat treatment at 14-21 d intervals if necessary
- Treatment of established disease will be less effective
- Evidence of effectiveness in bulb onions, shallots and leeks is limited

Restrictions
- Maximum total dose equivalent to 2 full doses on broad beans, field beans, cauliflowers, calabrese; 3 full doses on Brussels sprouts, leeks, onions and shallots

Crop-specific information
- HI 14 d for all crops

Environmental safety
- Dangerous for the environment
- Very toxic to aquatic organisms
- Buffer zone requirement 20 m in gooseberries [2]
- LERAP Category B

Hazard classification and safety precautions
Hazard Harmful, Dangerous for the environment [1, 2]; Harmful if inhaled, Very toxic to aquatic organisms [2]
Transport code 9
Packaging group III
UN Number 3082

SEE SECTION 3 FOR PRODUCTS ALSO REGISTERED

Risk phrases H315, H317, H319, H335, H351 [2]; R20, R36, R37, R38, R40, R43, R50, R53a [1]
Operator protection A, C, H; U05a, U09a, U15, U20a
Environmental protection E15a, E16a, E38 [1, 2]; H410 [2]
Consumer protection C02a (14 d)
Storage and disposal D01, D02, D05, D09a, D10c, D12a

81   chlorothalonil + penthiopyrad

A protectant and systemic fungicide mixture for disease control in wheat
FRAC mode of action code: 3 + 7

See also penthiopyrad

Products

1	Aylora	DuPont	250:100 g/l	SL	17905
2	Intellis Plus	DuPont	252.5:100.5 g/l	SL	17908
3	Treoris	DuPont	252.5:100.5 g/l	SL	17849
4	Trust	Generica	250:100 g/l	SL	16561

Uses

- Brown rust in **spring wheat, winter wheat** [1-4]
- Disease control in **spring barley, spring wheat, winter barley, winter wheat** [1, 2]
- Net blotch in **spring barley, winter barley** [1-4]
- Ramularia leaf spots in **spring barley, winter barley** [1-4]
- Rhynchosporium in **spring barley, winter barley** [1-4]
- Septoria leaf blotch in **spring wheat, winter wheat** [1-4]

Approval information

- Chlorothalonil and penthiopyrad are included in Annex 1 under EC Regulation 1107/2009

Restrictions

- To reduce the risk of resistance, only two applications of any SDHI fungicide may be made to a cereal crop.
- A maximum total dose of not more than 500 g/ha penthiopyrad may be applied in a two-year period to the same field.
- Do not apply by hand-held equipment

Environmental safety

- Buffer zone requirement 6 m [4]
- Buffer zone requirement 12 m [1-3]
- Low drift spraying equipment up to 30m from the top of the bank of any surface water bodies [1-4]
- To protect non target insects/arthropods respect an unsprayed buffer zone of 5 m to non-crop land [4]
- LERAP Category B

Hazard classification and safety precautions

Hazard Harmful, Dangerous for the environment [1, 3, 4]; Harmful if inhaled, Very toxic to aquatic organisms [1-3]
Transport code 9
Packaging group III
UN Number 3082
Risk phrases H317, H335, H351 [1-3]; R20, R37, R40b, R50, R53a, R70 [4]
Operator protection A, H; U02a, U05a, U09a, U11, U19a, U20a [1-4]; U09c [2]
Environmental protection E15b, E16a [1-4]; E34 [1, 3, 4]; E38 [1, 2]; H410 [1-3]
Storage and disposal D01, D02, D09a, D10c, D12a [1-4]; D06e, D12b [1, 2]
Medical advice M03a

82   chlorothalonil + propiconazole

A systemic and protectant fungicide for winter wheat and barley
FRAC mode of action code: M5 + 3

See also propiconazole

Products

Prairie	Syngenta	250:62.5 g/l	SC	13994

Uses
- Brown rust in **spring barley, spring wheat, winter barley, winter wheat**
- Glume blotch in **spring wheat, winter wheat**
- Rhynchosporium in **spring barley, winter barley**
- Septoria leaf blotch in **spring wheat, winter wheat**
- Yellow rust in **spring wheat, winter wheat**

Approval information
- Chlorothalonil and propiconazole included in Annex I under EC Regulation 1107/2009
- Accepted by BBPA for use on malting barley

Efficacy guidance
- On wheat apply from start of flag leaf emergence up to and including when ears just fully emerged (GS 37-59), on barley at any time to ears fully emerged (GS 59)
- Best results achieved from early treatment, especially if weather wet, or as soon as disease develops

Restrictions
- Maximum number of treatments 2 per crop or 1 per crop if other propiconazole based fungicide used in programme

Crop-specific information
- Latest use: up to and including emergence of ear just complete (GS 59).
- HI 42 d

Environmental safety
- Irritating to eyes, skin and respiratory system
- Risk of serious damage to eyes
- Dangerous to fish or other aquatic life. Do not contaminate surface waters or ditches with chemical or used container
- LERAP Category B

Hazard classification and safety precautions
Hazard Harmful, Dangerous for the environment, Very toxic to aquatic organisms
Transport code 9
Packaging group III
UN Number 3082
Risk phrases R20, R36, R37, R40, R43
Operator protection A, C, H; U02a, U05a, U11, U20b
Environmental protection E15b, E16a, E16b, E38, H410
Storage and disposal D01, D02, D09a, D10c, D12a

83   chlorothalonil + proquinazid

A chlorophenyl and quinazolinone fungicide mixture for powdery mildew control in cereals
FRAC mode of action code: M5 + U7

Products

Fielder SE	DuPont	500:25 g/l	SE	15120

Uses
- Powdery mildew in **spring barley, spring wheat, winter barley, winter wheat**

SEE SECTION 3 FOR PRODUCTS ALSO REGISTERED

SECTION 2

- Rhynchosporium in **spring barley** *(suppression only)*, **winter barley** *(suppression only)*
- Septoria leaf blotch in **spring wheat** *(suppression only)*, **winter wheat** *(suppression only)*

Approval information
- Chlorothalonil and proquinazid included in Annex 1 under EC Regulation 1107/2009

Efficacy guidance
- If disease has actively spread to new growth, application should be in tank-mixture with a fungicide that has an alternative mode of action with curative activity.
- If two applications are used consecutively in a crop, the second application should be in tank-mixture with a fungicide with an alternative mode of action against mildew, i.e. not a quinazolinone or quinoline fungicide.
- Should be considered as a member of the quinoline fungicides for resistance management.

Restrictions
- Maximum number of applications: 2 per crop.

Crop-specific information
- Latest use in wheat is before full flowering (GS65)
- Latest use in barley is before first awns visible

Environmental safety
- LERAP Category B

Hazard classification and safety precautions
  Hazard Harmful, Dangerous for the environment, Fatal if inhaled, Very toxic to aquatic organisms
  Transport code 9
  Packaging group III
  UN Number 3082
  Risk phrases H317, H318, H335, H351
  Operator protection A, C, H; U05a, U09a, U12, U14, U15, U20a
  Environmental protection E15a, E15b, E16a, E38, E40b, H410
  Storage and disposal D01, D02, D05, D09a, D10c, D12a
  Medical advice M05a

84   chlorothalonil + tebuconazole

A protectant and systemic fungicide mixture for winter wheat
FRAC mode of action code: M5 + 3

See also tebuconazole

Products

1	Cigal Plus	Rotam	375:125 g/l	SC	17229
2	Confucius	Rotam	375:125 g/l	SC	17257
3	Crafter	Nufarm UK	250:90 g/l	SC	15895
4	Fezan Plus	Sipcam	166:60 g/l	SC	15490
5	Pentangle	Nufarm UK	500:180 g/l	SC	16792
6	Timpani	Nufarm UK	250:90 g/l	SC	14651

Uses
- Brown rust in **spring wheat**, **winter wheat** [1, 2]
- Fusarium ear blight in **winter wheat** [4]
- Rhynchosporium in **winter barley** *(useful reduction)* [3, 6]
- Septoria leaf blotch in **spring wheat** [1, 2]; **winter wheat** [1-6]
- Sooty moulds in **winter wheat** [4]

Approval information
- Chlorothalonil and tebuconazole included in Annex I under EC Regulation 1107/2009

Efficacy guidance
- Latest time of application before anthesis (GS61)

FOR FULL CONDITIONS OF USE ALWAYS READ THE PRODUCT LABEL

- Maximum individual dose 1.0 l/ha. Where disease pressure is severe, a second application may be required but applications after flag leaf visible (GS39) may only result in a reduction of disease severity

Restrictions
- Newer authorisations for tebuconazole products require application to cereals only after GS 30 and applications to oilseed rape and linseed after GS20 - check label

Environmental safety
- Buffer zone requirement 8 m [5]
- Buffer zone requirement 6 m [1-3, 6]
- LERAP Category B

Hazard classification and safety precautions
Hazard Very toxic [1-3, 5, 6]; Harmful [4]; Dangerous for the environment [1-6]; Fatal if inhaled [3]; Harmful if inhaled [1, 2, 4-6]; Very toxic to aquatic organisms [1-3, 5]
Transport code 6.1 [5]; 9 [1-4, 6]
Packaging group II [5]; III [1-4, 6]
UN Number 2902 [5]; 3082 [1-4, 6]
Risk phrases H317, H335, H361 [1-3, 5, 6]; H318 [1, 2, 5]; H319 [3, 6]; H351 [1-6]; H360 [4]
Operator protection A, H [1-6]; C [1-3, 5, 6]; U02a, U09a, U10, U11, U13, U14, U15, U19b [1-3, 5, 6]; U05a, U19a [1-6]
Environmental protection E15a, E16b [1-3, 5, 6]; E15b [4]; E16a, E34, E38, H410 [1-6]
Storage and disposal D01, D02, D09a, D12a [1-6]; D05, D10c [4]; D10b [1-3, 5, 6]
Medical advice M03 [4]; M04a [1-3, 5, 6]

85   chlorotoluron

A contact and residual urea herbicide for cereals only available in mixtures
HRAC mode of action code: C2

86   chlorotoluron + diflufenican + pendimethalin

A herbicide mixture for weed control in cereals
HRAC mode of action code: C2 + F1 + K1

Products
Tower	Adama	250:40:300 g/l	SC	16586

Uses
- Annual meadow grass in *spring barley, spring wheat, triticale, winter barley, winter rye, winter wheat*
- Charlock in *spring barley, spring wheat, triticale, winter barley, winter rye, winter wheat*
- Chickweed in *spring barley, spring wheat, triticale, winter barley, winter rye, winter wheat*
- Cleavers in *spring barley, spring wheat, triticale, winter barley, winter rye, winter wheat*
- Crane's-bill in *spring barley, spring wheat, triticale, winter barley, winter rye, winter wheat*
- Dead nettle in *spring barley, spring wheat, triticale, winter barley, winter rye, winter wheat*
- Field pansy in *spring barley, spring wheat, triticale, winter barley, winter rye, winter wheat*
- Forget-me-not in *spring barley, spring wheat, triticale, winter barley, winter rye, winter wheat*
- Fumitory in *spring barley, spring wheat, triticale, winter barley, winter rye, winter wheat*
- Loose silky bent in *spring barley, spring wheat, triticale, winter barley, winter rye, winter wheat*
- Mayweeds in *spring barley, spring wheat, triticale, winter barley, winter rye, winter wheat*
- Penny cress in *spring barley, spring wheat, triticale, winter barley, winter rye, winter wheat*
- Poppies in *spring barley, spring wheat, triticale, winter barley, winter rye, winter wheat*
- Runch in *spring barley, spring wheat, triticale, winter barley, winter rye, winter wheat*

SEE SECTION 3 FOR PRODUCTS ALSO REGISTERED

- Shepherd's purse in *spring barley, spring wheat, triticale, winter barley, winter rye, winter wheat*
- Speedwells in *spring barley, spring wheat, triticale, winter barley, winter rye, winter wheat*
- Volunteer oilseed rape in *spring barley, spring wheat, triticale, winter barley, winter rye, winter wheat*

Approval information
- Chlorotoluron, diflufenican and pendimethalin included in Annex I under EC Regulation 1107/2009

Efficacy guidance
- Minimum water volume for full dose (2.0 l/ha) is 200 l/ha while reducing the dose to 1.0 l/ha allows a lowest water volume of 100 l/ha.

Restrictions
- Maximum number of treatments 1 per crop
- Use only on listed crop varieties when applying post-emergence to winter wheat. Do not apply to undersown crops
- Substances containing the chlorotoluron active agent may not be applied more than once per year on the same area
- Application must not be made after 31 October in the year of drilling
- Do not use if it is frosty
- DO NOT apply to soils with greater than 10% organic matter

Crop-specific information
- Following normal harvest, there are no restrictions. In the event of crop failure, plough before drilling the following crop in the spring. Only winter wheat can be re-drilled in the same autumn following crop failure

Environmental safety
- Buffer zone requirement 6 m [1]
- LERAP Category B

Hazard classification and safety precautions
Transport code 9
Packaging group III
UN Number 3082
Risk phrases H351, H361
Operator protection A; U02a, U04a, U20b
Environmental protection E15b, E16a, E34, E38, H410
Storage and disposal D01, D09a, D10b, D12a
Medical advice M03

87   chlorpropham

A residual carbamate herbicide and potato sprout suppressant
HRAC mode of action code: K2

Products

1	Aceto Chlorpropham 50M	Aceto	500 g/l	HN	14134
2	Aceto Sprout Nip Pellet	Aceto	100% w/w	HN	15397
3	Aceto Sprout Nip Ultra	Aceto	500 g/l	HN	15456
4	Cleancrop Amigo 2	UPL Europe	400 g/l	EC	15419
5	Gro-Stop 100	Certis	300 g/l	HN	14182
6	Gro-Stop Fog	Certis	300 g/l	HN	14183
7	Gro-Stop Ready	Certis	120 g/l	EW	15109
8	Gro-Stop Solid	Certis	100% w/w	BR	14103
9	Intruder	UPL Europe	400 g/l	EC	15076
10	Pro-Long	UPL Europe	500 g/l	HN	14389

FOR FULL CONDITIONS OF USE ALWAYS READ THE PRODUCT LABEL

Uses

- Annual dicotyledons in **baby leaf crops** [9]; **bulb onions, chard** *(off-label)*, **endives, herbs (see appendix 6)** *(off-label)*, **lettuce, ornamental plant production, salad onions, shallots, spinach** *(off-label)* [4, 9]; **forest nurseries** *(off-label)*, **leaf brassicas** *(off-label)*, **leeks** *(off-label)* [4]
- Annual grasses in **chard** *(off-label)*, **herbs (see appendix 6)** *(off-label)*, **spinach** *(off-label)* [9]
- Annual meadow grass in **baby leaf crops** [9]; **bulb onions, endives, lettuce, ornamental plant production, salad onions, shallots** [4, 9]; **chard** *(off-label)*, **forest nurseries** *(off-label)*, **herbs (see appendix 6)** *(off-label)*, **leaf brassicas** *(off-label)*, **spinach** *(off-label)* [4]
- Chickweed in **chard** *(off-label)*, **forest nurseries** *(off-label)*, **herbs (see appendix 6)** *(off-label)*, **leaf brassicas** *(off-label)*, **leeks** *(off-label)* [4]
- Polygonums in **chard** *(off-label)*, **forest nurseries** *(off-label)*, **herbs (see appendix 6)** *(off-label)*, **leaf brassicas** *(off-label)*, **leeks** *(off-label)* [4]
- Small nettle in **chard** *(off-label)*, **forest nurseries** *(off-label)*, **herbs (see appendix 6)** *(off-label)*, **leaf brassicas** *(off-label)*, **leeks** *(off-label)* [4]
- Sprout suppression in **potatoes** [1-3, 7]; **ware potatoes** [5]; **ware potatoes** *(thermal fog)* [6, 8, 10]
- Wild oats in **baby leaf crops** [9]; **bulb onions, endives, lettuce, ornamental plant production, salad onions, shallots** [4, 9]; **chard** *(off-label)*, **forest nurseries** *(off-label)*, **herbs (see appendix 6)** *(off-label)*, **leaf brassicas** *(off-label)*, **spinach** *(off-label)* [4]

Extension of Authorisation for Minor Use (EAMUs)

- **chard** *20120728* [4], *20110596* [9]
- **forest nurseries** *20120727* [4]
- **herbs (see appendix 6)** *20120728* [4], *20110596* [9]
- **leaf brassicas** *20120728* [4]
- **leeks** *20121841* [4]
- **spinach** *20120728* [4], *20110596* [9]

Approval information

- Chlorpropham included in Annex I under EC Regulation 1107/2009
- Some products are formulated for application by thermal fogging. See labels for details

Efficacy guidance

- For sprout suppression apply with suitable fogging or rotary atomiser equipment over dry tubers before sprouting commences. Repeat applications may be needed. See labels for details
- Best results on potatoes obtained in purpose-built box stores with suitable forced draft ventilation. Potatoes in bulk stores should not be stacked more than 3 m high. Positive ventilation systems are a requirement for use after 2017
- Blockage of air spaces between tubers prevents circulation of vapour and consequent loss of efficacy
- It is important to treat potatoes before the eyes open to obtain best results
- Effectiveness of fogging reduced in non-dedicated stores without proper insulation and temperature controls. Best results obtained at 5-10°C and 75-80% humidity
- When treating potatoes in store with CIPC best practice guidelines are available on-line at www. BeCIPCcompliant.co.uk - some potatoes have recently been found to exceed the MRL and if more are found then the registration of CIPC is in jeopardy.

Restrictions

- Only clean, mature, disease-free potatoes should be treated for sprout suppression. Use of chlorpropham can inhibit tuber wound healing and the severity of skin spot infection in store may be increased if damaged tubers are treated
- Do not fog potatoes with a high level of skin spot
- Do not use on potatoes for seed. Do not handle, dry or store seed potatoes or any other seed or bulbs in boxes or buildings in which potatoes are being or have been treated
- Do not remove treated potatoes from store for sale or processing for at least 21 d after application

Crop-specific information

- Cure potatoes according to label instructions before treatment and allow 2 days - 4 wks between completion of loading into store and first treatment - check labels.

SEE SECTION 3 FOR PRODUCTS ALSO REGISTERED

Following crops guidance
- There is a risk of damage to seed potatoes which are handled or stored in boxes or buildings previously treated with chlorpropham

Environmental safety
- Dangerous for the environment
- Toxic to aquatic organisms
- Keep unprotected persons out of treated stores for at least 24 h after application
- Keep in original container, tightly closed, in a safe place, under lock and key

Hazard classification and safety precautions
Hazard Harmful, Harmful if swallowed [4, 7, 9]; Dangerous for the environment [1-10]; Highly flammable liquid and vapour [1, 10]; Toxic if swallowed, Toxic in contact with skin, Toxic if inhaled [1]; Very toxic to aquatic organisms [7]

Transport code 6.1 [1, 5-7, 10]; 9 [3, 4, 8, 9]

Packaging group III [1, 3-10]

UN Number 2902 [1, 5-7, 10]; 3077 [3, 8]; 3082 [4, 9]

Risk phrases H314, H318 [7]; H315 [4-7, 9]; H317 [4, 7, 9]; H319 [4-6]; H320, H335 [9]; H336 [4, 9]; H351, H373 [1-10]; H370 [10]

Operator protection A, H [1-10]; C [1, 3, 4, 9, 10]; D, M [1, 3, 5, 6, 8, 10]; E [3, 5, 6, 8, 10]; J [3, 8, 10]; U01, U19b [8]; U02a [4-6, 8, 9]; U04a, U15, U20a [5, 6]; U05a [1, 3-6, 8-10]; U08 [1, 3, 5, 6, 8, 10]; U09a, U11, U16b [4, 9]; U14 [5-7]; U19a [1-6, 8-10]; U20b [1, 3, 4, 8-10]

Environmental protection E02a [1, 3, 10] (24 h); E15a [1, 3, 5, 6, 10]; E15b [4, 7-9]; E22c [4, 9]; E34 [1, 3-10]; E38 [1-4, 7-9]; H410 [7, 10]; H411 [1-4, 8, 9]; H412 [5, 6]

Consumer protection C02b, C12 [5, 6]; C02c [1, 3, 10]

Storage and disposal D01, D02 [1, 3-10]; D05 [1, 3, 5-8, 10]; D07 [4-6, 8, 9]; D09a, D10c [4, 9]; D09b [1, 3, 5, 6, 10]; D10a, D12b [1, 3, 10]; D11a [5-8]; D12a [2, 4-9]

Medical advice M03, M05b [4, 9]; M04a [1, 3, 5, 6, 8, 10]

---

88   chlorpyrifos

A contact and ingested organophosphorus insecticide with all uses except application to Brassica seedlings via gantry sprayers revoked March 2016
IRAC mode of action code: 1B

Products

1 Cyren	Headland	480 g/l	EC	17934
2 Dursban WG	Dow	75% w/w	WG	17932
3 Equity	Dow	480 g/l	EC	17933

Uses
- Cabbage root fly in *protected broccoli, protected brussels sprouts, protected cabbages, protected calabrese, protected cauliflowers*
- Insect pests in *protected broccoli, protected brussels sprouts, protected cabbages, protected calabrese, protected cauliflowers*

Approval information
- Chlorpyrifos included in Annex I under EC Regulation 1107/2009
- In 2006 CRD required that all products containing this active ingredient should carry the following warning in the main area of the container label: "Chlorpyrifos is an anticholinesterase organophosphate. Handle with care"
- In 2016 CRD revoked all uses except for treatment of brassica crops in peat blocks via gantry-mounted sprayers only.

Efficacy guidance
- Apply via gantry-mounted sprayer only.

Restrictions
- Contains an anticholinesterase organophosphate compound. Do not use if under medical advice not to work with such compounds

Environmental safety
- Dangerous for the environment
- Very toxic to aquatic organisms
- Flammable
- LERAP Category A

Hazard classification and safety precautions

Hazard Harmful, Dangerous for the environment [1-3]; Flammable, Flammable liquid and vapour [3]; Toxic if swallowed [1]; Harmful if swallowed [2, 3]; Harmful if inhaled [1, 3]

Transport code 6.1 [1, 3]; 9 [2]

Packaging group III

UN Number 3017 [3]; 3018 [1]; 3077 [2]

Risk phrases H304 [1]; H315, H319 [1, 3]; H335, H336 [3]

Operator protection A [1-3]; C, H [1]; P [3]; U05a [1, 3]; U08, U19a, U20b [1-3]; U11 [3]; U14 [1]

Environmental protection E06a [1-3] (14 d); E12a [1, 2]; E12c [3]; E12e, E16c, E16d, E34, H410 [1-3]; E13a [1]; E15a, E38 [2, 3]; E17a [1] (18 m)

Storage and disposal D01, D09a, D10b [1-3]; D02, D12a [2, 3]; D05, D12b [1]

Medical advice M01 [2, 3]; M03 [1-3]; M05b [1, 3]

89   chlorpyrifos-methyl

An organophosphorus insecticide and acaricide for grain store use
IRAC mode of action code: 1B

Products

1 Garrison	Pan Agriculture	225 g/l	EC	18174
2 Reldan 22	Dow	225 g/l	EC	16710

Uses
- Grain storage pests in **stored grain**
- Insect pests in **grain stores, stored grain**
- Mites in **grain stores, stored grain**
- Pre-harvest hygiene in **grain stores**

Approval information
- Chlorpyrifos-methyl included in Annex I under EC Regulation 1107/2009
- Accepted by BBPA for use in stores for malting barley
- In 2006 CRD required that all products containing this active ingredient should carry the following warning in the main area of the container label: "Chlorpyrifos-methyl is an anticholinesterase organophosphate. Handle with care"

Efficacy guidance
- Apply to grain after drying to moisture content below 14%, cooling and cleaning
- Insecticide may become depleted at grain surface if grain is being cooled by continuous extraction of air from the base leading to reduced control of grain store pests especially mites
- Resistance to organophosphorus compounds sometimes occurs in insect and mite pests of stored products

Restrictions
- Contains an anticholinesterase organophosphorus compound. Do not use if under medical advice not to work with such compounds
- Maximum number of treatments 1 per batch or 1 per store, prior to storage
- Only treat grain in good condition
- Do not treat grain intended for sowing

Crop-specific information
- May be applied pre-harvest to surfaces of empty store and grain handling machinery and as admixture with grain
- May be used in wheat, barley, oats, rye, triticale or oilseed rape stores

SEE SECTION 3 FOR PRODUCTS ALSO REGISTERED

Environmental safety
- Dangerous for the environment
- Very toxic to aquatic organisms

Hazard classification and safety precautions
>**Hazard** Harmful, Dangerous for the environment
>**Transport code** M6
>**Packaging group** III
>**UN Number** 3082
>**Risk phrases** H304, H315, H317, H336
>**Operator protection** A, C, H, M; U05a, U08, U19a, U20b
>**Environmental protection** E15a, E34, H410
>**Storage and disposal** D01, D02, D05, D09a, D10b
>**Medical advice** M01, M03, M05b

90   citronella oil

A natural plant extract herbicide

Products

Barrier H	Barrier	22.9% w/w	OD	17145

Uses
- Ragwort in *amenity grassland, land temporarily removed from production*

Approval information
- Plant oils such as citronella oil are included in Annex 1 under EC Regulation 1107/2009

Efficacy guidance
- Best results obtained from spot treatment of ragwort in the rosette stage, during dry still conditions
- Aerial growth of ragwort is rapidly destroyed. Longer term control depends on overall management strategy
- Check for regrowth after 28 d and re-apply as necessary

Crop-specific information
- Contact with grasses will result in transient scorch which is outgrown in good growing conditions

Environmental safety
- Apply away from bees
- Harmful to fish or other aquatic life. Do not contaminate surface waters or ditches with chemical or used container
- Keep livestock out of treated areas for at least 2 wk and until foliage of any poisonous weeds such as ragwort has died and become unpalatable

Hazard classification and safety precautions
>**Hazard** Irritant
>**UN Number** N/C
>**Risk phrases** H317
>**Operator protection** A, C; U05a, U20c
>**Environmental protection** E07b (2 wk); E12g, E13c
>**Storage and disposal** D01, D02, D05, D09a, D11a

91   clethodim

A post-emergence grass herbicide for use in listed broad-leaved crops
HRAC mode of action code: A

Products

1	Centurion Max	Interfarm	120 g/l	EC	16310
2	Chellist	Chem-Wise	120 g/l	EC	16263

FOR FULL CONDITIONS OF USE ALWAYS READ THE PRODUCT LABEL

Products – continued

3	Clayton Gatso	Clayton	120 g/l	EC	17698
4	Clayton Gatso	Clayton	120 g/l	EC	17732
5	Clayton Gatso	Clayton	120 g/l	EC	17746
6	Knight	Harvest	120 g/l	EC	18218
7	Mollivar	Generica	120 g/l	EC	17795
8	Mollivar	Generica	120 g/l	EC	17798
9	Spartan	Harvest	120 g/l	EC	17920

Uses

- Annual grasses in **broad beans** *(off-label)*, **brussels sprouts** *(off-label)*, **bulb onions** *(off-label)*, **cabbages** *(off-label)*, **carrots** *(off-label)*, **dwarf beans** *(off-label)*, **french beans** *(off-label)*, **garlic** *(off-label)*, **runner beans** *(off-label)*, **shallots** *(off-label)*, **vining peas** *(off-label)* [1]; **sugar beet**, **winter oilseed rape** [1-9]
- Annual meadow grass in **corn gromwell** *(off-label)*, **crambe** *(off-label)*, **linseed** *(off-label)* [1]; **sugar beet**, **winter oilseed rape** [1-9]
- Blackgrass in **corn gromwell** *(off-label)*, **crambe** *(off-label)*, **linseed** *(off-label)* [1]; **sugar beet**, **winter oilseed rape** [1-9]
- Couch in **broad beans** *(off-label)*, **brussels sprouts** *(off-label)*, **bulb onions** *(off-label)*, **cabbages** *(off-label)*, **carrots** *(off-label)*, **dwarf beans** *(off-label)*, **french beans** *(off-label)*, **garlic** *(off-label)*, **runner beans** *(off-label)*, **shallots** *(off-label)*, **vining peas** *(off-label)* [1]
- Volunteer barley in **corn gromwell** *(off-label)*, **crambe** *(off-label)*, **linseed** *(off-label)* [1]; **sugar beet**, **winter oilseed rape** [1-9]
- Volunteer wheat in **corn gromwell** *(off-label)*, **crambe** *(off-label)*, **linseed** *(off-label)* [1]; **sugar beet**, **winter oilseed rape** [1-9]
- Wild oats in **corn gromwell** *(off-label)*, **crambe** *(off-label)*, **linseed** *(off-label)* [1]

Extension of Authorisation for Minor Use (EAMUs)

- **broad beans** *20162041* [1]
- **brussels sprouts** *20162041* [1]
- **bulb onions** *20162041* [1]
- **cabbages** *20162041* [1]
- **carrots** *20162041* [1]
- **corn gromwell** *20151656* [1]
- **crambe** *20161841* [1]
- **dwarf beans** *20162041* [1]
- **french beans** *20162041* [1]
- **garlic** *20162041* [1]
- **linseed** *20152306* [1]
- **runner beans** *20162041* [1]
- **shallots** *20162041* [1]
- **vining peas** *20162041* [1]

Approval information

- Clethodim included in Annex I under EC Regulation 1107/2009

Efficacy guidance

- Optimum efficacy is achieved from applications made when grasses are actively growing and have 3 lvs up to beginning of tillering

Restrictions

- To avoid the build-up of resistance, do not apply products containing an ACCase inhibitor more than twice to any crop.
- Label requires a 14 day interval before and after application before other products are applied, particularly in oilseed rape.
- Do not apply in tank mix with other products.
- Do not apply to winter oilseed rape after end of October in year of sowing

Crop-specific information

- Consult processors before use on crops grown for processing.
- Do not apply to crops suffering from stress for any reason.

SEE SECTION 3 FOR PRODUCTS ALSO REGISTERED

Following crops guidance

- Broad-leaved crops can be sown at any time after application but when planting cereals, maize or grasses, it is advisable to wait at least 4 weeks after application and to cultivate to at least 20 cm deep before sowing.

Hazard classification and safety precautions

Hazard Harmful, Dangerous for the environment

Transport code 9

Packaging group III

UN Number 3082

Risk phrases H304, H336 [1, 3-9]; R22b, R51, R53a, R66, R67 [2]

Operator protection A, H; U02a, U04a, U05a, U08, U19a, U20a

Environmental protection E15b, E38 [1-9]; H410 [5, 7, 8]; H411 [1, 3, 4, 6, 9]

Storage and disposal D01, D02, D05, D09a, D10b, D12a

Medical advice M05b

92   clodinafop-propargyl

A foliar acting herbicide for annual grass weed control in wheat, triticale and rye
HRAC mode of action code: A

Products

1	Clodinastar	Life Scientific	240 g/l	EC	17404
2	Epee	Harvest	240 g/l	EC	17601
3	Kipota	Life Scientific	240 g/l	EC	17993
4	Ravena	Headland	240 g/l	EC	17313
5	Sword	Adama	240 g/l	EC	15099
6	Topik	Syngenta	240 g/l	EC	15123
7	Tuli	AgChem Access	240 g/l	EC	16401

Uses

- Blackgrass in *durum wheat*, *spring rye*, *spring wheat*, *triticale*, *winter rye*, *winter wheat* [1-7]; *grass seed crops* (off-label) [6]
- Rough-stalked meadow grass in *durum wheat*, *spring rye*, *spring wheat*, *triticale*, *winter rye*, *winter wheat* [1-7]; *grass seed crops* (off-label) [6]
- Wild oats in *durum wheat*, *spring rye*, *spring wheat*, *triticale*, *winter rye*, *winter wheat* [1-7]; *grass seed crops* (off-label) [6]

Extension of Authorisation for Minor Use (EAMUs)

- *grass seed crops* 20132310 [6]

Approval information

- Clodinafop-propargyl included in Annex I under EC Regulation 1107/2009

Efficacy guidance

- Spray when majority of weeds have germinated but before competition reduces yield
- Products contain a herbicide safener (cloquintocet-mexyl) that improves crop tolerance to clodinafop-propargyl
- Optimum control achieved when all grass weeds emerged. Wait for delayed germination on dry or cloddy seedbed
- A mineral oil additive is recommended to give more consistent control of very high blackgrass populations or for late season treatments. See label for details
- Weed control not affected by soil type, organic matter or straw residues
- Control may be reduced if rain falls within 1 h of treatment
- Clodinafop-propargyl is an ACCase inhibitor herbicide. To avoid the build up of resistance do not apply products containing an ACCase inhibitor herbicide more than twice to any crop. In addition do not use any product containing clodinafop-propargyl in mixture or sequence with any other product containing the same ingredient
- Use these products as part of a resistance management strategy that includes cultural methods of control and does not use ACCase inhibitors as the sole chemical method of grass weed control

FOR FULL CONDITIONS OF USE ALWAYS READ THE PRODUCT LABEL

- Applying a second product containing an ACCase inhibitor to a crop will increase the risk of resistance development; only use a second ACCase inhibitor to control different weeds at a different timing
- Always follow WRAG guidelines for preventing and managing herbicide resistant weeds. See Section 5 for more information

Restrictions
- Maximum number of treatments 1 per crop
- Do not use on barley or oats
- Do not treat crops under stress or suffering from waterlogging, pest attack, disease or frost
- Do not treat crops undersown with grass mixtures
- Do not mix with products containing MCPA, mecoprop-P, 2,4-D or 2,4-DB
- MCPA, mecoprop, 2,4-D or 2,4-DB should not be applied within 21 d before, or 7 d after, treatment

Crop-specific information
- Latest use: before second node detectable stage (GS 32) for durum wheat, triticale, rye; before flag leaf sheath extending (GS 41) for wheat
- Spray in autumn, winter or spring from 1 true leaf stage (GS 11) to before second node detectable (GS 32) on durum, rye, triticale; before flag leaf sheath extends (GS 41) on wheat

Following crops guidance
- Any broad leaved crop or cereal (except oats) may be sown after failure of a treated crop provided that at least 3 wk have elapsed between application and drilling a cereal
- After normal harvest of a treated crop any broad leaved crop or wheat, durum wheat, rye, triticale or barley should be sown. Oats and grass should not be sown until the following spring

Environmental safety
- Dangerous for the environment
- Very toxic to aquatic organisms

Hazard classification and safety precautions
Hazard Dangerous for the environment, Very toxic to aquatic organisms [1-7]; Harmful if swallowed [2, 5]
Transport code 9
Packaging group III
UN Number 3082
Risk phrases H304, H373
Operator protection A, H [1-7]; C, K [1-3, 5-7]; U02a, U05a, U09a, U20a [1-7]; U14 [1, 2, 4, 5]
Environmental protection E15a, E38, H410
Storage and disposal D01, D02, D05, D09a, D10c, D12a

93   clodinafop-propargyl + cloquintocet-mexyl

A grass herbicide for use in winter wheat and durum wheat
HRAC mode of action code: A

Products

1 Buguis	UPL Europe	100:25 g/l	EC	17151
2 Ciclope	Clayton	100:25 g/l	EC	17226

Uses
- Annual grasses in *durum wheat, winter wheat*
- Blackgrass in *durum wheat, winter wheat*

Approval information
- Clodinafop-propargyl included in Annex I under EC Regulation 1107/2009 but cloquintocet-mexyl not yet listed

Efficacy guidance
- Reduced doses require addition of an adjuvant - see label

SEE SECTION 3 FOR PRODUCTS ALSO REGISTERED

Restrictions

- Do not use on barley or oats.
- Do not spray crops under stress or crops suffering from waterlogging, pest attack, disease or frost
- Do not spray crops undersown with grass mixtures
- Rain within one hour after application may reduce grass weed control
- Avoid the use of hormone-containing herbicides in mixture or sequence

Following crops guidance

- Activity is not affected by soil type, organic matter or straw residues. In the event of crop failure, any broad-leaved crop may be sown or after an interval of 3 weeks any cereal may be sown. Following normal harvest of a treated crop, any broad-leaved or cereal crop maybe sown.

Hazard classification and safety precautions

Transport code 9
Packaging group III
UN Number 3082
Risk phrases H304, H317, H319, H373
Operator protection A, H
Environmental protection H411

94   clodinafop-propargyl + pinoxaden

A foliar acting herbicide mixture for winter wheat
HRAC mode of action code: A + A

See also pinoxaden

Products

Traxos	Syngenta	100:100 g/l	EC	12742

Uses

- Italian ryegrass in **winter wheat**
- Perennial ryegrass in **winter wheat** *(from seed)*
- Wild oats in **winter wheat**

Approval information

- Clodinafop-propargyl and pinoxaden included in Annex I under EC Regulation 1107/2009
- Approved for use on crops for brewing by BBPA

Efficacy guidance

- Products contain a herbicide safener (cloquintocet-mexyl) that improves crop tolerance to clodinafop-propargyl
- Best results obtained from treatment when all grass weeds have emerged. There is no residual activity
- Broad-leaved weeds are not controlled
- Treat before emerged weed competition reduces yield
- For ryegrass control use as part of a programme including other products with activity against ryegrass
- For blackgrass control must be used as part of an integrated control strategy
- Grass weed control may be reduced if rain falls within 1 hr of application
- Clodinafop-propargyl and pinoxaden are ACCase inhibitor herbicides. To avoid the build up of resistance do not apply products containing an ACCase inhibitor herbicide more than twice to any crop. In addition do not use any product containing clodinafop-propargyl or pinoxaden in mixture or sequence with any other product containing the same ingredient
- Use as part of a resistance management strategy that includes cultural methods of control and does not use ACCase inhibitors as the sole chemical method of grass weed control
- Applying a second product containing an ACCase inhibitor to a crop will increase the risk of resistance development; only use a second ACCase inhibitor to control different weeds at a different timing

- Always follow WRAG guidelines for preventing and managing herbicide resistant weeds. See Section 5 for more information

Restrictions
- Maximum number of treatments 1 per crop
- Product must always be used with specified adjuvant. See label
- Do not use on other cereals
- Do not treat crops under stress from any cause
- Do not treat crops undersown with grass or grass mixtures
- Avoid use of hormone-containing herbicides in mixture or in sequence. See label for restrictions on tank mixes

Crop-specific information
- Latest use: before flag leaf sheath extended (before GS 41) for winter wheat
- All varieties of winter wheat may be treated

Following crops guidance
- There are no restrictions on succeeding crops in a normal rotation
- In the event of failure of a treated crop ryegrass, maize, oats or any broad-leaved crop may be planted after a minimum interval of 4 wk from application

Environmental safety
- Dangerous for the environment
- Toxic to aquatic organisms

Hazard classification and safety precautions
Hazard Irritant, Dangerous for the environment
Transport code 9
Packaging group III
UN Number 3082
Risk phrases H304, H317, H319, H360
Operator protection A, C, H, K; U05a, U09a, U20b
Environmental protection E15b, E38, H411
Storage and disposal D01, D02, D05, D09a, D10c, D11a, D12a

95   clodinafop-propargyl + prosulfocarb

A foliar acting herbicide mixture for grass and broad-leaved weed control in winter wheat
HRAC mode of action code: A + N

See also prosulfocarb

Products
Grapple	Syngenta	10:800 g/l	EC	16523

Uses
- Annual dicotyledons in **winter wheat**
- Annual grasses in **winter wheat**

Approval information
- Clodinafop-propargyl and prosulfocarb included in Annex I under EC Regulation 1107/2009

Efficacy guidance
- To avoid the build up of resistance do not apply products containing an ACCase inhibitor herbicide more than twice to any crop. In addition, do not use this product in mixture or sequence with any other product containing clodinafop-propargyl.

Restrictions
- Do not use on barley or oats
- Do not spray crops undersown with grass mixtures

SECTION 2

- When it is applied first, leave 7 days before applying hormone herbicides. If mecoprop-P or 2,4-DB containing products are applied first, leave 14 days before it is applied. If MCPA or 2,4-D containing products are applied first, leave 21 days before it is applied.
- The cereal seed must be covered by 3 cm of soil and for best results apply to a firm, moist seedbed free of clods.

Crop-specific information
- Only one application per crop
- Latest use before 5 tiller stage (GS25)

Environmental safety
- LERAP Category B

Hazard classification and safety precautions
Hazard Harmful, Dangerous for the environment
Transport code 9
Packaging group III
UN Number 3082
Risk phrases H304, H319
Operator protection A, H; U02a, U05a, U08, U20c
Environmental protection E15b, E16a, E16b, E38, H410
Storage and disposal D01, D02, D05, D09a, D10c, D12a
Medical advice M05b

96   clofentezine

A selective ovicidal tetrazine acaricide for use in top fruit
IRAC mode of action code: 10A

Products
1	Apollo 50 SC	Adama	500 g/l	SC	17187
2	Clayton Sputnik	Clayton	500 g/l	SC	17344

Uses
- Conifer spinning mite in **forest nurseries** *(off-label)*, **ornamental plant production** *(off-label)* [1]
- Red spider mites in **apples**, **pears** [1, 2]
- Rust mite in **forest nurseries** *(off-label)*, **ornamental plant production** *(off-label)* [1]
- Spider mites in **forest nurseries** *(off-label)*, **ornamental plant production** *(off-label)* [1]

Extension of Authorisation for Minor Use (EAMUs)
- **forest nurseries** *20162082* [1]
- **ornamental plant production** *20162082* [1]

Approval information
- Clofentezine included in Annex I under EC Regulation 1107/2009

Efficacy guidance
- Acts on eggs and early motile stages of mites. For effective control total cover of plants is essential, particular care being needed to cover undersides of leaves

Restrictions
- Maximum number of treatments 1 per yr for apples, pears, cherries, plums

Crop-specific information
- HI apples, pears 28 d; cherries, plums 8 wk
- For red spider mite control spray apples and pears between bud burst and pink bud, plums and cherries between white bud and first flower. Rust mite is also suppressed
- On established infestations apply in conjunction with an adult acaricide

Environmental safety
- Harmful to aquatic organisms
- Product safe on predatory mites, bees and other predatory insects

FOR FULL CONDITIONS OF USE ALWAYS READ THE PRODUCT LABEL

Hazard classification and safety precautions
  UN Number N/C
  Operator protection U05a, U08, U20b
  Environmental protection E15a, H412
  Storage and disposal D01, D02, D05, D09a, D11a

97 clomazone

An isoxazolidinone residual herbicide for oilseed rape, field beans, combining and vining peas
HRAC mode of action code: F3

Products

1	Blanco	Adama	360 g/l	CS	16704
2	Centium 360 CS	Headland	360 g/l	CS	17327
3	Cirrus CS	Headland	371 g/l	CS	17328
4	Clayton Chrome CS	Clayton	360 g/l	CS	16383
5	Cleancrop Chicane	Headland	360 g/l	CS	17330
6	Cleancrop Covert	Headland	360 g/l	CS	17340
7	Clomate	Albaugh UK	360 g/l	CS	15565
8	Gamit 36 CS	Headland	360 g/l	CS	17314
9	Notion	AgChem Access	360 g/l	CS	16773
10	Sirtaki CS	Sipcam	360 g/l	CS	18032
11	Throne	Harvest	360 g/l	CS	17826

Uses

- Annual dicotyledons in *carrots, spring field beans* [7, 9]; *combining peas, potatoes, vining peas, winter field beans, winter oilseed rape* [1, 7, 9]; *field beans, spring oilseed rape* [1]
- Chickweed in *asparagus* (off-label), *cucumbers* (off-label), *gherkins* (off-label), *melons* (off-label), *okra* (off-label), *pumpkins* (off-label), *watermelons* (off-label), *winter squash* (off-label) [8]; *baby leaf crops* (off-label), *borage* (off-label), *broad beans* (off-label), *broccoli* (off-label), *brussels sprouts* (off-label), *cabbages* (off-label), *calabrese* (off-label), *cauliflowers* (off-label), *celeriac* (off-label), *celery (outdoor)* (off-label), *choi sum* (off-label), *collards* (off-label), *courgettes* (off-label), *fennel* (off-label), *french beans* (off-label), *herbs (see appendix 6)* (off-label), *kale* (off-label), *lupins* (off-label), *oriental cabbage* (off-label), *poppies for morphine production* (off-label), *rhubarb* (off-label), *runner beans* (off-label), *soya beans* (off-label), *spinach* (off-label), *spring cabbage* (off-label), *summer squash* (off-label), *swedes* (off-label), *sweet potato* (off-label) [6, 8]; *carrots* [4, 6, 8, 11]; *combining peas, spring field beans, vining peas* [2-5]; *forest nurseries* (off-label), *game cover* (off-label), *hops* (off-label), *ornamental plant production* (off-label) [2, 3, 6, 8]; *potatoes* [4, 6, 8, 10, 11]; *spring oilseed rape* [2, 3, 5, 10]; *winter field beans* [2, 3, 5]; *winter oilseed rape* [2-5, 10]
- Cleavers in *asparagus* (off-label), *cucumbers* (off-label), *gherkins* (off-label), *melons* (off-label), *okra* (off-label), *pumpkins* (off-label), *watermelons* (off-label), *winter squash* (off-label) [8]; *baby leaf crops* (off-label), *borage* (off-label), *broad beans* (off-label), *broccoli* (off-label), *brussels sprouts* (off-label), *cabbages* (off-label), *calabrese* (off-label), *cauliflowers* (off-label), *celeriac* (off-label), *celery (outdoor)* (off-label), *choi sum* (off-label), *collards* (off-label), *courgettes* (off-label), *fennel* (off-label), *french beans* (off-label), *herbs (see appendix 6)* (off-label), *kale* (off-label), *lupins* (off-label), *oriental cabbage* (off-label), *poppies for morphine production* (off-label), *rhubarb* (off-label), *runner beans* (off-label), *soya beans* (off-label), *spinach* (off-label), *spring cabbage* (off-label), *summer squash* (off-label), *swedes* (off-label), *sweet potato* (off-label) [6, 8]; *carrots* [4, 6-9, 11]; *combining peas, vining peas* [1-5, 7, 9]; *field beans* [1]; *forest nurseries* (off-label), *game cover* (off-label), *hops* (off-label), *ornamental plant production* (off-label) [2, 3, 6, 8]; *potatoes* [1, 4, 6-11]; *spring field beans* [2-5, 7, 9]; *spring oilseed rape* [1-3, 5, 10]; *winter field beans* [1-3, 5, 7, 9]; *winter oilseed rape* [1-5, 7, 9, 10]
- Fool's parsley in *asparagus* (off-label), *cucumbers* (off-label), *gherkins* (off-label), *melons* (off-label), *okra* (off-label), *pumpkins* (off-label), *watermelons* (off-label), *winter squash* (off-label) [8]; *baby leaf crops* (off-label), *borage* (off-label), *broad beans* (off-label), *broccoli* (off-label), *brussels sprouts* (off-label), *cabbages* (off-label), *calabrese* (off-label), *cauliflowers* (off-label), *celeriac* (off-label), *celery (outdoor)* (off-label), *choi sum* (off-label), *collards* (off-label),

courgettes *(off-label)*, **fennel** *(off-label)*, **french beans** *(off-label)*, **herbs (see appendix 6)** *(off-label)*, **kale** *(off-label)*, **lupins** *(off-label)*, **oriental cabbage** *(off-label)*, **poppies for morphine production** *(off-label)*, **rhubarb** *(off-label)*, **runner beans** *(off-label)*, **soya beans** *(off-label)*, **spinach** *(off-label)*, **spring cabbage** *(off-label)*, **summer squash** *(off-label)*, **swedes** *(off-label)*, **sweet potato** *(off-label)* [6, 8]; **carrots** [4, 6, 8, 11]; **combining peas**, **spring field beans**, **vining peas** [2-5]; **forest nurseries** *(off-label)*, **game cover** *(off-label)*, **hops** *(off-label)*, **ornamental plant production** *(off-label)* [2, 3, 6, 8]; **potatoes** [4, 6, 8, 10, 11]; **spring oilseed rape** [2, 3, 5, 10]; **winter oilseed rape** [4]

- Ivy-leaved speedwell in **baby leaf crops** *(off-label)*, **borage** *(off-label)*, **broad beans** *(off-label)*, **courgettes** *(off-label)*, **herbs (see appendix 6)** *(off-label)*, **lupins** *(off-label)*, **spinach** *(off-label)*, **summer squash** *(off-label)* [6, 8]; **broccoli** *(off-label)*, **brussels sprouts** *(off-label)*, **cabbages** *(off-label)*, **calabrese** *(off-label)*, **carrots**, **cauliflowers** *(off-label)*, **celeriac** *(off-label)*, **celery (outdoor)** *(off-label)*, **choi sum** *(off-label)*, **collards** *(off-label)*, **fennel** *(off-label)*, **forest nurseries** *(off-label)*, **french beans** *(off-label)*, **game cover** *(off-label)*, **hops** *(off-label)*, **kale** *(off-label)*, **oriental cabbage** *(off-label)*, **ornamental plant production** *(off-label)*, **poppies for morphine production** *(off-label)*, **potatoes**, **rhubarb** *(off-label)*, **runner beans** *(off-label)*, **soya beans** *(off-label)*, **spring cabbage** *(off-label)*, **swedes** *(off-label)*, **sweet potato** *(off-label)* [6]; **cucumbers** *(off-label)*, **gherkins** *(off-label)*, **melons** *(off-label)*, **okra** *(off-label)*, **pumpkins** *(off-label)*, **watermelons** *(off-label)*, **winter squash** *(off-label)* [8]
- Perennial weeds in **asparagus** *(off-label - spring-germinating)* [6, 8]
- Red dead-nettle in **asparagus** *(off-label)*, **cucumbers** *(off-label)*, **gherkins** *(off-label)*, **melons** *(off-label)*, **okra** *(off-label)*, **pumpkins** *(off-label)*, **watermelons** *(off-label)*, **winter squash** *(off-label)* [8]; **baby leaf crops** *(off-label)*, **borage** *(off-label)*, **broad beans** *(off-label)*, **broccoli** *(off-label)*, **brussels sprouts** *(off-label)*, **cabbages** *(off-label)*, **calabrese** *(off-label)*, **cauliflowers** *(off-label)*, **celeriac** *(off-label)*, **celery (outdoor)** *(off-label)*, **choi sum** *(off-label)*, **collards** *(off-label)*, **courgettes** *(off-label)*, **fennel** *(off-label)*, **french beans** *(off-label)*, **herbs (see appendix 6)** *(off-label)*, **kale** *(off-label)*, **lupins** *(off-label)*, **oriental cabbage** *(off-label)*, **poppies for morphine production** *(off-label)*, **rhubarb** *(off-label)*, **runner beans** *(off-label)*, **soya beans** *(off-label)*, **spinach** *(off-label)*, **spring cabbage** *(off-label)*, **summer squash** *(off-label)*, **swedes** *(off-label)*, **sweet potato** *(off-label)* [6, 8]; **carrots** [4, 6, 8, 11]; **combining peas**, **spring field beans**, **vining peas** [2-5]; **forest nurseries** *(off-label)*, **game cover** *(off-label)*, **hops** *(off-label)*, **ornamental plant production** *(off-label)* [2, 3, 6, 8]; **potatoes** [4, 6, 8, 10, 11]; **spring oilseed rape** [2, 3, 5, 10]; **winter field beans** [2, 3, 5]; **winter oilseed rape** [2-5, 10]
- Shepherd's purse in **asparagus** *(off-label)*, **cucumbers** *(off-label)*, **gherkins** *(off-label)*, **melons** *(off-label)*, **okra** *(off-label)*, **pumpkins** *(off-label)*, **watermelons** *(off-label)*, **winter squash** *(off-label)* [8]; **baby leaf crops** *(off-label)*, **borage** *(off-label)*, **broad beans** *(off-label)*, **broccoli** *(off-label)*, **brussels sprouts** *(off-label)*, **cabbages** *(off-label)*, **calabrese** *(off-label)*, **cauliflowers** *(off-label)*, **celeriac** *(off-label)*, **celery (outdoor)** *(off-label)*, **choi sum** *(off-label)*, **collards** *(off-label)*, **courgettes** *(off-label)*, **fennel** *(off-label)*, **french beans** *(off-label)*, **herbs (see appendix 6)** *(off-label)*, **kale** *(off-label)*, **lupins** *(off-label)*, **oriental cabbage** *(off-label)*, **poppies for morphine production** *(off-label)*, **rhubarb** *(off-label)*, **runner beans** *(off-label)*, **soya beans** *(off-label)*, **spinach** *(off-label)*, **spring cabbage** *(off-label)*, **summer squash** *(off-label)*, **swedes** *(off-label)*, **sweet potato** *(off-label)* [6, 8]; **carrots** [4, 6, 8, 11]; **combining peas**, **spring field beans**, **vining peas** [2-5]; **forest nurseries** *(off-label)*, **game cover** *(off-label)*, **hops** *(off-label)*, **ornamental plant production** *(off-label)* [2, 3, 6, 8]; **potatoes** [4, 6, 8, 10, 11]; **spring oilseed rape** [2, 3, 5, 10]; **winter field beans** [2, 3, 5]; **winter oilseed rape** [2-5, 10]
- Speedwells in **carrots**, **spring field beans** [7, 9]; **combining peas**, **potatoes**, **vining peas**, **winter field beans**, **winter oilseed rape** [1, 7, 9]; **field beans**, **spring oilseed rape** [1]

Extension of Authorisation for Minor Use (EAMUs)

- **asparagus** *(spring-germinating) 20152992* [6], *(spring-germinating) 20152817* [8], *20152817* [8]
- **baby leaf crops** *20160076* [6], *20152830* [8]
- **borage** *20152994* [6], *20152829* [8]
- **broad beans** *20152999* [6], *20152828* [8]
- **broccoli** *20152997* [6], *20152820* [8]
- **brussels sprouts** *20152997* [6], *20152820* [8]
- **cabbages** *20152997* [6], *20152820* [8]
- **calabrese** *20152997* [6], *20152820* [8]
- **cauliflowers** *20152997* [6], *20152820* [8]

- *celeriac* 20152993 [6], 20152815 [8]
- *celery (outdoor)* 20160074 [6], 20152811 [8]
- *choi sum* 20152997 [6], 20152820 [8]
- *collards* 20152997 [6], 20152820 [8]
- *courgettes* 20152990 [6], 20152831 [8]
- *cucumbers* 20152831 [8]
- *fennel* 20160072 [6], 20152810 [8]
- *forest nurseries* 20152924 [2], 20152921 [3], 20160075 [6], 20152813 [8]
- *french beans* 20152991 [6], 20152819 [8]
- *game cover* 20152924 [2], 20152921 [3], 20160075 [6], 20152813 [8]
- *gherkins* 20152831 [8]
- *herbs (see appendix 6)* 20160073 [6], 20152824 [8]
- *hops* 20152924 [2], 20152921 [3], 20160075 [6], 20152813 [8]
- *kale* 20152997 [6], 20152820 [8]
- *lupins* 20152998 [6], 20152827 [8]
- *melons* 20152831 [8]
- *okra* 20152831 [8]
- *oriental cabbage* 20152997 [6], 20152820 [8]
- *ornamental plant production* 20152924 [2], 20152921 [3], 20160075 [6], 20152813 [8]
- *poppies for morphine production* 20160070 [6], 20152821 [8]
- *pumpkins* 20152831 [8]
- *rhubarb* 20160077 [6], 20152816 [8]
- *runner beans* 20152985 [6], 20152818 [8]
- *soya beans* 20152982 [6], 20152823 [8]
- *spinach* 20160076 [6], 20152830 [8]
- *spring cabbage* 20152997 [6], 20152820 [8]
- *summer squash* 20152990 [6], 20152831 [8]
- *swedes* 20152925 [6], 20152814 [8]
- *sweet potato* 20160071 [6], 20152812 [8]
- *watermelons* 20152831 [8]
- *winter squash* 20152831 [8]

Approval information
- Clomazone included in Annex I under EC Regulation 1107/2009
- Approval expiry 02 Nov 2018 [1]

Efficacy guidance
- Best results obtained from application as soon as possible after sowing crop and before emergence of crop or weeds
- Uptake is via roots and shoots. Seedbeds should be firm, level and free from clods. Loose puffy seedbeds should be consolidated before spraying
- Efficacy is reduced on organic soils, on dry cloddy seedbeds and if prolonged dry weather follows application
- Clomazone acts by inhibiting synthesis of chlorophyll pigments. Susceptible weeds emerge but are chlorotic and die shortly afterwards
- Season-long control of weeds may not be achieved
- Always follow WRAG guidelines for preventing and managing herbicide resistant weeds. See Section 5 for more information

Restrictions
- Maximum number of treatments one per crop
- Crops must be covered by a minimum of 20 mm settled soil. Do not apply to broadcast crops. Direct-drilled crops should be harrowed across the slits to cover seed before spraying
- Do not use on compacted soils or soils of poor structure that may be liable to waterlogging
- Do not use on Sands or Very Light soils or those with more than 10% organic matter
- Do not treat two consecutive crops of carrots with clomazone in one calendar yr
- Consult manufacturer or your advisor before use on potato seed crops
- Do not overlap spray swaths. Crop plants emerged at time of treatment may be severely damaged
- Application must be made using a coarse spray quality.

SEE SECTION 3 FOR PRODUCTS ALSO REGISTERED

Crop-specific information
- Latest use: pre-emergence of crop
- Severe, but normally transient, crop damage may occur in overlaps on field beans
- Some transient crop bleaching may occur under certain climatic conditions and can be severe where heavy rain follows application. This is normally rapidly outgrown and has no effect on final crop yield. Overlapping spray swaths may cause severe damage to field beans

Following crops guidance
- Following normal harvest of a spring or autumn treated crop, cereals, oilseed rape, field beans, combining peas, potatoes, maize, turnips, linseed or sugar beet may be sown
- In the event of failure of an autumn treated crop, winter cereals or winter beans may be sown in the autumn if 6 wk have elapsed since treatment. In the spring following crop failure combining peas, field beans or potatoes may be sown if 6 wk have elapsed since treatment, and spring cereals, maize, turnips, onions, carrots or linseed may be sown if 7 mth have elapsed since treatment
- In the event of a failure of a spring treated crop a wide range of crops may be sown provided intervals of 6-9 wk have elapsed since treatment. See label for details
- Prior to resowing any listed replacement crop the soil should be ploughed and cultivated to 15 cm

Environmental safety
- Take extreme care to avoid drift outside the target area, or on to ponds, waterways or ditches as considerable damage may occur. Apply using a coarse quality spray
- To protect non-target plants respect an untreated buffer zone of 10 metres to non-crop land [1, 4, 7, 9, 10]

Hazard classification and safety precautions
Transport code 9 [1, 4]
Packaging group III [1, 4]
UN Number 3082 [4]; N/C [1-3, 5-11]
Risk phrases H334 [10]
Operator protection A [1-8, 10, 11]; H [1]; U05a, U14, U20b [2-8, 10, 11]; U05b [1]
Environmental protection E15b [2-8, 10, 11]; E39 [4, 7]; H410 [10]; H412 [1]; H413 [2, 3, 5, 6, 8, 11]
Storage and disposal D01, D02 [1-8, 10, 11]; D09a [2-8, 10, 11]; D10b [2, 3, 5, 6, 8, 10, 11]; D10c [4, 7]; D12a [1]

98   clomazone + dimethenamid-p + metazachlor

A pre-emergence herbicide mixture for use in winter oilseed rape
HRAC mode of action code: F3 + K3

See also clomazone + metazachlor
dimethenamid-p + metazachlor

Products
Nimbus Gold	BASF	40:200:200 g/l	CC	18192

Uses
- Annual dicotyledons in **winter oilseed rape**
- Annual grasses in **winter oilseed rape**

Approval information
- Clomazone, dimethenamid-P and metazachlor included in Annex I under EC Regulation 1107/2009

Restrictions
- Do not apply using hand-held equipment.
- To protect non-target plants respect an untreated buffer zone of 10 m to non-crop land.
- Applications shall be limited to a total dose of not more than 1.0 kg metazachlor/ha in a three year period on the same field.

FOR FULL CONDITIONS OF USE ALWAYS READ THE PRODUCT LABEL

- A maximum total dose of no more than 100 g/ha clomazone may be applied to the same field in a three year period.

Environmental safety
- LERAP Category B

Hazard classification and safety precautions
 Hazard Harmful, Irritant, Dangerous for the environment, Very toxic to aquatic organisms
 Transport code 9
 Packaging group III
 UN Number 3082
 Risk phrases H317, H351
 Operator protection A, H; U05a, U08, U14, U19a, U20b
 Environmental protection E07a, E15b, E16a, E19b, E38, H410
 Storage and disposal D01, D02, D08, D09a, D10c, D12a
 Medical advice M05a

99 clomazone + metribuzin

A mixture of isoxazolidinone and triazinone residual herbicides for use in potatoes
HRAC mode of action code: F3 + C1

See also metribuzin

Products

Metric	Belchim	60:233 g/l	CC	16720

Uses
- Annual dicotyledons in **potatoes**
- Annual meadow grass in **potatoes**

Approval information
- Clomazone and metribuzin included in Annex I under EC Regulation 1107/2009

Restrictions
- For use on specified varieties of potato only
- Safety to daughter tubers has not been tested; consult manufacturer before treating seed crops
- Application must be made using a coarse spray quality.

Following crops guidance
- Before drilling or planting any succeeding crop, soil MUST be mouldboard ploughed to a depth of at least 15 cm (6") taking care to ensure that the furrow slice is inverted. Ploughing should be carried out as soon as possible (preferably within 3-4 weeks) after lifting the potato crop, but certainly no later than the end of December.
- In the same year: Provided at least 16 weeks have elapsed after the application of the recommended rate cereals and winter beans may be grown as following crops.In the following year: Do not grow any vegetable brassica crop (including cauliflower, calabrese, Brussels sprout and cabbage), lettuce or radish on land treated in the previous year.
- Cereals, oilseed rape, field beans, combining peas, potatoes, maize, turnip, linseed and sugar beet may be sown from spring onwards in the year following use.

Environmental safety
- Buffer zone requirement 10 m [1]
- LERAP Category B

Hazard classification and safety precautions
 Hazard Dangerous for the environment, Very toxic to aquatic organisms
 Transport code 9
 Packaging group III
 UN Number 3082
 Operator protection A, H; U05a, U19a, U20b

SEE SECTION 3 FOR PRODUCTS ALSO REGISTERED

Environmental protection E15b, E16a, E16b, E34, E38, H410
Storage and disposal D01, D02, D09a, D10b, D12a
Medical advice M05a

100   clomazone + napropamide

A herbicide mixture for use in winter oilseed rape
HRAC mode of action code: F3 + K3

See also clomazone
         napropamide

Products

Altiplano DAMtec	Headland	3.5:40% w/w	WG	17189

Uses
- Annual dicotyledons in **winter oilseed rape**
- Annual grasses in **winter oilseed rape**
- Chickweed in **winter oilseed rape**
- Cleavers in **winter oilseed rape**
- Crane's-bill in **winter oilseed rape**
- Dead nettle in **winter oilseed rape**
- Mayweeds in **winter oilseed rape**
- Poppies in **winter oilseed rape**

Approval information
- Clomazone and napropamide included in Annex I under EC Regulation 1107/2009

Restrictions
- Do not use on sands, very light soils or on soils with more than 10% organic matter.
- Do not roll after application.
- Do not irrigate within 3 weeks of application.

Crop-specific information
- Ensure seed is covered by a minimum of 2 cm settled soil before application

Environmental safety
- LERAP Category B

Hazard classification and safety precautions
Hazard Dangerous for the environment, Very toxic to aquatic organisms
Transport code 9
Packaging group III
UN Number 3077
Operator protection A, H
Environmental protection E16a, E38, H410
Storage and disposal D12a

101   clomazone + pendimethalin

A residual herbicide mixture for weed control in combining peas and field beans
HRAC mode of action code: F3 + K1

Products

Stallion Sync TEC	Headland	30:333 g/l	CS	17243

Uses
- Annual dicotyledons in **combining peas**, **spring field beans**
- Cleavers in **combining peas**, **spring field beans**

FOR FULL CONDITIONS OF USE ALWAYS READ THE PRODUCT LABEL

Approval information
- Clomazone and pendimethalin included in Annex I under EC Regulation 1107/2009

Restrictions
- Application must be made using a coarse spray quality.

Environmental safety
- LERAP Category B

Hazard classification and safety precautions
Hazard Dangerous for the environment, Very toxic to aquatic organisms
Transport code 9
Packaging group III
UN Number 3082
Risk phrases H317
Operator protection A; U02a, U05a, U08, U13, U14, U19a, U20b
Environmental protection E15b, E16a, E38, H410
Storage and disposal D01, D02, D05, D09a, D10b, D12a

SECTION 2

102   clopyralid

A foliar translocated picolinic herbicide for a wide range of crops
HRAC mode of action code: O

See also 2,4-D + clopyralid + MCPA
bromoxynil + clopyralid

Products

1	Dow Shield 400	Dow	400 g/l	SL	14984
2	Glopyr 400	Globachem	400 g/l	SL	15009
3	Leash	Life Scientific	200 g/l	SL	17969
4	Vivendi 200	UPL Europe	200 g/l	SL	16966

Uses
- Annual dicotyledons in **broccoli, bulb onions, cabbages, calabrese, cauliflowers** [2]; **brussels sprouts, forage maize, linseed** [1, 2]; **fodder beet, mangels, red beet, spring oilseed rape, sugar beet, swedes, turnips, winter oilseed rape** [1-4]; **grassland, ornamental plant production, spring barley, spring oats, spring wheat, winter barley, winter oats, winter wheat** [1, 2, 4]
- Black bindweed in **forest** *(off-label)* [1]
- Clovers in **grass seed crops** *(off-label)* [1]
- Corn marigold in **broccoli, bulb onions, cabbages, calabrese, cauliflowers** [2]; **brussels sprouts, fodder beet, forage maize, grassland, linseed, mangels, ornamental plant production, red beet, spring barley, spring oats, spring oilseed rape, spring wheat, sugar beet, swedes, turnips, winter barley, winter oats, winter oilseed rape, winter wheat** [1, 2]; **forest** *(off-label)* [1]
- Creeping thistle in **broccoli, bulb onions, cabbages, calabrese, cauliflowers** [2]; **brussels sprouts, fodder beet, forage maize, grassland, linseed, mangels, ornamental plant production, red beet, spring barley, spring oats, spring oilseed rape, spring wheat, sugar beet, swedes, turnips, winter barley, winter oats, winter oilseed rape, winter wheat** [1, 2]; **forest** *(off-label)*, **sweetcorn** *(off-label)* [1]
- Dandelions in **forest** *(off-label)* [1]
- Groundsel in **all edible seed crops grown outdoors** *(off-label)*, **all non-edible seed crops grown outdoors** *(off-label)*, **bilberries** *(off-label)*, **blackcurrants** *(off-label)*, **blueberries** *(off-label)*, **chard** *(off-label)*, **corn gromwell** *(off-label)*, **cranberries** *(off-label)*, **elderberries** *(off-label)*, **farm forestry** *(off-label)*, **forest** *(off-label)*, **garlic** *(off-label)*, **gooseberries** *(off-label)*, **hemp grown for fibre production** *(off-label)*, **hops in propagation** *(off-label)*, **miscanthus** *(off-label)*, **mulberries** *(off-label)*, **nursery fruit trees** *(off-label)*, **outdoor leaf herbs** *(off-label)*, **redcurrants** *(off-label)*, **rose hips** *(off-label)*, **shallots** *(off-label)*, **spinach** *(off-label)*, **spinach beet** *(off-label)*, **strawberries** *(off-label)*, **sweetcorn** *(off-label)* [1]; **forest nurseries** *(off-label)*, **game cover** *(off-label)* [1, 4]

SEE SECTION 3 FOR PRODUCTS ALSO REGISTERED

- Mayweeds in *all edible seed crops grown outdoors* (off-label), *all non-edible seed crops grown outdoors* (off-label), *apples* (off-label), *bilberries* (off-label), *blackcurrants* (off-label), *blueberries* (off-label), *borage for oilseed production* (off-label), *canary flower (echium spp.)* (off-label), *chard* (off-label), *corn gromwell* (off-label), *crab apples* (off-label), *cranberries* (off-label), *durum wheat* (off-label), *elderberries* (off-label), *evening primrose* (off-label), *farm forestry* (off-label), *forest* (off-label), *garlic* (off-label), *gooseberries* (off-label), *hemp grown for fibre production* (off-label), *honesty* (off-label), *hops in propagation* (off-label), *leeks* (off-label), *miscanthus* (off-label), *mulberries* (off-label), *mustard* (off-label), *nursery fruit trees* (off-label), *outdoor leaf herbs* (off-label), *pears* (off-label), *poppies for morphine production* (off-label), *redcurrants* (off-label), *rose hips* (off-label), *salad onions* (off-label), *shallots* (off-label), *spinach* (off-label), *spinach beet* (off-label), *spring rye* (off-label), *strawberries* (off-label), *triticale* (off-label), *winter rye* (off-label) [1]; *broccoli, bulb onions, cabbages, calabrese, cauliflowers* [2]; *brussels sprouts, forage maize, linseed* [1, 2]; *fodder beet, mangels, red beet, spring oilseed rape, sugar beet, swedes, turnips, winter oilseed rape* [1-4]; *forest nurseries* (off-label), *game cover* (off-label) [1, 4]; *grassland, ornamental plant production, spring barley, spring oats, spring wheat, winter barley, winter oats, winter wheat* [1, 2, 4]
- Perennial dicotyledons in *leeks* (off-label), *salad onions* (off-label) [1]
- Ragwort in *forest* (off-label) [1]
- Sowthistle in *bilberries* (off-label), *blackcurrants* (off-label), *blueberries* (off-label), *cranberries* (off-label), *elderberries* (off-label), *forest* (off-label), *garlic* (off-label), *gooseberries* (off-label), *mulberries* (off-label), *poppies for morphine production* (off-label), *redcurrants* (off-label), *rose hips* (off-label), *shallots* (off-label) [1]; *forest nurseries* (off-label), *game cover* (off-label) [4]
- Thistles in *apples* (off-label), *asparagus* (off-label), *borage for oilseed production* (off-label), *canary flower (echium spp.)* (off-label), *chinese cabbage* (off-label), *choi sum* (off-label), *collards* (off-label), *corn gromwell* (off-label), *crab apples* (off-label), *durum wheat* (off-label), *evening primrose* (off-label), *forest* (off-label), *garlic* (off-label), *honesty* (off-label), *kale* (off-label), *leeks* (off-label), *mustard* (off-label), *pak choi* (off-label), *pears* (off-label), *poppies for morphine production* (off-label), *salad onions* (off-label), *shallots* (off-label), *spring greens* (off-label), *spring rye* (off-label), *strawberries* (off-label), *tatsoi* (off-label), *trees* (off-label), *triticale* (off-label), *winter rye* (off-label) [1]; *fodder beet, mangels, red beet, spring oilseed rape, sugar beet, swedes, turnips, winter oilseed rape* [3, 4]; *forest nurseries* (off-label) [1, 4]; *game cover* (off-label), *grassland, ornamental plant production, spring barley, spring oats, spring wheat, winter barley, winter oats, winter wheat* [4]
- Volunteer potatoes in *chinese cabbage* (off-label), *choi sum* (off-label), *collards* (off-label), *forest nurseries* (off-label), *garlic* (off-label), *kale* (off-label), *leeks* (off-label), *pak choi* (off-label), *salad onions* (off-label), *shallots* (off-label), *spring greens* (off-label), *sweetcorn* (off-label), *tatsoi* (off-label) [1]

Extension of Authorisation for Minor Use (EAMUs)
- *all edible seed crops grown outdoors* 20150007 [1]
- *all non-edible seed crops grown outdoors* 20150007 [1]
- *apples* 20102080 [1]
- *asparagus* 20102079 [1]
- *bilberries* 20161629 [1]
- *blackcurrants* 20161629 [1]
- *blueberries* 20161629 [1]
- *borage for oilseed production* 20102086 [1]
- *canary flower (echium spp.)* 20102086 [1]
- *chard* 20102081 [1]
- *chinese cabbage* 20131710 [1]
- *choi sum* 20131710 [1]
- *collards* 20131710 [1]
- *corn gromwell* 20121046 [1]
- *crab apples* 20102080 [1]
- *cranberries* 20161629 [1]
- *durum wheat* 20102085 [1]
- *elderberries* 20161629 [1]
- *evening primrose* 20102086 [1]

- **farm forestry** *20150007* [1]
- **forest** *20152633* [1]
- **forest nurseries** *20130514* [1], *20150007* [1], *20151529* [4]
- **game cover** *20150007* [1], *20151529* [4]
- **garlic** *20130292* [1]
- **gooseberries** *20161629* [1]
- **grass seed crops** *20102084* [1]
- **hemp grown for fibre production** *20122041* [1], *20150007* [1]
- **honesty** *20102086* [1]
- **hops in propagation** *20150007* [1]
- **kale** *20131710* [1]
- **leeks** *20140400* [1]
- **miscanthus** *20150007* [1]
- **mulberries** *20161629* [1]
- **mustard** *20102086* [1]
- **nursery fruit trees** *20150007* [1]
- **outdoor leaf herbs** *20113236* [1]
- **pak choi** *20131710* [1]
- **pears** *20102080* [1]
- **poppies for morphine production** *20102082* [1]
- **redcurrants** *20161629* [1]
- **rose hips** *20161629* [1]
- **salad onions** *20140400* [1]
- **shallots** *20130292* [1]
- **spinach** *20102081* [1]
- **spinach beet** *20102081* [1]
- **spring greens** *20131710* [1]
- **spring rye** *20102085* [1]
- **strawberries** *20131822* [1]
- **sweetcorn** *20152626* [1]
- **tatsoi** *20131710* [1]
- **trees** *20102083* [1]
- **triticale** *20102085* [1]
- **winter rye** *20102085* [1]

SECTION 2

Approval information
- Clopyralid included in Annex I under EC Regulation 1107/2009
- Accepted by BBPA for use on malting barley
- Approval expiry 03 Oct 2018 [3]

Efficacy guidance
- Best results achieved by application to young actively growing weed seedlings. Treat creeping thistle at rosette stage and repeat 3-4 wk later as directed
- High activity on weeds of Compositae family. For most crops recommended for use in tank mixes. See label for details

Restrictions
- Maximum total dose varies between the equivalent of one and two full dose treatments, depending on the crop treated. See labels for details
- Do not apply to cereals later than the second node detectable stage (GS 32)
- Do not apply when crop damp or when rain expected within 6 h
- Do not use straw from treated cereals in compost or any other form for glasshouse crops. Straw may be used for strawing down strawberries
- Straw from treated grass seed crops or linseed should be baled and carted away. If incorporated do not plant winter beans in same year
- Do not use on onions at temperatures above 20°C or when under stress
- Do not treat maiden strawberries or runner beds or apply to early leaf growth during blossom period or within 4 wk of picking. Aug or early Sep sprays may reduce yield
- Applications must be made earlier than 1 March in the year of harvest

SEE SECTION 3 FOR PRODUCTS ALSO REGISTERED

Crop-specific information

- Latest use: 7 d before cutting grass for hay or silage; before 3rd node detectable (GS 33) for cereals; before flower buds visible from above for oilseed rape, linseed
- HI grassland 7 d; apples, pears, strawberries 4 wk; maize, sweetcorn, onions, Brussels sprouts, broccoli, cabbage, cauliflowers, calabrese, kale, fodder rape, oilseed rape, swedes, turnips, sugar beet, red beet, fodder beet, mangels, sage, honesty 6 wk
- Timing of application varies with weed problem, crop and other ingredients of tank mixes. See labels for details
- Apply as directed spray in woody ornamentals, avoiding leaves, buds and green stems. Do not apply in root zone of families Compositae or Leguminosae

Following crops guidance

- Do not plant susceptible autumn-sown crops in same year as treatment. Do not apply later than Jul where susceptible crops are to be planted in spring. See label for details

Environmental safety

- Harmful to aquatic organisms [1, 2]
- Wash spray equipment thoroughly with water and detergent immediately after use. Traces of product can damage susceptible plants sprayed later
- Keep livestock out of treated areas for at least 7 d and until foliage of any poisonous weeds such as ragwort has died and become unpalatable
- Some pesticides pose a greater threat of contamination of water than others and clopyralid is one of these pesticides. Take special care when applying clopyralid near water and do not apply if heavy rain is forecast

Hazard classification and safety precautions

UN Number N/C
Operator protection A, C [1-4]; H [3]; U05a [3, 4]; U08, U19a, U20b [1-4]
Environmental protection E07a [1, 2]; E15a [1-4]; E34 [3, 4]; H412 [2]
Storage and disposal D01, D05, D09a, D10b [1-4]; D02 [3, 4]; D12a [1-3]

---

103   clopyralid + florasulam

A translocated herbicide mixture for weed control in cereals
HRAC mode of action code: O + B

Products

Gartrel	Dow	300:25 g/l	SC	16828

Uses

- Black bindweed in *durum wheat, spelt, spring barley, spring oats, spring wheat, triticale, winter barley, winter rye, winter wheat*
- Charlock in *durum wheat, spelt, spring barley, spring oats, spring wheat, triticale, winter barley, winter rye, winter wheat*
- Chickweed in *durum wheat, spelt, spring barley, spring oats, spring wheat, triticale, winter barley, winter rye, winter wheat*
- Cleavers in *durum wheat, spelt, spring barley, spring oats, spring wheat, triticale, winter barley, winter rye, winter wheat*
- Mayweeds in *durum wheat, spelt, spring barley, spring oats, spring wheat, triticale, winter barley, winter rye, winter wheat*
- Runch in *durum wheat, spelt, spring barley, spring oats, spring wheat, triticale, winter barley, winter rye, winter wheat*
- Shepherd's purse in *durum wheat, spelt, spring barley, spring oats, spring wheat, triticale, winter barley, winter rye, winter wheat*
- Volunteer oilseed rape in *durum wheat, spelt, spring barley, spring oats, spring wheat, triticale, winter barley, winter rye, winter wheat*
- Wild radish in *durum wheat, spelt, spring barley, spring oats, spring wheat, triticale, winter barley, winter rye, winter wheat*

Approval information

- Clopyralid and florasulam included in Annex I under EC Regulation 1107/2009

FOR FULL CONDITIONS OF USE ALWAYS READ THE PRODUCT LABEL

Efficacy guidance
- Best results obtained when weeds are small and actively growing
- Effectiveness may be reduced when soil is very dry
- Use adequate water volume to achieve complete spray coverage of the weeds
- Florasulam is a member of the ALS-inhibitor group of herbicides

Restrictions
- Do not roll or harrow for 7 d before or after application
- Do not use any treated plant material for composting or mulching
- Do not use manure from animals fed on treated crops for composting
- Specific restrictions apply to use in sequence or tank mixture with other sulfonylurea or ALS-inhibiting herbicides. See label for details

Following crops guidance
- Where residues of a treated crop have not completely decayed by the time of planting a succeeding crop, avoid planting peas, beans and other legumes, carrots and other Umbelliferae, potatoes, lettuce and other Compositae, glasshouse and protected crops
- Where the product has been used in mixture with certain named products (see label) only cereals or grass may be sown in the autumn following harvest. Otherwise cereals, oilseed rape, grass or vegetable brassicas as transplants may be sown as a following crop in the same calendar yr as treatment. Oilseed rape may show some temporary reduction of vigour after a dry summer, but yields are not affected
- In addition to the above, field beans, linseed, peas, sugar beet, potatoes, maize, clover (for use in grass/clover mixtures) or carrots may be sown in the calendar yr following treatment
- In the event of failure of a treated crop in spring, only spring wheat, spring barley, spring oats, maize or ryegrass may be sown

Environmental safety
- LERAP Category B

Hazard classification and safety precautions
Hazard Very toxic to aquatic organisms
Transport code 9
Packaging group III
UN Number 3082
Operator protection U05a, U19a, U20a
Environmental protection E15a, E16a, E34, H410
Storage and disposal D05, D09a, D10b, D12a
Medical advice M03

---

104   clopyralid + florasulam + fluroxypyr

A translocated herbicide mixture for cereals
HRAC mode of action code: O + B + O

See also florasulam
fluroxypyr

Products
1 Dakota	Dow	80:2.5:100 g/l	EC	16121
2 Galaxy	Dow	80:2.5:100 g/l	EC	14085
3 Leystar	Dow	80:2.5:100 g/l	EC	17921
4 Polax	Pan Amenity	80:2.5;100 g/l	EC	15915
5 Praxys	Dow	80:2.5;100 g/l	EC	13912

Uses
- Annual dicotyledons in **amenity grassland**, **lawns**, **managed amenity turf** [4, 5]; **forage maize**, **grass seed crops**, **grassland**, **newly sown grass leys** [3]; **spring barley**, **spring oats**, **spring wheat**, **winter barley**, **winter oats**, **winter wheat** [1, 2]
- Black medick in **amenity grassland** [5]

SEE SECTION 3 FOR PRODUCTS ALSO REGISTERED

- Bristly oxtongue in *amenity grassland* [4, 5]; *grass seed crops, grassland, newly sown grass leys* [3]; *lawns, managed amenity turf* [4]
- Buttercups in *amenity grassland* [4, 5]; *grass seed crops, grassland, newly sown grass leys* [3]; *lawns, managed amenity turf* [4]
- Chickweed in *forage maize* [3]; *spring barley, spring oats, spring wheat, winter barley, winter oats, winter wheat* [1, 2]
- Cleavers in *forage maize* [3]; *spring barley, spring oats, spring wheat, winter barley, winter oats, winter wheat* [1, 2]
- Common mouse-ear in *amenity grassland* [4, 5]; *grass seed crops, grassland, newly sown grass leys* [3]; *lawns, managed amenity turf* [4]
- Creeping thistle in *forage maize* [3]; *spring barley, spring oats, spring wheat, winter barley, winter oats, winter wheat* [1]
- Daisies in *amenity grassland* [4, 5]; *grass seed crops, grassland, newly sown grass leys* [3]; *lawns, managed amenity turf* [4]
- Dandelions in *amenity grassland* [4, 5]; *grass seed crops, grassland, newly sown grass leys* [3]; *lawns, managed amenity turf* [4]
- Mayweeds in *forage maize* [3]; *spring barley, spring oats, spring wheat, winter barley, winter oats, winter wheat* [1]
- Plantains in *amenity grassland* [4, 5]; *grass seed crops, grassland, newly sown grass leys* [3]; *lawns, managed amenity turf* [4]
- Self-heal in *amenity grassland* [5]
- Slender speedwell in *amenity grassland* [5]

Approval information
- Clopyralid, florasulam and fluroxypyr are all included in Annex I under EC Regulation 1107/2009

Efficacy guidance
- Best results obtained when weeds are small and actively growing
- Effectiveness may be reduced when soil is very dry
- Use adequate water volume to achieve complete spray coverage of the weeds
- Florasulam is a member of the ALS-inhibitor group of herbicides

Restrictions
- Maximum number of treatments 1 per yr for all crops
- Do not spray when crops are under stress from any cause
- Do not apply through CDA applicators
- Do not roll or harrow for 7 d before or after application
- Do not use any treated plant material for composting or mulching
- Do not use manure from animals fed on treated crops for composting
- Specific restrictions apply to use in sequence or tank mixture with other sulfonylurea or ALS-inhibiting herbicides. See label for details

Crop-specific information
- Latest use: before second node detectable for oats; before third node detectable for spring barley and spring wheat; before flag leaf detectable in winter barley and winter wheat.

Following crops guidance
- Where residues of a treated crop have not completely decayed by the time of planting a succeeding crop, avoid planting peas, beans and other legumes, carrots and other Umbelliferae, potatoes, lettuce and other Compositae, glasshouse and protected crops
- Where the product has been used in mixture with certain named products (see label) only cereals or grass may be sown in the autumn following harvest. Otherwise cereals, oilseed rape, grass or vegetable brassicas as transplants may be sown as a following crop in the same calendar yr as treatment. Oilseed rape may show some temporary reduction of vigour after a dry summer, but yields are not affected
- In addition to the above, field beans, linseed, peas, sugar beet, potatoes, maize, clover (for use in grass/clover mixtures) or carrots may be sown in the calendar yr following treatment
- In the event of failure of a treated crop in spring, only spring wheat, spring barley, spring oats, maize or ryegrass may be sown

FOR FULL CONDITIONS OF USE ALWAYS READ THE PRODUCT LABEL

Environmental safety
- Dangerous for the environment
- Very toxic to aquatic organisms
- Take extreme care to avoid drift outside the target area
- Some pesticides pose a greater threat of contamination of water than others and clopyralid is one of these pesticides. Take special care when applying clopyralid near water and do not apply if heavy rain is forecast

Hazard classification and safety precautions
Hazard Harmful, Dangerous for the environment [1-5]; Harmful if inhaled [1, 3, 4]
Transport code 9
Packaging group III
UN Number 3082
Risk phrases H304, H319 [1, 3]; H315 [1, 3, 4]; H317 [3]; H320 [4]; R20, R36, R38, R50, R53a [2, 5]
Operator protection A, H [1-5]; C [1-3]; U05a, U11, U19a, U20a
Environmental protection E15b, E34, E38 [1-5]; H410 [1, 3, 4]
Storage and disposal D01, D02, D09a, D10b, D12a [1-5]; D05 [4, 5]
Medical advice M03 [1-5]; M05a [1-3]

105 clopyralid + fluroxypyr + MCPA

A translocated herbicide mixture for use in sports and amenity turf
HRAC mode of action code: O + O + O

See also fluroxypyr
MCPA

Products
Greenor	Rigby Taylor	20:40:200 g/l	ME	15204

Uses
- Annual dicotyledons in **managed amenity turf**

Approval information
- Clopyralid, fluroxypyr and MCPA included in Annex I under EC Regulation 1107/2009

Efficacy guidance
- Best results achieved when weeds actively growing and turf grass competitive
- Treatment should normally be between Apr-Sep when the soil is moist
- Do not apply during drought unless irrigation is applied
- Allow 3 d before or after mowing established turf to ensure sufficient weed leaf surface present to allow uptake and movement

Restrictions
- Maximum number of treatments 2 per yr
- Do not treat grass under stress from frost, drought, waterlogging, trace element deficiency, disease or pest attack
- Do not treat if night temperatures are low, when frost is imminent or during prolonged cold weather

Crop-specific information
- Treat young turf only in spring when at least 2 mth have elapsed since sowing
- Allow 5 d after mowing young turf before treatment
- Product selective on a number of turf grass species (see label) but consultation or testing recommended before treatment of any cultivar

Environmental safety
- Dangerous for the environment
- Very toxic to aquatic organisms
- Wash spray equipment thoroughly with water and detergent immediately after use. Traces of product can damage susceptible plants sprayed later

SEE SECTION 3 FOR PRODUCTS ALSO REGISTERED

- Some pesticides pose a greater threat of contamination of water than others and clopyralid is one of these pesticides. Take special care when applying clopyralid near water and do not apply if heavy rain is forecast
- LERAP Category B

Hazard classification and safety precautions
    Hazard Irritant, Dangerous for the environment
    Transport code 9
    Packaging group III
    UN Number 3082
    Risk phrases H317, H320
    Operator protection A, C; U05a, U08, U14, U19a, U20b
    Environmental protection E15a, E16a, E38, H410
    Storage and disposal D01, D02, D05, D09a, D10b, D12a

106   clopyralid + picloram

A post-emergence herbicide mixture for oilseed rape
HRAC mode of action code: O + O

See also picloram

Products

1	Clopic	UPL Europe	267:67 g/l	SL	17387
2	Galera	Dow	267:67 g/l	SL	16413
3	Legara	AgChem Access	267:67 g/l	SL	16789

Uses
- Annual dicotyledons in **corn gromwell** *(off-label)*, **crambe** *(off-label)*, **forest nurseries** *(off-label)*, **game cover** *(off-label)*, **mustard** *(off-label)*, **ornamental plant production** *(off-label)* [2]
- Cleavers in **corn gromwell** *(off-label)*, **crambe** *(off-label)*, **forest nurseries** *(off-label)*, **game cover** *(off-label)*, **mustard** *(off-label)*, **ornamental plant production** *(off-label)*, **spring oilseed rape** *(off-label)* [2]; **winter oilseed rape** [1-3]
- Mayweeds in **corn gromwell** *(off-label)*, **crambe** *(off-label)*, **forest nurseries** *(off-label)*, **game cover** *(off-label)*, **mustard** *(off-label)*, **ornamental plant production** *(off-label)* [2]; **winter oilseed rape** [1-3]

Extension of Authorisation for Minor Use (EAMUs)
- **corn gromwell** *20160559* [2]
- **crambe** *20161622* [2]
- **forest nurseries** *20142720* [2]
- **game cover** *20142720* [2]
- **mustard** *20152939* [2]
- **ornamental plant production** *20150968* [2]
- **spring oilseed rape** *20140827* [2]

Approval information
- Clopyralid and picloram included in Annex I under EC Regulation 1107/2009

Efficacy guidance
- Best results obtained from treatment when weeds are small and actively growing
- Cleavers that germinate after treatment will not be controlled

Restrictions
- Maximum total dose equivalent to one full dose treatment
- Do not treat crops under stress from cold, drought, pest damage, nutrient deficiency or any other cause
- Do not roll or harrow for 7 d before or after spraying
- Do not apply through CDA applicators
- Do not use any treated plant material for composting or mulching
- Do not use manure from animals fed on treated crops for composting

FOR FULL CONDITIONS OF USE ALWAYS READ THE PRODUCT LABEL

- Chop and incorporate all treated plant remains in early autumn, or as soon as possible after harvest, to release any residues into the soil. Ensure that all treated plant remains have completely decayed before planting susceptible crops

Crop-specific information

- Latest use: before flower buds visible above crop canopy for winter oilseed rape

Following crops guidance

- Wheat, barley, oats, maize, or oilseed rape may be sown 120 days after application, all other crops should only be sown 12 months after application [2]
- Ploughing or thorough cultivation should be carried out before planting leguminous crops
- Do not attempt to plant peas, beans, other legumes, carrots, other umbelliferous crops, potatoes, lettuce, other Compositae, or any glasshouse or protected crops if treated crop remains have not fully decayed by the time of planting
- In the event of failure of an autumn treated crop only oilseed rape, wheat, barley, oats, maize or ryegrass may be sown in the spring and only after ploughing or thorough cultivation

Environmental safety

- Dangerous for the environment
- Toxic to aquatic organisms
- Take extreme care to avoid drift onto crops and non-target plants outside the target area
- Some pesticides pose a greater threat of contamination of water than others and clopyralid is one of these pesticides. Take special care when applying clopyralid near water and do not apply if heavy rain is forecast

Hazard classification and safety precautions

Hazard Dangerous for the environment
UN Number N/C
Operator protection A, C; U05a
Environmental protection E15a, E34 [1-3]; E38, H413 [1, 2]
Storage and disposal D01, D02, D05, D07, D09a [1-3]; D12a [1, 2]

107   clopyralid + triclopyr

A perennial and woody weed herbicide for use in grassland
HRAC mode of action code: O + O

See also triclopyr

Products

1 Blaster Pro	Headland Amenity	60:240 g/l	EC	18074
2 Grazon Pro	Dow	60:240 g/l	EC	15785
3 Prevail	Dow	200:200 g/l	SL	17395
4 Thistlex	Dow	200:200 g/l	SL	16123

Uses

- Annual and perennial weeds in **grassland** [2]
- Annual dicotyledons in **game cover** (off-label) [4]
- Brambles in **amenity grassland** [1]; **grassland** [1, 2]
- Broom in **amenity grassland** [1]; **grassland** [1, 2]
- Creeping thistle in **grassland** [2-4]; **rotational grass** [3]
- Docks in **amenity grassland** [1]; **grassland** [1, 2]
- Gorse in **amenity grassland** [1]; **grassland** [1, 2]
- Mayweeds in **game cover** (off-label) [4]
- Perennial dicotyledons in **amenity grassland**, **grassland** [1]
- Stinging nettle in **amenity grassland**, **grassland** [1]
- Thistles in **amenity grassland**, **grassland** [1]; **game cover** (off-label) [4]

Extension of Authorisation for Minor Use (EAMUs)

- **game cover** 20142512 [4]

Approval information
- Clopyralid and triclopyr included in Annex I under EC Regulation 1107/2009

Efficacy guidance
- Must be applied to actively growing weeds
- Correct timing crucial for good control. Spray stinging nettle before flowering, docks in rosette stage in spring, creeping thistle before flower stems 15-20 cm high, brambles, broom and gorse in Jun-Aug
- Allow 2-3 wk regrowth after grazing or mowing before spraying perennial weeds
- Where there is a large reservoir of weed seed in the soil further treatment in the following yr may be needed
- [4] available as a twin pack with Doxstar as Pas.Tor for control of docks, nettles and thistles in grassland

Restrictions
- Maximum number of treatments 1 per yr
- Only use on permanent pasture or rotational grassland established for at least 1 yr
- Do not apply where clover is an important constituent of sward
- Do not roll or harrow within 10 d before or 7 d after spraying
- Do not cut grass for 21 d before or 28 d after spraying
- Do not use any treated plant material for composting or mulching, and do not use manure for composting from animals fed on treated crops
- Do not apply by hand-held equipment
- Do not allow drift onto other crops, amenity plantings or gardens, ponds, lakes or water courses. All conifers, especially pine and larch, are very sensitive
- Maximum concentration must not exceed 60 mls product per 10 litres of water (6 ml product per litre of water) [1, 2]

Crop-specific information
- Latest use: 7 d before grazing or cutting grass
- Some transient yellowing of treated swards may occur but is quickly outgrown

Following crops guidance
- Residues in plant tissues which have not completely decayed may affect succeeding susceptible crops such as peas, beans, other legumes, carrots, parsnips, potatoes, tomatoes, lettuce, glasshouse and protected crops
- Do not plant susceptible autumn-sown crops (eg winter beans) in same year as treatment and allow at least 9 mth from treatment before planting a susceptible crop in the following yr
- Do not direct drill kale, swedes, turnips, grass or grass mixtures within 6 wk of spraying
- Do not spray after end Jul where susceptible crops are to be planted in the next spring

Environmental safety
- Dangerous for the environment
- Very toxic to aquatic organisms
- Keep livestock out of treated areas for at least 7 d after spraying and until foliage of any poisonous weeds such as ragwort or buttercup has died down and become unpalatable
- Some pesticides pose a greater threat of contamination of water than others and clopyralid is one of these pesticides. Take special care when applying clopyralid near water and do not apply if heavy rain is forecast
- LERAP Category B

Hazard classification and safety precautions
Hazard Harmful, Flammable liquid and vapour [1, 2]; Irritant [1, 3, 4]; Dangerous for the environment [1-4]
Transport code 3 [1, 2]
Packaging group III [1, 2]
UN Number 1993 [1, 2]; N/C [3, 4]
Risk phrases H304, H315, H317, H319, H336 [1, 2]; H318, H373 [3, 4]; H335 [1-4]
Operator protection A, C [1-4]; H, M [1, 2]; U02a, U05a, U11, U20b [1-4]; U08, U14, U19a, U23b [1, 2]; U15 [1, 3, 4]; U23a [3, 4]
Environmental protection E07a, E15a, E23, E34, E38, H411 [1, 2]; E07c [3, 4] (7 days); E16a [1-4]; H410 [3, 4]

FOR FULL CONDITIONS OF USE ALWAYS READ THE PRODUCT LABEL

Consumer protection C01 [1, 2]
Storage and disposal D01, D02, D05, D09a, D10b, D12a
Medical advice M03 [3, 4]; M05b [1]

108   clothianidin

A nitromethylene neonicotinoid insecticide seed treatment for control of virus vectors, aphids and a reduction of damage by slugs and wireworms in winter cereals
IRAC mode of action code: 4A

See also beta-cyfluthrin + clothianidin

Products

Deter	Bayer CropScience	250 g/l	FS	12411

Uses

- Leafhoppers in **durum wheat** *(seed treatment)*, **triticale** *(seed treatment)*, **winter barley** *(seed treatment)*, **winter oats** *(seed treatment)*, **winter rye** *(seed treatment)*, **winter wheat** *(seed treatment)*
- Slugs in **durum wheat** *(seed treatment)*, **triticale** *(seed treatment)*, **winter barley** *(seed treatment)*, **winter oats** *(seed treatment)*, **winter rye** *(seed treatment)*, **winter wheat** *(seed treatment)*
- Virus vectors in **durum wheat** *(seed treatment)*, **triticale** *(seed treatment)*, **winter barley** *(seed treatment)*, **winter oats** *(seed treatment)*, **winter rye** *(seed treatment)*, **winter wheat** *(seed treatment)*
- Wireworm in **durum wheat** *(seed treatment)*, **triticale** *(seed treatment)*, **winter barley** *(seed treatment)*, **winter oats** *(seed treatment)*, **winter rye** *(seed treatment)*, **winter wheat** *(seed treatment)*

Approval information

- Clothianidin included in Annex I under EC Regulation 1107/2009
- Accepted by BBPA for use on malting barley

Efficacy guidance

- May only be used in conjunction with manufacturer's approved seed treatment application equipment or by following procedures given in the operating instructions
- Treated seed should preferably be drilled in the same season
- Evenness of seed cover improved by simultaneous application with a small volume (1.5-3 litres/tonne) of water
- Calibrate drill for treated seed and drill at 2.5-4 cm into firm, well prepared seedbed
- Use minimum 125 kg treated seed per ha
- Seed should be drilled to a depth of 40 mm into a well prepared and firm seedbed. If seed is present on the surface, or if spills have occurred, the field should be harrowed and rolled if conditions are appropriate
- When aphid activity is unusually late, or is heavy and prolonged in areas of high risk, and mild weather predominates, follow-up foliar aphicide may be required
- Incidental suppression of leafhoppers in early spring (Mar/Apr) may be achieved but if a specific attack develops an additional foliar insecticide may be required

Restrictions

- Maximum number of seed treatments one per batch
- Product must be fully re-dispersed and homogeneous before use
- Do not use on seed with more than 16% moisture content, or on sprouted, cracked or skinned seed
- All seed batches should be tested to ensure they are suitable for treatment
- Do not sow crops treated with clothianidin between 1st January and 30th June

Crop-specific information

- Latest use: pre-drilling

Environmental safety
- Harmful to game and wildlife. Treated seed should not be left on the surface. Bury spillages.

Hazard classification and safety precautions
    Hazard Irritant, Very toxic to aquatic organisms
    UN Number N/C
    Risk phrases H317
    Operator protection A, H; U04a, U05a, U07, U13, U14, U20b
    Environmental protection E03, E15b, E34, E36a, H410
    Storage and disposal D01, D02, D05, D09a, D14
    Treated seed S01, S02, S03, S04b, S05, S06a, S06b, S07, S08
    Medical advice M03

109   clothianidin + prothioconazole

A combined fungicide and insecticide seed dressing for cereals
IRAC mode of action code: 4A + FRAC 3

See also prothioconazole

Products

Redigo Deter	Bayer CropScience	250:50 g/l	FS	12423

Uses
- Bunt in **durum wheat** *(seed treatment)*, **triticale** *(seed treatment)*, **winter rye** *(seed treatment)*, **winter wheat** *(seed treatment)*
- Covered smut in **winter barley** *(seed treatment)*
- Fusarium foot rot and seedling blight in **durum wheat** *(seed treatment)*, **triticale** *(seed treatment)*, **winter barley** *(seed treatment)*, **winter oats** *(seed treatment)*, **winter rye** *(seed treatment)*, **winter wheat** *(seed treatment)*
- Leaf stripe in **winter barley** *(seed treatment)*
- Leafhoppers in **durum wheat** *(seed treatment)*, **triticale** *(seed treatment)*, **winter barley** *(seed treatment)*, **winter oats** *(seed treatment)*, **winter rye** *(seed treatment)*, **winter wheat** *(seed treatment)*
- Loose smut in **durum wheat** *(seed treatment)*, **winter barley** *(seed treatment)*, **winter oats** *(seed treatment)*, **winter wheat** *(seed treatment)*
- Slugs in **durum wheat** *(seed treatment)*, **triticale** *(seed treatment)*, **winter barley** *(seed treatment)*, **winter oats** *(seed treatment)*, **winter rye** *(seed treatment)*, **winter wheat** *(seed treatment)*
- Virus vectors in **durum wheat** *(seed treatment)*, **triticale** *(seed treatment)*, **winter barley** *(seed treatment)*, **winter oats** *(seed treatment)*, **winter rye** *(seed treatment)*, **winter wheat** *(seed treatment)*
- Wireworm in **durum wheat** *(seed treatment)*, **triticale** *(seed treatment)*, **winter barley** *(seed treatment)*, **winter oats** *(seed treatment)*, **winter rye** *(seed treatment)*, **winter wheat** *(seed treatment)*

Approval information
- Clothianidin and prothioconazole included in Annex I under EC Regulation 1107/2009
- Accepted by BBPA for use on malting barley

Efficacy guidance
- May only be used in conjunction with manufacturer's approved seed treatment application equipment or by following procedures given in the operating instructions
- Treated wheat seed should preferably be drilled in the same season. Treated seed of other cereals must be used in the same season
- Evenness of seed cover improved by simultaneous application with a small volume (1.5-3 litres/tonne) of water
- Calibrate drill for treated seed and drill at 2.5-4 cm into firm, well prepared seedbed
- Use minimum 125 kg treated seed per ha
- When aphid activity unusually late, or is heavy and prolonged in areas of high risk, and mild weather predominates, follow-up foliar aphicide may be required

FOR FULL CONDITIONS OF USE ALWAYS READ THE PRODUCT LABEL

- Incidental suppression of leafhoppers in early spring (Mar/Apr) may be achieved but if a specific attack develops an additional foliar insecticide may be required
- Protection against slugs only applies to germinating seeds and not to aerial parts after emergence
- Where very high populations of aphids, wireworms or slugs occur additional specific control measures may be needed

Restrictions
- Maximum number of seed treatments one per batch for all cereals
- Product must be fully re-dispersed and homogeneous before use
- Do not use on seed with more than 16% moisture content, or on sprouted, cracked or skinned seed
- All seed batches should be tested to ensure they are suitable for treatment
- Do not sow treated crops between 1st Jan and 30th June

Crop-specific information
- Latest use: pre-drilling of all cereals

Environmental safety
- Harmful to aquatic organisms

Hazard classification and safety precautions
Hazard Irritant, Very toxic to aquatic organisms
UN Number N/C
Risk phrases H317
Operator protection A, H; U14, U19a, U20b
Environmental protection E15a, E34, E38, H410
Storage and disposal D01, D02, D09a, D11a, D12a
Treated seed S01, S02, S03, S04b, S05, S06a, S07, S08

110   Coniothyrium minitans

A fungal parasite of sclerotia in soil
FRAC mode of action code: 44

Products

Contans WG	Bayer CropScience	1000 IU/mg	WG	17985

Uses
- Sclerotinia in **all edible crops (outdoor)**, **all non-edible crops (outdoor)**

Approval information
- Coniothyrium minitans included in Annex 1 under EC Regulation 1107/2009

Efficacy guidance
- *C.minitans* is a soil acting biological fungicide with specific action against the resting bodies (sclerotia) of *Sclerotinia sclerotiorum* and *S.minor*
- Treat 3 mth before disease protection is required to allow time for the infective sclerotia in the soil to be reduced
- A post-harvest treatment of soil and debris prevents further contamination of the soil with sclerotia produced by the previous crop
- For best results the soil should be moist and the temperature between 12-20°C
- Application should be followed by soil incorporation into the surface layer to 10 cm with a rotovator or rotary harrow. Thorough spray coverage of the soil is essential to ensure uniform distribution
- If soil temperature drops below 0°C or rises above 27°C fungicidal activity is suspended but restarts when soil temperature returns within this range
- In glasshouses untreated areas at the margins should be covered by a film to avoid spread of spores from untreated sclerotia

SEE SECTION 3 FOR PRODUCTS ALSO REGISTERED

Restrictions
- Maximum number of treatments 1 per crop or situation
- Do not plough or cultivate after treatment
- Do not mix with other pesticides, acids, alkalines or any product that attacks organic material
- Store product under cool dry conditions away from direct sunlight and away from heat sources

Crop-specific information
- Latest use: pre-planting
- HI: zero

Following crops guidance
- There are no restrictions on following crops and no waiting interval is specified

Hazard classification and safety precautions
    UN Number N/C
    Operator protection A, C, D; U05a, U19a
    Environmental protection E15b, E34
    Storage and disposal D01, D02, D12a
    Medical advice M03

111   copper oxychloride

A protectant copper fungicide and bactericide
FRAC mode of action code: M1

Products

Cuprokylt	Certis	87% w/w (copper)	WP	17079

Uses
- Dothistroma needle blight in *forest nurseries* *(off-label)*
- Downy mildew in *table grapes*, *wine grapes*

Extension of Authorisation for Minor Use (EAMUs)
- *forest nurseries* *20152416*

Approval information
- Copper oxychloride included in Annex 1 under EC Regulation 1107/2009
- Accepted by BBPA for use on hops

Efficacy guidance
- Spray crops at high volume when foliage dry but avoid run off. Do not spray if rain expected soon
- Spray interval commonly 10-14 d but varies with crop, see label for details

Environmental safety
- Dangerous for the environment
- Very toxic to aquatic organisms
- Keep all livestock out of treated areas for at least 3 wks
- Buffer zone requirement 18 m in potatoes and ornamentals when using a horizontal sprayer, 50 m when using broadcast air-assisted sprayer on crops up to 1.2m height
- LERAP Category B

Hazard classification and safety precautions
    Hazard Dangerous for the environment, Harmful if swallowed, Harmful if inhaled, Very toxic to aquatic organisms
    Transport code 9
    Packaging group III
    UN Number 3077
    Operator protection U20a
    Environmental protection E06a (3 wk); E13c, E16a, E34, E38, H410, H411
    Storage and disposal D01, D09a, D10b

FOR FULL CONDITIONS OF USE ALWAYS READ THE PRODUCT LABEL

112   cyantraniliprole

An ingested and contact insecticide for insect pest control in brassica crops
IRAC mode of action code: 28

Products

Verimark 20 SC	DuPont	200 g/l	SC	17519

Uses

* Cabbage root fly in *broccoli*, *brussels sprouts*, *cabbages*, *cauliflowers*

Approval information

* Cyantraniliprole is included in Annex 1 under EC Regulation 1107/2009

Efficacy guidance

* May be used as part of an Integrated Pest Management (IPM) program

Restrictions

* To protect bees and pollinating insects avoid application to crops when in flower.
* Consult processor when the product is to be applied to crops grown for processing

Hazard classification and safety precautions

Hazard Very toxic to aquatic organisms
Transport code 9
Packaging group III
UN Number 3082
Operator protection U05a, U09a, U19a, U20a
Environmental protection E12e, E15b, H410
Storage and disposal D01, D02, D09a, D12a, D22
Medical advice M03

113   cyazofamid

A cyanoimidazole sulfonamide protectant fungicide for potatoes
FRAC mode of action code: 21

Products

1	Linford	AgChem Access	400 g/l	SC	13824
2	Ranman Top	Belchim	160 g/l	SC	14753
3	Swallow	AgChem Access	400 g/l	SC	14078

Uses

* Blight in *potatoes*

Approval information

* Cyazofamid included in Annex I under EC Regulation 1107/2009

Efficacy guidance

* Apply as a protectant treatment before blight enters the crop and repeat every 5 -10 d depending on severity of disease pressure
* Commence spray programme immediately the risk of blight in the locality occurs, usually when the crop meets along the rows
* Product must always be used with organosilicone adjuvant provided in the twin pack
* To minimise the chance of development of resistance no more than three applications should be made consecutively (out of a permissible total of six) in the blight control programme. For more information on Resistance Management see Section 5

Restrictions

* Maximum number of treatments 6 per crop (no more than three of which should be consecutive)
* Mixed product must not be allowed to stand overnight
* Consult processor before using on crops intended for processing

SECTION 2

SEE SECTION 3 FOR PRODUCTS ALSO REGISTERED

Crop-specific information
- HI 7 d

Environmental safety
- Dangerous for the environment
- Very toxic to aquatic organisms
- Do not empty into drains

Hazard classification and safety precautions

Hazard Harmful [1]; Harmful [3] (adjuvant); Irritant [2]; Dangerous for the environment [1-3]
Transport code 9
Packaging group III
UN Number 3082
Risk phrases H315 [2]; H319 [1]; R20, R36, R48 [3] (adjuvant); R41, R50, R53a [3]
Operator protection A, C; U02a, U05a, U11, U15 [1-3]; U19a [1]; U19a [3] (adjuvant); U20a [1, 3]; U20b [2]
Environmental protection E15a, E34 [1, 3]; E15b, H410 [2]; E19b, E38 [1-3]; H412 [1]
Storage and disposal D01, D02, D09a, D10c, D12a [1-3]; D05, D12b [2]
Medical advice M03

114 cycloxydim

A translocated post-emergence cyclohexanedione oxime herbicide for grass weed control
HRAC mode of action code: A

Products

Laser	BASF	200 g/l	EC	17339

Uses
- Agrostis spp. in **managed amenity turf** *(off-label)*
- Annual grasses in **angelica** *(off-label)*, **baby leaf crops** *(off-label)*, **beans without pods (fresh)** *(off-label)*, **broad beans** *(off-label)*, **broccoli** *(off-label)*, **calabrese** *(off-label)*, **caraway** *(off-label)*, **carrots**, **cauliflowers**, **celeriac** *(off-label)*, **celery leaves** *(off-label)*, **christmas trees**, **combining peas**, **coriander** *(off-label)*, **cress** *(off-label)*, **dill** *(off-label)*, **dwarf beans**, **early potatoes**, **edible podded peas** *(off-label)*, **endives** *(off-label)*, **fennel leaves** *(off-label)*, **fodder beet**, **forest**, **forest nurseries**, **herbs (see appendix 6)** *(off-label)*, **horseradish** *(off-label)*, **jerusalem artichokes** *(off-label)*, **lamb's lettuce** *(off-label)*, **land cress** *(off-label)*, **leeks**, **lentils** *(off-label)*, **lettuce** *(off-label)*, **linseed**, **lovage** *(off-label)*, **maincrop potatoes**, **mangels**, **ornamental plant production**, **ornamental plant production** *(off-label)*, **purslane** *(off-label)*, **red beet** *(off-label)*, **rocket** *(off-label)*, **runner beans** *(off-label)*, **salad burnet** *(off-label)*, **salad onions**, **salsify** *(off-label)*, **soya beans** *(off-label)*, **spinach** *(off-label)*, **spinach beet** *(off-label)*, **spring field beans**, **spring oilseed rape**, **strawberries**, **sugar beet**, **swedes**, **sweet ciceley** *(off-label)*, **turnips** *(off-label)*, **vining peas**, **watercress** *(off-label)*, **winter field beans**, **winter oilseed rape**
- Bent grasses in **managed amenity turf** *(off-label)*
- Black bent in **carrots**, **cauliflowers**, **combining peas**, **dwarf beans**, **early potatoes**, **fodder beet**, **leeks**, **linseed**, **maincrop potatoes**, **mangels**, **salad onions**, **spring field beans**, **spring oilseed rape**, **strawberries**, **sugar beet**, **swedes**, **vining peas**, **winter field beans**, **winter oilseed rape**
- Blackgrass in **carrots**, **cauliflowers**, **combining peas**, **dwarf beans**, **early potatoes**, **fodder beet**, **leeks**, **linseed**, **maincrop potatoes**, **mangels**, **salad onions**, **spring field beans**, **spring oilseed rape**, **strawberries**, **sugar beet**, **swedes**, **vining peas**, **winter field beans**, **winter oilseed rape**
- Couch in **carrots**, **cauliflowers**, **combining peas**, **dwarf beans**, **early potatoes**, **fodder beet**, **leeks**, **linseed**, **maincrop potatoes**, **mangels**, **salad onions**, **spring field beans**, **spring oilseed rape**, **strawberries**, **sugar beet**, **swedes**, **vining peas**, **winter field beans**, **winter oilseed rape**
- Creeping bent in **carrots**, **cauliflowers**, **combining peas**, **dwarf beans**, **early potatoes**, **fodder beet**, **leeks**, **linseed**, **maincrop potatoes**, **mangels**, **salad onions**, **spring field beans**, **spring oilseed rape**, **strawberries**, **sugar beet**, **swedes**, **vining peas**, **winter field beans**, **winter oilseed rape**

FOR FULL CONDITIONS OF USE ALWAYS READ THE PRODUCT LABEL

- Green cover in *land temporarily removed from production*
- Onion couch in *carrots, cauliflowers, combining peas, dwarf beans, early potatoes, fodder beet, leeks, linseed, maincrop potatoes, mangels, salad onions, spring field beans, spring oilseed rape, strawberries, sugar beet, swedes, vining peas, winter field beans, winter oilseed rape*
- Perennial grasses in *angelica* (off-label), *beans without pods (fresh)* (off-label), *broad beans* (off-label), *caraway* (off-label), *celeriac* (off-label), *celery leaves* (off-label), *christmas trees, coriander* (off-label), *cress* (off-label), *dill* (off-label), *edible podded peas* (off-label), *endives* (off-label), *fennel leaves* (off-label), *forest, forest nurseries, horseradish* (off-label), *jerusalem artichokes* (off-label), *lamb's lettuce* (off-label), *land cress* (off-label), *lentils* (off-label), *lettuce* (off-label), *lovage* (off-label), *ornamental plant production, ornamental plant production* (off-label), *purslane* (off-label), *red beet* (off-label), *rocket* (off-label), *runner beans* (off-label), *salad burnet* (off-label), *salsify* (off-label), *soya beans* (off-label), *spinach* (off-label), *spinach beet* (off-label), *sweet ciceley* (off-label), *turnips* (off-label)
- Volunteer cereals in *angelica* (off-label), *baby leaf crops* (off-label), *beans without pods (fresh)* (off-label), *broad beans* (off-label), *broccoli* (off-label), *calabrese* (off-label), *caraway* (off-label), *carrots, cauliflowers, celery leaves* (off-label), *combining peas, coriander* (off-label), *cress* (off-label), *dill* (off-label), *dwarf beans, early potatoes, edible podded peas* (off-label), *endives* (off-label), *fennel leaves* (off-label), *fodder beet, herbs (see appendix 6)* (off-label), *lamb's lettuce* (off-label), *land cress* (off-label), *leeks, lentils* (off-label), *lettuce* (off-label), *linseed, lovage* (off-label), *maincrop potatoes, mangels, purslane* (off-label), *rocket* (off-label), *runner beans* (off-label), *salad burnet* (off-label), *salad onions, soya beans* (off-label), *spinach* (off-label), *spinach beet* (off-label), *spring field beans, spring oilseed rape, strawberries, sugar beet, swedes, sweet ciceley* (off-label), *vining peas, watercress* (off-label), *winter field beans, winter oilseed rape*
- Wild oats in *carrots, cauliflowers, combining peas, dwarf beans, early potatoes, fodder beet, leeks, linseed, maincrop potatoes, mangels, salad onions, spring field beans, spring oilseed rape, strawberries, sugar beet, swedes, vining peas, winter field beans, winter oilseed rape*

Extension of Authorisation for Minor Use (EAMUs)
- *angelica* 20171266
- *baby leaf crops* 20171266
- *beans without pods (fresh)* 20171265
- *broad beans* 20171265
- *broccoli* 20171267
- *calabrese* 20171267
- *caraway* 20171266
- *celeriac* 20170438
- *celery leaves* 20171266
- *coriander* 20171266
- *cress* 20171266
- *dill* 20171266
- *edible podded peas* 20171265
- *endives* 20171266
- *fennel leaves* 20171266
- *herbs (see appendix 6)* 20171266
- *horseradish* 20170438
- *jerusalem artichokes* 20170438
- *lamb's lettuce* 20171266
- *land cress* 20171266
- *lentils* 20171265
- *lettuce* 20171266
- *lovage* 20171266
- *managed amenity turf* 20171715
- *ornamental plant production* 20170437
- *purslane* 20171266
- *red beet* 20170438
- *rocket* 20171266

SEE SECTION 3 FOR PRODUCTS ALSO REGISTERED

- ***runner beans*** *20171265*
- ***salad burnet*** *20171266*
- ***salsify*** *20170438*
- ***soya beans*** *20171265*
- ***spinach*** *20171266*
- ***spinach beet*** *20171266*
- ***sweet ciceley*** *20171266*
- ***turnips*** *20170438*
- ***watercress*** *20171266*

Approval information
- Cycloxydim included in Annex 1 under EC Regulation 1107/2009

Efficacy guidance
- Best results achieved when weeds small and have not begun to compete with crop. Effectiveness reduced by drought, cool conditions or stress. Weeds emerging after application are not controlled
- Foliage death usually complete after 3-4 wk but longer under cool conditions, especially late treatments to winter oilseed rape
- Perennial grasses should have sufficient foliage to absorb spray and should not be cultivated for at least 14 d after treatment
- On established couch pre-planting cultivation recommended to fragment rhizomes and encourage uniform emergence
- Split applications to volunteer wheat and barley at GS 12-14 will often give adequate control in winter oilseed rape. See label for details
- Apply to dry foliage when rain not expected for at least 2 h
- Cycloxydim is an ACCase inhibitor herbicide. To avoid the build up of resistance do not apply products containing an ACCase inhibitor herbicide more than twice to any crop. In addition do not use any product containing cycloxydim in mixture or sequence with any other product containing the same ingredient
- Use these products as part of a resistance management strategy that includes cultural methods of control and does not use ACCase inhibitors as the sole chemical method of grass weed control
- Applying a second product containing an ACCase inhibitor to a crop will increase the risk of resistance development; only use a second ACCase inhibitor to control different weeds at a different timing
- Always follow WRAG guidelines for preventing and managing herbicide resistant weeds. See Section 5 for more information

Restrictions
- Maximum number of treatments 1 per crop or yr in most situations. See label
- Must be used with authorised adjuvant oil. See label
- Do not apply to crops damaged or stressed by adverse weather, pest or disease attack or other pesticide treatment
- Prevent drift onto other crops, especially cereals and grass

Crop-specific information
- HI cabbage, cauliflower, calabrese, salad onions 4 wk; peas, dwarf beans 5 wk; bulb onions, carrots, parsnips, strawberries 6 wk; sugar and fodder beet, leeks, mangels, potatoes, field beans, swedes, Brussels sprouts, winter field beans 8 wk; oilseed rape, soya beans, linseed 12 wk
- Recommended time of application varies with crop. See label for details
- On peas a crystal violet wax test should be done if leaf wax likely to have been affected by weather conditions or other chemical treatment. The wax test is essential if other products are to be sprayed before or after treatment
- May be used on ornamental bulbs when crop 5-10 cm tall. Product has been used on tulips, narcissi, hyacinths and irises but some subjects may be more sensitive and growers advised to check tolerance on small number of plants before treating the rest of the crop

FOR FULL CONDITIONS OF USE ALWAYS READ THE PRODUCT LABEL

- May be applied to land temporarily removed from production where the green cover is made up predominantly of tolerant crops listed on label. Use on industrial crops of linseed and oilseed rape on land temporarily removed from production also permitted.

Following crops guidance

- Guideline intervals for sowing succeeding crops after failed treated crop: field beans, peas, sugar beet, rape, kale, swedes, radish, white clover, lucerne 1 wk; dwarf French beans 4 wk; wheat, barley, maize 8 wk
- Oats should not be sown after failure of a treated crop

Environmental safety

- Dangerous for the environment
- Toxic to aquatic organisms
- Harmful to fish or other aquatic life. Do not contaminate surface waters or ditches with chemical or used container

Hazard classification and safety precautions

Hazard Harmful, Dangerous for the environment
Transport code 9
Packaging group III
UN Number 3082
Risk phrases H304, H315, H318, H336, H361
Operator protection A, C, H; U05a, U08, U20b
Environmental protection E15a, E38, H411
Storage and disposal D01, D02, D09a, D10c, D12a
Medical advice M05b

115 Cydia pomonella GV

An baculovirus insecticide for codling moth control

Products

1	Carpovirusine	Arysta	1.0 x10^13 GV/l	SC	15243
2	Carpovirusine EVO 2	Arysta	1.0 x10^13 GV/l	SC	17565
3	Cyd-X Xtra	Certis	1% w/w virus	SC	17020

Uses

- Codling moth in *apples*, *pears*

Approval information

- Cydia pomonella included in Annex I under EC Regulation 1107/2009

Efficacy guidance

- Optimum results require a precise spraying schedule commencing soon after egg laying and before first hatch
- First generation of codling moth normally occurs in first 2 wk of Jun in UK
- Apply in sufficient water to achieve good coverage of whole tree
- Normally 3 applications at intervals of 8 sunny days are needed for each codling moth generation
- Fruit damage is reduced significantly in the first yr and the population of codling moth in the following year will be reduced

Restrictions

- Maximum number of treatments 3 per season
- Do not tank mix with copper or any products with a pH lower than 5 or higher than 8
- Do not use as the exclusive measure for codling moth control. Adopt an anti-resistance strategy. See Section 5 for more information
- Consult processor before use on crops for processing

Crop-specific information

- HI for apples and pears 1 day [1]
- Evidence of efficacy and crop safety on pears is limited [3]

SEE SECTION 3 FOR PRODUCTS ALSO REGISTERED

Environmental safety

- Product has no adverse effect on predatory insects and is fully compatible with IPM programmes

Hazard classification and safety precautions

Hazard Harmful [3]
UN Number N/C
Risk phrases H317 [2]; R42, R43 [3]
Operator protection A, D, H; U05a, U16b [1, 2]; U14, U19a [1-3]
Environmental protection E15a [3]; E15b, E34 [1, 2]
Storage and disposal D01, D02, D09a, D10b [1, 2]; D12a [1-3]
Medical advice M05a [1, 2]

116   cyflufenamid

An amidoxime fungicide for cereals
FRAC mode of action code: U6

Products

1	Clayton Roulette	Clayton	50 g/l	EW	18017
2	Cosine	Certis	50 g/l	EW	16404
3	Cyflamid	Certis	50 g/l	EW	12403
4	Takumi SC	Certis	100 g/l	SC	16000
5	Vegas	Certis	50 g/l	EW	15238

Uses

- Disease control in *courgettes, cucumbers, melons, summer squash* [4]
- Powdery mildew in *apples, pears, wine grapes* (off-label) [2]; *courgettes, cucumbers, melons, ornamental plant production* (off-label), *protected tomatoes* (off-label), *pumpkins* (off-label), *strawberries* (off-label), *summer squash, winter squash* (off-label) [4]; *durum wheat, spring barley, spring oats, spring wheat, triticale, winter barley, winter oats, winter rye, winter wheat* [1, 3, 5]; *forest nurseries* (off-label), *grass seed crops* (off-label), *hops* (off-label), *rye* (off-label), *soft fruit* (off-label) [3]

Extension of Authorisation for Minor Use (EAMUs)

- *forest nurseries* 20082915 [3]
- *grass seed crops* 20060602 [3]
- *hops* 20082915 [3]
- *ornamental plant production* 20131294 [4]
- *protected tomatoes* 20140800 [4]
- *pumpkins* 20162915 [4]
- *rye* 20060602 [3]
- *soft fruit* 20082915 [3]
- *strawberries* 20162055 [4]
- *wine grapes* 20170846 [2]
- *winter squash* 20162915 [4]

Approval information

- Cyflufenamid included in Annex 1 under EC Regulation 1107/2009
- Accepted by BBPA on malting barley

Efficacy guidance

- Best results obtained from treatment at the first visible signs of infection by powdery mildew
- Sustained disease pressure may require a second treatment
- Disease spectrum may be broadened by appropriate tank mixtures. See label
- Product is rainfast within 1 h
- Must be used as part of an Integrated Crop Management programme that includes alternating use, or mixture, with fungicides with a different mode of action effective against powdery mildew

Restrictions

- Maximum number of treatments 2 per crop on all recommended cereals

FOR FULL CONDITIONS OF USE ALWAYS READ THE PRODUCT LABEL

- Apply only in the spring
- Do not apply to crops under stress from drought, waterlogging, cold, pests or diseases, lime or nutrient deficiency or other factors affecting crop growth

Crop-specific information
- Latest use: before start of flowering of cereal crops

Following crops guidance
- No restrictions

Environmental safety
- Dangerous for the environment
- Very toxic to aquatic organisms
- Avoid drift onto ponds, waterways or ditches

Hazard classification and safety precautions
Hazard Harmful, Harmful if swallowed [2]; Dangerous for the environment [1-5]
Transport code 9 [1-3, 5]
Packaging group III [1-3, 5]
UN Number 3082 [1-3, 5]; N/C [4]
Operator protection A; U05a
Environmental protection E15b, E34, H411 [1-5]; E17a [2] (5 m); E38 [2]
Storage and disposal D01, D02, D05, D10c, D12b
Medical advice M05b [2]

---

117   cymoxanil

A cyanoacetamide oxime fungicide for potatoes
FRAC mode of action code: 27

See also chlorothalonil + cymoxanil
cyazofamid + cymoxanil

Products

1	Cymbal 45	Belchim	45% w/w	WG	16647
2	Danso Flow	Belchim	225 g/l	SC	17963
3	Dauphin 45	Clayton	45% w/w	WG	16434
4	Drum	Belchim	45% w/w	WG	16750
5	Krug Flow	Belchim	225 g/l	SC	17964
6	Option	DuPont	60% w/w	WG	16959
7	Sacron WG	UPL Europe	45% w/w	WG	16433
8	Sipcam C50 WG	Sipcam	50% w/w	WG	16743

Uses
- Blight in *early potatoes* [8]; *potatoes* [1-8]
- Downy mildew in *hops* *(off-label)*, *wine grapes* *(off-label)* [6]
- Late blight in *potatoes* [3]

Extension of Authorisation for Minor Use (EAMUs)
- *hops* *20161101* [6]
- *wine grapes* *20160542* [6]

Approval information
- Cymoxanil included in Annex I under EC Regulation 1107/2009
- Accepted by BBPA for use on hops

Efficacy guidance
- Product to be used in mixture with specified mixture partners in order to combine systemic and protective activity. See labels for details
- Commence spray programme as soon as weather conditions favourable for disease development occur and before infection appears. At latest the first treatment should be made as the foliage meets along the rows

SEE SECTION 3 FOR PRODUCTS ALSO REGISTERED

- Repeat treatments at 7-14 day intervals according to disease incidence and weather conditions
- Treat irrigated crops as soon as possible after irrigation and repeat at 10 d intervals

Restrictions
- Minimum spray interval 7 d
- Do not allow packs to become wet during storage and do not keep opened packs from one season to another
- Cymoxanil must be used in tank mixture with other fungicides

Crop-specific information
- HI for potatoes 1 d [6], 7 d [1, 4, 8], 14 d [7]
- Consult processor before treating potatoes grown for processing

Environmental safety
- Dangerous for the environment
- Very toxic to aquatic organisms
- LERAP Category B [3]

Hazard classification and safety precautions
Hazard Harmful [1-3, 5-8]; Irritant [4]; Dangerous for the environment [1-7]; Harmful if swallowed [1, 6-8]; Very toxic to aquatic organisms [6]
Transport code 9 [1-7]
Packaging group III [1-7]
UN Number 3077 [1, 3, 4, 6, 7]; 3082 [2, 5]; N/C [8]
Risk phrases H314 [2, 5]; H317 [1, 2, 4-8]; H319 [4]; H361, H373 [1, 2, 4, 5, 7]; R22a, R43, R48, R51, R53a, R62, R63 [3]
Operator protection A [1-8]; C [2, 5]; D [3]; H [1-5, 7]; U05a [1-8]; U08 [6, 8]; U09a [1-5, 7]; U14 [1-5, 7, 8]; U19a [6]; U20a [1-3, 5]; U20b [4, 6-8]
Environmental protection E13a, E34 [6]; E13c [8]; E15a [1-5, 7]; E16a [3]; E38 [1-7]; H410 [4, 6]; H411 [1, 2, 5, 7, 8]
Storage and disposal D01, D02, D09a, D11a [1-8]; D12a [1-7]
Medical advice M03 [6, 8]; M05a [1-3, 5, 7]

118   cymoxanil + fluazinam

A fungicide mixture for blight control in potatoes
FRAC mode of action code: 27 + 29

Products

1 Grecale	Sipcam	200:300 g/l	SC	17002
2 Kunshi	Belchim	25:37.5% w/w	WG	16991
3 Plexus	Headland	200:300 g/l	SC	17099
4 Tezuma	Belchim	25:37.5% w/w	WG	17396

Uses
- Blight in **potatoes**
- Late blight in **potatoes**
- Phytophthora in **potatoes**

Approval information
- Cymoxanil and fluazinam included in Annex 1 under EC Regulation 1107/2009

Restrictions
- Use as part of a programme with fungicides acting via a different mode of action to reduce the risk of resistance
- Use low drift spraying equipment up to 30m from the top of the bank of any surface water bodies [1, 2, 4]

Following crops guidance
- No restrictions on following crops

FOR FULL CONDITIONS OF USE ALWAYS READ THE PRODUCT LABEL

SECTION 2

Environmental safety
- Buffer zone requirement 6 m [1, 2], 7 m [3]
- To protect non target insects/arthropods respect an unsprayed buffer zone of 5 m to non-crop land [1, 3]
- Horizontal boom sprayers must be fitted with three star drift reduction technology for all uses [2, 4]
- LERAP Category B [3]

Hazard classification and safety precautions
Hazard Dangerous for the environment [1]; Harmful if swallowed [1, 3]; Very toxic to aquatic organisms [1-4]
Transport code 9
Packaging group III
UN Number 3077 [2, 4]; 3082 [1, 3]
Risk phrases H317, H361 [1-4]; H319 [2, 4]; H373 [1, 2, 4]
Operator protection A, C, H [1, 3]; U05a, U20d [3]
Environmental protection E15b, E16a, E38 [3]; H410 [1-4]
Medical advice M03 [3]

119   cymoxanil + fludioxonil + metalaxyl-M

A fungicide seed dressing for peas
FRAC mode of action code: 27 + 12 + 4

See also fludioxonil
metalaxyl-M

Products
Wakil XL	Syngenta	10:5:17.5% w/w	WS	17217

Uses
- Damping off in **carrots** *(off-label)*, **parsnips** *(off-label)*
- Downy mildew in **combining peas**, **vining peas**
- Pythium in **carrots** *(off-label)*, **parsnips** *(off-label)*

Extension of Authorisation for Minor Use (EAMUs)
- **carrots** *20170896*
- **parsnips** *20170896*

Approval information
- Cymoxanil, fludioxonil and metalaxyl-M included in Annex I under EC Regulation 1107/2009

Efficacy guidance
- Apply through continuous flow seed treaters which should be calibrated before use

Restrictions
- Max number of treatments 1 per seed batch
- Ensure moisture content of treated seed satisfactory and store in a dry place
- Check calibration of seed drill with treated seed before drilling and sow as soon as possible after treatment
- Consult before using on crops for processing

Crop-specific information
- Latest use: pre-drilling

Environmental safety
- Harmful to aquatic organisms
- Do not use treated seed as food or feed

Hazard classification and safety precautions
UN Number N/C
Risk phrases H361, H373
Operator protection A, D, H; U02a, U05a, U20b

SEE SECTION 3 FOR PRODUCTS ALSO REGISTERED

Environmental protection E03, E15a, H411
Storage and disposal D01, D02, D05, D07, D09a, D11a, D12a
Treated seed S02, S04b, S05, S07

120   cymoxanil + mancozeb

A protectant and systemic fungicide for potato blight control
FRAC mode of action code: 27 + M3

See also mancozeb

Products

1	Curzate M WG	DuPont	4.5:68% w/w	WG	11901
2	Cymozeb	Belchim	4.5:68% w/w	WG	18085
3	Nautile DG	UPL Europe	5:68% w/w	WG	16653
4	Palmas WP	Clayton	4.6:65% w/w	WP	16632
5	Profilux WG	Belchim	4.5:68% w/w	WG	16125
6	Video	UPL Europe	5:68% w/w	WG	16685
7	Zetanil WG	Sipcam	4.5:65% w/w	WG	15488

Uses

- Blight in **potatoes** [1-7]
- Disease control in **hops in propagation** *(off-label)* [1]
- Late blight in **potatoes** [2, 4, 5]

Extension of Authorisation for Minor Use (EAMUs)

- **hops in propagation** *20162919* [1]

Approval information

- Cymoxanil and mancozeb included in Annex I under EC Regulation 1107/2009

Efficacy guidance

- Apply immediately after blight warning or as soon as local conditions dictate and repeat at 7-14 d intervals until haulm dies down or is burnt off
- Spray interval should not be more than 7-10 d in irrigated crops (see product label). Apply treatment after irrigation
- To minimise the likelihood of development of resistance these products should be used in a planned Resistance Management strategy. See Section 5 for more information

Restrictions

- Not specified for most products
- Do not apply at less than 7 d intervals
- Do not allow packs to become wet during storage
- Use low drift spraying equipment up to 30m from the top of the bank of any surface water bodies [3, 6]
- Horizontal boom sprayers must be fitted with three star drift reduction technology for all uses [6]

Crop-specific information

- Destroy and remove any haulm that remains after harvest of early varieties to reduce blight pressure on neighbouring maincrop potatoes

Following crops guidance

- HI 7 days [5]

Environmental safety

- Dangerous for the environment
- Very toxic to aquatic organisms
- Keep product away from fire or sparks
- Buffer zone requirement 6m [3, 6]
- LERAP Category B [1, 2, 4, 5]

FOR FULL CONDITIONS OF USE ALWAYS READ THE PRODUCT LABEL

Hazard classification and safety precautions

Hazard Harmful [1, 2, 4, 5, 7]; Irritant [3, 6]; Dangerous for the environment [1-7]; Very toxic to aquatic organisms [1-6]

Transport code 9

Packaging group III

UN Number 3077

Risk phrases H317 [1-7]; H319 [4]; H335, H360 [7]; H361 [1-6]

Operator protection A [1-7]; C [1, 2, 4, 5, 7]; D [4]; H [3, 4, 6, 7]; U05a, U08, U14, U19a, U20b [1-7]; U11 [3, 4, 6, 7]

Environmental protection E13c [1, 2, 5]; E15a, E38 [1-7]; E16a [1, 2, 4, 5]; H410 [1, 3, 4, 6, 7]; H411 [2, 5]

Storage and disposal D01, D02, D09a [1-7]; D06b, D12a [3, 4, 6, 7]; D11a [1, 2, 5]

Medical advice M05a [6]

121   cymoxanil + mandipropamid

A fungicide mixture for use in potatoes
FRAC mode of action code: 27 + 40

See also cymoxanil

Products

Carial Flex	Syngenta	18:25% w/w	WG	16629

Uses

- Blight in *potatoes*
- Late blight in *potatoes*

Approval information

- Cymoxanil and mandipropamid are included in Annex 1 under EC Regulation 1107/2009

Efficacy guidance

- Where possible, use an alternating strategy using fungicides from different mode of action groups.
- Where CAA fungicides are applied as a mixture (co-formulated or as a tank mix) up to six applications (or max. of 50% of the total number of applications) may be made per crop or season.
- No more than 3 applications of any CAA fungicide should be made consecutively

Restrictions

- Application may only be made between 1 May and 31 August

Hazard classification and safety precautions

Hazard Harmful if swallowed

Transport code 9

Packaging group III

UN Number 3077

Risk phrases H361, H373, R53a

Operator protection A, H; U05a, U08, U20b

Environmental protection E15b, E34, E38, H410

Storage and disposal D01, D02, D09a, D10c, D12a, D19

Medical advice M03, M05a

122   cymoxanil + zoxamide

A protectant and systemic fungicide for blight control in potatoes
FRAC mode of action code: 27 + 22

See also cymoxanil
dimethomorph + zoxamide

Products

Lieto	Sipcam	33:33% w/w	WG	16703

Uses

- Blight in **potatoes**

Approval information

- Cymoxanil and zoxamide included in Annex I under EC Regulation 1107/2009

Environmental safety

- Buffer zone requirement 6 m [1]

Hazard classification and safety precautions

Hazard Harmful, Dangerous for the environment, Harmful if swallowed
Transport code 9
Packaging group III
UN Number 3077
Risk phrases H317
Operator protection A, C, H; U05a, U11, U14, U15, U19a
Environmental protection E38, H410
Storage and disposal D01, D02, D12a, D12b
Medical advice M05a

123   cypermethrin

A contact and stomach acting pyrethroid insecticide
IRAC mode of action code: 3

Products

1	Cythrin 500 EC	Arysta	500 g/l	EC	16993
2	Forester	Fargro	100 g/l	EW	13164
3	Vazor Cypermethrin 10	Killgerm	100 g/l	EW	H10085

Uses

- Aphids in **asparagus, broccoli, brussels sprouts, cabbages, calabrese, carrots, cauliflowers, celeriac, dwarf beans, edible podded peas, forage maize, french beans, grain maize, horseradish, linseed, lupins, mustard, ornamental plant production, parsley root, parsnips, radishes, runner beans, salsify, spelt, spring barley** (autumn sown), **spring oats, spring rye, spring wheat** (autumn sown), **triticale, vining peas, winter barley, winter oats, winter rye, winter wheat** [1]
- Barley yellow dwarf virus vectors in **spring barley** (autumn sown), **spring oats, spring rye, spring wheat** (autumn sown), **triticale, winter barley, winter oats, winter rye, winter wheat** [1]
- Beetles in **cut logs** [2]
- Bladder pod midge in **spring oilseed rape, winter oilseed rape** [1]
- Cabbage stem flea beetle in **winter oilseed rape** [1]
- Carrot fly in **carrots, celeriac, horseradish, parsley root, parsnips** [1]
- Caterpillars in **broccoli, brussels sprouts, cabbages, calabrese, cauliflowers** [1]
- Cutworms in **fodder beet, mangels, potatoes, red beet, sugar beet, swedes, turnips** [1]
- Flies in **agricultural premises, farm buildings, outdoor crops** [3]
- Insect pests in **asparagus, carrots, celeriac, dwarf beans, edible podded peas, french beans, grain maize, horseradish, linseed, lupins, mustard, ornamental plant production, parsley root, parsnips, radishes, runner beans, salsify, spelt** [1]
- Large pine weevil in **forest** [2]
- Moths in **agricultural premises, farm buildings, outdoor crops** [3]
- Pea and bean weevil in **combining peas, spring field beans, vining peas, winter field beans** [1]
- Pea moth in **vining peas** [1]
- Pollen beetle in **spring oilseed rape, winter oilseed rape** [1]
- Rape winter stem weevil in **winter oilseed rape** [1]
- Seed weevil in **spring oilseed rape, winter oilseed rape** [1]

FOR FULL CONDITIONS OF USE ALWAYS READ THE PRODUCT LABEL

- Wasps in *agricultural premises, farm buildings, outdoor crops* [3]
- Whitefly in *broccoli, brussels sprouts, cabbages, calabrese, cauliflowers* [1]
- Yellow cereal fly in *spring barley* (autumn sown), *spring oats, spring rye, spring wheat* (autumn sown), *triticale, winter barley, winter oats, winter rye, winter wheat* [1]

Approval information
- Cypermethrin included in Annex I under EC Regulation 1107/2009
- Accepted by BBPA for use on malting barley and hops

Efficacy guidance
- Products combine rapid action, good persistence, and high activity on Lepidoptera.
- As effect is mainly via contact, good coverage is essential for effective action. Spray volume should be increased on dense crops
- A repeat spray after 10-14 d is needed for some pests of outdoor crops
- Rates and timing of sprays vary with crop and pest. See label for details

Restrictions
- Maximum number of treatments varies with crop and product. See label or approval notice for details [2]
- Apply by knapsack or hand-held sprayer [2]
- Some new products advise using low drift spraying equipment up to 30m from the top of the bank of any surface water bodies - check label before use.

Crop-specific information
- HI vining peas 7 d; other crops 0 d
- Test spray sample of new or unusual ornamentals or trees before committing whole batches [2]
- Post-planting forestry applications should be made before damage is seen or at the onset of damage during the first 2 yrs after transplanting [2]

Environmental safety
- Dangerous for the environment
- Very toxic to aquatic organisms
- High risk to bees. Do not apply to crops in flower, or to those in which bees are actively foraging, except as directed. Do not apply when flowering weeds are present
- Give local beekeepers warning when using in flowering crops
- Do not spray cereals after 31 Mar within 6 m of the edge of the growing crop [2]
- Buffer zone requirement 18 m [1]
- To protect non target insects/arthropods respect an unsprayed buffer zone of 5 m to non-crop land [1]
- Horizontal boom sprayers must be fitted with three star drift reduction technology for all uses [1]
- Use low drift spraying equipment up to 30m from the top of the bank of any surface water bodies [1]
- LERAP Category A [1]

Hazard classification and safety precautions
Hazard Harmful, Dangerous for the environment [1, 2]; Flammable [1]; Harmful if swallowed [1-3]

Transport code 3 [1]; 9 [2, 3]

Packaging group III

UN Number 1993 [1]; 3082 [2, 3]

Risk phrases H315 [1, 2]; H317 [1-3]; H320, H335, H336, H373 [1]

Operator protection A [1-3]; C, D [3]; H [1, 2]; M [2]; U02a, U04a, U14, U16b [2]; U05a, U19a [1-3]; U08, U20b [1]; U10 [2, 3]

Environmental protection E12c, E12f, E34 [1, 2]; E13a, E16c, E16d, E16h [1]; E15a, E38 [2]; H410 [1-3]

Storage and disposal D01, D02, D05, D10b [1, 2]; D09a, D12a [1-3]

Medical advice M03 [2]; M03a [3]; M05b [1]

SEE SECTION 3 FOR PRODUCTS ALSO REGISTERED

124   cypermethrin + tetramethrin

A pyrethroid insecticide mixture for control of crawling and flying insects in a variety of situations
IRAC mode of action code: 3

See also cypermethrin

Products

Vazor Cypermax Plus	Killgerm	10% + 5% w/w	ME	H9988

Uses

- Ants in **agricultural premises, manure heaps, poultry houses, refuse tips**
- Bedbugs in **agricultural premises, manure heaps, poultry houses, refuse tips**
- Cockroaches in **agricultural premises, manure heaps, poultry houses, refuse tips**
- Mosquitoes in **agricultural premises, manure heaps, poultry houses, refuse tips**
- Moths in **agricultural premises, manure heaps, poultry houses, refuse tips**
- Silverfish in **agricultural premises, manure heaps, poultry houses, refuse tips**
- Wasps in **agricultural premises, manure heaps, poultry houses, refuse tips**

Approval information

- Cypermethrin is included but tetramethrin not included in Annex 1 under EC Regulation 1107/2009

Hazard classification and safety precautions

Transport code 9
Packaging group III
UN Number 3082
Operator protection A, C, D, H; U20b
Environmental protection E05c, E12b, H410
Storage and disposal D09a, D12a

125   cyproconazole + penthiopyrad

An SDHI and triazole mixture for disease control in wheat
FRAC mode of action code: 3 + 7

See also penthiopyrad

Products

Cielex	DuPont	60:150 g/l	SC	16540

Uses

- Septoria leaf blotch in **spring wheat, winter wheat**

Approval information

- Cyproconazole and penthiopyrad are included in Annex 1 under EC Regulation 1107/2009

Efficacy guidance

- Disease control is more effective when used as a protectant treatment or during the earliest stages of disease development

Restrictions

- To reduce the risk of resistance, only two applications of any SDHI fungicide may be made to a cereal crop.

Following crops guidance

- A period of a year must be observed prior to planting of leafy crops as a succeeding crop following treatment

Hazard classification and safety precautions

Transport code 9
Packaging group III
UN Number 3082

FOR FULL CONDITIONS OF USE ALWAYS READ THE PRODUCT LABEL

Risk phrases R51, R53a, R63
Operator protection A; U05a, U09b, U11, U20a
Environmental protection E15b, E38
Storage and disposal D01, D02, D09a, D09b, D10c, D12a, D12b
Medical advice M03a

126   cyproconazole + trifloxystrobin

A conazole and strobilurin fungicide mixture for disease control in sugar beet
FRAC mode of action code: 3 + 11

See also trifloxystrobin

Products

1 Clayton Cast	Clayton	160:375 g/l	SC	17727
2 Escolta	Bayer CropScience	160:375 g/l	SC	17397

Uses
- Cercospora leaf spot in *fodder beet*, *sugar beet* [1, 2]; *red beet* *(off-label)* [2]
- Powdery mildew in *fodder beet*, *sugar beet* [1, 2]; *red beet* *(off-label)* [2]
- Ramularia leaf spots in *fodder beet*, *sugar beet* [1, 2]; *red beet* *(off-label)* [2]
- Rust in *fodder beet*, *sugar beet* [1, 2]; *red beet* *(off-label)* [2]

Extension of Authorisation for Minor Use (EAMUs)
- *red beet* *20161089* [2]

Approval information
- Cyproconazole and trifloxystrobin included in Annex I under EC Regulation 1107/2009

Efficacy guidance
- Best results obtained from treatment at early stages of disease development. Further treatment may be needed if disease attack is prolonged
- Trifloxystrobin is a member of the QoI cross resistance group. Product should be used preventatively and not relied on for its curative potential
- Use product as part of an Integrated Crop Management strategy incorporating other methods of control, including where appropriate other fungicides with a different mode of action.
- Strains of barley powdery mildew resistant to QoIs are common in the UK

Restrictions
- Maximum total dose per crop equivalent to two full dose treatments
- First application not before BBCH 31 (10% ground cover); second application not before BBCH 40 (full canopy & start of harvestable root development); maintain at least 21 day interval between sprays.

Crop-specific information
- HI 35 d

Environmental safety
- Dangerous for the environment
- Very toxic to aquatic organisms
- LERAP Category B

Hazard classification and safety precautions
Hazard Harmful, Dangerous for the environment, Very toxic to aquatic organisms
Transport code 9
Packaging group III
UN Number 3082
Risk phrases H361
Operator protection A; U02a, U05a, U09a, U19a, U20a
Environmental protection E15b, E16a, E38, H410
Storage and disposal D01, D02, D05, D09a, D10b, D12a
Medical advice M03

SEE SECTION 3 FOR PRODUCTS ALSO REGISTERED

SECTION 2

127 cyprodinil

An anilinopyrimidine systemic broad spectrum fungicide for cereals
FRAC mode of action code: 9

See also cyproconazole + cyprodinil

Products

1	Coracle	AgChem Access	300 g/l	EC	15866
2	Kayak	Syngenta	300 g/l	EC	14847

Uses
- Disease control in **forest nurseries** *(off-label)* [2]
- Eyespot in **spring barley, winter barley** [1, 2]
- Net blotch in **spring barley, winter barley** [1, 2]
- Powdery mildew in **spring barley, winter barley** [1, 2]
- Rhynchosporium in **spring barley, winter barley** [1, 2]

Extension of Authorisation for Minor Use (EAMUs)
- **forest nurseries** *20122056* [2]

Approval information
- Cyprodinil included in Annex I under EC Regulation 1107/2009
- Accepted by BBPA for use on malting barley

Efficacy guidance
- Best results obtained from treatment at early stages of disease development
- For best control of eyespot spray before or during the period of stem extension in spring. Control may be reduced if very dry conditions follow treatment

Restrictions
- Maximum number of treatments 2 per crop on wheat and barley [2]

Crop-specific information
- Latest use: before first spikelet of inflorescence just visible stage for barley, before early milk stage for wheat; up to and including first awns visible stage (GS49) in barley [2]

Environmental safety
- Dangerous for the environment
- Very toxic to aquatic organisms
- LERAP Category B

Hazard classification and safety precautions
Hazard Irritant, Dangerous for the environment [1, 2]; Very toxic to aquatic organisms [2]
Transport code 9
Packaging group III
UN Number 3082
Risk phrases H317 [2]; R38, R43, R50, R53a [1]
Operator protection A, H; U05a, U09a, U20a
Environmental protection E15b, E16a, E34, E38 [1, 2]; H410 [2]
Storage and disposal D01, D02, D05, D07, D09a, D10c, D12a

128 cyprodinil + fludioxonil

A broad-spectrum fungicide mixture for fruit crops, legumes, some vegetables, forest nurseries and ornamentals
FRAC mode of action code: 9 + 12

See also fludioxonil

Products

1	Clayton Gear	Clayton	37.5:25% w/w	WG	17169
2	Switch	Syngenta	37.5:25 % w/w	WG	15129

FOR FULL CONDITIONS OF USE ALWAYS READ THE PRODUCT LABEL

Uses

- Alternaria in **apples**, **crab apples**, **pears**, **quinces** [1, 2]; **brassica leaves and sprouts** (off-label), **broccoli** (off-label), **brussels sprouts** (off-label), **cabbages** (off-label), **calabrese** (off-label), **cauliflowers** (off-label), **celery (outdoor)** (off-label), **collards** (off-label), **cress** (off-label), **endives** (off-label), **herbs (see appendix 6)** (off-label), **kale** (off-label), **lamb's lettuce** (off-label), **lettuce** (off-label), **protected baby leaf crops** (off-label), **protected cress** (off-label), **protected endives** (off-label), **protected herbs (see appendix 6)** (off-label), **protected lamb's lettuce** (off-label), **protected lettuce** (off-label), **protected radishes** (off-label), **protected red mustard** (off-label), **protected rocket** (off-label), **protected sorrel** (off-label), **protected spinach** (off-label), **protected spinach beet** (off-label), **red mustard** (off-label), **rocket** (off-label), **sorrel** (off-label), **spinach** (off-label), **spinach beet** (off-label), **spring cabbage** (off-label) [2]
- Alternaria blight in **carrots** (moderate control) [1, 2]
- Anthracnose in **protected blueberry** (off-label) [2]
- Ascochyta in **mange-tout peas**, **sugar snap peas**, **vining peas** [1, 2]
- Basal stem rot in **ornamental bulbs** (off-label) [2]
- Black spot in **protected strawberries** (qualified minor use), **strawberries** (qualified minor use) [1, 2]
- Blossom wilt in **apricots** (off-label), **cherries** (off-label), **peaches** (off-label), **plums** (off-label) [2]
- Botrytis in **apricots** (off-label), **asparagus** (off-label), **brassica leaves and sprouts** (off-label), **broccoli** (off-label), **brussels sprouts** (off-label), **bulb onions** (off-label), **cabbages** (off-label), **calabrese** (off-label), **cauliflowers** (off-label), **cherries** (off-label), **chicory** (off-label), **collards** (off-label), **cress** (off-label), **endives** (off-label), **garlic** (off-label), **grapevines** (off-label), **herbs (see appendix 6)** (off-label), **kale** (off-label), **lamb's lettuce** (off-label), **lettuce** (off-label), **lupins** (off-label), **peaches** (off-label), **plums** (off-label), **protected asparagus** (off-label), **protected aubergines** (off-label), **protected baby leaf crops** (off-label), **protected blueberry** (off-label), **protected courgettes** (off-label), **protected cress** (off-label), **protected cucumbers** (off-label), **protected endives** (off-label), **protected gherkins** (off-label), **protected herbs (see appendix 6)** (off-label), **protected lamb's lettuce** (off-label), **protected lettuce** (off-label), **protected peppers** (off-label), **protected pumpkins** (off-label), **protected red mustard** (off-label), **protected rocket** (off-label), **protected sorrel** (off-label), **protected spinach** (off-label), **protected spinach beet** (off-label), **protected summer squash** (off-label), **protected tomatoes** (off-label), **pumpkins** (off-label), **red mustard** (off-label), **rocket** (off-label), **salad onions** (off-label), **shallots** (off-label), **sorrel** (off-label), **spinach** (off-label), **spinach beet** (off-label), **spring cabbage** (off-label) [2]; **broad beans** (moderate control), **broad beans (dry-harvested)** (moderate control), **carrots** (moderate control), **celeriac** (moderate control), **combining peas** (moderate control), **forest nurseries**, **french beans (dry-harvested)** (moderate control), **green beans** (moderate control), **mange-tout peas** (moderate control), **ornamental plant production**, **protected forest nurseries**, **protected ornamentals**, **protected strawberries**, **runner beans** (moderate control), **strawberries**, **sugar snap peas** (moderate control), **vining peas** (moderate control) [1, 2]
- Botrytis fruit rot in **apples**, **blackberries**, **crab apples**, **pears**, **quinces**, **raspberries** [1, 2]
- Botrytis root rot in **bilberries** (qualified minor use recommendation), **blackcurrants** (qualified minor use recommendation), **blueberries** (qualified minor use recommendation), **cranberries** (qualified minor use recommendation), **gooseberries** (qualified minor use recommendation), **redcurrants** (qualified minor use recommendation), **whitecurrants** (qualified minor use recommendation) [1, 2]
- Fusarium in **protected cucumbers** (off-label) [2]
- Fusarium bulb rot in **ornamental bulbs** (off-label) [2]
- Fusarium diseases in **apples**, **crab apples**, **pears**, **quinces** [1, 2]
- Gloeosporium rot in **apples**, **crab apples**, **pears**, **quinces** [1, 2]
- Leaf spot in **celery (outdoor)** (off-label), **protected radishes** (off-label) [2]
- Mycosphaerella in **mange-tout peas**, **sugar snap peas**, **vining peas** [1, 2]
- Penicillium rot in **apples**, **crab apples**, **pears**, **quinces** [1, 2]
- Phoma in **brassica leaves and sprouts** (off-label), **broccoli** (off-label), **brussels sprouts** (off-label), **cabbages** (off-label), **calabrese** (off-label), **cauliflowers** (off-label), **collards** (off-label), **cress** (off-label), **endives** (off-label), **herbs (see appendix 6)** (off-label), **kale** (off-label), **lamb's lettuce** (off-label), **lettuce** (off-label), **protected baby leaf crops** (off-label), **protected cress** (off-label), **protected endives** (off-label), **protected herbs (see appendix 6)** (off-label), **protected lamb's lettuce** (off-label), **protected lettuce** (off-label), **protected red mustard** (off-label),

SEE SECTION 3 FOR PRODUCTS ALSO REGISTERED

 protected rocket *(off-label)*, *protected sorrel* *(off-label)*, *protected spinach* *(off-label)*,
 protected spinach beet *(off-label)*, *red mustard* *(off-label)*, *rocket* *(off-label)*, *sorrel* *(off-label)*,
 spinach *(off-label)*, *spinach beet* *(off-label)*, *spring cabbage* *(off-label)* [2]

- Rhizoctonia in *brassica leaves and sprouts* *(off-label)*, *broccoli* *(off-label)*, *brussels sprouts*
 (off-label), *cabbages* *(off-label)*, *calabrese* *(off-label)*, *cauliflowers* *(off-label)*, *collards* *(off-label)*,
 cress *(off-label)*, *endives* *(off-label)*, *herbs (see appendix 6)* *(off-label)*, *kale* *(off-label)*, *lamb's*
 lettuce *(off-label)*, *lettuce* *(off-label)*, *protected baby leaf crops* *(off-label)*, *protected cress* *(off-
 label)*, *protected endives* *(off-label)*, *protected herbs (see appendix 6)* *(off-label)*, *protected*
 lamb's lettuce *(off-label)*, *protected lettuce* *(off-label)*, *protected red mustard* *(off-label)*,
 protected rocket *(off-label)*, *protected sorrel* *(off-label)*, *protected spinach* *(off-label)*,
 protected spinach beet *(off-label)*, *red mustard* *(off-label)*, *rocket* *(off-label)*, *sorrel* *(off-label)*,
 spinach *(off-label)*, *spinach beet* *(off-label)*, *spring cabbage* *(off-label)* [2]
- Sclerotinia in *beetroot* *(off-label)*, *celery (outdoor)* *(off-label)*, *ginger* *(off-label)*, *herbs for*
 medicinal uses (see appendix 6) *(off-label)*, *horseradish* *(off-label)*, *parsley root* *(off-label)*,
 parsnips *(off-label)*, *protected radishes* *(off-label)*, *salsify* *(off-label)*, *spice roots* *(off-label)*,
 turmeric *(off-label)* [2]; *broad beans*, *broad beans (dry-harvested)*, *carrots* *(moderate control)*,
 combining peas, *french beans (dry-harvested)*, *green beans*, *mange-tout peas*, *runner*
 beans, *sugar snap peas*, *vining peas* [1, 2]

Extension of Authorisation for Minor Use (EAMUs)

- *apricots* *20103092* [2]
- *asparagus* *20103173* [2]
- *beetroot* *20103087* [2]
- *brassica leaves and sprouts* *20150085* [2]
- *broccoli* *20150085* [2]
- *brussels sprouts* *20150085* [2]
- *bulb onions* *20122285* [2]
- *cabbages* *20150085* [2]
- *calabrese* *20150085* [2]
- *cauliflowers* *20150085* [2]
- *celery (outdoor)* *20150865* [2]
- *cherries* *20103092* [2]
- *chicory* *20122285* [2]
- *collards* *20150085* [2]
- *cress* *20150085* [2]
- *endives* *20150085* [2]
- *garlic* *20122285* [2]
- *ginger* *20103087* [2]
- *grapevines* *20112098* [2]
- *herbs (see appendix 6)* *20150085* [2]
- *herbs for medicinal uses (see appendix 6)* *20103087* [2]
- *horseradish* *20103087* [2]
- *kale* *20150085* [2]
- *lamb's lettuce* *20150085* [2]
- *lettuce* *20150085* [2]
- *lupins* *20103170* [2]
- *ornamental bulbs* *20152274* [2]
- *parsley root* *20103087* [2]
- *parsnips* *20103087* [2]
- *peaches* *20103092* [2]
- *plums* *20103092* [2]
- *protected asparagus* *20170202* [2]
- *protected aubergines* *20103172* [2]
- *protected baby leaf crops* *20150085* [2]
- *protected blueberry* *20160648* [2]
- *protected courgettes* *20170202* [2]
- *protected cress* *20150085* [2]
- *protected cucumbers* *20103171* [2]
- *protected endives* *20150085* [2]

FOR FULL CONDITIONS OF USE ALWAYS READ THE PRODUCT LABEL

- **protected gherkins** *20170202* [2]
- **protected herbs (see appendix 6)** *20150085* [2]
- **protected lamb's lettuce** *20150085* [2]
- **protected lettuce** *20150085* [2]
- **protected peppers** *20103172* [2]
- **protected pumpkins** *20170202* [2]
- **protected radishes** *20150865* [2]
- **protected red mustard** *20150085* [2]
- **protected rocket** *20150085* [2]
- **protected sorrel** *20150085* [2]
- **protected spinach** *20150085* [2]
- **protected spinach beet** *20150085* [2]
- **protected summer squash** *20170202* [2]
- **protected tomatoes** *20110302* [2]
- **pumpkins** *20170202* [2]
- **red mustard** *20150085* [2]
- **rocket** *20150085* [2]
- **salad onions** *20103174* [2]
- **salsify** *20103087* [2]
- **shallots** *20122285* [2]
- **sorrel** *20150085* [2]
- **spice roots** *20103087* [2]
- **spinach** *20150085* [2]
- **spinach beet** *20150085* [2]
- **spring cabbage** *20150085* [2]
- **turmeric** *20103087* [2]

Approval information
- Cyprodinil and fludioxonil included in Annex I under EC Regulation 1107/2009

Efficacy guidance
- Best results obtained from application at the earliest stage of disease development or as a protective treatment following disease risk assessment
- Subsequent treatments may follow after a minimum of 10 d if disease pressure remains high
- To minimise the likelihood of development of resistance product should be used in a planned Resistance Management strategy. See Section 5 for more information

Restrictions
- Maximum number of treatments 2 per crop
- Consult processor before treating any crop for processing

Crop-specific information
- HI broad beans, green beans, mange-tout peas, runner beans, sugar snap peas, vining peas 14 d; protected strawberries, strawberries 3 d
- First signs of disease infection likely to be seen from early flowering for all crops
- Ensure peas are free from any stress before treatment. If necessary check wax with a crystal violet test

Environmental safety
- Dangerous for the environment
- Very toxic to aquatic organisms
- LERAP Category B

Hazard classification and safety precautions
Hazard Dangerous for the environment, Very toxic to aquatic organisms
Transport code 9
Packaging group III
UN Number 3077
Risk phrases H317

SEE SECTION 3 FOR PRODUCTS ALSO REGISTERED

SECTION 2

Operator protection A, H; U05a, U20a
Environmental protection E15b, E16a, E16b, E38, H410
Storage and disposal D01, D02, D05, D07, D09a, D10c, D12a

129 cyprodinil + isopyrazam

A fungicide mixture for disease control in barley
FRAC mode of action code: 7 + 9

See also cyprodinil

Products

1 Bontima	Syngenta	187.5:62.5 g/l	EC	17188
2 Cebara	Syngenta	187.5:62.5 g/l	EC	17286
3 Concorde	Syngenta	150:62.5 g/l	EC	17248

Uses

* Brown rust in *spring barley, winter barley*
* Net blotch in *spring barley, winter barley*
* Powdery mildew in *spring barley, winter barley*
* Ramularia leaf spots in *spring barley, winter barley*
* Rhynchosporium in *spring barley, winter barley*

Approval information

* Cyprodinil and isopyrazam are included in Annex 1 under EC Regulation 1107/2009
* Accepted by BBPA for use on malting barley before ear emergence only

Restrictions

* Isopyrazam is an SDH respiration inhibitor; Do not apply more than two foliar applications of products containing an SDH inhibitor to any cereal crop

Crop-specific information

* There are no restrictions on succeeding crops in a normal rotation

Environmental safety

* LERAP Category B

Hazard classification and safety precautions

Hazard Harmful, Dangerous for the environment, Harmful if inhaled [1-3]; Very toxic to aquatic organisms [3]
Transport code 9
Packaging group III
UN Number 3082
Risk phrases H304, H361 [1-3]; H351 [3]
Operator protection A, C, H; U05a, U09a, U20b [1-3]; U11 [1, 2]
Environmental protection E15b, E16a, E16b, E38, H410
Storage and disposal D01, D02, D05, D09a, D10c, D11a, D12a
Medical advice M05b [3]

130 2,4-D

A translocated phenoxycarboxylic acid herbicide for cereals, grass and amenity use
HRAC mode of action code: O

See also amitrole + 2,4-D + diuron
clopyralid + 2,4-D + MCPA

Products

1 Depitone Ultra	Nufarm UK	600 g/l	EC	17615
2 Depitox	Nufarm UK	500 g/l	SL	13258
3 Dioweed 50	UPL Europe	500 g/l	SL	13197

FOR FULL CONDITIONS OF USE ALWAYS READ THE PRODUCT LABEL

Products – continued

4	Headland Staff 500	Headland	500 g/l	SL	13196
5	Herboxone	Headland	500 g/l	SL	13958
6	HY-D Super	Agrichem	500 g/l	SL	13198

Uses

- Annual dicotyledons in **amenity grassland, managed amenity turf, spring barley, spring wheat, winter barley, winter oats, winter rye, winter wheat** [1-6]; **apple orchards, farm forestry** (off-label), **forest nurseries** (off-label), **game cover** (off-label), **ornamental plant production** (off-label), **pear orchards** [2]; **grassland** [1-3]; **miscanthus** (off-label), **undersown oats** [6]; **permanent grassland** [4-6]; **spring rye** [1, 2]; **undersown barley, undersown rye, undersown wheat** [2, 6]; **undersown spring cereals, undersown winter cereals** [3]
- Perennial dicotyledons in **amenity grassland, managed amenity turf, spring barley, spring wheat, winter barley, winter oats, winter rye, winter wheat** [1, 2, 4-6]; **apple orchards, farm forestry** (off-label), **forest nurseries** (off-label), **game cover** (off-label), **ornamental plant production** (off-label), **pear orchards** [2]; **grassland, spring rye** [1, 2]; **permanent grassland** [4-6]; **undersown barley, undersown rye, undersown wheat** [2, 6]; **undersown oats** [6]

Extension of Authorisation for Minor Use (EAMUs)

- **farm forestry** *20082843* [2]
- **forest nurseries** *20082843* [2]
- **game cover** *20082843* [2]
- **miscanthus** *20082914* [6]
- **ornamental plant production** *20082843* [2]

Approval information

- 2,4-D included in Annex I under EC Regulation 1107/2009
- Approved for aquatic weed control [2]. See Section 5 for information on use of herbicides in or near water
- Accepted by BBPA for use on malting barley

Efficacy guidance

- Best results achieved by spraying weeds in seedling to young plant stage when growing actively in a strongly competing crop
- Most effective stage for spraying perennials varies with species. See label for details

Restrictions

- Maximum number of treatments normally 1 per crop and in forestry, 1 per yr in grassland and 2 or 3 per yr in amenity turf. Check individual labels
- Do not use on newly sown leys containing clover
- Do not spray grass seed crops after ear emergence
- Do not spray within 6 mth of laying turf or sowing fine grass
- Do not dump surplus herbicide in water or ditch bottoms
- Do not plant conifers until at least 1 mth after treatment
- Do not spray crops stressed by cold weather or drought or if frost expected
- Do not roll or harrow within 7 d before or after spraying
- Do not spray if rain falling or imminent
- Do not mow or roll turf or amenity grass 4 d before or after spraying. The first 4 mowings after treatment must be composted for at least 6 mth before use
- Do not mow grassland or graze for at least 10 d after spraying

Crop-specific information

- Latest use: before 1st node detectable in cereals; end Aug for conifer plantations; before established grassland 25 cm high
- Spray winter cereals in spring when leaf-sheath erect but before first node detectable (GS 31), spring cereals from 5-leaf stage to before first node detectable (GS 15-31)
- Cereals undersown with grass and/or clover, but not with lucerne, may be treated
- Selective treatment of resistant conifers can be made in Aug when growth ceased and plants hardened off, spray must be directed if applied earlier. See label for details

Following crops guidance

- Do not use shortly before or after sowing any crop

SEE SECTION 3 FOR PRODUCTS ALSO REGISTERED

- Do not plant succeeding crops within 3 mth [4]
- Do not direct drill brassicas or grass/clover mixtures within 3 wk of application

Environmental safety
- Dangerous for the environment
- Very toxic to aquatic organisms [4]
- Dangerous to aquatic higher plants. Do not contaminate surface waters or ditches with chemical or used container
- 2,4-D is active at low concentrations. Take extreme care to avoid drift onto neighbouring crops, especially beet crops, brassicas, most market garden crops including lettuce and tomatoes under glass, pears and vines
- May be used to control aquatic weeds in presence of fish if used in strict accordance with directions for waterweed control and precautions needed for aquatic use [2]
- Keep livestock out of treated areas for at least 2 wk following treatment and until poisonous weeds, such as ragwort, have died down and become unpalatable
- Water containing the herbicide must not be used for irrigation purposes within 3 wk of treatment or until the concentration in water is below 0.05 ppm

Hazard classification and safety precautions
Hazard Harmful, Dangerous for the environment [3-6]; Irritant [1, 2]; Harmful if swallowed [2-6]
Transport code 9 [3-5]
Packaging group III [3-5]
UN Number 3082 [3-5]; N/C [1, 2, 6]
Risk phrases H315 [1]; H317 [1, 6]; H318 [2-6]
Operator protection A, C, H [1-6]; M [1, 2]; U05a, U11 [1-6]; U08, U20b [1, 2, 4-6]; U09a, U20a [3]; U14, U15 [6]
Environmental protection E07a, E34 [1-6]; E15a [1-3]; E15b, E38 [3-6]; H410 [4, 5]; H411 [1, 6]
Storage and disposal D01, D02, D05 [1-6]; D09a [1, 2, 4-6]; D10a [4-6]; D10b [1, 2]; D10c [3]; D12a [3-6]
Medical advice M03 [1-6]; M05a [3, 6]

---

131   2,4-D + dicamba

A translocated herbicide for use on turf
HRAC mode of action code: O + O

See also dicamba

Products
1 Magneto	Nufarm UK	344:120 g/l	SL	17444
2 Thrust	Nufarm UK	344:120 g/l	SL	15408

Uses
- Annual dicotyledons in **amenity grassland** [1, 2]; **grassland** [1]; **permanent grassland** [2]
- Perennial dicotyledons in **amenity grassland** [1, 2]; **grassland** [1]; **permanent grassland** [2]

Approval information
- 2,4-D and dicamba included in Annex I under EC Regulation 1107/2009
- Accepted by BBPA for use on malting barley

Efficacy guidance
- Best results achieved by application when weeds growing actively in spring or early summer (later with irrigation and feeding)
- More resistant weeds may need repeat treatment after 3 wk
- Improved control of some weeds can be obtained by use of specifed oil adjuvant [1, 2]
- Do not use during drought conditions

Restrictions
- Maximum number of treatments 2 per yr on amenity grass and established grassland [1, 2]
- Do not treat newly sown or turfed areas or grass less than 1 yr old
- Do not treat grass crops intended for seed production [1, 2]

FOR FULL CONDITIONS OF USE ALWAYS READ THE PRODUCT LABEL

- Do not treat grass suffering from drought, disease or other adverse factors [1, 2]
- Do not roll or harrow for 7 d before or after treatment [1, 2]
- Do not apply when grassland is flowering [1, 2]
- Avoid spray drift onto cultivated crops or ornamentals
- Do not graze grass for at least seven days after spraying [1, 2]
- Do not mow or roll four days before or after application. The first four mowings after treatment must be composted for at least six months before use [1, 2]

Crop-specific information
- Latest use: before grass 25 cm high for amenity grass, established grassland [2]
- The first four mowings after treatment must be composted for at least 6 mth before use

Environmental safety
- Dangerous for the environment [2]
- Toxic to aquatic organisms
- Keep livestock out of treated areas for at least two weeks following treatment and until poisonous weeds, such as ragwort, have died down and become unpalatable [1, 2]

Hazard classification and safety precautions
Hazard Harmful, Dangerous for the environment, Harmful if swallowed
Transport code 9
Packaging group III
UN Number 3082
Risk phrases H317, H318
Operator protection A, C, H, M; U05a, U11
Environmental protection E07a, E15a, E34, E38, H411 [1, 2]; E22c [1]
Storage and disposal D01, D02, D05, D09a, D10a, D12a
Medical advice M03

132   2,4-D + dicamba + fluroxypyr

A translocated and contact herbicide mixture for amenity turf
HRAC mode of action code: O + O + O

See also dicamba
fluroxypyr

Products

1 Holster XL	Barclay	285:52.5:105 g/l	EC	16407
2 Mascot Crossbar	Rigby Taylor	285:52.5:105 g/l	EC	16454
3 Speedster	Pan Amenity	285:52.5:105 g/l	EC	17025

Uses
- Annual dicotyledons in *amenity grassland, managed amenity turf* [1-3]
- Buttercups in *amenity grassland, managed amenity turf* [1]
- Clover in *amenity grassland, managed amenity turf* [1]
- Daisies in *amenity grassland, managed amenity turf* [1]
- Dandelions in *amenity grassland, managed amenity turf* [1]
- Yarrow in *amenity grassland, managed amenity turf* [1]

Approval information
- 2,4-D, dicamba and fluroxypyr included in Annex I under EC Regulation 1107/2009

Efficacy guidance
- Apply when weeds actively growing (normally between Apr and Sep) and when soil is moist
- Best results obtained from treatment in spring or early summer before weeds begin to flower
- Do not apply if turf is wet or if rainfall expected within 4 h of treatment. Both circumstances will reduce weed control

Restrictions
- Maximum number of treatments 2 per yr
- Do not mow for 3 d before or after treatment

SEE SECTION 3 FOR PRODUCTS ALSO REGISTERED

- Avoid overlapping or overdosing, especially on newly sown turf
- Do not spray in drought conditions or if turf under stress from frost, waterlogging, trace element deficiency, pest or disease attack

Crop-specific information
- Latest use: normally Sep for managed amenity turf
- New turf may be treated in spring provided at least 2 mth have elapsed since sowing

Environmental safety
- Dangerous for the environment
- Toxic to aquatic organisms
- Keep livestock out of treated areas for at least two weeks following treatment and until poisonous weeds, such as ragwort, have died down and become unpalatable
- LERAP Category B

Hazard classification and safety precautions
Hazard Harmful, Dangerous for the environment [1-3]; Harmful if swallowed [2, 3]; Harmful if inhaled [2]
Transport code 9
Packaging group III
UN Number 3082
Risk phrases H315, H320, H336 [2, 3]; H371 [3]; R22a, R22b, R36, R38, R51, R53a, R67 [1]
Operator protection A, C, H, M; U05a, U08, U11, U14, U15, U19a, U20b
Environmental protection E07a, E15a, E16a, E34, E38 [1-3]; H411 [2, 3]
Storage and disposal D01, D02, D09a, D10a
Medical advice M03, M05a, M05b

133   2,4-D + dicamba + MCPA + mecoprop-P

A translocated herbicide mixture for grassland
HRAC mode of action code: O + O + O + O

See also dicamba
MCPA
mecoprop-P

Products

1 Enforcer	Everris Ltd	70:20:70:42 g/l	SL	17274
2 Longbow	Bayer CropScience	70:20:70:42 g/l	SL	16528

Uses
- Buttercups in *managed amenity turf*
- Clovers in *managed amenity turf*
- Daisies in *managed amenity turf*
- Dandelions in *managed amenity turf*
- Self-heal in *managed amenity turf*

Approval information
- 2,4-D, dicamba, MCPA and mecoprop-P included in Annex I under EC Regulation 1107/2009

Environmental safety
- Some pesticides pose a greater threat of contamination of water than others and mecoprop-P is one of these pesticides. Take special care when applying mecoprop-P near water and do not apply if heavy rain is forecast

Hazard classification and safety precautions
UN Number N/C
Risk phrases H319
Operator protection A, C, H, M
Environmental protection E07a, E15a, H410
Storage and disposal D05, D09a, D10a, D12a

FOR FULL CONDITIONS OF USE ALWAYS READ THE PRODUCT LABEL

134   2,4-D + florasulam

A post-emergence herbicide mixture for control of broad leaved weeds in managed amenity turf
HRAC mode of action code: O + B

See also florasulam

Products

1	Esteron T	Everris Ltd	300:6.25 g/l	ME	17432
2	Junction	Rigby Taylor	452:6.25 g/l	ME	12493

Uses
- Clover in *amenity grassland* [1]; *managed amenity turf* [1, 2]
- Daisies in *amenity grassland* [1]; *managed amenity turf* [1, 2]
- Dandelions in *amenity grassland* [1]; *managed amenity turf* [1, 2]
- Plantains in *amenity grassland* [1]; *managed amenity turf* [1, 2]
- Sticky mouse-ear in *amenity grassland* [1]; *managed amenity turf* [1, 2]

Approval information
- 2,4-D and florasulam included in Annex I under EC Regulation 1107/2009

Efficacy guidance
- Best results obtained from treatment of actively growing weeds between Mar and Oct when soil is moist
- Do not apply when rain is imminent or during periods of drought unless irrigation is applied

Restrictions
- Maximum number of treatments on managed amenity turf: 1 per yr

Crop-specific information
- Avoid mowing turf 3 d before and after spraying
- Ensure newly sown turf has established before treating. Turf sown in late summer or autumn should not be sprayed until growth is resumed in the following spring

Following crops guidance
- An interval of 4 wk must elapse between application and re-seeding turf

Environmental safety
- Dangerous for the environment
- Toxic to aquatic organisms

Hazard classification and safety precautions
Hazard Harmful, Dangerous for the environment, Harmful if swallowed
Transport code 9
Packaging group III
UN Number 3082
Risk phrases H317
Operator protection A, H; U02a, U05a, U08, U14, U20a
Environmental protection E15a, E34, E38 [1, 2]; H410 [1]; H411 [2]
Storage and disposal D01, D02, D09a, D10b, D12a

135   2,4-D + glyphosate

A total herbicide for weed control in crops and on railway ballast
HRAC mode of action code: O + G

See also glyphosate

Products

Kyleo	Nufarm UK	160:240 g/l	SL	16567

Uses

- Annual and perennial weeds in *all edible crops (outdoor), all non-edible seed crops grown outdoors, amenity grassland, amenity vegetation, apple orchards, green cover on land temporarily removed from production, pear orchards, railway tracks*

Approval information

- 2,4-D and glyphosate included in Annex I under EC Regulation 1107/2009

Restrictions

- Do not use under polythene or glass
- Application of lime, fertiliser, farmyard manure and pesticides should be delayed until 5 days after application of Kyleo. After application large quantities of decaying foliage, stolons, roots or rhizomes should be dispersed or buried by thorough cultivation before crop drilling
- Do not use in or alongside hedges
- Do not apply by hand-held equipment
- Windfall (fruit fall to ground) must not be used as food or feeding stuff

Environmental safety

- LERAP Category B

Hazard classification and safety precautions

Transport code 9
Packaging group III
UN Number 3082
Risk phrases H317, H319
Operator protection A, C, H; U05a, U19a
Environmental protection E15a, E16a, E34, E38, H410
Storage and disposal D01, D09a, D10a, D12a, D12b

136   2,4-D + MCPA

A translocated herbicide mixture for cereals and grass
HRAC mode of action code: O + O

See also MCPA

Products

1 Cirran 360	Nufarm UK	360:315 g/l	SL	17346
2 Headland Polo	Headland	360:315 g/l	SL	14933
3 Lupo	Nufarm UK	360:315 g/l	SL	14931
4 PastureMaster	Nufarm UK	360:315 g/l	SL	17994

Uses

- Annual dicotyledons in *grassland, spring barley, spring wheat, winter barley, winter oats, winter wheat* [1-4]
- Perennial dicotyledons in *grassland* [1, 3, 4]; *spring barley, spring wheat, winter barley, winter oats, winter wheat* [1-4]

Approval information

- 2,4-D and MCPA included in Annex I under EC Regulation 1107/2009
- Accepted by BBPA for use on malting barley

Efficacy guidance

- Best results achieved by spraying weeds in seedling to young plant stage when growing actively in a strongly competing crop
- Most effective stage for spraying perennials varies with species. See label for details

Restrictions

- Maximum number of treatments 1 per crop in cereals, 2 per year in grassland and 3 per year in managed amenity turf [2]; 1 per crop in cereals and 1 per year in grassland [3]
- Do not spray if rain falling or imminent
- Do not cut grass or graze for at least 10 d after spraying

FOR FULL CONDITIONS OF USE ALWAYS READ THE PRODUCT LABEL

- Do not use on newly sown leys containing clover or other legumes
- Do not spray crops stressed by cold weather or drought or if frost expected
- Do not use shortly before or after sowing any crop
- Do not roll or harrow within 7 d before or after spraying

Crop-specific information
- Spray winter cereals in spring when leaf-sheath erect but before first node detectable (GS 31), spring cereals from 5-leaf stage to before first node detectable (GS 15-31)
- Latest use: before first node detectable (GS 31) in cereals;

Environmental safety
- Dangerous for the environment
- Toxic to aquatic organisms
- Keep livestock out of treated areas for at least two weeks following treatment and until poisonous weeds, such as ragwort, have died down and become unpalatable
- 2,4-D and MCPA are active at low concentrations. Take extreme care to avoid drift onto neighbouring crops, especially beet crops, brassicas, most market garden crops including lettuce and tomatoes under glass, pears and vines

Hazard classification and safety precautions
Hazard Harmful [1-4]; Dangerous for the environment [2]
Transport code 9 [2]
Packaging group III [2]
UN Number 3082 [2]; N/C [1, 3, 4]
Risk phrases H318
Operator protection A, C, H, M; U05a, U08, U11, U14, U20b [1-4]; U15 [2]
Environmental protection E07a, E15a, E34 [1-4]; E38 [1, 3, 4]
Storage and disposal D01, D02, D05, D09a, D10a [1-4]; D12a [1, 3, 4]
Medical advice M03 [1-4]; M05a [2]

137   daminozide

A hydrazide plant growth regulator for use in certain ornamentals

Products
1	B-Nine SG	Certis	85% w/w	SG	14435
2	Dazide Enhance	Fargro	85% w/w	SG	16092
3	Stature	Pure Amenity	85% w/w	SG	18064

Uses
- Growth regulation in **protected ornamentals**

Approval information
- Daminozide included in Annex I under EC Regulation 1107/2009

Efficacy guidance
- Best results obtained by application in late afternoon when glasshouse has cooled down
- Apply a fine spray using compressed air or power sprayers to give good coverage of dry foliage without run off
- Response to treatment differs widely depending on variety, stage of growth and physiological condition of the plant. It is recommended that any new variety is first tested on a small scale to observe if adverse effects occur

Restrictions
- Apply only to turgid, well watered plants. Do not water for 24 h after spraying
- Do not mix with other spray chemicals unless specifically recommended
- Do not store product in metal containers

Crop-specific information
- Do not use on chrysanthemum variety Fandango
- Evidence of effectiveness on Azaleas is limited

SEE SECTION 3 FOR PRODUCTS ALSO REGISTERED

- See label for guidance on a range of bedding plant species

Hazard classification and safety precautions
    UN Number N/C
    Operator protection A, H; U02a, U08, U19a, U20b [2, 3]; U05a [1-3]
    Environmental protection E15b
    Storage and disposal D01, D02, D09a [1-3]; D10b, D12a [2, 3]
    Medical advice M05a [2, 3]

138   dazomet

A methyl isothiocyanate releasing soil fumigant
HRAC mode of action code: Z

Products

Basamid	Certis	97% w/w	GR	11324

Uses
- Soil pests in **soil and compost**
- Soil-borne diseases in **soil and compost**

Approval information
- Dazomet included in Annex I under EC Regulation 1107/2009

Efficacy guidance
- Dazomet acts by releasing methyl isothiocyanate in contact with moist soil
- Soil sterilization is carried out after harvesting one crop and before planting the next
- The soil must be of fine tilth, free of clods and evenly moist to the depth of sterilization
- Soil moisture must not be less than 50% of water-holding capacity or oversaturated. If too dry, water at least 7-14 d before treatment
- In order to obtain short treatment times it is recommended to treat soils when soil temperature is above 7°C. Treatment should be used outdoors before winter rains make soil too wet to cultivate - usually early Nov
- For club root control treat only in summer when soil temperature above 10°C
- Where onion white rot is a problem it is unlikely to give effective control where inoculum level high or crop under stress
- Apply granules with suitable applicators, mix into soil immediately to desired depth and seal surface with polythene sheeting, by flooding or heavy rolling. See label for suitable application and incorporation machinery
- With 'planting through' technique polythene seal is left in place to form mulch into which new crop can be planted

Restrictions
- Maximum number of treatments 1 per batch of soil for protected crops or field crop situation
- Use is limited to one application every third year on the same area
- Crops must not be planted until the cress safety test has been carried out and germination found to be satisfactory
- Do not treat ground where water table may rise into treated layer

Crop-specific information
- Latest use: pre-planting

Following crops guidance
- With 'planting through' technique no gas release cultivations are made and safety test with cress is particularly important. Conduct cress test on soil samples from centre as well as edges of bed. Observe minimum of 30 d from application to cress test in soils at 10°C or above, at least 50 d in soils below 10°C. See label for details
- In all other situations 14-28 d after treatment cultivate lightly to allow gas to disperse and conduct cress test after a further 14-28 d (timing depends on soil type and temperature). Do not treat structures containing live plants or any ground within 1 m of live plants

FOR FULL CONDITIONS OF USE ALWAYS READ THE PRODUCT LABEL

Environmental safety
- Dangerous for the environment
- Very toxic to aquatic organisms

Hazard classification and safety precautions
> **Hazard** Harmful, Dangerous for the environment, Harmful if swallowed
> **Transport code** 9
> **Packaging group** III
> **UN Number** 3077
> **Risk phrases** H315, H317, H319, H335
> **Operator protection** A, M; U04a, U05a, U09a, U19a, U20a
> **Environmental protection** E15a, E34, E38, H410
> **Storage and disposal** D01, D02, D07, D09a, D11a, D12a
> **Medical advice** M03, M05a

139   2,4-DB

A translocated phenoxycarboxylic acid herbicide
HRAC mode of action code: O

Products

1	CloverMaster	Nufarm UK	400 g/l	SL	17927
2	DB Straight	UPL Europe	400 g/l	SL	13736
3	Headland Spruce	Headland	400 g/l	SL	15089

Uses
- Annual dicotyledons in *grassland, lucerne, red clover, spring barley, spring oats, spring wheat, white clover, winter barley, winter oats, winter wheat* [1-3]; *undersown barley, undersown oats, undersown wheat* [1]
- Buttercups in *grassland* [2, 3]
- Fat hen in *grassland* [2, 3]
- Penny cress in *grassland* [2, 3]
- Perennial dicotyledons in *grassland, lucerne, red clover, spring barley, spring oats, spring wheat, undersown barley, undersown oats, undersown wheat, white clover, winter barley, winter oats, winter wheat* [1]
- Shepherd's purse in *grassland* [2, 3]

Approval information
- 2,4-DB included in Annex I under EC Regulation 1107/2009
- Accepted by BBPA for use on malting barley

Efficacy guidance
- Best results achieved on young seedling weeds under good growing conditions. Treatment less effective in cold weather and dry soil conditions
- Rain within 12 h may reduce effectiveness

Restrictions
- Do not allow spray drift onto neighbouring crops

Crop-specific information
- Latest use: before first node detectable (GS 31) for undersown cereals; fourth trifoliate leaf for lucerne
- In direct sown lucerne spray when seedlings have reached first trifoliate leaf stage. Optimum time 3-4 trifoliate leaves
- Do not treat any lucerne after fourth trifoliate leaf
- In spring barley and spring oats undersown with lucerne spray from when cereal has 1 leaf unfolded and lucerne has first trifoliate leaf
- In spring wheat undersown with lucerne spray from when cereal has 3 leaves unfolded and lucerne has first trifoliate leaf

SEE SECTION 3 FOR PRODUCTS ALSO REGISTERED

Following crops guidance

- Do not sow any crop into soil treated with 2,4-DB for at least 3 months after application.

Environmental safety

- Dangerous for the environment
- Toxic to aquatic organisms
- Keep livestock out of treated areas for at least two weeks following treatment and until poisonous weeds, such as ragwort, have died down and become unpalatable

Hazard classification and safety precautions

Hazard Harmful, Dangerous for the environment, Harmful if swallowed
Transport code 9 [2, 3]
Packaging group III [2, 3]
UN Number 3082 [2, 3]; N/C [1]
Risk phrases H315 [3]; H318 [1-3]
Operator protection A, C, H; U05a, U09a, U11, U20a
Environmental protection E07a, E15a, E15b, E34, E38 [1-3]; H411 [3]
Storage and disposal D01, D02, D05, D10c, D12a
Medical advice M03, M05a

140   2,4-DB + MCPA

A translocated herbicide for cereals, clovers and leys
HRAC mode of action code: O + O

See also MCPA

Products

1	Clovermax	Nufarm UK	240:40 g/l	SL	15780
2	Redlegor	UPL Europe	240:40 g/l	SL	15783

Uses

- Annual dicotyledons in *grassland, spring wheat, winter wheat* [1, 2]; *oats, spring barley, undersown barley, undersown wheat, winter barley* [1]; *undersown spring cereals, undersown winter cereals* [2]
- Buttercups in *grassland, oats, spring barley, spring wheat, undersown barley, undersown wheat, winter barley, winter wheat* [1]
- Fat hen in *grassland, oats, spring barley, spring wheat, undersown barley, undersown wheat, winter barley, winter wheat* [1]
- Penny cress in *grassland, oats, spring barley, spring wheat, undersown barley, undersown wheat, winter barley, winter wheat* [1]
- Perennial dicotyledons in *grassland, spring wheat, undersown spring cereals, undersown winter cereals, winter wheat* [2]
- Plantains in *grassland* [1]
- Polygonums in *grassland, spring wheat, undersown spring cereals, undersown winter cereals, winter wheat* [2]
- Shepherd's purse in *grassland, oats, spring barley, spring wheat, undersown barley, undersown wheat, winter barley, winter wheat* [1]

Approval information

- 2,4-DB and MCPA included in Annex I under EC Regulation 1107/2009
- Accepted by BBPA for use on malting barley

Efficacy guidance

- Best results achieved on young seedling weeds under good growing conditions
- Spray thistles and other perennials when 10-20 cm high provided clover at correct stage
- Effectiveness may be reduced by rain within 12 h, by very cold conditions or drought

Restrictions

- Maximum number of treatments 1 per crop
- Do not spray established clover crops or lucerne

- Do not roll or harrow within 7 d before or after spraying
- Do not spray immediately before or after sowing any crop
- Do not graze crops within a week before or 2 weeks after applying
- Do not apply during drought, rain or if rain is expected
- Do not apply in very cold conditions as effectiveness may be reduced
- Do not use immediately before or after sowing any crop
- Do not use on cereals undersown with lucerne or in seed mixtures containing lucerne
- Do not spray in windy conditions as the spray drift may cause damage to neighbouring crops. The following crops are particularly susceptible: beet, Brassicae (e.g. turnips, swedes, oilseed rape), and most market garden crops including lettuce and tomatoes under glass, pears and vines

Crop-specific information
- Latest use: before first node detectable (GS 31) for cereals; 4th trifoliate leaf stage of clover for grass re-seeds
- Apply in spring to winter cereals from leaf sheath erect stage, to spring barley or oats from 2-leaf stage (GS 12), to spring wheat from 5-leaf stage (GS 15)
- Spray clovers as soon as possible after first trifoliate leaf, grasses after 2-3 leaf stage
- Red clover may suffer temporary distortion after treatment

Environmental safety
- Harmful to aquatic organisms
- Harmful to fish or other aquatic life. Do not contaminate surface waters or ditches with chemical or used container
- Keep livestock out of treated areas for at least two weeks following treatment and until poisonous weeds, such as ragwort, have died down and become unpalatable

Hazard classification and safety precautions
Hazard Harmful, Dangerous for the environment, Harmful if swallowed
UN Number N/C
Risk phrases H315 [2]; H318 [1, 2]
Operator protection A, C [1]; U05a, U20b [1, 2]; U08, U11, U13, U14, U15 [2]; U12, U19a [1]
Environmental protection E07a [1, 2]; E13c, H412 [2]; E15b, E34, E38, H411 [1]
Storage and disposal D01, D02, D05, D09a [1, 2]; D10b, D12a [1]; D10c [2]
Medical advice M03 [1, 2]; M05a [2]

141   deltamethrin

A pyrethroid insecticide with contact and residual activity
IRAC mode of action code: 3

Products

1 Bandu	Headland	25 g/l	EC	16153
2 Decis	Bayer CropScience	25 g/l	EC	16124
3 Decis Forte	Bayer CropScience	100 g/l	EC	16110
4 Decis Protech	Bayer CropScience	15 g/l	EW	16160
5 Grainstore	Pan Agriculture	25 g/l	EC	14932
6 K-Obiol EC 25	Bayer CropScience	25 g/l	EC	13573
7 K-Obiol ULV 6	Bayer CropScience	0.69% w/v	UL	13572

Uses
- American serpentine leaf miner in **gherkins** *(off-label)*, **summer squash** *(off-label)* [2]; **protected aubergines** *(off-label)*, **protected courgettes** *(off-label)* [1, 2]; **protected gherkins** *(off-label)*, **protected squashes** *(off-label)* [1]; **protected ornamentals** *(off-label)* [1-4]
- Aphids in **amenity vegetation**, **apples**, **cress** *(off-label)*, **durum wheat** *(off-label)*, **endives** *(off-label)*, **grass seed crops** *(off-label)*, **lamb's lettuce** *(off-label)*, **land cress** *(off-label)*, **leaves and shoots** *(off-label)*, **lettuce**, **protected peppers**, **rye** *(off-label)*, **spring barley**, **spring oats**, **spring wheat**, **triticale** *(off-label)*, **winter barley**, **winter oats**, **winter wheat** [1-4]; **brussels sprouts**, **cabbages**, **cauliflowers** [2, 3]; **celery (outdoor)** *(off-label)*, **fennel** *(off-label)*, **rhubarb** *(off-label)* [2]; **evening primrose** *(off-label)* [4]; **forest nurseries** *(off-label)*, **protected forest**

nurseries *(off-label)* [1, 2]; **ornamental plant production, protected cucumbers, protected ornamentals, protected tomatoes** [1-3]; **protected chilli peppers** [3]
- Apple sucker in **apples** [1-4]
- Barley yellow dwarf virus vectors in **winter barley, winter wheat** [1-4]
- Beet virus yellows vectors in **winter oilseed rape** [1-4]
- Bramble shoot moth in **blackberries** *(off-label)* [1-4]
- Cabbage moth in **collards** *(off-label - caterpillars)*, **kale** *(off-label - caterpillars)*, **oriental cabbage** *(off-label - caterpillars)*, **spring greens** *(off-label - caterpillars)* [2]
- Cabbage seed weevil in **mustard** [1]; **spring mustard, winter mustard** [2-4]; **spring oilseed rape, winter oilseed rape** [1-4]
- Cabbage stem flea beetle in **quinoa** *(off-label)* [2]; **winter oilseed rape** [1-4]
- Cabbage stem weevil in **mustard** [1]; **spring mustard, winter mustard** [2-4]; **spring oilseed rape, winter oilseed rape** [1-4]
- Cabbage white butterfly in **collards** *(off-label - caterpillars)*, **kale** *(off-label - caterpillars)*, **oriental cabbage** *(off-label - caterpillars)*, **spring greens** *(off-label - caterpillars)* [2]
- Capsids in **amenity vegetation, apples** [1-4]; **celery (outdoor)** *(off-label)*, **fennel** *(off-label)*, **rhubarb** *(off-label)* [2]; **ornamental plant production** [1-3]
- Carrot fly in **carrots** *(off-label)*, **horseradish** *(off-label)*, **mallow (althaea spp.)** *(off-label)*, **parsnips** *(off-label)* [1-4]
- Caterpillars in **amenity vegetation, apples, blackberries** *(off-label)*, **brussels sprouts, cabbages, cauliflowers, cress** *(off-label)*, **endives** *(off-label)*, **lamb's lettuce** *(off-label)*, **land cress** *(off-label)*, **leaves and shoots** *(off-label)*, **protected peppers, swedes, turnips** [1-4]; **celery (outdoor)** *(off-label)*, **fennel** *(off-label)*, **rhubarb** *(off-label)* [2]; **evening primrose** *(off-label)*, **lettuce, protected celery** *(off-label)*, **protected rhubarb** *(off-label)* [4]; **mustard** [1]; **ornamental plant production, protected cucumbers, protected ornamentals, protected tomatoes** [1-3]; **protected chilli peppers** [3]
- Cereal flea beetle in **durum wheat** *(off-label)*, **rye** *(off-label)*, **triticale** *(off-label)* [3]
- Codling moth in **apples** [1-4]
- Cutworms in **carrots** *(off-label)*, **horseradish** *(off-label)*, **leeks** *(off-label)*, **mallow (althaea spp.)** *(off-label)*, **parsnips** *(off-label)* [1-4]; **celery (outdoor)** *(off-label)*, **fennel** *(off-label)*, **rhubarb** *(off-label)* [2]; **lettuce** [1, 4]
- Diamond-back moth in **collards** *(off-label - caterpillars)*, **kale** *(off-label - caterpillars)*, **oriental cabbage** *(off-label - caterpillars)*, **spring greens** *(off-label - caterpillars)* [2]
- Flea beetle in **brussels sprouts, cabbages, cauliflowers, swedes, turnips** [2, 3]; **celery (outdoor)** *(off-label)*, **durum wheat** *(off-label - cereal)*, **fennel** *(off-label)*, **rhubarb** *(off-label)*, **rye** *(off-label - cereal)*, **triticale** *(off-label - cereal)* [2]; **cress** *(off-label)* [1, 3, 4]; **durum wheat** *(off-label)*, **rye** *(off-label)*, **triticale** *(off-label)* [1, 4]; **endives** *(off-label)*, **lamb's lettuce** *(off-label)*, **land cress** *(off-label)*, **leaves and shoots** *(off-label)*, **sugar beet** [1-4]; **lettuce** [4]; **mustard** [1]
- Gout fly in **durum wheat** *(off-label)*, **rye** *(off-label)*, **triticale** *(off-label)* [1-4]
- Insect pests in **blackberries** *(off-label)*, **carrots** *(off-label)*, **cress** *(off-label)*, **durum wheat** *(off-label)*, **endives** *(off-label)*, **grass seed crops** *(off-label)*, **horseradish** *(off-label)*, **lamb's lettuce** *(off-label)*, **land cress** *(off-label)*, **leaves and shoots** *(off-label)*, **leeks** *(off-label)*, **mallow (althaea spp.)** *(off-label)*, **parsnips** *(off-label)*, **protected ornamentals** *(off-label)*, **protected spring onions** *(off-label)*, **rye** *(off-label)*, **triticale** *(off-label)* [1]; **evening primrose** *(off-label)* [4]; **forest nurseries** *(off-label)*, **protected forest nurseries** *(off-label)* [1, 2]; **grain** *(off-label)* [7]; **grain stores** [5, 6]; **lettuce** [2, 3]; **stored grain, stored pulses** [5-7]
- Leaf miner in **protected spring onions** *(off-label)* [1-4]
- Leaf moth in **leeks** *(off-label)* [1-4]
- Leafhoppers in **lettuce** [4]
- Mealybugs in **amenity vegetation, protected peppers** [1-4]; **ornamental plant production, protected cucumbers, protected ornamentals, protected tomatoes** [1-3]; **protected chilli peppers** [3]
- Pea and bean weevil in **broad beans, combining peas, spring field beans, vining peas, winter field beans** [1-4]
- Pea midge in **combining peas, vining peas** [2-4]
- Pea moth in **combining peas, vining peas** [1-4]
- Pear sucker in **pears** [1-4]
- Pollen beetle in **lettuce** [4]; **mustard** [1]; **spring mustard, winter mustard** [2-4]; **spring oilseed rape, winter oilseed rape** [1-4]

FOR FULL CONDITIONS OF USE ALWAYS READ THE PRODUCT LABEL

- Raspberry beetle in **blackberries** *(off-label)*, **raspberries** [1-4]
- Saddle gall midge in **durum wheat** *(off-label)*, **rye** *(off-label)*, **triticale** *(off-label)* [1-4]
- Sawflies in **apples** [1-4]
- Scale insects in **amenity vegetation**, **protected peppers** [1-4]; **ornamental plant production**, **protected cucumbers**, **protected ornamentals**, **protected tomatoes** [1-3]; **protected chilli peppers** [3]
- Silver Y moth in **collards** *(off-label - caterpillars)*, **kale** *(off-label - caterpillars)*, **oriental cabbage** *(off-label - caterpillars)*, **spring greens** *(off-label - caterpillars)* [2]
- Spotted wing drosophila in **mirabelles** *(off-label)*, **plums** *(off-label)* [2]
- Thrips in **amenity vegetation**, **protected peppers** [1-4]; **lettuce** [4]; **ornamental plant production**, **protected cucumbers**, **protected tomatoes** [1-3]; **protected chilli peppers** [3]
- Tortrix moths in **apples** [1-4]
- Wessex flea beetle in **quinoa** *(off-label)* [2]
- Western flower thrips in **gherkins** *(off-label)*, **summer squash** *(off-label)* [2]; **protected aubergines** *(off-label)*, **protected courgettes** *(off-label)*, **strawberries** *(off-label)* [1, 2]; **protected chilli peppers** [3]; **protected cucumbers**, **protected tomatoes** [1-3]; **protected gherkins** *(off-label)*, **protected squashes** *(off-label)* [1]; **protected ornamentals** *(off-label)*, **protected peppers** [1-4]
- Whitefly in **amenity vegetation**, **protected peppers** [1-4]; **brussels sprouts**, **cabbages**, **cauliflowers** [2, 3]; **ornamental plant production**, **protected cucumbers**, **protected ornamentals**, **protected tomatoes** [1-3]; **protected chilli peppers** [3]
- Yellow cereal fly in **durum wheat** *(off-label)*, **rye** *(off-label)*, **triticale** *(off-label)* [1-4]

Extension of Authorisation for Minor Use (EAMUs)

- **blackberries** *20141106* [1], *20140905* [2], *20140917* [3], *20131673* [4]
- **carrots** *20141105* [1], *20140900* [2], *20140916* [3], *20131672* [4]
- **celery (outdoor)** *20171516* [2]
- **collards** *(caterpillars)* *20162782* [2]
- **cress** *20141104* [1], *20140901* [2], *20140911* [3], *20131670* [4]
- **durum wheat** *20141100* [1], *(cereal)* *20140908* [2], *20140908* [2], *20140921* [3], *20131674* [4]
- **endives** *20141104* [1], *20140901* [2], *20140911* [3], *20131670* [4]
- **evening primrose** *20142584* [4]
- **fennel** *20171516* [2]
- **forest nurseries** *20131856* [1], *20131857* [2]
- **gherkins** *20131663* [2]
- **grain** *20112491* [7]
- **grass seed crops** *20141107* [1], *20140909* [2], *20140920* [3], *20131671* [4]
- **horseradish** *20141105* [1], *20140900* [2], *20140916* [3], *20131672* [4]
- **kale** *(caterpillars)* *20162782* [2]
- **lamb's lettuce** *20141104* [1], *20140901* [2], *20140911* [3], *20131670* [4]
- **land cress** *20141104* [1], *20140901* [2], *20140911* [3], *20131670* [4]
- **leaves and shoots** *20141104* [1], *20140901* [2], *20140911* [3], *20131670* [4]
- **leeks** *20141103* [1], *20140906* [2], *20140901* [3], *20131665* [4]
- **mallow (althaea spp.)** *20141105* [1], *20140900* [2], *20140916* [3], *20131672* [4]
- **mirabelles** *20140492* [2]
- **oriental cabbage** *(caterpillars)* *20162782* [2]
- **parsnips** *20141105* [1], *20140900* [2], *20140916* [3], *20131672* [4]
- **plums** *20140492* [2]
- **protected aubergines** *20132523* [1], *20131661* [2]
- **protected celery** *20142585* [4]
- **protected courgettes** *20132524* [1], *20131663* [2]
- **protected forest nurseries** *20131856* [1], *20131857* [2]
- **protected gherkins** *20132524* [1]
- **protected ornamentals** *20141102* [1], *20141011* [2], *20141012* [3], *20131662* [4]
- **protected rhubarb** *20142585* [4]
- **protected spring onions** *20141101* [1], *20140902* [2], *20140914* [3], *20131664* [4]
- **protected squashes** *20132524* [1]
- **quinoa** *20151052* [2]
- **rhubarb** *20171516* [2]

SEE SECTION 3 FOR PRODUCTS ALSO REGISTERED

- **rye** *20141100* [1], *(cereal) 20140908* [2], *20140908* [2], *20140921* [3], *20131674* [4]
- **spring greens** *(caterpillars) 20162782* [2]
- **strawberries** *20132527* [1], *20131643* [2]
- **summer squash** *20131663* [2]
- **triticale** *20141100* [1], *(cereal) 20140908* [2], *20140908* [2], *20140921* [3], *20131674* [4]

Approval information
- Deltamethrin included in Annex I under EC Regulation 1107/2009
- Accepted by BBPA for use on malting barley and hops

Efficacy guidance
- A contact and stomach poison with 3-4 wk persistence, particularly effective on caterpillars and sucking insects
- Normally applied at first signs of damage with follow-up treatments where necessary at 10-14 d intervals. Rates, timing and recommended combinations with other pesticides vary with crop and pest. See label for details
- Spray is rainfast within 1 h
- May be applied in frosty weather provided foliage not covered in ice
- Temperatures above 35°C may reduce effectiveness or persistence

Restrictions
- Maximum number of treatments varies with crop and pest, 4 per crop for wheat and barley, only 1 application between 1 Apr and 31 Aug. See label for other crops
- Do not apply more than 1 aphicide treatment to cereals in summer
- Do not spray crops suffering from drought or other physical stress
- Consult processer before treating crops for processing
- Do not apply to a cereal crop if any product containing a pyrethroid insecticide or dimethoate has been applied to that crop after the start of ear emergence (GS 51)
- Do not spray cereals after 31 Mar in the year of harvest within 6 m of the outside edge of the crop
- Reduced volume spraying must not be used on cereals after 31 Mar in yr of harvest

Crop-specific information
- Latest use: early dough (GS 83) for barley, oats, wheat; before flowering for mustard, oilseed rape; before 31 Mar for grass seed crops [1, 2, 4]

Environmental safety
- Dangerous for the environment
- Toxic to aquatic organisms
- Flammable [1-4]
- Extremely dangerous to fish or other aquatic life. Do not contaminate surface waters or ditches with chemical or used container
- Dangerous to bees. Do not apply to crops in flower or to those in which bees are actively foraging. Do not apply when flowering weeds are present
- Do not apply in tank-mixture with a triazole-containing fungicide when bees are likely to be actively foraging in the crop
- To protect non target insects/arthropods respect an unsprayed buffer zone of 5 m to non-crop land [1-4]
- Buffer zone requirement 7 m [1-4]
- Broadcast air-assisted LERAP [4, 5] (30m); Broadcast air-assisted LERAP [2] (30m raspberry, 50m apple & pear; LERAP category A [1, 2, 4, 5]

Hazard classification and safety precautions
Hazard Harmful, Flammable, Harmful if swallowed, Harmful if inhaled [1-3, 5]; Dangerous for the environment [1-5]; Flammable liquid and vapour [1-3]; Very toxic to aquatic organisms [2, 4]
Transport code 3 [1-5]; 9 [6, 7]
Packaging group III
UN Number 1993 [1-5]; 3082 [6, 7]
Risk phrases H304, H335, H336 [1-3]; H315 [1, 2, 5]; H318 [1-3, 5]; H371 [5]; R20 [2]

FOR FULL CONDITIONS OF USE ALWAYS READ THE PRODUCT LABEL

Operator protection A, C, H [1-7]; D, M [5-7]; U04a, U05a, U19a [1-3, 5-7]; U08 [1-3, 5]; U09a, U20c [6, 7]; U10 [1]; U11, U19c [2, 3]; U14 [4]; U20b [1-5]

Environmental protection E12a [2, 3, 5]; E12c [1]; E12f [1] (cereals, oilseed rape, peas, beans); E12f [2-5] (cereals, oilseed rape, peas, beans - see label for guidance); E15a, E16d [1-3, 5]; E15b [4, 6, 7]; E16c [1, 4, 5]; E16j [2, 3]; E17b [2, 3] (30m (raspberry); 50m (apple & pear)); E17b [4, 5] (18 m); E22c [4]; E34 [1-3, 5-7]; E38 [1-5]; H410 [1, 3, 4]; H411 [5]

Storage and disposal D01, D02, D09a, D10c [1-7]; D05 [1-3, 5]; D12a [2-5]

Medical advice M03 [1-3, 5-7]; M05b [1-3, 5]

142   desmedipham

A contact phenyl carbamate herbicide available only in mixtures
HRAC mode of action code: C1

143   desmedipham + ethofumesate + lenacil + phenmedipham

A selective contact and residual herbicide mixture for weed control in beet
HRAC mode of action code: C1 + N + C1 + C1

Products

Betanal MaxxPro	Bayer CropScience	47:75:27:60 g/l	OD	15086

Uses

- Annual dicotyledons in *fodder beet*, *forest nurseries* (off-label), *mangels*, *sugar beet*
- Annual meadow grass in *beetroot* (off-label), *fodder beet*, *forest nurseries* (off-label), *mangels*, *sugar beet*
- Black bindweed in *beetroot* (off-label)
- Black nightshade in *beetroot* (off-label)
- Bugloss in *beetroot* (off-label)
- Charlock in *beetroot* (off-label)
- Chickweed in *beetroot* (off-label)
- Cleavers in *beetroot* (off-label)
- Field speedwell in *beetroot* (off-label)
- Fumitory in *beetroot* (off-label)

Extension of Authorisation for Minor Use (EAMUs)

- *beetroot* 20152038
- *forest nurseries* 20162138

Approval information

- Desmedipham, ethofumesate, lenacil and phenmedipham included in Annex I under EC Regulation 1107/2009

Efficacy guidance

- On soils with more than 5% organic matter content, residual activity may be reduced.

Restrictions

- The maximum total dose must not exceed 1.0 kg/ha ethofumesate in any 3 year period.

Following crops guidance

- Beet crops may be sown at any time after the use of Betanal MaxxPro. Any other crop may be sown 3 months after using Betanal MaxxPro. Ploughing (mould board) to a minimum depth of 15 cm should precede preparation of a new seed bed.
- If a crop is suffering from manganese deficiency it may be checked. To avoid crop check, manganese should ideally be applied to the crop first.
- Crops suffering from lime deficiency may also be checked. Growers should ensure that the lime status of the soil is satisfactory before drilling.
- When the temperature is, or is likely to be, above 21˚C (70˚F) on the day of spraying, application should be made after 5 pm otherwise crop check may occur. If crops are subjected to substantial

day to night temperature changes shortly before or after spraying, a check may occur from which the crop may not fully recover.

Hazard classification and safety precautions

Hazard Irritant, Dangerous for the environment, Very toxic to aquatic organisms
Transport code 9
Packaging group III
UN Number 3082
Risk phrases H317, H318
Operator protection A, C, H; U05a, U11, U14, U20b
Environmental protection E15b, E38, E39, H410
Storage and disposal D09a, D10b, D12a
Medical advice M03

---

144   desmedipham + ethofumesate + phenmedipham

A selective contact and residual herbicide for beet
HRAC mode of action code: C1 + N + C1

See also ethofumesate
phenmedipham

Products

1	Betanal Elite	Bayer CropScience	71:112:91 g/l	EC	17721
2	Betasana Trio SC	UPL Europe	15:115:75 g/l	SC	15551
3	Beta-Team	UPL Europe	25:150:75 g/l	SE	16975
4	Mundua	Agroquimicos	25:151:75 g/l	EC	16103
5	Sniper	Adama	50:200:150 g/l	SE	16133
6	Trilogy	UPL Europe	15:115:75 g/l	SC	15644

Uses

- Annual dicotyledons in *fodder beet* [1-3, 5, 6]; *mangels* [2, 3, 6]; *sugar beet* [1-6]
- Annual meadow grass in *fodder beet* [1, 3, 5]; *mangels* [3]; *sugar beet* [1, 3-5]
- Barnyard grass in *fodder beet*, *sugar beet* [1]
- Blackgrass in *fodder beet* *(moderately susceptible)*, *sugar beet* *(moderately susceptible)* [1]
- Bristle grasses in *fodder beet*, *sugar beet* [1]
- Gallant soldier in *fodder beet*, *sugar beet* [1]
- Loose silky bent in *fodder beet*, *sugar beet* [1]
- Thorn apple in *fodder beet*, *sugar beet* [1]
- Wild oats in *fodder beet* *(moderately susceptible)*, *sugar beet* *(moderately susceptible)* [1]

Approval information

- Desmedipham, ethofumesate and phenmedipham included in Annex I under EC Regulation 1107/2009

Efficacy guidance

- Product recommended for low-volume overall application in a planned spray programme
- Product acts mainly by contact action. A full programme also gives some residual control but this may be reduced on soils with more than 5% organic matter
- Best results achieved from treatments applied at fully expanded cotyledon stage of largest weeds present. Occasional larger weeds will usually be controlled from a full programme of sprays
- Where a pre-emergence band spray has been applied treatment must be timed according to size of the untreated weeds between the rows
- Susceptible weeds may not all be killed by the first spray. Repeat applications as each flush of weeds reaches cotyledon size normally necessary for season long control
- Sequential treatments should be applied when the previous one is still showing an effect on the weeds
- Various mixtures with other beet herbicides are recommended. See label for details

FOR FULL CONDITIONS OF USE ALWAYS READ THE PRODUCT LABEL

Restrictions
- Maximum total dose 3.9 l/ha [5], 4.5 l/ha [4] and 7.0 l/ha [2, 6]
- Do not spray crops stressed by nutrient deficiency, wind damage, pest or disease attack, or previous herbicide treatments. Stressed crops treated under conditions of high light intensity may be checked and not recover fully
- If temperature likely to exceed 21°C spray after 5 pm
- Crystallisation may occur if spray volume exceeds that recommended or spray mixture not used within 2 h, especially if the water temperature is below 5°C
- Before use, wash out sprayer to remove all traces of previous products, especially hormone and sulfonyl urea weedkillers.
- The maximum total dose must not exceed 1.0 kg/ha ethofumesate in any three year period

Crop-specific information
- Latest use: before crop meets between rows
- Apply first treatment when majority of crop plants have reached the fully expanded cotyledon stage
- Frost within 7 d of treatment may cause check from which the crop may not recover

Following crops guidance
- Beet crops may be sown at any time after treatment. Any other crop may be sown 3 mth after treatment following mouldboard ploughing to 15 cm minimum

Environmental safety
- Dangerous for the environment
- Toxic to aquatic organisms
- LERAP Category B [1, 3, 5]

Hazard classification and safety precautions
Hazard Irritant [2, 3, 6]; Dangerous for the environment [1-6]; Very toxic to aquatic organisms [1]
Transport code 9
Packaging group III
UN Number 3082
Risk phrases H317 [3]; H319 [2]; H320 [6]; R51, R53a [4]
Operator protection A, H [1, 3-5]; U08 [1-6]; U14 [3]; U20a [1, 3-5]
Environmental protection E15a [1, 3-5]; E15b [2, 6]; E16a [1, 3, 5]; E38 [1-6]; H410 [1, 5, 6]; H411 [2, 3]
Storage and disposal D05, D09a, D12a [1-6]; D10b [1, 3-5]; D10c [2, 6]
Treated seed S06a [2, 6]
Medical advice M03

145 desmedipham + phenmedipham

A mixture of contact herbicides for use in sugar beet
HRAC mode of action code: C1 + C1

See also phenmedipham

Products

1	Beetup Compact SC	UPL Europe	80:80 g/l	SC	15566
2	Betanal Turbo	Bayer CropScience	160:160 g/l	EC	15505
3	Rifle	Adama	160:160 g/l	SE	15705

Uses
- Amaranthus in *fodder beet*, *sugar beet* [1]
- Annual dicotyledons in *fodder beet*, *sugar beet* [1-3]; *mangels* [2]
- Black nightshade in *fodder beet*, *sugar beet* [1, 3]
- Chickweed in *fodder beet*, *sugar beet* [1, 3]
- Fat hen in *fodder beet*, *sugar beet* [1]
- Field pansy in *fodder beet*, *sugar beet* [1, 3]
- Field speedwell in *fodder beet*, *sugar beet* [3]
- Groundsel in *fodder beet*, *sugar beet* [3]

SEE SECTION 3 FOR PRODUCTS ALSO REGISTERED

- Ivy-leaved speedwell in *fodder beet*, *sugar beet* [3]
- Mayweeds in *fodder beet*, *sugar beet* [3]
- Penny cress in *fodder beet*, *sugar beet* [1, 3]
- Red dead-nettle in *fodder beet*, *sugar beet* [1]
- Shepherd's purse in *fodder beet*, *sugar beet* [1, 3]

Approval information
- Desmedipham and phenmedipham included in Annex I under EC Regulation 1107/2009

Efficacy guidance
- Product is recommended for low volume overall spraying in a planned programme involving pre- and/or post-emergence treatments at doses recommended for low volume programmes
- Best results obtained from treatment when earliest germinating weeds have reached cotyledon stage
- Further treatments must be applied as each flush of weeds reaches cotyledon stage but allowing a minimum of 7 d between each spray
- Where a pre-emergence band spray has been applied, the first treatment should be timed according to the size of the weeds in the untreated area between the rows
- Product is absorbed by leaves of emerged weeds which are killed by scorching action in 2-10 d. Apply overall as a fine spray to optimise weed cover and spray retention
- Various tank mixtures and sequences recommended to widen weed spectrum and add residual activity - see label for details

Restrictions
- Maximum total dose equivalent to three full dose treatments
- Do not spray crops stressed by nutrient deficiency, frost, wind damage, pest or disease attack, or previous herbicide treatments. Stressed crops may be checked and not recover fully
- If temperature likely to exceed 21°C spray after 5 pm
- Before use, wash out sprayer to remove all traces of previous products, especially hormone and sulfonyl urea weedkillers
- Crystallisation may occur if spray volume exceeds that recommended or spray mixture not used within 2 h, especially if the water temperature is below 5°C
- Product may cause non-reinforced PVC pipes and hoses to soften and swell. Wherever possible, use reinforced PVC or synthetic rubber hoses

Crop-specific information
- Latest use: before crop leaves meet between rows
- Product safe to use on all soil types

Environmental safety
- Dangerous for the environment
- Toxic to aquatic organisms
- Risk to certain non-target insects or other arthropods. Avoid spraying within 6 m of field boundary
- LERAP Category B

Hazard classification and safety precautions
Hazard Irritant [1]; Dangerous for the environment [1-3]; Very toxic to aquatic organisms [2]
Transport code 9
Packaging group III
UN Number 3082
Risk phrases H315, H317, H319 [1]
Operator protection A [1, 2]; U05a, U19a [1-3]; U08 [2]; U09a [1, 3]; U11, U14, U20b [1]; U20a [2, 3]
Environmental protection E15b, E16a, H410 [1-3]; E16b, E39 [2]; E38 [2, 3]
Storage and disposal D01, D02, D09a [1-3]; D05 [1, 3]; D10b, D12a [2, 3]; D10c [1]

FOR FULL CONDITIONS OF USE ALWAYS READ THE PRODUCT LABEL

146   dicamba

A translocated benzoic herbicide available in mixtures or formulated alone for weed control in maize
HRAC mode of action code: O

See also 2,4-D + dicamba
2,4-D + dicamba + dichlorprop-P
2,4-D + dicamba + fluroxypyr
2,4-D + dicamba + MCPA + mecoprop-P
2,4-D + dicamba + triclopyr

Products

1	Antuco	Ventura	70% w/w	SG	17409
2	Oceal	Rotam	70% w/w	SG	15618

Uses
- Amaranthus in **forage maize**, **grain maize** [1]; **maize** [2]
- Fat hen in **forage maize**, **grain maize** [1]; **maize** [2]
- Field pansy in **forage maize**, **grain maize** [1]; **maize** [2]
- Maple-leaved goosefoot in **forage maize**, **grain maize** [1]; **maize** [2]

Approval information
- Dicamba included in Annex I under EC Regulation 1107/2009

Restrictions
- Do not treat maize during cold or frosty conditions, during periods of high day/night temperature variations or during periods of very high temperatures.
- Avoid drift into greenhouses or onto other agricultural or horticultural crops, amenity plantings or gardens. No to be used in glasshouses. Do not allow spray applications to come into contact with desired broad-leaved trees. Beets, all brassicae (including oilseed rape), lettuce, peas, tomatoes, potatoes, all crops and ornamentals are particularly susceptible to dicamba and may be damaged by spray drift.
- Application may only be made between 1 May and 30 September

Following crops guidance
- In the event of crop failure, only redrill with maize but any crop may be sown after a normal harvest date

Hazard classification and safety precautions
Hazard Dangerous for the environment
Transport code 9
Packaging group III
UN Number 3077
Operator protection U20b
Environmental protection E15b, E38, H411
Storage and disposal D01, D02, D10c, D12b

147   dicamba + MCPA + mecoprop-P

A translocated herbicide for cereals, grassland, amenity grass and orchards
HRAC mode of action code: O + O + O

See also MCPA
mecoprop-P

Products

1	Hyprone P	UPL Europe	16:101:92 g/l	SL	17418
2	Hysward-P	Agrichem	16:101:92 g/l	SL	17609
3	Mircam Plus	Nufarm UK	19.5:245:43.3 g/l	SL	15868
4	T2 Green Pro	Nufarm UK	19:245:43.3 g/l	SL	16366
5	Turfmaster	Nufarm UK	19.5:245:43.3 g/l	SL	16344

SEE SECTION 3 FOR PRODUCTS ALSO REGISTERED

SECTION 2

Uses

- Annual dicotyledons in *amenity grassland* [2, 4]; *canary seed* *(off-label)*, *grass seed crops* [3]; *managed amenity turf* [2-5]; *spring barley, spring oats, spring wheat, winter barley, winter oats, winter wheat* [1, 3]
- Black bindweed in *amenity grassland* [4]; *managed amenity turf* [4, 5]
- Charlock in *amenity grassland* [4]; *managed amenity turf* [4, 5]
- Chickweed in *amenity grassland* [4]; *managed amenity turf* [4, 5]
- Clover in *amenity grassland* *(moderately susceptible)* [4]; *managed amenity turf* *(moderately susceptible)* [4, 5]
- Common orache in *amenity grassland* [4]; *managed amenity turf* [4, 5]
- Dandelions in *amenity grassland* *(moderately susceptible)* [4]; *managed amenity turf* *(moderately susceptible)* [4, 5]
- Docks in *amenity grassland, managed amenity turf* [2]
- Fat hen in *amenity grassland* [4]; *managed amenity turf* [4, 5]
- Forget-me-not in *amenity grassland* [4]; *managed amenity turf* [4, 5]
- Fumitory in *amenity grassland* [4]; *managed amenity turf* [4, 5]
- Groundsel in *amenity grassland* [4]; *managed amenity turf* [4, 5]
- Knotgrass in *amenity grassland* [4]; *managed amenity turf* [4, 5]
- Mayweeds in *amenity grassland* [4]; *managed amenity turf* [4, 5]
- Pale persicaria in *amenity grassland* [4]; *managed amenity turf* [4, 5]
- Penny cress in *amenity grassland* [4]; *managed amenity turf* [4, 5]
- Perennial dicotyledons in *amenity grassland* [2, 4]; *grass seed crops* [3]; *managed amenity turf* [2-5]; *spring barley, spring oats, spring wheat, winter barley, winter oats, winter wheat* [1, 3]
- Plantains in *amenity grassland* *(moderately susceptible)* [4]; *managed amenity turf* *(moderately susceptible)* [4, 5]
- Poppies in *amenity grassland* [4]; *managed amenity turf* [4, 5]
- Redshank in *amenity grassland* [4]; *managed amenity turf* [4, 5]
- Shepherd's purse in *amenity grassland* [4]; *managed amenity turf* [4, 5]
- Wild radish in *amenity grassland* [4]; *managed amenity turf* [4, 5]

Extension of Authorisation for Minor Use (EAMUs)

- *canary seed* 20160549 [3]

Approval information

- Dicamba, MCPA and mecoprop-P included in Annex I under EC Regulation 1107/2009
- Accepted by BBPA for use on malting barley

Efficacy guidance

- Treatment should be made when weeds growing actively. Weeds hardened by winter weather may be less susceptible
- For best results apply in fine warm weather, preferably when soil is moist. Do not spray if rain expected within 6 h or in drought
- Application of fertilizer 1-2 wk before spraying aids weed control in turf
- Where a second treatment later in the season is needed in amenity situations and on grass allow 4-6 wk between applications to permit sufficient foliage regrowth for uptake

Restrictions

- Maximum number of treatments (including other mecoprop-P products) or maximum total dose varies with crop and product. See label for details. The total amount of mecoprop-P applied in a single yr must not exceed the maximum total dose approved for any single product for the crop/situation
- Do not apply to cereals after the first node is detectable (GS 31), or to grass under stress from drought or cold weather
- Do not spray cereals undersown with clovers or legumes, to be undersown with grass or legumes or grassland where clovers or other legumes are important
- Do not spray leys established less than 18 mth or orchards established less than 3 yr
- Do not roll or harrow within 7 d before or after treatment, or graze for at least 7 d afterwards (longer if poisonous weeds present)
- Do not use on turf or grass in year of establishment. Allow 6-8 wk after treatment before seeding bare patches

FOR FULL CONDITIONS OF USE ALWAYS READ THE PRODUCT LABEL

- The first 4 mowings after use should not be used for mulching unless composted for 6 mth
- Turf should not be mown for 24 h before or after treatment (3-4 d for closely mown turf)
- Avoid drift onto all broad-leaved plants outside the target area

Crop-specific information
- Latest use: before first node detectable (GS 31) for cereals; 5-6 wk before head emergence for grass seed crops; mid-Oct for established grass
- HI 7-14 d before cutting or grazing for leys, permanent pasture
- Apply to winter cereals from the leaf sheath erect stage (GS 30), and to spring cereals from the 5 expanded leaf stage (GS 15)
- Spray grass seed crops 4-6 wk before flower heads begin to emerge (Timothy 6 wk)
- Turf containing bulbs may be treated once the foliage has died down completely [2]

Environmental safety
- Harmful to aquatic organisms
- Keep livestock out of treated areas for at least 2 wk following treatment and until poisonous weeds, such as ragwort, have died down and become unpalatable
- Harmful to fish or other aquatic life. Do not contaminate surface waters or ditches with chemical or used container
- Some pesticides pose a greater threat of contamination of water than others and mecoprop-P is one of these pesticides. Take special care when applying mecoprop-P near water and do not apply if heavy rain is forecast

Hazard classification and safety precautions
Hazard Harmful [4, 5]; Irritant [1-3]; Dangerous for the environment [3-5]
UN Number N/C
Risk phrases H315 [1, 2]; H318 [1-5]
Operator protection A, C, H [1-5]; M [1-3]; U05a, U08, U11, U20b [1-5]; U15 [3]; U19a [2]
Environmental protection E07a, H411 [1-5]; E13c [1, 2]; E15a, E38 [3]; E15b [4, 5]; E34 [3-5]
Storage and disposal D01, D02, D05, D09a [1-5]; D10b [3-5]; D10c [1, 2]; D12a [3]
Medical advice M03 [3-5]; M05a [1, 2]

148   dicamba + mecoprop-P

A translocated post-emergence herbicide for cereals and grassland
HRAC mode of action code: O + O

See also mecoprop-P

Products

1	Foundation	Headland	84:600 g/l	SL	16414
2	Headland Saxon	Headland	84:600 g/l	SL	11947
3	High Load Mircam	Nufarm UK	80:600 g/l	SL	11930
4	Hyban P	Agrichem	18.7:150 g/l	SL	16799
5	Hygrass P	Agrichem	18.7:150 g/l	SL	16802
6	Prompt	Headland	84:600 g/l	SL	11948
7	Quickfire	Headland Amenity	18.7:150 g/l	SL	17245

Uses
- Annual dicotyledons in *amenity grassland* [1-3, 6]; *durum wheat* (off-label), *rye* (off-label), *triticale* (off-label) [1, 6]; *managed amenity turf* [4, 5, 7]; *spring barley, spring oats, spring wheat, winter barley, winter oats, winter wheat* [1-4, 6]
- Chickweed in *amenity grassland* [3]; *durum wheat* (off-label), *rye* (off-label), *triticale* (off-label) [1]; *managed amenity turf* [4]; *spring barley, spring oats, spring wheat, winter barley, winter oats, winter wheat* [1-4, 6]
- Cleavers in *amenity grassland* [3]; *durum wheat* (off-label), *rye* (off-label), *triticale* (off-label) [1]; *managed amenity turf* [4]; *spring barley, spring oats, spring wheat, winter barley, winter oats, winter wheat* [1-4, 6]
- Docks in *managed amenity turf* [5, 7]

SEE SECTION 3 FOR PRODUCTS ALSO REGISTERED

- Mayweeds in **amenity grassland** [3]; **durum wheat** *(off-label)*, **rye** *(off-label)*, **triticale** *(off-label)* [1]; **managed amenity turf** [4]; **spring barley, spring oats, spring wheat, winter barley, winter oats, winter wheat** [1-4, 6]
- Perennial dicotyledons in **amenity grassland** [1, 2, 6]; **managed amenity turf** [4, 5, 7]; **spring barley, spring oats, spring wheat, winter barley, winter oats, winter wheat** [4]
- Plantains in **managed amenity turf, spring barley, spring oats, spring wheat, winter barley, winter oats, winter wheat** [4]
- Polygonums in **amenity grassland** [3]; **durum wheat** *(off-label)*, **rye** *(off-label)*, **triticale** *(off-label)* [1]; **managed amenity turf** [4]; **spring barley, spring oats, spring wheat, winter barley, winter oats, winter wheat** [1-4, 6]
- Thistles in **managed amenity turf** [5, 7]

Extension of Authorisation for Minor Use (EAMUs)
- **durum wheat** *20140666* [1], *20152161* [6]
- **rye** *20140666* [1], *20152161* [6]
- **triticale** *20140666* [1], *20152161* [6]

Approval information
- Dicamba and mecoprop-P included in Annex I under EC Regulation 1107/2009
- Accepted by BBPA for use on malting barley

Efficacy guidance
- Best results by application in warm, moist weather when weeds are actively growing

Restrictions
- Maximum number of treatments 1 per crop for cereals and 1 or 2 per yr on grass depending on label. The total amount of mecoprop-P applied in a single yr must not exceed the maximum total dose approved for any single product for the crop/situation
- Do not spray in cold or frosty conditions
- Do not spray if rain expected within 6 h
- Do not treat undersown grass until tillering begins
- Do not spray cereals undersown with clover or legume mixtures
- Do not roll or harrow within 7 d before or after spraying
- Do not treat crops suffering from stress from any cause
- Use product immediately following dilution; do not allow diluted product to stand before use [2]
- Avoid treatment when drift may damage neighbouring susceptible crops
- Do not use on new grass for at least 6 months after establishment [7]
- Do not cut grass for at least 1 day after treatment

Crop-specific information
- Latest use: before 1st node detectable for cereals; 7 d before cutting or 14 d before grazing grass
- Apply to winter sown crops from 5 expanded leaf stage (GS 15)
- Apply to spring sown cereals from 5 expanded leaf stage but before first node is detectable (GS 15-31)
- Treat grassland just before perennial weeds flower
- Transient crop prostration may occur after spraying but recovery is rapid

Following crops guidance
- The total amount of mecoprop-P applied in a single year must not exceed the maximum total dose approved for that crop for any single product

Environmental safety
- Dangerous for the environment [2, 3, 6]
- Toxic to aquatic organisms
- Harmful to fish or other aquatic life. Do not contaminate surface waters or ditches with chemical or used container
- Keep livestock out of treated areas for at least 2 wk and until foliage of poisonous weeds such as ragwort has died and become unpalatable
- Some pesticides pose a greater threat of contamination of water than others and mecoprop-P is one of these pesticides. Take special care when applying mecoprop-P near water and do not apply if heavy rain is forecast

FOR FULL CONDITIONS OF USE ALWAYS READ THE PRODUCT LABEL

Hazard classification and safety precautions

Hazard Harmful, Dangerous for the environment, Harmful if swallowed [1-3, 6]; Irritant [4, 5, 7]
UN Number N/C
Risk phrases H315 [1-3, 6, 7]; H318 [2-7]; H319 [1]
Operator protection A, C [1-7]; H, M [1, 2, 4-7]; U05a, U11, U20b [1-7]; U08 [1, 2, 4, 6]; U09a [3, 5, 7]; U15 [4, 5, 7]; U19a [5, 7]
Environmental protection E07a, E38 [1, 2, 4-7]; E13c [3-5, 7]; E15a [1, 2, 6]; E34 [1-7]; H411 [1, 3]; H412 [2, 4-7]
Storage and disposal D01, D02, D05, D09a [1-7]; D10b [1, 2, 6]; D10c [4, 5, 7]; D12a [1, 2, 4-7]
Medical advice M03 [1-7]; M05a [4, 5, 7]

149   dicamba + prosulfuron

A herbicide mixture for weed control in forage and grain maize
HRAC mode of action code: O + B

Products

1 Casper	Syngenta	50:5% w/w	WG	15573
2 Clayton Spook	Clayton	50:5% w/w	WG	15846
3 Rosan	Headland	50:5% w/w	WG	16591

Uses

- Annual dicotyledons in **forage maize**, **grain maize**
- Bindweeds in **forage maize**, **grain maize**
- Docks in **forage maize** *(seedlings only)*, **grain maize** *(seedlings only)*

Approval information

- Dicamba and prosulfuron included in Annex I under EC Regulation 1107/2009

Efficacy guidance

- Always apply in mixture with a non-ionic adjuvant.
- Do not apply in mixture with organo-phosphate insecticides

Following crops guidance

- Following normal harvest wheat, barley, rye, triticale and perennial ryegrass may be sown in the autumn, wheat, barley, rye, triticale, combining peas, maize, field beans, forage kale, broccoli and cauliflower may be sown the following spring but sugar beet, sunflowers or lucerne are not recommended.

Environmental safety

- LERAP Category B

Hazard classification and safety precautions

Hazard Dangerous for the environment [1-3]; Very toxic to aquatic organisms [3]
Transport code 9
Packaging group III
UN Number 3077
Risk phrases R50, R53a [1, 2]
Operator protection A, C, H; U02a, U05a, U14, U20b
Environmental protection E15b, E16a, E38 [1-3]; H410 [3]
Storage and disposal D01, D02, D09a, D10c, D12a

150   dichlorprop-P + ferrous sulphate + MCPA

A herbicide/fertilizer combination for moss and weed control in turf
HRAC mode of action code: O + O

See also ferrous sulphate
MCPA

SECTION 2

Products

SHL Granular Feed, Weed & Mosskiller	Sinclair	0.2:10.9:0.3% w/w	GR	10972

Uses

- Annual dicotyledons in *managed amenity turf*
- Buttercups in *managed amenity turf*
- Moss in *managed amenity turf*
- Perennial dicotyledons in *managed amenity turf*

Approval information

- Dichlorprop-P, ferrous sulphate and MCPA included in Annex I under EC Regulation 1107/2009

Efficacy guidance

- Apply between Mar and Sep when grass in active growth and soil moist
- A repeat treatment may be needed after 4-6 wk to control perennial weeds or if moss regrows
- Water in if rainfall does not occur within 48 h

Restrictions

- Do not treat newly sown grass for 6 mth after establishment [1]
- Do not apply during drought or freezing conditions or when rain imminent
- Avoid walking on treated areas until it has rained or turf has been watered
- Do not mow for 3-4 d before or after application
- Do not use first 4 mowings after treatment for mulching. Mowings should be composted for 6 mth before use
- Do not treat areas of fine turf such as golf or bowling greens [1]
- Avoid contact with tarmac surfaces as staining may occur

Environmental safety

- Avoid drift onto nearby plants and borders [1]

Hazard classification and safety precautions

UN Number N/C
Operator protection U20c
Storage and disposal D01, D09a, D11a

151　dichlorprop-P + MCPA + mecoprop-P

A translocated herbicide mixture for winter and spring cereals
HRAC mode of action code: O + O + O

See also MCPA
*　　　　mecoprop-P*

Products

1 Isomec Ultra	Nufarm UK	310:160:130 g/l	SL	16033
2 Optica Trio	Headland	310:160:130 g/l	SL	16113

Uses

- Annual dicotyledons in *durum wheat, spelt, spring barley, spring oats, spring wheat, winter barley, winter oats, winter wheat* [1, 2]; *rye, triticale* [2]
- Chickweed in *durum wheat, spelt, spring barley, spring oats, spring wheat, winter barley, winter oats, winter wheat* [1, 2]; *rye, triticale* [2]
- Cleavers in *durum wheat, spelt, spring barley, spring oats, spring wheat, winter barley, winter oats, winter wheat* [1, 2]; *rye, triticale* [2]
- Field pansy in *durum wheat, spelt, spring barley, spring oats, spring wheat, winter barley, winter oats, winter wheat* [1, 2]; *rye, triticale* [2]
- Mayweeds in *durum wheat, spelt, spring barley, spring oats, spring wheat, winter barley, winter oats, winter wheat* [1, 2]; *rye, triticale* [2]
- Poppies in *durum wheat, spelt, spring barley, spring oats, spring wheat, winter barley, winter oats, winter wheat* [1, 2]; *rye, triticale* [2]

FOR FULL CONDITIONS OF USE ALWAYS READ THE PRODUCT LABEL

SECTION 2

Approval information
- Dichlorprop-P, MCPA and mecoprop-P included in Annex I under EC Regulation 1107/2009
- Accepted by BBPA for use on malting barley

Efficacy guidance
- Best results obtained if application is made while majority of weeds are at seedling stage but not if temperatures are too low
- Optimum results achieved by spraying when temperature is above 10°C. If temperatures are lower delay spraying until growth becomes more active

Restrictions
- Maximum number of treatments 1 per crop
- Do not spray in windy conditions where spray drift may cause damage to neighbouring crops, especially sugar beet, oilseed rape, peas, turnips and most horticultural crops including lettuce and tomatoes under glass
- Do not apply before 1st March in year of application.

Crop-specific information
- Latest use: before second node detectable (GS 32) for all crops

Environmental safety
- Harmful to aquatic organisms
- Harmful to fish or other aquatic life. Do not contaminate surface waters or ditches with chemical or used container
- Some pesticides pose a greater threat of contamination of water than others and mecoprop-P is one of these pesticides. Take special care when applying mecoprop-P near water and do not apply if heavy rain is forecast

Hazard classification and safety precautions
Hazard Harmful, Dangerous for the environment, Harmful if swallowed
UN Number N/C
Risk phrases H315 [2]; H317, R51 [1]; H318 [1, 2]
Operator protection A, C, H, M; U05a, U08, U11, U20b [1, 2]; U14, U15 [1]
Environmental protection E13c, E34, E38 [1, 2]; H411 [2]
Storage and disposal D01, D02, D05, D09a, D10c [1, 2]; D12a [1]
Medical advice M03 [2]; M05a [1]

152   difenacoum

An anticoagulant coumarin rodenticide

Products

1	Difenag	Killgerm	0.005% w/w	RB	UK12-0497
2	Difenag Fresh Bait	Killgerm	0.005% w/w	RB	UK12-0436
3	Difenag Wax Bait	Killgerm	0.005% w/w	RB	UK14-0851
4	Difenag Wax Blocks	Killgerm	0.005% w/w	RB	UK12-0435
5	Neokil	BASF	0.005% w/w	RB	UK12-0298
6	Neosorexa Bait Blocks	BASF	0.005% w/w	RB	UK12-0360
7	Neosorexa Gold	BASF	0.005% w/w	RB	UK12-0304
8	Neosorexa Gold Ratpacks	BASF	0.005% w/w	RB	UK12-0306
9	Neosorexa Pasta Bait	BASF	0.005% w/w	RB	UK12-0365
10	Ratak Cut Wheat	Killgerm	0.005% w/w	RB	UK12-0313
11	Sakarat D Pasta Bait	Killgerm	0.005% w/w	RB	UK12-0371
12	Sakarat D Wax Bait	Killgerm	0.005% w/w	RB	UK12-0370
13	Sakarat D Whole Wheat	Killgerm	0.005% w/w	RB	UK12-0301
14	Sorexa D	BASF	0.005% w/w	RB	UK12-0319
15	Sorexa Gel	BASF	0.005% w/w	RB	UK12-0364

Uses
- Mice in *farm buildings* [5-15]; *farm buildings/yards* [1-4]; *farmyards* [5-13, 15]; *sewers* [5]

- Rats in **farm buildings** [5-10, 13, 14]; **farm buildings/yards** [1-4]; **farmyards** [5-10, 13]; **sewers** [5]

Approval information
- Difenacoum included in Annex I under EC Regulation 1107/2009
- Difenacoum approval was to expire 31/3/2015 but the EU has postponed expiry until 30/6/2018 to allow time for a decision to be reached on the applications for renewal.

Efficacy guidance
- Difenacoum is a chronic poison and rodents need to feed several times before accumulating a lethal dose. Effective against rodents resistant to other commonly used anticoagulants
- Best results achieved by placing baits at points between nesting and feeding places, at entry points, in holes and where droppings are seen
- A minimum of five baiting points normally required for a small infestation; more than 40 for a large infestation
- Inspect bait sites frequently and top up as long as there is evidence of feeding
- Product formulated for application through a skeleton or caulking gun [15]
- Maintain a few baiting points to guard against reinfestation after a successful control campaign
- Resistance to difenacoum in rats is now widespread in the UK. Take local professional advice before relying on these products to control infestations.

Restrictions
- Only for use by farmers, horticulturists and other professional users
- When working in rodent infested areas wear synthetic rubber/PVC gloves to protect against rodent-borne diseases

Environmental safety
- Harmful to wildlife
- Cover baits by placing in bait boxes, drain pipes or under boards to prevent access by children, animals or birds
- Products contain human taste deterrent

Hazard classification and safety precautions
    UN Number N/C
    Operator protection A [15]; U13, U20b
    Environmental protection E10b [5-9, 14]; E15a [15]
    Storage and disposal D09a, D11a
    Vertebrate/rodent control products V01a, V03a, V04a [15]; V01b, V04b [5-9, 11, 12, 14]; V02 [5-9, 11, 12, 14, 15]; V03b [11, 12]; V04c [5-9, 14]
    Medical advice M03

153   difenoconazole

A diphenyl-ether triazole protectant and curative fungicide
FRAC mode of action code: 3

See also azoxystrobin + difenoconazole

Products

1 Alternet	Belcrop	250 g/l	EC	17689
2 Difcon 250	Harvest	250 g/l	EC	17851
3 Difcor 250 EC	Belchim	250 g/l	EC	13917
4 Difend	Q-Chem	30 g/l	FS	16316
5 Difference	Belchim	250 g/l	EC	16129
6 Kix	Belchim	250 g/l	EC	17424
7 Narita	Belchim	250 g/l	EC	16210
8 Plover	Syngenta	250 g/l	EC	17288
9 Revel	Life Scientific	250 g/l	EC	17973

FOR FULL CONDITIONS OF USE ALWAYS READ THE PRODUCT LABEL

Uses

- Alternaria in **borage for oilseed production** *(off-label)*, **canary flower (echium spp.)** *(off-label)*, **durum wheat** *(off-label)*, **evening primrose** *(off-label)*, **honesty** *(off-label)*, **linseed** *(off-label)*, **mustard** *(off-label)*, **rye** *(off-label)*, **triticale** *(off-label)* [3]; **broccoli, brussels sprouts, cabbages, calabrese, cauliflowers, spring oilseed rape, winter oilseed rape** [2, 3, 8, 9]; **choi sum** *(off-label)* [8]; **collards, kale** [8, 9]
- Blight in **forest nurseries** *(off-label)* [3]
- Brown rust in **durum wheat** *(off-label)*, **rye** *(off-label)*, **triticale** *(off-label)* [3]; **winter wheat** [2, 3]
- Bunt in **winter wheat** [4]
- Cladosporium in **durum wheat** *(off-label)*, **rye** *(off-label)*, **triticale** *(off-label)* [3]
- Disease control in **potatoes** [1, 6, 7]
- Leaf spot in **choi sum** *(off-label)* [8]; **forest nurseries** *(off-label)* [3]
- Light leaf spot in **spring oilseed rape, winter oilseed rape** [2, 3, 8, 9]
- Powdery mildew in **borage for oilseed production** *(off-label)*, **canary flower (echium spp.)** *(off-label)*, **evening primrose** *(off-label)*, **honesty** *(off-label)*, **linseed** *(off-label)*, **mustard** *(off-label)* [3]
- Ring spot in **broccoli, brussels sprouts, cabbages, calabrese, cauliflowers** [2, 3, 8, 9]; **choi sum** *(off-label)* [8]; **collards, kale** [8, 9]
- Rust in **asparagus** *(off-label)* [8]; **forest nurseries** *(off-label)* [3]
- Scab in **apples, pears** [5]
- Sclerotinia in **borage for oilseed production** *(off-label)*, **canary flower (echium spp.)** *(off-label)*, **evening primrose** *(off-label)*, **honesty** *(off-label)*, **linseed** *(off-label)*, **mustard** *(off-label)* [3]
- Septoria leaf blotch in **durum wheat** *(off-label)*, **rye** *(off-label)*, **triticale** *(off-label)* [3]; **winter wheat** [2, 3]
- Stem canker in **spring oilseed rape, winter oilseed rape** [2, 3, 8, 9]
- Stinking smut in **winter wheat** [4]

Extension of Authorisation for Minor Use (EAMUs)

- **asparagus** *20171713* [8]
- **borage for oilseed production** *20140211* [3]
- **canary flower (echium spp.)** *20140211* [3]
- **choi sum** *20171714* [8]
- **durum wheat** *20140212* [3]
- **evening primrose** *20140211* [3]
- **forest nurseries** *20140205* [3]
- **honesty** *20140211* [3]
- **linseed** *20140211* [3]
- **mustard** *20140211* [3]
- **rye** *20140212* [3]
- **triticale** *20140212* [3]

Approval information

- Difenoconazole included in Annex I under EC Regulation 1107/2009

Efficacy guidance

- Improved control of established infections on oilseed rape achieved by mixture with carbendazim. See labels
- In brassicas a 3-spray programme should be used starting at the first sign of disease and repeated at 14-21 d intervals
- Product is fully rainfast 2 h after application
- For most effective control of Septoria, apply as part of a programme of sprays which includes a suitable flag leaf treatment
- Difenoconazole is a DMI fungicide. Resistance to some DMI fungicides has been identified in Septoria leaf blotch which may seriously affect performance of some products. For further advice contact a specialist advisor and visit the Fungicide Resistance Action Group (FRAG)-UK website

Restrictions

- Maximum number of treatments 3 per crop for brassicas; 2 per crop for oilseed rape; 1 per crop for wheat [3]

SEE SECTION 3 FOR PRODUCTS ALSO REGISTERED

- Maximum total dose equivalent to 3 full dose treatments on brassicas; 2 full dose treatments on oilseed rape; 1 full dose treatment on wheat [8]
- Apply to wheat any time from ear fully emerged stage but before early milk-ripe stage (GS 59-73)

Crop-specific information
- Latest use: before grain early milk-ripe stage (GS 73) for cereals; end of flowering for oilseed rape
- HI brassicas 21 d
- Treat oilseed rape in autumn from 4 expanded true leaf stage (GS 1,4). A repeat spray may be made in spring at the beginning of stem extension (GS 2,0) if visible symptoms develop

Environmental safety
- Dangerous for the environment
- Very toxic to aquatic organisms
- Buffer zone requirement 6 m [1]
- Broadcast air-assisted LERAP [5] (20m); LERAP Category B [1-3, 6-9]

Hazard classification and safety precautions
Hazard Harmful [5]; Irritant [2, 3]; Dangerous for the environment [1-9]; Harmful if swallowed [2, 3, 5-7]; Very toxic to aquatic organisms [9]
Transport code 9
Packaging group III
UN Number 3082
Risk phrases H304 [1-3, 6-9]; H315, H336 [1, 5]; H318 [1]; H319 [2, 3, 6, 7, 9]; H320 [5]
Operator protection A, H [1-9]; C [1-3, 5-9]; U05a [1-9]; U09a [1-3, 6-9]; U11 [2, 3, 5]; U20b [1-3, 5-9]
Environmental protection E15a, E16a [1-3, 6-9]; E15b [5]; E17b [5] (20m); E38 [1-9]; H410 [1, 8, 9]; H411 [2-7]
Storage and disposal D01, D02 [1-9]; D05 [1, 4, 6-9]; D07, D10c [1, 6-9]; D09a [1-3, 6-9]; D10b [2, 3]; D12a [1-4, 6-9]
Treated seed S01, S02, S03, S04a, S05, S06a, S07, S08 [4]
Medical advice M05b [5]

---

154 difenoconazole + fludioxonil

A triazole + phenylpyrrole seed treatment for use in cereals
FRAC mode of action code: 3 + 12

See also fludioxonil

Products
1 Celest Extra	Syngenta	25:25 g/l	FS	16630
2 Difend Extra	Globachem	25:25 g/l	FS	17739

Uses
- Bunt in **winter wheat** [1]
- Fusarium foot rot and seedling blight in **winter oats**, **winter wheat** [1]
- Microdochium nivale in **winter oats**, **winter wheat** [1]
- Pyrenophora leaf spot in **winter oats** [1]
- Seed-borne diseases in **winter wheat** [1, 2]
- Seedling blight and foot rot in **winter wheat** [1]
- Septoria seedling blight in **winter wheat** [1]
- Snow mould in **winter wheat** [1]
- Stripe smut in **winter rye** [1]

Approval information
- Difenoconazole and fludioxonil included in Annex I under EC Regulation 1107/2009

Efficacy guidance
- Effective against benzimidazole-resistant and benzimidazole-sensitive strains of *Microdochium nivale*

FOR FULL CONDITIONS OF USE ALWAYS READ THE PRODUCT LABEL

Crop-specific information
- Under adverse environmental or soil conditions, seed rates should be increased to compensate for a slight drop in germination capacity. Flow rates of treated seed should be checked before drilling commences.

Environmental safety
- Buffer zone requirement 10 m

Hazard classification and safety precautions
Hazard Harmful [1]; Dangerous for the environment [1, 2]
Transport code 9
Packaging group III
UN Number 3082
Risk phrases R20, R51, R53a [1]
Operator protection A, H; U05a [1, 2]; U20c [1]
Environmental protection E15a [1]; E38 [1, 2]; H410 [2]
Storage and disposal D01, D02, D12a [1, 2]; D05 [2]; D09a, D11a [1]
Treated seed S01, S02, S03, S04a, S05, S06a, S07, S08 [1, 2]; S04b, S09 [1]
Medical advice M05a [1]

155   difenoconazole + mandipropamid

A triazole/mandelamide fungicide mixture for blight control in potatoes
FRAC mode of action code: 3 + 40

Products

1 Amphore Plus	Syngenta	250:250 g/l	SL	16327
2 Carial Star	Syngenta	250:250 g/l	SL	16323

Uses
- Alternaria blight in **potatoes**
- Blight in **potatoes**
- Early blight in **potatoes**

Approval information
- Difenoconazole and mandipropamid included in Annex I under EC Regulation 1107/2009

Efficacy guidance
- Rainfast within 15 minutes
- Spray programme must start before blight enters the crop

Crop-specific information
- Can be used on all varieties of potato, including seed potatoes

Environmental safety
- LERAP Category B

Hazard classification and safety precautions
Transport code 9
Packaging group III
UN Number 3082
Operator protection A, H; U05a, U20b
Environmental protection E15b, E16a, E22c, E34, E38, H410
Storage and disposal D01, D09a, D10c
Medical advice M03

SEE SECTION 3 FOR PRODUCTS ALSO REGISTERED

156 difenoconazole + paclobutrazol

A triazole fungicide mixture for growth regulation and disease control in oilseed rape
FRAC mode of action code: 3

Products

Toprex	Syngenta	250:125 g/l	SC	16456

Uses
- Disease control in **winter oilseed rape**
- Kabatiella lini in **linseed** *(off-label)*
- Light leaf spot in **winter oilseed rape**
- Mildew in **linseed** *(off-label)*
- Phoma in **winter oilseed rape**
- Septoria leaf blotch in **linseed** *(off-label)*

Extension of Authorisation for Minor Use (EAMUs)
- **linseed** *20151633*

Approval information
- Difenoconazole and paclobutrazol included in Annex I under EC Regulation 1107/2009

Efficacy guidance
- Contains a growth regulator for height reduction and lodging control and a fungicide for disease control. It should only be used when both growth regulation and disease control are required. If this is not the case, use appropriate alternative products at the required timing

Hazard classification and safety precautions
  Hazard Dangerous for the environment, Very toxic to aquatic organisms
  Transport code 9
  Packaging group III
  UN Number 3082
  Risk phrases H361
  Operator protection A, H; U05a
  Environmental protection E38, H410
  Storage and disposal D01, D02, D12a
  Medical advice M05a

157 difenoconazole + propiconazole

A triazole fungicide mixture for disease control in fodder beet and sugar beet
FRAC mode of action code: 3

See also difenoconazole
propiconazole

Products

1	Armure	Syngenta	150:150 g/l	EC	16587
2	Difure Pro	Belchim	150:150 g/l	EC	17303

Uses
- Cercospora leaf spot in **fodder beet**, **sugar beet**
- Powdery mildew in **fodder beet**, **sugar beet**
- Ramularia leaf spots in **fodder beet** *(moderate control)*, **sugar beet** *(moderate control)*
- Rust in **fodder beet**, **sugar beet**

Approval information
- Difenoconazole and propiconazole included in Annex I under EC Regulation 1107/2009

Efficacy guidance
- Use preventatively or at the first sign of disease in the crop

FOR FULL CONDITIONS OF USE ALWAYS READ THE PRODUCT LABEL

Restrictions
- Do not apply before complete crop cover is achieved.
- Do not apply using hand-held equipment.

Crop-specific information
- Do not apply to beet before BBCH 40 growth stage

Environmental safety
- LERAP Category B

Hazard classification and safety precautions
Hazard Irritant, Dangerous for the environment [1, 2]; Harmful if swallowed [2]
Transport code 9
Packaging group III
UN Number 3082
Risk phrases H304 [1, 2]; H315, H318, H336, H373 [2]; H319, H335 [1]
Operator protection A, C; U04a, U09a, U20a
Environmental protection E15b, E16a, E34, E38, H410
Storage and disposal D01, D02, D05, D09a, D10c, D12a
Medical advice M03

158   diflubenzuron

A selective, persistent, contact and stomach acting insecticide
IRAC mode of action code: 15

Products

Dimilin Flo	Certis	480 g/l	SC	11056

Uses
- Browntail moth in *amenity vegetation, hedges, nursery stock, ornamental plant production*
- Bud moth in *apples, pears*
- Carnation tortrix moth in *amenity vegetation, hedges, nursery stock, ornamental plant production*
- Caterpillars in *broccoli, brussels sprouts, cabbages, calabrese, cauliflowers, chives* (off-label), *endives* (off-label), *frise* (off-label), *herbs (see appendix 6)* (off-label), *leaf spinach* (off-label), *lettuce* (off-label), *parsley* (off-label), *radicchio* (off-label), *salad brassicas* (off-label - for baby leaf production), *scarole* (off-label)
- Clouded drab moth in *apples, pears*
- Codling moth in *apples, pears*
- Fruit tree tortrix moth in *apples, pears*
- Houseflies in *livestock houses, manure heaps, refuse tips*
- Lackey moth in *amenity vegetation, hedges, nursery stock, ornamental plant production*
- Moths in *almonds* (off-label), *bilberries* (off-label), *blackberries* (off-label), *chestnuts* (off-label), *crab apples* (off-label), *cranberries* (off-label), *gooseberries* (off-label), *hazel nuts* (off-label), *quinces* (off-label), *redcurrants* (off-label), *walnuts* (off-label), *whitecurrants* (off-label)
- Oak leaf roller moth in *forest*
- Pear sucker in *pears*
- Phorid flies in *edible fungi* (off-label - other than mushrooms)
- Pine beauty moth in *forest*
- Pine looper in *forest*
- Plum fruit moth in *plums*
- Rust mite in *apples, pears, plums*
- Sciarid flies in *edible fungi* (off-label - other than mushrooms)
- Small ermine moth in *amenity vegetation, hedges, nursery stock, ornamental plant production*
- Tortrix moths in *plums*
- Winter moth in *amenity vegetation, apples, blackcurrants, forest, hedges, nursery stock, ornamental plant production, pears, plums*

SEE SECTION 3 FOR PRODUCTS ALSO REGISTERED

Extension of Authorisation for Minor Use (EAMUs)
- *almonds* 20060571
- *bilberries* 20060573
- *blackberries* 20060573
- *chestnuts* 20060571
- *chives* 20051321
- *crab apples* 20060572
- *cranberries* 20060573
- *edible fungi* (other than mushrooms) 20060574
- *endives* 20051321
- *frise* 20051321
- *gooseberries* 20060573
- *hazel nuts* 20060571
- *herbs (see appendix 6)* 20051321
- *leaf spinach* 20051321
- *lettuce* 20051321
- *parsley* 20051321
- *quinces* 20060572
- *radicchio* 20051321
- *redcurrants* 20060573
- *salad brassicas* (for baby leaf production) 20051321
- *scarole* 20051321
- *walnuts* 20060571
- *whitecurrants* 20060573

Approval information
- Approved for aerial application in forestry when average wind velocity does not exceed 18 knots and gusts do not exceed 20 knots. See Section 5 for more information
- Diflubenzuron included in Annex I under EC Regulation 1107/2009

Efficacy guidance
- Most active on young caterpillars and most effective control achieved by spraying as eggs start to hatch
- Dose and timing of spray treatments vary with pest and crop. See label for details
- Addition of wetter recommended for use on brassicas and for pear sucker control in pears

Restrictions
- Maximum number of treatments 3 per yr for apples, pears; 2 per yr for plums, blackcurrants; 2 per crop for brassicas; 1 per yr for forest
- Before treating ornamentals check varietal tolerance on a small sample
- Do not use as a compost drench or incorporated treatment on ornamental crops
- Do not spray protected plants in flower or with flower buds showing colour
- For use only on the food crops specified on the label

Crop-specific information
- HI apples, pears, plums, blackcurrants, brassicas 14 d

Environmental safety
- Dangerous for the environment
- Very toxic to aquatic organisms
- Broadcast air-assisted LERAP [1] (20 m when used in orchards; 10 m when used in blackcurrants, forestry or ornamentals); LERAP Category B [1]

Hazard classification and safety precautions
Hazard Dangerous for the environment, Very toxic to aquatic organisms
Transport code 9
Packaging group III
UN Number 3082
Operator protection U20c
Environmental protection E05a, E15a, E16a, E16b, E18, E38 [1]; E17b [1] (20 m when used in orchards; 10 m when used in blackcurrants, forestry or ornamentals)
Storage and disposal D05, D09a, D11a, D12b

FOR FULL CONDITIONS OF USE ALWAYS READ THE PRODUCT LABEL

159   diflufenican

A shoot absorbed pyridinecarboxamide herbicide for winter cereals
HRAC mode of action code: F1

See also bromoxynil + diflufenican
bromoxynil + diflufenican + ioxynil
chlorotoluron + diflufenican
chlorotoluron + diflufenican + pendimethalin
clodinafop-propargyl + diflufenican
clopyralid + diflufenican + MCPA

Products

1	Clayton El Nino	Clayton	500 g/l	SC	17337
2	Dican	Albaugh UK	50% w/w	WG	16922
3	Diflanil 500 SC	Q-Chem	500 g/l	SC	17024
4	Hurricane SC	Adama	500 g/l	SC	16027
5	Ossetia	Rotam	50% w/w	WG	17741
6	Sempra	UPL Europe	500 g/l	SC	16967
7	Sempra XL	UPL Europe	500 g/l	SC	17447
8	Solo 500 SC	Sipcam	500 g/l	SC	17622
9	Twister 500	Harvest	500 g/l	SC	17627

Uses

- Annual dicotyledons in **forest nurseries** *(off-label)*, **game cover** *(off-label)*, **ornamental plant production** *(off-label)*, **spring oats** *(off-label)*, **winter oats** *(off-label)* [4]; **grass seed crops** *(off-label)* [4, 6]; **oat seed crops** *(off-label)* [6]; **spring barley, triticale, winter barley, winter rye, winter wheat** [3, 6]
- Annual grasses in **forest nurseries** *(off-label)*, **game cover** *(off-label)*, **ornamental plant production** *(off-label)*, **spring oats** *(off-label)*, **winter oats** *(off-label)* [4]; **grass seed crops** *(off-label)* [4, 6]; **oat seed crops** *(off-label)*, **spring barley, triticale, winter barley, winter rye, winter wheat** [6]
- Annual meadow grass in **forest nurseries** *(off-label)*, **game cover** *(off-label)*, **ornamental plant production** *(off-label)*, **spring oats** *(off-label)*, **winter oats** *(off-label)* [4]; **grass seed crops** *(off-label)* [4, 7]; **oat seed crops** *(off-label)* [7]
- Black bindweed in **grass seed crops** *(off-label)*, **oat seed crops** *(off-label)* [7]
- Brassica spp. in **durum wheat, spring barley, spring wheat, triticale, winter barley, winter rye, winter wheat** [5]
- Charlock in **forest nurseries** *(off-label)*, **game cover** *(off-label)*, **grass seed crops** *(off-label)*, **oat seed crops** *(off-label)*, **ornamental plant production** *(off-label)* [7]
- Chickweed in **forest nurseries** *(off-label)*, **game cover** *(off-label)*, **grass seed crops** *(off-label)*, **oat seed crops** *(off-label)*, **ornamental plant production** *(off-label)* [7]; **spring barley, triticale, winter barley, winter rye, winter wheat** [3]
- Cleavers in **durum wheat, spring wheat** [1, 4, 5, 7, 9]; **rye** [8]; **spring barley, triticale** [1, 3-5, 7-9]; **winter barley, winter wheat** [1-5, 7-9]; **winter rye** [1, 3-5, 7, 9]
- Corn spurrey in **forest nurseries** *(off-label)*, **game cover** *(off-label)*, **grass seed crops** *(off-label)*, **oat seed crops** *(off-label)*, **ornamental plant production** *(off-label)* [7]
- Dead nettle in **forest nurseries** *(off-label)*, **game cover** *(off-label)*, **grass seed crops** *(off-label)*, **oat seed crops** *(off-label)*, **ornamental plant production** *(off-label)* [7]
- Field pansy in **durum wheat, spring wheat, winter rye** [1, 4, 5, 7, 9]; **forest nurseries** *(off-label)*, **game cover** *(off-label)*, **grass seed crops** *(off-label)*, **oat seed crops** *(off-label)*, **ornamental plant production** *(off-label)* [7]; **rye** [8]; **spring barley, triticale** [1, 4, 5, 7-9]; **winter barley, winter wheat** [1, 2, 4, 5, 7-9]
- Field speedwell in **durum wheat, spring wheat** [1, 4, 5, 7, 9]; **rye** [8]; **spring barley, triticale** [1, 3-5, 7-9]; **winter barley, winter wheat** [1-5, 7-9]; **winter rye** [1, 3-5, 7, 9]
- Flixweed in **forest nurseries** *(off-label)*, **game cover** *(off-label)*, **grass seed crops** *(off-label)*, **oat seed crops** *(off-label)*, **ornamental plant production** *(off-label)* [7]
- Forget-me-not in **forest nurseries** *(off-label)*, **game cover** *(off-label)*, **ornamental plant production** *(off-label)* [7]

SEE SECTION 3 FOR PRODUCTS ALSO REGISTERED

- Ivy-leaved speedwell in **durum wheat**, **spring wheat** [1, 4, 5, 7, 9]; **rye** [8]; **spring barley**, **triticale** [1, 3-5, 7-9]; **winter barley**, **winter wheat** [1-5, 7-9]; **winter rye** [1, 3-5, 7, 9]
- Knotgrass in **grass seed crops** (off-label), **oat seed crops** (off-label) [7]
- Mayweeds in **durum wheat**, **spring wheat**, **winter rye** [1, 4, 5, 7, 9]; **rye** [8]; **spring barley**, **triticale** [1, 4, 5, 7-9]; **winter barley**, **winter wheat** [1, 2, 4, 5, 7-9]
- Mouse-ear chickweed in **forest nurseries** (off-label), **game cover** (off-label), **grass seed crops** (off-label), **oat seed crops** (off-label), **ornamental plant production** (off-label) [7]
- Nipplewort in **grass seed crops** (off-label), **oat seed crops** (off-label) [7]
- Parsley-piert in **forest nurseries** (off-label), **game cover** (off-label), **grass seed crops** (off-label), **oat seed crops** (off-label), **ornamental plant production** (off-label) [7]
- Poppies in **durum wheat**, **spring wheat**, **winter rye** [1, 4, 5, 7, 9]; **forest nurseries** (off-label), **game cover** (off-label), **grass seed crops** (off-label), **oat seed crops** (off-label), **ornamental plant production** (off-label) [7]; **rye** [8]; **spring barley**, **triticale** [1, 4, 5, 7-9]; **winter barley**, **winter wheat** [1, 2, 4, 5, 7-9]
- Red dead-nettle in **durum wheat**, **spring wheat**, **winter rye** [1, 4, 7, 9]; **rye** [8]; **spring barley**, **triticale** [1, 4, 7-9]; **winter barley**, **winter wheat** [1, 2, 4, 7-9]
- Runch in **forest nurseries** (off-label), **game cover** (off-label), **grass seed crops** (off-label), **oat seed crops** (off-label), **ornamental plant production** (off-label) [7]
- Shepherd's purse in **forest nurseries** (off-label), **game cover** (off-label), **ornamental plant production** (off-label) [7]
- Sowthistle in **grass seed crops** (off-label), **oat seed crops** (off-label) [7]
- Speedwells in **forest nurseries** (off-label), **game cover** (off-label), **grass seed crops** (off-label), **oat seed crops** (off-label), **ornamental plant production** (off-label) [7]
- Treacle mustard in **forest nurseries** (off-label), **game cover** (off-label), **grass seed crops** (off-label), **oat seed crops** (off-label), **ornamental plant production** (off-label) [7]
- Volunteer oilseed rape in **forest nurseries** (off-label), **game cover** (off-label), **grass seed crops** (off-label), **oat seed crops** (off-label), **ornamental plant production** (off-label) [7]; **spring barley**, **triticale**, **winter barley**, **winter rye**, **winter wheat** [3]
- Wild radish in **forest nurseries** (off-label), **game cover** (off-label), **grass seed crops** (off-label), **oat seed crops** (off-label), **ornamental plant production** (off-label) [7]

Extension of Authorisation for Minor Use (EAMUs)
- **forest nurseries** *20130142* [4], *20170374* [7]
- **game cover** *20130142* [4], *20141325* [4], *20170374* [7]
- **grass seed crops** *20130141* [4], *20141324* [4], *20160023* [6], *20162935* [7]
- **oat seed crops** *20160022* [6], *20162935* [7]
- **ornamental plant production** *20130142* [4], *20170374* [7]
- **spring oats** *20130140* [4], *20141269* [4]
- **winter oats** *20130140* [4], *20141269* [4]

Approval information
- Diflufenican included in Annex I under EC Regulation 1107/2009
- Accepted by BBPA for use on malting barley

Efficacy guidance
- Best results achieved from treatment of small actively growing weeds in early autumn or spring
- Good weed control depends on efficient burial of trash or straw before or during seedbed preparation
- Loose or fluffy seedbeds should be rolled before application
- The final seedbed should be moist, fine and firm with clods no bigger than fist size
- Ensure good even spray coverage and increase spray volume for post-emergence treatments where the crop or weed foliage is dense
- Activity may be slow under cool conditions and final level of weed control may take some time to appear
- Where cleavers are a particular problem a separate specific herbicide treatment may be required
- Efficacy may be impaired on soils with a Kd factor greater than 6
- Always follow WRAG guidelines for preventing and managing herbicide resistant weeds. See Section 5 for more information

FOR FULL CONDITIONS OF USE ALWAYS READ THE PRODUCT LABEL

Restrictions
- Maximum number of treatments 1 per crop
- Do not treat broadcast crops [3, 4]
- Do not roll treated crops or harrow at any time after treatment
- Do not apply to soils with more than 10% organic matter or on Sands, or very stony or gravelly soils
- Do not treat after a period of cold frosty weather

Crop-specific information
- Latest use: before end of tillering (GS 29) for wheat and barley [3, 4], pre crop emergence for triticale and winter rye [3, 4]
- Treat only named varieties of rye or triticale [3, 4]

Following crops guidance
- Labels vary slightly but in general ploughing to 150 mm and thoroughly mixing the soil is recommended before drilling or planting any succeeding crops either after crop failure or after normal harvest
- In the event of crop failure only winter wheat or winter barley may be re-drilled immediately after ploughing. Spring crops of wheat, barley, oilseed rape, peas, field beans, sugar beet [3, 4], potatoes, carrots, edible brassicas or onions may be sown provided an interval of 12 wk has elapsed after ploughing
- After normal harvest of a treated crop winter cereals, oilseed rape, field beans, leaf brassicas, sugar beet seed crops and winter onions may be drilled in the following autumn. Other crops listed above for crop failure may be sown in the spring after normal harvest
- Successive treatments with any products containing diflufenican can lead to soil build-up and inversion ploughing to 150 mm must precede sowing any following non-cereal crop. Even where ploughing occurs some crops may be damaged

Environmental safety
- Dangerous for the environment
- Very toxic to aquatic organisms
- Buffer zone requirement 6m [1, 2, 4, 7, 8]
- Buffer zone requirement 7 m for spring barley, 8m for winter barley, winter rye, triticale and winter wheat [3, 6]
- Buffer zone requirement 8m for rye and triticale, 14m for spring barley, winter barley, durum wheat, spring wheat and winter wheat [5]
- LERAP Category B

Hazard classification and safety precautions
    Hazard Dangerous for the environment [1-9]; Very toxic to aquatic organisms [2, 4, 5, 8, 9]
    Transport code 9
    Packaging group III
    UN Number 3077 [2, 5]; 3082 [1, 3, 4, 6-9]
    Operator protection A [1-9]; C [1, 3, 4, 7-9]; H [2]; U05a [6, 8, 9]; U08, U13, U19a [1, 2, 4, 5, 7-9]; U15 [1, 2, 4, 5, 7]; U20a [1, 2, 4-9]
    Environmental protection E13a [6]; E13c [8, 9]; E15a [3]; E16a, H410 [1-9]; E16i [1, 4]; E34, E38 [1, 2, 4, 5, 7-9]
    Storage and disposal D01, D02 [1, 2, 4-9]; D05 [1-5, 7-9]; D09a, D10b [1-9]; D12a [1-7]
    Medical advice M05a [1, 2, 4, 5, 7]

160   diflufenican + florasulam

A herbicide mixture for weed control in cereals
HRAC mode of action code: B + F1

See also diflufenican
      florasulam

Products

Lector Delta	Headland	500:50 g/l	SC	16808

SEE SECTION 3 FOR PRODUCTS ALSO REGISTERED

SECTION 2

Uses

- Annual dicotyledons in *rye, spring barley, triticale, winter barley, winter wheat*
- Black bindweed in *rye, spring barley, triticale, winter barley, winter wheat*
- Charlock in *rye, spring barley, triticale, winter barley, winter wheat*
- Chickweed in *rye, spring barley, triticale, winter barley, winter wheat*
- Cleavers in *rye, spring barley, triticale, winter barley, winter wheat*
- Field pansy in *rye, spring barley, triticale, winter barley, winter wheat*
- Forget-me-not in *rye, spring barley, triticale, winter barley, winter wheat*
- Hemp-nettle in *rye, spring barley, triticale, winter barley, winter wheat*
- Mayweeds in *rye, spring barley, triticale, winter barley, winter wheat*
- Penny cress in *rye, spring barley, triticale, winter barley, winter wheat*
- Shepherd's purse in *rye, spring barley, triticale, winter barley, winter wheat*
- Volunteer oilseed rape in *rye, spring barley, triticale, winter barley, winter wheat*

Approval information

- Diflufenican and florasulam included in Annex I under EC Regulation 1107/2009

Restrictions

- Following application to grass seed crops, the treated grass must not be fed to livestock.
- The total amount of florasulam applied to a cereal crop must not exceed 7.5 g.a.i/ha

Following crops guidance

- In the event of crop failure in the spring, cultivate to 20 cms and then only plant spring wheat, spring barley or spring oats.
- Crops that can be sown in the same calendar year following treatment are cereals, oilseed rape, field beans, grass, peas, sugar beet, potatoes, maize and vegetable brassicas as transplants.

Environmental safety

- Buffer zone requirement 6 m for spring barley, 5 m for winter barley, winter wheat, winter rye and triticale [1]
- LERAP Category B

Hazard classification and safety precautions

Hazard Dangerous for the environment
Transport code 9
Packaging group III
UN Number 3082
Operator protection A
Environmental protection E15b, E16a, E38, H410
Storage and disposal D01, D02, D03, D05, D06a, D09a, D10c, D12a

161   diflufenican + flufenacet

A contact and residual herbicide mixture for cereals
HRAC mode of action code: F1 + K3

See also flufenacet

Products

1	Ascent	Agform	100:400 g/l	SC	17902
2	Clayton Vista	Clayton	200:400 g/l	SC	16705
3	Firestorm	Certis	100:400 g/l	SC	17631
4	Herold	Adama	200:400 g/l	SC	16195
5	Liberator	Bayer CropScience	100:400 g/l	SC	15206
6	Naceto	Globachem	200:400 g/l	SC	18063
7	Pincer	Agform	100:400 g/l	SC	17130
8	Regatta	Bayer CropScience	100:400 g/l	SC	15353
9	Terrane	Headland	100:400 g/l	SC	17619
10	Trumpet	Harvest	200:400 g/l	SC	18183

FOR FULL CONDITIONS OF USE ALWAYS READ THE PRODUCT LABEL

Uses

- Annual dicotyledons in **durum wheat** *(off-label)*, **rye** *(off-label)*, **triticale** *(off-label)* [4, 5]; **grass seed crops** *(off-label)* [5]; **spring barley** *(off-label)* [4, 5, 8]; **spring wheat** *(off-label)* [4]; **winter oats** *(off-label)* [4, 8]
- Annual grasses in **winter oats** *(off-label)* [5]
- Annual meadow grass in **durum wheat** *(off-label)*, **rye** *(off-label)*, **triticale** *(off-label)* [4, 5]; **grass seed crops** *(off-label)*, **spring wheat** [5]; **spring barley** *(off-label)* [4, 5, 8]; **spring wheat** *(off-label)* [4]; **triticale, winter rye** [6, 9]; **winter barley, winter wheat** [1-10]; **winter oats** *(off-label)* [4, 8]
- Blackgrass in **durum wheat** *(off-label)*, **grass seed crops** *(off-label)*, **rye** *(off-label)*, **spring wheat, triticale** *(off-label)* [5]; **spring barley** *(off-label)*, **winter oats** *(off-label)* [5, 8]; **triticale, winter rye** [6, 9]; **winter barley, winter wheat** [1-10]
- Chickweed in **spring wheat** [5]; **triticale, winter rye** [6, 9]; **winter barley, winter wheat** [1-10]
- Field pansy in **spring wheat** [5]; **triticale, winter rye** [6, 9]; **winter barley, winter wheat** [1-10]
- Field speedwell in **spring wheat** [5]; **triticale, winter rye** [6, 9]; **winter barley, winter wheat** [1-10]
- Ivy-leaved speedwell in **spring wheat** [5]; **triticale, winter rye** [6, 9]; **winter barley, winter wheat** [1-10]
- Mayweeds in **spring wheat** [5]; **triticale, winter rye** [6, 9]; **winter barley, winter wheat** [1-10]
- Red dead-nettle in **spring wheat** [5]; **winter barley, winter wheat** [1, 3, 5, 7, 8]
- Shepherd's purse in **triticale, winter rye** [6, 9]; **winter barley, winter wheat** [2, 4, 6, 9, 10]
- Volunteer oilseed rape in **triticale, winter rye** [6, 9]; **winter barley, winter wheat** [2, 4, 6, 9, 10]
- Wild oats in **winter oats** *(off-label)* [5]

Extension of Authorisation for Minor Use (EAMUs)

- **durum wheat** *20131493* [4], *20141699* [4], *20110297* [5]
- **grass seed crops** *20110297* [5]
- **rye** *20131493* [4], *20141699* [4], *20110297* [5]
- **spring barley** *20131492* [4], *20141698* [4], *20121010* [5], *20121905* [8]
- **spring wheat** *20131492* [4], *20141698* [4]
- **triticale** *20131493* [4], *20141699* [4], *20110297* [5]
- **winter oats** *20131491* [4], *20141693* [4], *20111550* [5], *20121904* [8]

Approval information

- Diflufenican and flufenacet included in Annex I under EC Regulation 1107/2009
- Accepted by BBPA for use on malting barley

Efficacy guidance

- Best results obtained when there is moist soil at and after application and rain falls within 7 d
- Residual control may be reduced under prolonged dry conditions
- Activity may be slow under cool conditions and final level of weed control may take some time to appear
- Good weed control depends on burying any trash or straw before or during seedbed preparation
- Established perennial grasses and broad-leaved weeds will not be controlled
- Do not use as a stand-alone treatment for blackgrass control. Always follow WRAG guidelines for preventing and managing herbicide resistant weeds. Section 5 for more information

Restrictions

- Maximum number of treatments 1 per crop
- Do not treat undersown cereals or those to be undersown
- Do not use on waterlogged soils or soils prone to waterlogging
- Do not use on Sands or Very Light soils, or very stony or gravelly soils, or on soils containing more than 10% organic matter
- Do not treat broadcast crops and treat shallow-drilled crops post-emergence only
- Do not incorporate into the soil or disturb the soil after application by rolling or harrowing
- Avoid treating crops under stress from whatever cause and avoid treating during periods of prolonged or severe frosts

Crop-specific information

- Latest use: before 31 Dec in yr of sowing and before 3rd tiller stage (GS 23) for wheat or 4th tiller stage (GS 24) for barley

SEE SECTION 3 FOR PRODUCTS ALSO REGISTERED

- For pre-emergence treatments the seed should be covered with a minimum of 32 mm settled soil

Following crops guidance
- In the event of crop failure wheat, barley or potatoes may be sown provided the soil is ploughed to 15 cm, and a minimum of 12 weeks elapse between treatment and sowing spring wheat or spring barley
- After normal harvest wheat, barley or potatoes my be sown without special cultivations. Soil must be ploughed or cultivated to 15 cm before sowing oilseed rape, field beans, peas, sugar beet, carrots, onions or edible brassicae
- Successive treatments with any products containing diflufenican can lead to soil build-up and inversion ploughing to 150 mm must precede sowing any following non-cereal crop. Even where ploughing occurs some crops may be damaged

Environmental safety
- Dangerous for the environment
- Very toxic to aquatic organisms
- Risk to non-target insects or other arthropods. Avoid spraying within 6 m of the field boundary to reduce the effects on non-target insects or other arthropods
- Buffer zone requirement 6 m [9]
- Buffer zone requirement 10 m
- Buffer zone requirement 12 m [6]
- LERAP Category B

Hazard classification and safety precautions
Hazard Harmful, Dangerous for the environment [1-10]; Harmful if swallowed, Very toxic to aquatic organisms [1-3, 5-8]
Transport code 9
Packaging group III
UN Number 3082
Risk phrases H317 [2, 6]; H373 [1-3, 5-9]; R22a, R48, R50, R53a [4, 10]; R43 [4, 9, 10]
Operator protection A, H; U05a, U14
Environmental protection E15a [2, 7, 8]; E15b [4, 6, 9, 10]; E16a, E34, E38 [1-10]; E22c [1-6, 9, 10]; H410 [1-3, 5-9]
Storage and disposal D01, D02, D09a, D10b, D12a
Medical advice M03, M05a

162   diflufenican + flufenacet + flurtamone

A herbicide mixture for use in wheat and barley
HRAC mode of action code: F1 + K3

Products

1 Movon	Bayer CropScience	90:240:120 g/l	SC	14784
2 Vigon	Bayer CropScience	60:240:120 g/l	SC	14785

Uses
- Annual meadow grass in *winter barley, winter wheat* [1, 2]
- Blackgrass in *winter wheat* *(moderately susceptible)* [1, 2]
- Chickweed in *winter barley, winter wheat* [1, 2]
- Field pansy in *winter barley, winter wheat* [1, 2]
- Field speedwell in *winter barley, winter wheat* [1, 2]
- Forget-me-not in *winter wheat* [1, 2]
- Groundsel in *winter wheat* [1, 2]
- Italian ryegrass in *winter wheat* *(moderately susceptible)* [1, 2]
- Ivy-leaved speedwell in *winter wheat* [1, 2]
- Mayweeds in *winter barley* [1]; *winter wheat* [1, 2]

Approval information
- Diflufenican, flufenacet and flurtamone included in Annex 1 under EC Regulation 1107/2009
- Accepted by BBPA for use on malting barley

FOR FULL CONDITIONS OF USE ALWAYS READ THE PRODUCT LABEL

Environmental safety
- Avoid spraying within 5 m of the field boundary to reduce effects on non-target insects or other arthropods
- LERAP Category B

Hazard classification and safety precautions
  Hazard Harmful, Dangerous for the environment, Harmful if swallowed, Very toxic to aquatic organisms
  Transport code 9
  Packaging group III
  UN Number 3082
  Risk phrases H373 [1, 2]; R22a, R50, R53a, R69 [2]
  Operator protection A; U05a
  Environmental protection E15b, E16a, E22c, E34, E38, E39, E40b, H410
  Storage and disposal D01, D02, D09a, D10b, D12a
  Medical advice M03, M05a

163   diflufenican + flurtamone

A contact and residual herbicide mixture for cereals
HRAC mode of action code: F1 + F1

See also flurtamone

Products

Bacara	Bayer CropScience	100:250 g/l	SC	16448

Uses
- Annual dicotyledons in *spring barley, winter barley, winter wheat*
- Annual meadow grass in *spring barley, winter barley, winter wheat*
- Blackgrass in *spring barley, winter barley, winter wheat*
- Loose silky bent in *spring barley, winter barley, winter wheat*
- Volunteer oilseed rape in *spring barley, winter barley, winter wheat*

Approval information
- Diflufenican and flurtamone included in Annex I under EC Regulation 1107/2009
- Accepted by BBPA for use on malting barley

Efficacy guidance
- Apply pre-emergence or from when crop has first leaf unfolded before susceptible weeds pass recommended size
- Best results obtained on firm, fine seedbeds with adequate soil moisture present at and after application. Increase water volume where crop or weed foliage is dense
- Good weed control requires ash, trash and burnt straw to be buried during seed bed preparation
- Loose fluffy seedbeds should be rolled before application and the final seed bed should be fine and firm without large clods
- Speed of control depends on weather conditions and activity can be slow under cool conditions
- Always follow WRAG guidelines for preventing and managing herbicide resistant weeds. See Section 5 for more information

Restrictions
- Maximum number of treatments 1 per crop
- Crops should be drilled to a normal depth of 25 mm and the seed well covered. Do not treat broadcast crops as uncovered seed may be damaged
- Do not treat spring sown cereals, durum wheat, oats, undersown cereals or those to be undersown
- Do not treat frosted crops or when frost is imminent. Severe frost after application, or any other stress, may lead to transient discoloration or scorch
- Do not use on Sands or Very Light soils or those that are very stony or gravelly
- Do not use on waterlogged soils, or on crops subject to temporary waterlogging by heavy rainfall, as there is risk of persistent crop damage which may result in yield loss

SEE SECTION 3 FOR PRODUCTS ALSO REGISTERED

- Do not use on soils with more than 10% organic matter
- Do not harrow at any time after application and do not roll autumn treated crops until spring

Crop-specific information
- Take particular care to match spray swaths otherwise crop discoloration and biomass reduction may occur which may lead to yield reduction

Following crops guidance
- In the event of crop failure winter wheat may be redrilled immediately after normal cultivation, and winter barley may be sown after ploughing. Fields must be ploughed to a depth of 15 cm and 20 wk must elapse before sowing spring crops of wheat, barley, oilseed rape, peas, field beans or potatoes
- After normal harvest autumn cereals can be drilled after ploughing. Thorough mixing of the soil must take place before drilling field beans, leaf brassicae or winter oilseed rape. For sugar beet seed crops and winter onions complete inversion of the furrow slice is essential
- Do not broadcast or direct drill oilseed rape or other brassica crops as a following crop on treated land. See label for detailed advice on preparing land for subsequent autumn cropping in the normal rotation
- Successive treatments with any products containing diflufenican can lead to soil build-up and inversion ploughing to 150 mm must precede sowing any following non-cereal crop. Even where ploughing occurs some crops may be damaged

Environmental safety
- Dangerous for the environment
- Very toxic to aquatic organisms
- Extremely dangerous to fish or other aquatic life. Do not contaminate surface waters or ditches with chemical or used container
- Buffer zone requirement 5 m for winter wheat and winter barley, 6 m for spring barley
- LERAP Category B

Hazard classification and safety precautions
Hazard Very toxic to aquatic organisms
Transport code 9
Packaging group III
UN Number 3082
Operator protection A; U20c
Environmental protection E15b, E16a, H410
Storage and disposal D09a, D10a

164   diflufenican + glyphosate

A foliar non-selective herbicide mixture
HRAC mode of action code: F1 + G

See also glyphosate

Products

1	Pistol	Bayer CropScience	40:250 g/l	SC	17451
2	Proshield	Everris Ltd	40:250 g/l	SC	17525

Uses
- Annual and perennial weeds in **hard surfaces** [1]; **natural surfaces not intended to bear vegetation, permeable surfaces overlying soil** [1, 2]

Approval information
- Diflufenican and glyphosate included in Annex I under EC Regulation 1107/2009

Efficacy guidance
- Treat when weeds actively growing from Mar to end Sep and before they begin to senesce
- Performance may be reduced if application is made to plants growing under stress, such as drought or water-logging

FOR FULL CONDITIONS OF USE ALWAYS READ THE PRODUCT LABEL

- Pre-emergence activity is reduced on soils containing more than 10% organic matter or where organic debris has collected
- Perennial weeds such as docks, perennial sowthistle and willowherb are best treated before flowering or setting seed
- Perennial weeds emerging from established rootstocks after treatment will not be controlled
- A rainfree period of at least 6 h (preferably 24 h) should follow spraying for optimum control
- For optimum control do not cultivate or rake after treatment

Restrictions
- Maximum number of treatments 1 per yr
- May only be used on porous surfaces overlying soil. Must not be used if an impermeable membrane lies between the porous surface and the soil, and must not be used on any non-porous man-made surfaces
- Do not add any wetting agent or adjuvant oil
- Do not spray in windy weather
- Use a knapsack sprayer giving a coarse spray via anti-drift nozzles for application [1]

Following crops guidance
- A period of at least 6 mth must be allowed after treatment of sites that are to be cleared and grubbed before sowing and planting. Soil should be ploughed or dug first to ensure thorough mixing and dilution of any herbicide residues

Environmental safety
- Dangerous for the environment
- Very toxic to aquatic organisms
- Avoid drift onto non-target plants
- Heavy rain after application may wash product onto sensitive areas such as newly sown grass or areas about to be planted
- Buffer zone requirement 2 m [1, 2]
- LERAP Category B

Hazard classification and safety precautions
Hazard Dangerous for the environment, Very toxic to aquatic organisms
Transport code 9
Packaging group III
UN Number 3082
Operator protection A, C [1, 2]; H [2]; U20b
Environmental protection E15a, E16a, E16b, E38, H410
Storage and disposal D01, D02, D09a, D11a, D12a

165   diflufenican + iodosulfuron-methyl-sodium + mesosulfuron-methyl

A contact and residual herbicide mixture containing sulfonyl ureas for winter wheat
HRAC mode of action code: F1 + B + B

See also iodosulfuron-methyl-sodium
mesosulfuron-methyl

Products

1	Hamlet	Bayer CropScience	50:2.5:7.5 g/l	OD	17370
2	Othello	Bayer CropScience	50:2.5:7.5 g/l	OD	16149

Uses
- Annual dicotyledons in **winter wheat**
- Annual meadow grass in **winter wheat**
- Cleavers in **winter wheat**
- Mayweeds in **winter wheat**
- Rough-stalked meadow grass in **winter wheat**
- Volunteer oilseed rape in **winter wheat**

SEE SECTION 3 FOR PRODUCTS ALSO REGISTERED

SECTION 2

Approval information
- Diflufenican, iodosulfuron-methyl-sodium and mesosulfuron-methyl included in Annex I under EC Regulation 1107/2009

Efficacy guidance
- Optimum control obtained when all weeds are emerged at spraying. Activity is primarily via foliar uptake and good spray coverage of the target weeds is essential
- Translocation occurs readily within the target weeds and growth is inhibited within hours of treatment but symptoms may not be apparent for up to 4 wk, depending on weed species, timing of treatment and weather conditions
- Iodosulfuron-methyl and mesosulfuron-methyl are both members of the ALS-inhibitor group of herbicides. To avoid the build up of resistance do not use any product containing an ALS-inhibitor herbicide with claims for control of grass weeds more than once on any crop
- Use this product as part of a Resistance Management Strategy that includes cultural methods of control and does not use ALS inhibitors as the sole chemical method of weed control in successive crops. See Section 5 for more information

Restrictions
- Maximum number of treatments 1 per crop
- Do not use on crops undersown with grasses, clover or other legumes or any other broad-leaved crop
- Do not use where annual grass weeds other than annual meadow grass and rough meadow grass are present
- Do not use as a stand-alone treatment for control of common chickweed or common poppy. Only use mixtures with non ALS-inhibitor herbicides for these weeds
- Do not use as the sole means of weed control in successive crops
- Specific restrictions apply to use in sequence or tank mixture with other sulfonylurea or ALS-inhibiting herbicides. See label for details
- Do not apply to crops under stress from any cause
- Do not apply when rain is imminent or during periods of frosty weather
- Specified adjuvant must be used. See label

Crop-specific information
- Latest use: before 2nd node detectable for winter wheat
- Transitory crop effects may occur, particularly on overlaps and after late season/spring applications. Recovery is normally complete and yield not affected

Following crops guidance
- In the event of crop failure winter or spring wheat may be drilled after normal cultivation and an interval of 6 wk
- Winter wheat, winter barley or winter oilseed rape may be drilled in the autumn following normal harvest of a treated crop. Spring wheat, spring barley, spring oilseed rape or sugar beet may be drilled in the following spring
- Where the product has been applied in sequence with a permitted ALS-inhibitor herbicide (see label) only winter or spring wheat or barley, or sugar beet may be sown as following crops
- Successive treatments with any products containing diflufenican can lead to soil build-up and inversion ploughing to 150 mm must precede sowing any following non-cereal crop. Even where ploughing occurs some crops may be damaged

Environmental safety
- Dangerous for the environment
- Very toxic to aquatic organisms
- Take extreme care to avoid drift outside the target area
- LERAP Category B

Hazard classification and safety precautions
Hazard Irritant, Dangerous for the environment, Very toxic to aquatic organisms
Transport code 9
Packaging group III
UN Number 3082
Risk phrases H319

FOR FULL CONDITIONS OF USE ALWAYS READ THE PRODUCT LABEL

Operator protection A, C; U05a, U11, U20b
Environmental protection E15b, E16a, E38, H410
Storage and disposal D01, D02, D05, D09a, D10a, D12a

166   diflufenican + metsulfuron-methyl

A contact and residual herbicide mixture containing sulfonyl ureas for cereals
HRAC mode of action code: F1 + B

Products

Pelican Delta	Headland	60:6% w/w	WG	14312

Uses

- Annual dicotyledons in **spring barley, winter barley, winter wheat**
- Charlock in **spring barley, winter barley, winter wheat**
- Corn spurrey in **spring barley, winter barley, winter wheat**
- Docks in **spring barley, winter barley, winter wheat**
- Hemp-nettle in **spring barley, winter barley, winter wheat**
- Mayweeds in **spring barley, winter barley, winter wheat**
- Pale persicaria in **spring barley, winter barley, winter wheat**
- Parsley-piert in **spring barley, winter barley, winter wheat**
- Redshank in **spring barley, winter barley, winter wheat**
- Scarlet pimpernel in **spring barley, winter barley, winter wheat**
- Shepherd's purse in **spring barley, winter barley, winter wheat**

Approval information

- Diflufenican and metsulfuron-methyl included in Annex I under EC Regulation 1107/2009

Efficacy guidance

- Do not apply to frosted crops or scorch may occur
- Can only be applied after 1st Feb up to latest growth stage stipulated

Restrictions

- Do not use on crops grown for seed
- Do not apply in mixture or sequence with any other ALS herbicide

Crop-specific information

- Do not use on soils containing more than 10% organic matter

Following crops guidance

- In the event of crop failure, only sow wheat within 3 months of application

Environmental safety

- Buffer zone requirement 7m [1]
- LERAP Category B

Hazard classification and safety precautions

Hazard Irritant, Dangerous for the environment
Transport code 9
Packaging group III
UN Number 3077
Risk phrases H319
Operator protection A, C; U05a, U08, U11, U19a, U20b
Environmental protection E15a, E16a, E34, E38, H410
Storage and disposal D01, D02, D05, D09a, D10b, D12a

167   diflufenican + pendimethalin

A mixture of pyridinecarboxamide and dinitroanaline herbicides, see Section 3
HRAC mode of action code: F1 + K1

See also pendimethalin

Products

1 Bulldog	Sipcam	15.6:313 g/l	SC	17538
2 Omaha 2	Adama	40:400 g/l	SC	16846

Uses
- Annual dicotyledons in *spring barley, spring wheat* [2]; *triticale, winter barley, winter rye, winter wheat* [1, 2]
- Annual grasses in *spring barley, spring wheat, triticale, winter barley, winter rye, winter wheat* [2]
- Annual meadow grass in *triticale, winter barley, winter rye, winter wheat* [1]

Approval information
- Diflufenican and pendimethalin included in Annex I under EC Regulation 1107/2009

Efficacy guidance
- Do not apply pre-emergence to winter crops drilled after 20th November
- Do not treat broadcast crops

Following crops guidance
- Plough to 150 mm and thoroughly mix the soil before planting any following crop

Environmental safety
- Buffer zone requirement 6 m [1, 2]

Hazard classification and safety precautions
  Hazard Dangerous for the environment [1, 2]; Very toxic to aquatic organisms [1]
  Transport code 9
  Packaging group III
  UN Number 3082
  Risk phrases H334 [1]; R50, R53a [2]
  Operator protection A; U05a, U08, U20a [1]
  Environmental protection E38 [1, 2]; H410 [1]
  Storage and disposal D01, D02, D09a, D10c [1]; D12a [1, 2]

168   dimethachlor

A residual anilide herbicide for use in oilseed rape
HRAC mode of action code: K3

Products

1 Clayton Dimethachlor	Clayton	500 g/l	EC	16188
2 Teridox	Syngenta	500 g/l	EC	15876

Uses
- Annual dicotyledons in *winter oilseed rape*
- Chickweed in *winter oilseed rape*
- Docks in *winter oilseed rape*
- Groundsel in *winter oilseed rape*
- Loose silky bent in *winter oilseed rape*
- Mayweeds in *winter oilseed rape*
- Red dead-nettle in *winter oilseed rape*
- Scarlet pimpernel in *winter oilseed rape*

Approval information
- Dimethachlor included in Annex 1 under EC Regulation 1107/2009

FOR FULL CONDITIONS OF USE ALWAYS READ THE PRODUCT LABEL

Efficacy guidance
- Optimum weed control requires moist soils.
- Heavy rain within a few days of application may lead to poor weed control

Restrictions
- Do not use on sands or very light soils, stony soils, organic soil, broadcast crops or late-drilled crops.
- Apply after drilling but before crop germination which can occur within 48 hours of drilling.

Crop-specific information
- High transpiration rates or persistent wet weather during the first weeks after treatment may reduce crop vigour and plant stand.
- Crops must be covered by at least 20mm of settled soil before application.
- Do not cultivate after application.

Following crops guidance
- In the event of autumn crop failure, winter oilseed rape may be re-drilled after cultivating the soil to at least 15 cms.
- If re-drilling the following spring, spring oilseed rape, field beans, peas, maize or potatoes may be sown. If planting spring wheat, barley, oats, spring linseed, sunflowers, sugar beet or grass crops the land should be ploughed before sowing.

Environmental safety
- Buffer zone requirement 10m [1, 2]
- LERAP Category B

Hazard classification and safety precautions
Hazard Irritant, Dangerous for the environment, Very toxic to aquatic organisms
Transport code 9
Packaging group III
UN Number 3082
Risk phrases H304, H315, H317
Operator protection A, C, H; U05a, U09a, U19a, U20c
Environmental protection E15b, E16a, E40b, H410
Storage and disposal D01, D02, D05, D09a, D10c, D12a
Medical advice M05a

169　dimethenamid-p

A chloroacetamide herbicide available only in mixtures
HRAC mode of action code: K3

See also clomazone + dimethenamid-p + metazachlor

170　dimethenamid-p + metazachlor

A soil acting herbicide mixture for oilseed rape
HRAC mode of action code: K3 + K3

See also metazachlor

Products

Springbok	BASF	200:200 g/l	EC	16786

Uses
- Annual dicotyledons in **broccoli** *(off-label)*, **brussels sprouts** *(off-label)*, **cabbages** *(off-label)*, **calabrese** *(off-label)*, **cauliflowers** *(off-label)*, **chinese leaf** *(off-label)*, **chives** *(off-label)*, **choi sum** *(off-label)*, **collards** *(off-label)*, **forest nurseries** *(off-label)*, **game cover** *(off-label)*, **kale** *(off-label)*, **kohlrabi** *(off-label)*, **leeks** *(off-label)*, **oriental cabbage** *(off-label)*, **ornamental plant production** *(off-label)*, **salad onions** *(off-label)*, **swedes** *(off-label)*, **tatsoi** *(off-label)*, **turnips** *(off-label)*, **winter oilseed rape**

SEE SECTION 3 FOR PRODUCTS ALSO REGISTERED

SECTION 2

- Annual grasses in **broccoli** *(off-label)*, **brussels sprouts** *(off-label)*, **cabbages** *(off-label)*, **calabrese** *(off-label)*, **cauliflowers** *(off-label)*, **chinese leaf** *(off-label)*, **chives** *(off-label)*, **choi sum** *(off-label)*, **collards** *(off-label)*, **forest nurseries** *(off-label)*, **game cover** *(off-label)*, **kale** *(off-label)*, **kohlrabi** *(off-label)*, **leeks** *(off-label)*, **oriental cabbage** *(off-label)*, **ornamental plant production** *(off-label)*, **salad onions** *(off-label)*, **swedes** *(off-label)*, **tatsoi** *(off-label)*, **turnips** *(off-label)*
- Annual meadow grass in **broccoli** *(off-label)*, **brussels sprouts** *(off-label)*, **cabbages** *(off-label)*, **calabrese** *(off-label)*, **cauliflowers** *(off-label)*, **chinese leaf** *(off-label)*, **chives** *(off-label)*, **choi sum** *(off-label)*, **collards** *(off-label)*, **kale** *(off-label)*, **kohlrabi** *(off-label)*, **leeks** *(off-label)*, **oriental cabbage** *(off-label)*, **ornamental plant production** *(off-label)*, **salad onions** *(off-label)*, **swedes** *(off-label)*, **tatsoi** *(off-label)*, **turnips** *(off-label)*
- Chickweed in **ornamental plant production** *(off-label)*, **winter oilseed rape**
- Cleavers in **ornamental plant production** *(off-label)*, **winter oilseed rape**
- Common storksbill in **winter oilseed rape**
- Crane's-bill in **chives** *(off-label)*, **forest nurseries** *(off-label)*, **game cover** *(off-label)*, **leeks** *(off-label)*, **ornamental plant production** *(off-label)*, **salad onions** *(off-label)*, **swedes** *(off-label)*, **turnips** *(off-label)*
- Fat hen in **broccoli** *(off-label)*, **brussels sprouts** *(off-label)*, **cabbages** *(off-label)*, **calabrese** *(off-label)*, **cauliflowers** *(off-label)*, **chinese leaf** *(off-label)*, **choi sum** *(off-label)*, **collards** *(off-label)*, **kale** *(off-label)*, **kohlrabi** *(off-label)*, **oriental cabbage** *(off-label)*, **tatsoi** *(off-label)*
- Field speedwell in **winter oilseed rape**
- Groundsel in **broccoli** *(off-label)*, **brussels sprouts** *(off-label)*, **cabbages** *(off-label)*, **calabrese** *(off-label)*, **cauliflowers** *(off-label)*, **chinese leaf** *(off-label)*, **choi sum** *(off-label)*, **collards** *(off-label)*, **kale** *(off-label)*, **kohlrabi** *(off-label)*, **oriental cabbage** *(off-label)*, **ornamental plant production** *(off-label)*, **tatsoi** *(off-label)*
- Mayweeds in **broccoli** *(off-label)*, **brussels sprouts** *(off-label)*, **cabbages** *(off-label)*, **calabrese** *(off-label)*, **cauliflowers** *(off-label)*, **chinese leaf** *(off-label)*, **choi sum** *(off-label)*, **collards** *(off-label)*, **kale** *(off-label)*, **kohlrabi** *(off-label)*, **oriental cabbage** *(off-label)*, **ornamental plant production** *(off-label)*, **tatsoi** *(off-label)*
- Poppies in **winter oilseed rape**
- Redshank in **broccoli** *(off-label)*, **brussels sprouts** *(off-label)*, **cabbages** *(off-label)*, **calabrese** *(off-label)*, **cauliflowers** *(off-label)*, **chinese leaf** *(off-label)*, **choi sum** *(off-label)*, **collards** *(off-label)*, **kale** *(off-label)*, **kohlrabi** *(off-label)*, **oriental cabbage** *(off-label)*, **tatsoi** *(off-label)*
- Scented mayweed in **winter oilseed rape**
- Shepherd's purse in **broccoli** *(off-label)*, **brussels sprouts** *(off-label)*, **cabbages** *(off-label)*, **calabrese** *(off-label)*, **cauliflowers** *(off-label)*, **chinese leaf** *(off-label)*, **chives** *(off-label)*, **choi sum** *(off-label)*, **collards** *(off-label)*, **forest nurseries** *(off-label)*, **game cover** *(off-label)*, **kale** *(off-label)*, **kohlrabi** *(off-label)*, **leeks** *(off-label)*, **oriental cabbage** *(off-label)*, **ornamental plant production** *(off-label)*, **salad onions** *(off-label)*, **swedes** *(off-label)*, **tatsoi** *(off-label)*, **turnips** *(off-label)*, **winter oilseed rape**
- Small nettle in **broccoli** *(off-label)*, **brussels sprouts** *(off-label)*, **cabbages** *(off-label)*, **calabrese** *(off-label)*, **cauliflowers** *(off-label)*, **chinese leaf** *(off-label)*, **choi sum** *(off-label)*, **collards** *(off-label)*, **kale** *(off-label)*, **kohlrabi** *(off-label)*, **oriental cabbage** *(off-label)*, **ornamental plant production** *(off-label)*, **tatsoi** *(off-label)*
- Sowthistle in **ornamental plant production** *(off-label)*

Extension of Authorisation for Minor Use (EAMUs)
- **broccoli** *20151540*
- **brussels sprouts** *20151540*
- **cabbages** *20151540*
- **calabrese** *20151540*
- **cauliflowers** *20151540*
- **chinese leaf** *20151540*
- **chives** *20143008*
- **choi sum** *20151540*
- **collards** *20151540*
- **forest nurseries** *20143006*
- **game cover** *20143006*
- **kale** *20151540*

FOR FULL CONDITIONS OF USE ALWAYS READ THE PRODUCT LABEL

- **kohlrabi** *20151540*
- **leeks** *20143008*
- **oriental cabbage** *20151540*
- **ornamental plant production** *20143006, 20152108*
- **salad onions** *20143008*
- **swedes** *20143007*
- **tatsoi** *20151540*
- **turnips** *20143007*

Approval information
- Dimethenamid-p and metazachlor included in Annex I under EC Regulation 1107/2009

Efficacy guidance
- Best results obtained from treatments to fine, firm and moist seedbeds
- Apply pre- or post-emergence of the crop and ideally before weed emergence
- Residual weed control may be reduced under prolonged dry conditions
- Weeds germinating from depth may not be controlled
- Dimethenamid-p is more active than metazachlor under dry soil conditions

Restrictions
- Maximum total dose on winter oilseed rape equivalent to one full dose treatment
- Do not disturb soil after application
- Do not treat broadcast crops until they have attained two fully expanded cotyledons
- Do not use on Sands, Very Light soils, or soils containing 10% organic matter
- Do not apply when heavy rain forecast or on soils waterlogged or prone to waterlogging
- Do not treat crops suffering from stress from any cause
- Do not use pre-emergence when crop seed has started to germinate or if not covered with 15 mm of soil
- Applications shall be limited to a total dose of not more than 1.0 kg metazachlor/ha in a three year period on the same field
- Do not apply in mixture with phosphate liquid fertilisers
- Metazachlor stewardship requires that all autumn applications should be made before the end of September to reduce the risk to water

Crop-specific information
- Latest use: before 7th true leaf for winter oilseed rape
- All varieties of winter oilseed rape may be treated

Following crops guidance
- Any crop may follow a normally harvested treated winter oilseed rape crop. Ploughing is not essential before a following cereal crop but is required for all other crops
- In the event of failure of a treated crop winter wheat (excluding durum) or winter barley may be drilled in the same autumn, and any cereal (excluding durum wheat), spring oilseed rape, peas or field beans may be sown in the following spring. Ploughing to at least 150 mm should precede in all cases

Environmental safety
- Dangerous for the environment
- Very toxic to aquatic organisms
- Take extreme care to avoid spray drift onto non-crop plants outside the target area
- Some pesticides pose a greater threat of contamination of water than others and metazachlor is one of these pesticides. Take special care when applying metazachlor near water and do not apply if heavy rain is forecast
- Metazachlor stewardship guidelines advise a maximum dose of 750 g.a.i/ha/annum. Applications to drained fields should be complete by 15th Oct but, if drains are flowing, complete applications by 1st Oct.
- LERAP Category B

Hazard classification and safety precautions
Hazard Harmful, Dangerous for the environment, Harmful if swallowed, Harmful if inhaled, Very toxic to aquatic organisms
Transport code 9

SEE SECTION 3 FOR PRODUCTS ALSO REGISTERED

Packaging group III
UN Number 3082
Risk phrases H304, H317, H319, H351
Operator protection A, C, H; U05a, U08, U20c
Environmental protection E15a, E16a, E34, E38
Storage and disposal D01, D02, D09a, D10c, D12a
Medical advice M03, M05a

171   dimethenamid-p + metazachlor + quinmerac

A soil-acting herbicide mixture for use in oilseed rape
HRAC mode of action code: K3 + K3 + O

See also metazachlor
*   quinmerac*

Products

1	Banastar	BASF	100:300:100 g/l	SE	16834
2	Elk	BASF	200:200:100 g/l	SE	16920
3	Katamaran Turbo	BASF	200:200:100 g/l	SE	16921
4	Shadow	BASF	200:200:100 g/l	SE	16804

Uses

- Annual dicotyledons in *winter oilseed rape* [1-4]
- Annual grasses in *winter oilseed rape* [2-4]
- Annual meadow grass in *winter oilseed rape* [1]
- Blackgrass in *winter corn gromwell* *(off-label)* [3]
- Chickweed in *winter corn gromwell* *(off-label)* [3]
- Cleavers in *winter corn gromwell* *(off-label)* [3]; *winter oilseed rape* [1-4]
- Crane's-bill in *winter corn gromwell* *(off-label)* [3]; *winter oilseed rape* [1]
- Fat hen in *winter corn gromwell* *(off-label)* [3]
- Field pansy in *winter corn gromwell* *(off-label)* [3]
- Forget-me-not in *winter corn gromwell* *(off-label)* [3]
- Groundsel in *winter corn gromwell* *(off-label)* [3]
- Ivy-leaved speedwell in *winter corn gromwell* *(off-label)* [3]
- Mayweeds in *winter corn gromwell* *(off-label)* [3]
- Poppies in *winter corn gromwell* *(off-label)* [3]; *winter oilseed rape* [1]
- Red dead-nettle in *winter corn gromwell* *(off-label)* [3]
- Shepherd's purse in *winter corn gromwell* *(off-label)* [3]
- Sowthistle in *winter corn gromwell* *(off-label)* [3]
- Speedwells in *winter oilseed rape* [1]

Extension of Authorisation for Minor Use (EAMUs)

- *winter corn gromwell* *20152318* [3]

Approval information

- Dimethenamid-p, metazachlor and quinmerac included in Annex I under EC Regulation 1107/ 2009

Efficacy guidance

- Dimethenamid-p is more active than metazachlor under dry soil conditions

Restrictions

- Applications shall be limited to a total dose of not more than 1.0 kg metazachlor/ha in a three year period on the same field
- Do not apply in mixture with phosphate liquid fertilisers
- Metazachlor stewardship requires that all autumn applications should be made before the end of September to reduce the risk to water

FOR FULL CONDITIONS OF USE ALWAYS READ THE PRODUCT LABEL

Following crops guidance

- For all situations following or rotational crops must not be planted until four months after application

Environmental safety

- Some pesticides pose a greater threat of contamination of water than others and metazachlor is one of these pesticides. Take special care when applying metazachlor near water and do not apply if heavy rain is forecast.
- Metazachlor stewardship guidelines advise a maximum dose of 750 g.a.i/ha/annum. Applications to drained fields should be complete by 15th Oct but, if drains are flowing, complete applications by 1st Oct.
- LERAP Category B

Hazard classification and safety precautions

Hazard Harmful [3, 4]; Irritant [2]; Dangerous for the environment, Very toxic to aquatic organisms [2-4]
Transport code 9
Packaging group III
UN Number 3082
Risk phrases H317, H351
Operator protection A, H; U05a, U08 [1-4]; U14, U20c [2-4]; U20a [1]
Environmental protection E07a, E38 [2-4]; E07d [1]; E15b, E16a, E34, H410 [1-4]
Storage and disposal D01, D02, D09a, D10c [1-4]; D05 [1]; D12a [2-4]
Medical advice M03, M05a [2-4]

172   dimethenamid-p + pendimethalin

A chloracetamide and dinitroaniline herbicide mixture for weed control in maize crops
HRAC mode of action code: K1 + K3

Products

1 Clayton Launch	Clayton	212.5:250 g/l	EC	16277
2 Dime	Pan Agriculture	212.5:250 g/l	EC	17398
3 Wing - P	BASF	212.5:250 g/l	EC	15425

Uses

- Annual dicotyledons in *baby leaf crops* (off-label), *broccoli* (off-label), *brussels sprouts* (off-label), *bulb onion sets* (off-label), *bulb onions* (off-label), *cabbages* (off-label), *calabrese* (off-label), *cauliflowers* (off-label), *chives* (off-label), *forage maize (under plastic mulches)* (off-label), *garlic* (off-label), *grain maize under plastic mulches* (off-label), *herbs (see appendix 6)* (off-label), *leeks* (off-label), *lettuce* (off-label), *ornamental plant production* (off-label), *salad onions* (off-label), *shallots* (off-label), *strawberries* (off-label) [3]; *forage maize, grain maize* [1-3]
- Annual grasses in *baby leaf crops* (off-label), *broccoli* (off-label), *brussels sprouts* (off-label), *bulb onion sets* (off-label), *bulb onions* (off-label), *cabbages* (off-label), *calabrese* (off-label), *cauliflowers* (off-label), *chives* (off-label), *forage maize (under plastic mulches)* (off-label), *garlic* (off-label), *grain maize under plastic mulches* (off-label), *herbs (see appendix 6)* (off-label), *leeks* (off-label), *lettuce* (off-label), *ornamental plant production* (off-label), *salad onions* (off-label), *shallots* (off-label), *strawberries* (off-label) [3]
- Annual meadow grass in *broccoli* (off-label), *brussels sprouts* (off-label), *calabrese* (off-label), *cauliflowers* (off-label) [3]; *forage maize, grain maize* [1-3]
- Wild oats in *ornamental plant production* (off-label) [3]

Extension of Authorisation for Minor Use (EAMUs)

- *baby leaf crops* 20170810 [3]
- *broccoli* 20131656 [3]
- *brussels sprouts* 20131656 [3]
- *bulb onion sets* 20122250 [3]
- *bulb onions* 20122250 [3]
- *cabbages* 20122251 [3]

SEE SECTION 3 FOR PRODUCTS ALSO REGISTERED

- *calabrese 20131656* [3]
- *cauliflowers 20131656* [3]
- *chives 20122249* [3]
- *forage maize (under plastic mulches) 20150101* [3]
- *garlic 20122250* [3]
- *grain maize (under plastic mulches) 20111841* [3]
- *herbs (see appendix 6) 20170810* [3]
- *leeks 20122248* [3]
- *lettuce 20170810* [3]
- *ornamental plant production 20130253* [3]
- *salad onions 20122249* [3]
- *shallots 20122250* [3]
- *strawberries 20160933* [3]

Approval information
- Dimethenamid-p and pendimethalin are included in Annex 1 under EC Regulation 1107/2009

Efficacy guidance
- Do not use on soil types with more than 10% organic matter
- Best results obtained when rain falls within 7 days of application

Crop-specific information
- Use on maize crops grown under plastic mulches is by EAMU only; do not use in greenhouses, covers or other forms of protection
- Risk of crop damage if heavy rain falls soon after application to stoney or gravelly soils
- Seed should be covered with at least 5 cm of settled soil

Environmental safety
- LERAP Category B

Hazard classification and safety precautions
Hazard Harmful, Dangerous for the environment [1-3]; Harmful if swallowed [2, 3]; Very toxic to aquatic organisms [3]
Transport code 9
Packaging group III
UN Number 3082
Risk phrases H304 [3]; H315, H317 [2, 3]; H371 [2]; R22a, R22b, R38, R43, R50, R53a [1]
Operator protection A, H; U05a, U14
Environmental protection E15b, E16a, E34, E38 [1-3]; H410 [2, 3]
Storage and disposal D01, D02, D09a, D10c, D12a
Medical advice M03, M05a

173   dimethenamid-p + quinmerac

A herbicide mixture for weed control in oilseed rape
HRAC mode of action code: K3 + O

See also dimethenamid-p + metazachlor + quinmerac
*   quinmerac*

Products
1	Tanaris	BASF	333 + 167 g/l	SE	17173
2	Topkat	BASF	333 + 167 g/l	SE	17356

Uses
- Annual dicotyledons in *winter oilseed rape*
- Cleavers in *winter oilseed rape*
- Crane's-bill in *winter oilseed rape*
- Dead nettle in *winter oilseed rape*
- Mayweeds in *winter oilseed rape*
- Poppies in *winter oilseed rape*

FOR FULL CONDITIONS OF USE ALWAYS READ THE PRODUCT LABEL

- Shepherd's purse in *winter oilseed rape*
- Sowthistle in *winter oilseed rape*
- Speedwells in *winter oilseed rape*

Approval information
- Dimethenamid-p and quinmerac included in Annex I under EC Regulation 1107/2009

Environmental safety
- LERAP Category B

Hazard classification and safety precautions
Hazard Very toxic to aquatic organisms
Transport code 9
Packaging group III
UN Number 3082
Risk phrases H317, H319
Operator protection A, C, H; U05a, U08, U20a
Environmental protection E15b, E16a, E34, H410
Storage and disposal D01, D02, D09a, D10c

174   dimethoate

A contact and systemic organophosphorus insecticide and acaricide
IRAC mode of action code: 1B

Products

Danadim Progress	Headland	400 g/l	EC	15890

Uses
- Aphids in *durum wheat, ornamental plant production, rye, spring wheat, triticale, winter wheat*
- Insect pests in *protected forest nurseries* (off-label)
- Leaf miner in *ornamental plant production*
- Red spider mites in *ornamental plant production*

Extension of Authorisation for Minor Use (EAMUs)
- *protected forest nurseries 20122130*

Approval information
- Dimethoate included in Annex I under EC Regulation 1107/2009
- In 2006 CRD required that all products containing this active ingredient should carry the following warning in the main area of the container label: "Dimethoate is an anticholinesterase organophosphate. Handle with care"

Efficacy guidance
- Chemical has quick knock-down effect and systemic activity lasts for up to 14 d
- With some crops, products differ in range of pests listed as controlled. Uses section above provides summary. See labels for details
- For most pests apply when pest first seen and repeat 2-3 wk later or as necessary. Timing and number of sprays varies with crop and pest. See labels for details
- Best results achieved when crop growing vigorously. Systemic activity reduced when crops suffering from drought or other stress
- In hot weather apply in early morning or late evening
- Where aphids or spider mites resistant to organophosphorus compounds occur control is unlikely to be satisfactory and repeat treatments may result in lower levels of control

Restrictions
- Contains an anticholinesterase organophosphorus compound. Do not use if under medical advice not to work with such compounds

SEE SECTION 3 FOR PRODUCTS ALSO REGISTERED

- Maximum number of treatments 6 per crop for Brussels sprouts, broccoli, cauliflower, calabrese, lettuce; 4 per crop for grass seed crops, cereals; 2 per crop for beet crops, triticale, rye; 1 per crop for mangel and sugar beet seed crops
- On beet crops only one treatment per crop may be made for control of leaf miners (max 84 g a.i./ ha) and black bean aphid (max 400 g a.i./ha)
- In beet crops resistant strains of peach-potato aphid (*Myzus persicae*) are common and dimethoate products must not be used to control this pest
- Test for varietal susceptibility on all unusual plants or new cultivars
- Do not tank mix with alkaline materials. See label for recommended tank-mixes
- Consult processor before spraying crops grown for processing
- Must not be applied to cereals if any product containing a pyrethroid insecticide or dimethoate has been sprayed after the start of ear emergence (GS 51)

Crop-specific information
- Latest use: flowering just complete stage (GS 30) for wheat, rye, triticale
- HI cereals 14 d; ornamental plant production 7 d

Environmental safety
- Dangerous for the environment
- Harmful to aquatic organisms
- Flammable
- Harmful to game, wild birds and animals
- Harmful to livestock. Keep all livestock out of treated areas for at least 7 d
- Likely to cause adverse effects on beneficial arthropods
- To protect non target insects/arthropods respect an unsprayed buffer zone of 5 m to non-crop land
- Dangerous to bees. Do not apply to crops in flower or to those in which bees are actively foraging. Do not apply when flowering weeds are present
- Surface residues may also cause bee mortality following spraying
- Dangerous to fish or other aquatic life. Do not contaminate surface waters or ditches with chemical or used container
- Keep in original container, tightly closed, in a safe place, under lock and key
- LERAP Category A

Hazard classification and safety precautions
>**Hazard** Harmful, Flammable, Flammable liquid and vapour, Harmful if swallowed, Harmful if inhaled
>**Transport code** 6.1
>**Packaging group** III
>**UN Number** 3017
>**Risk phrases** H304, H317
>**Operator protection** A, H, M; U02a, U04a, U08, U14, U15, U19a, U20b
>**Environmental protection** E06c (7 d); E10b, E12a, E12e, E13b, E16c, E16d, E22a, E34, H410
>**Storage and disposal** D01, D02, D05, D09b, D12b
>**Medical advice** M01, M03

175   dimethomorph

A cinnamic acid fungicide with translaminar activity
FRAC mode of action code: 40

See also ametoctradin + dimethomorph

Products

Paraat	BASF	50% w/w	WP	15445

Uses
- Crown rot in **protected strawberries** *(moderate control)*, **strawberries** *(moderate control)*
- Downy mildew in **brassica leaves and sprouts** *(off-label)*, **chard** *(off-label)*, **herbs (see appendix 6)** *(off-label)*, **lamb's lettuce** *(off-label)*, **lettuce** *(off-label)*, **protected baby leaf crops**

FOR FULL CONDITIONS OF USE ALWAYS READ THE PRODUCT LABEL

(off-label), **protected herbs (see appendix 6)** *(off-label)*, **protected lamb's lettuce** *(off-label)*, **protected lettuce** *(off-label)*, **protected ornamentals** *(off-label)*, **radishes** *(off-label)*, **rocket** *(off-label)*, **seedling brassicas** *(off-label)*, **spinach** *(off-label)*
- Root rot in **blackberries**, **protected blackberries** *(moderate control)*, **protected raspberries** *(moderate control)*, **raspberries** *(moderate control)*

Extension of Authorisation for Minor Use (EAMUs)
- **brassica leaves and sprouts** *20122244*
- **chard** *20122244*
- **herbs (see appendix 6)** *20122244*
- **lamb's lettuce** *20122244*
- **lettuce** *20122244*
- **protected baby leaf crops** *20112584*
- **protected herbs (see appendix 6)** *20112584*
- **protected lamb's lettuce** *20112584*
- **protected lettuce** *20112584*
- **protected ornamentals** *20112585*
- **radishes** *20160190*
- **rocket** *20122244*
- **seedling brassicas** *20130274*
- **spinach** *20122244*

Approval information
- Dimethomorph included in Annex I under EC Regulation 1107/2009

Restrictions
- Do not sow clover until the spring following application.
- Dimethomorph is a carboxylic acid amide fungicide. No more than three consecutive applications of CAA fungicides should be made.

Crop-specific information
- After application, allow 10 months before sowing clover [1]

Following crops guidance
- Cereal crops may be planted as part of a normal rotation following use. 120 days must elapse after the application before any other crops are planted.

Environmental safety
- LERAP Category B

Hazard classification and safety precautions
Transport code 9
Packaging group III
UN Number 3077
Operator protection A, D, H; U05a
Environmental protection E15b, E16a, E38
Storage and disposal D01, D02, D09a, D10a, D12a

176   dimethomorph + fluazinam

A mixture of cinnamic acid and dinitroaniline fungicides for blight control in potatoes
FRAC mode of action code: 29 + 40

See also fluazinam

Products

Hubble	Adama	200:200 g/l	SC	16089

Uses
- Blight in **potatoes**
- Late blight in **potatoes**

SECTION 2

Approval information
- Dimethomorph and fluazinam included in Annex I under EC Regulation 1107/2009

Environmental safety
- Buffer zone requirement 6m [1]
- LERAP Category B

Hazard classification and safety precautions
>**Hazard** Harmful, Dangerous for the environment
>**Transport code** 9
>**Packaging group** III
>**UN Number** 3082
>**Risk phrases** H361
>**Operator protection** A, H; U02a, U04a, U05a, U08, U14, U15, U20a
>**Environmental protection** E15b, E16a, E34, E38, H410
>**Storage and disposal** D01, D02, D05, D09a, D12a
>**Medical advice** M03

177   dimethomorph + mancozeb

A systemic and protectant fungicide for potato blight control
FRAC mode of action code: 40 + M3

See also mancozeb

Products

Invader	BASF	7.5:66.7% w/w	WG	15223

Uses
- Blight in **potatoes**
- Disease control in **chives** *(off-label)*, **cress** *(off-label)*, **forest nurseries** *(off-label)*, **frise** *(off-label)*, **herbs (see appendix 6)** *(off-label)*, **lamb's lettuce** *(off-label)*, **leaf brassicas** *(off-label)*, **leeks**, **lettuce** *(off-label)*, **ornamental plant production** *(off-label)*, **poppies for morphine production** *(off-label)*, **radicchio** *(off-label)*, **salad onions** *(off-label)*, **scarole** *(off-label)*, **shallots**
- Downy mildew in **bulb onions**, **chives** *(off-label)*, **cress** *(off-label)*, **forest nurseries** *(off-label)*, **frise** *(off-label)*, **garlic**, **herbs (see appendix 6)** *(off-label)*, **lamb's lettuce** *(off-label)*, **leaf brassicas** *(off-label)*, **lettuce** *(off-label)*, **ornamental plant production** *(off-label)*, **poppies for morphine production** *(off-label)*, **radicchio** *(off-label)*, **salad onions** *(off-label)*, **scarole** *(off-label)*
- White tip in **leeks** *(reduction)*

Extension of Authorisation for Minor Use (EAMUs)
- **chives** *20120111*
- **cress** *20120110*
- **forest nurseries** *20120109*
- **frise** *20120110*
- **herbs (see appendix 6)** *20120110*
- **lamb's lettuce** *20120110*
- **leaf brassicas** *20120110*
- **lettuce** *20120110*
- **ornamental plant production** *20120109*
- **poppies for morphine production** *20120112*
- **radicchio** *20120110*
- **salad onions** *20120111*
- **scarole** *20120110*

Approval information
- Dimethomorph and mancozeb included in Annex I under EC Regulation 1107/2009

Efficacy guidance
- Commence treatment as soon as there is a risk of blight infection

FOR FULL CONDITIONS OF USE ALWAYS READ THE PRODUCT LABEL

- In the absence of a warning treatment should start before the crop meets along the rows
- Repeat treatments every 7-14 d depending in the degree of infection risk
- Irrigated crops should be regarded as at high risk and treated every 7 d
- For best results good spray coverage of the foliage is essential
- To minimise the likelihood of development of resistance these products should be used in a planned Resistance Management strategy. See Section 5 for more information

Restrictions
- Maximum total dose equivalent to eight full dose treatments on potatoes

Crop-specific information
- HI 7 d for all crops

Environmental safety
- Dangerous for the environment
- Very toxic to aquatic organisms
- LERAP Category B

Hazard classification and safety precautions
Hazard Irritant, Dangerous for the environment, Very toxic to aquatic organisms
Transport code 9
Packaging group III
UN Number 3077
Risk phrases H317, H319, H335, H361
Operator protection A, C, D, H; U02a, U04a, U05a, U08, U13, U19a, U20b
Environmental protection E16a, E16b, E38, H410
Storage and disposal D01, D02, D05, D09a, D12a
Medical advice M05a

178 dimethomorph + pyraclostrobin

A fungicide mixture for disease control in onions, garlic and shallots
FRAC mode of action code: 11 + 40

Products

Cassiopeia	BASF	72:40 g/l	EC	16522

Uses
- Downy mildew in **bulb onions**, **garlic**, **shallots**

Approval information
- Dimethomorph and pyraclostrobin included in Annex I under EC Regulation 1107/2009

Efficacy guidance
- Use before the disease is established in the crop and use in a programme with other effective fungicides that have a different mode of action to reduce the risk of resistance

Hazard classification and safety precautions
Hazard Harmful, Dangerous for the environment
Transport code 9
Packaging group III
UN Number 3082
Risk phrases R20, R22a, R38, R50, R53a
Operator protection A, H; U05a
Environmental protection E38
Storage and disposal D01, D02, D12a, D12b
Medical advice M05b

SEE SECTION 3 FOR PRODUCTS ALSO REGISTERED

SECTION 2

179 dimethomorph + zoxamide

A fungicide mixture for blight control in potatoes
FRAC mode of action code: 40 + 22

See also dimethomorph
zoxamide

Products

Presidium	Gowan	180:180 g/l	SC	18119

Uses
- Blight in **potatoes**
- Late blight in **potatoes**

Approval information
- Dimethomorph and zoxamide included in Annex I under EC Regulation 1107/2009

Environmental safety
- Buffer zone requirement 6 m.

Hazard classification and safety precautions
Hazard Very toxic to aquatic organisms
Transport code 9
Packaging group III
UN Number 3082
Risk phrases H317
Operator protection A, H
Environmental protection H410

180 dimoxystrobin

A protectant strobilurin fungicide for cereals available only in mixtures
FRAC mode of action code: 11

See also boscalid + dimoxystrobin

181 dimoxystrobin + epoxiconazole

A protectant and curative fungicide mixture for cereals
FRAC mode of action code: 11 + 3

See also epoxiconazole

Products

Swing Gold	BASF	133:50 g/l	SC	16081

Uses
- Brown rust in **durum wheat, spring wheat, winter wheat**
- Fusarium ear blight in **durum wheat, spring wheat, winter wheat**
- Glume blotch in **durum wheat, spring wheat, winter wheat**
- Septoria leaf blotch in **durum wheat, spring wheat, winter wheat**
- Tan spot in **durum wheat, spring wheat, winter wheat**

Approval information
- Dimoxystrobin and epoxiconazole included in Annex I under EC Regulation 1107/2009

Efficacy guidance
- For best results apply from the start of ear emergence
- Dimoxystrobin is a member of the QoI cross resistance group. Product should be used preventatively and not relied on for its curative potential

FOR FULL CONDITIONS OF USE ALWAYS READ THE PRODUCT LABEL

- Use product as part of an Integrated Crop Management strategy incorporating other methods of control, including where appropriate other fungicides with a different mode of action. Do not apply more than two foliar applications of QoI containing products to any cereal crop
- There is a significant risk of widespread resistance occurring in *Septoria tritici* populations in UK. Failure to follow resistance management action may result in reduced levels of disease control
- Epoxiconazole is a DMI fungicide. Resistance to some DMI fungicides has been identified in Septoria leaf blotch which may seriously affect performance of some products. For further advice contact a specialist advisor and visit the Fungicide Resistance Action Group (FRAG)-UK website

Restrictions
- Maximum number of treatments 1 per crop
- Do not apply before the start of ear emergence

Crop-specific information
- Latest use: up to and including flowering (GS 69)

Environmental safety
- Dangerous for the environment
- Very toxic to aquatic organisms
- Avoid drift onto neighbouring crops
- LERAP Category B

Hazard classification and safety precautions

Hazard Harmful, Dangerous for the environment, Harmful if swallowed, Harmful if inhaled
Transport code 9
Packaging group III
UN Number 3082
Risk phrases H317, H351, H360
Operator protection A; U05a, U20b
Environmental protection E15a, E16a, E34, E38, H410
Storage and disposal D01, D02, D05, D09a, D10c, D12a
Medical advice M05a

182   diquat

A non-residual bipyridyl contact herbicide and crop desiccant
HRAC mode of action code: D

Products

1	Clayton Diquat	Clayton	200 g/l	SL	12739
2	Clayton Diquat 200	Clayton	200 g/l	SL	13942
3	Clayton IQ	Clayton	200 g/l	SL	14519
4	Dessicash 200	Sharda	200 g/l	SL	14848
5	Di Novia	Agroquimicos	200 g/l	SL	15840
6	Diqua	Sharda	200 g/l	SL	14032
7	Dragoon	Hermoo	200 g/l	SL	13927
8	Dragoon Gold	Belcrop	200 g/l	SL	14973
9	Life Scientific Diquat	Life Scientific	200 g/l	SL	15398
10	Mission 200 SL	UPL Europe	200 g/l	SL	16977
11	Quad-Glob 200 SL	Q-Chem	200 g/l	SL	14758
12	Quazar	Belchim	200 g/l	SL	15357
13	Quit	Belchim	200 g/l	SL	14874
14	Reglone	Syngenta	200 g/l	SL	10534
15	Retro	Syngenta	200 g/L	SL	13841
16	Roquat 20	Q-Chem	200 g/l	SL	15703
17	Woomera	Belcrop	200 g/l	SL	14972

SEE SECTION 3 FOR PRODUCTS ALSO REGISTERED

Uses

- Annual and perennial weeds in *all edible crops (outdoor)* (around only), *all non-edible crops (outdoor)* (around only), *bulb onions*, *chicory root*, *combining peas* (dry harvested only), *endives*, *linseed*, *ornamental plant production*, *potatoes*, *potatoes* (weed control or desiccation), *red clover* (seed crop), *spring field beans*, *spring oats* (animal feed only), *spring oilseed rape*, *strawberries* (around only), *sugar beet*, *white clover* (seed crop), *winter field beans*, *winter oats* (animal feed only), *winter oilseed rape* [10]; *forest nurseries* (off-label), *poppies for morphine production* (off-label) [11]; *hops* [8, 10]
- Annual dicotyledons in *all edible crops (outdoor)*, *all non-edible crops (outdoor)*, *potatoes* [12, 13]; *all edible crops (outdoor)* (around), *all non-edible crops (outdoor)* (around), *ornamental plant production*, *sugar beet* [1-9, 11, 14, 16, 17]; *fodder beet*, *mangels* [8]; *lucerne* (off-label) [14]
- Annual grasses in *all edible crops (outdoor)*, *all non-edible crops (outdoor)*, *potatoes* [12, 13]
- Basal defoliation in *hops* [3]
- Black bindweed in *all edible crops (outdoor)* [15]
- Chemical stripping in *hops* [11]
- Chickweed in *all edible crops (outdoor)*, *all non-edible crops (outdoor)*, *ornamental plant production*, *potatoes*, *sugar beet* [15]
- Crane's-bill in *all edible crops (outdoor)* [15]
- Desiccation in *barley seed crops* (off-label), *corn gromwell* (off-label), *crambe* (off-label), *hops* (off-label), *oat seed crops* (off-label), *potato dumps* (off-label) [14]; *borage for oilseed production* (off-label), *lupins* (off-label), *sweet potato* (off-label) [11, 14]; *combining peas*, *linseed*, *oats for animal feed*, *oilseed rape*, *red clover*, *spring field beans*, *white clover*, *winter field beans* [12, 13]; *dwarf beans* (off-label), *french beans* (off-label), *hemp for oilseed production* (off-label), *herbs (see appendix 6)* (off-label), *mustard* (off-label), *poppies for morphine production* (off-label), *poppies for oilseed production* (off-label), *sesame* (off-label), *sunflowers* (off-label) [11]; *hops* [8]; *potatoes* [10, 12, 13]
- Fat hen in *all edible crops (outdoor)* [15]
- General weed control in *chives* (off-label), *herbs (see appendix 6)* (off-label), *parsley* (off-label), *poppies for morphine production* (off-label) [14]; *potatoes* [1-9, 11, 14, 16]
- Grass weeds in *all edible crops (outdoor)* (around only), *all non-edible crops (outdoor)* (around only), *bulb onions*, *chicory root*, *combining peas* (dry harvested), *endives*, *hops*, *linseed*, *ornamental plant production*, *potatoes*, *potatoes* (weed control and desiccation), *red clover* (seed crop), *spring field beans*, *spring oats* (animal feed), *spring oilseed rape*, *strawberries* (around only), *sugar beet*, *white clover* (seed crop), *winter field beans*, *winter oats* (animal feed), *winter oilseed rape* [10]
- Growth regulation in *hops* [2]
- Haulm destruction in *potatoes* [1-9, 11, 14, 16]
- Perennial dicotyledons in *lucerne* (off-label) [14]
- Poppies in *all edible crops (outdoor)* [15]
- Pre-harvest desiccation in *broad beans* [1, 2, 4-6, 9, 14]; *brown mustard* (off-label), *french beans* (off-label), *hemp for oilseed production* (off-label), *poppies for morphine production* (off-label), *poppies for oilseed production* (off-label), *sesame* (off-label), *sunflowers* (off-label), *white mustard* (off-label) [14]; *clover seed crops*, *combining peas*, *laid oats* (stockfeed only), *linseed*, *spring field beans*, *spring oilseed rape*, *winter field beans*, *winter oilseed rape* [1-9, 11, 14, 16]
- Sowthistle in *all edible crops (outdoor)* [15]

Extension of Authorisation for Minor Use (EAMUs)

- *barley seed crops* 20111978 [14]
- *borage for oilseed production* 20093348 [11], 20140187 [14]
- *brown mustard* 20062387 [14]
- *chives* 20062388 [14]
- *corn gromwell* 20140187 [14]
- *crambe* 20082092 [14]
- *dwarf beans* 20093346 [11]
- *forest nurseries* 20093350 [11]
- *french beans* 20093346 [11], 20062389 [14]
- *hemp for oilseed production* 20093351 [11], 20062386 [14]

FOR FULL CONDITIONS OF USE ALWAYS READ THE PRODUCT LABEL

- **herbs (see appendix 6)** *20093345* [11], *20062388* [14]
- **hops** *20102019* [14]
- **lucerne** *20082842* [14]
- **lupins** *20093353* [11], *20073232* [14]
- **mustard** *20093347* [11]
- **oat seed crops** *20111978* [14]
- **parsley** *20062388* [14]
- **poppies for morphine production** *20093349* [11], *20062385* [14]
- **poppies for oilseed production** *20093347* [11], *20062387* [14]
- **potato dumps** *20111882* [14]
- **sesame** *20093347* [11], *20062387* [14]
- **sunflowers** *20093347* [11], *20062387* [14]
- **sweet potato** *20093352* [11], *20070734* [14]
- **white mustard** *20062387* [14]

Approval information

- Diquat included in Annex I under EC Regulation 1107/2009
- All approvals for aquatic weed control expired in 2004
- NOT accepted by BBPA for use on malting barley but accepted on hops on the basis of historical use

Efficacy guidance

- Acts rapidly on green parts of plants and rainfast in 15 min
- Best results for potato desiccation achieved by spraying in bright light and low humidity conditions
- For weed control in row crops apply as overall spray before crop emergence or before transplanting or as an inter-row treatment using a sprayer designed to prevent contamination of crop foliage by spray
- Use of authorised non-ionic wetter for improved weed control essential for most uses except potato desiccation

Restrictions

- Maximum number of treatments 1 per crop in most situations - see labels for details
- Do not apply potato haulm destruction treatment when soil dry. Tubers may be damaged if spray applied during or shortly after dry periods. See label for details of maximum allowable soil-moisture deficit and varietal drought resistance scores
- Do not add wetters to desiccant sprays for potatoes
- Consult processor before adding non-ionic or other wetter on peas. Staining may be increased
- Do not use treated straw or haulm as animal feed or bedding within 4 d of spraying
- Do not apply through ULV or mistblower equipment

Crop-specific information

- Latest use: before crop emergence for sugar beet and before transplanting ornamentals. Not specified for other crops where use varies according to situation
- HI zero when used as a crop desiccant but see label for advisory intervals
- For pre-harvest desiccation, apply to potatoes when tubers the desired size and to other crops when mature or approaching maturity. See label for details of timing and of period to be left before harvesting potatoes
- If potato tubers are to be stored leave 14 d after treatment before lifting

Environmental safety

- Dangerous for the environment
- Very toxic to aquatic organisms
- Keep all livestock out of treated areas and away from treated water for at least 24 h
- Do not feed treated straw or haulm to livestock within 4 days of spraying
- Do not use on crops if the straw is to be used as animal feed/bedding
- Do not dump surplus herbicide in water or ditch bottoms

SEE SECTION 3 FOR PRODUCTS ALSO REGISTERED

Hazard classification and safety precautions

Hazard Very toxic [1]; Toxic [2-17]; Dangerous for the environment [1-17]; Toxic if swallowed [8, 12, 13, 17]; Harmful if swallowed [4, 6, 7, 11-16]; Toxic if inhaled [4, 6-8, 11-17]; Very toxic to aquatic organisms [14, 15]

Transport code 8

Packaging group III

UN Number 1760

Risk phrases H290, H372 [14, 15]; H315, H373 [4, 6-8, 11-13, 16, 17]; H317, H320 [8, 11-13, 16, 17]; H335 [4, 6-8, 11-17]; R22a, R37, R38, R48, R50, R53a [1-3, 5, 9, 10]; R23a [2, 3, 5, 9, 10]; R25, R36, R69 [10]; R26 [1]; R43 [2, 10]

Operator protection A, M [1-17]; C [8, 10, 11, 16, 17]; H [1-10, 12-15, 17]; P [15]; U02a, U05a, U08, U19a, U20b [1-17]; U04a [1-11, 14-17]; U04c [10]; U10 [3-9, 11, 14-17]; U12, U12b [12, 13]; U15 [11-13, 16]; U19c [8, 17]

Environmental protection E06c [1-11] (24 h); E06c [12] (24 hrs); E06c [13] (24 hours); E06c [14-17] (24 h); E08a [11, 16] (4 d); E09 [1, 2] (4 d); E09 [3-10]; E09 [12] (4 d); E09 [13-15, 17]; E15a [10]; E15b [1-9, 11-14, 16, 17]; E34 [1-17]; E38 [3-11, 14-17]; H410 [4, 6-8, 11-17]

Storage and disposal D01, D02, D09a, D12a [1-17]; D05 [1-9, 11-14, 16, 17]; D10a [10]; D10b [1, 2]; D10c [3-9, 11-17]

Medical advice M04a

183   dithianon

A protectant and eradicant dicarbonitrile fungicide for scab control
FRAC mode of action code: M9

Products

Dithianon WG	BASF	70% w/w	WG	17018

Uses
- Scab in *apples*, *pears*

Approval information
- Dithianon included in Annex I under EC Regulation 1107/2009

Efficacy guidance
- Apply at bud-burst and repeat every 7-14 d until danger of scab infection ceases
- Application at high rate within 48 h of a Mills period prevents new infection
- Spray programme also reduces summer infection with apple canker

Restrictions
- Maximum number of treatments on apples and pears 6 per crop
- Do not use on Golden Delicious apples after green cluster
- Do not mix with lime sulphur or highly alkaline products

Crop-specific information
- HI 4 wk for apples, pears

Environmental safety
- Dangerous for the environment
- Very toxic to aquatic organisms

Hazard classification and safety precautions

Hazard Dangerous for the environment, Toxic if swallowed, Very toxic to aquatic organisms

Transport code 9

Packaging group III

UN Number 3077

Risk phrases H317, H318

Operator protection A, D, H; U05a, U11

Environmental protection E15a, E34, E38, H410

Storage and disposal D01, D02, D05, D12a

FOR FULL CONDITIONS OF USE ALWAYS READ THE PRODUCT LABEL

184   dithianon + potassium phosphonates

A fungicide mixture for use in apple and pears

See also dithianon
*   dithianon + pyraclostrobin*

Products

| Delan Pro | BASF | 125:561 g/l | SC | 17374 |

Uses

- Scab in **apples**, **pears**

Approval information

- Dithianon and potassium phosphonate included in Annex I under EC Regulation 1107/2009

Efficacy guidance

- Apply at bud-burst and repeat every 7-14 d until danger of scab infection ceases
- Application at high rate within 48 h of a Mills period prevents new infection
- Spray programme also reduces summer infection with apple canker

Restrictions

- When mixed with products containing carbonate or bicarbonate carbon dioxide may be released and foaming may occur.

Environmental safety

- Dangerous for the environment
- Very toxic to aquatic organisms
- LERAP Category B

Hazard classification and safety precautions

Hazard Very toxic to aquatic organisms
Transport code 9
Packaging group III
UN Number 3082
Risk phrases H317, H319, H351
Operator protection A, C, H; U19a, U20a
Environmental protection E16a, H410; E17a (30 m)
Storage and disposal D01, D02, D05

185   dithianon + pyraclostrobin

A protectant fungicide mixture for apples and pears
FRAC mode of action code: M9 + 11

See also pyraclostrobin

Products

| Maccani | BASF | 12:4% w/w | WG | 13227 |

Uses

- Powdery mildew in **apples**
- Scab in **apples**, **pears**

Approval information

- Pyraclostrobin and dithianon included in Annex I under EC Regulation 1107/2009

Efficacy guidance

- Best resullts obtained from treatments applied from bud burst before disease development and repeated at 10 d intervals
- Water volume should be adjusted according to size of tree and leaf area to ensure good coverage
- Pyraclostrobin is a member of the QoI cross resistance group. Product should be used preventatively at the full recommended dose and not relied on for its curative potential

SEE SECTION 3 FOR PRODUCTS ALSO REGISTERED

- Where field performance may be adversely affected apply QoI containing fungicides in mixtures in strict alternation with fungicides from a different cross-resistance group

Restrictions
- Maximum number of treatments 4 per crop. However where the total number of fungicide treatments is less than 12, the maximum number of applications of any QoI containing products should be three
- Consult processors before use on crops destined for fruit preservation

Crop-specific information
- HI 35 d for apples, pears
- Some apple varieties may be considered to be susceptible to russetting caused by the dithianon component during fruit development

Environmental safety
- Dangerous for the environment
- Very toxic to aquatic organisms
- Broadcast air-assisted LERAP (40 m)

Hazard classification and safety precautions
Hazard Harmful, Dangerous for the environment
Transport code 9
Packaging group III
UN Number 3077
Risk phrases R20, R22a, R36, R50, R53a
Operator protection A, C, D; U05a, U11, U20b
Environmental protection E15b, E34, E38; E17b (40 m)
Storage and disposal D01, D02, D05, D08, D11a, D12a
Medical advice M03, M05a

186   dodeca-8,10-dienyl acetate

An insect sex pheromone
FRAC mode of action code: 12 + IRAC 3

Products
Exosex CM	Exosect	0.1% w/w	DP	12103

Uses
- Codling moth in **apple orchards, pear orchards**

Approval information
- Dodeca-8,10-dienyl acetate is included in Annex 1 under EC Regulation 1107/2009

Efficacy guidance
- Store unopened sachets below 4°C for up to 24 months or store at ambient temperatures and use within 5 months.
- Allow at least 20 m between dispensers and use a minimum of 25 dispensers per hectare.
- Dispenser activity lasts 70 - 90 days depending upon temperatures.

Hazard classification and safety precautions
UN Number N/C
Operator protection A
Environmental protection E15a
Storage and disposal D01, D02
Medical advice M03

187 dodecadienol, tetradecenyl acetate and tetradecylacetate

An insect pheromone mixture

See also dodeca-8,10-dienyl acetate

Products

RAK 3+4	BASF	3.82 : 4.1 : 1.9 % w/w	VP	17824

Uses
- Codling moth in **apples**, **pears**
- Fruit tree tortrix moth in **apples**, **cherries**, **pears**

Approval information
- Dodecadienol, tetradecenyl acetate and tetradecylacetate are included in Annex 1 under EC Regulation 1107/2009

Efficacy guidance
- Use in conjunction with an insecticide if damage threshold is exceeded during treatment.

Restrictions
- Do not use in orchards less than 1 ha in area
- Efficacy impaired if high density of Tortrix moths in surrounding untreated area

Hazard classification and safety precautions
UN Number N/C
Risk phrases H315, H317
Operator protection A; U05a, U20b
Environmental protection E15b, E34, H411
Storage and disposal D01, D02, D09a

188 epoxiconazole

A systemic, protectant and curative triazole fungicide for use in cereals
FRAC mode of action code: 3

See also boscalid + epoxiconazole
boscalid + epoxiconazole + pyraclostrobin
chlorothalonil + epoxiconazole
dimoxystrobin + epoxiconazole
epoxiconazole + fluxapyroxad

Products

1	Bassoon EC	BASF	83 g/l	EC	15219
2	Cortez	Adama	125 g/l	SC	16280
3	Cube	Becesane	125 g/l	SC	16876
4	Epic	BASF	125 g/l	SC	16798
5	Mendoza	Adama	125 g/l	SC	17845
6	Rubric	Headland	125 g/l	SC	14118
7	Ruby	AgChem Access	125 g/l	SC	16809
8	Scholar	Becesane	125 g/l	SC	16402
9	Strand	Headland	125 g/l	SC	14261
10	Zenith	Becesane	125 g/l	SC	16403

Uses
- Brown rust in **durum wheat** [1, 8]; **rye** [1]; **spring barley**, **spring wheat**, **triticale**, **winter barley**, **winter wheat** [1-10]; **spring rye**, **winter rye** [2-10]
- Crown rust in **oats** [1]
- Disease control in **durum wheat** (off-label), **forest nurseries** (off-label) [2]; **grass seed crops** (off-label) [2, 6]
- Eyespot in **winter barley** (reduction), **winter wheat** (reduction) [2-10]

SEE SECTION 3 FOR PRODUCTS ALSO REGISTERED

- Fusarium ear blight in **durum wheat** *(good reduction)*, **winter wheat** *(good reduction)* [1]; **spring wheat** [4]; **spring wheat** *(good reduction)* [1, 2, 5]; **winter wheat** [2, 4, 5]
- Glume blotch in **durum wheat** [1, 8]; **spring wheat, triticale, winter wheat** [1-10]
- Leaf spot in **red beet** *(off-label)* [6]
- Net blotch in **spring barley** [3, 4, 6-10]; **spring barley** *(moderate control)* [1, 2, 5]; **winter barley** [2-10]; **winter barley** *(moderate control)* [1]
- Powdery mildew in **durum wheat** [8]; **fodder beet, sugar beet** [3, 6, 7, 9, 10]; **grass seed crops** *(off-label)* [6]; **oats, rye** [1]; **spring barley, winter barley** [1, 3, 4, 6-10]; **spring oats, spring rye, winter oats, winter rye, winter wheat** [2-10]; **spring wheat, triticale** [3, 4, 6-10]
- Ramularia leaf spots in **spring barley** *(reduction)*, **winter barley** *(reduction)* [1]
- Rhynchosporium in **rye** [1]; **spring barley, winter barley** [1-10]; **spring rye, winter rye** [2-10]
- Rust in **fodder beet, sugar beet** [3, 6, 7, 9, 10]; **grass seed crops** *(off-label)* [6]
- Septoria leaf blotch in **durum wheat** [1, 8]; **spring wheat, triticale, winter wheat** [1-10]
- Sooty moulds in **durum wheat** *(reduction)* [8]; **spring wheat** *(reduction)*, **winter wheat** *(reduction)* [2-10]
- Yellow rust in **durum wheat** [1, 8]; **rye** [1]; **spring barley, spring wheat, triticale, winter barley, winter wheat** [1-10]; **spring rye, winter rye** [2-10]

Extension of Authorisation for Minor Use (EAMUs)
- **durum wheat** *20131995* [2], *20141277* [2]
- **forest nurseries** *20131996* [2], *20141278* [2]
- **grass seed crops** *20131995* [2], *20141277* [2], *20122523* [6]
- **red beet** *20171513* [6]

Approval information
- Accepted by BBPA for use on malting barley (before ear emergence only)
- Epoxiconazole included in Annex I under EC Regulation 1107/2009

Efficacy guidance
- Apply at the start of foliar disease attack
- Optimum effect against eyespot achieved by spraying between leaf-sheath erect and second node detectable stages (GS 30-32)
- Best control of ear diseases of wheat obtained by treatment during ear emergence
- Mildew control improved by use of tank mixtures. See label for details
- For Septoria spray after third node detectable stage (GS 33) when weather favouring disease development has occurred
- Epoxiconazole is a DMI fungicide. Resistance to some DMI fungicides has been identified in Septoria leaf blotch which may seriously affect performance of some products. For further advice contact a specialist advisor and visit the Fungicide Resistance Action Group (FRAG)-UK website

Restrictions
- Maximum total dose equivalent to two full dose treatments
- Product may cause damage to broad-leaved plant species
- Avoid spray drift onto neighbouring crops

Crop-specific information
- Latest use: up to and including flowering just complete (GS 69) in wheat, rye, triticale; up to and including emergence of ear just complete (GS 59) in barley, oats

Environmental safety
- Dangerous for the environment
- Very toxic to aquatic organisms
- LERAP Category B

Hazard classification and safety precautions
Hazard Harmful, Dangerous for the environment [1-10]; Harmful if inhaled, Very toxic to aquatic organisms [4, 5]
Transport code 9
Packaging group III
UN Number 3082

FOR FULL CONDITIONS OF USE ALWAYS READ THE PRODUCT LABEL

Risk phrases H315 [4, 6, 9]; H351 [4-6, 9]; H360 [4, 5]; H361 [6, 9]; R20 [1]; R38 [1, 3, 7, 8, 10]; R40, R50, R53a [1-3, 7, 8, 10]; R62, R63 [1-3, 6-8, 10]
Operator protection A [1-10]; H [1, 4]; U05a, U20b
Environmental protection E15a, E16a, E38 [1-10]; H410 [4, 6, 9]; H411 [5]
Storage and disposal D01, D02, D05, D09a, D10c, D12a [1-10]; D19 [4]
Medical advice M05a

189   epoxiconazole + fenpropimorph

A systemic, protectant and curative fungicide mixture for cereals
FRAC mode of action code: 3 + 5

See also fenpropimorph

Products

1	Eclipse	BASF	84:250 g/l	SE	11731
2	Opus Team	BASF	84:250 g/l	SE	11759
3	Oracle Duet	AgChem Access	84:250 g/l	SE	16011

Uses
- Brown rust in **spring barley, spring rye, spring wheat, triticale, winter barley, winter rye, winter wheat** [1-3]
- Eyespot in **winter barley** *(reduction)*, **winter wheat** *(reduction)* [1-3]
- Foliar disease control in **durum wheat** *(off-label)*, **grass seed crops** *(off-label)* [1, 2]
- Fusarium ear blight in **spring wheat, winter wheat** [1-3]
- Glume blotch in **spring wheat, triticale, winter wheat** [1-3]
- Net blotch in **spring barley, winter barley** [1-3]
- Powdery mildew in **spring barley, spring oats, spring rye, spring wheat, triticale, winter barley, winter oats, winter rye, winter wheat** [1-3]
- Rhynchosporium in **spring barley, spring rye, winter barley, winter rye** [1-3]
- Septoria leaf blotch in **spring wheat, triticale, winter wheat** [1-3]
- Sooty moulds in **spring wheat** *(reduction)*, **winter wheat** *(reduction)* [1-3]
- Yellow rust in **spring barley, spring rye, spring wheat, triticale, winter barley, winter rye, winter wheat** [1-3]

Extension of Authorisation for Minor Use (EAMUs)
- **durum wheat** *20062644* [1], *20062645* [2]
- **grass seed crops** *20062644* [1], *20062645* [2]

Approval information
- Epoxiconazole and fenpropimorph included in Annex I under EC Regulation 1107/2009
- Accepted by BBPA for use on malting barley

Efficacy guidance
- Apply at the start of foliar disease attack
- Optimum effect against eyespot achieved by spraying between leaf-sheath erect and second node detectable stages (GS 30-32)
- Best control of ear diseases obtained by treatment during ear emergence
- For Septoria spray after third node detectable stage (GS 33) when weather favouring disease development has occurred
- Epoxiconazole is a DMI fungicide. Resistance to some DMI fungicides has been identified in Septoria leaf blotch which may seriously affect performance of some products. For further advice contact a specialist advisor and visit the Fungicide Resistance Action Group (FRAG)-UK website

Restrictions
- Maximum total dose equivalent to two full dose treatments
- Product may cause damage to broad-leaved plant species

SEE SECTION 3 FOR PRODUCTS ALSO REGISTERED

Crop-specific information
- Latest use: up to and including flowering just complete (GS 69) in wheat, rye, triticale; up to and including emergence of ear just complete (GS 59) in barley, oats

Environmental safety
- Dangerous for the environment
- Toxic to aquatic organisms
- Avoid spray drift onto neighbouring crops
- LERAP Category B

Hazard classification and safety precautions
> **Hazard** Harmful, Dangerous for the environment [1-3]; Harmful if inhaled [2]
> **Transport code** 9
> **Packaging group** III
> **UN Number** 3082
> **Risk phrases** H351, H360 [1, 2]; R20, R40, R43, R51, R53a, R62, R63 [3]
> **Operator protection** A, H; U05a, U20b
> **Environmental protection** E15a, E16a, E38 [1-3]; H411 [1, 2]
> **Storage and disposal** D01, D02, D05, D09a, D10c, D12a
> **Medical advice** M05a

190 epoxiconazole + fenpropimorph + kresoxim-methyl

A protectant, systemic and curative fungicide mixture for cereals
FRAC mode of action code: 3 + 5 + 11

See also fenpropimorph
kresoxim-methyl

Products

1	Asana	BASF	125:150:125 g/l	SE	11934
2	Mantra	BASF	125:150:125 g/l	SE	11728

Uses
- Brown rust in **spring barley**, **spring rye**, **spring wheat**, **triticale**, **winter barley**, **winter rye**, **winter wheat** [1, 2]
- Disease control in **grass seed crops** (off-label) [1]
- Eyespot in **spring oats** (reduction), **spring rye** (reduction), **triticale** (reduction), **winter barley** (reduction), **winter oats** (reduction), **winter rye** (reduction), **winter wheat** (reduction) [1, 2]
- Foliar disease control in **durum wheat** (off-label), **grass seed crops** (off-label) [2]
- Fusarium ear blight in **spring wheat** (reduction), **winter wheat** (reduction) [1, 2]
- Glume blotch in **spring wheat**, **triticale**, **winter wheat** [1, 2]
- Net blotch in **spring barley**, **winter barley** [1, 2]
- Powdery mildew in **spring barley**, **spring oats**, **spring rye**, **triticale**, **winter barley**, **winter oats**, **winter rye** [1, 2]
- Rhynchosporium in **spring barley**, **spring rye**, **winter barley**, **winter rye** [1, 2]
- Septoria leaf blotch in **spring wheat**, **triticale**, **winter wheat** [1, 2]
- Sooty moulds in **spring wheat** (reduction), **winter wheat** (reduction) [1, 2]
- Yellow rust in **spring barley**, **spring rye**, **spring wheat**, **triticale**, **winter barley**, **winter rye**, **winter wheat** [1, 2]

Extension of Authorisation for Minor Use (EAMUs)
- **durum wheat** 20061101 [2]
- **grass seed crops** 20122861 [1], 20061101 [2]

Approval information
- Kresoxim-methyl, epoxiconazole and fenpropimorph included in Annex I under EC Regulation 1107/2009
- Accepted by BBPA for use on malting barley (before ear emergence only)

Efficacy guidance
- For best results spray at the start of foliar disease attack and repeat if infection conditions persist
- Optimum effect against eyespot obtained by treatment between leaf sheaths erect and first node detectable stages (GS 30-32)
- For protection against ear diseases apply during ear emergence
- Epoxiconazole is a DMI fungicide. Resistance to some DMI fungicides has been identified in Septoria leaf blotch which may seriously affect performance of some products. For further advice contact a specialist advisor and visit the Fungicide Resistance Action Group (FRAG)-UK website
- Kresoxim-methyl is a member of the QoI cross resistance group. Product should be used preventatively and not relied on for its curative potential
- Use product as part of an Integrated Crop Management strategy incorporating other methods of control, including where appropriate other fungicides with a different mode of action. Do not apply more than two foliar applications of QoI containing products to any cereal crop
- There is a significant risk of widespread resistance occurring in *Septoria tritici* populations in UK. Failure to follow resistance management action may result in reduced levels of disease control
- Strains of barley powdery mildew resistant to QoIs are common in the UK

Restrictions
- Maximum total dose equivalent to two full dose treatments

Crop-specific information
- Latest use: mid flowering (GS 65) for wheat, rye, triticale; completion of ear emergence (GS 59) for barley, oats

Environmental safety
- Dangerous for the environment
- Very toxic to aquatic organisms
- Avoid spray drift onto neighbouring crops. Product may damage broad-leaved species
- LERAP Category B

Hazard classification and safety precautions
Hazard Harmful, Dangerous for the environment
Transport code 9
Packaging group III
UN Number 3082
Risk phrases R40, R50, R53a, R62, R63
Operator protection A, H; U05a, U14, U20b
Environmental protection E15a, E16a, E38
Storage and disposal D01, D02, D05, D09a, D10c, D12a
Medical advice M05a

191   epoxiconazole + fenpropimorph + metrafenone

A broad spectrum fungicide mixture for cereals
FRAC mode of action code: 3 + 5 + U8

See also fenpropimorph
metrafenone

Products

Capalo	BASF	62.5:200:75 g/l	SE	13170

Uses
- Brown rust in **spring barley**, **spring rye**, **spring wheat**, **triticale**, **winter barley**, **winter rye**, **winter wheat**
- Crown rust in **spring oats**, **winter oats**
- Fusarium ear blight in **spring wheat** *(reduction)*, **winter wheat** *(reduction)*
- Glume blotch in **spring wheat**, **triticale**, **winter wheat**
- Net blotch in **spring barley**, **winter barley**

SEE SECTION 3 FOR PRODUCTS ALSO REGISTERED

SECTION 2

- Powdery mildew in **spring barley**, **spring oats**, **spring rye**, **spring wheat**, **triticale**, **winter barley**, **winter oats**, **winter rye**, **winter wheat**
- Rhynchosporium in **spring barley**, **spring rye**, **winter barley**, **winter rye**
- Septoria leaf blotch in **spring wheat**, **triticale**, **winter wheat**
- Sooty moulds in **spring wheat** *(reduction)*, **winter wheat** *(reduction)*
- Yellow rust in **spring barley**, **spring rye**, **spring wheat**, **triticale**, **winter barley**, **winter rye**, **winter wheat**

Approval information
- Epoxiconazole, fenpropimorph and metrafenone included in Annex I under EC Regulation 1107/2009

Efficacy guidance
- For best results spray at the start of foliar disease attack and repeat if infection conditions persist
- Optimum effect against eyespot obtained by treatment between leaf sheaths erect and first node detectable stages (GS 30-32)
- For protection against ear diseases apply during ear emergence
- Treatment for control of Septoria diseases should normally be made after third node detectable stage if the disease is present and favourable weather for disease development has occurred
- Epoxiconazole is a DMI fungicide. Resistance to some DMI fungicides has been identified in Septoria leaf blotch which may seriously affect performance of some products. For further advice contact a specialist advisor and visit the Fungicide Resistance Action Group (FRAG)-UK website
- Should be used as part of a resistance management strategy that includes mixtures or sequences effective against mildew and non-chemical methods

Restrictions
- Maximum number of treatments 2 per crop
- Do not apply more than two applications of any products containing metrafenone to any one crop

Crop-specific information
- Latest use: up to and including beginning of flowering (GS 61) for rye, wheat, triticale; up to and including ear emergence complete (GS 59) for barley, oats

Following crops guidance
- Cereals. oilseed rape, sugar beet, linseed, maize, clover, field beans, peas, turnips, carrots, cauliflowers, onions, lettuce or potatoes may be sown following normal harvest of a treated crop

Environmental safety
- Dangerous for the environment
- Very toxic to aquatic organisms
- Avoid spray drift onto neighbouring crops. Broad-leaved species may be damaged
- LERAP Category B

Hazard classification and safety precautions
Hazard Harmful, Dangerous for the environment
Transport code 9
Packaging group III
UN Number 3082
Risk phrases R38, R40, R43, R50, R53a, R62, R63
Operator protection A, H; U05a, U08, U14, U15, U19a, U20b
Environmental protection E16a, E34, E38
Storage and disposal D01, D02, D05, D08, D09a, D10c, D12a
Medical advice M03

192   epoxiconazole + fluxapyroxad

An SDHI and triazole fungicide mixture for disease control in cereals
FRAC mode of action code: 3 + 7

See also epoxiconazole
fluxapyroxad

FOR FULL CONDITIONS OF USE ALWAYS READ THE PRODUCT LABEL

Products

1	Adexar	BASF	62.5:62.5 g/l	EC	17109
2	Clayton Index	Clayton	62.5:62.5 g/l	EC	17266
3	Pexan	BASF	62.5:59.4 g/l	EC	17110

Uses

- Brown rust in **durum wheat, rye, spring barley, spring wheat, triticale, winter barley, winter wheat**
- Crown rust in **oats**
- Eyespot in **durum wheat** *(moderate control)*, **oats** *(moderate control)*, **rye** *(moderate control)*, **triticale** *(moderate control)*, **winter wheat** *(moderate control)*
- Fusarium ear blight in **durum wheat** *(good reduction)*, **spring wheat** *(good reduction)*, **winter wheat** *(good reduction)*
- Glume blotch in **durum wheat, spring wheat, triticale, winter wheat**
- Net blotch in **spring barley, winter barley**
- Powdery mildew in **durum wheat** *(moderate control)*, **oats** *(moderate control)*, **rye** *(moderate control)*, **spring barley** *(moderate control)*, **spring wheat** *(moderate control)*, **triticale** *(moderate control)*, **winter barley** *(moderate control)*, **winter wheat** *(moderate control)*
- Ramularia leaf spots in **spring barley, winter barley**
- Rhynchosporium in **spring barley, winter barley**
- Septoria leaf blotch in **durum wheat, spring wheat, triticale, winter wheat**
- Sooty moulds in **durum wheat** *(reduction)*, **spring wheat** *(reduction)*, **winter wheat** *(reduction)*
- Tan spot in **durum wheat** *(moderate control)*, **spring wheat** *(moderate control)*, **winter wheat** *(moderate control)*
- Yellow rust in **durum wheat, rye, spring barley, spring wheat, triticale, winter barley, winter wheat**

Approval information

- Epoxiconazole and fluxapyroxad included in Annex I under EC Regulation 1107/2009
- Accepted by BBPA for use on malting barley up to GS 45 only (max. dose, 1 litre/ha)

Restrictions

- Avoid drift on to neighbouring crops since it may damage broad-leaved species.
- Do not apply more that two foliar applications of products containing SDHI fungicides to any cereal crop.

Crop-specific information

- After treating a cereal crop cabbage, carrot, clover, dwarf beans, field beans, lettuce, maize, oats, oilseed rape, onions, peas, potatoes, ryegrass, sugar beet, sunflower, winter barley and winter wheat may be sown as the following crop

Environmental safety

- LERAP Category B

Hazard classification and safety precautions

Hazard Harmful, Dangerous for the environment, Harmful if swallowed, Very toxic to aquatic organisms

Transport code 9

Packaging group III

UN Number 3082

Risk phrases H317, H319, H351, H360 [1-3]; R40 [3]; R43 [1, 2]

Operator protection A [1-3]; H [3]; U05a, U14, U20b

Environmental protection E15b, E16a, E38, H410

Storage and disposal D01, D02, D05, D10c, D12a, D19

Medical advice M05a

SEE SECTION 3 FOR PRODUCTS ALSO REGISTERED

193 epoxiconazole + fluxapyroxad + pyraclostrobin

A triazole + SDHI + strobilurin fungicide mixture for disease control in cereals
FRAC mode of action code: 3 + 7 + 11

Products

1 Ceriax	BASF	41.6:41.6:66.6 g/l	EC	17111
2 Vortex	BASF	41.6:41.6:61 g/l	EC	17112

Uses

- Brown rust in *durum wheat, rye, spring barley, spring wheat, triticale, winter barley, winter wheat*
- Crown rust in *oats*
- Eyespot in *durum wheat* (moderate control), *oats* (moderate control), *rye* (moderate control), *triticale* (moderate control), *winter wheat* (moderate control)
- Fusarium ear blight in *durum wheat* (good reduction), *spring wheat* (good reduction), *winter wheat* (good reduction)
- Glume blotch in *durum wheat, spring wheat, triticale, winter wheat*
- Net blotch in *spring barley, winter barley*
- Powdery mildew in *durum wheat* (moderate control), *oats* (moderate control), *rye* (moderate control), *spring barley* (moderate control), *spring wheat* (moderate control), *triticale* (moderate control), *winter barley* (moderate control), *winter wheat* (moderate control)
- Ramularia leaf spots in *spring barley, winter barley*
- Rhynchosporium in *rye, spring barley, winter barley*
- Septoria leaf blotch in *durum wheat, spring wheat, triticale, winter wheat*
- Sooty moulds in *durum wheat* (reduction), *spring wheat* (reduction), *winter wheat* (reduction)
- Tan spot in *durum wheat* (moderate control), *spring wheat* (moderate control), *winter wheat* (moderate control)
- Yellow rust in *durum wheat, rye, spring barley, spring wheat, triticale, winter barley, winter wheat*

Approval information

- Epoxiconazole, fluxapyroxad and pyraclostrobin included in Annex I under EC Regulation 1107/2009

Restrictions

- No more than two applications of any QoI fungicide may be applied to wheat, barley, oats, rye or triticale.
- Do not apply more that two foliar applications of products containing SDHI fungicides to any cereal crop.
- May cause damage to broad-leaved plant species.

Following crops guidance

- Cabbage, carrot, clover, dwarf French beans, lettuce, maize, oats, oilseed rape, onions, peas, potatoes, ryegrass, sugarbeet, sunflower, winter barley and winter wheat may be sown as a following crop but the effect on other crops has not been assessed.

Environmental safety

- LERAP Category B

Hazard classification and safety precautions

Hazard Harmful, Dangerous for the environment, Harmful if swallowed, Harmful if inhaled, Very toxic to aquatic organisms
Transport code 9
Packaging group III
UN Number 3082
Risk phrases H315, H317, H318, H351, H360
Operator protection A, H; U05a, U14, U20b
Environmental protection E15b, E16a, E38, H410
Storage and disposal D01, D02, D05, D07, D10c, D12a
Medical advice M05a

FOR FULL CONDITIONS OF USE ALWAYS READ THE PRODUCT LABEL

194   epoxiconazole + folpet

A systemic and protectant fungicide mixture for disease control in cereals
FRAC mode of action code: 3 + M4

See also epoxiconazole
folpet

Products

1	Manitoba	Adama	50:375 g/l	SC	16539
2	Scotia	Adama	50:375 g/l	SC	17415

Uses

- Brown rust in **spring wheat**, **triticale**, **winter wheat**
- Disease control in **spring barley**, **triticale**, **winter barley**
- Glume blotch in **spring wheat**, **triticale**, **winter wheat**
- Rhynchosporium in **spring barley**, **triticale**, **winter barley**
- Septoria leaf blotch in **spring wheat**, **triticale**, **winter wheat**
- Yellow rust in **spring wheat**, **triticale**, **winter wheat**

Approval information

- Epoxiconazole and folpet included in Annex I under EC Regulation 1107/2009

Environmental safety

- Dangerous for the environment
- Very toxic to aquatic organisms
- Buffer zone requirement 6m

Hazard classification and safety precautions

Hazard Harmful, Dangerous for the environment, Harmful if inhaled, Very toxic to aquatic organisms
Transport code 9
Packaging group III
UN Number 3082
Risk phrases H319, H351, H360
Operator protection A, H; U05a, U14
Environmental protection E38, H412
Storage and disposal D01, D12a

195   epoxiconazole + isopyrazam

A protectant and eradicant fungicide mixture for use in cereals
FRAC mode of action code: 3 + 7

Products

1	Clayton Kudos	Clayton	90:125 g/l	SC	16580
2	Keystone	Syngenta	99:125 g/l	SC	15493
3	Micaraz	Syngenta	90:125 g/l	SC	16106
4	Seguris	Syngenta	90:125 g/l	SC	15246

Uses

- Brown rust in **spring barley**, **triticale**, **winter barley**, **winter rye**, **winter wheat**
- Glume blotch in **winter wheat**
- Net blotch in **spring barley**, **winter barley**
- Ramularia leaf spots in **spring barley**, **winter barley**
- Rhynchosporium in **spring barley**, **winter barley**, **winter rye**
- Septoria leaf blotch in **triticale**, **winter wheat**
- Yellow rust in **winter wheat**

Approval information

- Epoxiconazole and isopyrazam are included in Annex I under EC Regulation 1107/2009

SEE SECTION 3 FOR PRODUCTS ALSO REGISTERED

- Approved by BBPA for use on malting barley before ear emergence only

Restrictions
- Isopyrazam is an SDH respiration inhibitor; Do not apply more than two foliar applications of products containing an SDH inhibitor to any cereal crop

Environmental safety
- Buffer zone requirement 6 m [3, 4]
- LERAP Category B

Hazard classification and safety precautions
> **Hazard** Harmful, Dangerous for the environment, Harmful if inhaled, Very toxic to aquatic organisms
> **Transport code** 9
> **Packaging group** III
> **UN Number** 3082
> **Risk phrases** H317, H319, H351, H361 [1-4]; H360 [2]
> **Operator protection** A, H; U09a, U14, U20b
> **Environmental protection** E15b, E16a, E16b, E38, H410
> **Storage and disposal** D01, D02, D05, D09a, D10c, D11a, D12a
> **Medical advice** M05a

196   epoxiconazole + metconazole

A triazole mixture for use in cereals
FRAC mode of action code: 3 + 3

See also metconazole

Products

1	Brutus	BASF	37.5:27.5 g/l	EC	14353
2	Osiris P	BASF	56.25:41.25 g/l	EC	15627

Uses
- Brown rust in *durum wheat*, *spring barley*, *spring wheat*, *triticale*, *winter barley*, *winter wheat* [1, 2]; *rye* [2]; *spring rye*, *winter rye* [1]
- Eyespot in *durum wheat* (reduction), *triticale* (reduction), *winter wheat* (reduction) [1, 2]; *rye* (reduction) [2]; *spring rye* (reduction), *winter barley* (reduction), *winter rye* (reduction) [1]
- Fusarium ear blight in *durum wheat* (good reduction), *spring wheat* (good reduction), *triticale* (good reduction), *winter wheat* (good reduction) [1]; *durum wheat* (moderate control), *spring wheat* (moderate control), *triticale* (moderate control), *winter wheat* (moderate control) [2]
- Glume blotch in *durum wheat*, *spring wheat*, *triticale*, *winter wheat* [1, 2]
- Net blotch in *spring barley*, *winter barley* [1, 2]
- Powdery mildew in *durum wheat* (moderate control), *spring barley* (moderate control), *spring wheat* (moderate control), *triticale* (moderate control), *winter barley* (moderate control), *winter wheat* (moderate control) [1, 2]; *rye* (moderate control) [2]; *spring rye* (moderate control), *winter rye* (moderate control) [1]
- Ramularia leaf spots in *spring barley* (moderate control), *winter barley* (moderate control) [1, 2]
- Rhynchosporium in *rye* (moderate control) [2]; *spring barley*, *winter barley* [1, 2]
- Septoria leaf blotch in *durum wheat*, *spring wheat*, *triticale*, *winter wheat* [1, 2]
- Sooty moulds in *durum wheat* (good reduction), *spring wheat* (good reduction), *winter wheat* (good reduction) [1]; *durum wheat* (moderate control), *spring wheat* (moderate control), *winter wheat* (moderate control) [2]
- Tan spot in *durum wheat* (moderate control), *spring wheat* (moderate control), *winter wheat* (moderate control) [1, 2]
- Yellow rust in *durum wheat*, *spring barley*, *spring wheat*, *triticale*, *winter barley*, *winter wheat* [1, 2]; *rye* [2]; *spring rye*, *winter rye* [1]

Approval information
- Epoxiconazole and metconazole included in Annex I under EC Regulation 1107/2009

FOR FULL CONDITIONS OF USE ALWAYS READ THE PRODUCT LABEL

Following crops guidance

- After treating a cereal crop, oilseed rape, cereals, sugar beet, linseed, maize, clover, beans, peas, carrots, potatoes, lettuce, cabbage, sunflower, ryegrass or onions may be sown as a following crop. The effect on other crops has not been evaluated.

Environmental safety

- LERAP Category B

Hazard classification and safety precautions

Hazard Harmful, Dangerous for the environment
Transport code 9
Packaging group III
UN Number 3082
Risk phrases R40, R43, R53a [1, 2]; R50, R62 [2]; R51, R63 [1]
Operator protection A, H; U05a, U14 [1]; U20b [1, 2]
Environmental protection E15b, E16a, E34 [1, 2]; E38 [2]
Storage and disposal D01, D02, D09a, D10c [1, 2]; D05, D12a, D19 [2]
Medical advice M05a

197   epoxiconazole + metrafenone

A broad spectrum fungicide mixture for cereals
FRAC mode of action code: 3 + U8

See also metrafenone

Products

1 Ceando	Dow	83:100 g/l	SC	16012
2 Cloister	Dow	83:100 g/l	SC	15994

Uses

- Brown rust in **durum wheat** [2]; **spring barley, spring rye, spring wheat, triticale, winter barley, winter rye, winter wheat** [1, 2]
- Crown rust in **spring oats, winter oats** [1, 2]
- Fusarium ear blight in **durum wheat** (reduction) [2]; **spring wheat** (reduction), **winter wheat** (reduction) [1, 2]
- Glume blotch in **durum wheat** [2]; **spring wheat, triticale, winter wheat** [1, 2]
- Net blotch in **spring barley, winter barley** [1, 2]
- Powdery mildew in **durum wheat** [2]; **spring barley, spring oats, spring rye, spring wheat, triticale, winter barley, winter oats, winter rye, winter wheat** [1, 2]
- Rhynchosporium in **spring barley, spring rye, winter barley, winter rye** [1, 2]
- Septoria leaf blotch in **durum wheat** [2]; **spring wheat, triticale, winter wheat** [1, 2]
- Sooty moulds in **durum wheat** (reduction) [2]; **spring wheat** (reduction), **winter wheat** (reduction) [1, 2]
- Yellow rust in **durum wheat** [2]; **spring barley, spring rye, spring wheat, triticale, winter barley, winter rye, winter wheat** [1, 2]

Approval information

- Epoxiconazole and metrafenone included in Annex I under EC Regulation 1107/2009

Efficacy guidance

- For best results spray at the start of foliar disease attack and repeat if infection conditions persist
- Optimum effect against eyespot obtained by treatment between leaf sheaths erect and first node detectable stages (GS 30-32)
- For protection against ear diseases apply during ear emergence
- Treatment for control of Septoria diseases should normally be made after third node detectable stage if the disease is present and favourable weather for disease development has occurred
- Epoxiconazole is a DMI fungicide. Resistance to some DMI fungicides has been identified in Septoria leaf blotch which may seriously affect performance of some products. For further advice contact a specialist advisor and visit the Fungicide Resistance Action Group (FRAG)-UK website

SEE SECTION 3 FOR PRODUCTS ALSO REGISTERED

- Should be used as part of a resistance management strategy that includes mixtures or sequences effective against mildew and non-chemical methods
- Where mildew well established at time of treatment product should be mixed with fenpropimorph. See label

Restrictions
- Maximum number of treatments 2 per crop
- Do not apply more than two applications of any products containing metrafenone to any one crop

Crop-specific information
- Latest use: up to and including beginning of flowering (GS 61) for rye, wheat, triticale; up to and including ear emergence complete (GS 59) for barley, oats

Following crops guidance
- Cereals. oilseed rape, sugar beet, linseed, maize, clover, field beans, peas, turnips, carrots, cauliflowers, onions, lettuce or potatoes may be sown following normal harvest of a treated crop

Environmental safety
- Dangerous for the environment
- Very toxic to aquatic organisms
- Avoid spray drift onto neighbouring crops. Broad-leaved species may be damaged
- LERAP Category B

Hazard classification and safety precautions
Hazard Harmful, Dangerous for the environment
UN Number N/C
Risk phrases H319, H351, H360
Operator protection A, H; U05a, U20b
Environmental protection E15a, E16a, E34, E38, H412
Storage and disposal D01, D02, D05, D08, D09a, D10c
Medical advice M03, M05a

198   epoxiconazole + pyraclostrobin

A protectant, systemic and curative fungicide mixture for cereals
FRAC mode of action code: 3 + 11

See also pyraclostrobin

Products

1 Envoy	BASF	62.5:85 g/l	SE	16297
2 Gemstone	BASF	62.5:80 g/l	SE	16298
3 Opera	BASF	50:133 g/l	SE	16420

Uses
- Brown rust in *durum wheat, rye, spring barley, spring wheat, triticale, winter barley, winter wheat* [1-3]
- Cercospora leaf spot in *sugar beet* [3]
- Crown rust in *spring oats, winter oats* [1-3]
- Disease control in *grass seed crops (off-label)* [3]
- Eyespot in *forage maize, forage maize (off-label), grain maize, grain maize (off-label)* [3]
- Fusarium ear blight in *durum wheat (good reduction), spring wheat (good reduction), winter wheat (good reduction)* [1-3]; *triticale (good reduction)* [3]
- Glume blotch in *durum wheat, spring wheat, triticale, winter wheat* [1-3]
- Leaf blight in *forage maize (off-label), grain maize (off-label)* [3]
- Net blotch in *spring barley, winter barley* [1-3]
- Northern leaf blight in *forage maize, grain maize* [3]
- Powdery mildew in *fodder beet (off-label), sugar beet* [3]; *rye, spring barley, spring oats, winter barley, winter oats* [1, 2]

- Ramularia leaf spots in *rye* *(moderate control)*, **spring barley** *(moderate control)*, **sugar beet**, **triticale** *(moderate control)*, **winter barley** *(moderate control)* [3]; **spring barley** *(reduction)*, **winter barley** *(reduction)* [1, 2]
- Rhynchosporium in *rye*, **spring barley**, **winter barley** [1-3]
- Rust in **fodder beet** *(off-label)*, **forage maize** *(off-label)*, **grain maize** *(off-label)*, **sugar beet** [3]
- Septoria leaf blotch in **durum wheat**, **spring wheat**, **triticale**, **winter wheat** [1, 2]; **durum wheat** *(moderate control)*, **spring wheat** *(moderate control)*, **triticale** *(moderate control)*, **winter wheat** *(moderate control)* [3]
- Sooty moulds in **spring wheat** *(reduction)*, **winter wheat** *(reduction)* [1, 2]
- Tan spot in **durum wheat**, **spring wheat**, **winter wheat** [1, 2]; **durum wheat** *(moderate control)*, **spring wheat** *(moderate control)*, **triticale** *(moderate control)*, **winter wheat** *(moderate control)* [3]
- Yellow rust in **durum wheat**, **rye**, **spring barley**, **spring wheat**, **triticale**, **winter barley**, **winter wheat** [1-3]

Extension of Authorisation for Minor Use (EAMUs)
- **fodder beet** *20140392* [3]
- **forage maize** *20141592* [3]
- **grain maize** *20141592* [3]
- **grass seed crops** *20140522* [3]

Approval information
- Epoxiconazole and pyraclostrobin included in Annex I under EC Regulation 1107/2009
- Accepted by BBPA for use on malting barley (before ear emergence only)

Efficacy guidance
- For best results apply at the start of foliar disease attack
- Good reduction of Fusarium ear blight can be obtained from treatment during flowering
- Yield response may be obtained in the absence of visual disease symptoms
- Epoxiconazole is a DMI fungicide. Resistance to some DMI fungicides has been identified in Septoria leaf blotch which may seriously affect performance of some products. For further advice contact a specialist advisor and visit the Fungicide Resistance Action Group (FRAG)-UK website
- Pyraclostrobin is a member of the QoI cross resistance group. Product should be used preventatively and not relied on for its curative potential
- Use product as part of an Integrated Crop Management strategy incorporating other methods of control, including where appropriate other fungicides with a different mode of action. Do not apply more than two foliar applications of QoI containing products to any cereal crop
- There is a significant risk of widespread resistance occurring in *Septoria tritici* populations in UK. Failure to follow resistance management action may result in reduced levels of disease control

Restrictions
- Maximum number of treatments 2 per crop
- Do not use on sugar beet crops grown for seed [3]

Crop-specific information
- Latest use: before grain watery ripe (GS 71) for wheat; up to and including emergence of ear just complete (GS 59) for barley and oats
- HI 6 wk for sugar beet [3]

Environmental safety
- Dangerous for the environment
- Very toxic to aquatic organisms
- LERAP Category B

Hazard classification and safety precautions
Hazard Harmful, Dangerous for the environment, Harmful if inhaled [1-3]; Toxic if swallowed [3]; Harmful if swallowed, Very toxic to aquatic organisms [1, 2]

Transport code 9
Packaging group III
UN Number 3082
Risk phrases H315, H317, H351 [1-3]; H360 [1, 3]

SEE SECTION 3 FOR PRODUCTS ALSO REGISTERED

Operator protection A; U05a, U14, U20b
Environmental protection E15a, E16a, E34, E38, H410 [1-3]; E16b [3]
Storage and disposal D01, D02, D09a, D10c, D12a [1-3]; D05 [3]
Medical advice M03, M05a

199   esfenvalerate

A contact and ingested pyrethroid insecticide
IRAC mode of action code: 3

Products

1 Clayton Cajole	Clayton	25 g/l	EC	14995
2 Clayton Slalom	Clayton	25 g/l	EC	15054
3 Clayton Vindicate	Clayton	25 g/l	EC	15055
4 Gocha	Belchim	25 g/l	EC	17320
5 Sumi-Alpha	Interfarm	25 g/l	EC	14023
6 Sven	Interfarm	25 g/l	EC	14859

Uses

- Aphids in *broccoli, brussels sprouts, cabbages, calabrese, cauliflowers, combining peas, edible podded peas, kale, kohlrabi, potatoes, spring barley, spring field beans, spring wheat, vining peas, winter barley, winter field beans, winter wheat* [1-6]; *chinese cabbage* [1-3, 5, 6]; *grass seed crops* (off-label); *protected ornamentals* [5]; *grassland, ornamental plant production* [1-3, 5]; *managed amenity turf* [1-3]; *oriental cabbage* [4]
- Bibionids in *grass seed crops* [6]; *grassland, managed amenity turf* [4, 6]
- Caterpillars in *broccoli, brussels sprouts, cabbages, calabrese, cauliflowers, kale, kohlrabi* [4, 6]; *chinese cabbage* [6]; *oriental cabbage* [4]
- Insect pests in *ornamental plant production* [4, 6]; *protected ornamentals* [6]
- Pea and bean weevil in *combining peas, edible podded peas, spring field beans, vining peas, winter field beans* [4, 6]

Extension of Authorisation for Minor Use (EAMUs)

- *grass seed crops* 20081266 [5]

Approval information

- Esfenvalerate included in Annex I under EC Regulation 1107/2009
- Accepted by BBPA for use on malting barley

Efficacy guidance

- For best reduction of spread of barley yellow dwarf virus winter sown crops at high risk (e.g. after grass or in areas with history of BYDV) should be treated when aphids first seen or by mid-Oct. Otherwise treat in late Oct-early Nov
- High risk winter sown crops will need a second treatment
- Spring sown crops should be treated from the 2-3 leaf stage if aphids are found colonising in the crop and a further application may be needed before the first node stage if they reinfest
- Product also recommended between onset of flowering and milky ripe stages (GS 61-73) for control of summer cereal aphids

Restrictions

- Maximum number of treatments 3 per crop of which 2 may be in autumn
- Do not use if another pyrethroid or dimethoate has been applied to crop after start of ear emergence (GS 51)
- Do not leave spray solution standing in spray tank

Crop-specific information

- Latest use: 31 Mar in yr of harvest (winter use dose); early milk stage for barley (summer use); late milk stage for wheat (summer use)

Environmental safety

- Dangerous for the environment
- Very toxic to aquatic organisms

FOR FULL CONDITIONS OF USE ALWAYS READ THE PRODUCT LABEL

- High risk to non-target insects or other arthropods. Do not spray within 5 m of the field boundary
- Give local beekeepers warning when using in flowering crops
- Flammable
- Store product in dark away from direct sunlight
- LERAP Category A [1-6]; LERAP Category B [2, 3]

Hazard classification and safety precautions

Hazard Harmful, Flammable, Dangerous for the environment [1-6]; Flammable liquid and vapour, Harmful if swallowed, Harmful if inhaled, Very toxic to aquatic organisms [1, 4]

Transport code 3

Packaging group II [1-3]; III [4-6]

UN Number 1993

Risk phrases H304, H317, H318, H373 [1, 4]; R20, R22a, R41, R43, R50, R53a [2, 3, 5, 6]

Operator protection A, C, H; U04a, U05a, U08, U11, U14, U19a, U20b

Environmental protection E15a, E16d, E22a, E34, E38 [1-6]; E16a [2, 3]; E16c [1, 4-6]; H410 [1, 4]

Storage and disposal D01, D02, D09a, D10c, D12a

Medical advice M03

200   ethephon

A plant growth regulator for cereals and various horticultural crops

See also chlormequat + 2-chloroethylphosphonic acid
chlormequat + 2-chloroethylphosphonic acid + imazaquin
chlormequat + 2-chloroethylphosphonic acid + mepiquat chloride

Products

1	Cerone	Nufarm UK	480 g/l	SL	15944
2	Coryx	SFP Europe	480 g/l	SL	17620
3	Ethe 480	Agroquimicos	480 g/l	SL	16887
4	Ipanema	SFP Europe	480 g/l	SL	15961
5	Padawan	SFP Europe	480 g/l	SL	15577
6	Telsee	Clayton	480 g/l	SL	16325

Uses

- Growth regulation in **apples** *(off-label)*, **durum wheat** *(off-label)*, **ornamental plant production** *(off-label)*, **protected tomatoes** *(off-label)* [1]
- Lodging control in **spring barley**, **winter barley**, **winter wheat** [1-6]; **triticale**, **winter rye** [1, 3]

Extension of Authorisation for Minor Use (EAMUs)

- **apples** *20122367* [1]
- **durum wheat** *20122368* [1]
- **ornamental plant production** *20122366* [1]
- **protected tomatoes** *20122369* [1]

Approval information

- Ethephon (2-chloroethylphosphonic acid) included in Annex I under EC Regulation 1107/2009
- All products containing ethephon carry the warning: 'ethephon is an anticholinesterase organophosphate. Handle with care'
- Accepted by BBPA for use on malting barley

Efficacy guidance

- Best results achieved on crops growing vigorously under conditions of high fertility
- Optimum timing varies between crops and products. See labels for details
- Do not spray crops when wet or if rain imminent

Restrictions

- Ethephon (2-chloroethylphosphonic acid) is an anticholinesterase organophosphorus compound. Do not use if under medical advice not to work with such compounds

SEE SECTION 3 FOR PRODUCTS ALSO REGISTERED

- Maximum number of treatments 1 per crop or yr
- Do not spray crops suffering from stress caused by any factor, during cold weather or period of night frost nor when soil very dry
- Do not apply to cereals within 10 d of herbicide or liquid fertilizer application
- Do not spray wheat or triticale where the leaf sheaths have split and the ear is visible

Crop-specific information
- Latest use: before 1st spikelet visible (GS 51) for spring barley, winter barley, winter rye; before flag leaf sheath opening (GS 47) for triticale, winter wheat
- HI cider apples, tomatoes 5 d

Environmental safety
- Harmful to aquatic organisms
- Avoid accidental deposits on painted objects such as cars, trucks, aircraft

Hazard classification and safety precautions
Hazard Harmful [1, 3]; Irritant [2, 4-6]; Harmful if inhaled [1]
Transport code 8
Packaging group III
UN Number 3265
Risk phrases H290, H315 [5]; H318 [1, 2, 5]; R20, R41, R52, R53a [3, 4, 6]; R38 [4, 6]
Operator protection A, C [1-6]; H [2]; U05a, U08, U11, U13, U19c, U20b
Environmental protection E13c, E38 [1-6]; H412 [1]
Storage and disposal D01, D02, D09a, D10b, D12a
Medical advice M01

201   ethephon + mepiquat chloride

A plant growth regulator for use in cereals

See also mepiquat chloride

Products
1 Clayton Proud	Clayton	155:305 g/l	SL	17067
2 Terpal	BASF	155:305 g/l	SL	16463

Uses
- Growth regulation in **forest nurseries** *(off-label)* [2]
- Increasing yield in **winter barley** *(low lodging situations)*, **winter wheat** *(low lodging situations)* [1, 2]
- Lodging control in **spring barley**, **triticale**, **winter barley**, **winter rye**, **winter wheat** [1, 2]

Extension of Authorisation for Minor Use (EAMUs)
- **forest nurseries** *20142725* [2]

Approval information
- Ethephon and mepiquat chloride included in Annex I under EC Regulation 1107/2009
- All products containing ethephon (2-chloroethylphosphonic acid) carry the warning: '2-chloroethylphosphonic acid is an anticholinesterase organophosphate. Handle with care'
- Accepted by BBPA for use on malting barley

Efficacy guidance
- Best results achieved on crops growing vigorously under conditions of high fertility
- Recommended dose and timing vary with crop, cultivar, growing conditions, previous treatment and desired degree of lodging control. See label for details
- May be applied to crops undersown with grass or clovers
- Do not apply to crops if wet or rain expected as efficacy will be impaired

Restrictions
- Ethephon is an anticholinesterase organophosphorus compound. Do not use if under medical advice not to work with such compounds
- Maximum number of treatments 2 per crop

FOR FULL CONDITIONS OF USE ALWAYS READ THE PRODUCT LABEL

- Add an authorised non-ionic wetter to spray solution. See label for recommended product and rate
- Do not treat crops damaged by herbicides or stressed by drought, waterlogging etc
- Do not treat crops on soils of low fertility unless adequately fertilized
- Do not apply to winter cultivars sown in spring or treat winter barley, triticale or winter rye on soils with more than 10% organic matter (winter wheat may be treated)
- Do not apply at temperatures above 21˚C

Crop-specific information
- Latest use: before ear visible (GS 49) for winter barley, spring barley, winter wheat and triticale; flag leaf just visible (GS 37) for winter rye
- Late tillering may be increased with crops subject to moisture stress and may reduce quality of malting barley

Environmental safety
- Do not use straw from treated cereals as a mulch or growing medium

Hazard classification and safety precautions
Hazard Harmful [1, 2]; Harmful if swallowed [2]
Transport code 8
Packaging group III
UN Number 3265
Risk phrases H290 [2]
Operator protection A, C; U20b
Environmental protection E15a [1, 2]; H413 [2]
Storage and disposal D01, D02, D08, D09a, D10c
Medical advice M01, M05a

202   ethofumesate

A benzofuran herbicide for grass weed control in various crops
HRAC mode of action code: N

See also chloridazon + ethofumesate
desmedipham + ethofumesate + lenacil + phenmedipham
desmedipham + ethofumesate + metamitron + phenmedipham
desmedipham + ethofumesate + phenmedipham

Products

1	Efeckt	UPL Europe	500 g/l	SC	17177
2	Ethefol	UPL Europe	500 g/l	SC	18224
3	Fumesate 500 SC	Hermoo	500 g/l	SC	14321
4	Oblix 500	UPL Europe	500 g/l	SC	16671
5	Palamares	Agroquimicos	500 g/l	SC	15653
6	Xerton	UPL Europe	417 g/l	SC	17335

Uses
- Annual dicotyledons in *amenity grassland* [5]; *fodder beet, red beet, sugar beet* [1-5]; *grass seed crops* [1, 4, 5]; *mangels* [1, 2, 4, 5]; *winter wheat* [6]
- Annual grasses in *fodder beet, red beet, sugar beet* [1, 3, 4]; *grass seed crops, mangels* [1, 4]; *winter wheat* [6]
- Annual meadow grass in *amenity grassland, fodder beet, grass seed crops, mangels, red beet, sugar beet* [5]
- Blackgrass in *amenity grassland, grass seed crops* [5]; *fodder beet, mangels, red beet, sugar beet* [2, 5]; *winter wheat* [6]
- Cleavers in *fodder beet, mangels, red beet, sugar beet* [2]

Approval information
- Ethofumesate included in Annex I under EC Regulation 1107/2009

SEE SECTION 3 FOR PRODUCTS ALSO REGISTERED

Efficacy guidance
- Most products may be applied pre- or post-emergence of crop or weeds but some restricted to pre-emergence or post-emergence use only. Check label
- Some products recommended for use only in mixtures. Check label
- Volunteer cereals not well controlled pre-emergence, weed grasses should be sprayed before fully tillered
- Grass crops may be sprayed during rain or when wet. Not recommended in very dry conditions or prolonged frost

Restrictions
- The maximum total dose must not exceed 1.0 kg/ha ethofumesate in any three year period
- Do not use on Sands or Heavy soils, Very Light soils containing a high percentage of stones, or soils with more than 5-10% organic matter (percentage varies according to label)
- Do not use on swards reseeded without ploughing
- Clovers will be killed or severely checked
- Do not graze or cut grass for 14 d after, or roll less than 7 d before or after spraying

Crop-specific information
- Latest use: before crops meet across rows for beet crops and mangels; not specified for other crops
- Apply in beet crops in tank mixes with other pre- or post-emergence herbicides. Recommendations vary for different mixtures. See label for details
- Safe timing on beet crops varies with other ingredient of tank mix. See label for details
- In grass crops apply to moist soil as soon as possible after sowing or post-emergence when crop in active growth, normally mid-Oct to mid-Dec. See label for details
- May be used in Italian, hybrid and perennial ryegrass, timothy, cocksfoot, meadow fescue and tall fescue. Apply pre-emergence to autumn-sown leys, post-emergence after 2-3 leaf stage. See label for details

Following crops guidance
- In the event of failure of a treated crop only sugar beet, red beet, fodder beet and mangels may be redrilled within 3 mth of application
- Any crop may be sown 3 mth after application of mixtures in beet crops following ploughing, 5 mth after application in grass crops

Environmental safety
- Dangerous for the environment
- Toxic to aquatic organisms
- Do not empty into drains

Hazard classification and safety precautions
Hazard Irritant [2]; Dangerous for the environment [5]
Transport code 9 [1, 2, 5]
Packaging group III [1, 2, 5]
UN Number 3082 [1, 2, 5]; N/C [3, 4, 6]
Risk phrases H317 [2, 6]
Operator protection A [2]; H [1, 2, 4, 6]; U05a [1, 4, 6]; U08, U20a [5]; U09a, U20b [2, 3]; U14 [2]; U19a [2, 3, 5]
Environmental protection E15a, E19b [5]; E15b [1-4, 6]; E23 [1, 4, 6]; E34 [1, 4-6]; E38 [2-4, 6]; H411 [1, 2, 4-6]; H412 [3]
Storage and disposal D01, D02 [1, 2, 4, 6]; D05 [2, 3, 5]; D09a [1-6]; D10a [1, 3, 4, 6]; D10b [5]; D10c [2]; D12a [1, 3-6]
Medical advice M05a [2]

203   ethofumesate + metamitron

A contact and residual herbicide mixture for beet crops
HRAC mode of action code: N + C1

See also metamitron

Products

1	Oblix MT	UPL Europe	150:350 g/l	SC	16487
2	Torero	Adama	150:350 g/l	SC	16484
3	Volcano	UPL Europe	150:350 g/l	SC	16532

Uses

- Annual dicotyledons in *fodder beet*, *mangels*, *sugar beet*
- Annual meadow grass in *fodder beet*, *mangels*, *sugar beet*

Approval information

- Ethofumesate and metamitron included in Annex I under EC Regulation 1107/2009

Efficacy guidance

- Best results obtained from a series of treatments applied as an overall fine spray commencing when earliest germinating weeds are no larger than fully expanded cotyledon and the majority of the crop at fully expanded cotyledon
- Apply subsequent sprays as each new flush of weeds reaches early cotyledon and continue until weed emergence ceases
- Product may be used on all soil types but residual activity may be reduced on those with more than 5% organic matter

Restrictions

- Maximum total dose equivalent to three full dose treatments
- The maximum total dose must not exceed 1.0 kg/ha ethofumesate in any three year period

Crop-specific information

- Latest use: before crop leaves meet between rows
- Crop tolerance may be reduced by stress caused by growing conditions, effects of pests, disease or other pesticides, nutrient deficiency etc

Following crops guidance

- Beet crops may be sown at any time after treatment. Any other crop may be sown after mouldboard ploughing to 15 cm and a minimum interval of 3 mth after treatment

Environmental safety

- Dangerous for the environment
- Very toxic to aquatic organisms
- Do not empty into drains

Hazard classification and safety precautions

Hazard Harmful [1]; Dangerous for the environment [1-3]
Transport code 9
Packaging group III
UN Number 3082
Risk phrases R22a, R51, R53a [1]
Operator protection A, H [1, 3]; U05a, U19a, U20a [1-3]; U08, U14, U15 [2, 3]
Environmental protection E13c, E34, H411 [2, 3]; E15a [1]; E19b, E38 [1-3]
Storage and disposal D01, D02, D09a, D10b, D12a [1-3]; D05 [2, 3]
Medical advice M05a [1]

204 ethofumesate + metamitron + phenmedipham

A contact and residual herbicide mixture for sugar beet
HRAC mode of action code: N + C1 + C1

See also metamitron
        phenmedipham

Products

Phemo	UPL Europe	51:153:51 g/l	SE	16984

Uses

- Annual dicotyledons in **fodder beet**, **mangels**, **red beet**, **sugar beet**
- Annual meadow grass in **fodder beet**, **mangels**, **red beet**, **sugar beet**

Approval information

- Ethofumesate, metamitron and phenmedipham included in Annex I under EC Regulation 1107/2009

Efficacy guidance

- Best results obtained from a series of treatments applied as an overall fine spray commencing when earliest germinating weeds are no larger than fully expanded cotyledon
- Apply subsequent sprays as each new flush of weeds reaches early cotyledon and continue until weed emergence ceases (maximum 3 sprays)
- Product must be applied with Actipron
- Product may follow certain pre-emergence treatments and be used in conjunction with other post-emergence sprays - see label for details
- Where a pre-emergence band spray has been applied, the first treatment should be timed according to the size of the weeds in the untreated area between the rows

Restrictions

- Maximum number of treatments 3 per crop; maximum total dose 6 kg product per ha
- Product may be used on all soil types but residual activity may be reduced on those with more than 5% organic matter
- Crop tolerance may be reduced by stress caused by growing conditions, effects of pests, disease or other pesticides, nutrient deficiency etc
- Only beet crops should be sown within 4 mth of last treatment; winter cereals may be sown after this interval
- Any spring crop may be sown in the year following use
- Mould-board ploughing to 150 mm followed by thorough cultivation recommended before planting any crop.
- The maximum total dose must not exceed 1.0 kg/ha ethofumesate in any three year period

Crop-specific information

- Latest use: before crop leaves meet between rows

Environmental safety

- Irritating to eyes
- Harmful to fish or other aquatic life. Do not contaminate surface waters or ditches with chemical or used container

Hazard classification and safety precautions

Hazard Dangerous for the environment
Transport code 9
Packaging group III
UN Number 3082
Operator protection A; U05a, U20c
Environmental protection E15a, E34, E38, H410
Storage and disposal D01, D02, D09a, D11a, D11b, D12a

FOR FULL CONDITIONS OF USE ALWAYS READ THE PRODUCT LABEL

205 ethofumesate + phenmedipham

A contact and residual herbicide for use in beet crops
HRAC mode of action code: N + C1

See also phenmedipham

Products

1	Brigida	Agroquimicos	200:200 g/l	SC	15658
2	Teamforce	UPL Europe	100:80 g/l	EC	16978
3	Twin	Adama	94:97 g/l	EC	13998

Uses
- Annual dicotyledons in *fodder beet*, *mangels*, *sugar beet*
- Annual meadow grass in *fodder beet*, *mangels*, *sugar beet*
- Blackgrass in *fodder beet*, *mangels*, *sugar beet*

Approval information
- Ethofumesate and phenmedipham included in Annex I under EC Regulation 1107/2009

Efficacy guidance
- Best results achieved by repeat applications to cotyledon stage weeds. Larger susceptible weeds not killed by first treatment usually checked and controlled by second application
- Apply on all soil types at 5-10 d intervals
- On soils with more than 5-10% organic matter residual activity may be reduced

Restrictions
- Maximum number of treatments normally 3 per crop or maximum total dose equivalent to three full dose treatments, or less - see labels for details
- Do not spray wet foliage or if rain imminent
- Spray in evening if daytime temperatures above 21°C expected
- Avoid or delay treatment if frost expected within 7 days
- Avoid or delay treating crops under stress from wind damage, manganese or lime deficiency, pest or disease attack etc.
- The maximum total dose must not exceed 1.0 kg/ha ethofumesate in any three year period

Crop-specific information
- Latest use: before crop foliage meets in the rows
- Check from which recovery may not be complete may occur if treatment made during conditions of sharp diurnal temperature fluctuation

Following crops guidance
- Beet crops may be sown at any time after treatment. Any other crop may be sown after mouldboard ploughing to 15 cm and a minimum interval of 3 mth after treatment

Environmental safety
- Dangerous for the environment [1, 3]
- Toxic to aquatic organisms
- Do not empty into drains
- Extra care necessary to avoid drift because product is recommended for use as a fine spray

Hazard classification and safety precautions

Hazard Harmful [2, 3]; Irritant [1, 2]; Dangerous for the environment [1-3]; Harmful if swallowed [2]

Transport code 9
Packaging group III
UN Number 3082
Risk phrases H312, H335, H351 [2]; H317 [1, 2]; R37, R40, R43, R51, R53a [3]
Operator protection A [1-3]; H [1, 2]; U05a, U08, U19a, U20a [1-3]; U14 [1]
Environmental protection E15a, E19b, E34 [1-3]; E38 [2]; H411 [1, 2]
Storage and disposal D01, D02, D09a, D10b, D12a [1-3]; D05 [1]
Medical advice M05a

SEE SECTION 3 FOR PRODUCTS ALSO REGISTERED

SECTION 2

206 ethylene

A gas used for fruit ripening and potato storage

Products

Biofresh Safestore	Biofresh	99.9% w/w	GA	15729

Uses
- Sprout suppression in **potatoes**

Approval information
- Ethylene is included in Annex 1 under EC Regulation 1107/2009

Efficacy guidance
- For use in stored fruit or potatoes after harvest

Restrictions
- Handling and release of ethylene must only be undertaken by operators suitably trained and competent to carry out the work
- Operators must vacate treated areas immediately after ethylene introduction
- Unprotected persons must be excluded from the treated areas until atmospheres have been thoroughly ventilated for 15 minutes minimum before re-entry
- Ambient atmospheric ethylene concentration must not exceed 1000 ppm. Suitable self-contained breathing apparatus must be worn in atmospheres containing ethylene in excess of 1000 ppm
- A minimum 3 d post treatment period is required before removal of treated crop from storage
- Ethylene treatment must only be undertaken in fully enclosed storage areas that are air tight with appropriate air circulation and venting facilities

Hazard classification and safety precautions
Hazard Extremely flammable
Transport code 2
UN Number 1962
Risk phrases H336
Operator protection A, D, H; U19a
Environmental protection E38, H412
Storage and disposal D06e, D06f, D06g
Medical advice M04a

207 etofenprox

An insecticide for use in oilseed rape
IRAC mode of action code: 3A

Products

Trebon 30 EC	Certis	287.5 g/l	EC	16666

Uses
- Pollen beetle in **spring oilseed rape** *(moderate control)*, **winter oilseed rape** *(moderate control)*

Approval information
- Etofenprox is included in Annex 1 under EC Regulation 1107/2009

Efficacy guidance
- Apply in a water volume of 100 - 300 l/ha when pollen beetle numbers exceed appropriate threshold levels. Follow up with a non-pyrethroid spray if above threshold numbers survive treatment. To control aphids, use an appropriate aphicide spray.

Restrictions
- Use low drift spraying equipment up to 30m from the top of the bank of any surface water bodies

FOR FULL CONDITIONS OF USE ALWAYS READ THE PRODUCT LABEL

Environmental safety
- Buffer zone requirement 12m
- Horizontal boom sprayers must be fitted with three star drift reduction technology for all uses
- LERAP Category B

Hazard classification and safety precautions
Hazard Irritant, Dangerous for the environment, Harmful if swallowed
Transport code 9
Packaging group III
UN Number 3082
Risk phrases H315, H320, H336, H373, R64
Operator protection A, C, H; U05a, U11, U14, U15, U20a
Environmental protection E12c, E12e, E15b, E16a, E22b, E38, H410
Storage and disposal D01, D02, D09a, D10b, D12a
Medical advice M05b

208   etoxazole

A mite growth inhibitor
IRAC mode of action code: 10B

Products

1	Borneo	Interfarm	110 g/l	SC	13919
2	Clayton Java	Clayton	110 g/l	SC	15155

Uses
- Mites in **protected aubergines**, **protected tomatoes**
- Spider mites in **protected ornamentals** *(off-label)*, **protected strawberries** *(off-label)*
- Two-spotted spider mite in **protected ornamentals** *(off-label)*, **protected strawberries** *(off-label)*

Extension of Authorisation for Minor Use (EAMUs)
- **protected ornamentals** *20081216* [1], *20111544* [2]
- **protected strawberries** *20092400* [1], *20111545* [2]

Approval information
- Etoxazole included in Annex I under EC Regulation 1107/2009

Hazard classification and safety precautions
Hazard Dangerous for the environment
Transport code 9
Packaging group III
UN Number 3082
Risk phrases R50, R53a [2]
Operator protection A; U05a, U20b
Environmental protection E15a, E34 [1, 2]; H410 [1]
Storage and disposal D01, D02, D09b, D11a, D12b

209   famoxadone

A strobilurin fungicide available only in mixtures
FRAC mode of action code: 11

See also cymoxanil + famoxadone

SECTION 2

210   fatty acids C7-C20

A soap concentrate insecticide and acaricide

Products

Flipper	Fargro	480 g/l	EW	17767

Uses
- Aphids in **protected aubergines** *(off-label)*, **protected chilli peppers** *(off-label)*, **protected cucumbers**, **protected peppers** *(off-label)*, **protected strawberries**, **protected tomatoes**
- Spider mites in **protected aubergines** *(off-label)*, **protected chilli peppers** *(off-label)*, **protected cucumbers**, **protected peppers** *(off-label)*, **protected strawberries**, **protected tomatoes**
- Whitefly in **protected aubergines** *(off-label)*, **protected chilli peppers** *(off-label)*, **protected cucumbers**, **protected peppers** *(off-label)*, **protected strawberries**, **protected tomatoes**

Extension of Authorisation for Minor Use (EAMUs)
- **protected aubergines** *20171761*
- **protected chilli peppers** *20171761*
- **protected peppers** *20171761*

Approval information
- Fatty acids C7 - C20 are included in Annex 1 under EC Regulation 1107/2009

Hazard classification and safety precautions
  Hazard Irritant
  UN Number N/C
  Risk phrases H315, H319, H335
  Operator protection A, C, H
  Environmental protection H412

211   fenamidone

A strobilurin fungicide available only in mixtures
FRAC mode of action code: 11

See also fenamidone + fosetyl-aluminium
fenamidone + mancozeb
fenamidone + propamocarb hydrochloride

212   fenamidone + fosetyl-aluminium

A strobilurin fungicide mixture
FRAC mode of action code: 11 + 33

See also fenamidone
fosetyl-aluminium

Products

Fenomenal	Bayer CropScience	60:600 g/l	WG	15494

Uses
- Collar rot in **strawberries** *(moderate control from foliar spray)*
- Crown rot in **strawberries** *(moderate control from foliar spray)*
- Downy mildew in **baby leaf crops** *(off-label)*, **celery leaves** *(off-label)*, **chives** *(off-label)*, **cress** *(off-label)*, **edible flowers** *(off-label)*, **endives** *(off-label)*, **herbs (see appendix 6)** *(off-label)*, **lamb's lettuce** *(off-label)*, **land cress** *(off-label)*, **lettuce** *(off-label)*, **ornamental plant production** *(off-label)*, **parsley** *(off-label)*, **protected ornamentals** *(off-label)*, **red mustard** *(off-label)*, **spinach** *(off-label)*, **spinach beet** *(off-label)*, **sweet ciceley** *(off-label)*, **wine grapes** *(off-label)*
- Phytophthora in **chicory** *(off-label)*, **ornamental plant production**, **strawberries**

- Pythium in *ornamental plant production, strawberries*
- Red core in *strawberries*
- Root rot in *lettuce* *(off-label)*, *ornamental plant production* *(off-label)*

Extension of Authorisation for Minor Use (EAMUs)
- *baby leaf crops* *20161308*
- *celery leaves* *20161308*
- *chicory* *20130448*
- *chives* *20161308*
- *cress* *20161308*
- *edible flowers* *20161308*
- *endives* *20161308*
- *herbs (see appendix 6)* *20160109, 20161308*
- *lamb's lettuce* *20161308*
- *land cress* *20161308*
- *lettuce* *20161308, 20171531*
- *ornamental plant production* *20131990, 20171531*
- *parsley* *20161308*
- *protected ornamentals* *20131990*
- *red mustard* *20161308*
- *spinach* *20161308*
- *spinach beet* *20161308*
- *sweet ciceley* *20161308*
- *wine grapes* *20130214*

Approval information
- Fenamidone and fosetyl-aluminium included in Annex I under EC Regulation 1107/2009

Efficacy guidance
- Best used preventatively. Do not rely on curative activity
- Hose down dirty plants before dipping
- Due to variations in growing conditions and the range of varieties when treating ornamentals, it is advisable to check the crop tolerance on a small sample before treating the whole crop
- Fenamidone is a member of the QoI cross resistance group. To minimise the likelihood of development of resistance these products should be used in a planned Resistance Management strategy. In addition before application consult and adhere to the latest FRAG-UK resistance guidance on application of QoI fungicides to potatoes. See Section 5 for more information

Crop-specific information
- There is only limited crop safety data on tulip bulbs
- Container-grown Viburnum, Pieris, Gaultheria, Chamaecyparis and Erica have been successfully treated in trials

Following crops guidance
- No limitations on choice of following crops

Environmental safety
- LERAP Category B

Hazard classification and safety precautions
Hazard Irritant, Dangerous for the environment, Very toxic to aquatic organisms
Transport code 9
Packaging group III
UN Number 3077
Risk phrases H319
Operator protection M; U05a, U11, U20b
Environmental protection E15b, E16a, E22c, E34, E38, H410
Storage and disposal D01, D02, D09a, D11a, D12a

SEE SECTION 3 FOR PRODUCTS ALSO REGISTERED

213   fenamidone + propamocarb hydrochloride

A systemic fungicide mixture for potatoes
FRAC mode of action code: 11 + 28

See also propamocarb hydrochloride

Products

1 Consento	Bayer CropScience	75:375 g/l	SC	15150
2 Prompto	Bayer CropScience	75:375 g/l	SC	15156

Uses

- Blight in **potatoes**
- Early blight in **potatoes**

Approval information

- Fenamidone and propamocarb hydrochloride included in Annex I under EC Regulation 1107/2009

Efficacy guidance

- Apply as soon as there is risk of blight infection or immediately after a blight warning
- Fenamidone is a member of the QoI cross resistance group. To minimise the likelihood of development of resistance these products should be used in a planned Resistance Management strategy. In addition before application consult and adhere to the latest FRAG-UK resistance guidance on application of QoI fungicides to potatoes. See Section 5 for more information

Restrictions

- Maximum number of treatments (including any other QoI containing fungicide) 6 per yr but no more than three should be applied consecutively. Use in alternation with fungicides from a different cross-resistance group
- Observe a 7 d interval between treatments
- Use only as a protective treatment. Do not use when blight has become readily visible (1% leaf area destroyed)

Crop-specific information

- HI 7 d for potatoes
- All varieties of early, main and seed crop potatoes may be sprayed [2]

Environmental safety

- Dangerous for the environment
- Very toxic to aquatic organisms
- Dangerous to fish or other aquatic life. Do not contaminate surface waters or ditches with chemical or used container
- LERAP Category B

Hazard classification and safety precautions

Hazard Irritant, Dangerous for the environment, Very toxic to aquatic organisms
Transport code 9
Packaging group III
UN Number 3082
Operator protection A, C; U05a, U08, U11 [1, 2]; U19a, U20b, U20c [2]; U20a [1]
Environmental protection E13b [1]; E15a [2]; E16a, E38, H410 [1, 2]
Storage and disposal D01, D02, D09a, D10b, D12a [1, 2]; D05 [1]

214   fenhexamid

A protectant hydroxyanilide fungicide for soft fruit and a range of horticultural crops
FRAC mode of action code: 17

Products

Teldor	Bayer CropScience	50% w/w	WG	11229

FOR FULL CONDITIONS OF USE ALWAYS READ THE PRODUCT LABEL

Uses

- Botrytis in **bilberries** *(off-label)*, **blackberries**, **blackcurrants**, **blueberries** *(off-label)*, **cherries** *(off-label)*, **frise** *(off-label)*, **gooseberries**, **herbs (see appendix 6)** *(off-label)*, **hops** *(off-label)*, **lamb's lettuce** *(off-label)*, **leaf brassicas** *(off-label - baby leaf production)*, **loganberries**, **plums** *(off-label)*, **protected aubergines** *(off-label)*, **protected blueberry** *(off-label)*, **protected chilli peppers** *(off-label)*, **protected courgettes** *(off-label)*, **protected cucumbers** *(off-label)*, **protected forest nurseries** *(off-label)*, **protected gherkins** *(off-label)*, **protected peppers** *(off-label)*, **protected squashes** *(off-label)*, **protected tomatoes** *(off-label)*, **raspberries**, **redcurrants**, **rocket** *(off-label)*, **rubus hybrids**, **scarole** *(off-label)*, **soft fruit** *(off-label)*, **strawberries**, **tomatoes (outdoor)** *(off-label)*, **whitecurrants**, **wine grapes**
- Grey mould in **protected aubergines** *(off-label)*, **protected chilli peppers** *(off-label)*, **protected courgettes** *(off-label)*, **protected cucumbers** *(off-label)*, **protected gherkins** *(off-label)*, **protected peppers** *(off-label)*, **protected squashes** *(off-label)*, **protected tomatoes** *(off-label)*

Extension of Authorisation for Minor Use (EAMUs)

- **bilberries** *20161214*
- **blueberries** *20161214*
- **cherries** *20031866*
- **frise** *20082062*
- **herbs (see appendix 6)** *20082062*
- **hops** *20082926*
- **lamb's lettuce** *20082062*
- **leaf brassicas** *(baby leaf production)* *20082062*
- **plums** *20031866*
- **protected aubergines** *20042087*
- **protected blueberry** *20161214*
- **protected chilli peppers** *20042086*
- **protected courgettes** *20042085*
- **protected cucumbers** *20042085*
- **protected forest nurseries** *20082926*
- **protected gherkins** *20042085*
- **protected peppers** *20042086*
- **protected squashes** *20042085*
- **protected tomatoes** *20042087*
- **rocket** *20082062*
- **scarole** *20082062*
- **soft fruit** *20082926*
- **tomatoes (outdoor)** *20042399*

Approval information

- Fenhexamid included in Annex I under EC Regulation 1107/2009

Efficacy guidance

- Use as part of a programme of sprays throughout the flowering period to achieve effective control of Botrytis
- To minimise possibility of development of resistance, no more than two sprays of the product may be applied consecutively. Other fungicides from a different chemical group should then be used for at least two consecutive sprays. If only two applications are made on grapevines, only one may include fenhexamid
- Complete spray cover of all flowers and fruitlets throughout the blossom period is essential for successful control of Botrytis
- Spray programmes should normally start at the start of flowering

Restrictions

- Maximum number of treatments 2 per yr on grapevines; 4 per yr on other listed crops but no more than 2 sprays may be applied consecutively

Crop-specific information

- HI 1 d for strawberries, raspberries, loganberries, blackberries, Rubus hybrids; 3 d for cherries; 7 d for blackcurrants, redcurrants, whitecurrants, gooseberries; 21d for outdoor grapes

SEE SECTION 3 FOR PRODUCTS ALSO REGISTERED

Environmental safety
- Dangerous for the environment
- Harmful to fish or other aquatic life. Do not contaminate surface waters or ditches with chemical or used container

Hazard classification and safety precautions
Hazard Dangerous for the environment
Transport code 9
Packaging group III
UN Number 3077
Risk phrases R53a
Operator protection U08, U19a, U20b
Environmental protection E13c, E38
Storage and disposal D05, D09a, D11a, D12a

215   fenoxaprop-P-ethyl

An aryloxyphenoxypropionate herbicide for use in wheat
HRAC mode of action code: A

See also diclofop-methyl + fenoxaprop-P-ethyl

Products

1	Foxtrot EW	Headland	69 g/l	EW	13243
2	Oskar	Headland	69 g/l	EW	13344
3	Polecat	Interfarm	69 g/l	EC	16112

Uses
- Awned canary grass in *spring wheat, winter wheat* [3]
- Blackgrass in *grass seed crops* (off-label), *spring barley, winter barley* [1, 2]; *spring wheat, winter wheat* [1-3]
- Canary grass in *grass seed crops* (off-label) [2]; *spring barley, spring wheat, winter barley, winter wheat* [1, 2]
- Rough-stalked meadow grass in *grass seed crops* (off-label), *spring barley, winter barley* [1, 2]; *spring wheat, winter wheat* [1-3]
- Wild oats in *grass seed crops* (off-label), *spring barley, winter barley* [1, 2]; *spring wheat, winter wheat* [1-3]

Extension of Authorisation for Minor Use (EAMUs)
- *grass seed crops* 20150680 [1], 20150866 [2]

Approval information
- Fenoxaprop-P-ethyl included in Annex I under EC Regulation 1107/2009

Efficacy guidance
- Treat weeds from 2 fully expanded leaves up to flag leaf ligule just visible; for awned canary-grass and rough meadow-grass from 2 leaves to the end of tillering
- A second application may be made in spring where susceptible weeds emerge after an autumn application
- Spray is rainfast 1 h after application
- Dry conditions resulting in moisture stress may reduce effectiveness
- Fenoxaprop-P-ethyl is an ACCase inhibitor herbicide. To avoid the build up of resistance do not apply products containing an ACCase inhibitor herbicide more than twice to any crop. In addition do not use any product containing fenoxaprop-P-ethyl in mixture or sequence with any other product containing the same ingredient
- Use these products as part of a resistance management strategy that includes cultural methods of control and does not use ACCase inhibitors as the sole chemical method of grass weed control
- Applying a second product containing an ACCase inhibitor to a crop will increase the risk of resistance development; only use a second ACCase inhibitor to control different weeds at a different timing

FOR FULL CONDITIONS OF USE ALWAYS READ THE PRODUCT LABEL

- Always follow WRAG guidelines for preventing and managing herbicide resistant weeds. See Section 5 for more information

Restrictions
- Maximum total dose equivalent to one or two full dose treatments depending on product used
- Do not apply to barley, durum wheat, undersown crops or crops to be undersown
- Do not roll or harrow within 1 wk of spraying
- Do not spray crops under stress, suffering from drought, waterlogging or nutrient deficiency or those grazed or if soil compacted
- Avoid spraying immediately before or after a sudden drop in temperature or a period of warm days/cold nights
- Do not mix with hormone weedkillers

Crop-specific information
- Latest use: before flag leaf sheath extending (GS 41)
- Treat from crop emergence to flag leaf fully emerged (GS 41).
- Product may be sprayed in frosty weather provided crop hardened off but do not spray wet foliage or leaves covered with ice
- Broadcast crops should be sprayed post-emergence after plants have developed well-established root system

Environmental safety
- Dangerous for the environment
- Very toxic to aquatic organisms

Hazard classification and safety precautions
> **Hazard** Irritant, Dangerous for the environment
> **Transport code** 9
> **Packaging group** III
> **UN Number** 3082
> **Risk phrases** H315 [1, 2]; H317 [1, 3]
> **Operator protection** A, C, H; U05a, U08, U14, U20b [1-3]; U11 [3]; U19a [1, 2]
> **Environmental protection** E15a, E38, H411 [1-3]; E34 [1, 2]
> **Storage and disposal** D01, D02, D05, D09a, D10b [1-3]; D12a [3]

216   fenoxycarb

> An insect specific growth regulator for top fruit
> IRAC mode of action code: 7B

Products

Insegar WG	Syngenta	25% w/w	WG	09789

Uses
- Summer-fruit tortrix moth in *apples*, *pears*

Approval information
- Fenoxycarb included in Annex I under EC Regulation 1107/2009

Efficacy guidance
- Best results from application at 5th instar stage before pupation. Product prevents transformation from larva to pupa
- Correct timing best identified from pest warnings
- Because of mode of action rapid knock-down of pest is not achieved and larvae continue to feed for a period after treatment
- Adequate water volume necessary to ensure complete coverage of leaves

Restrictions
- Maximum number of treatments 2 per crop
- Do not apply using knapsack sprayers
- Consult processors before use

SEE SECTION 3 FOR PRODUCTS ALSO REGISTERED

Crop-specific information
- HI 42 d for apples, pears
- Use on all varieties of apples and pears

Environmental safety
- Dangerous for the environment
- Toxic to aquatic organisms
- High risk to bees. Do not apply to crops in flower or to those in which bees are actively foraging. Do not apply when flowering weeds are present
- Risk to certain non-target insects or other arthropods. See directions for use
- Apply to minimise off-target drift to reduce effects on non-target organisms. Some margin of safety to beneficial arthropods is indicated.
- Broadcast air-assisted LERAP (8 m)

Hazard classification and safety precautions
Hazard Dangerous for the environment, Very toxic to aquatic organisms
Transport code 9
Packaging group III
UN Number 3077
Risk phrases H351
Operator protection A, C, H; U05a, U20a, U23a
Environmental protection E12a, E12e, E15a, E22b, E38, H410; E17b (8 m)
Storage and disposal D01, D02, D07, D09a, D11a, D12a

217   fenpropidin

A systemic, curative and protective piperidine (morpholine) fungicide only available in mixtures
FRAC mode of action code: 5

See also difenoconazole + fenpropidin
fenpropidin + prochloraz + tebuconazole

218   fenpropidin + prochloraz + tebuconazole

A contact and systemic fungicide mixture for cereals
FRAC mode of action code: 5 + 3 + 3

See also fenpropidin
fenpropidin + prochloraz
fenpropidin + tebuconazole
prochloraz
prochloraz + tebuconazole
tebuconazole

Products

Artemis	Adama	150:200:100 g/l	EC	14178

Uses
- Brown rust in **spring wheat, winter wheat**
- Ear diseases in **spring wheat, winter wheat**
- Eyespot in **spring wheat, winter wheat**
- Glume blotch in **spring wheat, winter wheat**
- Powdery mildew in **spring wheat, winter wheat**
- Septoria leaf blotch in **spring wheat, winter wheat**
- Yellow rust in **spring wheat, winter wheat**

Approval information
- Fenpropidin, prochloraz and tebuconazole included in Annex I under EC Regulation 1107/2009

FOR FULL CONDITIONS OF USE ALWAYS READ THE PRODUCT LABEL

Efficacy guidance
- Best disease control and yield benefit obtained when applied at early stage of disease development before infection spreads to new growth
- To protect the flag leaf and ear from Septoria diseases apply from flag leaf emergence to ear fully emerged (GS 37-59)
- Applications once foliar symptoms of *Septoria tritici* are already present on upper leaves will be less effective
- Resistance to some DMI fungicides has been identified in Septoria leaf blotch which may seriously affect the performance of some products.

Restrictions
- Newer authorisations for tebuconazole products require application to cereals only after GS 30 and applications to oilseed rape and linseed after GS20 - check label

Crop-specific information
- Do not apply when temperatures are high.

Hazard classification and safety precautions
Hazard Harmful, Dangerous for the environment, Harmful if inhaled, Very toxic to aquatic organisms
Transport code 9
Packaging group III
UN Number 3082
Risk phrases H315, H318, H361
Operator protection A, C, H; U02a, U04a, U05a, U09a, U10, U11, U15, U19a, U20b
Environmental protection E15a, E34, E38, H410
Storage and disposal D01, D02, D05, D12a
Medical advice M03, M05a

219   fenpropimorph

A contact and systemic morpholine fungicide
FRAC mode of action code: 5

See also azoxystrobin + fenpropimorph
epoxiconazole + fenpropimorph
epoxiconazole + fenpropimorph + kresoxim-methyl
epoxiconazole + fenpropimorph + metrafenone
epoxiconazole + fenpropimorph + pyraclostrobin

Products
1	Clayton Spigot	Clayton	750 g/l	EC	11560
2	Corbel	BASF	750 g/l	EC	00578
3	Crebol	AgChem Access	750 g/l	EC	13922

Uses
- Alternaria in **carrots** *(off-label)*, **horseradish** *(off-label)*, **parsley root** *(off-label)*, **parsnips** *(off-label)*, **salsify** *(off-label)* [2]
- Brown rust in **spring barley**, **spring wheat**, **triticale**, **winter barley**, **winter wheat** [1-3]
- Crown rot in **carrots** *(off-label)*, **horseradish** *(off-label)*, **parsley root** *(off-label)*, **parsnips** *(off-label)*, **salsify** *(off-label)* [1, 2]; **mallow (althaea spp.)** *(off-label)* [1]
- Foliar disease control in **durum wheat** *(off-label)* [2]; **grass seed crops** *(off-label)* [1]
- Powdery mildew in **bilberries** *(off-label)*, **blackcurrants** *(off-label)*, **blueberries** *(off-label)*, **carrots** *(off-label)*, **cranberries** *(off-label)*, **dewberries** *(off-label)*, **gooseberries** *(off-label)*, **hops** *(off-label)*, **horseradish** *(off-label)*, **loganberries** *(off-label)*, **parsley root** *(off-label)*, **parsnips** *(off-label)*, **raspberries** *(off-label)*, **redcurrants** *(off-label)*, **salsify** *(off-label)*, **strawberries** *(off-label)*, **tayberries** *(off-label)*, **whitecurrants** *(off-label)* [1, 2]; **durum wheat** *(off-label)*, **fodder beet** *(off-label)*, **mallow (althaea spp.)** *(off-label)*, **red beet** *(off-label)*, **sugar beet seed crops** *(off-label)* [1]; **spring barley**, **spring oats**, **spring wheat**, **winter barley**, **winter oats**, **winter rye**, **winter wheat** [1-3]
- Rhynchosporium in **spring barley**, **winter barley** [1-3]

SEE SECTION 3 FOR PRODUCTS ALSO REGISTERED

SECTION 2

- Rust in **fodder beet** *(off-label)*, **sugar beet seed crops** *(off-label)* [2]; **leeks** [1]; **red beet** *(off-label)* [1, 2]
- Yellow rust in **spring barley**, **spring wheat**, **triticale**, **winter barley**, **winter wheat** [1-3]

Extension of Authorisation for Minor Use (EAMUs)
- **bilberries** *20111798* [1], *20040804* [2]
- **blackcurrants** *20111798* [1], *20040804* [2]
- **blueberries** *20111798* [1], *20040804* [2]
- **carrots** *20111796* [1], *20023753* [2]
- **cranberries** *20111798* [1], *20040804* [2]
- **dewberries** *20111798* [1], *20040804* [2]
- **durum wheat** *20111799* [1], *20061460* [2]
- **fodder beet** *20111797* [1], *20023757* [2]
- **gooseberries** *20111798* [1], *20040804* [2]
- **grass seed crops** *20061060* [1]
- **hops** *20111794* [1], *20023759* [2]
- **horseradish** *20111796* [1], *20023753* [2]
- **loganberries** *20111798* [1], *20040804* [2]
- **mallow (althaea spp.)** *20111796* [1]
- **parsley root** *20111796* [1], *20023753* [2]
- **parsnips** *20111796* [1], *20023753* [2]
- **raspberries** *20111798* [1], *20040804* [2]
- **red beet** *20111795* [1], *20023751* [2]
- **redcurrants** *20111798* [1], *20040804* [2]
- **salsify** *20111796* [1], *20023753* [2]
- **strawberries** *20111798* [1], *20040804* [2]
- **sugar beet seed crops** *20111797* [1], *20023757* [2]
- **tayberries** *20111798* [1], *20040804* [2]
- **whitecurrants** *20111798* [1], *20040804* [2]

Approval information
- Accepted by BBPA for use on malting barley and hops
- Fenpropimorph included in Annex I under EC Regulation 1107/2009

Efficacy guidance
- On all crops spray at start of disease attack. See labels for recommended tank mixes. Follow-up treatments may be needed if disease pressure remains high
- Product rainfast after 2 h

Restrictions
- Maximum number of treatments 2 per crop for spring cereals; 3 or 4 per crop for winter cereals; 6 per crop for leeks
- Consult processors before using on crops for processing

Crop-specific information
- HI cereals 5 wk
- Scorch may occur on cereals if applied during frosty weather or in high temperatures
- Some leaf spotting may occur on undersown clovers
- Leeks should be treated every 2-3 wk as required, up to 6 applications

Environmental safety
- Dangerous for the environment
- Very toxic to aquatic organisms

Hazard classification and safety precautions
Hazard Harmful, Dangerous for the environment [1-3]; Harmful if inhaled, Very toxic to aquatic organisms [1]
Transport code 9
Packaging group III
UN Number 3082
Risk phrases H315, H361 [1]; R38, R50, R53a, R63 [2, 3]
Operator protection A [1-3]; C [1]; U05a, U08, U14, U15, U19a, U20b

FOR FULL CONDITIONS OF USE ALWAYS READ THE PRODUCT LABEL

Environmental protection E15a [1-3]; E38 [2, 3]; H410 [1]
Storage and disposal D01, D02, D09a, D10c, D12a [1-3]; D05 [1]; D08 [2, 3]
Medical advice M05a

220   fenpropimorph + pyraclostrobin

A protectant and curative fungicide mixture for cereals
FRAC mode of action code: 5 +11

See also pyraclostrobin

Products

Jenton	BASF	375:100 g/l	EC	11898

Uses
- Brown rust in *grass seed crops*, *spring barley*, *spring wheat*, *triticale*, *winter barley*, *winter wheat*
- Crown rust in *spring oats*, *winter oats*
- Disease control in *grass seed crops* *(off-label)*, *triticale* *(off-label)*
- Drechslera leaf spot in *grass seed crops*, *spring wheat*, *triticale*, *winter wheat*
- Glume blotch in *grass seed crops*, *grass seed crops* *(off-label)*, *spring wheat*, *triticale*, *triticale* *(off-label)*, *winter wheat*
- Net blotch in *spring barley*, *winter barley*
- Powdery mildew in *spring barley*, *spring oats*, *winter barley*, *winter oats*
- Rhynchosporium in *spring barley*, *winter barley*
- Rust in *grass seed crops* *(off-label)*, *triticale* *(off-label)*
- Septoria leaf blotch in *grass seed crops*, *grass seed crops* *(off-label)*, *spring wheat*, *triticale*, *triticale* *(off-label)*, *winter wheat*
- Yellow rust in *grass seed crops*, *spring barley*, *spring wheat*, *triticale*, *winter barley*, *winter wheat*

Extension of Authorisation for Minor Use (EAMUs)
- *grass seed crops* *20122661*
- *triticale* *20122661*

Approval information
- Fenpropimorph and pyraclostrobin included in Annex I under EC Regulation 1107/2009
- Accepted by BBPA for use on malting barley

Efficacy guidance
- Best results obtained from treatment at the start of foliar disease attack
- Yield response may be obtained in the absence of visual disease
- Pyraclostrobin is a member of the QoI cross resistance group. Product should be used preventatively and not relied on for its curative potential
- Use product as part of an Integrated Crop Management strategy incorporating other methods of control, including where appropriate other fungicides with a different mode of action. Do not apply more than two foliar applications of QoI containing products to any cereal crop
- There is a significant risk of widespread resistance occurring in *Septoria tritici* populations in UK. Strains of barley powdery mildew resistant to QoIs are common in UK. Failure to follow resistance management action may result in reduced levels of disease control

Restrictions
- Maximum number of treatments 2 per crop

Crop-specific information
- Latest use: up to and including emergence of ear just complete (GS 59) for barley

Environmental safety
- Dangerous for the environment
- Very toxic to aquatic organisms
- LERAP Category B

SEE SECTION 3 FOR PRODUCTS ALSO REGISTERED

Hazard classification and safety precautions

Hazard Harmful, Dangerous for the environment, Harmful if swallowed, Harmful if inhaled, Very toxic to aquatic organisms
Transport code 9
Packaging group III
UN Number 3082
Risk phrases H304, H315, H319, H361
Operator protection A, C; U05a, U14, U20b
Environmental protection E15a, E16a, E34, E38, H410
Storage and disposal D01, D02, D05, D09a, D10c, D12a
Medical advice M03, M05a

221   fenpyrazamine

A botriticide for protected crops
FRAC mode of action code: 17

Products

1 Clayton Vitis	Clayton	50% w/w	WG	16974
2 Empire	Pan Agriculture	50% w/w	WG	17357
3 Prolectus	Interfarm	50% w/w	WG	16607

Uses
* Botrytis in *protected aubergines, protected courgettes, protected cucumbers, protected peppers, protected tomatoes, strawberries, wine grapes* [1-3]; *protected strawberries* [2, 3]

Approval information
* Fenpyrazamine is included in Annex I under EC Regulation 1107/2009

Restrictions
* Consult processors before use on crops grown for processing
* No more than a third of the intended botryticide applications made per crop, per year, should contain 3-keto reductase (FRAC code 17) fungicides

Hazard classification and safety precautions

Hazard Dangerous for the environment
Transport code 6.1 [1]; 9 [2, 3]
Packaging group III
UN Number 3077
Operator protection U05a, U11, U13, U14, U15, U20b
Environmental protection E15b [2, 3]; E38 [1]; H410 [1-3]
Storage and disposal D01, D02, D09a, D11a [1-3]; D12a, D12b [2, 3]
Medical advice M03

222   ferric phosphate

A molluscicide bait for controlling slugs and snails

Products

1 Derrex	Certis	2.97% w/w	GB	15351
2 Ironmax Pro	De Sangosse	3% w/w	RB	17122
3 Iroxx	Certis	2.97% w/w	GB	16640
4 Sluxx HP	Certis	2.97% w/w	GB	16571

Uses
* Slugs and snails in *all edible crops (outdoor and protected), all non-edible crops (outdoor)* [1-4]; *amenity vegetation* [1, 3, 4]; *carrots, celeriac, fodder beet, grassland, managed amenity turf, radishes, red beet, sugar beet, turnips* [2]

FOR FULL CONDITIONS OF USE ALWAYS READ THE PRODUCT LABEL

Approval information
- Ferric phosphate included in Annex 1 under EC Regulation 1107/2009
- Accepted by BBPA for use on malting barley

Efficacy guidance
- Treat as soon as damage first seen preferably in early evening. Repeat as necessary to maintain control
- Best results obtained from moist soaked granules. This will occur naturally on moist soils or in humid conditions
- Ferric phosphate does not cause excessive slime secretion and has no requirement to collect moribund slugs from the soil surface
- Active ingredient is degraded by micro-organisms to beneficial plant nutrients

Restrictions
- Maximum total dose equivalent to four full dose treatments

Crop-specific information
- Latest use not specified for any crop

Hazard classification and safety precautions
  UN Number N/C
  Operator protection A, H [2]; U05a, U20a
  Environmental protection E15a [1, 3]; E34 [1-4]
  Storage and disposal D01, D09a

223   ferrous sulphate

A herbicide/fertilizer combination for moss control in turf

*See also dicamba + dichlorprop-P + ferrous sulphate + MCPA
    dichlorprop-P + ferrous sulphate + MCPA*

Products

1	Ferromex Mosskiller Concentrate	Omex	381.15 g/l	SL	16896
2	Greenmaster Autumn	Everris Ltd	16.3% w/w	GR	16762
3	Greentec Mosskiller Pro	Headland Amenity	16.78% w/w	GR	16728

Uses
- Moss in **amenity grassland** [1, 3]; **managed amenity turf** [1-3]

Approval information
- Ferrous sulphate included in Annex 1 under EC Regulation 1107/2009

Efficacy guidance
- For best results apply when turf is actively growing and the soil is moist
- Fertilizer component of most products encourages strong root growth and tillering
- Mow 3 d before treatment and do not mow for 3-4 d afterwards
- Water after 2 d if no rain
- Rake out dead moss thoroughly 7-14 d after treatment. Re-treatment may be necessary for heavy infestations

Restrictions
- Maximum number of treatments - see labels
- Do not apply during drought or when heavy rain expected
- Do not apply in frosty weather or when the ground is frozen
- Do not walk on treated areas until well watered

Crop-specific information
- If spilt on paving, concrete, clothes etc brush off immediately to avoid discolouration
- Observe label restrictions for interval before cutting after treatment

SEE SECTION 3 FOR PRODUCTS ALSO REGISTERED

Environmental safety

- Harmful to fish or other aquatic life. Do not contaminate surface waters or ditches with chemical or used container

Hazard classification and safety precautions

Hazard Harmful [3]
UN Number N/C
Risk phrases H315, H318 [3]; H319 [2]
Operator protection A [1]; U20b [1, 3]; U20c [2]
Environmental protection E13c [2]; E15a [3]; E15b [1]
Storage and disposal D01 [1, 3]; D09a [1-3]; D11a [2, 3]; D12a [1]

224   ferrous sulphate + MCPA + mecoprop-P

A translocated herbicide and moss killer mixture
HRAC mode of action code: O + O

See also MCPA
mecoprop-P

Products

1	Landscaper Pro Feed Weed Mosskiller	Everris Ltd	16.3:0.49:0.29% w/w	GR	17549
2	Renovator Pro	Everris Ltd	16.3:0.49:0.29% w/w	GR	16803

Uses

- Annual dicotyledons in *managed amenity turf*
- Moss in *managed amenity turf*
- Perennial dicotyledons in *managed amenity turf*

Approval information

- Ferrous sulphate, MCPA and mecoprop-P included in Annex I under EC Regulation 1107/2009

Efficacy guidance

- Apply from Apr to Sep when weeds are growing
- For best results apply when light showers or heavy dews are expected
- Apply with a suitable calibrated fertilizer distributor
- Retreatment may be necessary after 6 wk if weeds or moss persist
- For best control of moss scarify vigorously after 2 wk to remove dead moss
- Where regrowth of moss or weeds occurs a repeat treatment may be made after 6 wk
- Avoid treatment of wet grass or during drought. If no rain falls within 48 h water in thoroughly

Restrictions

- Maximum number of treatments 3 per yr
- Do not treat new turf until established for 6 mth
- The first 4 mowings after treatment should not be used to mulch cultivated plants unless composted at least 6 mth
- Avoid walking on treated areas until it has rained or they have been watered
- Do not re-seed or turf within 8 wk of last treatment
- Do not cut grass for at least 3 d before and at least 4 d after treatment
- Do not apply during freezing conditions or when rain imminent

Environmental safety

- Keep livestock out of treated areas for up to two weeks following treatment and until poisonous weeds, such as ragwort, have died down and become unpalatable
- Harmful to fish or other aquatic life. Do not contaminate surface waters or ditches with chemical or used container
- Do not empty into drains

FOR FULL CONDITIONS OF USE ALWAYS READ THE PRODUCT LABEL

- Some pesticides pose a greater threat of contamination of water than others and mecoprop-P is one of these pesticides. Take special care when applying mecoprop-P near water and do not apply if heavy rain is forecast

Hazard classification and safety precautions
> **UN Number** N/C
> **Risk phrases** H319
> **Operator protection** A, H [1]; U20b
> **Environmental protection** E07a, E13c, E19b
> **Storage and disposal** D01, D09a, D12a
> **Medical advice** M05a

225   flazasulfuron

A sulfonylurea herbicide for non-crop use
HRAC mode of action code: B

Products

1	Chikara Weed Control	Belchim	25% w/w	WG	14189
2	Clayton Apt	Clayton	25% w/w	WG	15157
3	Katana	Belchim	25% w/w	WG	16162
4	Pacaya	ProKlass	25% w/w	WG	16797
5	Paradise	Pan Agriculture	25% w/w	WG	16829
6	PureFlazasul	Pure Amenity	25% w/w	WG	15135
7	Sixte	Headland	25% w/w	WG	18142

Uses

- Annual and perennial weeds in *amenity vegetation, hard surfaces* [1, 3, 7]; *natural surfaces not intended to bear vegetation, permeable surfaces overlying soil* [1-7]; *railway tracks* [2, 4-6]

Approval information

- Flazasulfuron included in Annex 1 under EC Regulation 1107/2009

Environmental safety

- Buffer zone requirement 6 m [7]
- LERAP Category B

Hazard classification and safety precautions
> **Hazard** Dangerous for the environment [1-7]; Very toxic to aquatic organisms [1-4]
> **Transport code** 9
> **Packaging group** III
> **UN Number** 3077
> **Risk phrases** H318 [7]
> **Operator protection** A, H, M; U02a, U05a, U20b [1-7]; U08 [1, 3, 4, 6, 7]; U09a [2, 5]
> **Environmental protection** E15a, E16a, E16b, E38, H410
> **Storage and disposal** D01, D02, D09a, D11a, D12a

226   flocoumafen

A second generation anti-coagulant rodenticide

Products

1	Storm Mini Bits	BASF	0.005% w/w	RB	UK15-0915
2	Storm Pasta	BASF	0.005% w/w	RB	UK15-0852
3	Storm Secure	BASF	0.005% w/w	RB	UK15-0850

Uses

- Mice in *farm buildings, sewers*
- Rodents in *farm buildings, sewers*

SEE SECTION 3 FOR PRODUCTS ALSO REGISTERED

Approval information
- Flocoumafen is not included in Annex 1 under EC Regulation 1107/2009

Restrictions
- Not approved for outdoor use except in sewers and covered drains

Hazard classification and safety precautions
Hazard Harmful
UN Number N/C
Operator protection A; U05a, U20a
Environmental protection E38
Storage and disposal D01, D02, D09b, D12b
Medical advice M03

227   flonicamid

A selective feeding blocker aphicide
IRAC mode of action code: 9C

Products
1	Mainman	Belchim	50% w/w	WG	13123
2	Teppeki	Belchim	50% w/w	WG	12402

Uses
- Aphids in **apples, pears** [1]; **canary seed** *(off-label)*, **potatoes, winter wheat** [2]
- Damson-hop aphid in **hops** *(off-label)* [1]
- Glasshouse whitefly in **courgettes** *(off-label)*, **cucumbers** *(off-label)*, **protected courgettes** *(off-label)*, **protected cucumbers** *(off-label)*, **protected summer squash** *(off-label)*, **summer squash** *(off-label)* [1]
- Mealybugs in **protected aubergines** *(off-label)*, **protected tomatoes** *(off-label)* [1]
- Peach-potato aphid in **courgettes** *(off-label)*, **cucumbers** *(off-label)*, **protected courgettes** *(off-label)*, **protected cucumbers** *(off-label)*, **protected summer squash** *(off-label)*, **summer squash** *(off-label)* [1]
- Tobacco whitefly in **ornamental plant production** *(off-label)* [1]
- Whitefly in **ornamental plant production** *(off-label)*, **protected aubergines** *(off-label)*, **protected tomatoes** *(off-label)* [1]

Extension of Authorisation for Minor Use (EAMUs)
- **canary seed** *20170673* [2]
- **courgettes** *20142923* [1]
- **cucumbers** *20142923* [1]
- **hops** *20142225* [1]
- **ornamental plant production** *20130045* [1]
- **protected aubergines** *20160191* [1]
- **protected courgettes** *20142923* [1]
- **protected cucumbers** *20142923* [1]
- **protected summer squash** *20142923* [1]
- **protected tomatoes** *20160191* [1]
- **summer squash** *20142923* [1]

Approval information
- Flonicamid included in Annex 1 under EC Regulation 1107/2009
- Accepted by BBPA for use on hops

Efficacy guidance
- Apply when warning systems forecast significant aphid infestations
- Persistence of action is 21 d

Restrictions
- Maximum number of treatments 2 per crop for potatoes, winter wheat
- Must not be applied to winter wheat before 50% ear emerged stage (GS 53)

FOR FULL CONDITIONS OF USE ALWAYS READ THE PRODUCT LABEL

- Use a maximum of two consecutive applications of [1] in apples and pears. If further treatments are required, use an insecticide with a different mode of action before applying the final application of [1]

Crop-specific information
- HI: potatoes 14 d; winter wheat 28 d;
- HI: apples and pears 21 days [1]

Environmental safety
- Dangerous for the environment
- Harmful to aquatic organisms

Hazard classification and safety precautions
> **Hazard** Dangerous for the environment [2]
> **UN Number** N/C
> **Operator protection** A, H, M [1, 2]; C [2]; E [1]; U05a [1, 2]; U13, U14, U15 [1]
> **Environmental protection** E15a, H412
> **Storage and disposal** D01, D02, D12b [1, 2]; D03, D08 [1]; D07 [2]

228 florasulam

A triazolopyrimidine herbicide for cereals
HRAC mode of action code: B

See also 2,4-D + florasulam
clopyralid + florasulam
clopyralid + florasulam + fluroxypyr
diflufenican + florasulam

Products

1 Lector	Headland	50 g/l	SC	16565
2 Paramount	Headland	50 g/l	SC	16452
3 Sumir	Life Scientific	50 g/l	SC	17974

Uses
- Annual dicotyledons in *spring barley, spring oats, spring wheat, winter barley, winter oats, winter wheat* [2, 3]; *triticale, winter rye* [2]
- Charlock in *grass seed crops* (off-label), *ornamental plant production* (off-label) [1, 2]
- Chickweed in *grass seed crops* (off-label), *ornamental plant production* (off-label), *triticale, winter rye* [1, 2]; *spring barley, spring oats, spring wheat, winter barley, winter oats, winter wheat* [1-3]
- Cleavers in *grass seed crops* (off-label), *ornamental plant production* (off-label), *triticale, winter rye* [1, 2]; *spring barley, spring oats, spring wheat, winter barley, winter oats, winter wheat* [1-3]
- Groundsel in *ornamental plant production* (off-label) [1, 2]
- Hedge mustard in *grass seed crops* (off-label) [1, 2]; *spring barley, spring oats, spring wheat, triticale, winter barley, winter oats, winter rye, winter wheat* [1]
- Mayweeds in *grass seed crops* (off-label), *ornamental plant production* (off-label) [1, 2]; *spring barley, spring oats, spring wheat, winter barley, winter oats, winter wheat* [2, 3]; *triticale, winter rye* [2]
- Runch in *grass seed crops* (off-label) [1, 2]; *spring barley, spring oats, spring wheat, triticale, winter barley, winter oats, winter rye, winter wheat* [1]
- Scented mayweed in *spring barley, spring oats, spring wheat, triticale, winter barley, winter oats, winter rye, winter wheat* [1]
- Scentless mayweed in *spring barley, spring oats, spring wheat, triticale, winter barley, winter oats, winter rye, winter wheat* [1]
- Shepherd's purse in *grass seed crops* (off-label) [1, 2]; *spring barley, spring oats, spring wheat, triticale, winter barley, winter oats, winter rye, winter wheat* [1]
- Sowthistle in *ornamental plant production* (off-label) [1, 2]

- Volunteer oilseed rape in **grass seed crops** *(off-label)*, **ornamental plant production** *(off-label)*, **triticale**, **winter rye** [1, 2]; **spring barley**, **spring oats**, **spring wheat**, **winter barley**, **winter oats**, **winter wheat** [1-3]
- Wild radish in **grass seed crops** *(off-label)* [1, 2]; **spring barley**, **spring oats**, **spring wheat**, **triticale**, **winter barley**, **winter oats**, **winter rye**, **winter wheat** [1]

Extension of Authorisation for Minor Use (EAMUs)
- **grass seed crops** *20162960* [1], *20162916* [2]
- **ornamental plant production** *20141582 expires 24 Apr 2018* [1], *20141531 expires 30 Jun 2018* [2]

Approval information
- Florasulam included in Annex I under EC Regulation 1107/2009
- Accepted by BBPA for use on malting barley

Efficacy guidance
- Best results obtained from treatment of small actively growing weeds in good conditions
- Apply in autumn or spring once crop has 3 leaves
- Product is mainly absorbed by leaves of weeds and is effective on all soil types
- Florasulam is a member of the ALS-inhibitor group of herbicides

Restrictions
- Maximum total dose on any crop equivalent to one full dose treatment. The total amount of florasulam applied to a cereal crop must not exceed 7.5 g.a.i/ha [1]
- Do not roll or harrow within 7 d before or after application
- Do not spray when crops under stress from cold, drought, pest damage, nutrient deficiency or any other cause
- Specific restrictions apply to use in sequence or tank mixture with other sulfonylurea or ALS-inhibiting herbicides. See label for details

Crop-specific information
- Latest use: up to and including flag leaf just visible stage for all crops

Following crops guidance
- Where the product has been used in mixture with certain named products (see label) only cereals or grass may be sown in the autumn following harvest
- Unless otherwise restricted cereals, oilseed rape, field beans, grass or vegetable brassicas as transplants may be sown as a following crop in the same calendar yr as treatment. Oilseed rape may show some temporary reduction of vigour after a dry summer, but yields are not affected
- In addition to the above, linseed, peas, sugar beet, potatoes, maize, clover (for use in grass/clover mixtures) or carrots may be sown in the calendar yr following treatment
- In the event of failure of a treated crop in spring only spring wheat, spring barley, spring oats, maize or ryegrass may be sown

Environmental safety
- Dangerous for the environment
- Very toxic to aquatic organisms
- See label for detailed instructions on tank cleaning
- LERAP Category B [1, 2]

Hazard classification and safety precautions
Hazard Dangerous for the environment
Transport code 9
Packaging group III
UN Number 3082
Operator protection U05a
Environmental protection E15a, E34 [3]; E15b, E16a [1, 2]; E38, H410 [1-3]
Storage and disposal D01, D02, D05, D09a [1-3]; D06a, D10c, D12b [1, 2]; D10b, D12a [3]

229   florasulam + fluroxypyr

A post-emergence herbicide mixture for cereals or grass
HRAC mode of action code: B + O

See also fluroxypyr

Products

1	Cabadex	Headland Amenity	2.5:100 g/l	SE	13948
2	Cleave	Adama	2.5:100 g/l	SE	16774
3	Envy	Dow	2.5:100 g/l	SE	17901
4	Nevada	Dow	5:100 g/l	SE	17349
5	Slalom	Dow	2.5:100 g/l	SE	13772
6	Spitfire	Dow	5:100 g/l	SE	15101
7	Starane XL	Dow	2.5:100 g/l	SE	10921
8	Trafalgar	Pan Amenity	2.5:100 g/l	SE	14888

Uses

- Annual and perennial weeds in *amenity grassland, lawns, managed amenity turf* [1]
- Annual dicotyledons in *amenity grassland, lawns, managed amenity turf* [8]; *durum wheat* (off-label), *game cover* (off-label), *ornamental plant production* (off-label), *spring rye* (off-label), *triticale* (off-label), *winter rye* (off-label) [7]; *grass seed crops* (off-label) [6, 7]; *oats, rye* [4, 6]; *spring barley, spring wheat, winter barley, winter wheat* [2, 4-7]; *spring oats, winter oats* [2, 5, 7]; *triticale* [2, 4, 6]; *winter rye* [2]
- Black bindweed in *oats, rye, spring barley, spring wheat, triticale, winter barley, winter wheat* [4, 6]
- Black nightshade in *oats, rye, spring barley, spring wheat, triticale, winter barley, winter wheat* [4, 6]
- Buttercups in *amenity grassland, lawns, managed amenity turf* [1]
- Charlock in *grass seed crops, grassland, newly sown grass leys* [3]; *oats, rye, spring barley, spring wheat, triticale, winter barley, winter wheat* [4, 6]
- Chickweed in *amenity grassland, lawns, managed amenity turf* [8]; *grass seed crops, grassland, newly sown grass leys* [3]; *oats, rye* [4, 6]; *spring barley, spring wheat, winter barley, winter wheat* [2, 4-7]; *spring oats, winter oats* [2, 5, 7]; *triticale* [2, 4, 6]; *winter rye* [2]
- Cleavers in *grass seed crops, grassland, newly sown grass leys* [3]; *grass seed crops* (off-label) [6]; *oats, rye* [4, 6]; *spring barley, spring wheat, winter barley, winter wheat* [2, 4-7]; *spring oats, winter oats* [2, 5, 7]; *triticale* [2, 4, 6]; *winter rye* [2]
- Clover in *amenity grassland, lawns, managed amenity turf* [1]
- Common mouse-ear in *grassland* [3]
- Creeping buttercup in *grassland* [3]
- Daisies in *amenity grassland, lawns, managed amenity turf* [1]; *grassland* [3]
- Dandelions in *amenity grassland, lawns, managed amenity turf* [1]; *grassland* [3]
- Forget-me-not in *grass seed crops, grassland, newly sown grass leys* [3]; *oats, rye, spring barley, spring wheat, triticale, winter barley, winter wheat* [4, 6]
- Groundsel in *grass seed crops* (off-label) [6]
- Knotgrass in *grass seed crops, grassland, newly sown grass leys* [3]; *oats, rye, spring barley, spring wheat, triticale, winter barley, winter wheat* [4, 6]
- Mayweeds in *grass seed crops, grassland, newly sown grass leys* [3]; *grass seed crops* (off-label) [6]; *oats, rye* [4, 6]; *spring barley, spring wheat, winter barley, winter wheat* [2, 4-7]; *spring oats, winter oats* [2, 5, 7]; *triticale* [2, 4, 6]; *winter rye* [2]
- Plantains in *amenity grassland, lawns, managed amenity turf* [1]
- Poppies in *grass seed crops, grassland, newly sown grass leys* [3]; *oats, rye, spring barley, spring wheat, triticale, winter barley, winter wheat* [4, 6]
- Redshank in *oats, rye, spring barley, spring wheat, triticale, winter barley, winter wheat* [4, 6]
- Shepherd's purse in *grass seed crops, grassland, newly sown grass leys* [3]
- Sowthistle in *grass seed crops* (off-label) [6]
- Volunteer beans in *oats, rye, spring barley, spring wheat, triticale, winter barley, winter wheat* [4, 6]

SEE SECTION 3 FOR PRODUCTS ALSO REGISTERED

- Volunteer oilseed rape in **grass seed crops**, **grassland**, **newly sown grass leys** [3]; **oats**, **rye**, **spring barley**, **spring wheat**, **triticale**, **winter barley**, **winter wheat** [4, 6]

Extension of Authorisation for Minor Use (EAMUs)
- **durum wheat** *20072815* [7]
- **game cover** *20082904* [7]
- **grass seed crops** *20132321* [6], *20072815* [7]
- **ornamental plant production** *20082904* [7]
- **spring rye** *20072815* [7]
- **triticale** *20072815* [7]
- **winter rye** *20072815* [7]

Approval information
- Florasulam and fluroxypyr included in Annex I under EC Regulation 1107/2009
- Accepted by BBPA for use on malting barley

Efficacy guidance
- Best results obtained when weeds are small and growing actively
- Products are mainly absorbed through weed foliage. Cleavers emerging after application will not be controlled
- Florasulam is a member of the ALS-inhibitor group of herbicides

Restrictions
- Maximum total dose equivalent to one full dose treatment
- Do not roll or harrow 7 d before or after application
- Do not spray when crops are under stress from cold, drought, pest damage or nutrient deficiency
- Do not apply through CDA applicators
- Specific restrictions apply to use in sequence or tank mixture with other sulfonylurea or ALS-inhibiting herbicides. See label for details
- The total amount of florasulam applied to a cereal crop must not exceed 7.5 g.a.i/ha

Crop-specific information
- Latest use: before flag leaf sheath extended (before GS 41) for spring barley and spring wheat; before flag leaf sheath opening (before GS 47) for winter barley and winter wheat; before second node detectable (before GS 32) for winter oats

Following crops guidance
- Cereals, oilseed rape, field beans or grass may follow treated crops in the same yr. Oilseed rape may suffer temporary vigour reduction after a dry summer
- In addition to the above, linseed, peas, sugar beet, potatoes, maize or clover may be sown in the calendar yr following treatment
- In the event of failure of a treated crop in the spring, only spring cereals, maize or ryegrass may be planted
- May also be used on crops undersown with grass [4]

Environmental safety
- Dangerous for the environment
- Toxic to aquatic organisms
- Take extreme care to avoid drift onto non-target crops or plants
- Aquatic buffer zone of 5m
- LERAP Category B [2-4, 6]

Hazard classification and safety precautions
Hazard Irritant, Dangerous for the environment [1-8]; Flammable, Flammable liquid and vapour [4, 6]
Transport code 3 [4, 6]; 9 [1-3, 5, 7, 8]
Packaging group III
UN Number 1993 [4, 6]; 3082 [1-3, 5, 7, 8]
Risk phrases H304 [3, 4, 6]; H315, H336 [2-6, 8]; H317, H319 [2-6]; H320 [8]; H335 [3-6]; R36, R38, R53a, R67 [1, 7]; R50 [1]; R51 [7]
Operator protection A [1-8]; C [1-7]; H [3, 4, 6]; U05a, U11, U19a [1-8]; U08 [1, 3]; U09a [4, 6]; U14, U20b [1, 3, 4, 6]

FOR FULL CONDITIONS OF USE ALWAYS READ THE PRODUCT LABEL

Environmental protection E07c [3] (7 days); E15a [2, 4-8]; E15b [1]; E16a [2-4, 6]; E34, E38 [1-8]; H410 [2-6]; H411 [8]
Storage and disposal D01, D02, D09a, D10c, D12a [1-8]; D05 [3, 8]
Medical advice M05a [4, 6, 8]

230 florasulam + halauxifen-methyl

A herbicide mixture for control of broad-leaved weeds in cereals
HRAC mode of action code: B + O

See also florasulam
florasulam + fluroxypyr

Products

Zypar	Dow	5:6.25 g/l	OD	17938

Uses

- Annual dicotyledons in **durum wheat, rye, spelt, spring barley, spring wheat, triticale, winter barley, winter wheat**
- Chickweed in **durum wheat, rye, spelt, spring barley, spring wheat, triticale, winter barley, winter wheat**
- Cleavers in **durum wheat, rye, spelt, spring barley, spring wheat, triticale, winter barley, winter wheat**
- Crane's-bill in **durum wheat, rye, spelt, spring barley, spring wheat, triticale, winter barley, winter wheat**
- Fat hen in **durum wheat, rye, spelt, spring barley, spring wheat, triticale, winter barley, winter wheat**
- Fumitory in **durum wheat, rye, spelt, spring barley, spring wheat, triticale, winter barley, winter wheat**
- Groundsel in **durum wheat, rye, spelt, spring barley, spring wheat, triticale, winter barley, winter wheat**
- Mayweeds in **durum wheat, rye, spelt, spring barley, spring wheat, triticale, winter barley, winter wheat**
- Poppies in **durum wheat, rye, spelt, spring barley, spring wheat, triticale, winter barley, winter wheat**
- Volunteer oilseed rape in **durum wheat, rye, spelt, spring barley, spring wheat, triticale, winter barley, winter wheat**

Approval information

- Florasulam and halauxifen-methyl included in Annex I under EC Regulation 1107/2009

Efficacy guidance

- Rainfast 1 hour after application

Restrictions

- To avoid subsequent injury to crops other than cereals (wheat, durum wheat, spelt, barley, rye, triticale), all spraying equipment must be thoroughly cleaned both inside and out using All Clear Extra spray cleaner at 0.5 % v/v or bleach containing 5 % hypochlorite.

Following crops guidance

- There are no restrictions for sowing any succeeding crop after the cereal harvest but for sensitive species such as clover, lentils or sunflower ploughing is recommended prior to drilling.
- Where crop failure after an autumn application occurs, sowing the following spring crops is possible: 1 month after application (no cultivation restrictions): spring wheat, spring barley, maize, ryegrass; 3 months after application (after ploughing): spring oilseed rape, field beans, peas, sunflower
- Where crop failure after a spring application occurs, it is possible to sow spring wheat and spring barley 1 month after application with no need to cultivate while maize can be sown 2 months after application after ploughing.

SEE SECTION 3 FOR PRODUCTS ALSO REGISTERED

Environmental safety
- LERAP Category B

Hazard classification and safety precautions
    Transport code 9
    Packaging group III
    UN Number 3082
    Risk phrases H315, H317, H319
    Operator protection A, C, H; U05a, U08, U19a, U20c
    Environmental protection E15b, E16a, E34, H410
    Storage and disposal D09a, D10c

231   florasulam + pyroxsulam

A mixture of two triazolopirimidine sulfonamides for winter wheat
HRAC mode of action code: B

See also pyroxsulam

Products

1	Broadway Star	Dow	1.4:7.1% w/w	WG	14319
2	Palio	Dow	1.4:7.1% w/w	WG	17345

Uses
- Charlock in *triticale, winter rye, winter wheat*
- Chickweed in *triticale, winter rye, winter wheat*
- Cleavers in *triticale, winter rye, winter wheat*
- Field pansy in *triticale, winter rye, winter wheat*
- Field speedwell in *triticale, winter rye, winter wheat*
- Geranium species in *triticale, winter rye, winter wheat*
- Ivy-leaved speedwell in *triticale, winter rye, winter wheat*
- Mayweeds in *triticale, winter rye, winter wheat*
- Poppies in *triticale, winter rye, winter wheat*
- Ryegrass in *triticale, winter rye, winter wheat*
- Sterile brome in *triticale, winter rye, winter wheat*
- Volunteer oilseed rape in *triticale, winter rye, winter wheat*
- Wild oats in *triticale, winter rye, winter wheat*

Approval information
- Florasulam and pyroxsulam included in Annex I under EC Regulation 1107/2009.

Efficacy guidance
- Rainfast within 1 hour of application
- Requires an authorised adjuvant at application and recommended for use in a programme with herbicides employing a different mode of action.

Restrictions
- For autumn planted crops a maximum total dose of 3.75 g.a.i/ha of florasulam must be observed for applications made between crop emergence in the year of planting and February 1st in the year of harvest.
- The total amount of florasulam applied to a cereal crop must not exceed 7.5 g.a.i/ha

Crop-specific information
- Crop injury may occur if applied in tank mixture with plant growth regulators - allow a minimum interval of 7 days.
- Crop injury may occur if applied in tank mixture with OP insecticides, MCPB or dicamba - allow a minimum interval of 14 days

Following crops guidance
- Crop failure before 1st Feb - plough and allow 6 weeks to elapse and then drill spring wheat, spring barley, grass or maize.
- Crop failure after 1st Feb - plough and allow 6 weeks to elapse before drilling grass or maize.

FOR FULL CONDITIONS OF USE ALWAYS READ THE PRODUCT LABEL

Environmental safety
- Take extreme care to avoid drift on to susceptible crops, non-target plants or waterways
- LERAP Category B

Hazard classification and safety precautions
Hazard Dangerous for the environment
Transport code 9
Packaging group III
UN Number 3077
Operator protection U05a, U14, U15, U20a
Environmental protection E15b, E16a, E34, E38, E39, H410
Storage and disposal D01, D02, D09a, D12a

232   fluazifop-P-butyl

A phenoxypropionic acid grass herbicide for broadleaved crops
HRAC mode of action code: A

Products

1	Clayton Crowe	Clayton	125 g/l	EC	12572
2	Clayton Maximus	Clayton	125 g/l	EC	12543
3	Fusilade Max	Syngenta	125 g/l	EC	11519
4	Salvo	Generica	125 g/l	EC	16228

Uses
- Annual grasses in *almonds* (off-label), *apples* (off-label), *apricots* (off-label), *asparagus* (off-label), *bilberries* (off-label), *blackberries* (off-label), *blueberries* (off-label), *borage* (off-label), *broad beans (dry-harvested)* (off-label), *cabbages* (off-label), *canary flower (echium spp.)* (off-label), *cherries* (off-label), *chicory* (off-label), *choi sum* (off-label), *cob nuts* (off-label), *collards* (off-label), *cranberries* (off-label), *evening primrose* (off-label), *figs* (off-label), *filberts* (off-label), *fodder beet* (off-label), *garlic* (off-label), *globe artichoke* (off-label), *grapevines* (off-label), *haricot beans* (off-label), *hazel nuts* (off-label), *herbs (see appendix 6)* (off-label), *honesty* (off-label), *hybridberry* (off-label), *linseed* (off-label), *lucerne* (off-label), *lupins* (off-label), *mallow (althaea spp.)* (off-label), *mustard* (off-label), *navy beans* (off-label), *non-edible flowers* (off-label), *ornamental plant production* (off-label), *pak choi* (off-label), *parsley root* (off-label), *parsnips* (off-label), *peaches* (off-label), *pears* (off-label), *plums* (off-label), *protected non-edible flowers* (off-label), *protected ornamentals* (off-label), *quinces* (off-label), *red beet* (off-label), *redcurrants* (off-label), *shallots* (off-label), *spring field beans* (off-label), *spring greens* (off-label), *swedes* (off-label), *tic beans* (off-label), *turnips* (off-label), *walnuts* (off-label), *whitecurrants* (off-label), *winter field beans* (off-label) [3]; *blackcurrants, bulb onions, carrots, combining peas, farm forestry, fodder beet, gooseberries, hops, linseed, raspberries, spring field beans, spring oilseed rape, spring oilseed rape for industrial use, strawberries, sugar beet, swedes* (stockfeed only), *turnips* (stockfeed only), *vining peas, winter field beans, winter oilseed rape, winter oilseed rape for industrial use* [1-4]; *field margins, flax, flax for industrial use, linseed for industrial use* [1, 3, 4]; *kale* (stockfeed only) [2]
- Barren brome in *field margins* [1, 3, 4]
- Blackgrass in *hops, spring field beans, winter field beans* [1-4]
- Green cover in *land temporarily removed from production* [1-4]
- Perennial grasses in *almonds* (off-label), *apples* (off-label), *apricots* (off-label), *asparagus* (off-label), *bilberries* (off-label), *blackberries* (off-label), *blueberries* (off-label), *borage* (off-label), *broad beans (dry-harvested)* (off-label), *cabbages* (off-label), *canary flower (echium spp.)* (off-label), *cherries* (off-label), *chicory* (off-label), *choi sum* (off-label), *cob nuts* (off-label), *collards* (off-label), *cranberries* (off-label), *evening primrose* (off-label), *figs* (off-label), *filberts* (off-label), *fodder beet* (off-label), *garlic* (off-label), *globe artichoke* (off-label), *grapevines* (off-label), *haricot beans* (off-label), *hazel nuts* (off-label), *herbs (see appendix 6)* (off-label), *honesty* (off-label), *hybridberry* (off-label), *linseed* (off-label), *lucerne* (off-label), *lupins* (off-label), *mallow (althaea spp.)* (off-label), *mustard* (off-label), *navy beans* (off-label), *non-edible flowers* (off-label), *ornamental plant production* (off-label), *pak choi* (off-label), *parsley root*

(off-label), **parsnips** *(off-label)*, **peaches** *(off-label)*, **pears** *(off-label)*, **plums** *(off-label)*, **protected non-edible flowers** *(off-label)*, **protected ornamentals** *(off-label)*, **quinces** *(off-label)*, **red beet** *(off-label)*, **redcurrants** *(off-label)*, **shallots** *(off-label)*, **spring field beans** *(off-label)*, **spring greens** *(off-label)*, **swedes** *(off-label)*, **tic beans** *(off-label)*, **turnips** *(off-label)*, **walnuts** *(off-label)*, **whitecurrants** *(off-label)*, **winter field beans** *(off-label)* [3]; **blackcurrants, bulb onions, carrots, combining peas, farm forestry, fodder beet, gooseberries, hops, linseed, raspberries, spring field beans, spring oilseed rape, spring oilseed rape for industrial use, strawberries, sugar beet, swedes** *(stockfeed only)*, **turnips** *(stockfeed only)*, **vining peas, winter field beans, winter oilseed rape, winter oilseed rape for industrial use** [1-4]; **flax, flax for industrial use, linseed for industrial use** [1, 3, 4]; **kale** *(stockfeed only)* [2]

- Volunteer cereals in **blackcurrants, bulb onions, carrots, combining peas, farm forestry, fodder beet, gooseberries, hops, linseed, raspberries, spring field beans, spring oilseed rape, spring oilseed rape for industrial use, strawberries, sugar beet, swedes** *(stockfeed only)*, **turnips** *(stockfeed only)*, **vining peas, winter field beans, winter oilseed rape, winter oilseed rape for industrial use** [1-4]; **field margins, flax, flax for industrial use, linseed for industrial use** [1, 3, 4]; **kale** *(stockfeed only)* [2]

- Wild oats in **blackcurrants, bulb onions, carrots, combining peas, farm forestry, fodder beet, gooseberries, hops, linseed, raspberries, spring field beans, spring oilseed rape, spring oilseed rape for industrial use, strawberries, sugar beet, swedes** *(stockfeed only)*, **turnips** *(stockfeed only)*, **vining peas, winter field beans, winter oilseed rape, winter oilseed rape for industrial use** [1-4]; **field margins, flax, flax for industrial use, linseed for industrial use** [1, 3, 4]; **kale** *(stockfeed only)* [2]

Extension of Authorisation for Minor Use (EAMUs)

- **almonds** *20121261* [3]
- **apples** *20121261* [3]
- **apricots** *20121261* [3]
- **asparagus** *20121323* [3]
- **bilberries** *20121258* [3]
- **blackberries** *20121315* [3]
- **blueberries** *20121258* [3]
- **borage** *20121320* [3]
- **broad beans (dry-harvested)** *20121321* [3]
- **cabbages** *20121317* [3]
- **canary flower (echium spp.)** *20121320* [3]
- **cherries** *20121261* [3]
- **chicory** *20121322* [3]
- **choi sum** *20121321* [3]
- **cob nuts** *20121261* [3]
- **collards** *20121321* [3]
- **cranberries** *20121258* [3]
- **evening primrose** *20121320* [3]
- **figs** *20121261* [3]
- **filberts** *20121261* [3]
- **fodder beet** *20121256* [3]
- **garlic** *20121321* [3]
- **globe artichoke** *20121260* [3]
- **grapevines** *20121316* [3]
- **haricot beans** *20121321* [3]
- **hazel nuts** *20121261* [3]
- **herbs (see appendix 6)** *20121319* [3]
- **honesty** *20121320* [3]
- **hybridberry** *20121315* [3]
- **linseed** *20121320* [3]
- **lucerne** *20121318* [3]
- **lupins** *20121255* [3]
- **mallow (althaea spp.)** *20121259* [3]
- **mustard** *20121320* [3]
- **navy beans** *20121321* [3]

FOR FULL CONDITIONS OF USE ALWAYS READ THE PRODUCT LABEL

- **non-edible flowers** *20121321* [3]
- **ornamental plant production** *20121321* [3]
- **pak choi** *20121321* [3]
- **parsley root** *20121259* [3]
- **parsnips** *20121321* [3]
- **peaches** *20121261* [3]
- **pears** *20121261* [3]
- **plums** *20121261* [3]
- **protected non-edible flowers** *20121321* [3]
- **protected ornamentals** *20121321* [3]
- **quinces** *20121261* [3]
- **red beet** *20121321* [3]
- **redcurrants** *20121258* [3]
- **shallots** *20121257* [3]
- **spring field beans** *20121321* [3]
- **spring greens** *20121321* [3]
- **swedes** *20121321* [3]
- **tic beans** *20121321* [3]
- **turnips** *20121321* [3]
- **walnuts** *20121261* [3]
- **whitecurrants** *20121258* [3]
- **winter field beans** *20121321* [3]

Approval information
- Fluazifop-P-butyl included in Annex I under EC Regulation 1107/2009
- Accepted by BBPA for use on hops

Efficacy guidance
- Best results achieved by application when weed growth active under warm conditions with adequate soil moisture.
- Spray weeds from 2-expanded leaf stage to fully tillered, couch from 4 leaves when majority of shoots have emerged, with a second application if necessary
- Control may be reduced under dry conditions. Do not cultivate for 2 wk after spraying couch
- Annual meadow grass is not controlled
- May also be used to remove grass cover crops
- Fluazifop-P-butyl is an ACCase inhibitor herbicide. To avoid the build up of resistance do not apply products containing an ACCase inhibitor herbicide more than twice to any crop. In addition do not use any product containing fluazifop-P-butyl in mixture or sequence with any other product containing the same ingredient
- Use these products as part of a resistance management strategy that includes cultural methods of control and does not use ACCase inhibitors as the sole chemical method of grass weed control

Restrictions
- Maximum number of treatments 1 per crop or yr for all crops
- Do not sow cereals or grass crops for at least 8 wk after application of high rate or 2 wk after low rate
- Do not apply through CDA sprayer, with hand-held equipment or from air
- Avoid treatment before spring growth has hardened or when buds opening
- Do not treat bush and cane fruit or hops between flowering and harvest
- Consult processors before treating crops intended for processing
- Oilseed rape, linseed and flax for industrial use must not be harvested for human or animal consumption nor grazed
- Do not use for forestry establishment on land not previously under arable cultivation or improved grassland
- Treated vegetation in field margins, land temporarily removed from production etc, must not be grazed or harvested for human or animal consumption and unprotected persons must be kept out of treated areas for at least 24 h

SEE SECTION 3 FOR PRODUCTS ALSO REGISTERED

Crop-specific information

- Latest use: before 50% ground cover for swedes, turnips; before 5 leaf stage for spring oilseed rape; before flowering for blackcurrants, gooseberries, hops, raspberries, strawberries; before flower buds visible for field beans, peas, linseed, flax, winter oilseed rape; 2 wk before sowing cereals or grass for field margins, land temporarily removed from production
- HI beet crops, kale, carrots 8 wk; onions 4 wk; oilseed rape for industrial use 2 wk
- Apply to sugar and fodder beet from 1-true leaf to 50% ground cover
- Apply to winter oilseed rape from 1-true leaf to established plant stage
- Apply to spring oilseed rape from 1-true leaf but before 5-true leaves
- Apply in fruit crops after harvest. See label for timing details on other crops
- Before using on onions or peas use crystal violet test to check that leaf wax is sufficient

Environmental safety

- Dangerous for the environment
- Very toxic to aquatic organisms

Hazard classification and safety precautions

Hazard Harmful, Dangerous for the environment [1-4]; Very toxic to aquatic organisms [3]
Transport code 9
Packaging group III
UN Number 3082
Risk phrases H315 [2]; H361 [2, 3]; R38, R50, R53a, R63 [1, 4]
Operator protection A, C, H, M; U05a, U08, U20b
Environmental protection E15a [1-4]; E38 [1, 3, 4]; H410 [2, 3]
Storage and disposal D01, D02, D05, D09a, D12a [1-4]; D10b [2]; D10c [1, 3, 4]
Medical advice M05b [2]

233   fluazinam

A dinitroaniline fungicide for use in potatoes
FRAC mode of action code: 29

See also azoxystrobin + fluazinam
cymoxanil + fluazinam
dimethomorph + fluazinam

Products

1 Fluazinova	Barclay	500 g/l	SC	17625
2 Gando	Harvest	500 g/l	SC	18001
3 Nando 500SC	Nufarm UK	500 g/l	SC	16388
4 Shirlan	Syngenta	500 g/l	SC	16624
5 Tizca	Headland	500 g/l	SC	16289
6 Volley	Adama	500 g/l	SC	16451

Uses

- Blight in **potatoes** [1-6]
- Late blight in **potatoes** [1]
- Powdery scab in **seed potatoes** *(off-label)* [5]

Extension of Authorisation for Minor Use (EAMUs)

- **seed potatoes** *20170902* [5]

Approval information

- Fluazinam included in Annex I under EC Regulation 1107/2009

Efficacy guidance

- Commence treatment at the first blight risk warning (before blight enters the crop). Products are rainfast within 1 h
- In the absence of a warning, treatment should start before foliage of adjacent plants meets in the rows
- Spray at 5-14 d intervals depending on severity of risk (see label)

- Ensure complete coverage of the foliage and stems, increasing volume as haulm growth progresses, in dense crops and if blight risk increases

Restrictions
- Do not use with hand-held sprayers

Crop-specific information
- HI 0 - 10 d for potatoes. Check label
- Ensure complete kill of potato haulm before lifting and do not lift crops for storage while there is any green tissue left on the leaves or stem bases

Environmental safety
- Dangerous for the environment
- Very toxic to aquatic organisms
- Buffer zone requirement 6m [1, 6]
- Buffer zone requirement 7m [4, 5]
- Buffer zone requirement 8m [2, 3]
- LERAP Category B

Hazard classification and safety precautions
Hazard Harmful [2, 3]; Irritant [1, 4, 5]; Dangerous for the environment [1-6]; Very toxic to aquatic organisms [4, 6]
Transport code 9
Packaging group III
UN Number 3082
Risk phrases H315 [2, 3]; H317 [1-5]; H318 [5]; H361 [1-4, 6]; R43 [1]
Operator protection A, C, H [1, 4-6]; U02a, U04a, U08 [1-4, 6]; U05a, U14 [1-5]; U11 [5]; U15, U20a [1-4]; U20b [5, 6]
Environmental protection E15a, E16a, E38, H410 [1-6]; E16b, E34 [1-4, 6]
Storage and disposal D01, D02, D05, D10c [1-5]; D09a, D12a [1-6]; D10a [6]
Medical advice M03 [1-4, 6]; M05a [1-4]

234   fludioxonil

A phenylpyrrole fungicide seed treatment for wheat and barley
FRAC mode of action code: 12

See also chlorothalonil + fludioxonil + propiconazole
cymoxanil + fludioxonil + metalaxyl-M
cyprodinil + fludioxonil
difenoconazole + fludioxonil
difenoconazole + fludioxonil + tebuconazole

Products
1	Beret Gold	Syngenta	25 g/l	FS	16430
2	Emblem	Pure Amenity	125 g/l	SC	18066
3	Geoxe	Syngenta	50% w/w	WG	16596
4	Maxim 100FS	Syngenta	100 g/l	FS	15683
5	Medallion TL	Syngenta	125 g/l	SC	15287

Uses
- Alternaria in **apples** (reduction), **crab apples** (reduction), **pears** (reduction), **quinces** (reduction) [3]
- Anthracnose in **amenity grassland** (reduction), **managed amenity turf** (reduction) [2, 5]
- Black dot in **potatoes** (some reduction) [4]
- Black scurf in **potatoes** [4]
- Botrytis in **apples** (reduction), **crab apples** (reduction), **pears** (reduction), **quinces** (reduction) [3]
- Bunt in **spring wheat** (seed treatment), **winter wheat** (seed treatment) [1]
- Covered smut in **spring barley** (seed treatment), **winter barley** (seed treatment) [1]
- Disease control in **apples**, **crab apples**, **pears**, **quinces** [3]

- Drechslera leaf spot in **amenity grassland** *(useful levels of control)*, **managed amenity turf** *(useful levels of control)* [2, 5]
- Fusarium foot rot and seedling blight in **rye** *(seed treatment)*, **spring barley** *(seed treatment)*, **spring oats** *(seed treatment)*, **spring wheat** *(seed treatment)*, **triticale** *(seed treatment)*, **winter barley** *(seed treatment)*, **winter oats** *(seed treatment)*, **winter wheat** *(seed treatment)* [1]
- Fusarium patch in **amenity grassland**, **managed amenity turf** [2, 5]
- Leaf stripe in **spring barley** *(seed treatment - reduction)*, **winter barley** *(seed treatment - reduction)* [1]
- Microdochium nivale in **amenity grassland**, **managed amenity turf** [2, 5]; **rye** *(seed treatment)*, **triticale** [1]
- Monilinia spp. in **apples** *(reduction)*, **crab apples** *(reduction)*, **pears** *(reduction)*, **quinces** *(reduction)* [3]
- Nectria spp in **apples** *(reduction)*, **crab apples** *(reduction)*, **pears** *(reduction)*, **quinces** *(reduction)* [3]
- Penicillium rot in **apples** *(reduction)*, **crab apples** *(reduction)*, **pears** *(reduction)*, **quinces** *(reduction)* [3]
- Phlyctema vagabunda in **apples** *(reduction)*, **crab apples** *(reduction)*, **pears** *(reduction)*, **quinces** *(reduction)* [3]
- Pyrenophora leaf spot in **spring oats** *(seed treatment)*, **winter oats** *(seed treatment)* [1]
- Scab in **seed potatoes** *(off-label)* [4]
- Septoria seedling blight in **spring wheat** *(seed treatment)*, **winter wheat** *(seed treatment)* [1]
- Silver scurf in **potatoes** *(reduction)* [4]
- Snow mould in **spring barley** *(seed treatment)*, **spring oats** *(seed treatment)*, **spring wheat** *(seed treatment)*, **winter barley** *(seed treatment)*, **winter oats** *(seed treatment)*, **winter wheat** *(seed treatment)* [1]
- Stripe smut in **rye** [1]

Extension of Authorisation for Minor Use (EAMUs)
- **seed potatoes** *20160736* [4]

Approval information
- Fludioxonil included in Annex I under EC Regulation 1107/2009
- Accepted by BBPA for use on malting barley

Efficacy guidance
- Apply direct to seed using conventional seed treatment equipment. Continuous flow treaters should be calibrated using product before use
- Effective against benzimidazole-resistant strains of *Microdochium nivale*

Restrictions
- Maximum number of treatments 1 per seed batch
- Do not apply to cracked, split or sprouted seed
- Sow treated seed within 6 mth

Crop-specific information
- Latest use: before drilling
- Product may reduce flow rate of seed through drill. Recalibrate with treated seed before drilling

Environmental safety
- Dangerous for the environment
- Toxic to aquatic organisms
- Do not use treated seed as food or feed
- Treated seed harmful to game and wildlife
- LERAP Category B [2, 3, 5]

Hazard classification and safety precautions
Hazard Irritant, Very toxic to aquatic organisms [3]; Dangerous for the environment [1-5]
Transport code 9
Packaging group III
UN Number 3077 [3]; 3082 [1, 2, 4, 5]
Risk phrases H317 [3]

FOR FULL CONDITIONS OF USE ALWAYS READ THE PRODUCT LABEL

SECTION 2

Operator protection A [1, 3, 4]; H [1, 3]; U05a [1-5]; U14, U19a, U20a [3]; U20b [1, 2, 4, 5]
Environmental protection E03, E15a [1]; E15b [2-5]; E16a [2, 3, 5]; E16b [2, 5]; E34 [2, 4, 5]; E38 [1-5]; H410 [3]; H411 [1, 2, 4, 5]
Storage and disposal D01, D02, D09a, D12a [1-5]; D05, D11a [1, 4]; D10c [2-5]
Treated seed S01, S03 [4]; S02, S05 [1, 4]; S07 [1]

235   fludioxonil + metalaxyl-M

A seed dressing for use in forage maize
FRAC mode of action code: 12 + 4

Products

Maxim XL	Syngenta	25:9.69 g/l	FS	16599

Uses
- Damping off in *forage maize*
- Fusarium in *forage maize*
- Pythium in *forage maize*

Approval information
- Fludioxonil and metalaxyl-M included in Annex I under EC Regulation 1107/2009

Restrictions
- For advice on resistance management, refer to the latest Fungicide Resistance Action Group (FRAG) guidelines

Hazard classification and safety precautions
UN Number N/C
Operator protection A, H; U05a
Environmental protection E34, E38, H412
Storage and disposal D01, D02, D05, D09a, D10c, D11a, D12a
Treated seed S01, S02, S03, S04a, S05, S07

236   fludioxonil + sedaxane

A fungicide seed treatment for cereals
FRAC mode of action code: 12 + 7

See also sedaxane

Products

Vibrance Duo	Syngenta	25:25 g/l	FS	17838

Uses
- Bunt in *winter wheat*
- Fusarium diseases in *winter wheat*
- Fusarium ear blight in *winter wheat* *(moderate control)*
- Loose smut in *spring oats*, *winter wheat*
- Septoria leaf blotch in *winter wheat*
- Snow mould in *rye*, *triticale*, *winter wheat*
- Stripe smut in *rye*

Approval information
- Fuldioxonil and sedaxane included in Annex 1 under EC Regulation 1107/2009

Hazard classification and safety precautions
Hazard Harmful if inhaled
Transport code 9
Packaging group III
UN Number 3082
Risk phrases H317

SEE SECTION 3 FOR PRODUCTS ALSO REGISTERED

Operator protection A, H; U05a, U19a
Environmental protection E15b, E34, H410
Storage and disposal D01, D02, D05, D09a, D10c, D11a, D12a
Medical advice M03

237   fludioxonil + tebuconazole

A seed treatment for use in cereals
FRAC mode of action code: 3 + 12

See also fludioxonil
tebuconazole

Products

Fountain	Adama	50:10 g/l	FS	17708

Uses
• Seed-borne diseases in *triticale, winter barley, winter oats, winter rye, winter wheat*

Approval information
• Fludioxonil and tebuconazole included in Annex I under EC Regulation 1107/2009

Efficacy guidance
• Do not broadcast treated seed. Ensure that it is covered by at least 40mm settled soil.

Hazard classification and safety precautions
Transport code 9
Packaging group III
UN Number 3082
Operator protection A, H
Environmental protection H410

238   fludioxonil + tefluthrin

A fungicide and insecticide seed treatment mixture for cereals
FRAC mode of action code: 12 + IRAC 3

See also tefluthrin

Products

Austral Plus	Syngenta	10:40 g/l	FS	13314

Uses
• Bunt in *spring wheat, winter wheat*
• Covered smut in *spring barley, winter barley*
• Fusarium foot rot and seedling blight in *spring barley, spring oats, spring wheat, winter barley, winter oats, winter wheat*
• Leaf stripe in *spring barley* (partial control), *winter barley* (partial control)
• Pyrenophora leaf spot in *spring oats, winter oats*
• Seed-borne diseases in *triticale seed crop* (off-label)
• Septoria seedling blight in *spring wheat, winter wheat*
• Snow mould in *spring barley, spring wheat, winter barley, winter wheat*
• Wheat bulb fly in *spring barley, spring wheat, winter barley, winter wheat*
• Wireworm in *spring barley, spring oats, spring wheat, winter barley, winter oats, winter wheat*

Extension of Authorisation for Minor Use (EAMUs)
• *triticale seed crop* 20080982

Approval information
• Fludioxonil and tefluthrin included in Annex 1 under EC Regulation 1107/2009
• Accepted by BBPA for use on malting barley

FOR FULL CONDITIONS OF USE ALWAYS READ THE PRODUCT LABEL

Efficacy guidance
- Apply direct to seed using conventional seed treatment equipment. Continuous flow treaters should be calibrated using product before use
- Best results obtained from seed drilled into a firm even seedbed
- Tefluthrin is released into soil after drilling and repels or kills larvae of wheat bulb fly and wireworm attacking below ground. Pest control may be reduced by deep or shallow drilling
- Where egg counts of wheat bulb fly or population counts of wireworm indicate a high risk of severe attack follow up spray treatments may be needed
- Control of leaf stripe in barley may not be sufficient in crops grown for seed certification
- Effective against benzimidazole-resistant strains of *Microdochium nivale*
- Under adverse soil or environmental conditions seed rates should be increased to compensate for possible reduced germination capacity

Restrictions
- Maximum number of treatments 1 per batch
- Store treated seed in cool dry conditions and drill within 3 mth
- Do not use on seed above 16% moisture content or on sprouted, cracked or damaged seed

Crop-specific information
- Latest use: before drilling

Environmental safety
- Dangerous for the environment
- Very toxic to aquatic organisms

Hazard classification and safety precautions
Hazard Dangerous for the environment, Very toxic to aquatic organisms
Transport code 9
Packaging group III
UN Number 3082
Operator protection A, C, D, H; U02a, U04a, U05a, U07, U08, U14, U20b
Environmental protection E03, E15b, E34, E38, H410
Storage and disposal D01, D02, D05, D09b, D10c, D11a, D12a
Treated seed S02, S04b, S05, S07, S08

239 flufenacet

A broad spectrum oxyacetamide herbicide for weed control in winter cereals
HRAC mode of action code: K3

See also diflufenican + flufenacet
diflufenican + flufenacet + flurtamone
flufenacet + isoxaflutole
flufenacet + metribuzin
flufenacet + pendimethalin

Products

1	Fence	Albaugh UK	480 g/l	SL	17393
2	Gorgon	Sipcam	480 g/l	SL	17685
3	Macho	Albaugh UK	480 g/l	SL	18122
4	Sunfire	Certis	500 g/l	SL	16745
5	System 50	Certis	500 g/l	SL	16612

Uses
- Annual dicotyledons in ***ornamental plant production*** *(off-label)* [4]; ***winter barley, winter wheat*** [1-5]
- Annual meadow grass in ***ornamental plant production*** *(off-label)*, ***protected forest nurseries*** *(off-label)* [4]; ***rye*** *(off-label)*, ***triticale*** *(off-label)* [4, 5]
- Blackgrass in ***ornamental plant production*** *(off-label)* [4]; ***rye*** *(off-label)*, ***triticale*** *(off-label)* [4, 5]; ***winter barley, winter wheat*** [1-5]
- Chickweed in ***protected forest nurseries*** *(off-label)* [4]

SEE SECTION 3 FOR PRODUCTS ALSO REGISTERED

- Field pansy in **protected forest nurseries** *(off-label)* [4]
- Penny cress in **protected forest nurseries** *(off-label)* [4]
- Scented mayweed in **protected forest nurseries** *(off-label)* [4]
- Shepherd's purse in **protected forest nurseries** *(off-label)* [4]
- Sowthistle in **protected forest nurseries** *(off-label)* [4]
- Speedwells in **protected forest nurseries** *(off-label)* [4]

Extension of Authorisation for Minor Use (EAMUs)
- *ornamental plant production* *20171065* [4]
- *protected forest nurseries* *20170951* [4]
- *rye* *20170952* [4], *20162019* [5]
- *triticale* *20170952* [4], *20162019* [5]

Approval information
- Flufenacet included in Annex I under EC Regulation 1107/2009

Restrictions
- Do not apply by hand-held equipment [1-3]

Environmental safety
- LERAP Category B

Hazard classification and safety precautions
Hazard Harmful, Dangerous for the environment, Harmful if swallowed [1-5]; Very toxic to aquatic organisms [4, 5]
Transport code 9
Packaging group III
UN Number 3082
Risk phrases H317 [4, 5]; H373 [1-5]
Operator protection A, H
Environmental protection E16a, H410
Storage and disposal D01, D02 [1-3]

240   flufenacet + isoxaflutole

A residual herbicide mixture for maize
HRAC mode of action code: K3 + F2

See also isoxaflutole

Products
Amethyst	AgChem Access	48:10% w/w	WG	14079

Uses
- Annual dicotyledons in **forage maize**, **grain maize**
- Grass weeds in **forage maize**, **grain maize**

Approval information
- Flufenacet and isoxaflutole included in Annex I under EC Regulation 1107/2009

Efficacy guidance
- Best results obtained from applications to a fine, firm seedbed in the presence of some soil moisture
- Efficacy may be reduced on cloddy seedbeds or under prolonged dry conditions
- Established perennial grasses and broad-leaved weeds growing from rootstocks will not be controlled
- Always follow WRAG guidelines for preventing and managing herbicide resistant weeds. See Section 5 for more information

Restrictions
- Maximum number of treatments 1 per crop
- Do not use on soils both above 70% sand and less than 2% organic matter

FOR FULL CONDITIONS OF USE ALWAYS READ THE PRODUCT LABEL

- Do not use on maize crops intended for seed production

Crop-specific information
- Latest use: before crop emergence
- Ideally treatment should be made within 4 d of sowing, before the maize seeds have germinated

Following crops guidance
- In the event of failure of a treated crop maize, sweet corn or potatoes may be grown after surface cultivation or ploughing
- After normal harvest of a treated crop oats or barley may be sown after ploughing and wheat, rye, triticale, sugar beet, field beans, peas, soya, sorghum or sunflowers may be sown after surface cultivation or ploughing.

Environmental safety
- Dangerous for the environment
- Very toxic to aquatic organisms
- Take care to avoid drift over other crops. Beet crops and sunflowers are particularly sensitive
- LERAP Category B

Hazard classification and safety precautions
Hazard Harmful, Dangerous for the environment
Transport code 9
Packaging group III
UN Number 3077
Risk phrases R22a, R36, R43, R48, R50, R53a, R63
Operator protection A, C, H; U05a, U11, U20b
Environmental protection E15b, E16a, E34, E38
Storage and disposal D01, D02, D05, D09a, D10a, D12a
Medical advice M03

241   flufenacet + metribuzin

A herbicide mixture for potatoes
HRAC mode of action code: K3 + C1

See also metribuzin

Products
Artist	Bayer CropScience	24:17.5% w/w	WP	17049

Uses
- Annual dicotyledons in **bilberries** *(off-label)*, **blackcurrants** *(off-label)*, **cranberries** *(off-label)*, **early potatoes**, **gooseberries** *(off-label)*, **maincrop potatoes**, **redcurrants** *(off-label)*, **ribes species** *(off-label)*, **soya beans** *(off-label)*
- Annual grasses in **bilberries** *(off-label)*, **blackcurrants** *(off-label)*, **cranberries** *(off-label)*, **gooseberries** *(off-label)*, **redcurrants** *(off-label)*, **ribes species** *(off-label)*, **soya beans** *(off-label)*
- Annual meadow grass in **early potatoes**, **maincrop potatoes**
- Black bindweed in **soya beans** *(off-label)*

Extension of Authorisation for Minor Use (EAMUs)
- *bilberries* 20152968
- *blackcurrants* 20152968
- *cranberries* 20152968
- *gooseberries* 20152968
- *redcurrants* 20152968
- *ribes species* 20152968
- *soya beans* 20171098

Approval information
- Flufenacet and metribuzin included in Annex I under EC Regulation 1107/2009

SEE SECTION 3 FOR PRODUCTS ALSO REGISTERED

Efficacy guidance
- Product acts through root uptake and needs sufficient soil moisture at and shortly after application
- Effectiveness is reduced under dry soil conditions
- Residual activity is reduced on mineral soils with a high organic matter content and on peaty or organic soils
- Ensure application is made evenly to both sides of potato ridges
- Perennial weeds are not controlled

Restrictions
- Maximum total dose equivalent to one full dose treatment
- Potatoes must be sprayed before emergence of crop and weeds
- See label for list of tolerant varieties. Do not treat Maris Piper grown on Sands or Very Light soils
- Do not use on Sands
- On stony or gravelly soils there is risk of crop damage especially if heavy rain falls soon after application

Crop-specific information
- Latest use: before potato crop emergence
- Consult processor before use on crops for processing

Following crops guidance
- Before drilling or planting any succeeding crop soil must be mouldboard ploughed to at least 15 cm as soon as possible after lifting and no later than end Dec
- In W Cornwall on soils with more than 5% organic matter treated early potatoes may be followed by summer planted brassica crops 14 wk after treatment and after mouldboard ploughing. Elsewhere cereals or winter beans may be grown in the same year if at least 16 wk have elapsed since treatment
- In the yr following treatment any crop may be grown except lettuce or radish, or vegetable brassica crops on silt soils in Lincs

Environmental safety
- Dangerous for the environment
- Very toxic to aquatic organisms
- Take care to avoid spray drift onto neighbouring crops, especially lettuce or brassicas
- LERAP Category B

Hazard classification and safety precautions
Hazard Harmful, Dangerous for the environment, Harmful if swallowed
Transport code 9
Packaging group III
UN Number 3077
Risk phrases H317, H373
Operator protection A, C, D, H; U05a, U08, U13, U14, U19a, U20b
Environmental protection E15a, E16a, E38, H410
Storage and disposal D01, D02, D09a, D11a, D12a
Medical advice M03

242   flufenacet + pendimethalin

A broad spectrum residual and contact herbicide mixture for winter cereals
HRAC mode of action code: K3 + K1

See also pendimethalin

Products

1 Clayton Glacier	Clayton	60:300 g/l	EC	15920
2 Confluence	Pan Agriculture	60:300 g/l	EC	17199
3 Crystal	BASF	60:300 g/l	EC	13914
4 Frozen	Harvest	60:300 g/l	EC	17691
5 Shooter	BASF	60:300 g/l	EC	14106

FOR FULL CONDITIONS OF USE ALWAYS READ THE PRODUCT LABEL

Products – continued

6	Trooper	BASF	60:300 g/l	EC	13924

Uses

- Annual dicotyledons in **game cover** *(off-label)*, **spring barley** *(off-label)* [3]; **winter barley**, **winter wheat** [1-6]
- Annual grasses in **winter barley**, **winter wheat** [1-6]
- Annual meadow grass in **game cover** *(off-label)*, **spring barley** *(off-label)* [3]; **winter barley**, **winter wheat** [1-6]
- Blackgrass in **game cover** *(off-label)*, **spring barley** *(off-label)* [3]; **winter barley**, **winter wheat** [1-6]
- Chickweed in **winter barley**, **winter wheat** [1-6]
- Corn marigold in **winter barley**, **winter wheat** [1-6]
- Field speedwell in **winter barley**, **winter wheat** [1-6]
- Ivy-leaved speedwell in **winter barley**, **winter wheat** [1-6]
- Wild oats in **spring barley** *(off-label)* [3]

Extension of Authorisation for Minor Use (EAMUs)

- **game cover** *20090450* [3]
- **spring barley** *20130951* [3]

Approval information

- Flufenacet and pendimethalin included in Annex I under EC Regulation 1107/2009
- Accepted by BBPA for use on malting barley

Efficacy guidance

- Best results achieved when applied from pre-emergence of weeds up to 2 leaf stage but post emergence treatment is not recommended on clay soils
- Product requires some soil moisture to be activated ideally from rain within 7 d of application. Prolonged dry conditions may reduce residual control
- Product is slow acting and final level of weed control may take some time to appear
- For effective weed control seed bed preparations should ensure even incorporation of any trash, straw and ash to 15 cm
- Efficacy may be reduced on soils with more than 6% organic matter
- Always follow WRAG guidelines for preventing and managing herbicide resistant weeds. See Section 5 for more information

Restrictions

- Maximum total dose equivalent to one full dose treatment
- For pre-emergence treatments seed should be covered with at least 32 mm settled soil. Shallow drilled crops should be treated post-emergence only
- Do not treat undersown crops
- Avoid spraying during periods of prolonged or severe frosts
- Do not use on stony or gravelly soils or those with more than 10% organic matter
- Pre-emergence treatment may only be used on crops drilled before 30 Nov. All crops must be treated before 31 Dec in yr of planting.
- Concentrated or diluted product may stain clothing or skin

Crop-specific information

- Latest use: before third tiller stage (GS 23) and before 31 Dec in yr of planting
- Very wet weather before and after treatment may result in loss of crop vigour and reduced yield, particularly where soils become waterlogged

Following crops guidance

- Any crop may follow a failed or normally harvested treated crop provided ploughing to at least 15 cm is carried out beforehand

Environmental safety

- Dangerous for the environment
- Very toxic to aquatic organisms
- Risk to certain non-target insects or other arthropods - avoid spraying within 6 m of field boundary

SEE SECTION 3 FOR PRODUCTS ALSO REGISTERED

- Some products supplied in small volume returnable packs. Follow instructions for use
- LERAP Category B

Hazard classification and safety precautions

Hazard Harmful, Dangerous for the environment, Harmful if swallowed [1-6]; Very toxic to aquatic organisms [1, 3-6]

Transport code 3

Packaging group III

UN Number 1993 [3-6]; 3082 [1, 2]

Risk phrases H304 [1, 3-6]; H315 [1-6]; H351, H370 [2]

Operator protection A, C, H; U02a, U05a [1-6]; U14, U20b [5]; U20c [1-4, 6]

Environmental protection E15a [1-4, 6]; E15b [5]; E16a, E16b, E22b, E34, E38, H410 [1-6]

Storage and disposal D01, D02, D09a, D12a [1-6]; D08, D10c [5]; D10a [1-4, 6]

Medical advice M03 [1-6]; M05b [1-4, 6]

243   flufenacet + picolinafen

A herbicide mixture for weed control in winter cereals
HRAC mode of action code: K3 + F1

Products

1	Pontos	BASF	240:100 g/l	SC	17811
2	Quirinus	BASF	240:50 g/l	SC	17711

Uses

- Annual dicotyledons in *rye, triticale, winter barley, winter wheat*
- Annual grasses in *rye, triticale, winter barley, winter wheat*
- Annual meadow grass in *rye, triticale, winter barley, winter wheat*
- Charlock in *rye, triticale, winter barley, winter wheat*
- Chickweed in *rye, triticale, winter barley, winter wheat*
- Field pansy in *rye, triticale, winter barley, winter wheat*
- Loose silky bent in *rye, triticale, winter barley, winter wheat*
- Mayweeds in *rye, triticale, winter barley, winter wheat*
- Shepherd's purse in *rye, triticale, winter barley, winter wheat*
- Speedwells in *rye, triticale, winter barley, winter wheat*
- Volunteer oilseed rape in *rye, triticale, winter barley, winter wheat*

Approval information

- Flufenacet and picolinafen are included in Annex I under EC Regulation 1107/2009

Efficacy guidance

- Can be used on all varieties of winter crops of wheat, barley, rye and triticale

Restrictions

- Always follow WRAG guidelines for preventing and managing herbicide resistant weeds
- Use low drift spraying equipment up to 30m from the top of the bank of any surface water bodies [1, 2]
- Horizontal boom sprayers must be fitted with three star drift reduction technology for all uses

Following crops guidance

- There are no restrictions on following crops after the normal harvest.
- In the event of crop failure, winter wheat can be re-sown in the same autumn provided soil is cultivated to a minimum depth of 15cm. Any of the following crops may be sown provided there has been a minimum of 60 days after the application and the soil is cultivated to a minimum depth of 15cm; legumes, maize, sugar beet and sunflower. Oilseed rape can be re-sown after 90 days following a pre-emergence application or 60 days following a post emergence application and the soil is cultivated to a minimum depth of 15cm. Spring barley can be re-drilled 120 days following application and the soil is cultivated to a minimum depth of 15cm.

FOR FULL CONDITIONS OF USE ALWAYS READ THE PRODUCT LABEL

Environmental safety
- Buffer zone requirement 6 m [1, 2]
- LERAP Category B

Hazard classification and safety precautions
    Transport code 9
    Packaging group III
    UN Number 3082
    Risk phrases H373
    Operator protection A, H; U05a, U19a
    Environmental protection E16a, E34, H410
    Storage and disposal D01, D02, D05, D09a, D10c, D12b
    Medical advice M03

244　flumioxazin

A phenyphthalimide herbicide for winter wheat
HRAC mode of action code: E

Products

1	Digital	Interfarm	300 g/l	SC	13561
2	Sumimax	Interfarm	300 g/l	SC	13548

Uses
- Annual dicotyledons in **bulb onions** *(off-label)*, **carrots** *(off-label)*, **ornamental plant production** *(off-label)*, **parsnips** *(off-label)*, **vining peas** *(off-label)*, **winter oats** *(off-label)*, **winter wheat** [1, 2]; **farm forestry** *(off-label)*, **forest nurseries** *(off-label)* [1]; **hops** *(off-label)*, **soft fruit** *(off-label)*, **top fruit** *(off-label)* [2]
- Annual grasses in **bulb onions** *(off-label)*, **winter oats** *(off-label)* [1, 2]; **carrots** *(off-label)*, **hops** *(off-label)*, **ornamental plant production** *(off-label)*, **parsnips** *(off-label)*, **soft fruit** *(off-label)*, **top fruit** *(off-label)*, **vining peas** *(off-label)* [2]
- Annual meadow grass in **winter wheat** [1, 2]
- Blackgrass in **winter oats** *(off-label)* [1, 2]
- Chickweed in **winter wheat** [1, 2]
- Cleavers in **winter wheat** [1, 2]
- Groundsel in **bulb onions** *(off-label)*, **carrots** *(off-label)*, **ornamental plant production** *(off-label)*, **parsnips** *(off-label)*, **vining peas** *(off-label)* [1, 2]; **farm forestry** *(off-label)*, **forest nurseries** *(off-label)* [1]; **hops** *(off-label)*, **soft fruit** *(off-label)*, **top fruit** *(off-label)*, **winter oats** *(off-label)* [2]
- Loose silky bent in **winter wheat** [1, 2]
- Mayweeds in **winter wheat** [1, 2]
- Ryegrass in **winter oats** *(off-label)* [1, 2]
- Speedwells in **winter wheat** [1, 2]
- Volunteer oilseed rape in **bulb onions** *(off-label)*, **carrots** *(off-label)*, **ornamental plant production** *(off-label)*, **parsnips** *(off-label)*, **vining peas** *(off-label)*, **winter wheat** [1, 2]; **farm forestry** *(off-label)*, **forest nurseries** *(off-label)* [1]; **hops** *(off-label)*, **soft fruit** *(off-label)*, **top fruit** *(off-label)*, **winter oats** *(off-label)* [2]
- Volunteer potatoes in **bulb onions** *(off-label)*, **carrots** *(off-label)*, **ornamental plant production** *(off-label)*, **parsnips** *(off-label)*, **vining peas** *(off-label)* [1, 2]; **farm forestry** *(off-label)*, **forest nurseries** *(off-label)* [1]; **hops** *(off-label)*, **soft fruit** *(off-label)*, **top fruit** *(off-label)*, **winter oats** *(off-label)* [2]

Extension of Authorisation for Minor Use (EAMUs)
- **bulb onions** *20091114 expires 30 Jun 2018* [1], *20091108 expires 30 Jun 2018* [2]
- **carrots** *20091112 expires 30 Jun 2018* [1], *20091111 expires 30 Jun 2018* [2]
- **farm forestry** *20082844 expires 30 Jun 2018* [1]
- **forest nurseries** *20082844 expires 30 Jun 2018* [1]
- **hops** *20082881 expires 30 Jun 2018* [2]
- **ornamental plant production** *20082844 expires 30 Jun 2018* [1], *20082881 expires 30 Jun 2018* [2]

SEE SECTION 3 FOR PRODUCTS ALSO REGISTERED

- **parsnips** *20091112 expires 30 Jun 2018* [1], *20091111 expires 30 Jun 2018* [2]
- **soft fruit** *20082881 expires 30 Jun 2018* [2]
- **top fruit** *20082881 expires 30 Jun 2018* [2]
- **vining peas** *20091113 expires 30 Jun 2018* [1], *20091109 expires 30 Jun 2018* [2]
- **winter oats** *20093121 expires 30 Jun 2018* [1], *20093122 expires 30 Jun 2018* [2]

Approval information
- Flumioxazine is included in Annex 1 under EC Regulation 1107/2009

Efficacy guidance
- Best results obtained from applications to moist soil early post-emergence of the crop when weeds are germinating up to 1 leaf stage
- Flumioxazin is contact acting and relies on weeds germinating in moist soil and coming into sufficient contact with the herbicide
- Flumioxazin remains in the surface layer of the soil and does not migrate to lower layers
- Contact activity is long lasting during the autumn
- Seed beds should be firm, fine and free from clods. Crops should be drilled to 25 mm and well covered by soil
- Efficacy is reduced on soils with more than 10% organic matter
- Flumioxazin is a member of the protoporphyrin oxidase (PPO) inhibitor herbicides. To avoid the build up of resistance do not use PPO-inhibitor herbicides more than once on any crop
- Use as part of a resistance management strategy that includes cultural methods of control and does not use PPO-inhibitors as the sole chemical method of grass weed control

Restrictions
- Maximum number of treatments 1 per crop for winter wheat
- Only apply to crops that have been hardened by cool weather. Do not treat crops with lush or soft growth
- Do not apply to waterlogged soils or to soils with more than 10% organic matter
- Do not follow an application with another pesticide for 14 d
- Do not treat undersown crops
- Do not treat broadcast crops until they are past the 3 leaf stage
- Do not treat crops under stress for any reason, or during prolonged frosty weather
- Do not roll or harrow for 2 wk before treatment or at any time afterwards

Crop-specific information
- Latest use: before 5th true leaf stage (GS 15) for winter wheat
- Light discolouration of leaf margins can occur after treatment and treatment of soft crops will cause transient leaf bleaching

Following crops guidance
- Any crop may be sown following normal harvest of a treated crop
- In the event of failure of a treated crop spring cereals, spring oilseed rape, sugar beet, maize or potatoes may be redrilled after ploughing

Environmental safety
- Dangerous for the environment
- Very toxic to aquatic organisms
- Take extreme care to avoid drift onto plants outside the target area
- LERAP Category B

Hazard classification and safety precautions
Hazard Toxic, Dangerous for the environment
Transport code 9
Packaging group III
UN Number 3082
Risk phrases H360 [2]; H361 [1]
Operator protection A, H; U05a, U19a
Environmental protection E15b, E16a, E34, E38, H410
Storage and disposal D01, D02, D05, D07, D09a, D12b
Medical advice M04a

FOR FULL CONDITIONS OF USE ALWAYS READ THE PRODUCT LABEL

245   fluopicolide

An benzamide fungicide available only in mixtures
FRAC mode of action code: 43

246   fluopicolide + propamocarb hydrochloride

A protectant and systemic fungicide mixture for potato blight
FRAC mode of action code: 43 + 28

See also propamocarb hydrochloride

Products

Infinito	Bayer CropScience	62.5:625 g/l	SC	16335

Uses
* Blight in **potatoes**
* Downy mildew in **broccoli** *(off-label)*, **brussels sprouts** *(off-label)*, **bulb onions** *(off-label)*, **cabbages** *(off-label)*, **calabrese** *(off-label)*, **cauliflowers** *(off-label)*, **collards** *(off-label)*, **garlic** *(off-label)*, **herbs (see appendix 6)** *(off-label)*, **kale** *(off-label)*, **lamb's lettuce** *(off-label)*, **land cress** *(off-label)*, **leeks** *(off-label)*, **lettuce** *(off-label)*, **ornamental plant production** *(off-label)*, **protected broccoli** *(off-label)*, **protected brussels sprouts** *(off-label)*, **protected cabbages** *(off-label)*, **protected calabrese** *(off-label)*, **protected cauliflowers** *(off-label)*, **protected collards** *(off-label)*, **protected kale** *(off-label)*, **radishes** *(off-label)*, **red mustard** *(off-label)*, **rocket** *(off-label)*, **salad onions** *(off-label)*, **shallots** *(off-label)*, **spinach** *(off-label)*

Extension of Authorisation for Minor Use (EAMUs)
* *broccoli 20152557*
* *brussels sprouts 20152557*
* *bulb onions 20161552*
* *cabbages 20152557*
* *calabrese 20152557*
* *cauliflowers 20152557*
* *collards 20152557*
* *garlic 20161552*
* *herbs (see appendix 6) 20171301*
* *kale 20152557*
* *lamb's lettuce 20171301*
* *land cress 20171301*
* *leeks 20161552*
* *lettuce 20171301*
* *ornamental plant production 20142251*
* *protected broccoli 20152557*
* *protected brussels sprouts 20152557*
* *protected cabbages 20152557*
* *protected calabrese 20152557*
* *protected cauliflowers 20152557*
* *protected collards 20152557*
* *protected kale 20152557*
* *radishes 20171608*
* *red mustard 20171301*
* *rocket 20171301*
* *salad onions 20161552*
* *shallots 20161552*
* *spinach 20171301*

Approval information
* Fluopicolide and propamocarb hydrochloride included in Annex I under EC Regulation 1107/2009

SEE SECTION 3 FOR PRODUCTS ALSO REGISTERED

Efficacy guidance

- Commence spray programme before infection appears as soon as weather conditions favourable for disease development occur. At latest the first treatment should be made as the foliage meets along the rows
- Repeat treatments at 7-10 day intervals according to disease incidence and weather conditions
- Reduce spray interval if conditions are conducive to the spread of blight
- Spray as soon as possible after irrigation
- Increase water volume in dense crops
- When used from full canopy development to haulm desiccation as part of a full blight protection programme tubers will be protected from late blight after harvest and tuber blight incidence will be reduced
- To reduce the development of resistance product should be used in single or block applications with fungicides from a different cross-resistance group

Restrictions

- Maximum total dose on potatoes equivalent to four full dose treatments
- Do not apply if rainfall or irrigation is imminent. Product is rainfast in 1 h provided spray has dried on leaf
- Do not apply as a curative treatment when blight is present in the crop
- Do not apply more than 3 consecutive treatments of the product

Crop-specific information

- HI 7 d for potatoes
- All varieties of potatoes, including seed crops, may be treated

Environmental safety

- Dangerous for the environment
- Toxic to aquatic organisms

Hazard classification and safety precautions

Hazard Irritant, Dangerous for the environment, Very toxic to aquatic organisms
Transport code 9
Packaging group III
UN Number 3082
Risk phrases H317
Operator protection A, H; U05a, U08, U11, U19a, U20a
Environmental protection E15a, E38, H410
Storage and disposal D01, D02, D09a, D10b, D12a

247   fluopyram

An SDHI fungicide only available in mixtures
FRAC mode of action code: 7

See also bixafen + fluopyram + prothioconazole

248   fluopyram + prothioconazole

A foliar fungicide for use in oilseed rape
FRAC mode of action code: 3 + 7

Products

1 Clayton Repel	Clayton	125:125 g/l	SE	18037
2 Propulse	Bayer CropScience	125:125 g/l	SE	17837
3 Recital	Bayer CropScience	125:125 g/l	SE	17909

Uses

- Disease control in **spring oilseed rape, winter oilseed rape** [1-3]
- Light leaf spot in **winter oilseed rape** [1, 2]
- Phoma in **winter oilseed rape** [1, 2]

FOR FULL CONDITIONS OF USE ALWAYS READ THE PRODUCT LABEL

- Powdery mildew in *winter oilseed rape* [1, 2]
- Sclerotinia in *winter oilseed rape* [1, 2]
- Stem canker in *winter oilseed rape* [1, 2]

Approval information
- Fluopyram and prothioconazole included in Annex 1 under EC Regulation 1107/2009.

Environmental safety
- LERAP Category B

Hazard classification and safety precautions
Hazard Harmful [3]; Dangerous for the environment, Very toxic to aquatic organisms [1-3]
Transport code 9
Packaging group III
UN Number 3082
Operator protection A, H; U05a, U09b, U19a, U20c
Environmental protection E15b, E16a, E34, E38, H410
Storage and disposal D01, D02, D05, D09a, D10b, D12a
Medical advice M03

249   fluopyram + prothioconazole + tebuconazole

An SDHI and triazole mixture for seed treatment in winter barley
FRAC mode of action code: 3 + 7

Products

Raxil Star	Bayer CropScience	20:100:60 g/l	FS	17805

Uses
- Covered smut in *winter barley*
- Fusarium in *winter barley*
- Leaf stripe in *winter barley*
- Loose smut in *winter barley*
- Microdochium nivale in *winter barley*
- Net blotch in *winter barley* *(seed borne)*

Approval information
- Fluopyram, prothioconazole and tebuconazole included in Annex 1 under EC Regulation 1107/2009
- Accepted by BBPA for use on malting barley

Efficacy guidance
- Seed treatment products must be applied by manufacturer's recommended treatment application equipment
- Treated cereal seed should preferably be drilled in the same season
- Follow-up treatments will be needed later in the season to give protection against air-borne and splash-borne diseases
- Seed should be drilled to a depth of 40 mm into a well prepared and firm seedbed. If seed is present on the soil surface, or spills have occurred, then, if conditions are appropriate, the field should be harrowed then rolled to ensure good incorporation
- Do not use on seed with more than 16% moisture content or on sprouted, cracked, skinned or otherwise damaged seed

Hazard classification and safety precautions
Hazard Harmful, Dangerous for the environment
Transport code 9
Packaging group III
UN Number 3082
Risk phrases H361
Operator protection A, C, D, H; U05a, U12b, U20b, U20e

SEE SECTION 3 FOR PRODUCTS ALSO REGISTERED

Environmental protection E15b, E34, E38, H410
Storage and disposal D01, D02, D05, D09a, D10d, D12a, D14
Treated seed S01, S02, S04b, S05, S07, S08

250   fluopyram + trifloxystrobin

A fungicide mixture for disease control in protected strawberries
FRAC mode of action code: 7 + 11

See also fluopyram
       trifloxystrobin

Products

1	Exteris Stressgard	Bayer CropScience	12.5:12.5 g/l	SC	17825
2	Luna Sensation	Bayer CropScience	250:250 g/l	SC	15793

Uses

- Botrytis in **lettuce** *(off-label)*, **protected lettuce** *(off-label)* [2]
- Dollar spot in **managed amenity turf** [1]
- Grey mould in **protected strawberries** *(moderate control)* [2]
- Microdochium nivale in **managed amenity turf** [1]
- Powdery mildew in **lettuce** *(off-label)*, **protected lettuce** *(off-label)*, **protected strawberries** [2]
- Sclerotinia in **lettuce** *(off-label)*, **protected lettuce** *(off-label)* [2]

Extension of Authorisation for Minor Use (EAMUs)

- **lettuce** *20171179* [2]
- **protected lettuce** *20171179* [2]

Approval information

- Fluopyram and trifloxystrobin are included in Annex 1 under EC Regulation 1107/2009

Restrictions

- If more than two QoI products are to be used on the crop then the FRAC advice on treatment must be followed.

Environmental safety

- Buffer zone requirement 14 m
- LERAP Category B

Hazard classification and safety precautions

Hazard Harmful if swallowed [2]; Very toxic to aquatic organisms [1, 2]
Transport code 9
Packaging group III
UN Number 3082
Risk phrases H317 [1]
Operator protection A, H; U05a, U09a, U20a [2]
Environmental protection E15a [2]; E15b [1]; E16a, E34, H410 [1, 2]
Storage and disposal D01, D02, D09a, D10a
Medical advice M03 [2]

251   fluoxastrobin

A protectant stobilurin fungicide available in mixtures
FRAC mode of action code: 11

See also bixafen + fluoxastrobin + prothioconazole

252   fluoxastrobin + prothioconazole

A strobilurin and triazole fungicide mixture for cereals
FRAC mode of action code: 11 + 3

See also prothioconazole

Products

1	Clayton Edge	Clayton	100:100 g/l	EC	12911
2	Fandango	Bayer CropScience	100:100 g/l	EC	17318
3	Firefly 155	Bayer CropScience	45:110 g/l	EC	14818
4	Kurdi	AgChem Access	100:100 g/l	EC	15917
5	Unicur	Bayer CropScience	100:100 g/l	EC	17402

Uses

- Botrytis in **bulb onions** *(useful reduction)*, **shallots** *(useful reduction)* [5]
- Botrytis squamosa in **bulb onions** *(useful reduction)*, **shallots** *(useful reduction)* [5]
- Brown rust in **spring barley**, **winter barley** [1, 2, 4]; **spring wheat** [2-4]; **winter rye**, **winter wheat** [1-4]
- Crown rust in **spring oats**, **winter oats** [2, 4]
- Downy mildew in **bulb onions** *(control)*, **shallots** *(control)* [5]
- Eyespot in **spring barley** *(reduction)*, **winter barley** *(reduction)* [1, 2, 4]; **spring oats**, **winter oats** [2, 4]; **spring oats** *(reduction of incidence and severity)*, **winter oats** *(reduction of incidence and severity)* [3]; **spring wheat** *(reduction)* [2-4]; **winter rye** *(reduction)*, **winter wheat** *(reduction)* [1-4]
- Fusarium root rot in **spring wheat** *(reduction)*, **winter wheat** *(reduction)* [2, 4]
- Glume blotch in **spring wheat** [2-4]; **winter wheat** [1-4]
- Late ear diseases in **spring barley**, **winter barley**, **winter wheat** [1, 2, 4]; **spring wheat** [2, 4]
- Net blotch in **spring barley**, **winter barley** [1, 2, 4]
- Powdery mildew in **spring barley**, **winter barley** [1, 2, 4]; **spring oats**, **winter oats** [2, 4]; **spring wheat** [2-4]; **winter rye**, **winter wheat** [1-4]
- Rhynchosporium in **spring barley**, **winter barley** [1, 2, 4]; **winter rye** [1-4]
- Septoria leaf blotch in **spring wheat** [2-4]; **winter wheat** [1-4]
- Sharp eyespot in **spring wheat** *(reduction)*, **winter wheat** *(reduction)* [2, 4]
- Sooty moulds in **spring wheat** *(reduction)*, **winter wheat** *(reduction)* [2, 4]
- Take-all in **spring wheat** *(reduction)*, **winter barley** *(reduction)*, **winter wheat** *(reduction)* [2, 4]
- Tan spot in **spring wheat** [2, 4]; **winter wheat** [1, 2, 4]
- Yellow rust in **spring wheat** [2-4]; **winter wheat** [1-4]

Approval information

- Fluoxastrobin and prothioconazole included in Annex I under EC Regulation 1107/2009
- Accepted by BBPA for use on malting barley

Efficacy guidance

- Best results on foliar diseases obtained from treatment at early stages of disease development. Further treatment may be needed if disease attack is prolonged
- Foliar applications to established infections of any disease are likely to be less effective
- Best control of cereal ear diseases obtained by treatment during ear emergence
- Fluoxastrobin is a member of the QoI cross resistance group. Foliar product should be used preventatively and not relied on for its curative potential
- Use product as part of an Integrated Crop Management strategy incorporating other methods of control, including where appropriate other fungicides with a different mode of action. Do not apply more than two foliar applications of QoI containing products to any cereal crop
- There is a significant risk of widespread resistance occurring in *Septoria tritici* populations in UK. Failure to follow resistance management action may result in reduced levels of disease control
- Strains of wheat and barley powdery mildew resistant to QoIs are common in the UK. Control of wheat mildew can only be relied on from the triazole component
- Prothioconazole is a DMI fungicide. Resistance to some DMI fungicides has been identified in Septoria leaf blotch which may seriously affect performance of some products. For further

advice contact a specialist advisor and visit the Fungicide Resistance Action Group (FRAG)-UK website
- Where specific control of wheat mildew is required this should be achieved through a programme of measures including products recommended for the control of mildew that contain a fungicide from a different cross-resistance group and applied at a dose that will give robust control

Restrictions
- Maximum total dose of foliar sprays equivalent to two full dose treatments for the crop
- Do not apply by hand-held equipment, e.g Knapsack Sprayer [5]
- **Crop-specific information**
- Latest use: before grain milky ripe for spray treatments on wheat and rye; beginning of flowering for barley, oats
- Some transient leaf chlorosis may occur after treatment of wheat or barley but this has not been found to affect yield

Environmental safety
- Dangerous for the environment
- Toxic to aquatic organisms
- Risk to non-target insects or other arthropods. Avoid spraying within 6 m of the field boundary to reduce the effects on non-target insects or other arthropods
- LERAP Category B

Hazard classification and safety precautions
Hazard Dangerous for the environment, Very toxic to aquatic organisms
Transport code 9
Packaging group III
UN Number 3082
Risk phrases H318 [1, 2, 4]
Operator protection A [1-5]; H [2, 3, 5]; U05a [1-5]; U09a, U20b [3]; U09b, U20a [1, 2, 4, 5]
Environmental protection E15a, E16a, E34, H410 [1-5]; E22c [1, 2, 4, 5]; E38 [2-5]
Storage and disposal D01, D05, D09a, D10b, D12a [1-5]; D02 [1]
Medical advice M03

---

253   fluoxastrobin + prothioconazole + trifloxystrobin

A triazole and strobilurin fungicide mixture for cereals
FRAC mode of action code: 11 + 3 + 11

See also prothioconazole
trifloxystrobin

Products

Jaunt	Bayer CropScience	75:150:75 g/l	EC	12350

Uses
- Brown rust in *durum wheat, rye, spring barley, spring wheat, triticale, winter barley, winter wheat*
- Eyespot in *durum wheat* (reduction), *rye* (reduction), *spring barley* (reduction), *spring wheat* (reduction), *triticale* (reduction), *winter barley* (reduction), *winter wheat* (reduction)
- Glume blotch in *durum wheat, rye, spring wheat, triticale, winter wheat*
- Late ear diseases in *durum wheat, rye, spring barley, spring wheat, triticale, winter barley, winter wheat*
- Net blotch in *spring barley, winter barley*
- Powdery mildew in *durum wheat, rye, spring barley, spring wheat, triticale, winter barley, winter wheat*
- Rhynchosporium in *spring barley, winter barley*
- Septoria leaf blotch in *durum wheat, rye, spring wheat, triticale, winter wheat*
- Tan spot in *durum wheat, rye, spring wheat, triticale, winter wheat*
- Yellow rust in *durum wheat, rye, spring wheat, triticale, winter wheat*

FOR FULL CONDITIONS OF USE ALWAYS READ THE PRODUCT LABEL

Approval information
- Fluoxastrobin, prothioconazole and trifloxystrobin included in Annex I under EC Regulation 1107/2009
- Accepted by BBPA for use on malting barley

Efficacy guidance
- Best results obtained from treatment at early stages of disease development. Further treatment may be needed if disease attack is prolonged
- Applications to established infections of any disease are likely to be less effective
- Best control of cereal ear diseases obtained by treatment during ear emergence
- Fluoxastrobin and trifloxystrobin are members of the QoI cross resistance group. Product should be used preventatively and not relied on for its curative potential
- Use product as part of an Integrated Crop Management strategy incorporating other methods of control, including where appropriate other fungicides with a different mode of action. Do not apply more than two foliar applications of QoI containing products to any cereal crop
- There is a significant risk of widespread resistance occurring in *Septoria tritici* populations in UK. Failure to follow resistance management action may result in reduced levels of disease control
- Strains of wheat and barley powdery mildew resistant to QoIs are common in the UK. Control of wheat mildew can only be relied on from the triazole component
- Where specific control of wheat mildew is required this should be achieved through a programme of measures including products recommended for the control of mildew that contain a fungicide from a different cross-resistance group and applied at a dose that will give robust control
- Prothioconazole is a DMI fungicide. Resistance to some DMI fungicides has been identified in Septoria leaf blotch which may seriously affect performance of some products. For further advice contact a specialist advisor and visit the Fungicide Resistance Action Group (FRAG)-UK website

Restrictions
- Maximum total dose equivalent to two full dose treatments

Crop-specific information
- Latest use: before grain milky ripe for winter wheat; up to beginning of anthesis (GS 61) for barley

Environmental safety
- Dangerous for the environment
- Very toxic to aquatic organisms
- Risk to non-target insects or other arthropods. Avoid spraying within 6 m of the field boundary to reduce the effects on non-target insects or other arthropods
- LERAP Category B

Hazard classification and safety precautions
Hazard Irritant, Dangerous for the environment, Very toxic to aquatic organisms
Transport code 9
Packaging group III
UN Number 3082
Operator protection A, C, H; U05a, U09b, U19a, U20b
Environmental protection E15a, E16a, E22c, E34, E38, H410
Storage and disposal D01, D02, D05, D09a, D10b, D12a
Medical advice M03

SEE SECTION 3 FOR PRODUCTS ALSO REGISTERED

254   fluroxypyr

A post-emergence pyridinecarboxylic acid herbicide
HRAC mode of action code: O

See also 2,4-D + dicamba + fluroxypyr
aminopyralid + fluroxypyr
bromoxynil + fluroxypyr
bromoxynil + fluroxypyr + ioxynil
clopyralid + florasulam + fluroxypyr
clopyralid + fluroxypyr + MCPA
clopyralid + fluroxypyr + triclopyr
florasulam + fluroxypyr

Products

1	Cleancrop Gallifrey 3	Dow	333 g/l	EC	17399
2	Crescent	Certis	200 g/l	EC	17589
3	Flurostar 200	Globachem	200 g/l	EC	17438
4	GAL-GONE	Globachem	200 g/l	EC	17505
5	Hatchet Xtra	Certis	200 g/l	EC	17593
6	Hudson 200	Barclay	200 g/l	EC	17749
7	Hurler	Barclay	200 g/l	EC	17715
8	Minstrel	UPL Europe	200 g/l	EC	13745
9	Starane Hi-Load HL	Dow	333 g/l	EC	16557
10	Tandus	Nufarm UK	200 g/l	EC	18071

Uses

- Annual dicotyledons in **durum wheat, grassland, spring oats, spring wheat, triticale, winter barley, winter oats, winter wheat** [1-10]; **forage maize** [1-8, 10]; **grass seed crops, spelt** [1, 9]; **rye** [10]; **spring barley** [2-8, 10]; **spring rye, winter rye** [1-9]
- Black bindweed in **bulb onions** (off-label), **garlic** (off-label), **grass seed crops, leeks** (off-label), **salad onions** (off-label), **shallots** (off-label), **spelt** [1, 9]; **durum wheat, spring barley, spring oats, spring wheat, triticale, winter barley, winter oats, winter wheat** [1-10]; **forage maize** [1-8]; **grassland, spring rye, winter rye** [1-9]; **newly sown grass leys** (off-label) [9]; **rye** [10]
- Black nightshade in **forage maize** [1]
- Chickweed in **almonds** (off-label), **apples** (off-label), **bulb onions** (off-label), **chestnuts** (off-label), **garlic** (off-label), **grass seed crops, hazel nuts** (off-label), **leeks** (off-label), **pears** (off-label), **poppies for morphine production** (off-label), **salad onions** (off-label), **shallots** (off-label), **spelt, sweetcorn** (off-label), **walnuts** (off-label) [1, 9]; **canary seed** (off-label), **farm forestry** (off-label), **forest nurseries** (off-label), **game cover** (off-label), **miscanthus** (off-label), **newly sown grass leys** (off-label), **ornamental plant production** (off-label) [9]; **durum wheat, grassland, spring barley, spring oats, spring wheat, triticale, winter barley, winter oats, winter wheat** [1-10]; **forage maize** [1-8, 10]; **rye** [10]; **spring rye, winter rye** [1-9]
- Cleavers in **almonds** (off-label), **apples** (off-label), **bulb onions** (off-label), **chestnuts** (off-label), **garlic** (off-label), **grass seed crops, hazel nuts** (off-label), **leeks** (off-label), **pears** (off-label), **poppies for morphine production** (off-label), **salad onions** (off-label), **shallots** (off-label), **spelt, sweetcorn** (off-label), **walnuts** (off-label) [1, 9]; **canary seed** (off-label), **farm forestry** (off-label), **forest nurseries** (off-label), **game cover** (off-label), **miscanthus** (off-label), **newly sown grass leys** (off-label), **ornamental plant production** (off-label) [9]; **durum wheat, spring barley, spring oats, spring wheat, triticale, winter barley, winter oats, winter wheat** [1-10]; **forage maize** [1-8]; **grassland, spring rye, winter rye** [1-9]; **rye** [10]
- Corn spurrey in **bulb onions** (off-label), **garlic** (off-label), **leeks** (off-label), **salad onions** (off-label), **shallots** (off-label) [1, 9]; **newly sown grass leys** (off-label) [9]
- Docks in **durum wheat, spring barley, spring oats, spring wheat, triticale, winter barley, winter oats, winter wheat** [2-8, 10]; **forage maize, spring rye, winter rye** [2-8]; **grassland** [2-7]; **rye** [10]
- Forget-me-not in **bulb onions** (off-label), **garlic** (off-label), **grass seed crops, leeks** (off-label), **salad onions** (off-label), **shallots** (off-label), **spelt** [1, 9]; **durum wheat, spring barley, spring oats, spring wheat, triticale, winter barley, winter oats, winter wheat** [1-10]; **forage maize** [1-8]; **grassland, spring rye, winter rye** [1-9]; **newly sown grass leys** (off-label) [9]; **rye** [10]

FOR FULL CONDITIONS OF USE ALWAYS READ THE PRODUCT LABEL

- Fumitory in **bulb onions** *(off-label)*, **garlic** *(off-label)*, **leeks** *(off-label)*, **newly sown grass leys** *(off-label)*, **salad onions** *(off-label)*, **shallots** *(off-label)* [9]
- Groundsel in **bulb onions** *(off-label)*, **garlic** *(off-label)*, **leeks** *(off-label)*, **newly sown grass leys** *(off-label)*, **salad onions** *(off-label)*, **shallots** *(off-label)* [9]
- Hemp-nettle in **almonds** *(off-label)*, **apples** *(off-label)*, **bulb onions** *(off-label)*, **chestnuts** *(off-label)*, **garlic** *(off-label)*, **grass seed crops**, **hazel nuts** *(off-label)*, **leeks** *(off-label)*, **pears** *(off-label)*, **salad onions** *(off-label)*, **shallots** *(off-label)*, **spelt**, **sweetcorn** *(off-label)*, **walnuts** *(off-label)* [1, 9]; **durum wheat**, **spring barley**, **spring oats**, **spring wheat**, **triticale**, **winter barley**, **winter oats**, **winter wheat** [1-10]; **farm forestry** *(off-label)*, **forest nurseries** *(off-label)*, **game cover** *(off-label)*, **miscanthus** *(off-label)*, **newly sown grass leys** *(off-label)*, **ornamental plant production** *(off-label)* [9]; **forage maize** [1-8]; **grassland**, **spring rye**, **winter rye** [1-9]; **rye** [10]
- Knotgrass in **bulb onions** *(off-label)*, **garlic** *(off-label)*, **leeks** *(off-label)*, **newly sown grass leys** *(off-label)*, **salad onions** *(off-label)*, **shallots** *(off-label)* [9]
- Pale persicaria in **bulb onions** *(off-label)*, **garlic** *(off-label)*, **leeks** *(off-label)*, **salad onions** *(off-label)*, **shallots** *(off-label)* [1, 9]; **newly sown grass leys** *(off-label)* [9]
- Poppies in **poppies for morphine production** *(off-label)* [1, 9]
- Red dead-nettle in **bulb onions** *(off-label)*, **garlic** *(off-label)*, **leeks** *(off-label)*, **salad onions** *(off-label)*, **shallots** *(off-label)* [1, 9]; **newly sown grass leys** *(off-label)* [9]
- Redshank in **bulb onions** *(off-label)*, **garlic** *(off-label)*, **leeks** *(off-label)*, **salad onions** *(off-label)*, **shallots** *(off-label)* [1, 9]; **newly sown grass leys** *(off-label)* [9]
- Speedwells in **bulb onions** *(off-label)*, **garlic** *(off-label)*, **leeks** *(off-label)*, **newly sown grass leys** *(off-label)*, **salad onions** *(off-label)*, **shallots** *(off-label)* [9]
- Volunteer potatoes in **bulb onions** *(off-label)*, **farm forestry** *(off-label)*, **forest nurseries** *(off-label)*, **game cover** *(off-label)*, **garlic** *(off-label)*, **leeks** *(off-label)*, **miscanthus** *(off-label)*, **newly sown grass leys** *(off-label)*, **ornamental plant production** *(off-label)*, **salad onions** *(off-label)*, **shallots** *(off-label)* [9]; **poppies for morphine production** *(off-label)* [1]; **spring barley**, **spring oats**, **spring wheat** [10]; **sweetcorn** *(off-label)* [1, 9]; **winter barley**, **winter wheat** [2-8, 10]

Extension of Authorisation for Minor Use (EAMUs)
- **almonds** *20171043* [1], *20170227* [9]
- **apples** *20171043* [1], *20170227* [9]
- **bulb onions** *20171042* [1], *20142851* [9]
- **canary seed** *20171133* [9]
- **chestnuts** *20171043* [1], *20170227* [9]
- **farm forestry** *20171268* [9]
- **forest nurseries** *20171268* [9]
- **game cover** *20171268* [9]
- **garlic** *20171042* [1], *20142851* [9]
- **hazel nuts** *20171043* [1], *20170227* [9]
- **leeks** *20171042* [1], *20142851* [9]
- **miscanthus** *20171268* [9]
- **newly sown grass leys** *20142851* [9]
- **ornamental plant production** *20171268* [9]
- **pears** *20171043* [1], *20170227* [9]
- **poppies for morphine production** *20171040* [1], *20161040* [9]
- **salad onions** *20171042* [1], *20142851* [9]
- **shallots** *20171042* [1], *20142851* [9]
- **sweetcorn** *20171041* [1], *20170226* [9]
- **walnuts** *20171043* [1], *20170227* [9]

Approval information
- Fluroxypyr included in Annex I under EC Regulation 1107/2009
- Accepted by BBPA for use on malting barley

Efficacy guidance
- Best results achieved under good growing conditions in a strongly competing crop
- A number of tank mixtures with other herbicides are recommended for use in autumn and spring to extend range of species controlled. See label for details
- Spray is rainfast in 1 h

SEE SECTION 3 FOR PRODUCTS ALSO REGISTERED

Restrictions

- Maximum number of treatments 1 per crop or yr or maximum total dose equivalent to one full dose treatment
- Do not apply in any tank-mix on triticale or forage maize
- Do not use on crops undersown with clovers or other legumes
- Do not treat crops suffering stress caused by any factor
- Do not roll or harrow for 7 d before or after treatment
- Do not spray if frost imminent
- Straw from treated crops must not be returned directly to the soil but must be removed and used only for livestock bedding

Crop-specific information

- Latest use: before flag leaf sheath opening (GS 47) for winter wheat and barley; before flag leaf sheath extending (GS 41) for spring wheat and barley; before second node detectable (GS 32) for oats, rye, triticale and durum wheat; before 7 leaves unfolded and before buttress roots appear for maize
- Apply to new leys from 3 expanded leaf stage
- Timing varies in tank mixtures. See label for details
- Crops undersown with grass may be sprayed provided grasses are tillering

Following crops guidance

- Clovers, peas, beans and other legumes must not be sown for 12 mth following treatment at the highest dose
- In the event of crop failure, spring cereals, spring oilseed rape, maize, onions, poppies and new leys may be sown 5 weeks after application with no requirements for soil cultivation [9]

Environmental safety

- Dangerous for the environment
- Very toxic to aquatic organisms
- Flammable
- Keep livestock out of treated areas for at least 3 d following treatment and until poisonous weeds, such as ragwort, have died down and become unpalatable
- Wash spray equipment thoroughly with water and detergent immediately after use. Traces of product can damage susceptible plants sprayed later
- LERAP Category B [1-5, 8, 9]

Hazard classification and safety precautions

Hazard Harmful [2-5, 7, 8, 10]; Flammable [2-5, 7, 8]; Dangerous for the environment [1-5, 7-10]; Flammable liquid and vapour [2, 5-7, 10]; Very toxic to aquatic organisms [3, 4, 8]

Transport code 3 [1-3, 5, 7-9]; 9 [4, 6, 10]

Packaging group III

UN Number 1993 [1-5, 7-9]; 3082 [6, 10]

Risk phrases H304, H336 [2-8, 10]; H315 [2-7, 10]; H317 [1, 2, 5-7, 9, 10]; H318 [2, 5]; H319 [1, 3, 4, 6-10]; H335 [6-8, 10]

Operator protection A, H [1-10]; C [1-6, 8, 9]; U04c, U12 [1, 9]; U05a [1-5, 7, 9]; U08, U19a [1-10]; U11 [1-5, 9]; U14 [2-5, 7, 10]; U20b [1-9]

Environmental protection E06a [2] (3 d); E07a [3-6, 8, 10]; E07c [7] (14 d); E15a [6, 8]; E15b [1-5, 7, 9, 10]; E16a [1-5, 8, 9]; E34 [1-5, 7, 9]; E38, H410 [2-8, 10]; H411 [1, 9]

Storage and disposal D01, D02, D05 [1-5, 7, 9]; D09a [1-10]; D10a [7]; D10b [1-6, 8-10]; D12a [2-8, 10]; D12b [1, 9]

Treated seed S06a [10]

Medical advice M03 [2-5, 7]; M05b [2-8, 10]

---

255   fluroxypyr + halauxifen-methyl

A herbicide mixture for use in cereals
HRAC mode of action code: O

Products

1 Pixxaro EC	Dow	280:12 g/l	EC	17545
2 Whorl	Dow	280:12 g/l	EC	17819

FOR FULL CONDITIONS OF USE ALWAYS READ THE PRODUCT LABEL

Uses

- Chickweed in *durum wheat, rye, spelt, spring barley, spring wheat, triticale, winter barley, winter wheat*
- Cleavers in *durum wheat, rye, spelt, spring barley, spring wheat, triticale, winter barley, winter wheat*
- Crane's-bill in *durum wheat, rye, spelt, spring barley, spring wheat, triticale, winter barley, winter wheat*
- Fat hen in *durum wheat, rye, spelt, spring barley, spring wheat, triticale, winter barley, winter wheat*
- Fumitory in *durum wheat, rye, spelt, spring barley, spring wheat, triticale, winter barley, winter wheat*
- Poppies in *durum wheat, rye, spelt, spring barley, spring wheat, triticale, winter barley, winter wheat*

Approval information

- Fluroxypyr and halauxifen-methyl included in Annex I under EC Regulation 1107/2009

Efficacy guidance

- Rainfast one hour after application
- Addition of an adjuvant gives improved reliability against poppy, chickweed and volunteer potatoes

Following crops guidance

- After application of 0.5 l/ha up to BBCH 45, white mustard, pea, winter oilseed rape, phacelia, ryegrass, winter wheat and winter barley may be sown in the autumn of the same year after the cereal harvest; spring wheat, spring barley, spring oat, ryegrass, maize, spring oilseed rape, sugar beet, potato, field bean, tomato, sunflower, onion, clover, and pea may be sown in the spring of the year after the cereal harvest. Ploughing is recommended prior to drilling maize, alfalfa, broad bean, field bean, soybean and clover in the same year as application.
- In the event of crop failure spring wheat, spring barley, spring oats and ryegrass may be sown one month after application with no cultivation. Maize, spring oilseed rape, peas, broad beans and field beans may be sown two months after application and ploughing. Three months after application sorghum may be sown.

Environmental safety

- LERAP Category B

Hazard classification and safety precautions

Transport code 9
Packaging group III
UN Number 3082
Risk phrases H317, H319, H335
Operator protection A, C, H; U05a, U08, U19a, U20c
Environmental protection E15b, E16a, E34, H410
Storage and disposal D09a, D10c

256   fluroxypyr + metsulfuron-methyl + thifensulfuron-methyl

A herbicide mixture for cereals
HRAC mode of action code: O + B + B

*See also fluroxypyr + metsulfuron-methyl
thifensulfuron-methyl*

Products

Provalia LQM	DuPont	135:5:30 g/l	OD	17846

Uses

- Annual dicotyledons in *winter barley, winter wheat*
- Charlock in *winter barley, winter wheat*
- Chickweed in *winter barley, winter wheat*

SEE SECTION 3 FOR PRODUCTS ALSO REGISTERED

- Cleavers in *winter barley, winter wheat*
- Field pansy in *winter barley, winter wheat*
- Fumitory in *winter barley, winter wheat*
- Mayweeds in *winter barley, winter wheat*
- Poppies in *winter barley, winter wheat*
- Red dead-nettle in *winter barley, winter wheat*
- Shepherd's purse in *winter barley, winter wheat*

Approval information

- Fluroxypyr, metsulfuron-methyl and thifensulfuron-methyl included in Annex I under EC Regulation 1107/2009

Efficacy guidance

- Apply in a minimum of 150 litres of water per hectare.

Restrictions

- Do not apply to any crop suffering from stress as a result of drought, waterlogging, low temperatures, pest or disease attack, nutrient or lime deficiency or other factors reducing crop growth.
- Do not use on cereal crops undersown with grasses, clover or other legumes or any other broad-leaved crop.
- Take special care to avoid damage by drift onto broad-leaved plants outside the target area, or onto ponds, waterways or ditches. Thorough cleansing of equipment is also very important.
- Do not apply within 7 days of rolling the crop.

Following crops guidance

- Cereals, oilseed rape, field beans or grass may be sown in the same calendar year as harvest of a cereal crop. In case of crop failure for any reason, sow only wheat within three months of application. Before sowing, soil should be ploughed and cultivated to a depth of at least 15-cm.

Environmental safety

- LERAP Category B

Hazard classification and safety precautions

Transport code 9
Packaging group III
UN Number 3082
Risk phrases H317
Operator protection A, H; U05b, U09a, U11, U20a
Environmental protection E15b, E16a, H410
Storage and disposal D01, D02, D09a, D10c, D12a

257   fluroxypyr + triclopyr

A foliar acting herbicide for docks in grassland
HRAC mode of action code: O + O

See also triclopyr

Products

1 Doxstar Pro	Dow	150:150 g/l	EC	15664
2 Pivotal	Dow	150:150 g/l	EC	16943

Uses

- Docks in *amenity grassland, grassland* [2]; *established grassland* [1]

Approval information

- Fluroxypyr and triclopyr included in Annex I under EC Regulation 1107/2009

Efficacy guidance

- Seedling docks in established grass only are controlled up to 50 mm diameter. Apply in spring or autumn or, at lower dose, in spring and autumn on docks up to 200 mm. A second application in the subsequent yr may be needed

FOR FULL CONDITIONS OF USE ALWAYS READ THE PRODUCT LABEL

- Allow 2-3 wk after cutting or grazing to allow sufficient regrowth of docks to occur before spraying
- Control may be reduced if rain falls within 2 h of application
- To allow maximum translocation to the roots of docks do not cut grass for 28 d after spraying
- [1] available as a twin pack with Thistlex as Pas.Tor for control of docks, nettles and thistles in grassland

Restrictions
- Maximum total dose equivalent to one full dose treatment
- Do not roll or harrow for 10 d before or 7 d after spraying
- Do not spray in drought, very hot or very cold weather

Crop-specific information
- Latest use: 7 d before grazing or harvest of grass
- Grass less than one yr old may be treated at half dose from the third leaf visible stage
- Clover will be killed or severely checked by treatment

Following crops guidance
- Do not sow kale, turnips, swedes or grass mixtures containing clover by direct drilling or minimum cultivation techniques within 6 wk of application

Environmental safety
- Dangerous for the environment
- Toxic to aquatic organisms
- Keep livestock out of treated areas for at least 7 d following treatment and until poisonous weeds, such as ragwort, have died down and become unpalatable
- Do not allow drift to come into contact with crops, amenity plantings, gardens, ponds, lakes or watercourses
- Wash spray equipment thoroughly with water and detergent immediately after use. Traces of product can damage susceptible plants sprayed later

Hazard classification and safety precautions
Hazard Irritant, Dangerous for the environment
Transport code 3
Packaging group III
UN Number 1993
Risk phrases H317
Operator protection A; U14
Environmental protection E38, H410
Storage and disposal D01, D02, D05, D12a
Medical advice M05a

258   flurtamone

A carotenoid synthesis inhibitor available only in mixtures
HRAC mode of action code: F1

See also diflufenican + flufenacet + flurtamone
*   diflufenican + flurtamone*

259   flutolanil

An carboxamide fungicide for treatment of potato seed tubers
FRAC mode of action code: 7

Products
1 Rhino	Certis	460 g/l	SC	14311
2 Rhino DS	Certis	6% w/w	DS	12763

SEE SECTION 3 FOR PRODUCTS ALSO REGISTERED

Uses

- Black scurf in **potatoes** *(tuber treatment)*
- Stem canker in **potatoes** *(tuber treatment)*

Approval information

- Flutolanil included in Annex I under EC Regulation 1107/2009

Efficacy guidance

- Apply to clean tubers before chitting, prior to planting, or at planting
- Apply flowable concentrate through canopied, hydraulic or spinning disc equipment (with or without electrostatics) mounted on a rolling conveyor or table [1]
- Flowable concentrate may be diluted with water up to 2.0 l per tonne to improve tuber coverage. Disease in areas not covered by spray will not be controlled [1]
- Dry powder may be applied via an on-planter applicator [2]

Restrictions

- Maximum number of treatments 1 per batch of seed tubers
- Check with processor before use on crops for processing

Crop-specific information

- Latest use: at planting
- Seed tubers should be of good quality and free from bacterial rots, physical damage or virus infection, and should not be sprouted to such an extent that mechanical damage to the shoots will occur during treatment or planting

Environmental safety

- Harmful to aquatic organisms

Hazard classification and safety precautions

Hazard Irritant [1, 2]; Toxic to aquatic organisms [1]
Transport code 9 [1]
Packaging group III [1]
UN Number 3082 [1]; N/C [2]
Risk phrases H317 [1]; H319, R53a [2]
Operator protection A, H [1, 2]; C [2]; U04a, U05a, U19a, U20a [1, 2]; U14 [1]
Environmental protection E15a, E34, E38 [1, 2]; H412 [2]
Storage and disposal D01, D02, D09a, D12a [1, 2]; D05, D10a [1]; D11a [2]
Treated seed S01, S03, S04a, S05 [1, 2]; S02 [2]
Medical advice M03

260   flutriafol

A broad-spectrum triazole fungicide for cereals
FRAC mode of action code: 3

See also chlorothalonil + flutriafol
fludioxonil + flutriafol

Products

1 Consul	Headland	125 g/l	SC	12976
2 Pointer	Headland	125 g/l	SC	12975

Uses

- Brown rust in **spring barley, winter barley, winter wheat**
- Powdery mildew in **spring barley, winter barley, winter wheat**
- Rhynchosporium in **spring barley, winter barley**
- Septoria leaf blotch in **winter wheat**
- Yellow rust in **spring barley, winter barley, winter wheat**

Approval information

- Flutriafol included in Annex I under EC Regulation 1107/2009
- Accepted by BBPA for use on malting barley

FOR FULL CONDITIONS OF USE ALWAYS READ THE PRODUCT LABEL

Efficacy guidance
- Best results obtained from treatment in early stages of disease development. See label for detailed guidance on spray timing for specific diseases and the need for repeat treatments
- Tank mix options available on the label to broaden activity spectrum
- Good spray coverage essential for optimum performance
- Flutriafol is a DMI fungicide. Resistance to some DMI fungicides has been identified in Septoria leaf blotch which may seriously affect performance of some products. For further advice contact a specialist advisor and visit the Fungicide Resistance Action Group (FRAG)-UK website

Restrictions
- Maximum number of treatments 2 per crop (including other products containing flutriafol)

Crop-specific information
- Latest use: before early grain milky ripe stage (GS 73)
- Flag leaf tip scorch on wheat caused by stress may be increased by fungicide treatment

Environmental safety
- Harmful to aquatic organisms
- Harmful to fish or other aquatic life. Do not contaminate surface waters or ditches with chemical or used container

Hazard classification and safety precautions
Hazard Harmful
UN Number N/C
Risk phrases H317
Operator protection A, H; U05a, U09a, U19a, U20b
Environmental protection E13c, E38, H411
Storage and disposal D01, D02, D09a, D10c

---

261   fluxapyroxad

Also known as Xemium, it is an SDHI fungicide for use in cereals.
FRAC mode of action code: 7

See also chlorothalonil + fluxapyroxad
epoxiconazole + fluxapyroxad
epoxiconazole + fluxapyroxad + pyraclostrobin
fluxapyroxad + metconazole

Products
1	Allstar	BASF	300 g/l	SC	18138
2	Bugle	BASF	59.4 g/l	EC	17821
3	Imtrex	BASF	62.5 g/l	EC	17108
4	Sercadis	BASF	300 g/l	SC	17776

Uses
- Brown rust in **durum wheat, rye, spring barley, spring wheat, triticale, winter barley, winter wheat** [2, 3]
- Crown rust in **spring oats** [2]; **winter oats** [2, 3]
- Disease control in **potatoes** [1]
- Net blotch in **spring barley, winter barley** [2, 3]
- Powdery mildew in **apples, pears** [4]; **durum wheat** (reduction), **rye** (reduction), **spring wheat** (reduction), **triticale** (reduction), **winter barley** (reduction), **winter oats** (reduction), **winter wheat** (reduction) [3]
- Rhynchosporium in **rye, spring barley, winter barley** [2, 3]
- Scab in **apples, pears** [4]
- Septoria leaf blotch in **durum wheat, spring wheat, triticale, winter wheat** [2, 3]
- Tan spot in **durum wheat** (reduction), **spring wheat** (reduction), **winter wheat** (reduction) [3]
- Yellow rust in **durum wheat** (moderate control), **rye** (moderate control), **spring barley** (moderate control), **spring wheat** (moderate control), **triticale** (moderate control), **winter barley** (moderate control), **winter wheat** (moderate control) [2, 3]

SEE SECTION 3 FOR PRODUCTS ALSO REGISTERED

Approval information

- Fluxapyroxad included in Annex I under EC Regulation 1107/2009
- Accepted by BBPA for use on malting barley up to GS 45 only (max. dose, 1 litre/ha) [2, 3]

Restrictions

- Do not apply more that two foliar applications of products containing SDHI fungicides to any cereal crop.
- Do not apply by hand-held equipment

Following crops guidance

- Only cereals, cabbages, carrots, chicory, clover, dwarf french beans, field beans, leeks, lettuce, linseed, maize, oats, oilseed rape, onions, peas, potatoes, radishes, ryegrass, soyabean, spinach, sunflowers or sugar beet may be sown as following crops after treatment

Environmental safety

- Buffer zone requirement 15 m [4]
- LERAP Category B [4]

Hazard classification and safety precautions

Hazard Harmful, Dangerous for the environment [2-4]; Harmful if inhaled [2, 3]; Very toxic to aquatic organisms [1, 4]

Transport code 9

Packaging group III

UN Number 3082

Risk phrases H319 [2, 3]; H351 [1-4]

Operator protection A, H; U05a, U20b [2-4]; U23a [1]

Environmental protection E15b, E34, E38 [2-4]; E16a [4]; H410 [1, 4]; H411 [2, 3]

Storage and disposal D01, D02, D05, D09a, D10c, D12a [2-4]

Medical advice M05a [2-4]

262   fluxapyroxad + metconazole

An SDHI and triazole mixture for disease control in cereals
FRAC mode of action code: 3 + 7

Products

1 Clayton Tardis	Clayton	62.5:45 g/l	EC	17459
2 Librax	BASF	62.5:45 g/l	EC	17107
3 Wolverine	Dow	62.5:45 g/l	EC	17729

Uses

- Brown rust in **durum wheat, rye, spring barley, spring wheat, triticale, winter barley, winter wheat**
- Cladosporium in **durum wheat** *(moderate control)*, **spring wheat** *(moderate control)*, **winter wheat** *(moderate control)*
- Disease control in **durum wheat, rye, spring barley, spring wheat, triticale, winter barley, winter wheat**
- Eyespot in **durum wheat** *(good reduction)*, **rye** *(good reduction)*, **spring wheat** *(good reduction)*, **triticale** *(good reduction)*, **winter wheat** *(good reduction)*
- Fusarium ear blight in **durum wheat** *(good reduction)*, **spring wheat** *(good reduction)*, **winter wheat** *(good reduction)*
- Net blotch in **spring barley, winter barley**
- Powdery mildew in **durum wheat** *(moderate control)*, **rye** *(moderate control)*, **spring barley** *(moderate control)*, **spring wheat** *(moderate control)*, **triticale** *(moderate control)*, **winter barley** *(moderate control)*, **winter wheat** *(moderate control)*
- Ramularia leaf spots in **spring barley, winter barley**
- Rhynchosporium in **rye, spring barley, winter barley**
- Septoria leaf blotch in **durum wheat, spring wheat, triticale, winter wheat**
- Tan spot in **durum wheat** *(good reduction)*, **spring wheat** *(good reduction)*, **winter wheat** *(good reduction)*

FOR FULL CONDITIONS OF USE ALWAYS READ THE PRODUCT LABEL

- Yellow rust in *durum wheat*, *rye*, *spring barley*, *spring wheat*, *triticale*, *winter barley*, *winter wheat*

Approval information
- Fluxapyroxad and metconazole included in Annex I under EC Regulation 1107/2009

Efficacy guidance
- Metconazole is a DMI fungicide. Resistance to some DMI fungicides has been identified in Septoria leaf blotch which may seriously affect performance of some products. For further advice contact a specialist advisor and visit the Fungicide Resistance Action Group (FRAG)-UK website

Restrictions
- Do not apply more that two foliar applications of products containing SDHI fungicides to any cereal crop.

Following crops guidance
- Only cereals, cabbages, carrots, clover, dwarf french beans, field beans, lettuce, maize, oats, oilseed rape, onions, peas, potatoes, ryegrass, sunflowers or sugar beet may be sown as following crops after treatment

Environmental safety
- LERAP Category B

Hazard classification and safety precautions
Hazard Harmful, Dangerous for the environment, Harmful if inhaled, Very toxic to aquatic organisms
Transport code 9
Packaging group III
UN Number 3082
Risk phrases H317, H319, H351, H361
Operator protection A, C, H; U05a, U20b
Environmental protection E15b, E16a, E34, E38, H410
Storage and disposal D01, D02, D09a, D10c, D12a
Medical advice M03

263   fluxapyroxad + pyraclostrobin

A fungicide mixture for disease control in cereals
FRAC mode of action code: 7 + 11

Products

1	Priaxor EC	BASF	75:150 g/l	EC	17371
2	Serpent	BASF	75:150 g/l	EC	17427

Uses
- Brown rust in *durum wheat*, *rye*, *spring wheat*, *triticale*, *winter wheat* [1]; *spring barley*, *winter barley* [1, 2]
- Crown rust in *spring oats*, *winter oats* [1]
- Glume blotch in *durum wheat* (moderate control), *spring wheat* (moderate control), *winter wheat* (moderate control) [1]
- Net blotch in *spring barley*, *winter barley* [1, 2]
- Powdery mildew in *durum wheat* (moderate control), *rye* (moderate control), *spring oats* (moderate control), *spring wheat* (moderate control), *triticale* (moderate control), *winter oats* (moderate control), *winter wheat* (moderate control) [1]; *spring barley* (moderate control), *winter barley* (moderate control) [1, 2]
- Ramularia leaf spots in *spring barley*, *winter barley* [1, 2]
- Rhynchosporium in *rye* [1]; *spring barley*, *winter barley* [1, 2]
- Septoria leaf blotch in *durum wheat*, *spring wheat*, *triticale*, *winter wheat* [1]
- Tan spot in *durum wheat*, *spring wheat*, *winter wheat* [1]

SEE SECTION 3 FOR PRODUCTS ALSO REGISTERED

- Yellow rust in **durum wheat**, **rye**, **spring wheat**, **triticale**, **winter wheat** [1]; **spring barley**, **winter barley** [1, 2]

Approval information
- Fluxapyroxad and pyraclostrobin included in Annex I under EC Regulation 1107/2009

Environmental safety
- LERAP Category B

Hazard classification and safety precautions
Hazard Harmful if swallowed, Harmful if inhaled, Very toxic to aquatic organisms
Transport code 9
Packaging group III
UN Number 3082
Risk phrases H351
Operator protection A, H; U05a, U20a
Environmental protection E15b, E16a, H410
Storage and disposal D01, D02, D10c

---

264   folpet

A multi-site protectant fungicide for disease control in wheat and barley
FRAC mode of action code: M4

See also epoxiconazole + folpet

Products
1	Arizona	Adama	500 g/l	SC	15318
2	Phoenix	Adama	500 g/l	SC	15259

Uses
- Rhynchosporium in **spring barley** *(reduction)*, **winter barley** *(reduction)*
- Septoria leaf blotch in **spring wheat** *(reduction)*, **winter wheat** *(reduction)*

Approval information
- Folpet included in Annex I under EC Regulation 1107/2009
- Accepted by BBPA for use on cereals

Efficacy guidance
- Folpet is a protectant fungicide and so the first application must be made before the disease becomes established in the crop. A second application should be timed to protect new growth.
- Folpet only has contact activity and so good coverage of the target foliage is essential for good activity.

Environmental safety
- LERAP Category B

Hazard classification and safety precautions
Hazard Harmful, Dangerous for the environment
Transport code 9
Packaging group III
UN Number 3082
Risk phrases R40, R50, R53a
Operator protection A, H; U05a, U08, U11, U14, U15, U20a
Environmental protection E15b, E16a, E34, E38
Storage and disposal D01, D02, D09a, D12a
Medical advice M03

265   foramsulfuron + iodosulfuron-methyl-sodium

A sulfonylurea herbicide mixture for weed control in forage and grain maize.
HRAC mode of action code: B

See also iodosulfuron-methyl-sodium

Products

Maister WG	Bayer CropScience	0.3:0.01% w/w	SG	16116

Uses
- Annual dicotyledons in **forage maize**, **grain maize**
- Annual meadow grass in **forage maize**, **grain maize**
- Black nightshade in **forage maize**, **grain maize**
- Chickweed in **forage maize**, **grain maize**
- Cockspur grass in **forage maize**, **grain maize**
- Couch in **forage maize**, **grain maize**
- Fat hen in **forage maize**, **grain maize**
- Knotgrass in **forage maize**, **grain maize**
- Mayweeds in **forage maize**, **grain maize**
- Shepherd's purse in **forage maize**, **grain maize**

Approval information
- Foramsulfuron and iodosulfuron-methyl-sodium included in Annex 1 under EC Regulation 1107/2009

Efficacy guidance
- Do not spray if rain is imminent or if temperature is above 25°C.
- Do not use in a non-CRD approved tank-mixture or sequence with another ALS or sulfonyl urea herbicide.
- Do not use on crops suffering from stress.
- Do not use on crops undersown with grass or a broad-leaved crop.

Environmental safety
- Buffer zone requirement 11m [1]
- LERAP Category B

Hazard classification and safety precautions
Hazard Irritant, Dangerous for the environment, Very toxic to aquatic organisms
Transport code 9
Packaging group III
UN Number 3077
Risk phrases H317
Operator protection A, D, H; U05a, U14
Environmental protection E15b, E16a, E38, H410
Storage and disposal D01, D02, D10a, D12a

266   fosetyl-aluminium + propamocarb hydrochloride

A systemic and protectant fungicide mixture for use in horticulture
FRAC mode of action code: 33 + 28

See also propamocarb hydrochloride

Products

1	Pan Cradle	Pan Agriculture	310:530 g/l	SL	15923
2	Previcur Energy	Bayer CropScience	310:530 g/l	SL	15367

Uses
- Damping off in **herbs (see appendix 6)** (off-label), **protected aubergines** (off-label), **protected chilli peppers** (off-label), **protected courgettes** (off-label), **protected forest nurseries** (off-

label), **protected gherkins** *(off-label)*, **protected hops** *(off-label)*, **protected marrows** *(off-label)*, **protected pumpkins** *(off-label)*, **protected soft fruit** *(off-label)*, **protected squashes** *(off-label)*, **protected sweet peppers** *(off-label)*, **protected top fruit** *(off-label)*, **protected watermelon** *(off-label)* [2]; **protected broccoli**, **protected brussels sprouts**, **protected cabbages**, **protected calabrese**, **protected cauliflowers**, **protected chinese cabbage**, **protected collards**, **protected kale**, **protected melons**, **protected radishes** [1, 2]

- Downy mildew in **herbs (see appendix 6)** *(off-label)*, **ornamental plant production** *(off-label)*, **protected forest nurseries** *(off-label)*, **protected hops** *(off-label)*, **protected ornamentals** *(off-label)*, **protected soft fruit** *(off-label)*, **protected top fruit** *(off-label)*, **spinach** *(off-label)* [2]; **lettuce**, **protected broccoli**, **protected brussels sprouts**, **protected cabbages**, **protected calabrese**, **protected cauliflowers**, **protected chinese cabbage**, **protected collards**, **protected kale**, **protected lettuce**, **protected melons**, **protected radishes** [1, 2]
- Pythium in **lettuce**, **protected cucumbers**, **protected lettuce**, **protected tomatoes** [1, 2]; **protected ornamentals** *(off-label)* [2]

Extension of Authorisation for Minor Use (EAMUs)
- **herbs (see appendix 6)** *20130983* [2]
- **ornamental plant production** *20131845* [2]
- **protected aubergines** *20111553* [2]
- **protected chilli peppers** *20111553* [2]
- **protected courgettes** *20111556* [2]
- **protected forest nurseries** *20122045* [2]
- **protected gherkins** *20111556* [2]
- **protected hops** *20122045* [2]
- **protected marrows** *20111556* [2]
- **protected ornamentals** *20111557* [2], *20131845* [2]
- **protected pumpkins** *20111556* [2]
- **protected soft fruit** *20122045* [2]
- **protected squashes** *20111556* [2]
- **protected sweet peppers** *20111553* [2]
- **protected top fruit** *20122045* [2]
- **protected watermelon** *20111556* [2]
- **spinach** *20112452* [2]

Approval information
- Fosetyl-aluminium and propamocarb hydrochloride included in Annex I under EC Regulation 1107/2009

Hazard classification and safety precautions
Hazard Irritant
UN Number N/C
Risk phrases H317
Operator protection A, H, M; U04a, U05a, U08, U14, U19a, U20b
Environmental protection E15a
Storage and disposal D09a, D11a
Medical advice M03

267 fosthiazate

An organophosphorus contact nematicide for potatoes
IRAC mode of action code: 1B

Products

Nemathorin 10G	Syngenta	10% w/w	FG	11003

Uses
- Nematodes in **hops** *(off-label)*, **ornamental plant production** *(off-label)*, **soft fruit** *(off-label)*
- Potato cyst nematode in **potatoes**
- Spraing vectors in **potatoes** *(reduction)*
- Wireworm in **potatoes** *(reduction)*

FOR FULL CONDITIONS OF USE ALWAYS READ THE PRODUCT LABEL

Extension of Authorisation for Minor Use (EAMUs)
- *hops* 20082912
- *ornamental plant production* 20082912
- *soft fruit* 20082912

Approval information
- Fosthiazate included in Annex I under EC Regulation 1107/2009
- In 2006 CRD required that all products containing this active ingredient should carry the following warning in the main area of the container label: "Fosthiazate is an anticholinesterase organophosphate. Handle with care"

Efficacy guidance
- Application best achieved using equipment such as Horstine Farmery Microband Applicator, Matco or Stocks Micrometer applicators together with a rear mounted powered rotary cultivator
- Granules must not become wet or damp before use.
- For optimum efficacy incorporate evenly in to top 20 cm of soil.
- Apply as close as possible to time of planting.

Restrictions
- Contains an anticholinesterase organophosphorus compound. Do not use if under medical advice not to work with such compounds
- Maximum number of treatments 1 per crop
- Product must only be applied using tractor-mounted/drawn direct placement machinery. Do not use air assisted broadcast machinery other than that referenced on the product label
- Do not allow granules to stand overnight in the application hopper
- Do not apply more than once every four years on the same area of land
- Do not use on crops to be harvested less than 17 wk after treatment
- Consult before using on crops intended for processing

Crop-specific information
- Latest use: at planting
- HI 17 wk for potatoes

Environmental safety
- Dangerous for the environment
- Toxic to aquatic organisms
- Dangerous to game, wild birds and animals
- Dangerous to livestock. Keep all livestock out of treated areas for at least 13 wk
- Incorporation to 10-15 cm and ridging up of treated soil must be carried out immediately after application. Powered rotary cultivators are preferred implements for incorporation but discs, power, spring tine or Dutch harrows may be used provided two passes are made at right angles
- To protect birds and wild mammals remove spillages
- Failure completely to bury granules immediately after application is hazardous to wildlife
- To protect groundwater do not apply any product containing fosthiazate more than once every four yr

Hazard classification and safety precautions
Hazard Harmful, Dangerous for the environment, Toxic if swallowed, Very toxic to aquatic organisms
Transport code 9
Packaging group III
UN Number 3077
Risk phrases H317
Operator protection A, E, G, H, K, M; U02a, U04a, U05a, U09a, U13, U14, U19a, U20a
Environmental protection E06b (13 wk); E15b, E34, E38, H410
Storage and disposal D01, D02, D05, D09a, D11a, D14
Medical advice M01, M03, M05a

SEE SECTION 3 FOR PRODUCTS ALSO REGISTERED

268   garlic extract

A naturally occuring nematicide and animal repellent

Products

NEMguard DE	Certis	45% w/w	GR	16749

Uses
- Free-living nematodes in **fodder beet** *(off-label)*, **red beet** *(off-label)*
- Nematodes in **carrots**, **parsnips**
- Root nematodes in **bulb onions** *(off-label)*, **garlic** *(off-label)*, **leeks** *(off-label)*, **shallots** *(off-label)*
- Root-knot nematodes in **bulb onions** *(off-label)*, **garlic** *(off-label)*, **leeks** *(off-label)*, **shallots** *(off-label)*
- Stem and bulb nematodes in **bulb onions** *(off-label)*, **garlic** *(off-label)*, **leeks** *(off-label)*, **shallots** *(off-label)*

Extension of Authorisation for Minor Use (EAMUs)
- **bulb onions** *20151838 expires 12 Nov 2018*
- **fodder beet** *20151841 expires 12 Nov 2018*
- **garlic** *20151838 expires 12 Nov 2018*
- **leeks** *20151838 expires 12 Nov 2018*
- **red beet** *20151841 expires 12 Nov 2018*
- **shallots** *20151838 expires 12 Nov 2018*

Approval information
- Garlic extract included in Annex 1 under EC Regulation 1107/2009

Hazard classification and safety precautions
    Hazard Irritant, Dangerous for the environment
    UN Number N/C
    Risk phrases H315, H317, H320
    Operator protection A, H; U11, U14
    Environmental protection E38, H411

269   gibberellins

A plant growth regulator for use in top fruit and grassland

Products

1	Gibb 3	Globachem	10% w/w	TB	17013
2	Gibb Plus	Globachem	10 g/l	SL	17251
3	Novagib	Fine	10 g/l	SL	08954
4	Regulex 10 SG	Interfarm	10% w/w	SG	17158
5	Smartgrass	Interfarm	40% w/w	SG	15628

Uses
- Growth regulation in **grassland** [5]; **pears** *(off-label)* [3]; **protected rhubarb** *(off-label)* [1]
- Improved germination in **nothofagus** [4]
- Increasing fruit set in **pears** [1]
- Reducing fruit russeting in **apples** [2, 3]; **pears** *(off-label)*, **quinces** *(off-label)* [2]
- Russet reduction in **apples**, **pears** [4]

Extension of Authorisation for Minor Use (EAMUs)
- **pears** *20171152* [2], *20012756* [3]
- **protected rhubarb** *20162836* [1]
- **quinces** *20171152* [2]

Approval information
- Gibberellins included in Annex 1 under EC Regulation 1107/2009

Efficacy guidance

- For optimum results on apples spray under humid, slow drying conditions and ensure good spray cover [3]
- Treat apples and pears immediately after completion of petal fall and repeat as directed on the label [3]
- Fruit set in pears can be improved when blossom is spare, setting is poor or where frost has killed many flowers [1]

Restrictions

- Maximum total dose varies with crop and product. Check label
- Prepared spray solutions are unstable. Do not leave in the sprayer during meal breaks or overnight
- Return bloom may be reduced in yr following treatment
- Consult processor before treating crops grown for processing [1, 3]
- Avoid storage at temperatures above 32°C [1]
- Do not exceed 2.5 g per 5 litres when used as a seed soak in Nothofagus [4]

Crop-specific information

- HI zero for apples [2-4]
- Apply to apples at completion of petal fall and repeat 3 or 4 times at 7-10 d intervals. Number of sprays and spray interval depend on weather conditions and dose (see labels) [3]
- Good results achieved on apple varieties Cox's Orange Pippin, Discovery, Golden Delicious and Karmijn. For other cultivars test on a small number of trees [3]
- Split treatment on pears allows second spray to be omitted if pollinating conditions become very favourable [1]
- Pear variety Conference usually responds well to treatment; Doyenne du Comice can be variable [1]

Hazard classification and safety precautions

UN Number N/C
Operator protection U05a [4]; U08 [1]; U20b [3]; U20c [1, 2, 4]
Environmental protection E15a [1-4]
Storage and disposal D01 [3, 4]; D02 [4]; D03, D05, D07 [3]; D09a [1-4]; D10b [2-4]

270   Gliocladium catenulatum

A fungus that acts as an antagonist against other fungi
FRAC mode of action code: 44

Products

Prestop	Everris Ltd	32% w/w	WP	17223

Uses

- Botrytis in *almonds* (off-label), *angelica* (off-label), *apples* (off-label), *apricots* (off-label), *asparagus* (off-label), *baby leaf crops* (off-label), *beans without pods (fresh)* (off-label), *bilberries* (off-label), *blackcurrants* (off-label), *blueberries* (off-label), *broad beans* (off-label), *broccoli* (off-label), *brussels sprouts* (off-label), *bulb vegetables* (off-label), *cabbages* (off-label), *calabrese* (off-label), *cane fruit* (off-label), *carrots* (off-label), *cauliflowers* (off-label), *celeriac* (off-label), *celery (outdoor)* (off-label), *celery leaves* (off-label), *cherries* (off-label), *chervil* (off-label), *chestnuts* (off-label), *chicory root* (off-label), *chives* (off-label), *choi sum* (off-label), *collards* (off-label), *cranberries* (off-label), *cress* (off-label), *dwarf beans* (off-label), *edible flowers* (off-label), *edible podded peas* (off-label), *endives* (off-label), *fennel leaves* (off-label), *florence fennel* (off-label), *french beans* (off-label), *fruiting vegetables* (off-label), *globe artichoke* (off-label), *gooseberries* (off-label), *hazel nuts* (off-label), *herbs (see appendix 6)* (off-label), *hops* (off-label), *horseradish* (off-label), *hyssop* (off-label), *jerusalem artichokes* (off-label), *kale* (off-label), *kohlrabi* (off-label), *lamb's lettuce* (off-label), *land cress* (off-label), *leeks* (off-label), *lentils* (off-label), *lettuce* (off-label), *lovage* (off-label), *marjoram* (off-label), *medlar* (off-label), *mint* (off-label), *nectarines* (off-label), *oregano* (off-label), *oriental cabbage* (off-label), *ornamental plant production* (off-label), *parsley* (off-label), *parsley root* (off-label), *parsnips* (off-label), *peaches* (off-label), *pears* (off-label), *plums* (off-label), *quinces* (off-label),

> *radishes (off-label), **red beet** (off-label), **redcurrants** (off-label), **rhubarb** (off-label), **rocket** (off-label), **rosemary** (off-label), **runner beans** (off-label), **sage** (off-label), **salad burnet** (off-label), **salsify** (off-label), **savory** (off-label), **seakale** (off-label), **soya beans** (off-label), **spinach** (off-label), **spinach beet** (off-label), **swedes** (off-label), **sweet ciceley** (off-label), **table grapes** (off-label), **tarragon** (off-label), **thyme** (off-label), **turnips** (off-label), **vining peas** (off-label), **walnuts** (off-label), **wine grapes** (off-label)*

- Didymella in **all protected edible crops** *(moderate control)*, **all protected non-edible crops** *(moderate control)*, **strawberries** *(moderate control)*
- Fusarium in **all protected edible crops** *(moderate control)*, **all protected non-edible crops** *(moderate control)*, **almonds** *(off-label)*, **angelica** *(off-label)*, **apples** *(off-label)*, **apricots** *(off-label)*, **asparagus** *(off-label)*, **baby leaf crops** *(off-label)*, **beans without pods (fresh)** *(off-label)*, **bilberries** *(off-label)*, **blackcurrants** *(off-label)*, **blueberries** *(off-label)*, **broad beans** *(off-label)*, **broccoli** *(off-label)*, **brussels sprouts** *(off-label)*, **bulb vegetables** *(off-label)*, **cabbages** *(off-label)*, **calabrese** *(off-label)*, **cane fruit** *(off-label)*, **carrots** *(off-label)*, **cauliflowers** *(off-label)*, **celeriac** *(off-label)*, **celery (outdoor)** *(off-label)*, **celery leaves** *(off-label)*, **cherries** *(off-label)*, **chervil** *(off-label)*, **chestnuts** *(off-label)*, **chicory root** *(off-label)*, **chives** *(off-label)*, **choi sum** *(off-label)*, **collards** *(off-label)*, **cranberries** *(off-label)*, **cress** *(off-label)*, **dwarf beans** *(off-label)*, **edible flowers** *(off-label)*, **edible podded peas** *(off-label)*, **endives** *(off-label)*, **fennel leaves** *(off-label)*, **florence fennel** *(off-label)*, **french beans** *(off-label)*, **fruiting vegetables** *(off-label)*, **globe artichoke** *(off-label)*, **gooseberries** *(off-label)*, **hazel nuts** *(off-label)*, **herbs (see appendix 6)** *(off-label)*, **hops** *(off-label)*, **horseradish** *(off-label)*, **hyssop** *(off-label)*, **jerusalem artichokes** *(off-label)*, **kale** *(off-label)*, **kohlrabi** *(off-label)*, **lamb's lettuce** *(off-label)*, **land cress** *(off-label)*, **leeks** *(off-label)*, **lentils** *(off-label)*, **lettuce** *(off-label)*, **lovage** *(off-label)*, **marjoram** *(off-label)*, **medlar** *(off-label)*, **mint** *(off-label)*, **nectarines** *(off-label)*, **oregano** *(off-label)*, **oriental cabbage** *(off-label)*, **ornamental plant production** *(off-label)*, **parsley** *(off-label)*, **parsley root** *(off-label)*, **parsnips** *(off-label)*, **peaches** *(off-label)*, **pears** *(off-label)*, **plums** *(off-label)*, **quinces** *(off-label)*, **radishes** *(off-label)*, **red beet** *(off-label)*, **redcurrants** *(off-label)*, **rhubarb** *(off-label)*, **rocket** *(off-label)*, **rosemary** *(off-label)*, **runner beans** *(off-label)*, **sage** *(off-label)*, **salad burnet** *(off-label)*, **salsify** *(off-label)*, **savory** *(off-label)*, **seakale** *(off-label)*, **soya beans** *(off-label)*, **spinach** *(off-label)*, **spinach beet** *(off-label)*, **strawberries** *(moderate control)*, **swedes** *(off-label)*, **sweet ciceley** *(off-label)*, **table grapes** *(off-label)*, **tarragon** *(off-label)*, **thyme** *(off-label)*, **turnips** *(off-label)*, **vining peas** *(off-label)*, **walnuts** *(off-label)*, **wine grapes** *(off-label)*
- Phytophthora in **all protected edible crops** *(moderate control)*, **all protected non-edible crops** *(moderate control)*, **almonds** *(off-label)*, **angelica** *(off-label)*, **apples** *(off-label)*, **apricots** *(off-label)*, **asparagus** *(off-label)*, **baby leaf crops** *(off-label)*, **beans without pods (fresh)** *(off-label)*, **bilberries** *(off-label)*, **blackcurrants** *(off-label)*, **blueberries** *(off-label)*, **broad beans** *(off-label)*, **broccoli** *(off-label)*, **brussels sprouts** *(off-label)*, **bulb vegetables** *(off-label)*, **cabbages** *(off-label)*, **calabrese** *(off-label)*, **cane fruit** *(off-label)*, **carrots** *(off-label)*, **cauliflowers** *(off-label)*, **celeriac** *(off-label)*, **celery (outdoor)** *(off-label)*, **celery leaves** *(off-label)*, **cherries** *(off-label)*, **chervil** *(off-label)*, **chestnuts** *(off-label)*, **chicory root** *(off-label)*, **chives** *(off-label)*, **choi sum** *(off-label)*, **collards** *(off-label)*, **cranberries** *(off-label)*, **cress** *(off-label)*, **dwarf beans** *(off-label)*, **edible flowers** *(off-label)*, **edible podded peas** *(off-label)*, **endives** *(off-label)*, **fennel leaves** *(off-label)*, **florence fennel** *(off-label)*, **french beans** *(off-label)*, **fruiting vegetables** *(off-label)*, **globe artichoke** *(off-label)*, **gooseberries** *(off-label)*, **hazel nuts** *(off-label)*, **herbs (see appendix 6)** *(off-label)*, **hops** *(off-label)*, **horseradish** *(off-label)*, **hyssop** *(off-label)*, **jerusalem artichokes** *(off-label)*, **kale** *(off-label)*, **kohlrabi** *(off-label)*, **lamb's lettuce** *(off-label)*, **land cress** *(off-label)*, **leeks** *(off-label)*, **lentils** *(off-label)*, **lettuce** *(off-label)*, **lovage** *(off-label)*, **marjoram** *(off-label)*, **medlar** *(off-label)*, **mint** *(off-label)*, **nectarines** *(off-label)*, **oregano** *(off-label)*, **oriental cabbage** *(off-label)*, **ornamental plant production** *(off-label)*, **parsley** *(off-label)*, **parsley root** *(off-label)*, **parsnips** *(off-label)*, **peaches** *(off-label)*, **pears** *(off-label)*, **plums** *(off-label)*, **quinces** *(off-label)*, **radishes** *(off-label)*, **red beet** *(off-label)*, **redcurrants** *(off-label)*, **rhubarb** *(off-label)*, **rocket** *(off-label)*, **rosemary** *(off-label)*, **runner beans** *(off-label)*, **sage** *(off-label)*, **salad burnet** *(off-label)*, **salsify** *(off-label)*, **savory** *(off-label)*, **seakale** *(off-label)*, **soya beans** *(off-label)*, **spinach** *(off-label)*, **spinach beet** *(off-label)*, **strawberries** *(moderate control)*, **swedes** *(off-label)*, **sweet ciceley** *(off-label)*, **table grapes** *(off-label)*, **tarragon** *(off-label)*, **thyme** *(off-label)*, **turnips** *(off-label)*, **vining peas** *(off-label)*, **walnuts** *(off-label)*, **wine grapes** *(off-label)*

FOR FULL CONDITIONS OF USE ALWAYS READ THE PRODUCT LABEL

- Pythium in **all protected edible crops** *(moderate control)*, **all protected non-edible crops** *(moderate control)*, **strawberries** *(moderate control)*
- Rhizoctonia in **all protected edible crops** *(moderate control)*, **all protected non-edible crops** *(moderate control)*, **almonds** *(off-label)*, **angelica** *(off-label)*, **apples** *(off-label)*, **apricots** *(off-label)*, **asparagus** *(off-label)*, **baby leaf crops** *(off-label)*, **beans without pods (fresh)** *(off-label)*, **bilberries** *(off-label)*, **blackcurrants** *(off-label)*, **blueberries** *(off-label)*, **broad beans** *(off-label)*, **broccoli** *(off-label)*, **brussels sprouts** *(off-label)*, **bulb vegetables** *(off-label)*, **cabbages** *(off-label)*, **calabrese** *(off-label)*, **cane fruit** *(off-label)*, **carrots** *(off-label)*, **cauliflowers** *(off-label)*, **celeriac** *(off-label)*, **celery (outdoor)** *(off-label)*, **celery leaves** *(off-label)*, **cherries** *(off-label)*, **chervil** *(off-label)*, **chestnuts** *(off-label)*, **chicory root** *(off-label)*, **chives** *(off-label)*, **choi sum** *(off-label)*, **collards** *(off-label)*, **cranberries** *(off-label)*, **cress** *(off-label)*, **dwarf beans** *(off-label)*, **edible flowers** *(off-label)*, **edible podded peas** *(off-label)*, **endives** *(off-label)*, **fennel leaves** *(off-label)*, **florence fennel** *(off-label)*, **french beans** *(off-label)*, **fruiting vegetables** *(off-label)*, **globe artichoke** *(off-label)*, **gooseberries** *(off-label)*, **hazel nuts** *(off-label)*, **herbs (see appendix 6)** *(off-label)*, **hops** *(off-label)*, **horseradish** *(off-label)*, **hyssop** *(off-label)*, **jerusalem artichokes** *(off-label)*, **kale** *(off-label)*, **kohlrabi** *(off-label)*, **lamb's lettuce** *(off-label)*, **land cress** *(off-label)*, **leeks** *(off-label)*, **lentils** *(off-label)*, **lettuce** *(off-label)*, **lovage** *(off-label)*, **marjoram** *(off-label)*, **medlar** *(off-label)*, **mint** *(off-label)*, **nectarines** *(off-label)*, **oregano** *(off-label)*, **oriental cabbage** *(off-label)*, **ornamental plant production** *(off-label)*, **parsley** *(off-label)*, **parsley root** *(off-label)*, **parsnips** *(off-label)*, **peaches** *(off-label)*, **pears** *(off-label)*, **plums** *(off-label)*, **quinces** *(off-label)*, **radishes** *(off-label)*, **red beet** *(off-label)*, **redcurrants** *(off-label)*, **rhubarb** *(off-label)*, **rocket** *(off-label)*, **rosemary** *(off-label)*, **runner beans** *(off-label)*, **sage** *(off-label)*, **salad burnet** *(off-label)*, **salsify** *(off-label)*, **savory** *(off-label)*, **seakale** *(off-label)*, **soya beans** *(off-label)*, **spinach** *(off-label)*, **spinach beet** *(off-label)*, **strawberries** *(moderate control)*, **swedes** *(off-label)*, **sweet ciceley** *(off-label)*, **table grapes** *(off-label)*, **tarragon** *(off-label)*, **thyme** *(off-label)*, **turnips** *(off-label)*, **vining peas** *(off-label)*, **walnuts** *(off-label)*, **wine grapes** *(off-label)*

Extension of Authorisation for Minor Use (EAMUs)
- **almonds** *20152773*
- **angelica** *20152773*
- **apples** *20152773*
- **apricots** *20152773*
- **asparagus** *20152773*
- **baby leaf crops** *20152773*
- **beans without pods (fresh)** *20152773*
- **bilberries** *20152773*
- **blackcurrants** *20152773*
- **blueberries** *20152773*
- **broad beans** *20152773*
- **broccoli** *20152773*
- **brussels sprouts** *20152773*
- **bulb vegetables** *20152773*
- **cabbages** *20152773*
- **calabrese** *20152773*
- **cane fruit** *20152773*
- **carrots** *20152773*
- **cauliflowers** *20152773*
- **celeriac** *20152773*
- **celery (outdoor)** *20152773*
- **celery leaves** *20152773*
- **cherries** *20152773*
- **chervil** *20152773*
- **chestnuts** *20152773*
- **chicory root** *20152773*
- **chives** *20152773*
- **choi sum** *20152773*
- **collards** *20152773*
- **cranberries** *20152773*
- **cress** *20152773*

SEE SECTION 3 FOR PRODUCTS ALSO REGISTERED

- *dwarf beans* 20152773
- *edible flowers* 20152773
- *edible podded peas* 20152773
- *endives* 20152773
- *fennel leaves* 20152773
- *florence fennel* 20152773
- *french beans* 20152773
- *fruiting vegetables* 20152773
- *globe artichoke* 20152773
- *gooseberries* 20152773
- *hazel nuts* 20152773
- *herbs (see appendix 6)* 20152773
- *hops* 20152773
- *horseradish* 20152773
- *hyssop* 20152773
- *jerusalem artichokes* 20152773
- *kale* 20152773
- *kohlrabi* 20152773
- *lamb's lettuce* 20152773
- *land cress* 20152773
- *leeks* 20152773
- *lentils* 20152773
- *lettuce* 20152773
- *lovage* 20152773
- *marjoram* 20152773
- *medlar* 20152773
- *mint* 20152773
- *nectarines* 20152773
- *oregano* 20152773
- *oriental cabbage* 20152773
- *ornamental plant production* 20152773
- *parsley* 20152773
- *parsley root* 20152773
- *parsnips* 20152773
- *peaches* 20152773
- *pears* 20152773
- *plums* 20152773
- *quinces* 20152773
- *radishes* 20152773
- *red beet* 20152773
- *redcurrants* 20152773
- *rhubarb* 20152773
- *rocket* 20152773
- *rosemary* 20152773
- *runner beans* 20152773
- *sage* 20152773
- *salad burnet* 20152773
- *salsify* 20152773
- *savory* 20152773
- *seakale* 20152773
- *soya beans* 20152773
- *spinach* 20152773
- *spinach beet* 20152773
- *swedes* 20152773
- *sweet ciceley* 20152773
- *table grapes* 20152773
- *tarragon* 20152773
- *thyme* 20152773
- *turnips* 20152773

- **vining peas** *20152773*
- **walnuts** *20152773*
- **wine grapes** *20152773*

Approval information
- Gliocladium catenulatum included in Annex 1 under EC Regulation 1107/2009

Hazard classification and safety precautions
UN Number N/C
Operator protection A, D, H; U05a, U10, U14, U15, U20a
Environmental protection E15b, E34
Storage and disposal D01, D02, D09a, D10a
Medical advice M03

271   glufosinate-ammonium

A non-selective, non-residual phosphinic acid contact herbicide
HRAC mode of action code: H

Products

1	Amenity Whippet	Harvest	150 g/l	SL	17931
2	Finale 150	Bayer CropScience	150 g/l	SL	16733
3	Glam	Pan Amenity	150 g/l	SL	17707
4	Halta	Belcrop	150 g/l	SL	17315
5	Tailout	ProKlass	150 g/l	SL	17854
6	Whippet	AgChem Access	150 g/l	SL	17669

Uses
- Annual and perennial weeds in **hard surfaces, managed amenity turf, natural surfaces not intended to bear vegetation, permeable surfaces overlying soil** [1-3, 5]
- Desiccation in **potatoes** [4, 6]

Approval information
- Glufosinate-ammonium included in Annex I under EC Regulation 1107/2009
- Accepted by BBPA for use on malting barley and hops

Efficacy guidance
- Activity quickest under warm, moist conditions. Light rainfall 3-4 h after application will not affect activity. Do not spray wet foliage or if rain likely within 6 h
- For weed control uses treat when weeds growing actively. Deep rooted weeds may require second treatment
- On uncropped headlands apply in May/Jun to prevent weeds invading field
- Glufosinate kills all green tissue but does not harm mature bark

Restrictions
- Do not desiccate potatoes in exceptionally wet weather or in saturated soil. See label for details
- Do not spray hedge bottoms or on broadcast crops
- Must be applied only between 1st March and 30th September
- Applications must be made as band or spot treatments using low drift nozzles and/or spray shield [6]

Crop-specific information
- Crops can normally be sown/planted immediately after spraying or sprayed post-drilling. On sand, very light or immature peat soils allow at least 3 d before sowing/planting or expected emergence
- For grass destruction apply before winter dormancy occurs. Heavily grazed fields should show active regrowth. Plough from the day after spraying
- For potato haulm desiccation apply to listed varieties (not seed crops) at onset of senescence, 14-21 d before harvest

Environmental safety
- Harmful to fish or other aquatic life. Do not contaminate surface waters or ditches with chemical or used container
- Keep livestock out of treated areas until foliage of any poisonous weeds such as ragwort has died and become unpalatable

Hazard classification and safety precautions

Hazard Harmful [4-6]; Harmful if swallowed, Toxic in contact with skin [1-6]; Harmful if inhaled [4]
Transport code 6.1
Packaging group III
UN Number 2902
Risk phrases H312 [4]; H318, H360, H373 [1-6]
Operator protection A, C [1-6]; H [3, 4, 6]; U04a [2, 4-6]; U04b [3]; U05a, U13, U14, U15, U19a [2-6]; U08 [3-6]; U09a [2]; U11, U20a [4-6]
Environmental protection E07a, E22c [4-6]; E15a [2-6]; E22b [2, 3]; E34 [3-6]; H411 [4]
Storage and disposal D01, D02, D09a, D10b [3-6]; D05 [3]; D12a [4-6]
Medical advice M03 [3-6]

272   glyphosate

A translocated non-residual glycine derivative herbicide
HRAC mode of action code: G

See also 2,4-D + glyphosate
diflufenican + glyphosate
diuron + glyphosate
flufenacet + glyphosate + metosulam
glyphosate + pyraflufen-ethyl
glyphosate + sulfosulfuron

Products

1	Amega Duo	Nufarm UK	540 g/l	SL	13358
2	Amega Duo	Nufarm UK	540 g/l	SL	14131
3	Ardee XL	Barclay	360 g/l	SL	17515
4	Asteroid	Headland	360 g/l	SL	11118
5	Asteroid Pro	Headland	360 g/l	SL	15373
6	Asteroid Pro 450	Headland	450 g/l	SL	16384
7	Azural	Monsanto	360 g/l	SL	16239
8	Barbarian XL	Barclay	360 g/l	SL	17665
9	Barclay Gallup Biograde 360	Barclay	360 g/l	SL	17612
10	Barclay Gallup Biograde Amenity	Barclay	360 g/l	SL	12716
11	Barclay Gallup Biograde Amenity	Barclay	360 g/l	SL	17674
12	Barclay Gallup Hi-Aktiv	Barclay	490 g/l	SL	17646
13	Barclay Glyde 144	Barclay	144 g/l	SL	15362
14	Buggy XTG	Sipcam	36% w/w	SG	12951
15	CDA Vanquish Biactive	Bayer CropScience	120 g/l	UL	12586
16	Clinic UP	Nufarm UK	360 g/l	SL	17893
17	Clipper	Adama	360 g/l	SL	14820
18	Credit	Nufarm UK	540 g/l	SL	16775
19	Crestler	Nufarm UK	540 g/l	SL	16555
20	Crestler	Nufarm UK	540 g/l	SL	16663
21	Discman Biograde	Barclay	216 g/l	SL	12856
22	Envision	Headland	450 g/l	SL	10569
23	Gallup Hi-Aktiv Amenity	Barclay	360 g/l	SL	17681
24	Gallup XL	Barclay	360 g/l	SL	17663
25	Garryowen XL	Barclay	360 g/l	SL	17508

FOR FULL CONDITIONS OF USE ALWAYS READ THE PRODUCT LABEL

Products – continued

26	Glyfos Dakar	Headland	68% w/w	SG	13054
27	Glyfos Dakar Pro	Headland Amenity	68% w/w	SG	13147
28	Hilite	Nomix Enviro	144 g/l	RC	13871
29	Klean G	Ventura	360 g/l	SL	17645
30	Landmaster 360 TF	Albaugh UK	360 g/l	SL	17683
31	Liaison	Monsanto	360 g/l	SL	17089
32	Mascot Hi-Aktiv Amenity	Rigby Taylor	490 g/l	SL	15178
33	Master Gly	Generica	360 g/l	SL	17594
34	Mentor	Monsanto	360 g/l	SL	16508
35	Monsanto Amenity Glyphosate	Monsanto	360 g/l	SL	16382
36	Monsanto Amenity Glyphosate XL	Monsanto	360 g/l	SL	17997
37	Motif	Monsanto	360 g/l	SL	16509
38	Nomix Conqueror Amenity	Nomix Enviro	144 g/l	RC	16039
39	Palomo Green	ProKlass	360 g/l	SL	17081
40	Proliance Quattro	Syngenta	360 g/l	SL	13670
41	Rattler	Nufarm UK	540 g/l	SL	15522
42	Rattler	Nufarm UK	540 g/l	SL	15692
43	Reaper Green	AgChem Access	360 g/l	SL	15935
44	Rodeo	Monsanto	360 g/l	SL	16242
45	Rosate 360 TF	Albaugh UK	360 g/l	SL	17682
46	Rosate Green	Albaugh UK	360 g/l	SL	15122
47	Roundup Biactive GL	Monsanto	360 g/l	SL	17348
48	Roundup Energy	Monsanto	450 g/l	SL	12945
49	Roundup Flex	Monsanto	480 g/l	SL	15541
50	Roundup Metro XL	Monsanto	360 g/l	SL	17684
51	Roundup ProVantage	Monsanto	480 g/l	SL	15534
52	Roundup Sonic	Monsanto	450 g/l	SL	17152
53	Roundup Star	Monsanto	450 g/l	SL	15470
54	Roundup Vista Plus	Monsanto	450 g/l	SL	18002
55	Rustler Pro-Green	ChemSource	360 g/l	SL	15798
56	Samurai	Monsanto	360 g/l	SL	16238
57	Snapper	Nufarm UK	550 g/l	SL	15489
58	Snapper	Nufarm UK	550 g/l	SL	15695
59	Surrender T/F	Agrovista	360 g/l	SL	17575
60	Tanker	Nufarm UK	540 g/l	SL	15016
61	Tanker	Nufarm UK	540 g/l	SL	15694
62	Touchdown Quattro	Syngenta	360 g/l	SL	10608
63	Trustee Amenity	Barclay	450 g/l	SL	12897
64	Trustee Amenity	Barclay	450 g/l	SL	17697
65	Vesuvius Green	Ventura	360 g/l	SL	17650

SECTION 2

Uses

- Annual and perennial weeds in *all edible crops (outdoor and protected)* [1, 2, 16, 18-20, 33, 60, 61]; *all edible crops (outdoor)* [16, 33, 41, 42, 47, 49, 51, 57, 58]; *all edible crops (outdoor) (before planting)*, *all non-edible crops (outdoor) (before planting)* [3, 8, 9, 12, 17, 24-26, 29, 30, 39, 43, 45, 46, 50, 55, 59, 62, 65]; *all edible crops (stubble)*, *all non-edible crops (stubble)* [3, 8, 9, 12, 29, 47, 65]; *all edible crops (stubble) (before planting)*, *all non-edible crops (stubble) (before planting)* [24, 25]; *all non-edible crops (outdoor)* [1, 2, 16, 18-20, 33, 41, 42, 47, 49, 51, 57, 58, 60, 61]; *all non-edible seed crops grown outdoors* [16, 33]; *almonds (off-label)*, *apples (off-label)*, *apricots (off-label)*, *cherries (off-label)*, *peaches (off-label)*, *pears (off-label)*, *plums (off-label)*, *quinces (off-label)* [4, 47, 48]; *amenity vegetation* [1, 2, 5-7, 10, 11, 13, 15, 17-21, 23, 28, 31, 34-38, 41, 42, 44, 46, 49-61, 63, 64]; *apple orchards*, *pear orchards* [1-6, 8, 9, 12, 14, 16-20, 22, 24-26, 29, 30, 33, 39, 41-43, 45, 46, 48-55, 57-61, 65]; *apples*, *managed amenity turf (pre-establishment only)*, *ornamental plant production*, *pears* [38]; *asparagus* [3, 8, 9, 12, 16-18, 24, 25, 29, 30, 33, 39, 45, 49-51, 55, 59, 65]; *asparagus (off-*

SEE SECTION 3 FOR PRODUCTS ALSO REGISTERED

label), **crab apples** *(off-label)* [4, 47]; **beetroot** *(off-label)*, **carrots** *(off-label)*, **edible podded peas** *(off-label)*, **garlic** *(off-label)*, **horseradish** *(off-label)*, **leeks** *(off-label)*, **lupins** *(off-label)*, **miscanthus** *(off-label)*, **onions** *(off-label)*, **parsley root** *(off-label)*, **parsnips** *(off-label)*, **rye** *(off-label)*, **salad onions** *(off-label)*, **salsify** *(off-label)*, **shallots** *(off-label)*, **swedes** *(off-label)*, **turnips** *(off-label)*, **vining peas** *(off-label)* [48, 49]; **bilberries** *(off-label)*, **blackcurrants** *(off-label)*, **blueberries** *(off-label)*, **cranberries** *(off-label)*, **gooseberries** *(off-label)*, **redcurrants** *(off-label)* [4, 48]; **blackberries** *(off-label)*, **buckwheat** *(off-label)*, **canary seed** *(off-label)*, **farm forestry** *(off-label)*, **game cover** *(off-label)*, **hemp** *(off-label)*, **hops** *(off-label)*, **loganberries** *(off-label)*, **medlar** *(off-label)*, **millet** *(off-label)*, **nectarines** *(off-label)*, **quinoa** *(off-label)*, **raspberries** *(off-label)*, **rubus hybrids** *(off-label)*, **soft fruit** *(off-label)*, **sorghum** *(off-label)*, **strawberries** *(off-label)*, **top fruit** *(off-label)*, **woad** *(off-label)* [48]; **bulb onions** [1, 2, 16-20, 33, 41, 42, 49, 51, 57, 58, 60, 61]; **cherries, plums** [1-9, 12, 14, 16-20, 22, 24-26, 29-31, 33-35, 37-39, 41-46, 48-61, 65]; **chestnuts** *(off-label)*, **hazel nuts** *(off-label)*, **walnuts** *(off-label)* [47, 48]; **combining peas** [1, 2, 4, 16, 18-20, 33, 41, 42, 47, 49, 51, 54, 57, 58, 60-62]; **cultivated land/soil, fruit trees** *(off-label)*, **grapevines** *(off-label)*, **rhubarb** *(off-label)*, **vaccinium spp.** *(off-label)*, **whitecurrants** *(off-label)* [4]; **damsons** [2, 4, 14, 19, 20, 22, 26, 48, 50, 55, 60, 61]; **durum wheat** [1, 2, 4, 16, 18-20, 33, 41, 42, 49, 51, 57, 58, 60, 61]; **enclosed waters** [5-7, 10, 11, 23, 31, 32, 34, 37, 38, 44, 46, 52-54, 56, 63, 64]; **farm forestry** [7, 31, 34, 37, 44, 52-54, 56]; **forest** [1-3, 5-14, 16-25, 28-31, 33-35, 37-46, 49-61, 63-65]; **forest nurseries** [1, 17, 35, 36, 40, 46, 59]; **forestry plantations** [1, 2, 18-20, 41, 42, 57, 58, 60, 61]; **grassland** [1, 2, 7, 16, 18-20, 31, 33-35, 37, 41, 42, 44, 52-54, 56-58, 60, 61]; **green cover on land temporarily removed from production** [1-3, 7-9, 12, 16, 18-20, 28, 29, 31, 33-35, 37, 39, 41-44, 47, 49, 51, 56-58, 60, 61, 65]; **hard surfaces** [1, 3, 5-13, 16-18, 21, 23-25, 28-47, 49-59, 63-65]; **kentish cobnuts** *(off-label)*, **rotational grass** [47]; **land immediately adjacent to aquatic areas** [3, 5-13, 21, 23, 29-34, 37, 38, 44-47, 50, 52-54, 56, 63-65]; **land not intended to bear vegetation** [14, 15, 22]; **leeks, sugar beet, vining peas** [1, 2, 7, 16-20, 31, 33-35, 37, 41, 42, 44, 49, 51, 56-58, 60, 61]; **linseed** [1, 2, 4, 7, 16, 18-20, 31, 33-35, 37, 41, 42, 44, 49, 51, 56-58, 60-62]; **managed amenity turf** *(pre-establishment)* [28]; **mustard** [1, 2, 16, 18-20, 33, 41, 42, 49, 51, 57, 58, 60-62]; **natural surfaces not intended to bear vegetation, permeable surfaces overlying soil** [1-3, 5-13, 16-21, 23-25, 28-47, 49-61, 63-65]; **oilseed rape** [7, 18, 31, 34, 35, 37, 44, 56]; **open waters** [7, 10, 11, 23, 31, 32, 34, 37, 38, 44, 46, 52-54, 56, 63, 64]; **permanent grassland** [14, 47, 48]; **poppies for morphine production** *(off-label)* [49]; **potatoes** [47, 49, 51]; **spring barley, spring field beans, spring wheat, winter barley, winter field beans, winter wheat** [1, 2, 4, 16, 18-20, 33, 41, 42, 47, 49, 51, 57, 58, 60-62]; **spring oats, winter oats** [1, 2, 4, 16, 18-20, 33, 41, 42, 47, 49, 51, 57, 58, 60, 61]; **spring oilseed rape, winter oilseed rape** [1, 2, 4, 16, 18-20, 33, 41, 42, 49, 51, 57, 58, 60-62]; **stubbles** [4, 7, 14, 30, 31, 34, 35, 37, 39, 43-46, 48, 50, 52-56, 59, 62]; **swedes, turnips** [1, 2, 7, 16-20, 31, 33-35, 37, 41, 42, 44, 47, 49, 51, 56-58, 60, 61]]

- Annual dicotyledons in **bulb onions** [3, 7-9, 12, 24-26, 29-31, 34, 35, 37, 39, 43-46, 48, 50, 52-56, 59, 65]; **combining peas** [3, 8, 9, 12, 17, 24-26, 29, 30, 39, 43, 45, 46, 50, 55, 59, 65]; **durum wheat** [3, 8, 9, 12, 17, 24-26, 29, 30, 35, 39, 43, 45, 46, 48, 50, 52-55, 59, 65]; **grassland, potatoes** [52-54]; **leeks, sugar beet, swedes** [3, 8, 9, 12, 24-26, 29, 30, 39, 43, 45, 46, 48, 50, 52-55, 59, 65]; **linseed, mustard, spring field beans, spring oilseed rape, winter field beans, winter oilseed rape** [3, 8, 9, 12, 17, 24-26, 29, 30, 39, 43, 45, 46, 48, 50, 52-55, 59, 65]; **permanent grassland** [48]; **spring barley, spring oats, spring wheat, winter barley, winter oats, winter wheat** [3, 8, 9, 12, 17, 22, 24-26, 29, 30, 39, 43, 45, 46, 48, 50, 52-55, 59, 65]; **turnips** [3, 8, 9, 12, 24-26, 29, 30, 39, 43, 45, 46, 48, 52-54, 59, 65]; **vining peas** [3, 8, 9, 12, 24, 25, 29, 30, 39, 43, 45, 46, 48, 50, 52-55, 59, 65]
- Annual grasses in **bulb onions** [7, 26, 31, 34, 35, 37, 44, 48, 52-54, 56]; **combining peas** [26]; **cultivated land/soil** [22]; **durum wheat** [26, 35, 48, 52-54]; **grassland, potatoes** [52-54]; **leeks, linseed, mustard, spring barley, spring field beans, spring oats, spring oilseed rape, spring wheat, sugar beet, swedes, turnips, winter barley, winter field beans, winter oats, winter oilseed rape, winter wheat** [26, 48, 52-54]; **permanent grassland** [48]; **vining peas** [48, 52-54]
- Aquatic weeds in **enclosed waters, open waters** [7, 23, 31, 34, 37, 44, 46, 47, 49, 51, 56, 63, 64]; **land immediately adjacent to aquatic areas** [3, 7-9, 12, 23, 29-31, 34, 37, 44-47, 49-51, 56, 63-65]
- Biennial weeds in **beetroot** *(off-label)*, **carrots** *(off-label)*, **garlic** *(off-label)*, **horseradish** *(off-label)*, **leeks** *(off-label)*, **onions** *(off-label)*, **parsley root** *(off-label)*, **parsnips** *(off-label)*, **salad**

onions (off-label), *salsify* (off-label), *shallots* (off-label), *swedes* (off-label), *turnips* (off-label) [49]

- Black bent in *combining peas, linseed, spring field beans, spring oilseed rape, stubbles, winter field beans, winter oilseed rape* [22]
- Bolters in *sugar beet* (wiper application) [26, 48]
- Bracken in *amenity vegetation* [17]; *forest* [3, 8, 9, 12, 14, 17, 22, 24, 25, 27, 29, 30, 39, 43, 45, 46, 49-51, 55, 59, 65]; *forest nurseries* [17, 46, 59]
- Canary grass in *aquatic areas* [26]; *enclosed waters, land immediately adjacent to aquatic areas* [27]
- Chemical thinning in *farm forestry* [7, 31, 34, 37, 44, 52-54, 56]; *forest* [3, 5-9, 12, 22, 24, 25, 27, 29-31, 34-37, 39, 40, 43-46, 49-56, 59, 65]; *forest* (stump treatment) [14]; *forest nurseries* [35, 40, 46, 59]
- Couch in *combining peas* [4, 7, 14, 22, 26, 31, 34, 35, 37, 44, 48, 52, 53, 56, 62]; *durum wheat* [4, 7, 26, 31, 34, 35, 37, 44, 48, 52-54, 56]; *forest* [27]; *linseed, spring oilseed rape, stubbles, winter oilseed rape* [4, 14, 22, 26, 48, 52-54, 62]; *mustard* [7, 14, 31, 34, 35, 37, 44, 48, 52-54, 56, 62]; *oats* [7, 31, 34, 35, 37, 44, 56]; *permanent grassland* [14]; *spring barley, spring wheat, winter barley, winter wheat* [4, 7, 14, 26, 31, 34, 35, 37, 44, 48, 52-54, 56, 62]; *spring field beans, winter field beans* [4, 7, 14, 22, 26, 31, 34, 35, 37, 44, 48, 52-54, 56, 62]; *spring oats, winter oats* [4, 14, 26, 48, 52-54]
- Creeping bent in *aquatic areas* [26]; *combining peas, linseed, spring field beans, spring oilseed rape, stubbles, winter field beans, winter oilseed rape* [22]; *enclosed waters, land immediately adjacent to aquatic areas* [27]
- Desiccation in *amenity vegetation* [10, 11, 46]; *buckwheat* (off-label), *canary seed* (off-label), *hemp* (off-label), *millet* (off-label), *quinoa* (off-label), *sorghum* (off-label) [48]; *combining peas* [47, 49, 51, 54]; *durum wheat, linseed, mustard, spring barley, spring oats, spring oilseed rape, spring wheat, winter barley, winter oats, winter oilseed rape, winter wheat* [49, 51]; *lupins* (off-label), *poppies for morphine production* (off-label), *rye* (off-label) [48, 49]; *spring field beans, winter field beans* [47, 49, 51]
- Destruction of crops in *all edible crops (outdoor and protected)* [1, 2, 18, 60, 61]; *all edible crops (outdoor)* [7, 16, 31, 33-35, 37, 41, 42, 44, 47, 56-58]; *all edible crops (stubble), all non-edible crops (stubble), permanent grassland, rotational grass* [47]; *all non-edible crops (outdoor)* [7, 18, 31, 34, 35, 37, 41, 42, 44, 47, 56-58]; *all non-edible seed crops grown outdoors* [16, 33]; *amenity vegetation* [21]; *grassland* [1, 2, 18, 41, 42, 49, 51, 57, 58, 60, 61]; *green cover on land temporarily removed from production* [7, 31, 34, 35, 37, 44, 47, 56]
- Destruction of short term leys in *grassland* [28]
- Grass weeds in *forest, hard surfaces, land immediately adjacent to aquatic areas, natural surfaces not intended to bear vegetation, permeable surfaces overlying soil* [21]
- Green cover in *land immediately adjacent to aquatic areas* [24, 25]; *land temporarily removed from production* [4, 14, 17, 22, 24-26, 30, 39, 43, 45, 46, 48, 50, 52-55, 59, 62]
- Growth suppression in *tree stumps* [2, 16, 41, 42, 57, 58]
- Harvest management/desiccation in *combining peas* [1-3, 7-9, 12, 14, 16-18, 24, 25, 29-31, 33-35, 37, 39, 41-46, 48, 50, 52, 53, 55-61, 65]; *durum wheat* [1-3, 7-9, 12, 16-18, 24-26, 29-31, 33-35, 37, 39, 41-46, 48, 50, 52-61, 65]; *grassland, oats, oilseed rape* [7, 31, 34, 35, 37, 44, 56]; *linseed* [1-3, 7-9, 12, 14, 16-18, 24, 25, 29-31, 33-35, 37, 39, 41-46, 48, 50, 52-61, 65]; *mustard* [1-3, 7-9, 12, 14, 16-18, 24, 25, 29-31, 33-35, 37, 39, 41-46, 48, 50, 52-62, 65]; *spring barley, spring wheat, winter barley, winter wheat* [1-3, 7-9, 12, 14, 16-18, 22, 24-26, 29-31, 33-35, 37, 39, 41-46, 48, 50, 52-61, 65]; *spring field beans, winter field beans* [3, 7-9, 12, 14, 17, 24, 25, 29-31, 34, 35, 37, 39, 41-46, 48, 50, 52-59, 65]; *spring oats, winter oats* [1-3, 8, 9, 12, 14, 16-18, 22, 24-26, 29, 30, 33, 39, 41-43, 45, 46, 48, 50, 52-55, 57-61, 65]; *spring oilseed rape, winter oilseed rape* [1-4, 8, 9, 12, 14, 16-18, 22, 24-26, 29, 30, 33, 39, 41-43, 45, 46, 48, 50, 52-55, 57-62, 65]
- Heather in *forest* [3, 8, 9, 12, 14, 22, 24, 25, 27, 29, 30, 39, 43, 45, 46, 50, 55, 59, 65]; *forest nurseries* [46, 59]
- Perennial dicotyledons in *combining peas, linseed, spring field beans, spring oilseed rape, winter field beans, winter oilseed rape* [22, 26]; *grassland* (wiper application) [52-54]; *permanent grassland* (wiper application) [48]
- Perennial grasses in *aquatic areas* [4-6, 14, 22]; *combining peas, linseed, spring field beans, stubbles, winter field beans* [26]

- Pre-harvest desiccation in *combining peas, linseed, spring barley, spring field beans, spring wheat, winter barley, winter field beans, winter wheat* [62]
- Reeds in *aquatic areas* [4-6, 14, 22, 26]; *enclosed waters, land immediately adjacent to aquatic areas* [27, 49, 51]; *open waters* [49, 51]
- Rhododendrons in *forest* [3, 8, 9, 12, 14, 22, 24, 25, 27, 29, 30, 39, 43, 45, 46, 50, 55, 59, 65]; *forest nurseries* [46, 59]
- Rushes in *aquatic areas* [4-6, 14, 22, 26]; *enclosed waters, land immediately adjacent to aquatic areas* [27, 49, 51]; *open waters* [49, 51]
- Sedges in *aquatic areas* [4-6, 14, 22, 26]; *enclosed waters, land immediately adjacent to aquatic areas* [27, 49, 51]; *open waters* [49, 51]
- Sprout suppression in *tree stumps* [23, 49, 51-54]
- Sucker control in *apple orchards, cherries, pear orchards, plums* [7, 31, 34, 35, 37, 44, 48, 52-54, 56]; *damsons* [48]
- Sucker inhibition in *forest* [5, 6]; *tree stumps* [1, 7, 18, 31, 34, 35, 37, 44, 56, 60, 61]
- Sward destruction in *amenity vegetation* [23, 63, 64]; *grassland* [3, 8, 9, 12, 17, 24, 25, 29, 30, 39, 43, 45, 46, 50, 52-54, 59, 65]; *permanent grassland* [4, 14, 22, 26, 48, 55, 62]; *rotational grass* [55, 62]
- Total vegetation control in *all edible crops (outdoor)* (pre-sowing/planting), *all non-edible crops (outdoor)* (pre-sowing/planting) [48, 52-54]; *amenity vegetation* [15]; *hard surfaces, natural surfaces not intended to bear vegetation, permeable surfaces overlying soil* [27]; *land not intended to bear vegetation* [15, 62]
- Volunteer cereals in *bulb onions* [7, 26, 31, 34, 35, 37, 44, 48, 52-54, 56]; *combining peas* [26]; *cultivated land/soil* [22]; *durum wheat* [35, 48, 52-54]; *leeks, linseed, mustard, spring field beans, sugar beet, swedes, turnips, winter field beans* [26, 48, 52-54]; *potatoes* [52-54]; *spring barley, spring oats, spring oilseed rape, spring wheat, vining peas, winter barley, winter oats, winter oilseed rape, winter wheat* [48, 52-54]; *stubbles* [4, 14, 22, 26, 48, 52-54, 62]
- Volunteer potatoes in *stubbles* [4, 14, 26, 48, 52-54, 62]
- Waterlilies in *aquatic areas* [4-6, 14, 22, 26]; *enclosed waters* [27, 49, 51]; *land immediately adjacent to aquatic areas, open waters* [49, 51]
- Weed beet in *sugar beet* (wiper application) [26]
- Woody weeds in *forest* [3, 8, 9, 12, 14, 22, 24, 25, 27, 29, 30, 39, 43, 45, 46, 50, 55, 59, 65]; *forest nurseries* [46, 59]

Extension of Authorisation for Minor Use (EAMUs)
- *almonds* 20102476 [4], 20170234 [47], 20082888 [48]
- *apples* 20102476 [4], 20170234 [47], 20082888 [48]
- *apricots* 20102476 [4], 20170234 [47], 20082888 [48]
- *asparagus* 20071476 [4], 20171123 [47]
- *beetroot* 20130354 [48], 20132528 expires 30 Jun 2018 [49]
- *bilberries* 20071475 [4], 20082888 [48]
- *blackberries* 20082888 [48]
- *blackcurrants* 20071475 [4], 20082888 [48]
- *blueberries* 20071475 [4], 20082888 [48]
- *buckwheat* 20071839 [48]
- *canary seed* 20071839 [48]
- *carrots* 20130354 [48], 20132528 expires 30 Jun 2018 [49]
- *cherries* 20102476 [4], 20170234 [47], 20082888 [48]
- *chestnuts* 20170234 [47], 20082888 [48]
- *crab apples* 20102476 [4], 20170234 [47]
- *cranberries* 20071475 [4], 20082888 [48]
- *edible podded peas* 20141672 [48], 20141671 expires 30 Jun 2018 [49]
- *farm forestry* 20082888 [48]
- *fruit trees* 20102476 [4]
- *game cover* 20082888 [48]
- *garlic* 20130354 [48], 20132528 expires 30 Jun 2018 [49]
- *gooseberries* 20071475 [4], 20082888 [48]
- *grapevines* 20071479 [4]
- *hazel nuts* 20170234 [47], 20082888 [48]

- **hemp** *20071840* [48]
- **hops** *20082888* [48]
- **horseradish** *20130354* [48], *20132528 expires 30 Jun 2018* [49]
- **kentish cobnuts** *20170234* [47]
- **leeks** *20130354* [48], *20132528 expires 30 Jun 2018* [49]
- **loganberries** *20082888* [48]
- **lupins** *20071279* [48], *20131667 expires 30 Jun 2018* [49]
- **medlar** *20082888* [48]
- **millet** *20071839* [48]
- **miscanthus** *20082888* [48], *20132131 expires 30 Jun 2018* [49]
- **nectarines** *20082888* [48]
- **onions** *20130354* [48], *20132528 expires 30 Jun 2018* [49]
- **parsley root** *20130354* [48], *20132528 expires 30 Jun 2018* [49]
- **parsnips** *20130354* [48], *20132528 expires 30 Jun 2018* [49]
- **peaches** *20102476* [4], *20170234* [47], *20082888* [48]
- **pears** *20102476* [4], *20170234* [47], *20082888* [48]
- **plums** *20102476* [4], *20170234* [47], *20082888* [48]
- **poppies for morphine production** *20081058* [48], *20131668 expires 30 Jun 2018* [49]
- **quinces** *20102476* [4], *20170234* [47], *20082888* [48]
- **quinoa** *20071840* [48]
- **raspberries** *20082888* [48]
- **redcurrants** *20071475* [4], *20082888* [48]
- **rhubarb** *20071478* [4]
- **rubus hybrids** *20082888* [48]
- **rye** *20111701* [48], *20131666 expires 30 Jun 2018* [49]
- **salad onions** *20130354* [48], *20132528 expires 30 Jun 2018* [49]
- **salsify** *20130354* [48], *20132528 expires 30 Jun 2018* [49]
- **shallots** *20130354* [48], *20132528 expires 30 Jun 2018* [49]
- **soft fruit** *20082888* [48]
- **sorghum** *20071839* [48]
- **strawberries** *20082888* [48]
- **swedes** *20130354* [48], *20132528 expires 30 Jun 2018* [49]
- **top fruit** *20082888* [48]
- **turnips** *20130354* [48], *20132528 expires 30 Jun 2018* [49]
- **vaccinium spp.** *20071475* [4]
- **vining peas** *20141672* [48], *20141671 expires 30 Jun 2018* [49]
- **walnuts** *20170234* [47], *20082888* [48]
- **whitecurrants** *20071475* [4]
- **woad** *20082888* [48]

Approval information
- Glyphosate included in Annex I under EC Regulation 1107/2009
- Accepted by BBPA for use on malting barley and hops

Efficacy guidance
- For best results apply to actively growing weeds with enough leaf to absorb chemical
- For most products a rainfree period of at least 6 h (preferably 24 h) should follow spraying
- Adjuvants are obligatory for some products and recommended for some uses with others. See labels
- Mixtures with other pesticides or fertilizers may lead to reduced control.
- Products are formulated as isopropylamine, ammonium, potassium, or trimesium salts of glyphosate and may vary in the details of efficacy claims. See individual product labels
- Some products in ready-to-use formulations for use through hand-held applicators. See label for instructions [15, 28].
- If using to treat hard surfaces, ensure spraying takes place only when weeds are actively growing (normally March to October) and is confined only to visible weeds including those in the 30cm swath covering the kerb edge and road gulley – do not overspray drains.
- When applying products through rotary atomisers the spray droplet spectra must have a minimum Volume Median Diameter (VMD) of 200 microns
- With wiper application weeds should be at least 10 cm taller than crop

SEE SECTION 3 FOR PRODUCTS ALSO REGISTERED

- Annual weed grasses should have at least 5 cm of leaf and annual broad-leaved weeds at least 2 expanded true leaves
- Perennial grass weeds should have 4-5 new leaves and be at least 10 cm long when treated. Perennial broad-leaved weeds should be treated at or near flowering but before onset of senescence
- Volunteer potatoes and polygonums are not controlled by harvest-aid rates
- Bracken must be treated at full frond expansion
- Fruit tree suckers best treated in late spring
- Chemical thinning treatment can be applied as stump spray or stem injection
- In order to allow translocation, do not cultivate before spraying and do not apply other pesticides, lime, fertilizer or farmyard manure within 5 d of treatment
- Recommended intervals after treatment and before cultivation vary. See labels

Restrictions
- Maximum total dose per crop or season normally equivalent to one full dose treatment on field and edible crops and no restriction for non-crop uses. However some older labels indicate a maximum number of treatments. Check for details
- Do not treat cereals grown for seed or undersown crops
- Consult grain merchant before treating crops grown on contract or intended for malting
- Do not use treated straw as a mulch or growing medium for horticultural crops
- For use in nursery stock, shrubberies, orchards, grapevines and tree nuts care must be taken to avoid contact with the trees. Do not use in orchards established less than 2 yr and keep off low-lying branches
- Certain conifers may be sprayed overall in dormant season. See label for details
- Use a tree guard when spraying in established forestry plantations
- Do not spray root suckers in orchards in late summer or autumn
- Do not use under glass or polythene as damage to crops may result
- Do not mix, store or apply in galvanised or unlined mild steel containers or spray tanks
- Do not leave diluted chemical in spray tanks for long periods and make sure that tanks are well vented

Crop-specific information
- Harvest intervals: 4 wk for blackcurrants, blueberries; 14 d for linseed, oilseed rape; 8 d for mustard; 5-7 d for all other edible crops. Check label for exact details
- Latest use: for most products 2-14 d before cultivating, drilling or planting a crop in treated land; after harvest (post-leaf fall) but before bud formation in the following season for nuts and most fruit and vegetable crops; before fruit set for grapevines. See labels for details

Following crops guidance
- Decaying remains of plants killed by spraying must be dispersed before direct drilling
- Crops may be drilled 48 h after application. Trees and shrubs may be planted 7 d after application.

Environmental safety
- Products differ in their hazard and environmental safety classification. See labels
- Do not dump surplus herbicide in water or ditch bottoms or empty into drains
- Check label for maximum permitted concentration in treated water
- The Environment Agency or Local River Purification Authority must be consulted before use in or near water
- Take extreme care to avoid drift and possible damage to neighbouring crops or plants
- Treated poisonous plants must be removed before grazing or conserving
- Do not use in covered areas such as greenhouses or under polythene
- For field edge treatment direct spray away from hedge bottoms
- Some products require livestock to be excluded from treated areas and do not permit treated forage to be used for hay, silage or bedding. Check label for details

Hazard classification and safety precautions
Hazard Harmful [33, 59]; Irritant [3, 8, 14, 16, 17, 23-25, 30, 35, 36, 41, 42, 45, 46, 50, 57, 58]; Dangerous for the environment [3, 7, 8, 16, 17, 23-25, 28, 30, 31, 33-38, 41, 42, 44-46, 49-51, 56-59, 62]; Harmful if inhaled [50]
Transport code 9 [3, 7, 8, 17, 19, 20, 24, 25, 28, 30, 31, 33-37, 41, 42, 44-46, 50, 51, 56-59]

FOR FULL CONDITIONS OF USE ALWAYS READ THE PRODUCT LABEL

Packaging group III [3, 7, 8, 17, 19, 20, 24, 25, 28, 30, 31, 33-37, 41, 42, 44-46, 50, 51, 56-59]
UN Number 3082 [3, 7, 8, 17, 19, 20, 24, 25, 28, 30, 31, 33-37, 41, 42, 44-46, 50, 51, 56-59]; N/C [1, 2, 4-6, 9-16, 18, 21-23, 26, 27, 29, 32, 38-40, 43, 47-49, 52-55, 60-65]
Risk phrases H318 [19, 20, 41, 42, 50, 57, 58]; H319 [3, 7, 25, 31, 34, 36, 37, 44, 52-54]; R36 [56]; R41 [14, 35]; R50 [38]; R51 [35, 56]; R52 [26, 27, 46, 48, 52]; R53a [26, 27, 35, 38, 46, 48, 49, 52, 56]
Operator protection A [1-65]; C [3-12, 14, 16, 17, 22, 24, 25, 29-37, 39, 41-45, 48, 50, 52-59, 65]; D [4-6, 22]; H [1-6, 8-15, 17-30, 32, 38, 39, 41-43, 45, 47, 49-51, 55, 57-65]; M [1-15, 17-31, 34-39, 41-43, 45, 47-61, 63-65]; U02a [1, 2, 9-13, 16, 18-21, 23, 26-29, 33, 35, 36, 38, 39, 41-43, 55, 57, 58, 60, 61, 63-65]; U05a [1-3, 8, 17-20, 24, 25, 28, 30, 32, 38, 41, 42, 45, 46, 50, 57-62]; U08 [9-12, 17, 23, 26, 27, 29, 39, 43, 55, 63-65]; U09a [13, 21, 28, 38]; U11 [3, 7, 8, 14, 16, 17, 24-27, 30-37, 44-46, 50, 56, 59]; U12; U15 [41, 42, 57, 58]; U19a [9-13, 15, 17, 21, 23, 26-29, 38, 39, 43, 55, 63-65]; U20a [3, 8, 21, 24, 25, 30, 32, 45-47, 50, 59, 62]; U20b [1, 2, 4-7, 9-13, 15-20, 22, 23, 26-29, 31, 33-44, 48, 49, 51-58, 60, 61, 63-65]
Environmental protection E06a [40] (2 weeks); E06d, E08b [62]; E07d [60, 61]; E13c [12, 28, 38, 63, 64]; E15a [4-6, 15, 22, 47]; E15b [1-3, 7-11, 13, 14, 16, 18-21, 23-31, 33-46, 48-62, 65]; E19a [4-6, 12, 22, 63, 64]; E34 [15, 18-20, 60, 61]; E36a [18-20, 60, 61]; E38 [3, 7, 8, 14, 16, 24-27, 30, 31, 33-37, 41, 42, 44-46, 49-51, 56-59, 62]; H410 [28, 41, 42]; H411 [3, 7, 19, 20, 25, 31, 32, 34, 37, 50, 57, 58, 62]; H412 [18, 53]; H413 [1, 2, 16, 51, 60, 61]
Storage and disposal D01 [1-11, 13-15, 17-21, 23-32, 34, 37-39, 41-62, 65]; D02 [1-3, 7-11, 13-15, 17-21, 23-32, 34, 37-39, 41-62, 65]; D05 [1-3, 7-13, 15, 16, 18-21, 23-25, 28-31, 33-39, 41-65]; D09a [3-17, 21-31, 33-59, 62-65]; D10a [9-11, 15, 23, 29, 39, 43, 55, 65]; D10b [4-6, 12, 17, 22, 63, 64]; D10c [1-3, 7, 8, 16, 18-20, 24, 25, 30, 31, 33-37, 40-42, 44-54, 56-62]; D11a [13, 14, 21, 26-28, 38]; D12a [3, 7, 8, 16, 24-28, 30, 31, 33-38, 44-46, 49-51, 56, 59, 62]; D14 [18-20, 60, 61]
Medical advice M03 [62]; M05a [7, 16, 17, 31-37, 44, 56]

273   glyphosate + pyraflufen-ethyl

A herbicide mixture for non-selective weed control around amenity plants and on areas not intended to bear vegetation
HRAC mode of action code: G + E

See also glyphosate
        pyraflufen-ethyl

Products

Hammer	Everris Ltd	240:0.67 g/l	SE	16498

Uses

- Annual and perennial weeds in **amenity vegetation**, **hard surfaces**, **land not intended to bear vegetation**, **natural surfaces not intended to bear vegetation**, **ornamental plant production**, **permeable surfaces overlying soil**

Approval information

- Glyphosate and pyraflufen-ethyl included in Annex I under EC Regulation 1107/2009

Hazard classification and safety precautions

Hazard Dangerous for the environment
Transport code 9
Packaging group III
UN Number 3082
Risk phrases R51, R53a
Operator protection A, H, M; U02a, U08, U20b
Environmental protection E15b, E38
Storage and disposal D01, D02, D05, D09a, D10c, D12a

SEE SECTION 3 FOR PRODUCTS ALSO REGISTERED

274   glyphosate + sulfosulfuron

A translocated non-residual glycine derivative herbicide + a sulfonyl urea herbicide
HRAC mode of action code: G + B

See also sulfosulfuron

Products

Nomix Dual	Nomix Enviro	120:2.22 g/l	RC	13420

Uses
- Annual and perennial weeds in **amenity vegetation**, **hard surfaces**, **natural surfaces not intended to bear vegetation**, **permeable surfaces overlying soil**

Approval information
- Glyphosate and sulfosulfuron included in Annex I under EC Regulation 1107/2009

Hazard classification and safety precautions
Hazard Dangerous for the environment
Transport code 9
Packaging group III
UN Number 3082
Risk phrases R50, R53a
Operator protection A, H, M; U02a, U09a, U19a, U20b
Environmental protection E15b
Storage and disposal D01, D05, D09a, D11a

275   halauxifen-methyl

A herbicide for use in cereal crops only available in mixtures
HRAC mode of action code: O

See also aminopyralid + halauxifen-methyl
*   florasulam + halauxifen-methyl*
*   fluroxypyr + halauxifen-methyl*

276   hymexazol

A systemic heteroaromatic fungicide for pelleting sugar beet seed
FRAC mode of action code: 32

Products

Tachigaren 70 WP	Sumi Agro	70% w/w	WP	17977

Uses
- Black leg in **sugar beet** *(seed treatment)*

Approval information
- Hymexazol included in Annex 1 under EC Regulation 1107/2009

Efficacy guidance
- Incorporate into pelleted seed using suitable seed pelleting machinery

Restrictions
- Maximum number of treatments 1 per batch of seed
- Do not use treated seed as food or feed

Crop-specific information
- Latest use: before planting sugar beet seed

Environmental safety
- Harmful to aquatic organisms

- Harmful to fish or other aquatic life. Do not contaminate surface waters or ditches with chemical or used container
- Treated seed harmful to game and wildlife

Hazard classification and safety precautions
   Hazard Flammable solid
   Transport code 3
   Packaging group III
   UN Number 1325
   Risk phrases H317, H318, H361
   Operator protection A, C, F; U05a, U11, U20b
   Environmental protection E03, E13c, H411
   Storage and disposal D01, D02, D09a, D11a
   Treated seed S01, S02, S03, S04a, S05

277　imazalil

A systemic and protectant imidazole fungicide
FRAC mode of action code: 3

See also guazatine + imazalil

Products

Gavel	Certis	100 g/l	SL	17586

Uses
- Dry rot in **seed potatoes**
- Gangrene in **seed potatoes**
- Silver scurf in **seed potatoes**
- Skin spot in **seed potatoes**

Approval information
- Imazalil included in Annex I under EC Regulation 1107/2009

Efficacy guidance
- For best control of skin and wound diseases of ware potatoes treat as soon as possible after harvest, preferably within 7-10 d, before any wounds have healed

Restrictions
- Maximum number of treatments 1 per batch of ware tubers; 2 per batch of seed tubers;
- Consult processor before treating potatoes for processing

Crop-specific information
- Latest use: during storage and before chitting for seed potatoes
- Apply to clean soil-free potatoes post-harvest before putting into store, or at first grading. A further treatment may be applied in early spring before planting
- Apply through canopied hydraulic or spinning disc equipment preferably diluted with up to two litres water per tonne of potatoes to obtain maximum skin cover and penetration
- Use on ware potatoes subject to discharges of imazalil from potato washing plants being within emission limits set by the UK monitoring authority

Environmental safety
- Dangerous for the environment
- Toxic to aquatic organisms
- Do not empty into drains
- Personal protective equipment requirements may vary for each pack size. Check label

Hazard classification and safety precautions
   Hazard Irritant
   UN Number N/C
   Risk phrases H318, H351
   Operator protection A, C, H; U04a, U05a, U11, U14, U19a, U20a

SEE SECTION 3 FOR PRODUCTS ALSO REGISTERED

Environmental protection E15a, E19b, E34, H410
Storage and disposal D01, D02, D05, D09a, D10c, D12a
Treated seed S01, S02, S03, S04a, S05, S06a

278   imazamox

An imidazolinone contact and residual herbicide available only in mixtures
HRAC mode of action code: B

See also imazamox + metazachlor
imazamox + pendimethalin

279   imazamox + metazachlor

A contact and residual herbicide mixture for weed control in oilseed rape
HRAC mode of action code: B + K3

Products

Cleranda	BASF	17.5:375 g/l	SC	15036

Uses
- Annual dicotyledons in *winter oilseed rape*
- Annual grasses in *winter oilseed rape*

Approval information
- Imazamox and metazachlor are included in Annex I under EC Regulation 1107/2009

Efficacy guidance
- Must only be used for weed control in CLEARFIELD oilseed rape hybrids. Treatment of an oilseed rape variety that is not a 'CLEARFIELD Hybrid' will result in complete crop loss.

Restrictions
- Applications shall be limited to a total dose of not more than 1.0 kg metazachlor/ha in a three year period on the same field
- To avoid the buildup of resistance, do not apply this or any other product containing an ALS inhibitor herbicide with claims for grass weed control more than once to any crop.
- Do not apply in mixture with phosphate liquid fertilisers
- Metazachlor stewardship requires that all autumn applications should be made before the end of September to reduce the risk to water

Following crops guidance
- Wheat, barley, oats, oilseed rape, field beans, combining peas and sugar beet can follow normally harvested oilseed rape treated with [1].
- In the event of crop failure CLEARFIELD oilseed rape may be drilled 4 weeks after application after thorough mixing of the soil to distribute residues. If winter field beans are to be planted, allow 10 weeks after application and plough before planting. Wheat and barley also require ploughing before planting but may be drilled 8 weeks after application.

Environmental safety
- Metazachlor stewardship guidelines advise a maximum dose of 750 g.a.i/ha/annum. Applications to drained fields should be complete by 15th Oct but, if drains are flowing, complete applications by 1st Oct.
- LERAP Category B

Hazard classification and safety precautions
Hazard Irritant, Dangerous for the environment
Transport code 9
Packaging group III
UN Number 3082
Operator protection A, H; U05a
Environmental protection E07d, E15b, E16a, E40b
Storage and disposal D01, D02, D09a, D10c

FOR FULL CONDITIONS OF USE ALWAYS READ THE PRODUCT LABEL

280 imazamox + pendimethalin

A pre-emergence broad-spectrum herbicide mixture for legumes
HRAC mode of action code: B + K1

See also pendimethalin

Products

Nirvana	BASF	16.7:250 g/l	EC	14256

Uses
- Annual dicotyledons in **broad beans** *(off-label)*, **combining peas**, **hops** *(off-label)*, **ornamental plant production** *(off-label)*, **soft fruit** *(off-label)*, **spring field beans**, **top fruit** *(off-label)*, **vining peas**, **winter field beans**
- Black bindweed in **soya beans** *(off-label)*
- Charlock in **soya beans** *(off-label)*
- Chickweed in **soya beans** *(off-label)*
- Common orache in **soya beans** *(off-label)*
- Fat hen in **soya beans** *(off-label)*
- Field speedwell in **soya beans** *(off-label)*
- Fumitory in **soya beans** *(off-label)*
- Poppies in **soya beans** *(off-label)*

Extension of Authorisation for Minor Use (EAMUs)
- **broad beans** *20092891*
- **hops** *20092894*
- **ornamental plant production** *20092894*
- **soft fruit** *20092894*
- **soya beans** *20152425*
- **top fruit** *20092894*

Approval information
- Imazamox and pendimethalin included in Annex I under EC Regulation 1107/2009

Efficacy guidance
- Best results obtained from applications to fine firm seedbeds in the presence of adequate moisture
- Weed control may be reduced on cloddy seedbeds and on soils with over 6% organic matter
- Residual control may be reduced under prolonged dry conditions

Restrictions
- Maximum number of treatments 1 per crop
- Seed must be drilled to at least 2.5 cm of settled soil
- Do not use on soils containing more than 10% organic matter
- Do not apply to soils that are waterlogged or are prone to waterlogging
- Do not apply if heavy rain is forecast
- Do not soil incorporate the product or disturb the soil after application
- Consult processors before use on crops destined for processing
- To avoid the build-up of resistance do not apply this or any other product containing an ALS inhibitor herbicide with claims for control of grass-weeds more than once to any crop

Crop-specific information
- Latest use: pre-crop emergence for all crops
- Inadequately covered seed may result in cupping of the leaves after application from which recovery is normally complete
- Crop damage may occur on stony or gravelly soils especially if heavy rain follows treatment
- Winter oilseed rape and other brassica crops should not be drilled as the following crop.

Following crops guidance
- Winter wheat or winter barley may be drilled as a following crop provided 3 mth have elapsed since treatment and the land has been at least cultivated by a non-inversion technique such as discing

SEE SECTION 3 FOR PRODUCTS ALSO REGISTERED

SECTION 2

- Winter oilseed rape or other brassica crops should not be drilled as following crops
- A minimum of 12 mth must elapse between treatment and sowing red beet, sugar beet or spinach

Environmental safety

- Dangerous for the environment
- Very toxic to aquatic organisms
- LERAP Category B

Hazard classification and safety precautions

Hazard Irritant, Dangerous for the environment
Transport code 9
Packaging group III
UN Number 3082
Risk phrases R38, R43, R50, R53a
Operator protection A, H; U05a, U14, U20b
Environmental protection E15b, E16a, E38
Storage and disposal D01, D02, D09a, D10c, D12a
Medical advice M03

281   imazamox + quinmerac

A herbicide mixture for weed control in CLEARFIELD oilseed rape only
HRAC mode of action code: B + O

Products

Cleravo	BASF	35:250 g/l	SC	16877

Uses

- Annual dicotyledons in *spring oilseed rape*, *winter oilseed rape*
- Charlock in *spring oilseed rape*, *winter oilseed rape*
- Cleavers in *spring oilseed rape*, *winter oilseed rape*
- Crane's-bill in *spring oilseed rape* (cut-leaved), *winter oilseed rape* (cut-leaved)
- Dead nettle in *spring oilseed rape*, *winter oilseed rape*
- Flixweed in *spring oilseed rape*, *winter oilseed rape*
- Volunteer oilseed rape in *spring oilseed rape*, *winter oilseed rape*

Approval information

- Imazamox and quinmerac included in Annex I under EC Regulation 1107/2009

Efficacy guidance

- Transient crop scorch may occur if applied under frosty conditions.
- Soil moisture is required for efficacy via root uptake.
- Maximum activity is from application before or shortly after weed emergence. Uptake is via cotyledons, roots and shoots.

Restrictions

- Must only be applied to CLEARFIELD oilseed rape hybrids.
- Do not apply if heavy rain is forecast.
- Do not use on sands, very light soils or soils with more than 10% organic matter.
- Use on stony soils may result in a reduction in plant vigour or crop stand.
- To avoid the build-up of resistance do not apply this or any other product containing an ALS inhibitor herbicide with claims for control of grass-weeds more than once to any crop.

Crop-specific information

- In the event of crop failure, do not re-drill with non CLEARFIELD oilseed rape or with sugar beet. After ploughing Clearfield rape, maize or peas may be sown 4 weeks after application, wheat, barley or oats 8 weeks after application, field beans 10 weeks after application or sunflower 15 weeks after application.

Hazard classification and safety precautions
    Hazard Very toxic to aquatic organisms
    Transport code 9
    Packaging group III
    UN Number 3082
    Operator protection U05a, U20a
    Environmental protection E15b, E38, H410
    Storage and disposal D01, D02, D09a, D10c, D12b

282   imazaquin

An imidazolinone herbicide and plant growth regulator available only in mixtures
HRAC mode of action code: B

See also chlormequat + 2-chloroethylphosphonic acid + imazaquin
    chlormequat + imazaquin

283   imazosulfuron

Herbicide for use in winter cereals
HRAC mode of action code: B

Products

Brazzos	Interfarm	50% w/w	WG	16932

Uses
- Charlock in *triticale, winter barley, winter rye, winter wheat*
- Mayweeds in *triticale, winter barley, winter rye, winter wheat*
- Shepherd's purse in *triticale, winter barley, winter rye, winter wheat*
- Volunteer oilseed rape in *triticale, winter barley, winter rye, winter wheat*

Approval information
- Imazosulfuron included in Annex I under EC Regulation 1107/2009

Efficacy guidance
- Seedbeds should be moist, firm, fine and free of clods.
- Can be used on all soil types.

Restrictions
- Do not treat broadcast crops until they are past the 3 leaf stage.
- Do not roll or harrow crops 2 weeks before or after treatment.
- Do not disturb the soil after application.
- Do not apply to crops with lush or soft growth.
- Crops should be drilled to 25mm and well covered by soil.
- Extreme care should be taken to avoid damage by drift onto plants outside the target area.
- Do not apply to undersown crops.

Crop-specific information
- Brazzos can be used on all varieties of winter wheat, winter barley, winter rye and winter triticale.

Following crops guidance
- In the year of harvest, winter wheat, winter barley, winter rye, winter triticale and winter beans can be sown following a cereal crop. In the following year after harvest, spring cereals, winter oilseed rape, spring oilseed rape, peas, sugar beet, onions, carrots, lettuce and maize can be sown. Crops other than winter wheat, winter barley, winter rye and winter triticale may only be sown after the field has been ploughed.
- In the event of crop failure for any reason, only winter wheat, winter barley, winter rye and winter triticale may be planted.

Environmental safety
- LERAP Category B

SEE SECTION 3 FOR PRODUCTS ALSO REGISTERED

SECTION 2

Hazard classification and safety precautions
    Hazard Very toxic to aquatic organisms
    Transport code 9
    Packaging group III
    UN Number 3077
    Environmental protection E15b, E16a, E34, E38, H410
    Storage and disposal D01, D02, D09a, D10c, D12a, D12b

284   imidacloprid

A neonicotinoid insecticide for seed, soil, peat or foliar treatment with a 2 year ban from Dec 2013 on use in many crops due to concerns over effects on bees.
IRAC mode of action code: 4A

See also beta-cyfluthrin + imidacloprid
           bitertanol + fuberidazole + imidacloprid
           fuberidazole + imidacloprid + triadimenol

Products

1	Imidasect 5GR	Fargro	5 % w/w	GR	16746
2	Merit Forest	Bayer CropScience	70% w/w	WG	17242

Uses
- Aphids in **protected ornamentals** [1]
- Glasshouse whitefly in **protected ornamentals** [1]
- Hylastes spp. in **protected forest nurseries** *(reduction in damage)* [2]
- Hylobius abietes in **protected forest nurseries** [2]
- Sciarid flies in **protected ornamentals** [1]
- Tobacco whitefly in **protected ornamentals** [1]
- Vine weevil in **protected ornamentals** [1]

Approval information
- Imidacloprid is included in Annex I under EC Regulation 1107/20
- Accepted by BBPA for use on hops

Efficacy guidance
- Treated seed should be drilled within 18 months of purchase
- Product must only be used incorporated into compost using suitable automated equipment [1]

Restrictions
- Maximum number of treatments 1 per batch of seed
- Do not use treated seed as food or feed
- Avoid puddling or run-off of irrigation water after treatment
- Plants grown in treated media may not be put outside until after they have finished flowering

Crop-specific information
- Latest use: before sowing or planting bedding plants, hardy ornamental nursery stock, pot plants

Environmental safety
- High risk to bees. Do not apply to crops in flower or to those in which bees are actively foraging. Do not apply when flowering weeds are present
- Treated seed harmful to game and wildlife

Hazard classification and safety precautions
    Hazard Irritant, Harmful if swallowed [2]; Dangerous for the environment, Very toxic to aquatic organisms [1]
    Transport code 9 [1]
    Packaging group III [1]
    UN Number 3077 [1]; N/C [2]
    Operator protection A [1, 2]; H [2]; U05a [1, 2]; U20b [1]; U20c [2]
    Environmental protection E15a [2]; E15b [1]; H410 [1, 2]
    Storage and disposal D01, D02, D09a, D11a

FOR FULL CONDITIONS OF USE ALWAYS READ THE PRODUCT LABEL

285   4-indol-3-ylbutyric acid

A plant growth regulator promoting the rooting of cuttings

Products

1	Chryzoplus Grey 0.8%	Fargro	0.8% w/w	DP	17569
2	Chryzopon Rose 0.1%	Fargro	0.1% w/w	DP	17566
3	Chryzotek Beige 0.4%	Fargro	0.4% w/w	DP	17568
4	Chryzotop Green 0.25%	Fargro	0.25% w/w	DP	17567
5	Rhizopon AA Powder (0.5%)	Fargro	0.5% w/w	DP	17570
6	Rhizopon AA Powder (1%)	Fargro	1% w/w	DP	17571
7	Rhizopon AA Powder (2%)	Fargro	2% w/w	DP	17572
8	Rhizopon AA Tablets	Fargro	50 mg a.i.	WT	17573

Uses
- Rooting of cuttings in ***ornamental plant production***

Approval information
- 4-indol-3-ylbutyric acid not included in Annex 1 under EC Regulation 1107/2009

Efficacy guidance
- Dip base of cuttings into powder immediately before planting
- Powders or solutions of different concentration are required for different types of cutting. Lowest concentration for softwood, intermediate for semi-ripe, highest for hardwood
- See label for details of concentration and timing recommended for different species
- Use of planting holes recommended for powder formulations to ensure product is not removed on insertion of cutting. Cuttings should be watered in if necessary

Restrictions
- Maximum number of treatments 1 per situation
- Use of too strong a powder or solution may cause injury to cuttings
- No unused moistened powder should be returned to container

Crop-specific information
- Latest use: before cutting insertion for ornamental specimens

Hazard classification and safety precautions
  UN Number N/C
  Operator protection U14, U15 [1-4]; U19a [1-8]; U20a [1-4, 6-8]; U20b [5]
  Environmental protection E15a
  Storage and disposal D09a, D11a

286   indoxacarb

An oxadiazine insecticide for caterpillar control in a range of crops
IRAC mode of action code: 22A

Products

1	Explicit	DuPont	30% w/w	WG	15359
2	Rumo	DuPont	30% w/w	WG	14883
3	Steward	DuPont	30% w/w	WG	13149

Uses
- Cabbage leafroller in ***chinese cabbage*** (off-label), ***choi sum*** (off-label), ***collards*** (off-label), ***kale*** (off-label), ***kohlrabi*** (off-label), ***pak choi*** (off-label), ***protected baby leaf crops*** (off-label), ***protected herbs (see appendix 6)*** (off-label), ***protected kohlrabi*** (off-label), ***spring greens*** (off-label), ***tatsoi*** (off-label) [1-3]

SEE SECTION 3 FOR PRODUCTS ALSO REGISTERED

SECTION 2

- Cabbage moth in *chinese cabbage* (off-label), *choi sum* (off-label), *collards* (off-label), *kale* (off-label), *kohlrabi* (off-label), *pak choi* (off-label), *spring greens* (off-label), *tatsoi* (off-label) [1-3]; *protected baby leaf crops* (off-label), *protected herbs (see appendix 6)* (off-label), *protected kohlrabi* (off-label) [1, 2]
- Cabbage white butterfly in *chinese cabbage* (off-label), *choi sum* (off-label), *collards* (off-label), *kale* (off-label), *kohlrabi* (off-label), *pak choi* (off-label), *protected baby leaf crops* (off-label), *protected herbs (see appendix 6)* (off-label), *protected kohlrabi* (off-label), *spring greens* (off-label), *tatsoi* (off-label) [1-3]
- Capsids in *apricots* (off-label), *nectarines* (off-label), *peaches* (off-label) [1]; *strawberries* (off-label) [1, 3]
- Caterpillars in *all edible seed crops grown outdoors* (off-label), *all non-edible seed crops grown outdoors* (off-label), *blackberries* (off-label), *blueberries* (off-label), *endives* (off-label), *lettuce* (off-label), *protected blackberries* (off-label), *protected raspberries* (off-label), *raspberries* (off-label) [3]; *apples*, *cherries* (off-label), *forest nurseries* (off-label), *grapevines* (off-label), *hops* (off-label), *ornamental plant production* (off-label), *pears*, *protected forest nurseries* (off-label), *protected hops* (off-label), *protected soft fruit* (off-label), *protected top fruit* (off-label), *soft fruit* (off-label), *top fruit* (off-label) [1, 3]; *broccoli*, *brussels sprouts* (off-label), *cabbages*, *cauliflowers*, *protected aubergines*, *protected courgettes*, *protected cucumbers*, *protected marrows*, *protected melons*, *protected ornamentals*, *protected peppers*, *protected pumpkins*, *protected squashes*, *protected tomatoes*, *sweetcorn* (off-label) [1-3]; *chinese cabbage* (off-label), *choi sum* (off-label), *collards* (off-label), *kale* (off-label), *kohlrabi* (off-label) [1, 2]; *pak choi* (off-label), *protected all edible seed crops* (off-label), *protected all non-edible seed crops* (off-label), *protected baby leaf crops* (off-label), *protected herbs (see appendix 6)* (off-label), *protected kohlrabi* (off-label), *spring greens* (off-label), *tatsoi* (off-label) [1]
- Diamond-back moth in *chinese cabbage* (off-label), *choi sum* (off-label), *collards* (off-label), *kale* (off-label), *kohlrabi* (off-label), *pak choi* (off-label), *protected baby leaf crops* (off-label), *protected herbs (see appendix 6)* (off-label), *protected kohlrabi* (off-label), *spring greens* (off-label), *tatsoi* (off-label) [1-3]; *endives* (off-label), *lettuce* (off-label) [1, 2]; *sweetcorn* (off-label) [3]
- Flax tortrix moth in *endives* (off-label), *lettuce* (off-label) [1, 2]
- Garden pebble moth in *chinese cabbage* (off-label), *choi sum* (off-label), *collards* (off-label), *kale* (off-label), *kohlrabi* (off-label), *pak choi* (off-label), *protected baby leaf crops* (off-label), *protected herbs (see appendix 6)* (off-label), *protected kohlrabi* (off-label), *spring greens* (off-label), *tatsoi* (off-label) [1-3]
- Ghost moth in *endives* (off-label), *lettuce* (off-label) [1, 2]
- Insect pests in *all edible seed crops grown outdoors* (off-label), *all non-edible seed crops grown outdoors* (off-label), *forest nurseries* (off-label), *hops* (off-label), *ornamental plant production* (off-label), *protected forest nurseries* (off-label), *protected hops* (off-label) [2]; *blackberries* (off-label), *blueberries* (off-label), *protected blackberries* (off-label), *protected raspberries* (off-label), *raspberries* (off-label) [1, 3]
- Peach moth in *apricots* (off-label), *nectarines* (off-label), *peaches* (off-label) [1, 3]; *strawberries* (off-label) [1]
- Pollen beetle in *mustard* (off-label) [1, 3]; *spring oilseed rape*, *winter oilseed rape* [1-3]
- Silver Y moth in *endives* (off-label), *lettuce* (off-label), *protected baby leaf crops* (off-label), *protected herbs (see appendix 6)* (off-label), *protected kohlrabi* (off-label) [1-3]
- Swift moth in *endives* (off-label), *lettuce* (off-label) [1, 2]
- Winter moth in *apricots* (off-label), *nectarines* (off-label), *peaches* (off-label) [1, 3]; *strawberries* (off-label) [1]

Extension of Authorisation for Minor Use (EAMUs)
- *all edible seed crops grown outdoors* 20101537 [2], 20082905 [3]
- *all non-edible seed crops grown outdoors* 20101537 [2], 20082905 [3]
- *apricots* 20141468 [1], 20141031 [3]
- *blackberries* 20131369 [1], 20130988 [3]
- *blueberries* 20131369 [1], 20130988 [3]
- *brussels sprouts* 20112093 [1], 20101537 [2], 20081485 [3]
- *cherries* 20112092 [1], 20092300 [3]
- *chinese cabbage* 20141466 [1], 20141858 [1], 20141465 [2], 20141859 [2], 20141690 [3]

- *choi sum* *20141858* [1], *20141859* [2], *20141690* [3]
- *collards* *20141858* [1], *20141859* [2], *20141690* [3]
- *endives* *20161959* [1], *20161960* [2], *20161566* [3]
- *forest nurseries* *20112090* [1], *20101538* [2], *20082905* [3]
- *grapevines* *20112091* [1], *20072289* [3]
- *hops* *20112090* [1], *20101537* [2], *20082905* [3]
- *kale* *20141466* [1], *20141858* [1], *20141465* [2], *20141859* [2], *20141690* [3]
- *kohlrabi* *20141466* [1], *20141858* [1], *20141465* [2], *20141859* [2], *20141690* [3]
- *lettuce* *20161959* [1], *20161960* [2], *20161566* [3]
- *mustard* *20141836* [1], *20151918* [3]
- *nectarines* *20141468* [1], *20141031* [3]
- *ornamental plant production* *20112090* [1], *20101537* [2], *20082905* [3]
- *pak choi* *20141466* [1], *20141858* [1], *20141465* [2], *20141690* [3]
- *peaches* *20141468* [1], *20141031* [3]
- *protected all edible seed crops* *20112090* [1]
- *protected all non-edible seed crops* *20112090* [1]
- *protected baby leaf crops* *20141467* [1], *20141464* [2], *20141030* [3]
- *protected blackberries* *20131369* [1], *20130988* [3]
- *protected forest nurseries* *20112090* [1], *20101537* [2], *20082905* [3]
- *protected herbs (see appendix 6)* *20141467* [1], *20141464* [2], *20141030* [3]
- *protected hops* *20112090* [1], *20101537* [2], *20082905* [3]
- *protected kohlrabi* *20141467* [1], *20141464* [2], *20141030* [3]
- *protected raspberries* *20131369* [1], *20130988* [3]
- *protected soft fruit* *20112090* [1], *20082905* [3]
- *protected top fruit* *20112090* [1], *20082905* [3]
- *raspberries* *20131369* [1], *20130988* [3]
- *soft fruit* *20112090* [1], *20082905* [3]
- *spring greens* *20141466* [1], *20141858* [1], *20141465* [2], *20141690* [3]
- *strawberries* *20141468* [1], *20141031* [3]
- *sweetcorn* *20112089* [1], *20101536* [2], *20072288* [3]
- *tatsoi* *20141466* [1], *20141858* [1], *20141465* [2], *20141690* [3]
- *top fruit* *20112090* [1], *20082905* [3]

Approval information
- Indoxacarb included in Annex I under EC Regulation 1107/2009

Efficacy guidance
- Best results in brassica and protected crops obtained from treatment when first caterpillars are detected, or when damage first seen, or 7-10 d after trapping first adults in pheromone traps
- In apples and pears apply at egg-hatch
- Subsequent treatments in all crops may be applied at 8-14 d intervals
- Indoxacarb acts by ingestion and contact. Only larval stages are controlled but there is some ovicidal action against some species

Restrictions
- Maximum number of treatments 3 per crop or yr for apples, pears, brassica crops; 6 per yr for protected crops and 1 per crop for oilseed rape
- Do not apply to any crop suffering from stress from any cause
- When applying to protected crops the maximum concentration must not exceed 12.5 g of product per 100 litre water

Crop-specific information
- HI: brassica crops, protected crops 1 d; apples, pears 7 d

Following crops guidance
- When treating protected crops, observe the maximum concentration permitted.

Environmental safety
- Dangerous for the environment
- Toxic to aquatic organisms

SEE SECTION 3 FOR PRODUCTS ALSO REGISTERED

- In accordance with good agricultural practice apply in early morning or late evening when bees are less active
- Broadcast air-assisted LERAP [1, 3] (15m); LERAP Category B [1, 2]

Hazard classification and safety precautions
Hazard Harmful, Dangerous for the environment, Harmful if swallowed
Transport code 9
Packaging group III
UN Number 3077
Risk phrases H371
Operator protection A, D; U04a, U05a
Environmental protection E12f, E15b, E34, E38, H410 [1-3]; E16a [1, 2]; E17b [1, 3] (15m)
Storage and disposal D01, D02, D09a, D12a
Medical advice M03

287   iodosulfuron-methyl-sodium + mesosulfuron-methyl

A sulfonyl urea herbicide mixture for winter wheat
HRAC mode of action code: B + B

See also mesosulfuron-methyl

Products

1	Atlantis OD	Bayer CropScience	2:10 g/l	OD	18100
2	Atlantis WG	Bayer CropScience	0.6:3.0% w/w	WG	12478
3	Cintac	Life Scientific	1.0:3.0% w/w	WG	18222
4	Greencrop Doonbeg 2	Clayton	0.6:3.0 % w/w	WG	13025
5	Hatra	Bayer CropScience	2:10 g/l	OD	16190
6	Horus	Bayer CropScience	2:10 g/l	OD	16216
7	Lanista	Life Scientific	0.6:3.0% w/w	WG	18118
8	Nemo	AgChem Access	0.6:3.0% w/w	WG	14140
9	Pacifica	Bayer CropScience	1.0:3.0% w/w	WG	12049

Uses
- Annual grasses in *durum wheat* (off-label), *rye* (off-label), *triticale* (off-label) [2, 9]; *spring wheat* (off-label) [2]
- Annual meadow grass in *durum wheat* (off-label), *rye* (off-label), *spring wheat* (off-label), *triticale* (off-label) [2]; *winter wheat* [1-9]
- Blackgrass in *durum wheat* (off-label), *rye* (off-label), *triticale* (off-label) [2, 9]; *spring wheat* (off-label) [2]; *winter wheat* [1-9]
- Chickweed in *winter wheat* [1-9]
- Italian ryegrass in *winter wheat* [1-9]
- Mayweeds in *winter wheat* [1-9]
- Perennial ryegrass in *winter wheat* [1-9]
- Rough-stalked meadow grass in *winter wheat* [1-9]
- Wild oats in *winter wheat* [1-9]

Extension of Authorisation for Minor Use (EAMUs)
- *durum wheat* 20170897 [2], 20170898 [9]
- *rye* 20170897 [2], 20170898 [9]
- *spring wheat* 20170899 [2]
- *triticale* 20170897 [2], 20170898 [9]

Approval information
- Iodosulfuron-methyl-sodium and mesosulfuron-methyl included in Annex I under EC Regulation 1107/2009

Efficacy guidance
- Optimum grass weed control obtained when all grass weeds are emerged at spraying. Activity is primarily via foliar uptake and good spray coverage of the target weeds is essential

FOR FULL CONDITIONS OF USE ALWAYS READ THE PRODUCT LABEL

- Translocation occurs readily within the target weeds and growth is inhibited within hours of treatment but symptoms may not be apparent for up to 4 wk, depending on weed species, timing of treatment and weather conditions
- Residual activity is important for best results and is optimised by treatment on fine moist seedbeds. Avoid application under very dry conditions
- Residual efficacy may be reduced by high soil temperatures and cloddy seedbeds
- Iodosulfuron-methyl and mesosulfuron-methyl are both members of the ALS-inhibitor group of herbicides. To avoid the build up of resistance do not use any product containing an ALS-inhibitor herbicide with claims for control of grass weeds more than once on any crop
- Use these products as part of a Resistance Management Strategy that includes cultural methods of control and does not use ALS inhibitors as the sole chemical method of grass weed control. See Section 5 for more information

Restrictions

- Maximum number of treatments 1 per crop with a maximum total dose equivalent to one full dose treatment
- Do not use on crops undersown with grasses, clover or other legumes or any other broad-leaved crop
- Do not use as a stand-alone treatment for blackgrass, ryegrass or chickweed control
- Do not use as the sole means of weed control in successive crops
- Do not use in mixture or in sequence with any other ALS-inhibitor herbicide except those (if any) specified on the label
- Do not apply earlier than 1 Feb in the yr of harvest [3, 9]
- Do not apply to crops under stress from any cause
- Do not apply when rain is imminent or during periods of frosty weather
- Specified adjuvant must be used. See label

Crop-specific information

- Latest use: flag leaf just visible (GS 39)
- Winter wheat may be treated from the two-leaf stage of the crop
- Safety to crops grown for seed not established
- Avoid application before a severe frost or transient yellowing of the crop may occur.

Following crops guidance

- In the event of crop failure sow only winter wheat in the same cropping season
- Only winter wheat or winter barley (or winter oilseed rape [2]) may be sown in the year of harvest of a treated crop but ploughing before drilling oilseed rape is advisable, especially in a season where applications are made later than usual.
- Spring wheat, spring barley, sugar beet or spring oilseed rape may be drilled in the following spring. Plough before drilling oilseed rape

Environmental safety

- Dangerous for the environment
- Very toxic to aquatic organisms
- Dangerous to fish or other aquatic life. Do not contaminate surface waters or ditches with chemical or used container
- Take extreme care to avoid drift onto plants outside the target area or on to ponds, waterways or ditches
- LERAP Category B

Hazard classification and safety precautions

Hazard Irritant, Dangerous for the environment [1-9]; Very toxic to aquatic organisms [1-3, 8, 9]
Transport code 9
Packaging group III
UN Number 3077 [2-4, 7-9]; 3082 [1, 5, 6]
Risk phrases H315 [2, 7, 8]; H317 [7]; H318 [2, 3, 7-9]; H319 [1]; R36, R38 [5, 6]; R41 [4]; R50, R53a [4-6]
Operator protection A, C, H; U05a, U11, U20b
Environmental protection E15a [1-3, 5-9]; E15b [4]; E16a, E38 [1-9]; H410 [1-3, 8, 9]; H411 [7]
Storage and disposal D01, D02, D05, D09a, D10a, D12a

SEE SECTION 3 FOR PRODUCTS ALSO REGISTERED

SECTION 2

288   ipconazole

A triazole fungicide for disease control in cereals
FRAC mode of action code: 3

See also imazalil + ipconazole

Products

Rancona 15 ME	Certis	15.1 g/l	MS	16500

Uses
* Seed-borne diseases in **spring barley, spring wheat, winter barley, winter wheat**
* Soil-borne diseases in **spring barley, spring wheat, winter barley, winter wheat**

Approval information
* Ipconazole included in Annex I under EC Regulation 1107/2009
* Accepted by BBPA for use on malting barley

Hazard classification and safety precautions
  UN Number N/C
  Operator protection A, H; U05a, U09a, U20b
  Environmental protection E03, E15b, E34, H410
  Storage and disposal D01, D02, D09a, D10a
  Treated seed S01, S02, S03, S04d, S05, S07
  Medical advice M03

289   iprodione

A protectant dicarboximide fungicide with some eradicant activity
FRAC mode of action code: 2

See also carbendazim + iprodione

Products

1	Cavron Green	Belcrop	250 g/l	SC	15340
2	Chipco Green	Bayer CropScience	255 g/l	SC	13843
3	Grisu	Sipcam	500 g/l	SC	15375
4	Mascot Rayzor	Rigby Taylor	255 g/l	SC	14903
5	Panto	Pan Amenity	255 g/l	SC	14307
6	Pure Turf	Pure Amenity	255 g/l	SC	14851
7	Rovral WG	BASF	75% w/w	WG	13811
8	Surpass Pro	Headland Amenity	250 g/l	SC	15368

Uses
* Alternaria in **brassica seed crops, lettuce, oilseed rape, ornamental plant production, protected lettuce, protected strawberries, protected tomatoes, raspberries, salad onions, strawberries** [3]; **brassica seed crops** (pre-storage only), **dwarf beans** (off-label), **french beans** (off-label), **garlic** (off-label), **grapevines** (off-label), **mange-tout peas** (off-label), **protected herbs (see appendix 6)** (off-label), **runner beans** (off-label), **shallots** (off-label), **spring oilseed rape, winter oilseed rape** [7]; **brussels sprouts, cauliflowers** [3, 7]
* Anthracnose in **managed amenity turf** [2]
* Botrytis in **brassica seed crops, brussels sprouts, cauliflowers, oilseed rape** [3]; **bulb onions, cabbages** (off-label), **dwarf beans** (off-label), **forest nurseries** (off-label), **french beans** (off-label), **garlic** (off-label), **grapevines** (off-label), **hops** (off-label), **mange-tout peas** (off-label), **ornamental plant production, pears** (off-label), **protected aubergines** (off-label), **protected forest nurseries** (off-label), **protected herbs (see appendix 6)** (off-label), **protected hops** (off-label), **protected ornamentals** (off-label), **protected rhubarb** (off-label), **protected soft fruit** (off-label), **protected top fruit** (off-label), **red beet** (off-label), **runner beans** (off-label), **shallots** (off-label), **soft fruit** (off-label), **spring oilseed rape, top fruit** (off-label), **winter oilseed rape** [7]; **lettuce, ornamental plant production, protected lettuce, protected strawberries, protected tomatoes, raspberries, salad onions, strawberries** [3, 7]

FOR FULL CONDITIONS OF USE ALWAYS READ THE PRODUCT LABEL

- Brown patch in **amenity grassland** [6]; **managed amenity turf** [1, 6, 8]
- Collar rot in **bulb onions**, **garlic** *(off-label)*, **salad onions**, **shallots** *(off-label)* [7]
- Disease control in **managed amenity turf** [4, 5]; **ornamental plant production** *(off-label)* [7]
- Dollar spot in **amenity grassland** [2, 6]; **managed amenity turf** [1, 2, 6, 8]
- Fusarium patch in **amenity grassland** [2, 6]; **managed amenity turf** [1, 2, 6, 8]
- Melting out in **amenity grassland** [2, 6]; **managed amenity turf** [1, 2, 6, 8]
- Red thread in **amenity grassland** [2, 6]; **managed amenity turf** [1, 2, 6, 8]
- Rust in **managed amenity turf** [2]
- Sclerotinia in **oilseed rape** *(moderate control)* [3]
- Snow mould in **amenity grassland** [2, 6]; **managed amenity turf** [1, 2, 6, 8]

Extension of Authorisation for Minor Use (EAMUs)
- **cabbages** *20091845* [7]
- **dwarf beans** *20081760* [7]
- **forest nurseries** *20082902* [7]
- **french beans** *20081760* [7]
- **garlic** *20081768* [7]
- **grapevines** *20081758* [7]
- **hops** *20082902* [7]
- **mange-tout peas** *20081760* [7]
- **ornamental plant production** *20113200* [7]
- **pears** *20081761* [7]
- **protected aubergines** *20081767* [7]
- **protected forest nurseries** *20082902* [7]
- **protected herbs (see appendix 6)** *20081757* [7]
- **protected hops** *20082902* [7]
- **protected ornamentals** *20113200* [7]
- **protected rhubarb** *20081763* [7]
- **protected soft fruit** *20082902* [7]
- **protected top fruit** *20082902* [7]
- **red beet** *20081762* [7]
- **runner beans** *20081760* [7]
- **shallots** *20081768* [7]
- **soft fruit** *20082902* [7]
- **top fruit** *20082902* [7]

Approval information
- Iprodione included in Annex I under EC Regulation 1107/2009

Efficacy guidance
- Many diseases require a programme of 2 or more sprays at intervals of 2-4 wk. Recommendations vary with disease and crop - see label for details
- Use as a drench to control cabbage storage diseases. Spray ornamental pot plants and cucumbers to run-off
- Apply turf and amenity grass treatments after mowing to dry grass, free of dew. Do not mow again for at least 24 h [2]

Restrictions
- Maximum number of treatments or maximum total dose equivalent to 1 per batch for seed treatments; 1 per crop on cereals, cabbage (as drench), 2 per crop on field beans and stubble turnips; 3 per crop on brassicas (including seed crops), oilseed rape, protected winter lettuce (Oct-Feb); 4 per crop on strawberries, grapevines, salad onions, cucumbers; 5 per crop on raspberries; 6 per crop on bulb onions, tomatoes, turf; 7 per crop on lettuce (Mar-Sep)
- A minimum of 3 wk must elapse between treatments on leaf brassicas
- Treated brassica seed crops not to be used for human or animal consumption
- See label for pot plants showing good tolerance. Check other species before applying on a large scale
- Do not treat oilseed rape seed that is cracked or broken or of low viability
- Do not treat oats

SECTION 2

SEE SECTION 3 FOR PRODUCTS ALSO REGISTERED

Crop-specific information

- Latest use: pre-planting for seed treatments; at planting for potatoes; before grain watery ripe (GS 69) for wheat and barley; 21 d before consumption by livestock for stubble turnips
- HI strawberries, protected tomatoes 1 d; outdoor tomatoes 2 d; bulb onions, salad onions, raspberries, lettuce 7 d; brassicas, brassica seed crops, oilseed rape, mustard, field beans, stubble turnips 21 d
- Turf and amenity grass may become temporarily yellowed if frost or hot weather follows treatment
- Personal protective equipment requirements may vary for each pack size. Check label

Environmental safety

- Dangerous for the environment
- Very toxic to aquatic organisms
- See label for guidance on disposal of spent drench liquor
- Treatment harmless to *Encarsia* or *Phytoseiulus* being used for integrated pest control
- To protect non target insects/arthropods respect an unsprayed buffer zone of 5 m to non-crop land [3]
- Broadcast air-assisted LERAP [3, 7] (15 m); LERAP Category B [1-8]

Hazard classification and safety precautions

Hazard Harmful, Dangerous for the environment [1-8]; Very toxic to aquatic organisms [2, 6]
Transport code 9
Packaging group III
UN Number 3077 [7]; 3082 [1-6, 8]
Risk phrases H315 [1, 4, 5]; H320 [4, 5]; H351 [1-6]; R36, R40, R50, R53a [7, 8]; R38 [8]
Operator protection A [1-8]; C [1, 2, 4-8]; H [3, 7]; U04a, U08 [1, 2, 4-6, 8]; U05a [1-8]; U11 [3, 7]; U16b, U19a, U20b [7]; U20c [1-6, 8]
Environmental protection E15a [1-6, 8]; E15b [7]; E16a, E34, E38 [1-8]; E17b [3, 7] (15 m); H410 [1-6]
Storage and disposal D01, D02, D09a, D12a [1-8]; D05, D10b [1, 2, 4-6, 8]; D11a [7]
Treated seed S01, S02, S03, S04a, S05 [3]
Medical advice M03 [1-6, 8]; M05a [3, 7]

---

290   iprodione + thiophanate-methyl

A protectant and systemic fungicide for oilseed rape
FRAC mode of action code: 2 + 1

See also thiophanate-methyl

Products

Compass	BASF	167:167 g/l	SC	16801

Uses

- Alternaria in **spring oilseed rape, winter oilseed rape**
- Grey mould in **spring oilseed rape, winter oilseed rape**
- Light leaf spot in **spring oilseed rape, winter oilseed rape**
- Sclerotinia stem rot in **spring oilseed rape, winter oilseed rape**
- Stem canker in **spring oilseed rape, winter oilseed rape**

Approval information

- Iprodione and thiophanate-methyl included in Annex I under EC Regulation 1107/2009

Efficacy guidance

- Timing of sprays on oilseed rape varies with disease, see label for details
- For season-long control product should be applied as part of a disease control programme

Restrictions

- Maximum number of treatments (refers to total sprays containing benomyl, carbendazim or thiophanate methyl) 2 per crop for oilseed rape, field beans, meadowfoam; 3 per crop for carrots, horseradish, parsnips

FOR FULL CONDITIONS OF USE ALWAYS READ THE PRODUCT LABEL

Crop-specific information
- Latest use: before end of flowering for oilseed rape, field beans
- HI meadowfoam, field beans, oilseed rape 3 wk; carrots, horseradish, parsnips 28 d
- Treatment may extend duration of green leaf in winter oilseed rape

Environmental safety
- Dangerous for the environment
- Very toxic to aquatic organisms
- LERAP Category B

Hazard classification and safety precautions
Hazard Harmful, Dangerous for the environment
Transport code 9
Packaging group III
UN Number 3082
Risk phrases R40, R43, R50, R53a, R68
Operator protection A, C, H, M; U05a, U20c
Environmental protection E15a, E16a, E38
Storage and disposal D01, D02, D08, D09a, D10b, D12a
Medical advice M05a

291   iprodione + trifloxystrobin

A fungicide for use in managed amenity turf
FRAC mode of action code: 2 + 11

Products

| Interface | Bayer CropScience | 256:16 g/l | SC | 16060 |

Uses
- Anthracnose in *managed amenity turf (moderate control)*
- Disease control in *managed amenity turf*
- Dollar spot in *managed amenity turf*
- Fusarium patch in *managed amenity turf*
- Leaf spot in *managed amenity turf*
- Red thread in *managed amenity turf*
- Rust in *managed amenity turf*

Approval information
- Iprodione and trifloxystrobin included in Annex I under EC Regulation 1107/2009

Efficacy guidance
- Do not apply during drought or when turf is frozen
- Do not mow for at least 48 hours after application
- Allow at least 4 weeks between applications

Restrictions
- Do not use on commercial, estate or residential lawns

Environmental safety
- LERAP Category B

Hazard classification and safety precautions
Hazard Very toxic to aquatic organisms
Transport code 9
Packaging group III
UN Number 3082
Risk phrases H351
Operator protection A
Environmental protection E15a, E16a, E16b, E34, H410
Storage and disposal D01, D09a, D10b

SEE SECTION 3 FOR PRODUCTS ALSO REGISTERED

SECTION 2

292　isopyrazam

An SDHI fungicide available for disease control in barley.
FRAC mode of action code: 7

See also azoxystrobin + isopyrazam
*　　　cyprodinil + isopyrazam*
*　　　epoxiconazole + isopyrazam*

Products

1	Reflect	Syngenta	125 g/l	EC	17228
2	Zulu	Syngenta	125 g/l	EC	16913

Uses

- Alternaria in **red beet** *(off-label)* [1]
- Alternaria blight in **carrots** [1]
- Canker in **parsnips** *(off-label)* [1]
- Cercospora leaf spot in **red beet** *(off-label)* [1]
- Disease control in **protected melons, protected peppers, protected watermelon** [1]; **spring barley, winter barley** [2]
- Grey mould in **protected hemp** *(off-label)* [1]
- Mildew in **red beet** *(off-label)* [1]
- Powdery mildew in **carrots, ornamental plant production** *(off-label)*, **protected aubergines, protected chilli peppers, protected courgettes, protected cucumbers, protected hemp** *(off-label)*, **protected melons, protected ornamentals** *(off-label)*, **protected peppers, protected summer squash, protected tomatoes, protected watermelon, swedes** *(off-label)*, **turnips** *(off-label)* [1]
- Ramularia leaf spots in **red beet** *(off-label)* [1]
- Sclerotinia in **celeriac** *(off-label)*, **protected hemp** *(off-label)* [1]
- Septoria seedling blight in **celeriac** *(off-label)* [1]

Extension of Authorisation for Minor Use (EAMUs)

- **celeriac** *20160798* [1]
- **ornamental plant production** *20161602* [1]
- **parsnips** *20160798* [1]
- **protected hemp** *20170870 expires 31 Dec 2018* [1], *20171302* [1]
- **protected ornamentals** *20161602* [1]
- **red beet** *20160798* [1]
- **swedes** *20160798* [1]
- **turnips** *20160798* [1]

Approval information

- Isopyrazam is included in Appendix 1 under EC Regulation 1107/2009

Efficacy guidance

- Use only in mixture with fungicides with a different mode of action that are effective against the target diseases.

Restrictions

- Isopyrazam is an SDH respiration inhibitor; Do not apply more than two foliar applications of products containing an SDH inhibitor to any cereal crop

Crop-specific information

- There are no restrictions on succeeding crops in a normal rotation

Environmental safety

- LERAP Category B

Hazard classification and safety precautions

Hazard Harmful, Irritant, Dangerous for the environment, Harmful if inhaled, Very toxic to aquatic organisms [2]; Harmful if swallowed [1, 2]
Transport code 9

FOR FULL CONDITIONS OF USE ALWAYS READ THE PRODUCT LABEL

Packaging group III
UN Number 3082
Risk phrases H319, H361 [1, 2]; H351 [1]
Operator protection A, H [1, 2]; C [1]
Environmental protection E16a, H410 [1, 2]; E38 [2]
Storage and disposal D01, D02, D12a, D12b [2]

293 isoxaben

A soil-acting benzamide herbicide for use in cereals, grass and fruit
HRAC mode of action code: L

Products

Flexidor	Landseer	500 g/l	SC	18042

Uses
- Annual dicotyledons in **amenity vegetation, apples, asparagus, blackberries, blackcurrants, cherries, forest, forest nurseries, gooseberries, hops, ornamental plant production, pears, plums, raspberries, redcurrants, strawberries, triticale, winter barley, winter oats, winter rye, winter wheat**

Approval information
- Isoxaben included in Annex 1 under EC Regulation 1107/2009
- Accepted by BBPA for use on hops

Efficacy guidance
- When used alone apply pre-weed emergence
- Effectiveness is reduced in dry conditions. Weed seeds germinating at depth are not controlled
- Activity reduced on soils with more than 10% organic matter. Do not use on peaty soils
- Various tank mixtures are recommended for early post-weed emergence treatment (especially for grass weeds). See label for details

Restrictions
- Maximum number of treatments 1 per crop for all edible crops; 2 per yr on amenity vegetation and non-edible crops

Crop-specific information
- Latest use: before 1 Apr in yr of harvest for edible crops

Following crops guidance
- See label for details of crops which may be sown in the event of failure of a treated crop

Environmental safety
- Keep all livestock out of treated areas for at least 50 d

Hazard classification and safety precautions
Hazard Very toxic to aquatic organisms
Transport code 9
Packaging group III
UN Number 3082
Operator protection A, C, H, M; U05a, U20a
Environmental protection E06a (50 days); E15b, H410
Storage and disposal D01, D02, D09a, D11a

294 isoxaflutole

An isoxazole herbicide available only in mixtures
HRAC mode of action code: F2

See also flufenacet + isoxaflutole

SEE SECTION 3 FOR PRODUCTS ALSO REGISTERED

295   kresoxim-methyl

A protectant strobilurin fungicide for apples
FRAC mode of action code: 11

See also epoxiconazole + fenpropimorph + kresoxim-methyl
epoxiconazole + kresoxim-methyl
epoxiconazole + kresoxim-methyl + pyraclostrobin
fenpropimorph + kresoxim-methyl

Products

Stroby WG	BASF	50% w/w	WG	17316

Uses

- American gooseberry mildew in **blueberries** *(off-label)*, **cranberries** *(off-label)*, **gooseberries** *(off-label)*, **whitecurrants** *(off-label)*
- Black rot in **wine grapes** *(off-label)*
- Powdery mildew in **apple for cider making** *(reduction)*, **apples** *(reduction)*, **blackcurrants**, **blueberries** *(off-label)*, **cranberries** *(off-label)*, **gooseberries** *(off-label)*, **ornamental plant production**, **protected strawberries**, **redcurrants**, **strawberries**, **whitecurrants** *(off-label)*, **wine grapes** *(off-label)*
- Scab in **apple for cider making**, **apples**

Extension of Authorisation for Minor Use (EAMUs)

- **blueberries** *20160089*
- **cranberries** *20160089*
- **gooseberries** *20160089*
- **whitecurrants** *20160089*
- **wine grapes** *20160960*

Approval information

- Kresoxim-methyl included in Annex I under EC Regulation 1107/2009
- Approved for use in ULV systems

Efficacy guidance

- Activity is protectant. Best results achieved from treatments prior to disease development. See label for timing details on each crop. Treatments should be repeated at 10-14 d intervals but note limitations below
- To minimise the likelihood of development of resistance to strobilurin fungicides these products should be used in a planned Resistance Management strategy. See Section 5 for more information
- Product may be applied in ultra low volumes (ULV) but disease control may be reduced

Restrictions

- Maximum number of treatments 4 per yr on apples; 3 per yr on other crops. See notes in Efficacy about limitations on consecutive treatments
- Consult before using on crops intended for processing

Crop-specific information

- HI 14 d for blackcurrants, protected strawberries, strawberries; 35 d for apples
- On apples do not spray product more than twice consecutively and separate each block of two consecutive treatments with at least two applications from a different cross-resistance group. For all other crops do not apply consecutively and use a maximum of once in every three fungicide sprays
- Product should not be used as final spray of the season on apples

Environmental safety

- Dangerous for the environment
- Very toxic to aquatic organisms
- Harmless to ladybirds and predatory mites

FOR FULL CONDITIONS OF USE ALWAYS READ THE PRODUCT LABEL

- Harmless to honey bees and may be applied during flowering. Nevertheless local beekeepers should be notified when treatment of orchards in flower is to occur
- Broadcast air-assisted LERAP (5 m)

Hazard classification and safety precautions

Hazard Harmful, Dangerous for the environment, Very toxic to aquatic organisms
Transport code 9
Packaging group III
UN Number 3077
Risk phrases H351
Operator protection U05a, U20b
Environmental protection E15a, E38, H410; E17b (5 m)
Storage and disposal D01, D02, D08, D09a, D10c, D12a
Medical advice M05a

296   lambda-cyhalothrin

A quick-acting contact and ingested pyrethroid insecticide
IRAC mode of action code: 3

Products

1	Clayton Lanark	Clayton	100 g/l	CS	12942
2	Clayton Sparta	Clayton	50 g/l	EC	13457
3	Dalda 5	Globachem	50 g/l	EC	13688
4	Hallmark with Zeon Technology	Syngenta	100 g/l	CS	12629
5	Karate 2.5WG	Syngenta	2.5% w/w	GR	14060
6	Karis 10 CS	Headland	100 g/l	CS	16747
7	Lambda-C 100	Becesane	100 g/l	CS	15987
8	Lambda-C 50	Becesane	50 g/l	EC	15986
9	Lambdastar	Life Scientific	100 g/l	CS	17406
10	Markate 50	Agrovista	50 g/l	EC	13529
11	Seal Z	AgChem Access	100 g/l	CS	14201
12	Sparviero	Sipcam	100 g/l	CS	15687

Uses

- Aphids in *all edible seed crops grown outdoors* (off-label), *all non-edible seed crops grown outdoors* (off-label), *crambe* (off-label), *farm forestry* (off-label), *forest nurseries* (off-label), *hops* (off-label), *miscanthus* (off-label), *ornamental plant production* (off-label), *protected aubergines* (off-label), *protected forest nurseries* (off-label), *protected ornamentals* (off-label), *protected pak choi* (off-label), *protected peppers* (off-label), *protected tatsoi* (off-label), *protected tomatoes* (off-label), *soft fruit* (off-label), *top fruit* (off-label), *willow (short rotation coppice)* (off-label) [4]; *broccoli, brussels sprouts, calabrese, kale* [6]; *chestnuts* (off-label), *hazel nuts* (off-label), *walnuts* (off-label) [4, 10]; *cob nuts* (off-label), *lupins* (off-label) [10]; *combining peas, vining peas* [2, 3, 8, 10]; *durum wheat* [1, 3-8, 10, 11]; *edible podded peas, spring field beans, spring oilseed rape, winter field beans, winter oilseed rape* [3, 8, 10]; *fodder beet* (off-label) [1]; *lettuce, pears, rye, triticale* [3]; *potatoes, spring wheat* [1-4, 6-12]; *spring barley, spring oats* [1, 3, 4, 6, 7, 9, 11, 12]; *sugar beet* [3, 6, 8, 10]; *winter barley, winter oats* [1, 3-7, 9, 11, 12]; *winter wheat* [1-12]
- Barley yellow dwarf virus vectors in *durum wheat* [1, 4-7, 11]; *winter barley, winter oats, winter wheat* [1, 4-7, 9, 11, 12]
- Beet leaf miner in *sugar beet* [1, 4, 6, 7, 9, 11, 12]
- Beet virus yellows vectors in *spring oilseed rape* [1, 4, 6, 7, 9, 11, 12]; *winter oilseed rape* [1, 4-7, 9, 11, 12]
- Beetles in *combining peas, durum wheat, edible podded peas, potatoes, spring field beans, spring oilseed rape, spring wheat, sugar beet, vining peas, winter field beans, winter oilseed rape, winter wheat* [3, 8, 10]; *rye, spring barley, spring oats, triticale, winter barley, winter oats* [3]
- Brassica pod midge in *spring oilseed rape, winter oilseed rape* [2]

- Cabbage seed weevil in **spring oilseed rape**, **winter oilseed rape** [1, 2, 4, 6, 7, 9, 11, 12]
- Cabbage stem flea beetle in **spring oilseed rape** [1, 4, 6, 7, 9, 11, 12]; **winter oilseed rape** [1, 4-7, 9, 11, 12]
- Cabbage stem weevil in **celeriac** *(off-label)*, **radishes** *(off-label)* [1, 10]
- Capsids in **blackberries** *(off-label)*, **dewberries** *(off-label)*, **raspberries** *(off-label)*, **rubus hybrids** *(off-label)* [1, 4]; **grapevines** *(off-label)* [1, 4, 10]; **table grapes** *(off-label)*, **wine grapes** *(off-label)* [1]
- Carrot fly in **carrots** *(off-label)*, **parsnips** *(off-label)* [10]; **celeriac** *(off-label)*, **radishes** *(off-label)* [1, 10]; **celery (outdoor)** *(off-label)*, **horseradish** *(off-label)*, **mallow (althaea spp.)** *(off-label)*, **parsley root** *(off-label)* [1, 4, 10]; **fennel** *(off-label)* [4]
- Caterpillars in **beetroot** *(off-label)* [1, 4]; **broccoli**, **brussels sprouts**, **cabbages**, **calabrese**, **cauliflowers** [1, 4, 7, 9, 11]; **celery (outdoor)** *(off-label)*, **french beans** *(off-label)*, **navy beans** *(off-label)* [1, 4, 10]; **combining peas**, **durum wheat**, **edible podded peas**, **potatoes**, **spring field beans**, **spring oilseed rape**, **spring wheat**, **sugar beet**, **vining peas**, **winter field beans**, **winter oilseed rape**, **winter wheat** [3, 8, 10]; **dwarf beans** *(off-label)*, **protected pak choi** *(off-label)*, **protected tatsoi** *(off-label)*, **red beet** *(off-label)* [10]; **pears**, **rye**, **spring barley**, **spring oats**, **triticale**, **winter barley**, **winter oats** [3]; **runner beans** *(off-label)* [1, 10]; **swedes** *(off-label)*, **turnips** *(off-label)* [4, 10]; **table grapes** *(off-label)*, **wine grapes** *(off-label)* [1]
- Clay-coloured weevil in **blackberries** *(off-label)*, **dewberries** *(off-label)*, **raspberries** *(off-label)*, **rubus hybrids** *(off-label)* [1, 4]
- Common green capsid in **hops** *(off-label)* [4]
- Cutworms in **beetroot** *(off-label)* [1, 4]; **carrots**, **lettuce** [1, 4, 6, 7, 9, 11, 12]; **celeriac** *(off-label)*, **radishes** *(off-label)* [1, 10]; **chicory** *(off-label)*, **fennel** *(off-label)* [4]; **fodder beet** *(off-label)*, **red beet** *(off-label)* [10]; **parsnips** [4, 6, 7, 9, 11, 12]; **sugar beet** [1, 2, 4, 6, 7, 9, 11, 12]
- Flea beetle in **corn gromwell** *(off-label)*, **hops** *(off-label)* [4]; **crambe** *(off-label)*, **fodder beet** *(off-label)*, **protected pak choi** *(off-label)*, **protected tatsoi** *(off-label)* [10]; **poppies for morphine production** *(off-label)* [1, 4, 10]; **spring oilseed rape**, **sugar beet** [1, 2, 4, 6, 7, 9, 11, 12]; **winter oilseed rape** [1, 2, 4-7, 9, 11, 12]
- Frit fly in **sweetcorn** *(off-label)* [1, 4, 10]
- Insect pests in **all edible seed crops grown outdoors** *(off-label)*, **all non-edible seed crops grown outdoors** *(off-label)*, **celeriac** *(off-label)*, **celery (outdoor)** *(off-label)*, **chestnuts** *(off-label)*, **farm forestry** *(off-label)*, **forest nurseries** *(off-label)*, **hazel nuts** *(off-label)*, **hops** *(off-label)*, **lupins** *(off-label)*, **miscanthus** *(off-label)*, **ornamental plant production** *(off-label)*, **protected forest nurseries** *(off-label)*, **protected ornamentals** *(off-label)*, **radishes** *(off-label)*, **soft fruit** *(off-label)*, **spring rye** *(off-label)*, **top fruit** *(off-label)*, **walnuts** *(off-label)*, **winter rye** *(off-label)* [4, 10]; **beetroot** *(off-label)*, **cherries** *(off-label)*, **crambe** *(off-label)*, **mirabelles** *(off-label)*, **plums** *(off-label)*, **poppies for morphine production** *(off-label)*, **protected aubergines** *(off-label)*, **protected pak choi** *(off-label)*, **protected peppers** *(off-label)*, **protected strawberries** *(off-label)*, **protected tatsoi** *(off-label)*, **protected tomatoes** *(off-label)*, **strawberries** *(off-label)*, **willow (short rotation coppice)** *(off-label)* [4]; **blackberries** *(off-label)*, **dewberries** *(off-label)*, **frise** *(off-label)*, **lamb's lettuce** *(off-label)*, **leaf brassicas** *(off-label)*, **raspberries** *(off-label)*, **rubus hybrids** *(off-label)*, **scarole** *(off-label)* [1, 4]; **blackcurrants** *(off-label)*, **borage for oilseed production** *(off-label)*, **bulb onions** *(off-label)*, **canary flower (echium spp.)** *(off-label)*, **evening primrose** *(off-label)*, **fodder beet** *(off-label)*, **garlic** *(off-label)*, **gooseberries** *(off-label)*, **grass seed crops** *(off-label)*, **herbs (see appendix 6)** *(off-label)*, **honesty** *(off-label)*, **horseradish** *(off-label)*, **leeks** *(off-label)*, **mallow (althaea spp.)** *(off-label)*, **mustard** *(off-label)*, **navy beans** *(off-label)*, **parsley root** *(off-label)*, **redcurrants** *(off-label)*, **runner beans** *(off-label)*, **salad onions** *(off-label)*, **shallots** *(off-label)*, **spring linseed** *(off-label)*, **triticale** *(off-label)*, **whitecurrants** *(off-label)*, **winter linseed** *(off-label)* [1, 4, 10]; **broad beans** *(off-label)*, **carrots** *(off-label)*, **cob nuts** *(off-label)*, **dwarf beans** *(off-label)*, **land cress** *(off-label)*, **parsnips** *(off-label)*, **protected all edible seed crops** *(off-label)*, **protected all non-edible seed crops** *(off-label)*, **red beet** *(off-label)*, **short rotation coppice willow** *(off-label)* [10]; **french beans** *(off-label)*, **sweetcorn** *(off-label)* [1, 10]; **grapevines** *(off-label)*, **rye** *(off-label)* [1]
- Leaf midge in **blackcurrants** *(off-label)*, **gooseberries** *(off-label)*, **redcurrants** *(off-label)*, **whitecurrants** *(off-label)* [1, 4, 10]
- Leaf miner in **fodder beet** *(off-label)* [10]; **sugar beet** [2]
- Mangold fly in **sugar beet** [6]
- Pea and bean weevil in **combining peas**, **spring field beans**, **vining peas**, **winter field beans** [1, 2, 4, 6, 7, 9, 11, 12]; **edible podded peas** [1, 4, 6, 7, 9, 11, 12]

FOR FULL CONDITIONS OF USE ALWAYS READ THE PRODUCT LABEL

- Pea aphid in **combining peas**, **edible podded peas**, **vining peas** [1, 4, 6, 7, 9, 11, 12]
- Pea midge in **combining peas**, **edible podded peas**, **vining peas** [1, 4, 7, 9, 11, 12]
- Pea moth in **combining peas**, **vining peas** [1, 2, 4, 6, 7, 9, 11, 12]; **edible podded peas** [1, 4, 6, 7, 9, 11, 12]
- Pear midge in **combining peas**, **edible podded peas**, **vining peas** [6]
- Pear sucker in **pears** [1, 2, 4, 6, 7, 9, 11, 12]; **winter wheat** [2]
- Pod midge in **spring oilseed rape**, **winter oilseed rape** [1, 4, 6, 7, 9, 11, 12]
- Pollen beetle in **corn gromwell** *(off-label)* [4]; **crambe** *(off-label)* [10]; **poppies for morphine production** *(off-label)* [1, 4, 10]; **spring oilseed rape**, **winter oilseed rape** [1, 2, 4, 6, 7, 9, 11, 12]
- Rosy rustic moth in **hops** *(off-label)* [4]
- Sawflies in **blackcurrants** *(off-label)*, **gooseberries** *(off-label)*, **redcurrants** *(off-label)*, **whitecurrants** *(off-label)* [1, 4, 10]; **short rotation coppice willow** *(off-label)* [10]
- Seed beetle in **broad beans** *(off-label)* [1, 4, 10]
- Silver Y moth in **celery (outdoor)** *(off-label)*, **dwarf beans** *(off-label)*, **navy beans** *(off-label)*, **red beet** *(off-label)* [10]; **french beans** *(off-label)*, **runner beans** *(off-label)* [4, 10]
- Spotted wing drosophila in **bilberries** *(off-label)*, **blueberries** *(off-label)*, **cranberries** *(off-label)*, **ribes hybrids** *(off-label)* [4]
- Thrips in **bilberries** *(off-label)*, **blueberries** *(off-label)*, **cranberries** *(off-label)*, **ribes hybrids** *(off-label)* [4]; **broad beans** *(off-label)*, **bulb onions** *(off-label)*, **garlic** *(off-label)*, **leeks** *(off-label)*, **salad onions** *(off-label)*, **shallots** *(off-label)* [1, 4]
- Wasps in **grapevines** *(off-label)* [1, 4, 10]; **table grapes** *(off-label)*, **wine grapes** *(off-label)* [1]
- Weevils in **combining peas**, **durum wheat**, **edible podded peas**, **potatoes**, **spring field beans**, **spring oilseed rape**, **spring wheat**, **sugar beet**, **vining peas**, **winter field beans**, **winter oilseed rape**, **winter wheat** [3, 8, 10]; **rye**, **spring barley**, **spring oats**, **triticale**, **winter barley**, **winter oats** [3]
- Wheat-blossom midge in **winter wheat** [1, 4, 6, 7, 9, 11, 12]
- Whitefly in **broccoli**, **brussels sprouts**, **calabrese** [1, 4, 6, 7, 9, 11]; **cabbages**, **cauliflowers** [1, 4, 7, 9, 11]; **kale** [6]
- Willow aphid in **short rotation coppice willow** *(off-label)* [10]
- Willow beetle in **short rotation coppice willow** *(off-label)* [10]
- Willow sawfly in **short rotation coppice willow** *(off-label)* [10]
- Yellow cereal fly in **winter wheat** [1, 2, 4, 6, 7, 9, 11, 12]

Extension of Authorisation for Minor Use (EAMUs)
- **all edible seed crops grown outdoors** *20082944 expires 31 Dec 2018* [4], *20102880 expires 31 Dec 2018* [10]
- **all non-edible seed crops grown outdoors** *20082944 expires 31 Dec 2018* [4], *20102880 expires 31 Dec 2018* [10]
- **beetroot** *20063761 expires 31 Dec 2018* [1], *20060743 expires 31 Dec 2018* [4]
- **bilberries** *20140521 expires 31 Dec 2018* [4]
- **blackberries** *20063755 expires 31 Dec 2018* [1], *20060728 expires 31 Dec 2018* [4]
- **blackcurrants** *20063752 expires 31 Dec 2018* [1], *20060727 expires 31 Dec 2018* [4], *20073269 expires 31 Dec 2018* [10]
- **blueberries** *20140521 expires 31 Dec 2018* [4]
- **borage for oilseed production** *20063748 expires 31 Dec 2018* [1], *20060634 expires 31 Dec 2018* [4], *20073258 expires 31 Dec 2018* [10]
- **broad beans** *20063764 expires 31 Dec 2018* [1], *20060753 expires 31 Dec 2018* [4], *20073234 expires 31 Dec 2018* [10]
- **bulb onions** *20063756 expires 31 Dec 2018* [1], *20060730 expires 31 Dec 2018* [4], *20073256 expires 31 Dec 2018* [10]
- **canary flower (echium spp.)** *20063748 expires 31 Dec 2018* [1], *20060634 expires 31 Dec 2018* [4], *20073258 expires 31 Dec 2018* [10]
- **carrots** *20080201 expires 31 Dec 2018* [10]
- **celeriac** *20063757 expires 31 Dec 2018* [1], *20060731 expires 31 Dec 2018* [4], *20080204 expires 31 Dec 2018* [10]
- **celery (outdoor)** *20063762 expires 31 Dec 2018* [1], *20060744 expires 31 Dec 2018* [4], *20073257 expires 31 Dec 2018* [10]
- **cherries** *20131273 expires 31 Dec 2018* [4]

SEE SECTION 3 FOR PRODUCTS ALSO REGISTERED

- **chestnuts** *20060742 expires 31 Dec 2018* [4], *20080206 expires 31 Dec 2018* [10]
- **chicory** *20060740 expires 31 Dec 2018* [4]
- **cob nuts** *20080206 expires 31 Dec 2018* [10]
- **corn gromwell** *20121047 expires 31 Dec 2018* [4]
- **crambe** *20081046 expires 31 Dec 2018* [4], *20102876 expires 31 Dec 2018* [10]
- **cranberries** *20140521 expires 31 Dec 2018* [4]
- **dewberries** *20063755 expires 31 Dec 2018* [1], *20060728 expires 31 Dec 2018* [4]
- **dwarf beans** *20080202 expires 31 Dec 2018* [10]
- **evening primrose** *20063664 expires 31 Dec 2018* [1], *20060634 expires 31 Dec 2018* [4], *20073258 expires 31 Dec 2018* [10]
- **farm forestry** *20082944 expires 31 Dec 2018* [4], *20102880 expires 31 Dec 2018* [10]
- **fennel** *20060733 expires 31 Dec 2018* [4]
- **fodder beet** *20063665 expires 31 Dec 2018* [1], *20060637 expires 31 Dec 2018* [4], *20073233 expires 31 Dec 2018* [10]
- **forest nurseries** *20082944 expires 31 Dec 2018* [4], *20102880 expires 31 Dec 2018* [10]
- **french beans** *20063758 expires 31 Dec 2018* [1], *20060739 expires 31 Dec 2018* [4], *20080202 expires 31 Dec 2018* [10]
- **frise** *20063751 expires 31 Dec 2018* [1], *20060636 expires 31 Dec 2018* [4]
- **garlic** *20063756 expires 31 Dec 2018* [1], *20060730 expires 31 Dec 2018* [4], *20073256 expires 31 Dec 2018* [10]
- **gooseberries** *20063752 expires 31 Dec 2018* [1], *20060727 expires 31 Dec 2018* [4], *20073269 expires 31 Dec 2018* [10]
- **grapevines** *20063747 expires 31 Dec 2018* [1], *20060266 expires 31 Dec 2018* [4], *20080205 expires 31 Dec 2018* [10]
- **grass seed crops** *20063749 expires 31 Dec 2018* [1], *20060624 expires 31 Dec 2018* [4], *20073260 expires 31 Dec 2018* [10]
- **hazel nuts** *20060742 expires 31 Dec 2018* [4], *20080206 expires 31 Dec 2018* [10]
- **herbs (see appendix 6)** *20063751 expires 31 Dec 2018* [1], *20060636 expires 31 Dec 2018* [4], *20073259 expires 31 Dec 2018* [10]
- **honesty** *20063748 expires 31 Dec 2018* [1], *20060634 expires 31 Dec 2018* [4], *20073258 expires 31 Dec 2018* [10]
- **hops** *20082944 expires 31 Dec 2018* [4], *20131719 expires 31 Dec 2018* [4], *20102880 expires 31 Dec 2018* [10]
- **horseradish** *20071307 expires 31 Dec 2018* [1], *20071301 expires 31 Dec 2018* [4], *20080201 expires 31 Dec 2018* [10]
- **lamb's lettuce** *20063751 expires 31 Dec 2018* [1], *20060636 expires 31 Dec 2018* [4]
- **land cress** *20102878 expires 31 Dec 2018* [10]
- **leaf brassicas** *20063751 expires 31 Dec 2018* [1], *20060636 expires 31 Dec 2018* [4]
- **leeks** *20063756 expires 31 Dec 2018* [1], *20060730 expires 31 Dec 2018* [4], *20073256 expires 31 Dec 2018* [10]
- **lupins** *20060635 expires 31 Dec 2018* [4], *20102877 expires 31 Dec 2018* [10]
- **mallow (althaea spp.)** *20071307 expires 31 Dec 2018* [1], *20071301 expires 31 Dec 2018* [4], *20080201 expires 31 Dec 2018* [10]
- **mirabelles** *20131273 expires 31 Dec 2018* [4]
- **miscanthus** *20082944 expires 31 Dec 2018* [4], *20102880 expires 31 Dec 2018* [10]
- **mustard** *20063664 expires 31 Dec 2018* [1], *20060634 expires 31 Dec 2018* [4], *20073258 expires 31 Dec 2018* [10]
- **navy beans** *20063758 expires 31 Dec 2018* [1], *20060739 expires 31 Dec 2018* [4], *20080202 expires 31 Dec 2018* [10]
- **ornamental plant production** *20082944 expires 31 Dec 2018* [4], *20102880 expires 31 Dec 2018* [10]
- **parsley root** *20071307 expires 31 Dec 2018* [1], *20071301 expires 31 Dec 2018* [4], *20080201 expires 31 Dec 2018* [10]
- **parsnips** *20080201 expires 31 Dec 2018* [10]
- **plums** *20131273 expires 31 Dec 2018* [4]
- **poppies for morphine production** *20063763 expires 31 Dec 2018* [1], *20060749 expires 31 Dec 2018* [4], *20102875 expires 31 Dec 2018* [10]
- **protected all edible seed crops** *20102880 expires 31 Dec 2018* [10]
- **protected all non-edible seed crops** *20102880 expires 31 Dec 2018* [10]

FOR FULL CONDITIONS OF USE ALWAYS READ THE PRODUCT LABEL

- **protected aubergines** *20121994 expires 31 Dec 2018* [4]
- **protected forest nurseries** *20082944 expires 31 Dec 2018* [4], *20102880 expires 31 Dec 2018* [10]
- **protected ornamentals** *20082944 expires 31 Dec 2018* [4], *20102880 expires 31 Dec 2018* [10]
- **protected pak choi** *20081263 expires 31 Dec 2018* [4], *20102879 expires 31 Dec 2018* [10]
- **protected peppers** *20121994 expires 31 Dec 2018* [4]
- **protected strawberries** *20111705 expires 31 Dec 2018* [4]
- **protected tatsoi** *20081263 expires 31 Dec 2018* [4], *20102879 expires 31 Dec 2018* [10]
- **protected tomatoes** *20121994 expires 31 Dec 2018* [4]
- **radishes** *20063757 expires 31 Dec 2018* [1], *20060731 expires 31 Dec 2018* [4], *20080204 expires 31 Dec 2018* [10]
- **raspberries** *20063755 expires 31 Dec 2018* [1], *20060728 expires 31 Dec 2018* [4]
- **red beet** *20073254 expires 31 Dec 2018* [10]
- **redcurrants** *20063752 expires 31 Dec 2018* [1], *20060727 expires 31 Dec 2018* [4], *20073269 expires 31 Dec 2018* [10]
- **ribes hybrids** *20140521 expires 31 Dec 2018* [4]
- **rubus hybrids** *20063755 expires 31 Dec 2018* [1], *20060728 expires 31 Dec 2018* [4]
- **runner beans** *20063758 expires 31 Dec 2018* [1], *20060739 expires 31 Dec 2018* [4], *20080202 expires 31 Dec 2018* [10]
- **rye** *20063749 expires 31 Dec 2018* [1]
- **salad onions** *20063756 expires 31 Dec 2018* [1], *20060730 expires 31 Dec 2018* [4], *20073256 expires 31 Dec 2018* [10]
- **scarole** *20063751 expires 31 Dec 2018* [1], *20060636 expires 31 Dec 2018* [4]
- **shallots** *20063756 expires 31 Dec 2018* [1], *20060730 expires 31 Dec 2018* [4], *20073256 expires 31 Dec 2018* [10]
- **short rotation coppice willow** *20073268 expires 31 Dec 2018* [10]
- **soft fruit** *20082944 expires 31 Dec 2018* [4], *20102880 expires 31 Dec 2018* [10]
- **spring linseed** *20063748 expires 31 Dec 2018* [1], *20060634 expires 31 Dec 2018* [4], *20073258 expires 31 Dec 2018* [10]
- **spring rye** *20060624 expires 31 Dec 2018* [4], *20073260 expires 31 Dec 2018* [10]
- **strawberries** *20111705 expires 31 Dec 2018* [4]
- **swedes** *20101856 expires 31 Dec 2018* [4], *20102911 expires 31 Dec 2018* [10]
- **sweetcorn** *20063760 expires 31 Dec 2018* [1], *20060732 expires 31 Dec 2018* [4], *20080203 expires 31 Dec 2018* [10]
- **table grapes** *20063747 expires 31 Dec 2018* [1]
- **top fruit** *20082944 expires 31 Dec 2018* [4], *20102880 expires 31 Dec 2018* [10]
- **triticale** *20063749 expires 31 Dec 2018* [1], *20060624 expires 31 Dec 2018* [4], *20073260 expires 31 Dec 2018* [10]
- **turnips** *20101856 expires 31 Dec 2018* [4], *20102911 expires 31 Dec 2018* [10]
- **walnuts** *20060742 expires 31 Dec 2018* [4], *20080206 expires 31 Dec 2018* [10]
- **whitecurrants** *20063752 expires 31 Dec 2018* [1], *20060727 expires 31 Dec 2018* [4], *20073269 expires 31 Dec 2018* [10]
- **willow (short rotation coppice)** *20060748 expires 31 Dec 2018* [4]
- **wine grapes** *20063747 expires 31 Dec 2018* [1]
- **winter linseed** *20063748 expires 31 Dec 2018* [1], *20060634 expires 31 Dec 2018* [4], *20073258 expires 31 Dec 2018* [10]
- **winter rye** *20060624 expires 31 Dec 2018* [4], *20073260 expires 31 Dec 2018* [10]

Approval information

- Lambda-cyhalothrin included in Annex I under EC Regulation 1107/2009
- Accepted by BBPA for use on malting barley and hops

Efficacy guidance

- Best results normally obtained from treatment when pest attack first seen. See label for detailed recommendations on each crop
- Timing for control of barley yellow dwarf virus vectors depends on specialist assessment of the level of risk in the area
- Repeat applications recommended in some crops where prolonged attack occurs, up to maximum total dose. See label for details
- Where strains of aphids resistant to lambda-cyhalothrin occur control is unlikely to be satisfactory

SEE SECTION 3 FOR PRODUCTS ALSO REGISTERED

- Addition of wetter recommended for control of certain pests in brassicas and oilseed rape
- Use of sufficient water volume to ensure thorough crop penetration recommended for optimum results
- Use of drop-legged sprayer gives improved results in crops such as Brussels sprouts

Restrictions
- Maximum number of applications or maximum total dose per crop varies - see labels
- Do not apply to a cereal crop if any product containing a pyrethroid insecticide or dimethoate has been applied to the crop after the start of ear emergence (GS 51)
- Do not spray cereals in the spring/summer (i.e. after 1 Apr) within 6 m of edge of crop

Crop-specific information
- Latest use before late milk stage on cereals; before end of flowering for winter oilseed rape
- HI 3 d for radishes, red beet; 7 d for lettuce [1, 4]; 14 d for carrots and parsnips; 25 d for peas, field beans; 6 wk for spring oilseed rape; 8 wk for sugar beet

Environmental safety
- Dangerous for the environment
- Very toxic to aquatic organisms
- Flammable [2]
- To protect non-target arthropods respect an untreated buffer zone of 5 m to non-crop land
- Since there is a high risk to non-target insects or other anthropods, do not spray cereals in spring/ summer i.e. after 1st April within 6m of the field boundary [5]
- Dangerous to bees
- Broadcast air-assisted LERAP [1, 3, 4, 7-11] (25 m); Broadcast air-assisted LERAP [2, 12] (38 m); LERAP Category A [2, 5, 12]; LERAP Category B [1, 3, 4, 6-11]

Hazard classification and safety precautions
Hazard Harmful, Dangerous for the environment [1-12]; Corrosive [3, 8, 10]; Flammable, Flammable liquid and vapour [2]; Toxic if swallowed [3]; Harmful if swallowed [2, 4, 6, 9]; Harmful if inhaled [2-4, 6, 9]; Very toxic to aquatic organisms [2, 4-6, 9]
Transport code 3 [2]; 8 [3, 8, 10]; 9 [1, 4-7, 9, 11, 12]
Packaging group II [3]; III [1, 2, 4-12]
UN Number 1760 [3, 8, 10]; 1993 [2]; 3077 [5]; 3082 [1, 4, 6, 7, 9, 11, 12]
Risk phrases H304, H336 [2, 3]; H314 [3]; H317 [4]; H319 [2]; R20, R22a [1, 5, 7, 8, 10-12]; R22b, R34, R51 [8, 10]; R36, R38 [5]; R43 [1, 5, 7, 11, 12]; R50 [1, 7, 11, 12]; R53a [1, 7, 8, 10-12]
Operator protection A, H [1-12]; C [2, 3, 5, 8, 10]; J, K, M [5]; U02a [1, 3-12]; U04a [3, 8, 10]; U05a [1-12]; U08 [1-4, 6-12]; U09a [5]; U11 [3, 5, 8, 10]; U14, U20b [1, 4-7, 9, 11, 12]; U19a, U20a [2, 3, 8, 10]
Environmental protection E12a [3, 8, 10]; E12c [1, 2, 4, 5, 7, 9, 11, 12]; E15b [2, 3, 5, 8, 10]; E16a [1, 3, 4, 6-11]; E16c [2, 5, 12]; E17b [1] (25 m); E17b [2] (38 m); E17b [3, 4, 7-11] (25 m); E17b [12] (38 m); E22a [5] (after 1st April); E22b [3, 4, 9, 10]; E34 [1-4, 6-12]; E38 [1, 3-12]; H410 [2-6, 9]
Storage and disposal D01, D02 [1-12]; D05, D07 [2]; D09a [1-4, 6-12]; D10b [2, 3, 8, 10]; D10c [1, 4, 6, 7, 9, 11, 12]; D12a [1, 2, 4-7, 9, 11, 12]
Medical advice M03 [1, 4-7, 9, 11, 12]; M04a [3, 8, 10]; M05b [2, 3, 8, 10]

297   Lecanicillium muscarium

A fungal pathogen of whitefly
FRAC mode of action code: 44

Products

Mycotal	Koppert	4.8% w/w	WP	16644

Uses
- Spider mites in *leaf brassicas* (off-label), *protected bilberries* (off-label), *protected blackberries* (off-label), *protected blackcurrants* (off-label), *protected blueberry* (off-label), *protected cayenne peppers* (off-label), *protected cranberries* (off-label), *protected gooseberries* (off-label), *protected herbs (see appendix 6)* (off-label), *protected hops* (off-

FOR FULL CONDITIONS OF USE ALWAYS READ THE PRODUCT LABEL

label), **protected loganberries** *(off-label)*, **protected nursery fruit trees** *(off-label)*, **protected raspberries** *(off-label)*, **protected redcurrants** *(off-label)*, **protected ribes hybrids** *(off-label)*, **protected rubus hybrids** *(off-label)*, **protected table grapes** *(off-label)*, **protected wine grapes** *(off-label)*

- Thrips in **leaf brassicas** *(off-label)*, **protected bilberries** *(off-label)*, **protected blackberries** *(off-label)*, **protected blackcurrants** *(off-label)*, **protected blueberry** *(off-label)*, **protected cayenne peppers** *(off-label)*, **protected cranberries** *(off-label)*, **protected gooseberries** *(off-label)*, **protected herbs (see appendix 6)** *(off-label)*, **protected hops** *(off-label)*, **protected loganberries** *(off-label)*, **protected nursery fruit trees** *(off-label)*, **protected raspberries** *(off-label)*, **protected redcurrants** *(off-label)*, **protected ribes hybrids** *(off-label)*, **protected rubus hybrids** *(off-label)*, **protected table grapes** *(off-label)*, **protected wine grapes** *(off-label)*
- Whitefly in **leaf brassicas** *(off-label)*, **protected bilberries** *(off-label)*, **protected blackberries** *(off-label)*, **protected blackcurrants** *(off-label)*, **protected blueberry** *(off-label)*, **protected cayenne peppers** *(off-label)*, **protected cranberries** *(off-label)*, **protected cucumbers**, **protected gooseberries** *(off-label)*, **protected herbs (see appendix 6)** *(off-label)*, **protected hops** *(off-label)*, **protected loganberries** *(off-label)*, **protected nursery fruit trees** *(off-label)*, **protected ornamentals**, **protected peppers**, **protected raspberries** *(off-label)*, **protected redcurrants** *(off-label)*, **protected ribes hybrids** *(off-label)*, **protected rubus hybrids** *(off-label)*, **protected strawberries**, **protected table grapes** *(off-label)*, **protected tomatoes**, **protected wine grapes** *(off-label)*

Extension of Authorisation for Minor Use (EAMUs)
- **leaf brassicas** *20142679*
- **protected bilberries** *20142685*
- **protected blackberries** *20142684*
- **protected blackcurrants** *20142685*
- **protected blueberry** *20142685*
- **protected cayenne peppers** *20142680*
- **protected cranberries** *20142685*
- **protected gooseberries** *20142685*
- **protected herbs (see appendix 6)** *20142679*
- **protected hops** *20142681*
- **protected loganberries** *20142684*
- **protected nursery fruit trees** *20142681*
- **protected raspberries** *20142684*
- **protected redcurrants** *20142685*
- **protected ribes hybrids** *20142685*
- **protected rubus hybrids** *20142684*
- **protected table grapes** *20142685*
- **protected wine grapes** *20142685*

Approval information
- *Lecanicillium* included in Annex I under EC Regulation 1107/2009

Efficacy guidance
- *Lecanicillium muscarium* is a pathogenic fungus that infects the target pests and destroys them
- Apply spore powder as spray as part of biological control programme keeping the spray liquid well agitated
- Treat before infestations build to high levels and repeat as directed on the label
- Spray during late afternoon and early evening directing spray onto underside of leaves and to growing points
- Best results require minimum 80% relative humidity (70% if applied with the adjuvant Addit) and 18°C within the crop canopy

Restrictions
- Never use in tank mixture
- Do not use a fungicide within 3 d of treatment. Pesticides containing captan, chlorothalonil, fenarimol, dichlofluanid, imazalil, maneb, prochloraz, quinomethionate, thiram or tolylfluanid may not be used on the same crop
- Keep in a refrigerated store at 2-6°C but do not freeze

SEE SECTION 3 FOR PRODUCTS ALSO REGISTERED

Environmental safety
- Products have negligible effects on commercially available natural predators or parasites but consult manufacturer before using with a particular biological control agent for the first time

Hazard classification and safety precautions
  UN Number N/C
  Operator protection A, D, H; U19a, U20b
  Environmental protection E15a
  Storage and disposal D09a, D11a

298   lenacil

A residual, soil-acting uracil herbicide for beet crops
HRAC mode of action code: C1

See also desmedipham + ethofumesate + lenacil + phenmedipham

Products

1	Lenazar Flow 500 SC	Belcrop	500 g/l	SC	18124
2	Venzar 500 SC	DuPont	500 g/l	SC	17743

Uses
- Annual dicotyledons in *farm woodland* (off-label) [2]; *sugar beet* [1, 2]
- Annual meadow grass in *farm woodland* (off-label) [2]; *sugar beet* [1, 2]

Extension of Authorisation for Minor Use (EAMUs)
- *farm woodland* 20171654 [2]

Approval information
- Lenacil included in Annex I under EC Regulation 1107/2009

Efficacy guidance
- Best results, especially from pre-emergence treatments, achieved on fine, even, firm and moist soils free from clods. Continuing presence of moisture from rain or irrigation gives improved residual control of later germinating weeds. Effectiveness may be reduced by dry conditions
- On beet crops may be used pre- or post-emergence, alone or in mixture to broaden weed spectrum
- Apply overall or as band spray to beet crops pre-drilling incorporated, pre- or post-emergence
- All labels have limitations on soil types that may be treated. Residual activity reduced on soils with high OM content

Restrictions
- Do not use any other residual herbicide within 3 mth of the initial application to fruit or ornamental crops
- Do not treat crops under stress from drought, low temperatures, nutrient deficiency, pest or disease attack, or waterlogging
- Must only be applied after BBCH10 growth stage
- May only be applied once in every three years

Crop-specific information
- Latest use: pre-emergence for red beet, fodder beet, spinach, spinach beet, mangels; before leaves meet over rows when used on these crops post-emergence; 24 h after planting new strawberry runners or before flowering for established strawberry crops, blackcurrants, gooseberries, raspberries
- Heavy rain after application to beet crops may cause damage especially if followed by very hot weather
- Reduction in beet stand may occur where crop emergence or vigour is impaired by soil capping or pest attack
- Strawberry runner beds to be treated should be level without depressions around the roots
- New soft fruit cuttings should be planted at least 15 cm deep and firmed before treatment
- Check varietal tolerance of ornamentals before large scale treatment

FOR FULL CONDITIONS OF USE ALWAYS READ THE PRODUCT LABEL

Following crops guidance

- Succeeding crops should not be planted or sown for at least 4 mth (6 mth on organic soils) after treatment following ploughing to at least 150 mm.
- Only beet crops, mangels or strawberries may be sown within 4 months of treatment and no further applications of lenacil should be made for at least 4 months

Environmental safety

- Dangerous for the environment
- Very toxic to aquatic organisms
- LERAP Category B

Hazard classification and safety precautions

Hazard Dangerous for the environment
Transport code 9
Packaging group III
UN Number 3082
Risk phrases H351
Operator protection A, C; U05a, U08, U19a, U20a
Environmental protection E13c, E16a, E38, H410
Storage and disposal D01, D02, D05, D09a, D10a, D12a

299 lenacil + triflusulfuron-methyl

A foliar and residual herbicide mixture for sugar beet
HRAC mode of action code: C1 + B

See also triflusulfuron-methyl

Products

Safari Lite WSB	DuPont	71.4:5.4% w/w	WG	17954

Uses

- Annual dicotyledons in **sugar beet**
- Volunteer oilseed rape in **sugar beet**

Approval information

- Lenacil and triflusulfuron-methyl included in Annex I under EC Regulation 1107/2009

Efficacy guidance

- Best results obtained when weeds are small and growing actively
- Product recommended for use in a programme of treatments in tank mixture with a suitable herbicide partner to broaden the weed spectrum
- Ensure good spray cover of weeds. Apply when first weeds have emerged provided crop has reached cotyledon stage
- Susceptible plants cease to grow almost immediately after treatment and symptoms can be seen 5-10 d later
- Weed control may be reduced in very dry soil conditions
- Triflusulfuron-methyl is a member of the ALS-inhibitor group of herbicides

Restrictions

- Maximum number of treatments on sugar beet 3 per crop and do not apply more than 4 applications of any product containing triflusulfuron-methyl
- Do not apply to crops suffering from stress caused by drought, water-logging, low temperatures, pest or disease attack, nutrient deficiency or any other factors affecting crop growth
- Do not use on Sands, stony or gravelly soils or on soils with more than 10% organic matter
- Do not apply when temperature above or likely to exceed 21°C on day of spraying or under conditions of high light intensity

Crop-specific information

- Latest use: before crop leaves meet between rows for sugar beet

Following crops guidance
- Only cereals may be sown in the same calendar yr as a treated sugar beet crop. Any crop may be sown in the following spring
- In the event of crop failure sow only sugar beet within 4 mth of treatment

Environmental safety
- Dangerous for the environment
- Very toxic to aquatic organisms
- Take extreme care to avoid drift onto broad-leaved plants outside the target area or onto surface waters or ditches, or land intended for cropping
- Spraying equipment should not be drained or flushed onto land planted, or to be planted, with trees or crops other than cereals and should be thoroughly cleansed after use - see label for instructions
- LERAP Category B

Hazard classification and safety precautions
Hazard Irritant, Dangerous for the environment
Transport code 9
Packaging group III
UN Number 3077
Risk phrases H351
Operator protection A, C; U05a, U08, U11, U19a, U20b, U22a
Environmental protection E15a, E16a, E38, H410
Storage and disposal D01, D02, D09a, D12a

300 magnesium phosphide

A phosphine generating compound used to control insect pests in stored commodities

Products
Degesch Plates	Rentokil	56% w/w	GE	17105

Uses
- Insect pests in *carob, cocoa, coffee, crop handling & storage structures, herbal infusions, processed consumable products, stored dried spices, stored grain, stored linseed, stored nuts, stored oilseed rape, stored pulses, tea, tobacco*

Approval information
- Magnesium phosphide included in Annex I under EC Regulation 1107/2009
- Accepted by BBPA for use in stores for malting barley

Efficacy guidance
- Product acts as fumigant by releasing poisonous hydrogen phosphide gas on contact with moisture in the air
- Place plates on the floor or wall of the building or on the surface of the commodity. Exposure time varies depending on temperature and pest. See label

Restrictions
- Magnesium phosphide is subject to the Poisons Rules 1982 and the Poisons Act 1972. See Section 5 for more information
- Only to be used by professional operators trained in the use of magnesium phosphide and familiar with the precautionary measures to be observed. See label for full precautions

Environmental safety
- Highly flammable
- Prevent access to buildings under fumigation by livestock, pets and other non-target mammals and birds
- Dangerous to fish or other aquatic life. Do not contaminate surface waters or ditches with chemical or used container
- Keep in original container, tightly closed, in a safe place, under lock and key

- Do not allow plates or their spent residues to come into contact with food other than raw cereal grains
- Remove used plates after treatment. Do not bulk spent plates and residues: spontaneous ignition could result
- Keep livestock out of treated areas

Hazard classification and safety precautions

Hazard Very toxic, Highly flammable, Dangerous for the environment, In contact with water releases flammable gases which may ignite spontaneously, Fatal if swallowed, Fatal if inhaled, Very toxic to aquatic organisms
Transport code 4.3
Packaging group I
UN Number 2011
Risk phrases H312
Operator protection A, D, H; U01, U05b, U07, U13, U19a, U20a
Environmental protection E02a (4 h min); E02b, E13b, E34
Storage and disposal D01, D02, D05, D07, D09b, D11b
Medical advice M04a

301   maleic hydrazide

A pyridazine plant growth regulator suppressing sprout and bud growth

See also fatty acids + maleic hydrazide

Products

1 Clayton Stun	Clayton	60% w/w	SG	17194
2 Fazor	Arysta	60% w/w	SG	13617
3 Fazor	Arysta	60% w/w	SG	13679

Uses

- Growth regulation in **amenity grassland** *(off-label)*, **amenity vegetation** *(off-label)* [3]; **bilberries** *(off-label)*, **blackberries** *(off-label)*, **blackcurrants** *(off-label)*, **blueberries** *(off-label)*, **carrots** *(off-label)*, **cranberries** *(off-label)*, **garlic** *(off-label)*, **gooseberries** *(off-label)*, **hops** *(off-label)*, **loganberries** *(off-label)*, **ornamental plant production** *(off-label)*, **parsnips** *(off-label)*, **raspberries** *(off-label)*, **redcurrants** *(off-label)*, **rubus hybrids** *(off-label)*, **shallots** *(off-label)*, **strawberries** *(off-label)* [2, 3]; **bulb onions**, **potatoes** [1-3]

Extension of Authorisation for Minor Use (EAMUs)

- **amenity grassland** *20160184* [3]
- **amenity vegetation** *20160185* [3]
- **bilberries** *20160183* [3], *20161834* [2]
- **blackberries** *20160183* [3], *20161834* [2]
- **blackcurrants** *20160183* [3], *20161834* [2]
- **blueberries** *20160183* [3], *20161834* [2]
- **carrots** *20160186* [3], *20161835* [2]
- **cranberries** *20160183* [3], *20161834* [2]
- **garlic** *20160187* [3], *20161836* [2]
- **gooseberries** *20160183* [3], *20161834* [2]
- **hops** *20160183* [3], *20161834* [2]
- **loganberries** *20160183* [3], *20161834* [2]
- **ornamental plant production** *20160183* [3], *20161834* [2]
- **parsnips** *20160186* [3], *20161835* [2]
- **raspberries** *20160183* [3], *20161834* [2]
- **redcurrants** *20160183* [3], *20161834* [2]
- **rubus hybrids** *20160183* [3], *20161834* [2]
- **shallots** *20160187* [3], *20161836* [2]
- **strawberries** *20160183* [3], *20161834* [2]

SEE SECTION 3 FOR PRODUCTS ALSO REGISTERED

Approval information
- Maleic hydrazide included in Annex I under EC Regulation 1107/2009

Efficacy guidance
- Apply to grass at any time of yr when growth active, best when growth starting in Apr-May and repeated when growth recommences
- Uniform coverage and dry weather necessary for effective results
- Accurate timing essential for good results on potatoes but rain or irrigation within 24 h may reduce effectiveness on onions and potatoes
- Mow 2-3 d before and 5-10 d after spraying for best results. Need for mowing reduced for up to 6 wk
- When used for suppression of volunteer potatoes treatment may also give some suppression of sprouting in store but separate treatment will be necessary if sprouting occurs

Restrictions
- Maximum number of treatments 2 per yr on amenity grass, land not intended for cropping and land adjacent to aquatic areas; 1 per crop on onions, potatoes and on or around tree trunks
- Do not apply in drought or when crops are suffering from pest, disease or herbicide damage. Do not treat fine turf or grass seeded less than 8 mth previously
- Do not treat potatoes within 3 wk of applying a haulm desiccant or if temperatures above 26°C
- Consult processor before use on potato crops for processing

Crop-specific information
- Latest use: 3 wk before haulm destruction for potatoes; before 50% necking for onions
- HI onions 1 wk; potatoes 3 wk
- Apply to onions at 10% necking and not later than 50% necking stage when the tops are still green
- Only treat onions in good condition and properly cured, and do not treat more than 2 wk before maturing. Treated onions may be stored until Mar but must then be removed to avoid browning
- Apply to second early or maincrop potatoes at least 3 wk before haulm destruction
- Only treat potatoes of good keeping quality; not on seed, first earlies or crops grown under polythene

Environmental safety
- Only apply to grass not to be used for grazing
- Do not use treated water for irrigation purposes within 3 wk of treatment or until concentration in water falls below 0.02 ppm
- Maximum permitted concentration in water 2 ppm
- Do not dump surplus product in water or ditch bottoms
- Avoid drift onto nearby vegetables, flowers or other garden plants

Hazard classification and safety precautions
Transport code 9 [1, 3]
Packaging group III [1, 3]
UN Number 3077 [1, 3]
Operator protection A, H; U08, U20b
Environmental protection E15a, H411
Storage and disposal D09a, D11a

302   maleic hydrazide + pelargonic acid

A pyridazine plant growth regulator suppressing sprout and bud growth + a naturally occuring 9 carbon chain acid
HRAC mode of action code: Not classified

Products
Finalsan Plus	Certis	30:186.7 g/l	SL	17928

Uses
- Algae in *amenity vegetation, natural surfaces not intended to bear vegetation, ornamental plant production, permeable surfaces overlying soil*

FOR FULL CONDITIONS OF USE ALWAYS READ THE PRODUCT LABEL

- Annual and perennial weeds in *amenity vegetation, natural surfaces not intended to bear vegetation, ornamental plant production, permeable surfaces overlying soil*
- Moss in *amenity vegetation, natural surfaces not intended to bear vegetation, ornamental plant production, permeable surfaces overlying soil*

Approval information
- Maleic hydrazide and pelargonic acid included in Annex I under EC Regulation 1107/2009

Environmental safety
- LERAP Category B

Hazard classification and safety precautions
  UN Number N/C
  Risk phrases H319
  Operator protection A, C, H; U05a, U12, U15
  Environmental protection E15a, E16a
  Storage and disposal D01, D02
  Medical advice M05a

303   maltodextrin

A polysaccharide used as a food additive and with activity against red spider mites
FRAC mode of action code: 12 + IRAC 3

Products

1	Eradicoat	Certis	598 g/l	SL	17121
2	Majestik	Certis	598 g/l	SL	17240

Uses
- Aphids in *all edible crops (outdoor), all non-edible crops (outdoor), all protected edible crops, all protected non-edible crops*
- Spider mites in *all edible crops (outdoor), all non-edible crops (outdoor), all protected edible crops, all protected non-edible crops*
- Whitefly in *all edible crops (outdoor), all non-edible crops (outdoor), all protected edible crops, all protected non-edible crops*

Approval information
- Maltodextrin is included in Annex 1 under EC Regulation 1107/2009

Hazard classification and safety precautions
  Hazard Irritant
  UN Number N/C
  Risk phrases H319
  Operator protection A, H; U05a, U09a, U14, U19a, U20b
  Environmental protection E12a [2]; E12e [1]; H412 [1, 2]
  Storage and disposal D01, D02, D09a, D10b, D12a

304   mancozeb

A protective dithiocarbamate fungicide for potatoes and other crops
FRAC mode of action code: M3

See also ametoctradin + mancozeb
          benalaxyl + mancozeb
          benthiavalicarb-isopropyl + mancozeb
          chlorothalonil + mancozeb
          cymoxanil + mancozeb
          dimethomorph + mancozeb
          fenamidone + mancozeb

SEE SECTION 3 FOR PRODUCTS ALSO REGISTERED

Products

1	Dithane 945	Interfarm	80% w/w	WP	14621
2	Dithane NT Dry Flowable	Interfarm	75% w/w	WG	14704
3	Karamate Dry Flo Neotec	Landseer	75% w/w	WG	14632
4	Laminator 75 WG	Interfarm	75% w/w	WG	15667
5	Laminator Flo	Interfarm	455 g/l	SC	15345
6	Malvi	AgChem Access	75% w/w	WG	14981
7	Manzate 75 WG	UPL Europe	75% w/w	WG	15052
8	Mewati	AgChem Access	80% w/w	WP	15763
9	Penncozeb 80 WP	UPL Europe	80% w/w	WP	16953
10	Penncozeb WDG	UPL Europe	75% w/w	WG	16885
11	Quell Flo	Interfarm	455 g/l	SC	15237

Uses

- Alternaria blight in **carrots**, **parsnips** [2]
- Blight in **potatoes** [1, 2, 4-11]; **tomatoes (outdoor)** [4, 6]
- Botrytis in **flower bulbs** [7]
- Brown rust in **durum wheat** *(useful control)*, **spring wheat** *(useful control)*, **winter wheat** *(useful control)* [1, 2, 8]; **spring wheat**, **winter wheat** [5, 11]
- Disease control in **amenity vegetation**, **bulb onions**, **carrots**, **courgettes**, **hops** *(off-label)*, **parsnips**, **shallots**, **soft fruit** *(off-label)*, **top fruit** *(off-label)* [3]; **apples**, **table grapes** [4, 6]; **forest nurseries** *(off-label)* [1, 3, 10]; **ornamental plant production** [3, 10]; **ornamental plant production** *(off-label)* [1]; **wine grapes** [3, 4, 6]
- Downy mildew in **baby leaf crops** *(off-label)*, **lettuce** *(off-label)* [3]; **bulb onions** [7, 9, 10]; **forest nurseries** *(off-label)*, **ornamental plant production** [10]
- Early blight in **potatoes** [1, 2, 8]
- Rust in **bulb onions**, **shallots** [2]
- Scab in **apples** [3, 7, 9, 10]; **pears** [3]
- Septoria leaf blotch in **durum wheat** *(reduction)* [1, 2, 8]; **spring wheat**, **winter wheat** [5, 9-11]; **spring wheat** *(reduction)*, **winter wheat** *(reduction)* [1, 2, 7, 8]
- Sooty moulds in **spring wheat**, **winter wheat** [5, 11]
- Yellow rust in **winter wheat** [5, 11]

Extension of Authorisation for Minor Use (EAMUs)

- **baby leaf crops** *20101082* [3]
- **forest nurseries** *20152652* [1], *20122047* [3], *20170900* [10]
- **hops** *20122047* [3]
- **lettuce** *20101082* [3]
- **ornamental plant production** *20152652* [1]
- **soft fruit** *20122047* [3]
- **top fruit** *20122047* [3]

Approval information

- Mancozeb included in Annex I under EC Regulation 1107/2009
- Accepted by BBPA for use on malting barley

Efficacy guidance

- Mancozeb is a protectant fungicide and will give moderate control, suppression or reduction of the cereal diseases listed if treated before they are established but in many cases mixture with carbendazim is essential to achieve satisfactory results. See labels for details
- May be recommended for suppression or control of mildew in cereals depending on product and tank mix. See label for details

Restrictions

- Maximum number of treatments varies with crop and product used - check labels for details
- Check labels for minimum interval that must elapse between treatments
- Avoid treating wet cereal crops or those suffering from drought or other stress
- Keep dry formulations away from fire and sparks
- Use dry formulations immediately. Do not store

Crop-specific information

- Latest use: before early milk stage (GS 73) for cereals;

FOR FULL CONDITIONS OF USE ALWAYS READ THE PRODUCT LABEL

- HI potatoes 7 d; apples, blackcurrants 28 d
- Apply to potatoes before haulm meets across rows (usually mid-Jun) or at earlier blight warning, and repeat every 7-14 d depending on conditions and product used (see label)
- May be used on potatoes up to desiccation of haulm
- Apply to cereals from 4-leaf stage to before early milk stage (GS 71). Recommendations vary, see labels for details

Environmental safety
- Dangerous for the environment
- Very toxic to aquatic organisms
- Harmful to fish or other aquatic life. Do not contaminate surface waters or ditches with chemical or used container
- Do not empty into drains
- LERAP Category B

Hazard classification and safety precautions
Hazard Harmful [4, 9, 10]; Irritant [1-3, 5-8, 11]; Dangerous for the environment [1-4, 6-11]; Very toxic to aquatic organisms [1-3]

Transport code 9

Packaging group III

UN Number 3077 [1-4, 6-10]; 3082 [5, 11]

Risk phrases H317 [1-4, 7, 9-11]; H319 [2, 3]; H335 [4, 7, 11]; H360 [10]; H361 [1-4]; R36 [6]; R37 [8]; R43 [5, 8]; R50, R53a [6, 8]; R63 [9]

Operator protection A [1-8, 10, 11]; C [6]; D [1-4, 7, 8, 10]; H [1, 8, 10]; U05a [1-8, 10, 11]; U08 [5, 6, 11]; U11 [6]; U14 [1-4, 7-11]; U19a [6, 9, 10]; U20b [1-8, 11]

Environmental protection E13c [5]; E15a [1-4, 6-8, 11]; E15b [9, 10]; E16a [1-11]; E19b [10]; E34 [5, 6, 11]; E38 [1-4, 6-11]; H410 [1, 2, 4, 7, 9-11]; H411 [3]

Storage and disposal D01, D12a [1-4, 6-11]; D02 [1-11]; D05 [1-8, 11]; D07 [5, 6]; D09a [1-8, 10, 11]; D10a [6]; D11a [1-5, 7, 8, 11]

Medical advice M05a [9, 10]

305   mancozeb + metalaxyl-M

A systemic and protectant fungicide mixture
FRAC mode of action code: M3 + 4

See also metalaxyl-M

Products

1	Clayton Mohawk	Clayton	64:4% w/w	WG	18086
2	Clayton Mohawk	Clayton	64:4% w/w	WG	18133
3	Ensis	AgChem Access	64:4% w/w	WG	15919
4	Fubol Gold WG	Syngenta	64:4% w/w	WG	14605

Uses

- Blight in **potatoes** [1-4]
- Disease control in **bilberries** *(off-label)*, **blackberries** *(off-label)*, **blackcurrants** *(off-label)*, **blueberries** *(off-label)*, **cranberries** *(off-label)*, **forest nurseries** *(off-label)*, **gooseberries** *(off-label)*, **hops** *(off-label)*, **loganberries** *(off-label)*, **ornamental plant production** *(off-label)*, **protected bilberries** *(off-label)*, **protected blackberries** *(off-label)*, **protected blackcurrants** *(off-label)*, **protected blueberry** *(off-label)*, **protected cranberries** *(off-label)*, **protected forest nurseries** *(off-label)*, **protected gooseberries** *(off-label)*, **protected hops** *(off-label)*, **protected loganberries** *(off-label)*, **protected ornamentals** *(off-label)*, **protected raspberries** *(off-label)*, **protected redcurrants** *(off-label)*, **protected rubus hybrids** *(off-label)*, **protected strawberries** *(off-label)*, **raspberries** *(off-label)*, **redcurrants** *(off-label)*, **rubus hybrids** *(off-label)*, **strawberries** *(off-label)* [4]
- Downy mildew in **baby leaf crops** *(off-label)*, **herbs (see appendix 6)** *(off-label)*, **lettuce** *(off-label)*, **poppies for morphine production** *(off-label)*, **protected herbs (see appendix 6)** *(off-label)*, **rhubarb** *(off-label)*, **salad onions** *(off-label)* [4]; **bulb onions** *(useful control)*, **shallots** *(useful control)* [1-4]

SEE SECTION 3 FOR PRODUCTS ALSO REGISTERED

- Fungus diseases in *apple orchards* *(off-label)* [4]
- Pink rot in *potatoes* *(reduction)* [1, 2, 4]
- White blister in *cabbages* *(off-label)* [4]

Extension of Authorisation for Minor Use (EAMUs)

- *apple orchards* *20132282* [4]
- *baby leaf crops* *20132285* [4]
- *bilberries* *20132288* [4]
- *blackberries* *20132288* [4]
- *blackcurrants* *20132288* [4]
- *blueberries* *20132288* [4]
- *cabbages* *20132281* [4]
- *cranberries* *20132288* [4]
- *forest nurseries* *20132288* [4]
- *gooseberries* *20132288* [4]
- *herbs (see appendix 6)* *20132285* [4]
- *hops* *20132288* [4]
- *lettuce* *20132285* [4]
- *loganberries* *20132288* [4]
- *ornamental plant production* *20132288* [4]
- *poppies for morphine production* *20132286* [4]
- *protected bilberries* *20132288* [4]
- *protected blackberries* *20132288* [4]
- *protected blackcurrants* *20132288* [4]
- *protected blueberry* *20132288* [4]
- *protected cranberries* *20132288* [4]
- *protected forest nurseries* *20132288* [4]
- *protected gooseberries* *20132288* [4]
- *protected herbs (see appendix 6)* *20132287* [4]
- *protected hops* *20132288* [4]
- *protected loganberries* *20132288* [4]
- *protected ornamentals* *20132288* [4]
- *protected raspberries* *20132288* [4]
- *protected redcurrants* *20132288* [4]
- *protected rubus hybrids* *20132288* [4]
- *protected strawberries* *20132288* [4]
- *raspberries* *20132288* [4]
- *redcurrants* *20132288* [4]
- *rhubarb* *20132283* [4]
- *rubus hybrids* *20132288* [4]
- *salad onions* *20132284* [4]
- *strawberries* *20132288* [4]

Approval information

- Mancozeb and metalaxyl-M included in Annex I under EC Regulation 1107/2009

Efficacy guidance

- Commence potato blight programme before risk of infection occurs as crops begin to meet along the rows and repeat every 7-14 d according to blight risk. Do not exceed a 14 d interval between sprays
- If infection risk conditions occur earlier than the above growth stage commence spraying potatoes immediately
- Complete the potato blight programme using a protectant fungicide starting no later than 10 d after the last phenylamide spray. At least 2 such sprays should be applied
- To minimise the likelihood of development of resistance these products should be used in a planned Resistance Management strategy. See Section 5 for more information

Crop-specific information

- Latest use: before end of active potato haulm growth or before end Aug, whichever is earlier
- HI 7 d for potatoes

FOR FULL CONDITIONS OF USE ALWAYS READ THE PRODUCT LABEL

- After treating early potatoes destroy and remove any remaining haulm after harvest to minimise blight pressure on neighbouring maincrop potatoes

Environmental safety
- Dangerous for the environment
- Very toxic to aquatic organisms
- Do not harvest crops for human consumption for at least 7 d after final application
- LERAP Category B

Hazard classification and safety precautions
Hazard Harmful, Dangerous for the environment [1-4]; Very toxic to aquatic organisms [4]
Transport code 9
Packaging group III
UN Number 3077
Risk phrases H317, H361 [4]; R37, R43, R50, R53a, R63 [1-3]
Operator protection A; U05a, U08, U20b
Environmental protection E15a, E16a, E34, E38 [1-4]; H410 [4]
Consumer protection C02a (7 d)
Storage and disposal D01, D02, D05, D09a, D11a, D12a

306 mancozeb + zoxamide

A protectant fungicide mixture for potatoes
FRAC mode of action code: M3 + 22

See also zoxamide

Products

1	Electis 75WG	Gowan	66.7:8.3% w/w	WG	17871
2	Roxam 75WG	Gowan	66.7:8.3% w/w	WG	17869
3	Unikat 75WG	Gowan	66.7:8.3% w/w	WG	17868

Uses
- Blight in **potatoes** [1-3]
- Disease control in **ornamental plant production** *(off-label)* [1]; **wine grapes** [1-3]

Extension of Authorisation for Minor Use (EAMUs)
- *ornamental plant production 20170462* [1]

Approval information
- Mancozeb and zoxamide included in Annex I under EC Regulation 1107/2009

Efficacy guidance
- Apply as protectant spray on potatoes immediately risk of blight in district or as crops begin to meet along the rows and repeat every 7-10 d according to blight risk
- Do not use if potato blight present in crop. Products are not curative
- Spray irrigated potato crops as soon as possible after irrigation once the crop leaves are dry

Restrictions
- Maximum number of treatments 8 per crop for potatoes

Crop-specific information
- HI 7 d for potatoes

Environmental safety
- Dangerous for the environment
- Very toxic to aquatic organisms
- Keep away from fire and sparks
- LERAP Category B

Hazard classification and safety precautions
Hazard Irritant, Dangerous for the environment, Very toxic to aquatic organisms
Transport code 9

SEE SECTION 3 FOR PRODUCTS ALSO REGISTERED

Packaging group III
UN Number 3077
Risk phrases H317, H361
Operator protection A, D, H; U05a, U14, U20b [1-3]; U19a [1, 2]
Environmental protection E15a, E16a, E16b, E38, H410
Storage and disposal D01, D02, D05, D09a, D11a, D12a

307   mandipropamid

A mandelamide fungicide for the control of potato blight
FRAC mode of action code: 40

*See also cymoxanil + mandipropamid
        difenoconazole + mandipropamid*

Products

1 Mandoprid	Generica	250 g/l	SC	18182
2 Revus	Syngenta	250 g/l	SC	17443

Uses

- Disease control in **baby leaf crops, celery leaves, chives, cress, endives, herbs (see appendix 6), lamb's lettuce, land cress, lettuce, protected baby leaf crops, protected chives, protected cress, protected endives, protected herbs (see appendix 6), protected lamb's lettuce, protected land cress, protected lettuce, protected red mustard, protected spinach, protected spinach beet, red mustard, spinach, spinach beet** [1, 2]
- Downy mildew in **baby leaf crops, celery leaves, chives, cress, endives, herbs (see appendix 6), lamb's lettuce, land cress, lettuce, protected baby leaf crops, protected chives, protected cress, protected endives, protected herbs (see appendix 6), protected lamb's lettuce, protected land cress, protected lettuce, protected red mustard, protected spinach, protected spinach beet, red mustard, spinach, spinach beet** [1, 2]; **hops** (off-label), **ornamental plant production** (off-label), **protected ornamentals** (off-label) [2]

Extension of Authorisation for Minor Use (EAMUs)

- **hops** *20162024* [2]
- **ornamental plant production** *20162763* [2]
- **protected ornamentals** *20162763* [2]

Approval information

- Mandipropamid is included in Annex 1 under EC Regulation 1107/2009

Efficacy guidance

- Mandipropamid acts preventatively by preventing spore germination and inhibiting mycelial growth during incubation. Apply immediately after blight warning or as soon as local conditions favour disease development but before blight enters the crop
- Spray at 7-10 d intervals reducing the interval as blight risk increases
- Spray programme should include a complete haulm desiccant to prevent tuber infection at harvest
- Eliminate other potential infection sources
- To minimise the likelihood of development of resistance this product should be used in a planned Resistance Management strategy. See Section 5 for more information
- See label for details of tank mixtures that may be used as part of a resistance management strategy
- Rainfast in potatoes within 15 minutes of application

Restrictions

- Maximum number of treatments 4 per crop. Do not apply more than 3 treatments of this, or any other fungicide in the same resistance category, consecutively

Crop-specific information

- HI 3 d for potatoes

FOR FULL CONDITIONS OF USE ALWAYS READ THE PRODUCT LABEL

Environmental safety
- Buffer zone requirement 10 m in hops [2]

Hazard classification and safety precautions
UN Number N/C
Operator protection A, H; U05a, U20b
Environmental protection E15b, E34, E38, H411
Storage and disposal D01, D02, D05, D09a, D10c, D12a
Medical advice M03, M05a

308   MCPA

A translocated phenoxycarboxylic acid herbicide for cereals and grassland
HRAC mode of action code: O

See also 2,4-D + clopyralid + MCPA
2,4-D + dicamba + MCPA + mecoprop-P
2,4-D + dichlorprop-P + MCPA + mecoprop-P
2,4-D + MCPA
2,4-DB + linuron + MCPA
2,4-DB + MCPA
bentazone + MCPA + MCPB
bifenox + MCPA + mecoprop-P
clopyralid + 2,4-D + MCPA
clopyralid + diflufenican + MCPA
clopyralid + fluroxypyr + MCPA
dicamba + dichlorprop-P + ferrous sulphate + MCPA
dicamba + dichlorprop-P + MCPA
dicamba + MCPA + mecoprop-P
dichlorprop-P + ferrous sulphate + MCPA
dichlorprop-P + MCPA
dichlorprop-P + MCPA + mecoprop-P
ferrous sulphate + MCPA + mecoprop-P

Products

1	Agritox	Nufarm UK	500 g/l	SL	14894
2	Agroxone	Headland	500 g/l	SL	14909
3	Easel	Nufarm UK	750 g/l	SL	15548
4	Headland Spear	Headland	500 g/l	SL	14910
5	HY-MCPA	Agrichem	500 g/l	SL	14927
6	Larke	Nufarm UK	750 g/l	SL	14914

Uses
- Annual and perennial weeds in **grass seed crops**, **grassland** [1, 3, 6]
- Annual dicotyledons in **farm forestry** *(off-label)*, **game cover** *(off-label)* [2]; **grass seed crops**, **grassland** [2, 4, 5]; **miscanthus** *(off-label)* [4]; **spring barley**, **spring oats**, **spring wheat**, **winter barley**, **winter oats**, **winter wheat** [1-6]; **spring rye**, **winter rye** [1, 3, 6]; **undersown barley**, **undersown oats**, **undersown rye**, **undersown wheat** [1, 3, 5, 6]; **undersown barley** *(red clover or grass)*, **undersown wheat** *(red clover or grass)* [2, 4]
- Charlock in **spring barley**, **spring oats**, **spring wheat**, **winter barley**, **winter oats**, **winter wheat** [1-6]; **spring rye**, **winter rye** [1, 3, 6]; **undersown barley** *(red clover or grass)*, **undersown wheat** *(red clover or grass)* [2, 4]
- Fat hen in **spring barley**, **spring oats**, **spring wheat**, **winter barley**, **winter oats**, **winter wheat** [1-6]; **spring rye**, **winter rye** [1, 3, 6]; **undersown barley** *(red clover or grass)*, **undersown wheat** *(red clover or grass)* [2, 4]
- Hemp-nettle in **spring barley**, **spring oats**, **spring wheat**, **winter barley**, **winter oats**, **winter wheat** [1-6]; **spring rye**, **winter rye** [1, 3, 6]; **undersown barley** *(red clover or grass)*, **undersown wheat** *(red clover or grass)* [2, 4]
- Perennial dicotyledons in **farm forestry** *(off-label)*, **game cover** *(off-label)* [2]; **grass seed crops**, **grassland**, **undersown barley** *(red clover or grass)*, **undersown wheat** *(red clover or grass)* [2,

4]; *miscanthus (off-label)* [4]; *spring barley, spring oats, spring wheat, winter barley, winter oats, winter wheat* [1-6]; *spring rye, winter rye* [1, 3, 6]
- Wild radish in *spring barley, spring oats, spring wheat, winter barley, winter oats, winter wheat* [1-6]; *spring rye, winter rye* [1, 3, 6]; *undersown barley* (red clover or grass), *undersown wheat* (red clover or grass) [2, 4]

Extension of Authorisation for Minor Use (EAMUs)
- *farm forestry* 20122061 [2]
- *game cover* 20122061 [2]
- *miscanthus* 20122050 [4]

Approval information
- MCPA included in Annex I under EC Regulation 1107/2009
- Accepted by BBPA for use on malting barley

Efficacy guidance
- Best results achieved by application to weeds in seedling to young plant stage under good growing conditions when crop growing actively
- Spray perennial weeds in grassland before flowering. Most susceptible growth stage varies between species. See label for details
- Do not spray during cold weather, drought, if rain or frost expected or if crop wet

Restrictions
- Maximum number of treatments normally 1 per crop or yr except grass (2 per yr) for some products. See label
- Do not treat grass within 3 mth of germination and preferably not in the first yr of a direct sown ley or after reseeding
- Do not use on cereals before undersowing
- Do not roll, harrow or graze for a few days before or after spraying; see label
- Do not use on grassland where clovers are an important part of the sward
- Do not use on any crop suffering from stress or herbicide damage
- Avoid spray drift onto nearby susceptible crops

Crop-specific information
- Latest use: before 1st node detectable (GS 31) for cereals; 4-6 wk before heading for grass seed crops; before crop 15-25 cm high for linseed
- Apply to winter cereals in spring from fully tillered, leaf sheath erect stage to before first node detectable (GS 31)
- Apply to spring barley and wheat from 5-leaves unfolded (GS 15), to oats from 1-leaf unfolded (GS 11) to before first node detectable (GS 31)
- Apply to cereals undersown with grass after grass has 2-3 leaves unfolded
- Recommendations for crops undersown with legumes vary. Red clover may withstand low doses after 2-trifoliate leaf stage, especially if shielded by taller weeds but white clover is more sensitive. See label for details
- Apply to grass seed crops from 2-3 leaf stage to 5 wk before head emergence
- Temporary wilting may occur on linseed but without long term effects

Following crops guidance
- Do not direct drill brassicas or legumes within 6 wk of spraying grassland

Environmental safety
- Harmful to aquatic organisms
- MCPA is active at low concentrations. Take extreme care to avoid drift onto neighbouring crops, especially beet crops, brassicas, most market garden crops including lettuce and tomatoes under glass, pears and vines
- Keep livestock out of treated areas for at least 2 wk and until foliage of poisonous weeds such as ragwort has died and become unpalatable
- LERAP Category B

Hazard classification and safety precautions
Hazard Harmful [1-6]; Dangerous for the environment [2-5]; Harmful if swallowed [1, 3-6]
UN Number N/C

FOR FULL CONDITIONS OF USE ALWAYS READ THE PRODUCT LABEL

Risk phrases H318 [1, 3-6]; H335 [4, 5]; R20, R21, R50, R53a [2]; R22a [2, 6]
Operator protection A, C; U05a, U08, U11, U20b [1-6]; U09a [2]; U12 [3]; U15 [4]
Environmental protection E06a [5] (2 wk); E07a [1-4, 6]; E15a [1, 2, 5, 6]; E15b, E38 [3, 4]; E16a, E34 [1-6]; E40b [4]; H410 [4, 5]
Storage and disposal D01, D02, D05, D09a [1-6]; D10a [1-3, 5, 6]; D10b [4]; D12a [3]
Medical advice M03 [1-6]; M05a [5]

309   MCPA + mecoprop-P

A translocated selective herbicide for amenity grass
HRAC mode of action code: O + O

See also mecoprop-P

Products

Cleanrun Pro	Everris Ltd	0.49:0.29% w/w	GR	15828

Uses

- Annual dicotyledons in **managed amenity turf**
- Perennial dicotyledons in **managed amenity turf**

Approval information

- MCPA and mecoprop-P included in Annex I under EC Regulation 1107/2009

Efficacy guidance

- Apply from Apr to Sep, when weeds growing actively and have large leaf area available for chemical absorption

Restrictions

- The total amount of mecoprop-P applied in a single year must not exceed the maximum total dose approved for any single product for use on turf
- Avoid contact with cultivated plants
- Do not use first 4 mowings as compost or mulch unless composted for 6 mth
- Do not treat newly sown or turfed areas for at least 6 mth
- Do not reseed bare patches for 8 wk after treatment
- Do not apply when heavy rain expected or during prolonged drought. Irrigate after 1-2 d unless rain has fallen
- Do not mow within 2-3 d of treatment
- Treat areas planted with bulbs only after the foliage has died down
- Avoid walking on treated areas until it has rained or irrigation has been applied

Crop-specific information

- Granules contain NPK fertilizer to encourage grass growth

Environmental safety

- Take extreme care to avoid drift onto neighbouring crops, especially beet crops, brassicas, most market garden crops including lettuce and tomatoes under glass, pears and vines
- Harmful to fish or other aquatic life. Do not contaminate surface waters or ditches with chemical or used container
- Keep livestock out of treated areas for at least 2 wk and until foliage of any poisonous weeds such as ragwort has died and become unpalatable
- Some pesticides pose a greater threat of contamination of water than others and mecoprop-P is one of these pesticides. Take special care when applying mecoprop-P near water and do not apply if heavy rain is forecast

Hazard classification and safety precautions

UN Number N/C
Operator protection A, C, H, M; U20b
Environmental protection E07a, E13c, E19b
Storage and disposal D01, D09a, D12a
Medical advice M05a

SEE SECTION 3 FOR PRODUCTS ALSO REGISTERED

310 MCPB

A translocated phenoxycarboxylic acid herbicide
HRAC mode of action code: O

See also bentazone + MCPA + MCPB
bentazone + MCPB
MCPA + MCPB

Products

1 Bellmac Straight	UPL Europe	400 g/l	SL	14448
2 Butoxone	Headland	400 g/l	SL	14406
3 Tropotox	Nufarm UK	400 g/l	SL	14450

Uses

- Annual dicotyledons in **combining peas**, **vining peas** [1-3]; **game cover** *(off-label)* [1, 2]
- Docks in **combining peas**, **vining peas** [3]
- Perennial dicotyledons in **combining peas**, **game cover** *(off-label)*, **vining peas** [1, 2]
- Thistles in **combining peas**, **vining peas** [3]

Extension of Authorisation for Minor Use (EAMUs)

- **game cover** *20122052* [1], *20122053* [2]

Approval information

- MCPB included in Annex I under EC Regulation 1107/2009

Efficacy guidance

- Best results achieved by spraying young seedling weeds in good growing conditions
- Best results on perennials by spraying before flowering
- Effectiveness may be reduced by rain within 12 h, by very cold or dry conditions

Restrictions

- Maximum number of treatments 1 per crop or yr.
- Do not roll or harrow for 7-10 d before or after treatment (check label)

Crop-specific information

- Latest use: first node detectable stage (GS 31) for cereals; before flower buds appear in terminal leaf (GS 201) for peas; before flower buds form for clover
- Apply to undersown cereals from 2-leaves unfolded to first node detectable (GS 12-31), and after first trifoliate leaf stage of clover
- Red clover seedlings may be temporarily damaged but later growth is normal
- Apply to white clover seed crops in Mar to early Apr, not after mid-May, and allow 3 wk before cutting and closing up for seed
- Apply to peas from 3-6 leaf stage but before flower bud detectable (GS 103-201). Consult PGRO (see Appendix 2) or label for information on susceptibility of cultivars.
- Do not use on leguminous crops not mentioned on the label
- Apply to cane and bush fruit after harvest and after shoot growth ceased but before weeds are damaged by frost, usually in late Aug or Sep; direct spray onto weeds as far as possible

Environmental safety

- Harmful to aquatic organisms
- Harmful to fish or other aquatic life. Do not contaminate surface waters or ditches with chemical or used container
- Keep livestock out of treated areas until foliage of any poisonous weeds such as ragwort has died and become unpalatable
- Take extreme care to avoid drift onto neighbouring sensitive crops

Hazard classification and safety precautions

Hazard Harmful, Harmful if swallowed [1-3]; Dangerous for the environment [2, 3]
UN Number N/C
Risk phrases H315, H318
Operator protection A, C; U05a, U08, U20b [1-3]; U11, U14, U15 [1, 2]; U19a [3]

FOR FULL CONDITIONS OF USE ALWAYS READ THE PRODUCT LABEL

Environmental protection E07a, E34 [1-3]; E13c [1, 2]; E15a, E38, H411 [3]
Storage and disposal D01, D02, D09a [1-3]; D05 [1, 2]; D10b [2, 3]; D10c [1]
Medical advice M03, M05a

311   mecoprop-P

A translocated phenoxycarboxylic acid herbicide for cereals and grassland
HRAC mode of action code: O

See also 2,4-D + dicamba + MCPA + mecoprop-P
2,4-D + dichlorprop-P + MCPA + mecoprop-P
2,4-D + mecoprop-P
bifenox + MCPA + mecoprop-P
bromoxynil + ioxynil + mecoprop-P
carfentrazone-ethyl + mecoprop-P
dicamba + MCPA + mecoprop-P
dicamba + mecoprop-P
dichlorprop-P + MCPA + mecoprop-P
diflufenican + mecoprop-P
ferrous sulphate + MCPA + mecoprop-P
fluroxypyr + mecoprop-P
MCPA + mecoprop-P

Products

1	Compitox Plus	Nufarm UK	600 g/l	SL	14390
2	Duplosan KV	Nufarm UK	600 g/l	SL	13971
3	Headland Charge	Headland	600 g/l	SL	14394
4	Optica	Headland	600 g/l	SL	14373

Uses

- Annual dicotyledons in **amenity grassland**, **durum wheat** *(off-label)*, **spring durum wheat** *(off-label)*, **spring rye** *(off-label)*, **spring triticale** *(off-label)*, **triticale** *(off-label)*, **winter rye** *(off-label)* [1, 2]; **game cover** *(off-label)*, **miscanthus** *(off-label)* [2, 4]; **grass seed crops, managed amenity turf, spring barley, spring oats, spring wheat, winter barley, winter oats, winter wheat** [1-4]
- Chickweed in **amenity grassland** [1, 2]; **grass seed crops, managed amenity turf, spring barley, spring oats, spring wheat, winter barley, winter oats, winter wheat** [1-4]
- Cleavers in **amenity grassland** [1, 2]; **grass seed crops, managed amenity turf, spring barley, spring oats, spring wheat, winter barley, winter oats, winter wheat** [1-4]
- Perennial dicotyledons in **amenity grassland, grass seed crops, managed amenity turf, spring barley, spring oats, spring wheat, winter barley, winter oats, winter wheat** [1, 2]; **game cover** *(off-label)*, **miscanthus** *(off-label)* [2]

Extension of Authorisation for Minor Use (EAMUs)

- **durum wheat** *20093214* [1], *20093211* [2]
- **game cover** *20111129* [2], *20111125* [4]
- **miscanthus** *20111129* [2], *20111125* [4]
- **spring durum wheat** *20093214* [1], *20093211* [2]
- **spring rye** *20093214* [1], *20093211* [2]
- **spring triticale** *20093214* [1], *20093211* [2]
- **triticale** *20093214* [1], *20093211* [2]
- **winter rye** *20093214* [1], *20093211* [2]

Approval information

- Mecoprop-P included in Annex I under EC Regulation 1107/2009
- Accepted by BBPA for use on malting barley

Efficacy guidance

- Best results achieved by application to seedling weeds which have not been frost hardened, when soil warm and moist and expected to remain so for several days

SEE SECTION 3 FOR PRODUCTS ALSO REGISTERED

SECTION 2

Restrictions
- Maximum number of treatments normally 1 per crop for spring cereals and 1 per yr for newly sown grass; 2 per crop or yr for winter cereals and grass crops. Check labels for details
- The total amount of mecoprop-P applied in a single yr must not exceed the maximum total dose approved for any single product for the crop/situation
- Do not spray cereals undersown with clovers or legumes or to be undersown with legumes or grasses
- Do not spray grass seed crops within 5 wk of seed head emergence
- Do not spray crops suffering from herbicide damage or physical stress
- Do not spray during cold weather, periods of drought, if rain or frost expected or if crop wet
- Do not roll or harrow for 7 d before or after treatment

Crop-specific information
- Latest use: generally before 1st node detectable (GS 31) for spring cereals and before 3rd node detectable (GS 33) for winter cereals, but individual labels vary; 5 wk before emergence of seed head for grass seed crops
- Spray winter cereals from 1 leaf stage in autumn up to and including first node detectable in spring (GS 10-31) or up to second node detectable (GS 32) if necessary. Apply to spring cereals from first fully expanded leaf stage (GS 11) but before first node detectable (GS 31)
- Spray cereals undersown with grass after grass starts to tiller
- Spray newly sown grass leys when grasses have at least 3 fully expanded leaves and have begun to tiller. Any clovers will be damaged

Environmental safety
- Harmful to aquatic organisms
- Harmful to fish or other aquatic life. Do not contaminate surface waters or ditches with chemical or used container
- Keep livestock out of treated areas for at least 2 wk and until foliage of any poisonous weeds, such as ragwort, has died and become unpalatable
- Take extreme care to avoid drift onto neighbouring crops, especially beet crops, brassicas, most market garden crops including lettuce and tomatoes under glass, pears and vines
- Some pesticides pose a greater threat of contamination of water than others and mecoprop-P is one of these pesticides. Take special care when applying mecoprop-P near water and do not apply if heavy rain is forecast

Hazard classification and safety precautions
Hazard Harmful, Dangerous for the environment, Harmful if swallowed
Transport code 9 [3, 4]
Packaging group III [3, 4]
UN Number 3082 [3, 4]; N/C [1, 2]
Risk phrases H315, H318
Operator protection A, C [1-4]; H [1, 2]; P [3, 4]; U05a, U08, U11, U20b [1-4]; U15 [3, 4]
Environmental protection E07a, E13c, H411 [3, 4]; E15b [1, 2]; E34, E38 [1-4]
Storage and disposal D01, D02 [1-4]; D05, D10b [3, 4]; D09a, D10c [1, 2]
Medical advice M03

312 mepanipyrim

An anilinopyrimidine fungicide for use in horticulture
FRAC mode of action code: 9

Products

Frupica SC	Certis	450 g/l	SC	12067

Uses
- Botrytis in *courgettes* (off-label), *forest nurseries* (off-label), *protected strawberries*, *strawberries*

Extension of Authorisation for Minor Use (EAMUs)
- *courgettes* 20093235
- *forest nurseries* 20082853

Approval information
- Mepanipyrim included in Annex I under EC Regulation 1107/2009

Efficacy guidance
- Product is protectant and should be applied as a preventative spray when conditions favourable for Botrytis development occur
- To maintain Botrytis control use as part of a programme with other fungicides that control the disease
- To minimise the possibility of development of resistance adopt resistance management procedures by using products from different chemical groups as part of a mixed spray programme

Restrictions
- Maximum number of treatments 2 per crop (including other anilinopyrimidine products)
- Consult processor before use on crops for processing
- Use spray mixture immediately after preparation

Crop-specific information
- HI 3 d

Environmental safety
- Dangerous for the environment
- Very toxic to aquatic organisms
- LERAP Category B

Hazard classification and safety precautions
Hazard Dangerous for the environment
Transport code 9
Packaging group III
UN Number 3082
Operator protection A, H; U20c
Environmental protection E16a, E16b, E34, H410
Storage and disposal D01, D02, D05, D10c, D11a, D12b

313   mepiquat chloride

A quaternary ammonium plant growth regulator available only in mixtures

See also ethephon + mepiquat chloride
mepiquat chloride + metconazole
mepiquat chloride + prohexadione-calcium

314   mepiquat chloride + metconazole

A PGR mixture for growth control in winter oilseed rape

Products

Caryx	BASF	210:30 g/l	SL	16100

Uses
- Growth regulation in **winter oilseed rape**
- Lodging control in **winter oilseed rape**

Approval information
- Metconazole and mepiquat chloride included in Annex I under EC Regulation 1107/2009

Efficacy guidance
- Apply in 200 - 400 l/ha water. Use at lower water volumes has not been evaluated.
- Apply from the beginning of stem extension once the crop is actively growing.

SEE SECTION 3 FOR PRODUCTS ALSO REGISTERED

Following crops guidance

- Beans, cabbage, carrots, cereals, clover, lettuce, linseed, maize, oilseed rape, onions, peas, potatoes, ryegrass, sugar beet or sunflowers may be sown as a following crop but the effect on other crops has not been tested.

Environmental safety

- LERAP Category B

Hazard classification and safety precautions

Hazard Harmful, Dangerous for the environment, Harmful if swallowed, Harmful if inhaled
Transport code 9
Packaging group III
UN Number 3082
Risk phrases H317, H318
Operator protection A, C, H; U02a, U05a, U11, U14, U15
Environmental protection E15b, E16a, E38, H411
Storage and disposal D01, D02, D05, D09a, D10c, D19
Medical advice M03, M05a

---

315   mepiquat chloride + prohexadione-calcium

A growth regulator mixture for cereals

See also prohexadione-calcium

Products

Canopy	BASF	300:50 g/l	SC	16314

Uses

- Increasing yield in **oats**, **spring barley**, **triticale**, **winter barley**, **winter rye**, **winter wheat**
- Lodging control in **oats**, **spring barley**, **triticale**, **winter barley**, **winter rye**, **winter wheat**

Approval information

- Mepiquat chloride and prohexadione-calcium included in Annex I under EC Regulation 1107/2009
- Accepted by BBPA for use on malting barley

Efficacy guidance

- Best results obtained from treatments applied to healthy crops from the beginning of stem extension

Restrictions

- Maximum total dose equivalent to one full dose treatment on all crops
- Do not apply to any crop suffering from physical stress caused by waterlogging, drought or other conditions
- Do not treat on soils with a substantial moisture deficit
- Consult grain merchant or processor before use on crops for bread making or brewing. Effects on these processes have not been tested

Crop-specific information

- Latest use: before flag leaf fully emerged on wheat and barley

Following crops guidance

- Any crop may follow a normally harvested treated crop. Ploughing is not essential.

Environmental safety

- Harmful to aquatic organisms
- Avoid spray drift onto neighbouring crops

Hazard classification and safety precautions

Hazard Harmful, Harmful if swallowed
UN Number N/C
Operator protection A; U05a, U20b

FOR FULL CONDITIONS OF USE ALWAYS READ THE PRODUCT LABEL

Environmental protection E15a, E34, E38, H412
Storage and disposal D01, D02, D05, D09a, D10c
Medical advice M05a

316   meptyldinocap

A protectant dinitrophenyl fungicide for powdery mildew control
FRAC mode of action code: 29

Products

Kindred	Landseer	350 g/l	EC	13891

Uses
• Powdery mildew in *apples* (off-label), *crab apples* (off-label), *pears* (off-label), *protected strawberries*, *quinces* (off-label), *table grapes*, *wine grapes*

Extension of Authorisation for Minor Use (EAMUs)
• *apples* 20092664
• *crab apples* 20092664
• *pears* 20092664
• *quinces* 20092664

Approval information
• Meptyldinocap is included in Annex 1 under EC Regulation 1107/2009

Environmental safety
• Broadcast air-assisted LERAP (50 m); LERAP Category B

Hazard classification and safety precautions
Hazard Harmful, Flammable, Dangerous for the environment, Flammable liquid and vapour, Harmful if swallowed, Harmful if inhaled
Transport code 3
Packaging group III
UN Number 1993
Risk phrases H317, H319, H336
Operator protection A, C, H; U04a, U05a, U11, U14, U19a, U20a
Environmental protection E15b, E16a, E34, E38, H410; E17b (50 m)
Storage and disposal D01, D02, D09a, D10a, D12a
Medical advice M05a

317   mesosulfuron-methyl

A sulfonyl urea herbicide for cereals available only in mixtures
HRAC mode of action code: B

See also amidosulfuron + iodosulfuron-methyl-sodium + mesosulfuron-methyl
diflufenican + iodosulfuron-methyl-sodium + mesosulfuron-methyl
iodosulfuron-methyl-sodium + mesosulfuron-methyl

318   mesosulfuron-methyl + propoxycarbazone-sodium

A herbicide mixture for use in winter cereals
HRAC mode of action code: B + B

Products

Monolith	Bayer CropScience	4.5:6.75% w/w	WG	17687

Uses
• Annual dicotyledons in *durum wheat*, *spelt*, *triticale*, *winter rye*, *winter wheat*
• Annual meadow grass in *durum wheat*, *spelt*, *triticale*, *winter rye*, *winter wheat*

SEE SECTION 3 FOR PRODUCTS ALSO REGISTERED

- Blackgrass in *durum wheat, spelt, triticale, winter rye, winter wheat*
- Brome grasses in *durum wheat, spelt, triticale, winter rye, winter wheat*
- Chickweed in *durum wheat, spelt, triticale, winter rye, winter wheat*
- Loose silky bent in *durum wheat, spelt, triticale, winter rye, winter wheat*
- Mayweeds in *durum wheat, spelt, triticale, winter rye, winter wheat*
- Ryegrass in *durum wheat, spelt, triticale, winter rye, winter wheat*
- Wild oats in *durum wheat, spelt, triticale, winter rye, winter wheat*

Approval information
- Mesosulfuron-methyl and propoxycarbazone-sodium included in Annex I under EC Regulation 1107/2009

Efficacy guidance
- Apply as early as possible and before GS 29 of grass weeds

Restrictions
- This product must only be applied between 1 February in the year of harvest and the specified latest time of application.
- To avoid the build-up of resistance do not apply this or any other product containing an ALS herbicide with claims of control of grass-weeds more than once to any crop.
- This product must not be applied via hand-held equipment.
- DO NOT use on crops undersown with grasses, clover or other legumes or any other broad-leaved crop.
- Do not use on cereal crops grown for seed as effects on germination have not been established.
- Do not use as the sole means of grass weed or broad-leaved weed control in successive crops

Crop-specific information
- Clean the spray tank with clean water and add a liquid sprayer cleaner specifically formulated for sulfonylurea herbicides after use.

Following crops guidance
- Winter wheat, winter barley, winter oilseed rape, mustard, lupins and phacelia may be sown in the year of harvest to succeed a treated cereal crop. Spray overlaps in the treated cereal crop should be avoided in order to reduce the risk of localised adverse effects on following crops of winter oilseed rape and mustard.
- Spring wheat, spring barley, maize, spring oilseed rape, sugar beet, Italian rye-grass, peas and sunflowers may be drilled in the spring following harvest of a treated cereal crop.

Environmental safety
- LERAP Category B

Hazard classification and safety precautions
Hazard Very toxic to aquatic organisms
Transport code 9
Packaging group III
UN Number 3077
Risk phrases H319
Operator protection A, C, H; U05a, U11, U20b
Environmental protection E15a, E16a, H410
Storage and disposal D01, D02, D05, D09a, D10a, D12a, D12b

319   mesotrione

A foliar applied triketone herbicide for maize
HRAC mode of action code: F2

Products

1	Barracuda	Albaugh UK	100 g/l	SC	17433
2	Basilico	Life Scientific	100 g/l	SC	18028
3	Callisto	Syngenta	100 g/l	SC	12323
4	Clayton Goldcob	Clayton	100 g/l	SC	16299

FOR FULL CONDITIONS OF USE ALWAYS READ THE PRODUCT LABEL

Products – continued

5	Cuter	Clayton	100 g/l	SC	17668
6	Daneva	Rotam	100 g/l	SC	18029
7	Greencrop Goldcob	Clayton	100 g/l	SC	12863
8	Kideka	Nufarm UK	100 g/l	SC	18033
9	Mesostar	Life Scientific	100 g/l	SC	17410
10	Raikiri	Sipcam	100 g/l	SC	17670
11	Temsa SC	Belchim	100 g/l	SC	17234

Uses

- Annual dicotyledons in **asparagus** *(off-label)*, **game cover** *(off-label)*, **linseed** *(off-label)*, **poppies for morphine production** *(off-label)*, **sweetcorn** *(off-label)* [3]; **forage maize** [1-11]; **grain maize** [6]; **sweetcorn** [8]
- Annual grasses in **game cover** *(off-label)*, **sweetcorn** *(off-label)* [3]
- Annual meadow grass in **grain maize** [1-5, 8-11]; **linseed** *(off-label)* [3]
- Black bindweed in **rhubarb** *(off-label)* [3]
- Charlock in **rhubarb** *(off-label)* [3]
- Cleavers in **rhubarb** *(off-label)* [3]
- Cockspur grass in **forage maize**, **grain maize** [4]
- Field pansy in **rhubarb** *(off-label)* [3]
- Groundsel in **rhubarb** *(off-label)* [3]
- Himalayan balsam in **rhubarb** *(off-label)* [3]
- Mayweeds in **rhubarb** *(off-label)* [3]
- Volunteer oilseed rape in **forage maize** [1-11]; **grain maize** [1-6, 8-11]; **linseed** *(off-label)* [3]; **sweetcorn** [8]

Extension of Authorisation for Minor Use (EAMUs)

- **asparagus** *20111113* [3]
- **game cover** *20082830* [3]
- **linseed** *20071706* [3]
- **poppies for morphine production** *20113148* [3]
- **rhubarb** *20161761* [3]
- **sweetcorn** *20051893* [3]

Approval information

- Mesotrione included in Annex I under EC Regulation 1107/2009

Efficacy guidance

- Best results obtained from treatment of young actively growing weed seedlings in the presence of adequate soil moisture
- Treatment in poor growing conditions or in dry soil may give less reliable control
- Activity is mostly by foliar uptake with some soil uptake
- To minimise the possible development of resistance where continuous maize is grown the product should not be used for more than two consecutive seasons

Restrictions

- Maximum number of treatments 1 per crop of forage maize
- Do not use on seed crops or on sweetcorn varieties
- Do not spray when crop foliage wet or when excessive rainfall is expected to follow application
- Do not treat crops suffering from stress from cold or drought conditions, or when wide temperature fluctuations are anticipated

Crop-specific information

- Latest use: 8 leaves unfolded stage (GS 18) for forage maize
- Treatment under adverse conditions may cause mild to moderate chlorosis. The effect is transient and does not affect yield

Following crops guidance

- Winter wheat, durum wheat, winter barley or ryegrass may follow a normally harvested treated crop of maize. Oilseed rape may be sown provided it is preceded by deep ploughing to more than 15 cm

SEE SECTION 3 FOR PRODUCTS ALSO REGISTERED

- In the spring following application only forage maize, ryegrass, spring wheat or spring barley may be sown
- In the event of crop failure maize may be re-seeded immediately. Some slight crop effects may be seen soon after emergence but these are normally transient

Environmental safety
- Dangerous for the environment
- Very toxic to aquatic organisms
- Take extreme care to avoid drift onto all plants outside the target area
- LERAP Category B [1-5, 7-11]

Hazard classification and safety precautions
Hazard Irritant, Dangerous for the environment [1-11]; Very toxic to aquatic organisms [1-3, 5, 6, 8-10]
Transport code 9 [3]
Packaging group III [3]
UN Number 3082 [3]; N/C [1, 2, 4-11]
Risk phrases H317 [1, 5, 10, 11]; H318 [1, 5, 6, 8, 10, 11]; H319 [2, 3, 9]; R36, R50, R53a [4, 7]
Operator protection A, C; U05a, U20b [1-11]; U08, U19a [7]; U09a [1-6, 8-11]
Environmental protection E15b [1-11]; E16a, E16b [1-5, 7-11]; E38 [1-6, 8-11]; H410 [1-3, 5, 6, 8-11]
Storage and disposal D01, D02, D05, D09a, D12a [1-11]; D10b [7]; D10c, D11a [1-6, 8-11]
Medical advice M05a [7]

320   mesotrione + nicosulfuron

A herbicide for use in forage and grain maize
HRAC mode of action code: B + F2

Products

1 Clayton Aspen	Clayton	75:30 g/l	OD	16114
2 Elumis	Syngenta	75:30 g/l	OD	15800

Uses
- Amaranthus in *forage maize, grain maize*
- Annual dicotyledons in *forage maize, grain maize*
- Annual meadow grass in *forage maize, grain maize*
- Black bindweed in *forage maize, grain maize*
- Black nightshade in *forage maize, grain maize*
- Chickweed in *forage maize, grain maize*
- Cleavers in *forage maize, grain maize*
- Cockspur grass in *forage maize, grain maize*
- Fat hen in *forage maize, grain maize*
- Field pansy in *forage maize, grain maize*
- Field speedwell in *forage maize, grain maize*
- Fumitory in *forage maize, grain maize*
- Hemp-nettle in *forage maize, grain maize*
- Mayweeds in *forage maize, grain maize*
- Pale persicaria in *forage maize, grain maize*
- Red dead-nettle in *forage maize, grain maize*
- Redshank in *forage maize, grain maize*
- Ryegrass in *forage maize, grain maize*
- Shepherd's purse in *forage maize, grain maize*
- Volunteer oilseed rape in *forage maize, grain maize*

Approval information
- Mesotrione and nicosulfuron included in Annex 1 under EC Regulation 1107/2009

Efficacy guidance
- Optimum efficacy requires soils to be moist at application.

FOR FULL CONDITIONS OF USE ALWAYS READ THE PRODUCT LABEL

- To minimise the possible development of resistance where continuous maize is grown the product should not be used for more than two consecutive seasons.

Following crops guidance
- In the event of crop failure, only forage or grain maize may be sown as a replacement after ploughing.
- After normal harvest wheat or barley may be sown the same autumn after ploughing to 15 cm and after 4 months has elapsed since application. The following spring maize, wheat, barley or ryegrass may be planted after ploughing but other crops are not recommended.

Environmental safety
- LERAP Category B

Hazard classification and safety precautions
Hazard Irritant, Dangerous for the environment [1, 2]; Very toxic to aquatic organisms [2]
Transport code 9
Packaging group III
UN Number 3082
Risk phrases R38, R50, R53a [1]
Operator protection A, H; U02a, U05a, U14, U19a, U20a
Environmental protection E15b, E16a, E23 [1, 2]; H410 [2]
Storage and disposal D01, D02, D05, D09a, D10c, D12a
Medical advice M05a, M05d

321 mesotrione + terbuthylazine

A foliar and soil acting herbicide mixture for maize
HRAC mode of action code: F2 + C1

See also terbuthylazine

Products

1 Calaris	Syngenta	70:330 g/l	SC	12405
2 Clayton Faize	Clayton	70:330 g/l	SC	13810
3 Destiny	AgChem Access	70:330 g/l	SC	14159
4 Pan Theta	Pan Agriculture	70:330 g/l	SC	14501

Uses
- Annual dicotyledons in **forage maize**, **grain maize** [1-4]; **sweetcorn** *(off-label)* [1]
- Annual meadow grass in **forage maize**, **grain maize** [1-4]; **sweetcorn** *(off-label)* [1]

Extension of Authorisation for Minor Use (EAMUs)
- **sweetcorn** *20051892* [1]

Approval information
- Mesotrione and terbuthylazine included in Annex 1 under EC Regulation 1107/2009

Efficacy guidance
- Best results obtained from treatment of young actively growing weed seedlings in the presence of adequate soil moisture
- Treatment in poor growing conditions or in dry soil may give less reliable control
- Residual weed control is reduced on soils with more than 10% organic matter
- To minimise the possible development of resistance where continuous maize is grown the product should not be used for more than two consecutive seasons

Restrictions
- Maximum number of treatments 1 per crop of forage maize
- Do not use on seed crops or on sweetcorn varieties
- Do not spray when crop foliage wet or when excessive rainfall is expected to follow application
- Do not treat crops suffering from stress from cold or drought conditions, or when wide temperature fluctuations are anticipated
- Do not apply on Sands or Very Light soils

SEE SECTION 3 FOR PRODUCTS ALSO REGISTERED

SECTION 2

Crop-specific information
- Latest use: 8 leaves unfolded stage (GS 18) for forage maize
- Treatment under adverse conditions may cause mild to moderate chlorosis. The effect is transient and does not affect yield

Following crops guidance
- Winter wheat, durum wheat, winter barley or ryegrass may follow a normally harvested treated crop of maize. Oilseed rape may be sown provided it is preceded by deep ploughing to more than 15 cm
- In the spring following application forage maize, ryegrass, spring wheat or spring barley may be sown
- Spinach, beet crops, peas, beans, lettuce and cabbages must not be sown in the yr following application
- In the event of crop failure maize may be re-seeded immediately. Some slight crop effects may be seen soon after emergence but these are normally transient

Environmental safety
- Dangerous for the environment
- Very toxic to aquatic organisms
- Take extreme care to avoid drift onto all plants outside the target area
- LERAP Category B

Hazard classification and safety precautions
Hazard Harmful, Dangerous for the environment [1-4]; Harmful if swallowed [1, 4]; Very toxic to aquatic organisms [1]

Transport code 9
Packaging group III
UN Number 3082
Risk phrases R22a, R50, R53a [2, 3]
Operator protection A; U05a
Environmental protection E15a, E16a, E16b, E34, E38 [1-4]; H410 [1, 4]
Storage and disposal D01, D02, D05, D09a, D10c, D12a

322 metalaxyl-M

A phenylamide systemic fungicide
FRAC mode of action code: 4

See also chlorothalonil + metalaxyl-M
cymoxanil + fludioxonil + metalaxyl-M
fluazinam + metalaxyl-M
fludioxonil + metalaxyl-M
fludioxonil + metalaxyl-M + thiamethoxam
mancozeb + metalaxyl-M

Products

1	Apron XL	Syngenta	339.2 g/l	ES	14654
2	Caveo	Pan Agriculture	465.2 g/l	SL	16361
3	Clayton Tine	Clayton	465 g/l	SL	14072
4	Floreo	Pan Agriculture	465.2 g/l	SL	16358
5	Hobson	AgChem Access	465.2 g/l	SL	16529
6	SL 567A	Syngenta	465.2 g/l	SL	12380
7	Subdue	Fargro	465 g/l	SL	12503

Uses
- Cavity spot in **carrots** [2, 5, 6]; **carrots** *(reduction only)*, **parsnips** *(off-label - reduction)* [3]; **parsnips** *(off-label)* [6]
- Crown rot in **protected water lilies** *(off-label)*, **water lilies** *(off-label)* [6]
- Damping off in **watercress** *(off-label)* [6]
- Downy mildew in **asparagus** *(off-label)*, **baby leaf crops** *(off-label)*, **blackberries** *(off-label)*, **broccoli** *(off-label)*, **calabrese** *(off-label)*, **collards** *(off-label)*, **grapevines** *(off-label)*, **herbs (see**

FOR FULL CONDITIONS OF USE ALWAYS READ THE PRODUCT LABEL

appendix 6) (off-label), **hops** *(off-label),* **horseradish** *(off-label),* **kale** *(off-label),* **protected baby leaf crops** *(off-label),* **protected cucumbers** *(off-label),* **protected herbs (see appendix 6)** *(off-label),* **protected soft fruit** *(off-label),* **protected spinach** *(off-label),* **protected spinach beet** *(off-label),* **protected water lilies** *(off-label),* **raspberries** *(off-label),* **rubus hybrids** *(off-label),* **salad onions** *(off-label),* **soft fruit** *(off-label),* **spinach** *(off-label),* **spinach beet** *(off-label),* **spring field beans** *(off-label),* **water lilies** *(off-label),* **watercress** *(off-label),* **winter field beans** *(off-label)* [6]

- Phytophthora in **amenity vegetation** *(off-label),* **forest nurseries** *(off-label)* [7]; **ornamental plant production, protected ornamentals** [4]
- Phytophthora root rot in **ornamental plant production, protected ornamentals** [7]
- Pythium in **beetroot** *(off-label),* **brassica leaves and sprouts** *(off-label),* **broccoli, brussels sprouts, bulb onions, cabbages, calabrese, cauliflowers, chard** *(off-label),* **chinese cabbage, herbs (see appendix 6)** *(off-label),* **kohlrabi, ornamental plant production** *(off-label),* **radishes** *(off-label),* **shallots** *(off-label),* **spinach** [1]; **ornamental plant production, protected ornamentals** [4, 7]
- Root malformation disorder in **red beet** *(off-label)* [6]
- Storage rots in **cabbages** *(off-label)* [6]
- White blister in **horseradish** *(off-label)* [6]

Extension of Authorisation for Minor Use (EAMUs)
- **amenity vegetation** *20120383* [7]
- **asparagus** *20051502* [6]
- **baby leaf crops** *20051507* [6]
- **beetroot** *20102539* [1]
- **blackberries** *20072195* [6]
- **brassica leaves and sprouts** *20120526* [1]
- **broccoli** *20112502* [6]
- **cabbages** *20062117 expires 30 Jun 2018* [6]
- **calabrese** *20112502* [6]
- **chard** *20120526* [1]
- **collards** *20112048* [6]
- **forest nurseries** *20120383* [7]
- **grapevines** *20051504* [6]
- **herbs (see appendix 6)** *20102539* [1], *20051507* [6]
- **hops** *20051500* [6]
- **horseradish** *20051499* [6], *20061040* [6]
- **kale** *20112048* [6]
- **ornamental plant production** *20102539* [1]
- **parsnips** *(reduction) 20111033* [3], *20051508* [6]
- **protected baby leaf crops** *20051507* [6]
- **protected cucumbers** *20051503* [6]
- **protected herbs (see appendix 6)** *20051507* [6]
- **protected soft fruit** *20082937* [6]
- **protected spinach** *20051507* [6]
- **protected spinach beet** *20051507* [6]
- **protected water lilies** *20051501* [6]
- **radishes** *20102539* [1]
- **raspberries** *20072195* [6]
- **red beet** *20051307* [6]
- **rubus hybrids** *20072195* [6]
- **salad onions** *20072194* [6]
- **shallots** *20121635* [1]
- **soft fruit** *20082937* [6]
- **spinach** *20051507* [6]
- **spinach beet** *20051507* [6]
- **spring field beans** *20130917* [6]
- **water lilies** *20051501* [6]
- **watercress** *20072193* [6]
- **winter field beans** *20130917* [6]

SEE SECTION 3 FOR PRODUCTS ALSO REGISTERED

Approval information
- Metalaxyl-M included in Annex I under EC Regulation 1107/2009
- Accepted by BBPA for use on hops

Efficacy guidance
- Best results achieved when applied to damp soil or potting media
- Treatments to ornamentals should be followed immediately by irrigation to wash any residues from the leaves and allow penetration to the rooting zone [7]
- Efficacy may be reduced in prolonged dry weather [6]
- Results may not be satisfactory on soils with high organic matter content [6]
- Control of cavity spot on carrots overwintered in the ground or lifted in winter may be lower than expected [6]
- Use in an integrated pest management strategy and, where appropriate, alternate with products from different chemical groups
- Product should ideally be used preventatively and the number of phenylamide applications should be limited to 1-2 consecutive treatments
- Always follow FRAG guidelines for preventing and managing fungicide resistance. See Section 5 for more information

Restrictions
- Maximum number of treatments on ornamentals 1 per situation for media treatment. See label for details of drench treatment of protected ornamentals [7]
- Maximum total dose on carrots equivalent to one full dose treatment [6]
- Do not use where carrots have been grown on the same site within the previous eight yrs [6]
- Consult before use on crops intended for processing [6]
- Do not re-use potting media from treated plants for subsequent crops [7]
- Disinfect pots thoroughly prior to re-use [7]

Crop-specific information
- Latest use: 6 wk after drilling for carrots [6]
- Because of the large number of species and ornamental cultivars susceptibility should be checked before large scale treatment [7]
- Treatment of *Viburnum* and *Prunus* species not recommended [7]

Environmental safety
- Harmful to aquatic organisms
- Limited evidence suggests that metalaxyl-M is not harmful to soil dwelling predatory mites

Hazard classification and safety precautions
Hazard Harmful, Harmful if swallowed
UN Number N/C
Risk phrases H319 [2-7]; H335 [7]
Operator protection A [1-7]; C [7]; D [1]; H [1, 7]; U02a, U05a, U20b [1-7]; U04a, U10, U19a [2-7]
Environmental protection E15b, E34, E38, H412
Storage and disposal D01, D02, D05, D09a, D12a [1-7]; D07 [3, 5-7]; D10c [2-7]; D11a [1]
Treated seed S02, S04d, S05, S07, S08, S09 [1]
Medical advice M03 [2-7]; M05a [1]

323   metaldehyde

A molluscicide bait for controlling slugs and snails

Products

1 Allure	Chiltern	1.5% w/w	PT	12651
2 Appeal	Chiltern	1.5% w/w	PT	12713
3 Attract	Chiltern	1.5% w/w	PT	12712
4 Carakol 3	Adama	3% w/w	PT	14309
5 Certis Metaldehyde 3	Certis	3% w/w	PT	14337
6 Condor 3	Certis	3% w/w	PT	14324
7 Desire	Chiltern	1.5% w/w	PT	14048
8 Enzo	Adama	3% w/w	PT	14306

FOR FULL CONDITIONS OF USE ALWAYS READ THE PRODUCT LABEL

Products – continued

9	ESP3	De Sangosse	3% w/w	RB	16088
10	Gusto 3	Adama	3% w/w	PT	14308
11	Lynx H	De Sangosse	3% w/w	RB	14426
12	Osarex W	De Sangosse	3% w/w	RB	14428
13	Prowler	De Sangosse	3% w/w	RB	16087
14	Super 3	Certis	3% w/w	PT	14370
15	TDS Major	De Sangosse	4% w/w	CB	13462
16	TDS Metarex Amba	De Sangosse	4% w/w	CB	13461
17	Tempt	Chiltern	1.5% w/w	PT	14227
18	Trigger 3	Certis	3% w/w	GB	14304
19	Trounce	Chiltern	1.5% w/w	PT	14222

SECTION 2

Uses

- Slugs in *all edible crops (outdoor)*, *protected crops* [16]; *all edible crops except potatoes and cauliflowers*, *cauliflowers*, *potatoes* [2, 3, 7, 15]; *all non-edible crops (outdoor)* [2, 3, 7, 15, 16]; *natural surfaces not intended to bear vegetation* [15, 16]
- Slugs and snails in *all edible crops (outdoor and protected)* [1]; *all edible crops (outdoor)* [6]; *all edible crops except potatoes and cauliflowers* [4, 5, 8-14, 17-19]; *all non-edible crops (outdoor)* [1, 4, 6, 8-14, 17, 19]; *amenity grassland*, *managed amenity turf* [5, 18]; *cauliflowers* [1, 4, 8-13, 17, 19]; *natural surfaces not intended to bear vegetation* [5, 6, 9, 11-14, 18]; *potatoes* [1, 4, 5, 8-14, 17-19]
- Snails in *all edible crops (outdoor)*, *protected crops* [16]; *all edible crops except potatoes and cauliflowers*, *cauliflowers*, *potatoes* [2, 3, 7, 15]; *all non-edible crops (outdoor)* [2, 3, 7, 15, 16]; *natural surfaces not intended to bear vegetation* [15, 16]

Approval information

- Metaldehyde included in Annex I under EC Regulation 1107/2009
- Accepted by BBPA for use on malting barley and hops

Efficacy guidance

- Apply pellets by hand, fiddle drill, fertilizer distributor, by air (check label) or in admixture with seed. See labels for rates and timing.
- Best results achieved from an even spread of granules applied during mild, damp weather when slugs and snails most active. May be applied in standing crops
- To establish the need for pellet application on winter wheat or winter oilseed rape, monitor for slug activity. Where bait traps are used, use a foodstuff attractive to slugs e.g. chicken layer's mash
- Varieties of oilseed rape low in glucosinolates can be more acceptable to slugs than "single low" varieties and control may not be as good
- To prevent slug build up apply at end of season to brassicas and other leafy crops
- To reduce tuber damage in potatoes apply twice in Jul and Aug
- For information on slug trapping and damage risk assessment refer to AHDB Information Sheet 02 - Integrated slug control, available from the AHDB website (www.ahdb.org.uk/slugcontrol)

Restrictions

- Do not apply when rain imminent or water glasshouse crops within 4 d of application
- Take care to avoid lodging of pellets in the foliage when making late applications to edible crops.
- The maximum total dose of metaldehyde must not exceed 700 g active substance/ha/year.
- The risk to ground water on drained fields or sloping sites can be avoided by substituting metaldehyde products for ferric phosphate products

Crop-specific information

- Put slug traps out before cultivation, when the soil surface is visibly moist and the weather mild (5-25°C) (see label for guidance)
- For winter wheat, a catch of 4 or more slugs/trap indicates a possible risk, where soil and weather conditions favour slug activity
- For winter oilseed rape a catch of 4 or more slugs in standing cereals, or 1 or more in cereal stubble, if other conditions were met, would indicate possible risk of damage

SEE SECTION 3 FOR PRODUCTS ALSO REGISTERED

Environmental safety

- Dangerous to game, wild birds and animals
- Some products contain proprietary cat and dog deterrent
- Keep poultry out of treated areas for at least 7 d
- Do not use slug pellets in traps in winter wheat or winter oilseed rape since they are a potential hazard to wildlife and pets
- Some pesticides pose a greater threat of contamination of water than others and metaldehyde is one of these pesticides. Take special care when applying metaldehyde near water and do not apply if heavy rain is forecast
- No pellets should fall within 10 metres of a field boundary or watercourse.

Hazard classification and safety precautions

UN Number N/C

Operator protection A, H; U05a [1-13, 15-19]; U20a [4, 8, 10]; U20b [5, 6, 9, 11-13, 18]; U20c [1-3, 7, 14-17, 19]

Environmental protection E05b [1-5] (7 d); E05b [6, 7] (7 days); E05b [8] (7 d); E05b [9] (7 days); E05b [10] (7 d); E05b [11-13] (7 days); E05b [14-19] (7 d); E06b [7] (7 d); E07a, E34 [6, 9, 11-13]; E10a [1-4, 6-17, 19]; E10c [5, 18]; E15a [1-19]

Storage and disposal D01, D09a [1-19]; D02 [1-13, 15-19]; D05 [5, 18]; D07 [1-3, 5-7, 9, 11-19]; D11a [1-14, 17-19]; D11b [15, 16]

Treated seed S04a [6, 9, 11-13, 15, 16]

Medical advice M04a [5, 18]; M05a [6, 9, 11-13, 15, 16]

324 metamitron

A contact and residual triazinone herbicide for use in beet crops and a fruit thinner for apples and pears.
HRAC mode of action code: C1

See also chloridazon + chlorpropham + metamitron
chloridazon + metamitron
chlorpropham + metamitron
desmedipham + ethofumesate + metamitron + phenmedipham
ethofumesate + metamitron
ethofumesate + metamitron + phenmedipham

Products

1	Beetron 700	AgChem Access	700 g/l	SC	17458
2	Bettix Flo	UPL Europe	700 g/l	SC	16559
3	Brevis	Adama	15% w/w	SG	17479
4	Defiant	UPL Europe	700 g/l	SC	18216
5	Glotron 700 SC	Belchim	700 g/l	SC	17308
6	Goltix 70 SC	Adama	700 g/l	SC	16638
7	Mitron 700 SC	Belcrop	700 g/l	SC	16908
8	Target SC	UPL Europe	700 g/l	SC	16969
9	Tronix 700 SC	Belchim	700 g/l	SC	18137

Uses

- Annual dicotyledons in **fodder beet, mangels, red beet, sugar beet** [1, 2, 4-9]; **horseradish** *(off-label)*, **parsnips** *(off-label)*, **strawberries** *(off-label)* [6]
- Annual grasses in **fodder beet, mangels, red beet, sugar beet** [1, 2, 4, 7]
- Annual meadow grass in **fodder beet, mangels, red beet, sugar beet** [1, 2, 4-9]; **horseradish** *(off-label)*, **ornamental plant production** *(off-label)*, **parsnips** *(off-label)*, **protected ornamentals** *(off-label)*, **strawberries** *(off-label)* [6]
- Fat hen in **fodder beet, mangels, red beet, sugar beet** [1, 2, 4, 7]; **ornamental plant production** *(off-label)*, **protected ornamentals** *(off-label)* [6]
- Fruit thinning in **apples, pears** [3]
- Groundsel in **ornamental plant production** *(off-label)*, **protected ornamentals** *(off-label)* [6]
- Growth regulation in **apples, pears** [3]
- Mayweeds in **ornamental plant production** *(off-label)*, **protected ornamentals** *(off-label)* [6]

FOR FULL CONDITIONS OF USE ALWAYS READ THE PRODUCT LABEL

- Red dead-nettle in *ornamental plant production* (off-label), *protected ornamentals* (off-label) [6]
- Scarlet pimpernel in *ornamental plant production* (off-label), *protected ornamentals* (off-label) [6]
- Small nettle in *ornamental plant production* (off-label), *protected ornamentals* (off-label) [6]

Extension of Authorisation for Minor Use (EAMUs)
- *horseradish* 20142918 [6]
- *ornamental plant production* 20151175 [6]
- *parsnips* 20142918 [6]
- *protected ornamentals* 20151175 [6]
- *strawberries* 20142919 [6]

Approval information
- Metamitron included in Annex I under EC Regulation 1107/2009

Efficacy guidance
- May be used pre-emergence alone or post-emergence in tank mixture or with an authorised adjuvant oil
- Low dose programme (LDP). Apply a series of low-dose post-weed emergence sprays, including adjuvant oil, timing each treatment according to weed emergence and size. See label for details and for recommended tank mixes and sequential treatments. On mineral soils the LDP should be preceded by pre-drilling or pre-emergence treatment
- Traditional application. Apply either pre-drilling before final cultivation with incorporation to 8-10 cm, or pre-crop emergence at or soon after drilling into firm, moist seedbed to emerged weeds from cotyledon to first true leaf stage
- On emerged weeds at or beyond 2-leaf stage addition of adjuvant oil advised
- For control of wild oats and certain other weeds, tank mixes with other herbicides or sequential treatments are recommended. See label for details
- When used for fruit thinning in apples and pears, apply post blossom if fruit set is excessive. Apply in temperatures between 10 and 25°C [3]

Restrictions
- Maximum total dose equivalent to three full dose treatments for most products. Check label
- Using traditional method post-crop emergence on mineral soils do not apply before first true leaves have reached 1 cm long

Crop-specific information
- Latest use: before crop foliage meets across rows for beet crops
- HI herbs 6 wk
- HI for apples and pears 60 days
- Crop tolerance may be reduced by stress caused by growing conditions, effects of pests, disease or other pesticides, nutrient deficiency etc

Following crops guidance
- Only sugar beet, fodder beet or mangels may be drilled within 4 mth after treatment. Winter cereals may be sown in same season after ploughing, provided 16 wk passed since last treatment

Environmental safety
- Dangerous for the environment
- Very toxic to aquatic organisms
- Dangerous to fish or other aquatic life. Do not contaminate surface waters or ditches with chemical or used container
- Do not empty into drains

Hazard classification and safety precautions
Hazard Harmful [1, 4-9]; Dangerous for the environment [1-9]; Harmful if swallowed [2-5, 7-9]
Transport code 9
Packaging group III
UN Number 3077 [3]; 3082 [1, 2, 4-9]
Risk phrases H317 [2, 4]; H318 [3]; R22a, R53a [1, 6]; R43 [4]; R50 [1]; R51 [6]

SEE SECTION 3 FOR PRODUCTS ALSO REGISTERED

SECTION 2

Operator protection A [1-7]; C [3]; H [5, 6]; U02a, U04a, U11 [3]; U05a [3, 7-9]; U08, U19a [7-9]; U09a, U20c [2, 4]; U14 [2, 4, 7-9]; U20a [7]; U20b [1, 3, 8, 9]
Environmental protection E13b [2, 4]; E15a [1, 7-9]; E15b [3]; E19b [1]; E34 [1, 3, 4, 8, 9]; E38 [2, 4-9]; H410 [5, 9]; H411 [2-4, 7, 8]
Storage and disposal D01, D02, D12a [1-9]; D05 [2-4]; D09a [1-4, 7-9]; D10c [2, 4]; D11a [1, 3, 8, 9]
Medical advice M03 [1, 2, 4, 8, 9]; M05a [1, 2, 4-7]

325   metam-sodium

A methyl isothiocyanate producing sterilant for glasshouse, nursery and outdoor soils
HRAC mode of action code: Z

Products

Metam 510	Certis	510 g/l	LI	16079

Uses

- Brown rot in *potting soils*, *soils*
- Millipedes in *potting soils*, *soils*
- Potato cyst nematode in *potting soils*, *soils*
- Root rot in *potting soils*, *soils*
- Root-knot nematodes in *potting soils*, *soils*
- Symphylids in *potting soils*, *soils*
- Weed seeds in *potting soils*, *soils*
- Wireworm in *potting soils*, *soils*

Approval information

- Metam-sodium is not included in Annex 1 under EC Regulation 1107/2009

Efficacy guidance

- Metam-sodium is a partial soil sterilant and acts by breaking down in contact with soil to release methyl isothiocyanate (MIT)
- Apply to glasshouse soils as a drench, or inject undiluted to 20 cm at 30 cm intervals and seal immediately, or apply to surface and rotavate
- May also be used by mixing into potting soils
- Apply when soil temperatures exceed 7˚C, preferably above 10˚C, between 1 Apr and 31 Oct. Soil must be of fine tilth, free from debris and with 'potting moisture' content. If soil is too dry postpone treatment and water soil

Restrictions

- No plants must be present during treatment
- Crops must not be planted until a cress germination test has been completed satisfactorily
- Do not treat glasshouses within 2 m of growing crops. Fumes are damaging to all plants
- Avoid using in equipment incorporating natural rubber parts

Crop-specific information

- Latest use: pre-planting of crop

Following crops guidance

- Do not plant until soil is entirely free of fumes

Environmental safety

- Dangerous for the environment
- Very toxic to aquatic organisms
- Keep unprotected persons, livestock and pets out of treated areas for at least 24 h following treatment
- When diluted breakdown commences almost immediately. Only quantities for immediate use should be made up
- Divert or block drains which could carry solution under untreated glasshouses

FOR FULL CONDITIONS OF USE ALWAYS READ THE PRODUCT LABEL

- After treatment allow sufficient time (several weeks) for residues to dissipate and aerate soil by forking. Time varies with soil and season. Soils with high clay or organic matter content will retain gas longer than lighter soils

Hazard classification and safety precautions

Hazard Corrosive, Dangerous for the environment, Harmful if swallowed, Harmful if inhaled
Transport code 8
Packaging group III
UN Number 3267
Risk phrases H317, R31, R34
Operator protection A, C, H, M; U02a, U04a, U05a, U08, U11, U16a, U19a, U20c
Environmental protection E02a (24 hours); E15a, E34, E38, H410
Storage and disposal D01, D02, D09b, D10a, D12b
Medical advice M03, M04a

326   Metarhizium anisopliae

A naturally occurring insect parasitic fungus, Metarhizium anisopliae var. anisopliae strain F52
IRAC mode of action code: 11

Products

1	Met52 granular bioinsecticide	Fargro	2% w/w	GR	15168
2	Met52 OD	Fargro	104.8 g/l	OD	17367

Uses

- Cabbage root fly in *vegetable brassicas (off-label)* [1]
- Insect pests in *bulb onions, garlic, leeks, protected aubergines, protected chilli peppers, protected courgettes, protected cucumbers, protected melons, protected nursery fruit trees, protected ornamentals, protected peppers, protected pumpkins, protected strawberries, protected summer squash, protected tomatoes, protected winter squash, salad onions, shallots* [2]
- Leatherjackets in *bilberries (off-label), cranberries (off-label), loganberries (off-label), ribes hybrids (off-label), rubus hybrids (off-label), table grapes (off-label), wine grapes (off-label)* [1]
- Lettuce root aphid in *herbs (see appendix 6) (off-label), leafy vegetables (off-label)* [1]
- Midges in *amenity vegetation (off-label), bilberries (off-label), blueberries (off-label), container-grown ornamentals (off-label), cranberries (off-label), herbs (see appendix 6) (off-label), leafy vegetables (off-label), loganberries (off-label), ornamental plant production (off-label), ribes hybrids (off-label), rubus hybrids (off-label), table grapes (off-label), top fruit (off-label), wine grapes (off-label)* [1]
- Sciarid flies in *amenity vegetation (off-label), bilberries (off-label), blueberries (off-label), container-grown ornamentals (off-label), cranberries (off-label), edible fungi (off-label), herbs (see appendix 6) (off-label), leafy vegetables (off-label), loganberries (off-label), ornamental plant production (off-label), ribes hybrids (off-label), rubus hybrids (off-label), table grapes (off-label), top fruit (off-label), wine grapes (off-label)* [1]
- Thrips in *amenity vegetation (off-label), bilberries (off-label), blueberries (off-label), container-grown ornamentals (off-label), cranberries (off-label), herbs (see appendix 6) (off-label), leafy vegetables (off-label), loganberries (off-label), ornamental plant production (off-label), ribes hybrids (off-label), rubus hybrids (off-label), table grapes (off-label), top fruit (off-label), wine grapes (off-label)* [1]
- Vine weevil in *amenity vegetation (off-label), blackberries, blackcurrants, blueberries, gooseberries, ornamental plant production, raspberries, redcurrants, strawberries, top fruit (off-label)* [1]

Extension of Authorisation for Minor Use (EAMUs)

- *amenity vegetation 20111568* [1], *20111997* [1]
- *bilberries 20111568* [1], *20111997* [1]
- *blueberries 20111997* [1]
- *container-grown ornamentals 20111997* [1]

- **cranberries** *20111568* [1], *20111997* [1]
- **edible fungi** *20111568* [1]
- **herbs (see appendix 6)** *20111568* [1], *20111997* [1]
- **leafy vegetables** *20111568* [1], *20111997* [1]
- **loganberries** *20111568* [1], *20111997* [1]
- **ornamental plant production** *20111997* [1]
- **ribes hybrids** *20111568* [1], *20111997* [1]
- **rubus hybrids** *20111568* [1], *20111997* [1]
- **table grapes** *20111568* [1], *20111997* [1]
- **top fruit** *20111997* [1]
- **vegetable brassicas** *20111568* [1]
- **wine grapes** *20111568* [1], *20111997* [1]

Approval information

- Metarhizium anisopliae included in Annex 1 under EC Regulation 1107/2009

Efficacy guidance

- Incorporate into growing media at any growth stage

Hazard classification and safety precautions

UN Number N/C
Risk phrases R70
Operator protection A, D, H; U14, U20c
Environmental protection E15b, E34, E38
Storage and disposal D01, D09a, D10a, D12a, D20
Medical advice M03, M05a

---

327   metazachlor

A residual anilide herbicide for use in brassicas, nurseries and forestry
HRAC mode of action code: K3

See also aminopyralid + metazachlor + picloram
clomazone + dimethenamid-p + metazachlor
clomazone + metazachlor
clomazone + metazachlor + napropamide
dimethenamid-p + metazachlor
dimethenamid-p + metazachlor + quinmerac
imazamox + metazachlor
imazamox + metazachlor + quinmerac

Products

1 Clayton Buzz	Clayton	500 g/l	SC	16928
2 Rapsan 500 SC	Belchim	500 g/l	SC	16592
3 Stalwart	UPL Europe	500 g/l	SC	17405
4 Sultan 50 SC	Adama	500 g/l	SC	16680
5 Taza 500	Becesane	500 g/l	SC	17284

Uses

- Annual dicotyledons in **borage for oilseed production** *(off-label)*, **canary flower (echium spp.)** *(off-label)*, **evening primrose** *(off-label)*, **honesty** *(off-label)*, **mustard** *(off-label)*, **spring linseed** *(off-label)*, **winter linseed** *(off-label)* [4]; **broccoli, brussels sprouts, cabbages, calabrese, cauliflowers, nursery fruit trees, ornamental plant production** [3-5]; **spring oilseed rape, winter oilseed rape** [1-5]
- Annual meadow grass in **borage for oilseed production** *(off-label)*, **canary flower (echium spp.)** *(off-label)*, **evening primrose** *(off-label)*, **honesty** *(off-label)*, **mustard** *(off-label)*, **spring linseed** *(off-label)*, **winter linseed** *(off-label)* [4]; **broccoli, brussels sprouts, cabbages, calabrese, cauliflowers, nursery fruit trees, ornamental plant production** [3-5]; **spring oilseed rape, winter oilseed rape** [1-5]

- Blackgrass in **broccoli, brussels sprouts, cabbages, calabrese, cauliflowers, nursery fruit trees, ornamental plant production** [3-5]; **spring oilseed rape, winter oilseed rape** [2-5]
- Chickweed in **choi sum** *(off-label)*, **collards** *(off-label)*, **kale** *(off-label)*, **kohlrabi** *(off-label)*, **oriental cabbage** *(off-label)*, **tatsoi** *(off-label)* [4]; **spring oilseed rape, winter oilseed rape** [1]
- Groundsel in **choi sum** *(off-label)*, **collards** *(off-label)*, **kale** *(off-label)*, **kohlrabi** *(off-label)*, **oriental cabbage** *(off-label)*, **tatsoi** *(off-label)* [4]
- Mayweeds in **choi sum** *(off-label)*, **collards** *(off-label)*, **kale** *(off-label)*, **kohlrabi** *(off-label)*, **oriental cabbage** *(off-label)*, **tatsoi** *(off-label)* [4]; **spring oilseed rape, winter oilseed rape** [1]
- Shepherd's purse in **choi sum** *(off-label)*, **collards** *(off-label)*, **kale** *(off-label)*, **kohlrabi** *(off-label)*, **oriental cabbage** *(off-label)*, **tatsoi** *(off-label)* [4]

Extension of Authorisation for Minor Use (EAMUs)
- *borage for oilseed production* 20142381 [4]
- *canary flower (echium spp.)* 20142381 [4]
- *choi sum* 20161213 [4]
- *collards* 20161213 [4]
- *evening primrose* 20142381 [4]
- *honesty* 20142381 [4]
- *kale* 20161213 [4]
- *kohlrabi* 20161213 [4]
- *mustard* 20142381 [4]
- *oriental cabbage* 20161213 [4]
- *spring linseed* 20142381 [4]
- *tatsoi* 20161213 [4]
- *winter linseed* 20142381 [4]

Approval information
- Metazachlor is included in Annex I under EC Regulation 1107/2009

Efficacy guidance
- Activity is dependent on root uptake. For pre-emergence use apply to firm, moist, clod-free seedbed
- Some weeds (chickweed, mayweed, blackgrass etc) susceptible up to 2- or 4-leaf stage. Moderate control of cleavers achieved provided weeds not emerged and adequate soil moisture present
- Split pre- and post-emergence treatments recommended for certain weeds in winter oilseed rape on light and/or stony soils
- Effectiveness is reduced on soils with more than 10% organic matter
- Always follow WRAG guidelines for preventing and managing herbicide resistant weeds. See Section 5 for more information

Restrictions
- Maximum number of treatments 1 per crop for spring oilseed rape, swedes, turnips and brassicas; 2 per crop for winter oilseed rape (split dose treatment); 3 per yr for ornamentals, nursery stock, nursery fruit trees, forestry and farm forestry
- Do not use on sand, very light or poorly drained soils
- Do not treat protected crops or spray overall on ornamentals with soft foliage
- Do not spray crops suffering from wilting, pest or disease
- Do not spray broadcast crops or if a period of heavy rain forecast
- When used on nursery fruit trees any fruit harvested within 1 yr of treatment must be destroyed
- Applications shall be limited to a total dose of not more than 1.0 kg metazachlor/ha in a three year period on the same field
- Do not apply in mixture with phosphate liquid fertilisers
- Metazachlor stewardship requires that all autumn applications should be made before the end of September to reduce the risk to water

Crop-specific information
- Latest use: pre-emergence for swedes and turnips; before 10 leaf stage for spring oilseed rape; before end of Jan for winter oilseed rape
- HI brassicas 6 wk

SEE SECTION 3 FOR PRODUCTS ALSO REGISTERED

- On winter oilseed rape may be applied pre-emergence from drilling until seed chits, post-emergence after fully expanded cotyledon stage (GS 1,0) or by split dose technique depending on soil and weeds. See label for details
- On spring oilseed rape may also be used pre-weed-emergence from cotyledon to 10-leaf stage of crop (GS 1,0-1,10)
- With pre-emergence treatment ensure seed covered by 15 mm of well consolidated soil. Harrow across slits of direct-drilled crops
- Ensure brassica transplants have roots well covered and are well established. Direct drilled brassicas should not be treated before 3 leaf stage
- In ornamentals and hardy nursery stock apply after plants established and hardened off as a directed spray or, on some subjects, as an overall spray. See label for list of tolerant subjects. Do not treat plants in containers

Following crops guidance
- Any crop can follow normally harvested treated winter oilseed rape. See label for details of crops which may be planted after spring treatment and in event of crop failure

Environmental safety
- Dangerous for the environment
- Very toxic to aquatic organisms
- Keep livestock out of treated areas until foliage of any poisonous weeds such as ragwort has died and become unpalatable
- Keep livestock out of treated areas of swede and turnip for at least 5 wk following treatment
- Some pesticides pose a greater threat of contamination of water than others and metazachlor is one of these pesticides. Take special care when applying metazachlor near water and do not apply if heavy rain is forecast.
- Metazachlor stewardship guidelines advise a maximum dose of 750 g.a.i/ha/annum. Applications to drained fields should be complete by 15th Oct but, if drains are flowing, complete applications by 1st Oct.
- LERAP Category B [1, 3-5]

Hazard classification and safety precautions
Hazard Harmful, Dangerous for the environment [1-5]; Harmful if swallowed [1, 3-5]; Very toxic to aquatic organisms [3]
Transport code 9
Packaging group III
UN Number 3082
Risk phrases H317, H351
Operator protection A, H [1-5]; C, M [2-5]; U05a, U19a [1-5]; U08, U20b [1, 3-5]; U09a, U20a [2]; U14, U15 [3-5]
Environmental protection E06a [1-3, 5] (5 wk for swedes, turnips); E07a [1-3, 5]; E15a, E34, H410 [1-5]; E16a [1, 3-5]; E38 [1]
Consumer protection C02a [2] (5 weeks for swedes and turnips)
Storage and disposal D01, D02, D05, D09a, D10c [1-5]; D12a [1]; D12b [3-5]
Medical advice M03 [1-5]; M05a [1]

328 metazachlor + quinmerac

A residual herbicide mixture for oilseed rape
HRAC mode of action code: K3 + O

See also quinmerac

Products

1 Legion	Adama	375:125 g/l	SC	16249
2 Metazerac	Becesane	375:125 g/l	SC	17744
3 Naspar Extra	Belchim	375:125 g/l	SC	17038
4 Parsan Extra	Belchim	375:125 g/l	SC	17050

FOR FULL CONDITIONS OF USE ALWAYS READ THE PRODUCT LABEL

Uses

- Annual dicotyledons in **spring oilseed rape** [3, 4]; **winter oilseed rape** [1-4]
- Annual meadow grass in **spring oilseed rape** [3, 4]; **winter oilseed rape** [1-4]
- Blackgrass in **corn gromwell** *(off-label)* [1]; **spring oilseed rape** [3, 4]; **winter oilseed rape** [1-4]
- Chickweed in **corn gromwell** *(off-label)* [1]
- Cleavers in **corn gromwell** *(off-label)* [1]; **spring oilseed rape** [3, 4]; **winter oilseed rape** [1-4]
- Mayweeds in **corn gromwell** *(off-label)* [1]; **spring oilseed rape** [3, 4]; **winter oilseed rape** [1-4]
- Poppies in **corn gromwell** *(off-label)* [1]; **spring oilseed rape** [3, 4]; **winter oilseed rape** [1-4]
- Speedwells in **corn gromwell** *(off-label)* [1]

Extension of Authorisation for Minor Use (EAMUs)

- *corn gromwell 20162047* [1]

Approval information

- Metazachlor and quinmerac included in Annex I under EC Regulation 1107/2009

Efficacy guidance

- Activity is dependent on root uptake. Pre-emergence treatments should be applied to firm moist seedbeds. Applications to dry soil do not become effective until after rain has fallen
- Maximum activity achieved from treatment before weed emergence for some species
- Weed control may be reduced if excessive rain falls shortly after application especially on light soils
- May be used on all soil types except Sands, Very Light Soils, and soils containing more than 10% organic matter. Crop vigour and/or plant stand may be reduced on brashy and stony soils

Restrictions

- Maximum total dose equivalent to one full dose treatment
- Damage may occur in waterlogged conditions. Do not use on poorly drained soils
- Do not treat stressed crops. In frosty conditions transient scorch may occur
- Applications shall be limited to a total dose of not more than 1.0 kg metazachlor/ha in a three year period on the same field
- Do not apply in mixture with phosphate liquid fertilisers
- Metazachlor stewardship requires that all autumn applications should be made before the end of September to reduce the risk to water

Crop-specific information

- Latest use: end Jan in yr of harvest
- To ensure crop safety it is essential that crop seed is well covered with soil to 15 mm. Loose or puffy seedbeds must be consolidated before treatment. Do not use on broadcast crops
- Crop vigour and possibly plant stand may be reduced if excessive rain falls shortly after treatment especially on light soils

Following crops guidance

- In the event of crop failure after use, wheat or barley may be sown in the autumn after ploughing to 15 cm. Spring cereals or brassicas may be planted after ploughing in the spring
- For all situations following or rotational crops must not be planted until four months after application

Environmental safety

- Dangerous for the environment
- Very toxic to aquatic organisms
- Keep livestock out of treated areas until foliage of any poisonous weeds such as ragwort has died and become unpalatable
- To reduce risk of movement to water do not apply to dry soil or if heavy rain is forecast. On clay soils create a fine consolidated seedbed
- Some pesticides pose a greater threat of contamination of water than others and metazachlor is one of these pesticides. Take special care when applying metazachlor near water and do not apply if heavy rain is forecast.
- Metazachlor stewardship guidelines advise a maximum dose of 750 g.a.i/ha/annum. Applications to drained fields should be complete by 15th Oct but, if drains are flowing, complete applications by 1st Oct.
- LERAP Category B

SEE SECTION 3 FOR PRODUCTS ALSO REGISTERED

Hazard classification and safety precautions

Hazard Irritant, Dangerous for the environment [1-4]; Very toxic to aquatic organisms [3, 4]
Transport code 9
Packaging group III
UN Number 3082
Risk phrases H317, R43 [3, 4]; H351 [1-4]
Operator protection A; U05a, U08, U14, U19a, U20b
Environmental protection E07a, E15a, E16a, E38, H410
Storage and disposal D01, D02, D08, D09a, D10c, D12a
Medical advice M05a

329 metconazole

A triazole fungicide for cereals and oilseed rape
FRAC mode of action code: 3

See also boscalid + metconazole
epoxiconazole + metconazole
fluxapyroxad + metconazole
mepiquat chloride + metconazole

Products

1	Ambarac	Life Scientific	60 g/l	EC	17971
2	Caramba 90	BASF	90 g/l	EC	15524
3	Gringo	AgChem Access	60 g/l	EC	15774
4	Juventus	BASF	90 g/l	EC	15528
5	Metcostar	Life Scientific	60 g/l	EC	17446
6	Redstar	AgChem Access	90 g/l	SL	15938
7	Sirena	Belchim	60 g/l	EC	17103
8	Sunorg Pro	BASF	90 g/l	EC	15433

Uses

- Alternaria in *spring oilseed rape*, *winter oilseed rape* [1-8]
- Ascochyta in *combining peas* (reduction), *lupins* (qualified minor use), *vining peas* (reduction) [1-8]
- Botrytis in *combining peas* (reduction), *lupins* (qualified minor use), *vining peas* (reduction) [1-8]
- Brown rust in *durum wheat*, *rye*, *spring barley*, *triticale*, *winter barley*, *winter wheat* [1-8]; *spring wheat* [2, 4, 6, 8]
- Disease control in *broad beans* (off-label) [8]; *spring linseed* (off-label), *winter linseed* (off-label) [4, 8]
- Fusarium ear blight in *durum wheat* (reduction) [1, 3-5, 7, 8]; *rye* (reduction) [1-3, 5-8]; *spring wheat* (reduction) [2, 4, 6, 8]; *triticale* (reduction), *winter wheat* (reduction) [1-8]
- Growth regulation in *winter oilseed rape* [8]
- Light leaf spot in *spring oilseed rape*, *winter oilseed rape* [1-3, 5-8]; *spring oilseed rape* (reduction), *winter oilseed rape* (reduction) [4]
- Mycosphaerella in *combining peas* (reduction), *vining peas* (reduction) [1-8]
- Net blotch in *durum wheat* (reduction) [2]; *spring barley* (reduction), *winter barley* (reduction) [1-8]
- Phoma in *spring oilseed rape* (reduction), *winter oilseed rape* (reduction) [1-8]
- Powdery mildew in *durum wheat* (moderate control), *rye* (moderate control), *spring barley* (moderate control), *triticale* (moderate control), *winter wheat* (moderate control) [1-8]; *spring wheat* (moderate control) [2, 4, 6, 8]; *winter barley* [1-3, 5-8]; *winter barley* (moderate control) [4]
- Rhynchosporium in *rye* (moderate control), *spring barley* (reduction), *winter barley* (reduction) [4]; *spring barley*, *winter barley* [1-3, 5-8]
- Rust in *combining peas*, *lupins* (qualified minor use), *spring field beans*, *vining peas*, *winter field beans* [1-8]

- Septoria leaf blotch in **durum wheat** [1, 3-5, 7, 8]; **rye** [1-3, 5-8]; **spring wheat** [2, 4, 6, 8]; **triticale**, **winter wheat** [1-8]
- Yellow rust in **durum wheat** [1, 3-5, 7, 8]; **rye**, **triticale**, **winter wheat** [1-8]; **spring barley**, **winter barley** [4]; **spring wheat** [2, 4, 6, 8]

Extension of Authorisation for Minor Use (EAMUs)
- **broad beans** *20111928* [8]
- **spring linseed** *20130387* [4], *20111929* [8]
- **winter linseed** *20130387* [4], *20111929* [8]

Approval information
- Metconazole included in Annex I under EC Regulation 1107/2009
- Accepted by BBPA for use on malting barley

Efficacy guidance
- Best results from application to healthy, vigorous crops when disease starts to develop
- Good spray cover of the target is essential for best results. Spray volume should be increased to improve spray penetration into dense crops
- Metconazole is a DMI fungicide. Resistance to some DMI fungicides has been identified in Septoria leaf blotch which may seriously affect performance of some products. For further advice contact a specialist advisor and visit the Fungicide Resistance Action Group (FRAG)-UK website

Restrictions
- Maximum total dose equivalent to two full dose treatments on all crops
- Do not apply to oilseed rape crops that are damaged or stressed from previous treatments, adverse weather, nutrient deficiency or pest attack. Spring application may lead to reduction of crop height
- The addition of adjuvants is neither advised nor necessary and can lead to enhanced growth regulatory effects on stressed crops of oilseed rape
- Ensure sprayer is free from residues of previous treatments that may harm the crop, especially oilseed rape. Use of a detergent cleaner is advised before and after use
- Do not apply with pyrethroid insecticides on oilseed rape at flowering
- Maintain an interval of at least 14 days between applications to oilseed rape, peas, beans and lupins, 21 days between applications to cereals [4]
- To protect birds, only one application is allowed on cereals before end of tillering (GS29) [4]

Crop-specific information
- Latest use: up to and including milky ripe stage (GS 71) for cereals; 10% pods at final size for oilseed rape
- HI: 14 d for peas, field beans, lupins
- Spring treatments on oilseed rape can reduce the height of the crop
- Treatment for Septoria leaf spot in wheat should be made before second node detectable stage and when weather favouring development of the disease has occurred. If conditions continue to favour disease development a follow-up treatment may be needed
- Treat mildew infections in cereals before 3% infection on any green leaf. A specific mildewicide will improve control of established infections
- Treat yellow rust on wheat before 1% infection on any leaf or as preventive treatment after GS 39
- For brown rust in cereals spray susceptible varieties before any of top 3 leaves has more than 2% infection
- Peas should be treated at the start of flowering and repeat 3-4 wk later if required
- Field beans and lupins should be treated at petal fall and repeat 3-4 wk later if required

Following crops guidance
- Only cereals, oilseed rape, sugar beet, linseed, maize, clover, beans, peas, carrots, potatoes or onions may be sown as following crops after treatment

Environmental safety
- Dangerous for the environment
- Very toxic to aquatic organisms

SEE SECTION 3 FOR PRODUCTS ALSO REGISTERED

- Avoid treatment close to field boundary, even if permitted by LERAP assessment, to reduce effects on non-target insects or other arthropods
- LERAP Category B

Hazard classification and safety precautions

Hazard Harmful [1, 3-5, 7]; Irritant [2, 6, 8]; Flammable [1, 3, 5, 7]; Dangerous for the environment [1-8]; Flammable liquid and vapour [1, 5, 7]; Very toxic to aquatic organisms [1, 5]
Transport code 3 [1, 3, 5, 7]; 9 [2, 4, 6, 8]
Packaging group III
UN Number 1993 [1, 3, 5, 7]; 3082 [2, 4, 6, 8]
Risk phrases H304, H315, H317, H318, H335 [1, 5, 7]; H319, H373 [2, 4, 8]; H361 [1, 2, 4, 5, 7, 8]; R22b, R38, R41, R43, R50 [3]; R36, R51 [6]; R53a [1, 3, 6]; R63 [3, 6]
Operator protection A, C [1-8]; H [1, 3-5, 7]; M [1]; U02a, U05a, U20c [1-8]; U11, U14 [1, 3, 5, 7]
Environmental protection E15a, E16a, E34, E38 [1-8]; E16b [1, 3-5, 7]; H410 [1, 5, 7]; H411 [2, 4, 8]
Storage and disposal D01, D02, D05, D09a, D12a [1-8]; D06c [2, 4, 6, 8]; D10a [1-3, 5-8]; D10c [4]
Medical advice M03, M05a [4]; M05b [1, 3, 5, 7]

330   methiocarb

A stomach acting carbamate molluscicide and insecticide
IRAC mode of action code: 1A

Products

Mesurol	Bayer CropScience	500 g/l	FS	15311

Uses

- Frit fly in *fodder maize, grain maize, sweetcorn*

Approval information

- Methiocarb included in Annex I under EC Regulation 1107/2009
- In 2006 CRD required that all products containing this active ingredient should carry the following warning in the main area of the container label: "Methiocarb is an anticholinesterase carbamate. Handle with care"
- Accepted by BBPA for use on malting barley

Efficacy guidance

- See label for details of suitable application equipment

Restrictions

- This product contains an anticholinesterase carbamate compound. Do not use if under medical advice not to work with such compounds

Environmental safety

- Dangerous for the environment
- Very toxic to aquatic organisms
- Harmful to aquatic organisms
- Dangerous to game, wild birds and animals
- Admixed seed should be drilled and not broadcast, and may not be applied from the air

Hazard classification and safety precautions

Hazard Toxic, Dangerous for the environment, Toxic if swallowed, Very toxic to aquatic organisms
Transport code 6.1
Packaging group III
UN Number 2992
Operator protection A, D, H; U05a, U14, U19d
Environmental protection E10a, E15a, E22b, H410
Storage and disposal D01, D02, D09a, D10d, D12b
Treated seed S01, S02, S03, S04a, S04b, S05, S06a, S07, S08
Medical advice M04a

FOR FULL CONDITIONS OF USE ALWAYS READ THE PRODUCT LABEL

331   methoxyfenozide

A moulting accelerating diacylhydrazine insecticide
IRAC mode of action code: 18A

Products

| 1 | Agrovista Trotter | Agrovista | 240 g/l | SC | 14236 |
| 2 | Runner | Landseer | 240 g/l | SC | 15629 |

Uses
- Caterpillars in *aubergines* *(off-label)*, *chillies* *(off-label)*, *peppers* *(off-label)*, *tomatoes* *(off-label)* [2]
- Codling moth in *apples*, *pears* [1, 2]
- Insect pests in *aubergines* *(off-label)*, *chillies* *(off-label)*, *hops* *(off-label)*, *ornamental plant production* *(off-label)*, *peppers* *(off-label)*, *soft fruit* *(off-label)*, *tomatoes* *(off-label)* [2]
- Plum fruit moth in *plums* *(off-label)* [2]
- Tortrix moths in *apples*, *pears* [1, 2]; *table grapes* *(off-label)*, *wine grapes* *(off-label)* [2]
- Winter moth in *apples*, *pears* [1, 2]

Extension of Authorisation for Minor Use (EAMUs)
- *aubergines* *20170901* [2]
- *chillies* *20170901* [2]
- *hops* *20120448* [2]
- *ornamental plant production* *20120448* [2]
- *peppers* *20170901* [2]
- *plums* *20161694* [2]
- *soft fruit* *20120448* [2]
- *table grapes* *20161694* [2]
- *tomatoes* *20170901* [2]
- *wine grapes* *20161694* [2]

Approval information
- Methoxyfenozide included in Annex 1 under EC Regulation 1107/2009

Efficacy guidance
- To achieve best results uniform coverage of the foliage and full spray penetration of the leaf canopy is important, particularly when spraying post-blossom
- For maximum effectiveness on winter moth and tortrix spray pre-blossom when first signs of active larvae are seen, followed by a further spray in June if larvae of the summer generation are present
- For codling moth spray post-blossom to coincide with early to peak egg deposition. Follow-up treatments will normally be needed
- Methoxyfenozide is a moulting accelerating compound (MAC) and may be used in an anti-resistance strategy with other top fruit insecticides (including chitin biosynthesis inhibitors and juvenile hormones) which have a different mode of action
- To reduce further the likelihood of resistance development use at full recommended dose in sufficient water volume to achieve required spray penetration

Restrictions
- Maximum number of treatments 3 per yr but no more than two should be sprayed consecutively

Crop-specific information
- HI 14 d for apples, pears

Environmental safety
- Risk to non-target insects or other arthropods
- Broadcast air-assisted LERAP (5 m)

Hazard classification and safety precautions
UN Number N/C
Operator protection U20b
Environmental protection E15b, E16b, E22c; E17b (5 m)
Storage and disposal D05, D09a, D11a

SEE SECTION 3 FOR PRODUCTS ALSO REGISTERED

332   1-methylcyclopropene

An inhibitor of ethylene production for use in stored apples

Products
SmartFresh	Landseer	3.3% w/w	SP	11799

Uses
- Ethylene inhibition in **apples** *(post-harvest use)*, **pears** *(off-label - post harvest only)*
- Scald in **apples** *(post-harvest use)*, **pears** *(off-label - post harvest only)*
- Storage rots in **broccoli** *(off-label)*, **brussels sprouts** *(off-label)*, **cabbages** *(off-label)*, **calabrese** *(off-label)*, **cauliflowers** *(off-label)*

Extension of Authorisation for Minor Use (EAMUs)
- **broccoli** *20111502*
- **brussels sprouts** *20111503*
- **cabbages** *20111503*
- **calabrese** *20111502*
- **cauliflowers** *20111502*
- **pears** *(post harvest only) 20102424*

Approval information
- Methylcyclopropene included in Annex 1 under EC Regulation 1107/2009

Efficacy guidance
- Best results obtained from treatment of fruit in good condition and of proper quality for long-term storage
- Effects may be reduced in fruit that is in poor condition or ripe prior to storage or harvested late
- Product acts by releasing vapour into store when mixed with water
- Apply as soon as possible after harvest
- Treatment controls superficial scald and maintains fruit firmness and acid content for 3-6 mth in normal air, and 6-9 mth in controlled atmosphere storage
- Ethylene production recommences after removal from storage

Restrictions
- Maximum number of treatments 1 per batch of apples
- Must only be used by suitably trained and competent persons in fumigation operations
- Consult processors before treatment of fruit destined for processing or cider making
- Do not apply in mixture with other products
- Ventilate all areas thoroughly with all refrigeration fans operating at maximum power for at least 15 min before re-entry

Crop-specific information
- Latest use: 7 d after harvest of apples
- Product tested on Granny Smith, Gala, Jonagold, Bramley and Cox. Consult distributor or supplier before treating other varieties

Environmental safety
- Unprotected persons must be kept out of treated stores during the 24 h treatment period
- Prior to application ensure that the store can be properly and promptly sealed

Hazard classification and safety precautions
UN Number N/C
Operator protection U05a
Environmental protection E02a (24 h); E15a, E34
Consumer protection C12
Storage and disposal D01, D02, D07, D14

FOR FULL CONDITIONS OF USE ALWAYS READ THE PRODUCT LABEL

333   metobromuron

A pre-emergence herbicide for weed control in potatoes
HRAC mode of action code: C2

Products

1	Inigo	Belchim	500 g/l	SC	16933
2	Praxim	Belchim	500 g/l	SC	16871
3	Soleto	Belchim	500 g/l	SC	16935

Uses
- Annual dicotyledons in *potatoes*
- Annual meadow grass in *potatoes*
- Charlock in *potatoes*
- Chickweed in *potatoes*
- Fat hen in *potatoes*
- Groundsel in *potatoes*
- Mayweeds in *potatoes*
- Shepherd's purse in *potatoes*

Approval information
- Metobromuron included in Annex I under EC Regulation 1107/2009

Restrictions
- Heavy rain after application may cause transient discolouration of the crop.

Crop-specific information
- May be used on all varieties of potato

Following crops guidance
- Following normal harvest of a treated crop Brassicas (oilseed rape, turnip, brassica vegetables) and sugar beet may be sown, providing the soil has been ploughed. No restrictions apply when other crops are sown.
- In the event of crop failure maize, beans, peas and carrots may be sown after ploughing. It is not recommended to sow Brassicas (oilseed rape, turnip, brassica vegetables) or sugar beet as a replacement crop after crop failure. Potatoes can be re-planted after minimal cultivation.

Hazard classification and safety precautions
Transport code 9
Packaging group III
UN Number 3082
Risk phrases H351, H373
Operator protection A, H, M; U05a, U09a, U19a, U20a
Environmental protection E15b, E34, E38, H410
Storage and disposal D01, D02, D05, D09a, D12b
Medical advice M05a

334   metrafenone

A benzophenone protectant and curative fungicide for cereals
FRAC mode of action code: U8

See also epoxiconazole + fenpropimorph + metrafenone
epoxiconazole + metrafenone
fenpropimorph + metrafenone

Products

1	Flexity	BASF	300 g/l	SC	11775
2	Lexi	AgChem Access	300 g/l	SC	14890
3	Vivando	BASF	500 g/l	SC	18026

SEE SECTION 3 FOR PRODUCTS ALSO REGISTERED

Uses

- Cobweb in **mushroom boxes** [3]
- Disease control/foliar feed in **ornamental plant production** *(off-label)* [1]
- Eyespot in **spring wheat** *(reduction)*, **winter wheat** *(reduction)* [1, 2]
- Powdery mildew in **ornamental plant production** [1]; **spring barley**, **spring oats** *(evidence of mildew control on oats is limited)*, **spring wheat**, **winter barley**, **winter oats** *(evidence of mildew control on oats is limited)*, **winter wheat** [1, 2]

Extension of Authorisation for Minor Use (EAMUs)

- **ornamental plant production** *20082850* [1]

Approval information

- Metrafenone included in Annex I under EC Regulation 1107/2009
- Accepted by BBPA for use on malting barley

Efficacy guidance

- Best results obtained from treatment at the start of foliar disease attack
- Activity against mildew in wheat is mainly protectant with moderate curative control in the latent phase; activity is entirely protectant in barley
- Useful reduction of eyespot in wheat is obtained if treatment applied at GS 30-32
- Should be used as part of a resistance management strategy that includes mixtures or sequences effective against mildew and non-chemical methods

Restrictions

- Maximum number of treatments 2 per crop
- Avoid the use of sequential applications of Flexity unless it is used in tank mixture with other products active against powdery mildew employing a different mode of action

Crop-specific information

- Latest use: beginning of flowering (GS 31) for wheat, barley, oats

Following crops guidance

- Cereals, oilseed rape, sugar beet, linseed, maize, clover, field beans, peas, turnips, carrots, cauliflowers, onions, lettuce or potatoes may follow a treated cereal crop

Environmental safety

- Dangerous for the environment
- Toxic to aquatic organisms

Hazard classification and safety precautions

Hazard Irritant, Dangerous for the environment
Transport code 9 [3]
Packaging group III [3]
UN Number 3082 [3]; N/C [1, 2]
Risk phrases H317, H319 [1]; R43, R51, R53a [2]
Operator protection A, H; U05a, U20b
Environmental protection E15a, E34 [1-3]; H411 [1, 3]
Storage and disposal D01, D02, D05, D09a, D10c
Medical advice M03

---

335   metribuzin

A contact and residual triazinone herbicide for use in potatoes
HRAC mode of action code: C1

See also clomazone + metribuzin
diflufenican + metribuzin
flufenacet + metribuzin
mecoprop-P + metribuzin

Products

1 Cleancrop Frizbee	Agrii	70% w/w	WG	16642

Products – continued

2	Sencorex Flow	Nufarm UK	600 g/l	SL	16167
3	Shotput	Adama	70% w/w	WG	15968

Uses

- Annual dicotyledons in **asparagus** *(off-label)*, **carrots** *(off-label)*, **early potatoes, maincrop potatoes, mallow (althaea spp.)** *(off-label)*, **parsnips** *(off-label)*, **sweet potato** *(off-label)* [1-3]; **celeriac** *(off-label)*, **ornamental plant production** *(off-label)* [2]
- Annual grasses in **asparagus** *(off-label)*, **carrots** *(off-label)*, **mallow (althaea spp.)** *(off-label)*, **parsnips** *(off-label)*, **sweet potato** *(off-label)* [2, 3]; **celeriac** *(off-label)*, **ornamental plant production** *(off-label)* [2]; **early potatoes, maincrop potatoes** [1-3]
- Annual meadow grass in **carrots** *(off-label)*, **mallow (althaea spp.)** *(off-label)*, **parsnips** *(off-label)*, **sweet potato** *(off-label)* [3]; **maincrop potatoes** [1, 3]; **ornamental plant production** *(off-label)*, **protected ornamentals** *(off-label)* [2]
- Charlock in **ornamental plant production** *(off-label)*, **protected ornamentals** *(off-label)* [2]
- Cleavers in **ornamental plant production** *(off-label)*, **protected ornamentals** *(off-label)* [2]
- Common orache in **ornamental plant production** *(off-label)*, **protected ornamentals** *(off-label)* [2]
- Fat hen in **ornamental plant production** *(off-label)*, **protected ornamentals** *(off-label)* [2]
- Field pansy in **ornamental plant production** *(off-label)*, **protected ornamentals** *(off-label)* [2]
- Fool's parsley in **asparagus** *(off-label)*, **sweet potato** *(off-label)* [1, 2]; **carrots** *(off-label)*, **mallow (althaea spp.)** *(off-label)*, **parsnips** *(off-label)* [1-3]; **celeriac** *(off-label)* [2]
- Groundsel in **ornamental plant production** *(off-label)*, **protected ornamentals** *(off-label)* [2]
- Himalayan balsam in **rhubarb** *(off-label)* [2]
- Knotgrass in **ornamental plant production** *(off-label)*, **protected ornamentals** *(off-label)* [2]
- Mayweeds in **ornamental plant production** *(off-label)*, **protected ornamentals** *(off-label)* [2]
- Pale persicaria in **ornamental plant production** *(off-label)*, **protected ornamentals** *(off-label)* [2]
- Perennial dicotyledons in **asparagus** *(off-label)*, **asparagus** *(off-label - from seed)* [3]
- Perennial weeds in **asparagus** *(off-label)*, **sweet potato** *(off-label)* [1]
- Redshank in **ornamental plant production** *(off-label)*, **protected ornamentals** *(off-label)* [2]
- Shepherd's purse in **ornamental plant production** *(off-label)*, **protected ornamentals** *(off-label)* [2]
- Small nettle in **ornamental plant production** *(off-label)*, **protected ornamentals** *(off-label)* [2]
- Volunteer oilseed rape in **asparagus** *(off-label)*, **carrots** *(off-label)*, **celeriac** *(off-label)*, **mallow (althaea spp.)** *(off-label)*, **ornamental plant production** *(off-label)*, **parsnips** *(off-label)*, **protected ornamentals** *(off-label)* [2]; **early potatoes, maincrop potatoes** [1-3]; **sweet potato** *(off-label)* [2, 3]
- Wild mignonette in **asparagus** *(off-label)*, **sweet potato** *(off-label)* [1, 2]; **carrots** *(off-label)*, **mallow (althaea spp.)** *(off-label)*, **parsnips** *(off-label)* [1-3]; **celeriac** *(off-label)* [2]
- Willowherb in **ornamental plant production** *(off-label)*, **protected ornamentals** *(off-label)* [2]

Extension of Authorisation for Minor Use (EAMUs)

- **asparagus** *20151018* [1], *20150917* [2], *(from seed) 20142321* [3], *20142230 expires 31 Aug 2018* [3], *20142321* [3]
- **carrots** *20151017* [1], *20150916* [2], *20142229 expires 31 Aug 2018* [3], *20142320* [3]
- **celeriac** *20150916* [2]
- **mallow (althaea spp.)** *20151017* [1], *20150916* [2], *20142229 expires 31 Aug 2018* [3], *20142320* [3]
- **ornamental plant production** *20131867* [2], *20171732* [2]
- **parsnips** *20151017* [1], *20150916* [2], *20142229 expires 31 Aug 2018* [3], *20142320* [3]
- **protected ornamentals** *20171732* [2]
- **rhubarb** *20152218* [2]
- **sweet potato** *20151019* [1], *20150915* [2], *20142228 expires 31 Aug 2018* [3], *20142252* [3]

Approval information

- Metribuzin included in Annex I under EC Regulation 1107/2009

Efficacy guidance

- Best results achieved on weeds at cotyledon to 1-leaf stage

SEE SECTION 3 FOR PRODUCTS ALSO REGISTERED

- May be applied pre- or post-emergence of crop on named maincrop varieties and cv. Marfona; pre-emergence only on named early varieties
- Apply to moist soil with well-rounded ridges and few clods
- Activity reduced by dry conditions and on soils with high organic matter content
- On fen and moss soils pre-planting incorporation to 10-15 cm gives increased activity. Incorporate thoroughly and evenly
- With named maincrop and second early potato varieties on soils with more than 10% organic matter shallow pre- or post-planting incorporation may be used. See label for details
- Effective control using a programme of reduced doses is made possible by using a spray of smaller droplets, thus improving retention

Restrictions
- Maximum total dose equivalent to one full dose treatment on early potato varieties and one and a third full doses on maincrop varieties
- Only certain varieties may be treated. Apply pre-emergence only on named first earlies, pre- or post-emergence on named second earlies. On named maincrop varieties apply pre-emergence (except for certain varieties on Sands or Very Light soils) or post-emergence. See label
- All post-emergence treatments must be carried out before longest shoots reach 15 cm
- Do not cultivate after treatment
- Some recommended varieties may be sensitive to post-emergence treatment if crop under stress

Crop-specific information
- Latest use: pre-crop emergence for named early potato varieties; before most advanced shoots have reached 15 cm for post-emergence treatment of potatoes
- On stony or gravelly soils there is risk of crop damage, especially if heavy rain falls soon after application
- When days are hot and sunny delay spraying until evening

Following crops guidance
- Ryegrass, cereals or winter beans may be sown in same season provided at least 16 wk elapsed after treatment and ground ploughed to 15 cm and thoroughly cultivated as soon as possible after harvest and no later than end Dec
- In W Cornwall on soil with more than 5% organic matter early potatoes treated as recommended may be followed by summer planted brassica crops provided the soil has been ploughed, spring rainfall has been normal and at least 14 wk have elapsed since treatment
- Do not grow any vegetable brassicas, lettuces or radishes on land treated the previous yr. Other crops may be sown normally in spring of next yr

Environmental safety
- Dangerous for the environment
- Very toxic to aquatic organisms
- Do not empty into drains
- LERAP Category B

Hazard classification and safety precautions
Hazard Harmful, Dangerous for the environment [1-3]; Very toxic to aquatic organisms [1, 3]
Transport code 9
Packaging group III
UN Number 3077 [1, 3]; 3082 [2]
Risk phrases R48, R50, R53a [2]
Operator protection A, H; U05a, U08, U13, U19a [1-3]; U14, U20b [1, 3]; U20a [2]
Environmental protection E15a, E16b, E19b, H410 [1, 3]; E16a [1-3]; E38 [2]
Storage and disposal D01, D02, D09a, D11a, D12a
Medical advice M03 [2]; M05a [1-3]

336 metsulfuron-methyl

A contact and residual sulfonylurea herbicide used in cereals, linseed and set-aside
HRAC mode of action code: B

See also carfentrazone-ethyl + metsulfuron-methyl
diflufenican + metsulfuron-methyl
flupyrsulfuron-methyl + metsulfuron-methyl
fluroxypyr + metsulfuron-methyl
fluroxypyr + metsulfuron-methyl + thifensulfuron-methyl
mecoprop-P + metsulfuron-methyl

SECTION 2

Products

1	Alias SX	DuPont	20% w/w	SG	13398
2	Answer SX	DuPont	20% w/w	SG	15632
3	Deft Premium	Rotam	20% w/w	SG	16272
4	Finy	UPL Europe	20% w/w	WG	16960
5	Gropper SX	DuPont	20% w/w	SG	16290
6	Jubilee SX	DuPont	20% w/w	SG	12203
7	Laya	Life Scientific	20% w/w	SG	17984
8	Lorate	DuPont	20% w/w	SG	12743
9	Metmeth 200	Becesane	20% w/w	SG	15993
10	Savvy	Rotam	20% w/w	WG	14266
11	Savvy Premium	Rotam	20% w/w	SG	14782
12	Simba SX	DuPont	20% w/w	SG	15503
13	Snicket	AgChem Access	20% w/w	SG	15681

Uses

- Amsinkia in **green cover on land temporarily removed from production, linseed, spring barley, spring oats, spring wheat, triticale, winter barley, winter oats, winter wheat** [6]
- Annual dicotyledons in **farm forestry** *(off-label)*, **forest nurseries** *(off-label)*, **ornamental plant production** *(off-label)* [6]; **game cover** *(off-label)*, **miscanthus** *(off-label)* [6, 8]; **green cover on land temporarily removed from production** [4, 6]; **linseed, spring barley, spring oats, spring wheat, triticale, winter barley, winter oats, winter wheat** [1-13]
- Charlock in **green cover on land temporarily removed from production** [6]; **linseed, spring barley, spring oats, spring wheat, triticale, winter barley, winter oats, winter wheat** [3, 6]
- Chickweed in **green cover on land temporarily removed from production** [6]; **linseed, spring barley, spring oats, spring wheat, triticale, winter barley, winter oats, winter wheat** [1-3, 5-13]
- Docks in **green cover on land temporarily removed from production, linseed, spring barley, spring oats, spring wheat, triticale, winter barley, winter oats, winter wheat** [6]
- Green cover in **land temporarily removed from production** [1-3, 5, 7-13]
- Groundsel in **green cover on land temporarily removed from production, linseed, spring barley, spring oats, spring wheat, triticale, winter barley, winter oats, winter wheat** [6]
- Hemp-nettle in **green cover on land temporarily removed from production, linseed, spring barley, spring oats, spring wheat, triticale, winter barley, winter oats, winter wheat** [6]
- Mayweeds in **green cover on land temporarily removed from production** [6]; **linseed, spring barley, spring oats, spring wheat, triticale, winter barley, winter oats, winter wheat** [1-3, 5-13]
- Mouse-ear chickweed in **green cover on land temporarily removed from production, linseed, spring barley, spring oats, spring wheat, triticale, winter barley, winter oats, winter wheat** [6]
- Scarlet pimpernel in **green cover on land temporarily removed from production, linseed, spring barley, spring oats, spring wheat, triticale, winter barley, winter oats, winter wheat** [6]
- Venus looking-glass in **green cover on land temporarily removed from production, linseed, spring barley, spring oats, spring wheat, triticale, winter barley, winter oats, winter wheat** [6]

- Volunteer oilseed rape in *green cover on land temporarily removed from production, linseed, spring barley, spring oats, spring wheat, triticale, winter barley, winter oats, winter wheat* [6]
- Volunteer sugar beet in *green cover on land temporarily removed from production, linseed, spring barley, spring oats, spring wheat, triticale, winter barley, winter oats, winter wheat* [6]

Extension of Authorisation for Minor Use (EAMUs)
- *farm forestry 20082859* [6]
- *forest nurseries 20082859* [6]
- *game cover 20082859* [6], *20082860* [8]
- *miscanthus 20082859* [6], *20082860* [8]
- *ornamental plant production 20082859* [6]

Approval information
- Metsulfuron-methyl included in Annex I under EC Regulation 1107/2009
- Accepted by BBPA for use on malting barley

Efficacy guidance
- Best results achieved on small, actively growing weeds up to 6-true leaf stage. Good spray cover is important
- Weed control may be reduced in dry soil conditions
- Commonly used in tank-mixture on wheat and barley with other cereal herbicides to improve control of resistant dicotyledons (cleavers, fumitory, ivy-leaved speedwell), larger weeds and grasses. See label for recommended mixtures
- Metsulfuron-methyl is a member of the ALS-inhibitor group of herbicides and products should be used in a planned Resistance Management strategy. See Section 5 for more information

Restrictions
- Maximum number of treatments 1 per crop or per yr (for set-aside)
- Product must only be used after 1 Feb
- Do not apply within 7 d of rolling
- Do not use on cereal crops undersown with grass or legumes
- Do not use in tank mixture on oats, triticale or linseed. On linseed allow at least 7 d before or after other treatments
- Do not tank mix with chlorpyrifos. Allow at least 14 d before or after chlorpyrifos treatments
- Consult contract agents before use on a cereal crop for seed
- Do not use on any crop suffering stress from drought, waterlogging, frost, deficiency, pest or disease attack
- Specific restrictions apply to use in sequence or tank mixture with other sulfonylurea or ALS-inhibiting herbicides. See label for details
- Spraying equipment should not be drained or flushed onto land planted, or to be planted, with trees or crops other than cereals and should be thoroughly cleansed after use - see label for instructions
- Must only be applied from 1 February in the year of harvest until the specified latest time of application and must only be applied to green cover on land not being used for crop production where a full green cover is established [10]

Crop-specific information
- Latest use: before flag leaf sheath extending stage for cereals (GS 41); before flower buds visible or crop 30 cm tall for linseed [1, 2, 5, 6, 8, 12]; before 1 Aug in yr of treatment for land not being used for crop production [1, 2, 5, 6, 8, 12]
- Apply after 1 Feb to wheat, oats and triticale from 2-leaf (GS 12), and to barley from 3-leaf (GS 13) until flag-leaf sheath extending (GS 41)
- On linseed allow at least 7 d (10 d if crop is growing poorly or under stress) before or after other treatments
- Use in set-aside when a full green cover is established and made up predominantly of grassland, wheat, barley, oats or triticale. Do not use on seedling grasses

FOR FULL CONDITIONS OF USE ALWAYS READ THE PRODUCT LABEL

Following crops guidance

- Only cereals, oilseed rape, field beans or grass may be sown in same calendar year after treating cereals with the product alone. Other restrictions apply to tank mixtures. See label for details
- Only cereals should be planted within 16 mth of applying to a linseed crop or set-aside
- In the event of failure of a treated crop sow only wheat within 3 mth after treatment

Environmental safety

- Dangerous for the environment
- Very toxic to aquatic organisms
- Take extreme care to avoid damage by drift onto broad-leaved plants outside the target area, onto surface waters or ditches or onto land intended for cropping
- A range of broad leaved species will be fully or partially controlled when used in land temporarily removed from production, hence product may not be suitable where wild flower borders or other forms of conservation headland are being developed
- Before use on land temporarily removed from production as part of grant-aided scheme, ensure compliance with the management rules
- Green cover on land temporarily removed from production must not be grazed by livestock or harvested for human or animal consumption or used for animal bedding [6]
- LERAP Category B [2-6, 9-13]

Hazard classification and safety precautions

Hazard Dangerous for the environment [1-13]; Very toxic to aquatic organisms [1, 2, 5, 6, 8, 9, 12, 13]

Transport code 9

Packaging group III

UN Number 3077

Risk phrases H315, H318 [3, 11]

Operator protection A, C, H [3]; U04b, U11 [3]; U05a [4]; U19a [10]; U20b [3, 10]

Environmental protection E15a [10, 11]; E15b [1-9, 12, 13]; E16a [2-6, 9-13]; E16b [2]; E23 [4]; E38, H410 [1-13]

Storage and disposal D01, D02 [4]; D09a [2-4, 10, 11]; D10b [3, 10, 11]; D11a [2, 4]; D12a [1-3, 5-13]; D12b [3, 4]

337   metsulfuron-methyl + thifensulfuron-methyl

A contact residual and translocated sulfonylurea herbicide mixture for use in cereals
HRAC mode of action code: B + B

See also thifensulfuron-methyl

Products

1	Assuage	Becesane	6.8:68.2% w/w	SG	15801
2	Avro SX	DuPont	2.9:42.9 % w/w	SG	14313
3	Chimera SX	DuPont	2.9:42.9% w/w	SG	13171
4	Concert SX	DuPont	4:40% w/w	SG	12288
5	Ergon	Rotam	6.8:68.2% w/w	WG	14382
6	Finish SX	DuPont	6.7:33.3% w/w	SG	12259
7	Harmony M SX	DuPont	4:40% w/w	SG	12258
8	Mozaic SX	DuPont	4:40% w/w	SG	16108
9	Pennant	Headland	4:40% w/w	SG	13350
10	Presite SX	DuPont	6.7:33.3% w/w	SG	12291
11	Refine Max SX	DuPont	6.7:33.3% w/w	SG	15622

Uses

- Annual dicotyledons in *durum wheat* (off-label) [10]; *miscanthus* (off-label) [3, 7]; *spring barley, spring wheat, winter wheat* [1-11]; *spring oats, triticale, winter rye* [5]; *winter barley* [2, 3, 6, 10, 11]; *winter oats* [6, 10, 11]
- Black bindweed in *spring barley, spring wheat, winter barley, winter wheat* [2]
- Charlock in *spring barley, spring wheat, winter wheat* [1, 2, 5]; *spring oats, triticale, winter rye* [5]; *winter barley* [2]

SEE SECTION 3 FOR PRODUCTS ALSO REGISTERED

- Chickweed in **durum wheat** *(off-label)* [10]; **spring barley, spring wheat, winter wheat** [1-11]; **spring oats, triticale, winter rye** [5]; **winter barley** [2, 3, 6, 10, 11]; **winter oats** [6, 10, 11]
- Creeping thistle in **spring barley, spring wheat, winter barley, winter wheat** [2, 3]
- Fat hen in **spring barley, spring wheat, winter barley, winter wheat** [2]
- Field pansy in **durum wheat** *(off-label)* [10]; **spring barley, spring wheat, winter barley, winter wheat** [2, 6, 10, 11]; **winter oats** [6, 10, 11]
- Field speedwell in **durum wheat** *(off-label)* [10]; **spring barley, spring wheat, winter barley, winter wheat** [2, 6, 10, 11]; **winter oats** [6, 10, 11]
- Forget-me-not in **spring barley, spring wheat, winter barley, winter wheat** [2]
- Hemp-nettle in **spring barley, spring wheat, winter wheat** [1, 2, 5]; **spring oats, triticale, winter rye** [5]; **winter barley** [2]
- Ivy-leaved speedwell in **durum wheat** *(off-label)* [10]; **spring barley, spring wheat, winter barley, winter oats, winter wheat** [6, 10, 11]
- Knotgrass in **durum wheat** *(off-label)* [10]; **spring barley, spring wheat, winter barley, winter wheat** [2, 6, 10, 11]; **winter oats** [6, 10, 11]
- Mayweeds in **durum wheat** *(off-label)* [10]; **spring barley, spring wheat, winter wheat** [1-11]; **spring oats, triticale, winter rye** [5]; **winter barley** [2, 3, 6, 10, 11]; **winter oats** [6, 10, 11]
- Parsley-piert in **spring barley, spring wheat, winter wheat** [1, 5]; **spring oats, triticale, winter rye** [5]
- Polygonums in **spring barley, spring wheat, winter wheat** [3, 4, 7-9]; **winter barley** [3]
- Poppies in **spring barley, spring wheat, winter barley, winter wheat** [2]
- Red dead-nettle in **spring barley, spring wheat, winter wheat** [1-5, 7-9]; **spring oats, triticale, winter rye** [5]; **winter barley** [2, 3]
- Redshank in **spring barley, spring wheat, winter wheat** [1, 2, 5]; **spring oats, triticale, winter rye** [5]; **winter barley** [2]
- Shepherd's purse in **spring barley, spring wheat, winter wheat** [1-5, 7-9]; **spring oats, triticale, winter rye** [5]; **winter barley** [2, 3]

Extension of Authorisation for Minor Use (EAMUs)
- **durum wheat** *20070452* [10]
- **miscanthus** *20082835* [3], *20082916* [7]

Approval information
- Metsulfuron-methyl and thifensulfuron-methyl included in Annex I under EC Regulation 1107/2009
- Accepted by BBPA for use on malting barley

Efficacy guidance
- Best results by application to small, actively growing weeds up to 6-true leaf stage
- Ensure good spray cover
- Susceptible weeds stop growing almost immediately but symptoms may not be visible for about 2 wk
- Effectiveness may be reduced by heavy rain or if soil conditions very dry
- Metsulfuron-methyl and thifensulfuron-methyl are members of the ALS-inhibitor group of herbicides and products should be used in a planned Resistance Management strategy. See Section 5 for more information

Restrictions
- Maximum number of treatments 1 per crop
- Products may only be used after 1 Feb
- Do not use on any crop suffering stress from drought, waterlogging, frost, deficiency, pest or disease attack or any other cause
- Do not use on crops undersown with grasses, clover or legumes, or any other broad leaved crop
- Specific restrictions apply to use in sequence or tank mixture with other sulfonylurea or ALS-inhibiting herbicides. See label for details
- Do not apply within 7 d of rolling
- Consult contract agents before use on a cereal crop grown for seed

Crop-specific information
- Latest use: before flag leaf sheath extending (GS 39) for barley and wheat, before 2nd node (GS32) for winter oats

FOR FULL CONDITIONS OF USE ALWAYS READ THE PRODUCT LABEL

Following crops guidance
- Only cereals, oilseed rape, field beans or grass may be sown in same calendar year after treatment
- Additional constraints apply after use of certain tank mixtures. See label
- In the event of crop failure sow only winter wheat within 3 mth after treatment and after ploughing and cultivating to a depth of at least 15 cm

Environmental safety
- Dangerous for the environment
- Very toxic to aquatic organisms
- Take extreme care to avoid damage by drift onto broad-leaved plants outside the target area, or onto ponds, waterways or ditches
- Spraying equipment should not be drained or flushed onto land planted, or to be planted, with trees or crops other than cereals and should be thoroughly cleansed after use - see label for instructions
- LERAP Category B

Hazard classification and safety precautions
Hazard Dangerous for the environment [1-11]; Very toxic to aquatic organisms [2-4, 6-11]
Transport code 9
Packaging group III
UN Number 3077
Risk phrases R50, R53a [1]
Operator protection U05a [6, 10, 11]; U08, U20b [2-4, 6-11]; U19a [1-11]; U20c [1, 5]
Environmental protection E15a [6, 10, 11]; E15b [1, 4, 5, 7-9]; E16a [1-11]; E38 [2-4, 6-11]; H410 [2-11]
Storage and disposal D01, D02 [6, 10, 11]; D09a [1-11]; D10b [1, 5]; D11a [4, 6-11]; D12a [2-4, 6-11]

---

338 metsulfuron-methyl + tribenuron-methyl

A sulfonylurea herbicide mixture for cereals
HRAC mode of action code: B + B

See also tribenuron-methyl

Products

1	Ally Max SX	DuPont	14.3:14.3% w/w	SG	14835
2	BiPlay SX	DuPont	11.1:22.2% w/w	SG	14836
3	Boudha	Rotam	25:25% w/w	WG	15158
4	DP911 SX	DuPont	11.1:22.2% w/w	SG	17493
5	Traton SX	DuPont	11.1:22.2% w/w	SG	14837

Uses
- Annual dicotyledons in *durum wheat (off-label)*, *rye (off-label)*, *spring rye (off-label)* [1]; *spring barley, spring wheat, triticale, winter barley, winter oats, winter wheat* [1-5]; *spring oats* [1, 3]; *winter rye* [3]
- Chickweed in *spring barley, spring wheat, triticale, winter barley, winter oats, winter wheat* [1-3, 5]; *spring oats* [1, 3]; *winter rye* [3]
- Field pansy in *spring barley, spring wheat, triticale, winter barley, winter oats, winter wheat* [1, 2, 5]
- Field speedwell in *spring oats* [1]
- Hemp-nettle in *spring barley, spring wheat, triticale, winter barley, winter oats, winter wheat* [1, 2, 5]; *spring oats* [1]
- Mayweeds in *spring barley, spring wheat, triticale, winter barley, winter oats, winter wheat* [1-3, 5]; *spring oats* [1, 3]; *winter rye* [3]
- Poppies in *spring barley, spring oats, spring wheat, triticale, winter barley, winter oats, winter rye, winter wheat* [3]
- Red dead-nettle in *spring barley, spring wheat, triticale, winter barley, winter oats, winter wheat* [1-3, 5]; *spring oats* [1, 3]; *winter rye* [3]

SEE SECTION 3 FOR PRODUCTS ALSO REGISTERED

- Shepherd's purse in **spring barley**, **spring oats**, **spring wheat**, **triticale**, **winter barley**, **winter oats**, **winter rye**, **winter wheat** [3]
- Volunteer oilseed rape in **spring barley**, **spring oats**, **spring wheat**, **triticale**, **winter barley**, **winter oats**, **winter wheat** [1, 3]; **winter rye** [3]
- Volunteer sugar beet in **spring barley**, **spring oats**, **spring wheat**, **triticale**, **winter barley**, **winter oats**, **winter wheat** [1]

Extension of Authorisation for Minor Use (EAMUs)
- **durum wheat** *20101080* [1]
- **rye** *20101080* [1]
- **spring rye** *20101080* [1]

Approval information
- Metsulfuron-methyl and tribenuron-methyl included in Annex I under EC Regulation 1107/2009
- Accepted by BBPA for use on malting barley

Efficacy guidance
- Best results obtained when applied to small actively growing weeds
- Product acts by foliar and root uptake. Good spray cover essential but performance may be reduced when soil conditions are very dry and residual effects may be reduced by heavy rain
- Weed growth inhibited within hours of treatment and many show marked colour changes as they die back. Full effects may not be apparent for up to 4 wk
- Metsulfuron-methyl and tribenuron-methyl are members of the ALS-inhibitor group of herbicides and products should be used in a planned Resistance Management strategy. See Section 5 for more information

Restrictions
- Maximum number of treatments 1 per crop
- Product must only be used after 1 Feb and after crop has three leaves
- Do not apply to a crop suffering from drought, waterlogging, low temperatures, pest or disease attack, nutrient deficiency, soil compaction or any other stress
- Do not use on crops undersown with grasses, clover or other legumes
- Do not apply within 7 d of rolling
- Specific restrictions apply to use in sequence or tank mixture with other sulfonylurea or ALS-inhibiting herbicides. See label for details

Crop-specific information
- Latest use: before flag leaf sheath extending

Following crops guidance
- Only cereals, field beans, grass or oilseed rape may be sown in the same calendar yr as harvest of a treated crop [1, 2, 5]
- In the event of failure of a treated crop only winter wheat may be sown within 3 mth of treatment, and only after ploughing and cultivation to 15 cm minimum

Environmental safety
- Dangerous for the environment
- Very toxic to aquatic organisms
- Some non-target crops are highly sensitive. Take extreme care to avoid drift outside the target area, or onto ponds, waterways or ditches
- Spraying equipment should be thoroughly cleaned in accordance with manufacturer's instructions
- LERAP Category B [2-5]

Hazard classification and safety precautions
Hazard Irritant [1, 2, 4, 5]; Dangerous for the environment [1-5]; Very toxic to aquatic organisms [4]
Transport code 9
Packaging group III
UN Number 3077
Risk phrases R43, R50, R53a [1, 2, 5]
Operator protection A, H [1, 2, 5]; U08, U19a [1-5]; U20b [1, 2, 4, 5]

FOR FULL CONDITIONS OF USE ALWAYS READ THE PRODUCT LABEL

Environmental protection E15a [1, 2, 4, 5]; E15b, E39 [3]; E16a [2-5]; E38 [1-3, 5]; H410 [3, 4]
Storage and disposal D01, D02, D09a, D11a, D12a

339 1-naphthylacetic acid

A plant growth regulator to promote rooting of cuttings

See also 4-indol-3-ylbutyric acid + 1-naphthylacetic acid

Products

Fixor	Belcrop	100 g/l	SL	17428

Uses
• Fruit thinning in **apples**

Approval information
• 1-naphthylacetic acid included in Annex I under EC Regulation 1107/2009

Efficacy guidance
• One application should be made when the fruit have a diameter between 8 and 12 mm (BBCH 71)
• Dose rate: Max 150 mL product/ha in 1000L water
• Adapt the spray volume to the tree size and canopy density to ensure a thorough fruit and leaf coverage

Hazard classification and safety precautions
UN Number N/C
Risk phrases H318, H361
Operator protection U20c
Environmental protection E15a
Storage and disposal D05, D09a, D11a

340 napropamide

A soil applied alkanamide herbicide for oilseed rape, fruit and woody ornamentals
HRAC mode of action code: K3

See also clomazone + metazachlor + napropamide
clomazone + napropamide

Products

1	AC 650	UPL Europe	450 g/l	SC	17966
2	Colzamid	UPL Europe	450 g/l	SC	17967
3	Devrinol	UPL Europe	450 g/l	SC	17853

Uses
• Amaranthus in **baby leaf crops** *(off-label)*, **celeriac** *(off-label)*, **choi sum** *(off-label)*, **collards** *(off-label)*, **horseradish** *(off-label)*, **lamb's lettuce** *(off-label)*, **oriental cabbage** *(off-label)*, **protected baby leaf crops** *(off-label)*, **protected lamb's lettuce** *(off-label)*, **protected red mustard** *(off-label)*, **protected rocket** *(off-label)*, **red mustard** *(off-label)*, **rocket** *(off-label)*, **swedes** *(off-label)*, **turnips** *(off-label)* [3]
• Annual dicotyledons in **broccoli**, **brussels sprouts**, **cabbages**, **calabrese**, **cauliflowers**, **kale** [2, 3]; **winter oilseed rape** [1-3]
• Annual grasses in **broccoli**, **brussels sprouts**, **cabbages**, **calabrese**, **cauliflowers**, **kale** [2, 3]; **winter oilseed rape** [1-3]
• Annual meadow grass in **baby leaf crops** *(off-label)*, **borage for oilseed production** *(off-label)*, **celeriac** *(off-label)*, **choi sum** *(off-label)*, **collards** *(off-label)*, **evening primrose** *(off-label)*, **game cover** *(off-label)*, **horseradish** *(off-label)*, **lamb's lettuce** *(off-label)*, **linseed** *(off-label)*, **mustard** *(off-label)*, **oriental cabbage** *(off-label)*, **protected baby leaf crops** *(off-label)*, **protected lamb's lettuce** *(off-label)*, **protected red mustard** *(off-label)*, **protected rocket** *(off-label)*, **red mustard** *(off-label)*, **rocket** *(off-label)*, **swedes** *(off-label)*, **turnips** *(off-label)* [3]; **winter oilseed rape** [2, 3]

SEE SECTION 3 FOR PRODUCTS ALSO REGISTERED

- Barnyard grass in **baby leaf crops** *(off-label)*, **celeriac** *(off-label)*, **choi sum** *(off-label)*, **collards** *(off-label)*, **horseradish** *(off-label)*, **lamb's lettuce** *(off-label)*, **oriental cabbage** *(off-label)*, **protected baby leaf crops** *(off-label)*, **protected lamb's lettuce** *(off-label)*, **protected red mustard** *(off-label)*, **protected rocket** *(off-label)*, **red mustard** *(off-label)*, **rocket** *(off-label)*, **swedes** *(off-label)*, **turnips** *(off-label)* [3]
- Black bindweed in **baby leaf crops** *(off-label)*, **celeriac** *(off-label)*, **choi sum** *(off-label)*, **collards** *(off-label)*, **horseradish** *(off-label)*, **lamb's lettuce** *(off-label)*, **oriental cabbage** *(off-label)*, **protected baby leaf crops** *(off-label)*, **protected lamb's lettuce** *(off-label)*, **protected red mustard** *(off-label)*, **protected rocket** *(off-label)*, **red mustard** *(off-label)*, **rocket** *(off-label)*, **swedes** *(off-label)*, **turnips** *(off-label)* [3]
- Blackgrass in **borage for oilseed production** *(off-label)*, **evening primrose** *(off-label)*, **game cover** *(off-label)*, **linseed** *(off-label)*, **mustard** *(off-label)* [3]; **winter oilseed rape** [2, 3]
- Chickweed in **baby leaf crops** *(off-label)*, **borage for oilseed production** *(off-label)*, **celeriac** *(off-label)*, **choi sum** *(off-label)*, **collards** *(off-label)*, **evening primrose** *(off-label)*, **game cover** *(off-label)*, **horseradish** *(off-label)*, **lamb's lettuce** *(off-label)*, **linseed** *(off-label)*, **mustard** *(off-label)*, **oriental cabbage** *(off-label)*, **protected baby leaf crops** *(off-label)*, **protected lamb's lettuce** *(off-label)*, **protected red mustard** *(off-label)*, **protected rocket** *(off-label)*, **red mustard** *(off-label)*, **rocket** *(off-label)*, **swedes** *(off-label)*, **turnips** *(off-label)* [3]
- Cleavers in **baby leaf crops** *(off-label)*, **celeriac** *(off-label)*, **choi sum** *(off-label)*, **collards** *(off-label)*, **horseradish** *(off-label)*, **lamb's lettuce** *(off-label)*, **oriental cabbage** *(off-label)*, **protected baby leaf crops** *(off-label)*, **protected lamb's lettuce** *(off-label)*, **protected red mustard** *(off-label)*, **protected rocket** *(off-label)*, **red mustard** *(off-label)*, **rocket** *(off-label)*, **swedes** *(off-label)*, **turnips** *(off-label)* [3]; **broccoli, brussels sprouts, cabbages, calabrese, cauliflowers, kale** [2, 3]
- Common purslane in **baby leaf crops** *(off-label)*, **celeriac** *(off-label)*, **choi sum** *(off-label)*, **collards** *(off-label)*, **horseradish** *(off-label)*, **lamb's lettuce** *(off-label)*, **oriental cabbage** *(off-label)*, **protected baby leaf crops** *(off-label)*, **protected lamb's lettuce** *(off-label)*, **protected red mustard** *(off-label)*, **protected rocket** *(off-label)*, **red mustard** *(off-label)*, **rocket** *(off-label)*, **swedes** *(off-label)*, **turnips** *(off-label)* [3]
- Fat hen in **baby leaf crops** *(off-label)*, **borage for oilseed production** *(off-label)*, **celeriac** *(off-label)*, **choi sum** *(off-label)*, **collards** *(off-label)*, **evening primrose** *(off-label)*, **horseradish** *(off-label)*, **lamb's lettuce** *(off-label)*, **linseed** *(off-label)*, **mustard** *(off-label)*, **oriental cabbage** *(off-label)*, **protected baby leaf crops** *(off-label)*, **protected lamb's lettuce** *(off-label)*, **protected red mustard** *(off-label)*, **protected rocket** *(off-label)*, **red mustard** *(off-label)*, **rocket** *(off-label)*, **swedes** *(off-label)*, **turnips** *(off-label)* [3]
- Field pansy in **borage for oilseed production** *(off-label)*, **evening primrose** *(off-label)*, **game cover** *(off-label)*, **linseed** *(off-label)*, **mustard** *(off-label)* [3]
- Field speedwell in **baby leaf crops** *(off-label)*, **borage for oilseed production** *(off-label)*, **celeriac** *(off-label)*, **choi sum** *(off-label)*, **collards** *(off-label)*, **evening primrose** *(off-label)*, **game cover** *(off-label)*, **horseradish** *(off-label)*, **lamb's lettuce** *(off-label)*, **linseed** *(off-label)*, **mustard** *(off-label)*, **oriental cabbage** *(off-label)*, **protected baby leaf crops** *(off-label)*, **protected lamb's lettuce** *(off-label)*, **protected red mustard** *(off-label)*, **protected rocket** *(off-label)*, **red mustard** *(off-label)*, **rocket** *(off-label)*, **swedes** *(off-label)*, **turnips** *(off-label)* [3]
- Forget-me-not in **borage for oilseed production** *(off-label)*, **evening primrose** *(off-label)*, **game cover** *(off-label)*, **linseed** *(off-label)*, **mustard** *(off-label)* [3]
- Fumitory in **borage for oilseed production** *(off-label)*, **evening primrose** *(off-label)*, **game cover** *(off-label)*, **linseed** *(off-label)*, **mustard** *(off-label)* [3]
- Groundsel in **baby leaf crops** *(off-label)*, **celeriac** *(off-label)*, **choi sum** *(off-label)*, **collards** *(off-label)*, **horseradish** *(off-label)*, **lamb's lettuce** *(off-label)*, **oriental cabbage** *(off-label)*, **protected baby leaf crops** *(off-label)*, **protected lamb's lettuce** *(off-label)*, **protected red mustard** *(off-label)*, **protected rocket** *(off-label)*, **red mustard** *(off-label)*, **rocket** *(off-label)*, **swedes** *(off-label)*, **turnips** *(off-label)* [3]; **broccoli, brussels sprouts, cabbages, calabrese, cauliflowers, kale** [2, 3]
- Ivy-leaved speedwell in **borage for oilseed production** *(off-label)*, **evening primrose** *(off-label)*, **game cover** *(off-label)*, **linseed** *(off-label)*, **mustard** *(off-label)* [3]
- Knotgrass in **baby leaf crops** *(off-label)*, **celeriac** *(off-label)*, **choi sum** *(off-label)*, **collards** *(off-label)*, **horseradish** *(off-label)*, **lamb's lettuce** *(off-label)*, **oriental cabbage** *(off-label)*, **protected baby leaf crops** *(off-label)*, **protected lamb's lettuce** *(off-label)*, **protected red mustard** *(off-*

label), **protected rocket** *(off-label)*, **red mustard** *(off-label)*, **rocket** *(off-label)*, **swedes** *(off-label)*, **turnips** *(off-label)* [3]

- Mayweeds in **baby leaf crops** *(off-label)*, **borage for oilseed production** *(off-label)*, **celeriac** *(off-label)*, **choi sum** *(off-label)*, **collards** *(off-label)*, **evening primrose** *(off-label)*, **game cover** *(off-label)*, **horseradish** *(off-label)*, **lamb's lettuce** *(off-label)*, **linseed** *(off-label)*, **mustard** *(off-label)*, **oriental cabbage** *(off-label)*, **protected baby leaf crops** *(off-label)*, **protected lamb's lettuce** *(off-label)*, **protected red mustard** *(off-label)*, **protected rocket** *(off-label)*, **red mustard** *(off-label)*, **rocket** *(off-label)*, **swedes** *(off-label)*, **turnips** *(off-label)* [3]
- Pale persicaria in **borage for oilseed production** *(off-label)*, **evening primrose** *(off-label)*, **game cover** *(off-label)*, **linseed** *(off-label)*, **mustard** *(off-label)* [3]
- Poppies in **borage for oilseed production** *(off-label)*, **evening primrose** *(off-label)*, **game cover** *(off-label)*, **linseed** *(off-label)*, **mustard** *(off-label)* [3]
- Redshank in **baby leaf crops** *(off-label)*, **borage for oilseed production** *(off-label)*, **celeriac** *(off-label)*, **choi sum** *(off-label)*, **collards** *(off-label)*, **evening primrose** *(off-label)*, **game cover** *(off-label)*, **horseradish** *(off-label)*, **lamb's lettuce** *(off-label)*, **linseed** *(off-label)*, **mustard** *(off-label)*, **oriental cabbage** *(off-label)*, **protected baby leaf crops** *(off-label)*, **protected lamb's lettuce** *(off-label)*, **protected red mustard** *(off-label)*, **protected rocket** *(off-label)*, **red mustard** *(off-label)*, **rocket** *(off-label)*, **swedes** *(off-label)*, **turnips** *(off-label)* [3]
- Shepherd's purse in **baby leaf crops** *(off-label)*, **celeriac** *(off-label)*, **choi sum** *(off-label)*, **collards** *(off-label)*, **horseradish** *(off-label)*, **lamb's lettuce** *(off-label)*, **oriental cabbage** *(off-label)*, **protected baby leaf crops** *(off-label)*, **protected lamb's lettuce** *(off-label)*, **protected red mustard** *(off-label)*, **protected rocket** *(off-label)*, **red mustard** *(off-label)*, **rocket** *(off-label)*, **swedes** *(off-label)*, **turnips** *(off-label)* [3]
- Small nettle in **baby leaf crops** *(off-label)*, **celeriac** *(off-label)*, **choi sum** *(off-label)*, **collards** *(off-label)*, **horseradish** *(off-label)*, **lamb's lettuce** *(off-label)*, **oriental cabbage** *(off-label)*, **protected baby leaf crops** *(off-label)*, **protected lamb's lettuce** *(off-label)*, **protected red mustard** *(off-label)*, **protected rocket** *(off-label)*, **red mustard** *(off-label)*, **rocket** *(off-label)*, **swedes** *(off-label)*, **turnips** *(off-label)* [3]

Extension of Authorisation for Minor Use (EAMUs)
- **baby leaf crops** *20171231* [3]
- **borage for oilseed production** *20170447* [3]
- **celeriac** *20171126* [3]
- **choi sum** *20171127* [3]
- **collards** *20171127* [3]
- **evening primrose** *20170447* [3]
- **game cover** *20170446* [3]
- **horseradish** *20171126* [3]
- **lamb's lettuce** *20171231* [3]
- **linseed** *20170447* [3]
- **mustard** *20170447* [3]
- **oriental cabbage** *20171127* [3]
- **protected baby leaf crops** *20171231* [3]
- **protected lamb's lettuce** *20171231* [3]
- **protected red mustard** *20171231* [3]
- **protected rocket** *20171231* [3]
- **red mustard** *20171231* [3]
- **rocket** *20171231* [3]
- **swedes** *20171126* [3]
- **turnips** *20171126* [3]

Approval information
- Napropamide included in Annex I under EC Regulation 1107/2009

Efficacy guidance
- Best results obtained from treatment pre-emergence of weeds but product may be used in conjunction with contact herbicides for control of emerged weeds. Otherwise remove existing weeds before application
- Weed control may be reduced where spray is mixed too deeply in the soil
- Manure, crop debris or other organic matter may reduce weed control

SEE SECTION 3 FOR PRODUCTS ALSO REGISTERED

- Napropamide broken down by sunlight, so application during such conditions not recommended. Most crops recommended for treatment between Nov and end-Feb
- Apply to winter oilseed rape pre-emergence or as pre-drilling treatment in tank-mixture with trifluralin and incorporate within 30 min
- Post-emergence use of specific grass weedkilller recommended where volunteer cereals are a serious problem
- Increase water volume to ensure adequate dampening of compost when treating containerised nursery stock with dense leaf canopy

Restrictions
- Maximum number of treatments 1 per crop or yr
- Do not use on Sands
- Do not use on soils with more than 10% organic matter
- Strawberries, blackcurrants, gooseberries and raspberries must be treated between 1 Nov and end Feb
- Applications to strawberries must only be made where the mature foliage has been previously removed
- Applications to ornamentals and forest nurseries must be made after 1 Nov and before end Apr
- Do not treat ornamentals in containers of less than 1 litre
- Some phytotoxicity seen on yellow and golden varieties of conifers and container grown alpines. On any ornamental variety treat only a small number of plants in first season
- Consult processors before use on any crop for processing

Crop-specific information
- Latest use: before transplanting for brassicas; pre-emergence for winter oilseed rape; before end Feb for strawberries, bush and cane fruit; before end of Apr for field and container grown ornamental trees and shrubs
- Where minimal cultivation used to establish oilseed rape, tank-mixture may be applied directly to stubble and mixed into top 25 mm as part of surface cultivations
- Apply up to 14 d prior to drilling winter oilseed rape
- Apply to strawberries established for at least one season or to maiden crops as long as planted carefully and no roots exposed, between 1 Nov and end Feb. Do not treat runners of poor vigour or with shallow roots, or runner beds
- Newly planted ornamentals should have no roots exposed. Do not treat stock of poor vigour or with shallow roots. Treatments made in Mar and Apr must be followed by 25 mm irrigation within 24 h
- Bush and cane fruit must be established for at least 10 mth before spraying and treated between 1 Nov and end Feb

Following crops guidance
- After use in fruit or ornamentals no crop can be drilled within 7 mth of treatment. Leaf, flowerhead, root and fodder brassica crops may be drilled after 7 mth; potatoes, maize, peas or dwarf beans after 9 mth; autumn sown wheat or grass after 18 mth; any crop after 2 yr
- After use in oilseed rape only oilseed rape, swedes, fodder turnips, brassicas or potatoes should be sown within 12 mth of application
- Soil should be mould-board ploughed to a depth of at least 200 mm before drilling or planting any following crop

Environmental safety
- Dangerous for the environment
- Toxic to aquatic organisms
- LERAP Category B

Hazard classification and safety precautions
Hazard Irritant, Dangerous for the environment, Very toxic to aquatic organisms
Transport code 9
Packaging group III
UN Number 3082
Risk phrases H315, H320 [2, 3]
Operator protection A [1-3]; C [2, 3]; H [1]; U05a, U09a, U11, U20b

---

FOR FULL CONDITIONS OF USE ALWAYS READ THE PRODUCT LABEL

Environmental protection E15a [2, 3]; E16a, E34, E38, H410 [1-3]
Storage and disposal D01, D02, D05, D09a, D10c [1-3]; D12a [1]; D12b [2, 3]

341   nicosulfuron

A sulfonylurea herbicide for maize
HRAC mode of action code: B

See also mesotrione + nicosulfuron

Products

1	Accent	DuPont	75% w/w	WG	15392
2	Crew	Nufarm UK	40 g/l	OD	15770
3	Entail	Headland	240 g/l	OD	15671
4	Fornet 6 OD	Belchim	60 g/l	OD	15891
5	Milagro 240 OD	Syngenta	240 g/l	OD	15957
6	Nico Pro 4SC	Belchim	40 g/l	OD	16424
7	Pampa 4 SC	Belchim	40 g/l	OD	17521
8	Primero	Rotam	40.8 g/l	OD	16444
9	Samson Extra 6%	Belchim	60 g/l	OD	16418
10	Templier	Rotam	75% w/w	WG	16866

Uses

- Amaranthus in *forage maize* [2, 6, 7]; *grain maize* [2]
- Annual dicotyledons in *forage maize* [1, 3-5, 8-10]; *forest nurseries* (off-label), *game cover* (off-label), *ornamental plant production* (off-label), *top fruit* (off-label) [4, 6, 9]; *grain maize* [3-5, 8-10]; *sweetcorn* (off-label), *sweetcorn under plastic mulches* (off-label) [4, 9]
- Annual grasses in *forest nurseries* (off-label), *game cover* (off-label), *ornamental plant production* (off-label), *top fruit* (off-label) [6]
- Annual meadow grass in *forage maize* [1, 2, 4, 6-8, 10]; *forest nurseries* (off-label), *game cover* (off-label), *ornamental plant production* (off-label), *sweetcorn* (off-label), *sweetcorn under plastic mulches* (off-label), *top fruit* (off-label) [4, 6, 9]; *grain maize* [2, 4, 8, 10]
- Black nightshade in *sweetcorn* (off-label), *sweetcorn under plastic mulches* (off-label) [4]
- Blackgrass in *forage maize* [1, 10]; *grain maize* [10]
- Cockspur grass in *forage maize* [1, 10]; *grain maize* [10]
- Common amaranth in *sweetcorn* (off-label), *sweetcorn under plastic mulches* (off-label) [6]
- Couch in *forage maize* [1, 10]; *grain maize* [10]
- Grass weeds in *forage maize, grain maize* [3, 5, 9]; *forest nurseries* (off-label), *game cover* (off-label), *ornamental plant production* (off-label), *top fruit* (off-label) [9]
- Groundsel in *forage maize* [2, 6, 7]; *grain maize* [2]; *sweetcorn* (off-label), *sweetcorn under plastic mulches* (off-label) [6]
- Mayweeds in *forage maize* [2, 6, 7]; *grain maize* [2]
- Ryegrass in *forage maize* [1, 2, 6, 7, 10]; *forest nurseries* (off-label), *game cover* (off-label), *ornamental plant production* (off-label), *top fruit* (off-label) [4, 9]; *grain maize* [2, 10]; *sweetcorn* (off-label), *sweetcorn under plastic mulches* (off-label) [4, 6, 9]
- Shepherd's purse in *forage maize* [2, 6, 7]; *grain maize* [2]; *sweetcorn* (off-label), *sweetcorn under plastic mulches* (off-label) [6]
- Volunteer cereals in *forage maize* [1, 10]; *grain maize* [10]
- Wild oats in *forage maize* [1, 10]; *grain maize* [10]

Extension of Authorisation for Minor Use (EAMUs)

- *forest nurseries* 20140942 [4], 20140752 [6], 20140490 [9], 20141054 [9]
- *game cover* 20140942 [4], 20140752 [6], 20140490 [9], 20141054 [9]
- *ornamental plant production* 20140942 [4], 20140752 [6], 20140490 [9], 20141054 [9]
- *sweetcorn* 20140895 [4], 20140941 [6], 20140940 [9]
- *sweetcorn under plastic mulches* 20140895 [4], 20140941 [6], 20140940 [9]
- *top fruit* 20140942 [4], 20140752 [6], 20140490 [9], 20141054 [9]

Approval information

- Nicosulfuron included in Annex I under EC Regulation 1107/2009

SEE SECTION 3 FOR PRODUCTS ALSO REGISTERED

Efficacy guidance
- Product should be applied post-emergence between 2 and 8 crop leaf stage and to emerged weeds from the 2-leaf stage
- Product acts mainly by foliar activity. Ensure good spray cover of the weeds
- Nicosulfuron is a member of the ALS-inhibitor group of herbicides. To avoid the build up of resistance do not use any product containing an ALS-inhibitor herbicide with claims for control of grass weeds more than once on any crop
- Use these products as part of a resistance management strategy that includes cultural methods of control and does not use ALS inhibitors as the sole chemical method of grass weed control

Restrictions
- Maximum number of treatments 1 per crop
- Do not use if an organophosphorus soil insecticide has been used on the same crop
- Do not mix with foliar or liquid fertilisers or specified herbicides. See label for details
- Do not apply in mixture, or sequence, with any other sulfonyl-urea containing product
- Do not treat crops under stress
- Do not apply if rainfall is forecast to occur within 6 h of application

Crop-specific information
- Latest use: up to and including 8 true leaves for maize
- Some transient yellowing may be seen from 1-2 wk after treatment
- Only healthy maize crops growing in good field conditions should be treated

Following crops guidance
- Winter wheat (not undersown) may be sown 4 mth after treatment; maize may be sown in the spring following treatment; all other crops may be sown from the next autumn
- In normal crop rotation, after ploughing winter wheat, winter barley, winter rye and triticale can be sown and all other crops can be sown the following spring. In the event of crop failure maize or soybeans can be sown after ploughing [4, 6, 9]

Environmental safety
- Dangerous for the environment
- Very toxic to aquatic organisms
- LERAP Category B

Hazard classification and safety precautions
Hazard Harmful [4, 9]; Irritant [2, 3, 5-7]; Dangerous for the environment [1-10]; Harmful if inhaled [2]; Very toxic to aquatic organisms [1, 4, 5, 8, 9]
Transport code 9
Packaging group III
UN Number 3077 [1, 10]; 3082 [2-9]
Risk phrases H315 [2, 3, 5-7]; H317 [2-5, 9]; H319 [4, 9]; R50, R53a [10]
Operator protection A [2-7, 9]; C [3, 9]; H [2, 3, 5-7, 9]; U02a, U20b [3, 5, 9]; U05a [1-7, 9]; U10, U11 [3-5, 9]; U14, U19a [2-7, 9]; U15 [2, 6, 7]
Environmental protection E15b [1-3, 5-7, 9]; E16a [1-10]; E16b [3, 5, 9]; E34 [1, 2, 6, 7]; E38 [1-7, 9, 10]; E40b [1]; H410 [1-9]
Storage and disposal D01, D02 [1-9]; D06a [4]; D09a [1, 2, 6, 7]; D09b [8]; D10c [2, 3, 5-7, 9]; D11a [1]; D12a [1, 3-5, 9, 10]; D12b [2, 4, 6, 7]
Medical advice M03 [2, 6, 7]; M05a [3-5, 9]

342 oxamyl

A soil-applied, systemic carbamate nematicide and insecticide
IRAC mode of action code: 1A

Products

Vydate 10G	DuPont	10% w/w	GR	16595

Uses
- American serpentine leaf miner in *ornamental plant production* (off-label)
- Docking disorder vectors in *fodder beet*, *sugar beet*

FOR FULL CONDITIONS OF USE ALWAYS READ THE PRODUCT LABEL

- Free-living nematodes in *fodder beet*, *potatoes*, *sugar beet*
- Leaf miner in *ornamental plant production* *(off-label)*
- Potato cyst nematode in *potatoes*
- Silver leaf in *ornamental plant production* *(off-label)*
- South American leaf miner in *ornamental plant production* *(off-label)*
- Stem and bulb nematodes in *bulb onions* *(off-label)*, *garlic* *(off-label)*, *ornamental plant production* *(off-label)*, *shallots* *(off-label)*
- Stem nematodes in *carrots*, *parsnips*
- Whitefly in *ornamental plant production* *(off-label)*

Extension of Authorisation for Minor Use (EAMUs)
- *bulb onions* 20141637
- *garlic* 20141637
- *ornamental plant production* 20141636
- *shallots* 20141637

Approval information
- Oxamyl included in Annex I under EC Regulation 1107/2009
- In 2006 CRD required that all products containing this active ingredient should carry the following warning in the main area of the container label: "Oxamyl is an anticholinesterase carbamate. Handle with care"

Efficacy guidance
- Apply granules with suitable applicator before drilling or planting. See label for details of recommended machines
- In potatoes incorporate thoroughly to 10 cm and plant within 1-2 d
- In sugar beet apply in seed furrow at drilling

Restrictions
- Oxamyl is subject to the Poisons Rules 1982 and the Poisons Act 1972. See Section 5 for more information
- Contains an anticholinesterase carbamate compound. Do not use if under medical advice not to work with such compounds
- Maximum number of treatments 1 per crop or yr
- Keep in original container, tightly closed, in a safe place, under lock and key
- Wear protective gloves if handling treated compost or soil within 2 wk after treatment
- Allow at least 12 h, followed by at least 1 h ventilation, before entry of unprotected persons into treated glasshouses

Crop-specific information
- Latest use: at drilling/planting for vegetables; before drilling/planting for potatoes.
- HI: A minimum of 80 days must elapse between application and initial haulm destruction in potatoes

Environmental safety
- Dangerous for the environment
- Toxic to aquatic organisms
- Dangerous to fish or other aquatic life. Do not contaminate surface waters or ditches with chemical or used container
- Dangerous to game, wild birds and animals. Bury spillages

Hazard classification and safety precautions
Hazard Toxic, Fatal if swallowed, Toxic if inhaled
Transport code 6.1
Packaging group II
UN Number 2757
Operator protection A, B, H, K, M; C (or D); U02a, U04a, U05a, U08, U13, U19a, U20a, U23a
Environmental protection E10a, E13b, E15b, H412
Storage and disposal D01, D02, D06e, D09a, D11b
Medical advice M04a, M04c, M05a

343   paclobutrazol

A triazole plant growth regulator for ornamentals and fruit
FRAC mode of action code: 3

See also difenoconazole + paclobutrazol

Products

1 Bonzi	Syngenta	4 g/l	SC	17576
2 Pirouette	Fargro	4 g/l	SC	17203

Uses
- Growth regulation in **azaleas, bedding plants, kalanchoes, mini roses, pelargoniums, poinsettias** [2]; **protected ornamentals** [1]
- Increasing flowering in **azaleas, bedding plants, kalanchoes, mini roses, ornamental plant production, ornamental plant production** (off-label)**, pelargoniums, poinsettias** [2]
- Stem shortening in **ornamental plant production, ornamental plant production** (off-label) [2]

Extension of Authorisation for Minor Use (EAMUs)
- **ornamental plant production** *20171269* [2]

Approval information
- Paclobutrazol included in Annex I under EC Regulation 1107/2009

Efficacy guidance
- Chemical is active via both foliage and root uptake. For best results apply in dull weather when relative humidity not high

Restrictions
- Maximum number of treatments 1 per specimen for some species [2]

Following crops guidance
- Chemical has residual soil activity which can affect growth of following crops. Do not use treated pots or soil for other crops

Environmental safety
- Harmful to aquatic organisms
- Do not use on food crops [2]
- Keep livestock out of treated areas for at least 2 years after treatment

Hazard classification and safety precautions
   Transport code 9 [2]
   Packaging group III [2]
   UN Number 3082 [2]; N/C [1]
   Operator protection A [2]; U05a, U08, U20a
   Environmental protection E13c [2]; E15a, H411 [1, 2]
   Consumer protection C01
   Storage and disposal D01, D02, D05, D09a [1, 2]; D10b [2]; D10c [1]

344   pelargonic acid

A naturally occuring 9 carbon acid terminating in a carboxylic acid
HRAC mode of action code: Not classified

See also maleic hydrazide + pelargonic acid

Products

Katoun Gold	Belchim	500 g/l	EC	17879

Uses
- Annual dicotyledons in **amenity vegetation**

Approval information
- Pelargonic acid included in Annex 1 under EC Regulation 1107/2009

Hazard classification and safety precautions
   UN Number N/C
   Risk phrases H319
   Operator protection A, C

345   penconazole

A protectant triazole fungicide with antisporulant activity
FRAC mode of action code: 3

See also captan + penconazole

Products

1	Pan Penco	Pan Amenity	100 g/l	EC	18180
2	Topas	Syngenta	100 g/l	EC	16765
3	Topenco	Globachem	100 g/l	EC	17539
4	Topenco 100 EC	Globachem	100 g/l	EC	17161

Uses
- Powdery mildew in **apples, blackcurrants, crab apples, pears, protected strawberries, redcurrants, strawberries, table grapes, wine grapes** [1-3]; **courgettes** *(off-label),* **cucumbers** *(off-label),* **gherkins** *(off-label),* **globe artichoke** *(off-label),* **hops in propagation** *(off-label),* **nursery fruit trees** *(off-label),* **protected chilli peppers** *(off-label),* **protected courgettes** *(off-label),* **protected cucumbers** *(off-label),* **protected gherkins** *(off-label),* **protected hops** *(off-label),* **protected melons** *(off-label),* **protected nursery fruit trees** *(off-label),* **protected peppers** *(off-label),* **protected pumpkins** *(off-label),* **protected summer squash** *(off-label),* **protected tomatoes** *(off-label),* **protected watermelon** *(off-label),* **protected winter squash** *(off-label),* **summer squash** *(off-label)* [2]; **protected courgettes, protected cucumbers, protected summer squash** [4]

Extension of Authorisation for Minor Use (EAMUs)
- **courgettes** *20152336* [2]
- **cucumbers** *20152336* [2]
- **gherkins** *20152336* [2]
- **globe artichoke** *20152901* [2]
- **hops in propagation** *20152338* [2]
- **nursery fruit trees** *20152338* [2]
- **protected chilli peppers** *20152335* [2]
- **protected courgettes** *20152336* [2]
- **protected cucumbers** *20152336* [2]
- **protected gherkins** *20152336* [2]
- **protected hops** *20152338* [2]
- **protected melons** *20152337* [2]
- **protected nursery fruit trees** *20152338* [2]
- **protected peppers** *20152335* [2]
- **protected pumpkins** *20152337* [2]
- **protected summer squash** *20152336* [2]
- **protected tomatoes** *20152335* [2]
- **protected watermelon** *20152337* [2]
- **protected winter squash** *20152337* [2]
- **summer squash** *20152336* [2]

Approval information
- Accepted by BBPA for use on hops
- Penconazole included in Annex I under EC Regulation 1107/2009

Efficacy guidance
- Use as a protectant fungicide by treating at the earliest signs of disease

SEE SECTION 3 FOR PRODUCTS ALSO REGISTERED

- Treat crops every 10-14 d (every 7-10 d in warm, humid weather) at first sign of infection or as a protective spray ensuring complete coverage. See label for details of timing
- Increase dose and volume with growth of hops but do not exceed 2000 l/ha. Little or no activity will be seen on established powdery mildew
- Antisporulant activity reduces development of secondary mildew in apples

Restrictions
- Maximum number of treatments 10 per yr for apples, 6 per yr for hops, 4 per yr for blackcurrants
- Check for varietal susceptibility in roses. Some defoliation may occur after repeat applications on Dearest

Crop-specific information
- HI apples, hops 14 d; currants, bilberries, blueberries, cranberries, gooseberries 4 wk

Environmental safety
- Dangerous for the environment
- Toxic to aquatic organisms

Hazard classification and safety precautions
Hazard Irritant, Dangerous for the environment
Transport code 3
Packaging group III
UN Number 1915
Risk phrases H319, H361 [1-4]; R51 [4]
Operator protection A, C; U02a, U05a, U20b [1-4]; U08 [1-3]; U11 [4]
Environmental protection E15a [1-4]; E22c, H410 [4]; E38, H411 [1-3]
Storage and disposal D01, D02, D05, D09a [1-4]; D07, D10b [4]; D10c, D12a [1-3]
Medical advice M05b

346   pencycuron

A non-systemic urea fungicide for use on seed potatoes
FRAC mode of action code: 20

See also imazalil + pencycuron

Products

1 Monceren DS	Bayer CropScience	12.5% w/w	DS	11292
2 Solaren	AgChem Access	12.5% w/w	DS	15912

Uses
- Black scurf in **potatoes** *(tuber treatment)*
- Stem canker in **potatoes** *(tuber treatment)*

Approval information
- Pencycuron included in Annex I under EC Regulation 1107/2009

Efficacy guidance
- Provides control of tuber-borne disease and gives some reduction of stem canker
- Seed tubers should be of good quality and vigour, and free from soil deposits when treated
- Dust formulations should be applied immediately before, or during, the planting process by treating seed tubers in chitting trays, in bulk bins immediately before planting or in hopper at planting. Liquid formulations may be applied by misting equipment at any time, into or out of store but treatment over a roller table at the end of grading out, or at planting, is usually most convenient
- Whichever method is chosen, complete and even distribution on the tubers is essential for optimum efficacy
- Apply in accordance with detailed guidelines in manufacturer's literature
- If rain interrupts planting, cover dust treated tubers in hopper

Restrictions
- Maximum number of treatments 1 per batch of seed potatoes

FOR FULL CONDITIONS OF USE ALWAYS READ THE PRODUCT LABEL

- Treated tubers must be used only as seed and not for human or animal consumption
- Do not use on tubers previously treated with a dry powder seed treatment or hot water
- To prevent rapid absorption through damaged 'eyes' seed tubers should only be treated before the 'eyes' begin to open or immediately before or during planting
- Tubers removed from cold storage must be allowed to attain a temperature of at least 8°C before treatment
- Use of suitable dust mask is mandatory when applying dust, filling the hopper or riding on planter
- Treated tubers in bulk bins must be allowed to dry before being stored, especially when intended for cold storage
- Some internal staining of boxes used to store treated tubers may occur. Ware tubers should not be stored or supplied to packers in such stained boxes

Crop-specific information
- Latest use: at planting

Environmental safety
- Do not use treated seed as food or feed
- Treated seed harmful to game and wildlife

Hazard classification and safety precautions
UN Number N/C
Operator protection A, F; U20b
Environmental protection E03, E15a [1, 2]; H411 [1]
Storage and disposal D09a, D11a
Treated seed S01, S02, S03, S04a, S05, S06a

347   pendimethalin

A residual dinitroaniline herbicide for cereals and other crops
HRAC mode of action code: K1

See also bentazone + pendimethalin
chlorotoluron + diflufenican + pendimethalin
chlorotoluron + pendimethalin
clomazone + pendimethalin
diflufenican + pendimethalin
dimethenamid-p + pendimethalin
flufenacet + pendimethalin
imazamox + pendimethalin
isoproturon + pendimethalin

Products

1	Anthem	Adama	400 g/l	SC	15761
2	Claymore	BASF	400 g/l	SC	13441
3	Cleancrop National	Agrii	400 g/l	SC	17094
4	Most Micro	Sipcam	365 g/l	CS	16063
5	Nighthawk 330 EC	AgChem Access	330 g/l	EC	14083
6	Stomp 400 SC	BASF	400 g/l	SC	13405
7	Stomp Aqua	BASF	455 g/l	CS	14664
8	Trample	AgChem Access	330 g/l	EC	14173
9	Yomp	Becesane	330 g/l	EC	15609

Uses
- Annual dicotyledons in **almonds** *(off-label)*, **asparagus** *(off-label)*, **bilberries** *(off-label)*, **blueberries** *(off-label)*, **celery (outdoor)** *(off-label)*, **chervil** *(off-label)*, **chestnuts** *(off-label)*, **coriander** *(off-label)*, **cranberries** *(off-label)*, **cress** *(off-label)*, **dill** *(off-label)*, **evening primrose** *(off-label)*, **farm forestry** *(off-label)*, **fennel** *(off-label)*, **forest** *(off-label)*, **forest nurseries** *(off-label)*, **frise** *(off-label)*, **grass seed crops** *(off-label)*, **hazel nuts** *(off-label)*, **herbs (see appendix 6)** *(off-label)*, **lamb's lettuce** *(off-label)*, **leeks** *(off-label)*, **lettuce** *(off-label)*, **miscanthus** *(off-label)*, **ornamental plant production** *(off-label)*, **parsley root** *(off-label)*, **protected forest**

nurseries (off-label), *quinces* (off-label), *radicchio* (off-label), *redcurrants* (off-label), *rhubarb* (off-label), *salad onions* (off-label), *scarole* (off-label), *top fruit* (off-label), *walnuts* (off-label), *whitecurrants* (off-label) [6, 7]; *apple orchards, pear orchards* [2, 5, 6]; *apples, forage maize (under plastic mulches), grain maize, grain maize under plastic mulches, pears* [1, 3, 7, 9]; *blackberries, blackcurrants, broccoli, brussels sprouts, cabbages, calabrese, carrots, cauliflowers, cherries, gooseberries, leeks, loganberries, parsnips, plums, raspberries, rubus hybrids, strawberries* [1-3, 5-7, 9]; *bog myrtle* (off-label), *broccoli* (off-label), *brussels sprouts* (off-label), *cabbages* (off-label), *calabrese* (off-label), *carrots* (off-label), *cauliflowers* (off-label), *collards* (off-label), *french beans, hops* (off-label), *horseradish* (off-label), *kale* (off-label), *leaf brassicas* (off-label), *mallow (althaea spp.)* (off-label), *navy beans, parsley* (off-label), *parsnips* (off-label), *sweetcorn* (off-label), *sweetcorn under plastic mulches* (off-label) [7]; *broad beans* (off-label), *game cover* (off-label) [1, 3, 6, 7]; *bulb onion sets* (off-label) [1, 6]; *bulb onions* [1-3, 6, 7, 9]; *bulb onions* (off-label), *garlic* (off-label), *shallots* (off-label), *spring wheat* (off-label) [1, 6, 7]; *bulb onions* (pre + post emergence treatment), *chives* (off-label), *hops, navy beans* (off-label), *parsley, salad brassicas* (off-label - for baby leaf production), *sweetcorn* (off-label - under covers) [6]; *combining peas, forage maize* [1-3, 5-9]; *durum wheat, potatoes, spring barley, sunflowers, triticale, winter barley, winter rye, winter wheat* [1-9]; *dwarf beans* (off-label), *french beans* (off-label) [4, 6]; *edible podded peas* (off-label), *vining peas* (off-label) [2, 4, 5, 7]; *field beans* (off-label), *redcurrants* [3]; *forage soya bean* (off-label) [4, 7]; *lupins* (off-label), *runner beans* (off-label) [2, 4, 6, 7]; *spring field beans* (off-label) [1, 2, 4, 6, 7]; *winter field beans* (off-label) [1, 2, 4, 7]

- Annual grasses in *apples, blackberries, blackcurrants, broccoli, brussels sprouts, bulb onions, cabbages, calabrese, carrots, cauliflowers, cherries, forage maize (under plastic mulches), gooseberries, grain maize, grain maize under plastic mulches, leeks, loganberries, parsnips, pears, plums, raspberries, strawberries* [1, 3, 7, 9]; *bog myrtle* (off-label), *broccoli* (off-label), *brussels sprouts* (off-label), *calabrese* (off-label), *cauliflowers* (off-label), *celery (outdoor)* (off-label), *collards* (off-label), *fennel* (off-label), *grass seed crops* (off-label), *kale* (off-label), *rhubarb* (off-label) [7]; *broad beans* (off-label), *game cover* (off-label) [1, 3, 6]; *bulb onion sets* (off-label), *bulb onions* (off-label), *garlic* (off-label), *shallots* (off-label) [1, 6]; *chervil* (off-label), *coriander* (off-label), *dill* (off-label), *forest nurseries* (off-label), *leeks* (off-label), *miscanthus* (off-label), *navy beans* (off-label), *ornamental plant production* (off-label), *protected forest nurseries* (off-label), *salad onions* (off-label), *top fruit* (off-label) [6]; *combining peas, forage maize* [1, 3, 7-9]; *durum wheat, potatoes, spring barley, sunflowers, triticale, winter barley, winter rye, winter wheat* [1, 3, 4, 7-9]; *dwarf beans* (off-label), *french beans* (off-label) [4, 6]; *edible podded peas* (off-label), *vining peas* (off-label) [4, 5]; *field beans* (off-label), *redcurrants* [3]; *forage soya bean* (off-label), *lupins* (off-label), *runner beans* (off-label) [4]; *rubus hybrids* [1, 3, 5, 7, 9]; *spring field beans* (off-label), *winter field beans* (off-label) [1, 4, 5]; *spring wheat* (off-label) [1, 6, 7]
- Annual meadow grass in *apple orchards, blackberries, blackcurrants, broccoli, brussels sprouts, cabbages, calabrese, carrots, cauliflowers, cherries, combining peas, forage maize, gooseberries, leeks, loganberries, parsnips, pear orchards, plums, raspberries, strawberries* [2, 5, 6]; *broad beans* (off-label), *dwarf beans* (off-label), *french beans* (off-label) [4]; *bulb onion sets* (off-label), *bulb onions* (off-label), *garlic* (off-label), *shallots* (off-label) [1, 6]; *bulb onions, rubus hybrids* [2, 6]; *cabbages* (off-label), *lettuce* (off-label) [7]; *durum wheat, potatoes, spring barley, sunflowers, triticale, winter barley, winter rye, winter wheat* [2, 4-6]; *edible podded peas* (off-label), *lupins* (off-label), *runner beans* (off-label), *spring field beans* (off-label), *vining peas* (off-label), *winter field beans* (off-label) [2, 4]; *forage soya bean* (off-label) [4, 7]; *hops, leeks* (off-label), *parsley, salad onions* (off-label) [6]
- Blackgrass in *apple orchards, blackberries, blackcurrants, broccoli, brussels sprouts, cabbages, calabrese, carrots, cauliflowers, cherries, combining peas, durum wheat, gooseberries, leeks, loganberries, parsnips, pear orchards, plums, raspberries, spring barley, strawberries, sunflowers, triticale, winter barley, winter rye, winter wheat* [2, 5, 6]; *broad beans* (off-label), *game cover* (off-label), *spring field beans* (off-label), *spring wheat* (off-label), *winter field beans* (off-label) [1]; *bulb onions, rubus hybrids* [2, 6]; *forage soya bean* (off-label) [7]; *hops, parsley* [6]
- Chickweed in *broad beans* (off-label) [4]; *celeriac* (off-label), *lettuce* (off-label) [7]
- Cleavers in *broad beans* (off-label), *spring field beans* (off-label) [1]; *winter field beans* (off-label) [1, 5, 6]
- Common orache in *broad beans* (off-label) [4]

FOR FULL CONDITIONS OF USE ALWAYS READ THE PRODUCT LABEL

- Corn marigold in **broad beans** *(off-label)* [4]
- Dead nettle in **broad beans** *(off-label)* [4]
- Fat hen in **broad beans** *(off-label)* [4]; **celeriac** *(off-label)* [7]
- Field pansy in **broad beans** *(off-label)* [4]
- Forget-me-not in **broad beans** *(off-label)* [4]
- Fumitory in **broad beans** *(off-label)* [4]
- Groundsel in **celeriac** *(off-label)* [7]
- Hemp-nettle in **broad beans** *(off-label)* [4]
- Knotgrass in **edible podded peas** *(off-label)*, **vining peas** *(off-label)* [5, 6]
- Mayweeds in **celeriac** *(off-label)* [7]
- Parsley-piert in **broad beans** *(off-label)* [4]
- Polygonums in **celeriac** *(off-label)* [7]
- Poppies in **broad beans** *(off-label)* [4]
- Red dead-nettle in **lettuce** *(off-label)* [7]
- Rough-stalked meadow grass in **apple orchards, blackberries, blackcurrants, broccoli, brussels sprouts, cabbages, calabrese, carrots, cauliflowers, cherries, combining peas, durum wheat, forage maize, gooseberries, leeks, loganberries, parsnips, pear orchards, plums, potatoes, raspberries, spring barley, strawberries, sunflowers, triticale, winter barley, winter rye, winter wheat** [2, 5, 6]; **broad beans** *(off-label)* [4]; **bulb onions, rubus hybrids** [2, 6]; **hops, parsley** [6]
- Scarlet pimpernel in **broad beans** *(off-label)* [4]
- Shepherd's purse in **broad beans** *(off-label)* [4]
- Small nettle in **broad beans** *(off-label)* [4]; **celeriac** *(off-label)* [7]
- Sowthistle in **broad beans** *(off-label)* [4]; **celeriac** *(off-label)* [7]
- Speedwells in **broad beans** *(off-label)* [4]; **lettuce** *(off-label)* [7]
- Volunteer oilseed rape in **apple orchards, pear orchards, spring barley, sunflowers** [2, 5, 6]; **apples, forage maize (under plastic mulches), grain maize, grain maize under plastic mulches, pears** [1, 3, 9]; **blackberries, blackcurrants, broccoli, brussels sprouts, cabbages, calabrese, carrots, cauliflowers, cherries, gooseberries, leeks, loganberries, parsnips, plums, raspberries, rubus hybrids** [1-3, 5, 6, 9]; **broad beans** *(off-label)*, **field beans** *(off-label)*, **game cover** *(off-label)*, **redcurrants** [3]; **bulb onions** [1-3, 6, 9]; **combining peas, durum wheat, forage maize, potatoes, triticale, winter barley, winter rye, winter wheat** [1-3, 5, 6, 8, 9]; **hops, parsley** [6]; **strawberries** [2, 5, 6, 9]
- Wild oats in **durum wheat, triticale, winter barley, winter rye, winter wheat** [2, 5, 6]

Extension of Authorisation for Minor Use (EAMUs)

- *almonds 20071442* [6], *20092916* [7]
- *asparagus 20071431* [6], *20092920* [7]
- *bilberries 20071444* [6], *20092907* [7]
- *blueberries 20071444* [6], *20092907* [7]
- *bog myrtle 20120323* [7]
- *broad beans 20141259* [1], *20152446* [3], *20140230* [4], *20081016* [6], *20092922* [7]
- *broccoli 20112451* [7]
- *brussels sprouts 20112451* [7]
- *bulb onion sets 20171415* [1], *20171389* [6]
- *bulb onions 20171415* [1], *20171389* [6], *20092914* [7]
- *cabbages 20100650* [7]
- *calabrese 20112451* [7]
- *carrots 20093526* [7]
- *cauliflowers 20112451* [7]
- *celeriac 20131625* [7]
- *celery (outdoor) 20071430* [6], *20092924* [7], *20111287* [7]
- *chervil 20081808* [6], *20092913* [7]
- *chestnuts 20071442* [6], *20092916* [7]
- *chives 20071432* [6]
- *collards 20112451* [7]
- *coriander 20081808* [6], *20092913* [7]
- *cranberries 20071444* [6], *20092907* [7]
- *cress 20071432* [6], *20092921* [7]

SEE SECTION 3 FOR PRODUCTS ALSO REGISTERED

- **dill** _20081808_ [6], _20092913_ [7]
- **dwarf beans** _20140232_ [4], _20080053_ [6]
- **edible podded peas** _20152445_ [2], _20140233_ [4], _20111007_ [5], _20071437_ [6], _20092909_ [7]
- **evening primrose** _20071434_ [6], _20092915_ [7]
- **farm forestry** _20071436_ [6], _20092918_ [7]
- **fennel** _20071430_ [6], _20092924_ [7], _20111287_ [7]
- **field beans** _20152446_ [3]
- **forage soya bean** _20140231_ [4], _20121634_ [7]
- **forest** _20071436_ [6], _20092918_ [7]
- **forest nurseries** _20082923_ [6], _20092919_ [7]
- **french beans** _20140232_ [4], _20080053_ [6], _20092908_ [7]
- **frise** _20071432_ [6], _20092921_ [7]
- **game cover** _20141260_ [1], _20152496_ [3], _20082923_ [6], _20092919_ [7]
- **garlic** _20171415_ [1], _20171389_ [6], _20092914_ [7]
- **grass seed crops** _20071429_ [6], _20092923_ [7]
- **hazel nuts** _20071442_ [6], _20092916_ [7]
- **herbs (see appendix 6)** _20071432_ [6], _20092921_ [7]
- **hops** _20092919_ [7]
- **horseradish** _20093526_ [7]
- **kale** _20112451_ [7]
- **lamb's lettuce** _20071432_ [6], _20092921_ [7]
- **leaf brassicas** _20092921_ [7]
- **leeks** _20171389_ [6], _20092914_ [7]
- **lettuce** _20071432_ [6], _20092921_ [7], _20170372_ [7], _20170375_ [7]
- **lupins** _20152445_ [2], _20140233_ [4], _20071437_ [6], _20092909_ [7]
- **mallow (althaea spp.)** _20092921_ [7]
- **miscanthus** _20082923_ [6], _20092919_ [7]
- **navy beans** _20080053_ [6], _20092908_ [7]
- **ornamental plant production** _20082923_ [6], _20092919_ [7]
- **parsley** _20092913_ [7]
- **parsley root** _20071433_ [6], _20092911_ [7], _20093526_ [7]
- **parsnips** _20093526_ [7]
- **protected forest nurseries** _20082923_ [6], _20092919_ [7]
- **quinces** _20071443_ [6], _20092917_ [7]
- **radicchio** _20071432_ [6], _20092921_ [7]
- **redcurrants** _20071444_ [6], _20092907_ [7]
- **rhubarb** _20071430_ [6], _20092924_ [7], _20111287_ [7]
- **runner beans** _20152445_ [2], _20140233_ [4], _20071437_ [6], _20092909_ [7]
- **salad brassicas** _(for baby leaf production)_ _20071432_ [6]
- **salad onions** _20171389_ [6], _20092914_ [7]
- **scarole** _20071432_ [6], _20092921_ [7]
- **shallots** _20171415_ [1], _20171389_ [6], _20092914_ [7]
- **spring field beans** _20141259_ [1], _20152445_ [2], _20140233_ [4], _20092551_ [5], _20071437_ [6], _20092909_ [7]
- **spring wheat** _20141261_ [1], _20082947_ [6], _20092910_ [7]
- **sweetcorn** _(under covers)_ _20071435_ [6], _20092912_ [7]
- **sweetcorn under plastic mulches** _20092912_ [7]
- **top fruit** _20082923_ [6], _20092919_ [7]
- **vining peas** _20152445_ [2], _20140233_ [4], _20111007_ [5], _20071437_ [6], _20092909_ [7]
- **walnuts** _20071442_ [6], _20092916_ [7]
- **whitecurrants** _20071444_ [6], _20092907_ [7]
- **winter field beans** _20141259_ [1], _20152445_ [2], _20140233_ [4], _20092552_ [5], _20071437_ [6], _20092909_ [7]

Approval information
- Pendimethalin included in Annex I under EC Regulation 1107/2009
- Accepted by BBPA for use on malting barley and hops

FOR FULL CONDITIONS OF USE ALWAYS READ THE PRODUCT LABEL

Efficacy guidance

- Apply as soon as possible after drilling. Weeds are controlled as they germinate and emerged weeds will not be controlled by use of the product alone
- For effective blackgrass control apply not more than 2 d after final cultivation and before weed seeds germinate
- Best results by application to fine firm, moist, clod-free seedbeds when rain follows treatment. Effectiveness reduced by prolonged dry weather after treatment
- Effectiveness reduced on soils with more than 6% organic matter. Do not use where organic matter exceeds 10%
- Any trash, ash or straw should be incorporated evenly during seedbed preparation
- Do not disturb soil after treatment
- Apply to potatoes as soon as possible after planting and ridging in tank-mix with metribuzin but note that this will restrict the varieties that may be treated
- Always follow WRAG guidelines for preventing and managing herbicide resistant weeds. See Section 5 for more information

Restrictions

- Maximum number of treatments 1 per crop or yr
- Maximum total dose equivalent to one full dose treatment on most crops; 2 full dose treatments on leeks
- May be applied pre-emergence of cereal crops sown before 30 Nov provided seed covered by at least 32 mm soil, or post-emergence to early tillering stage (GS 23)
- Do not undersow treated crops
- Do not use on crops suffering stress due to disease, drought, waterlogging, poor seedbed conditions or chemical treatment or on soils where water may accumulate
- Do not apply with hand-held equipment [1]

Crop-specific information

- Latest use: pre-emergence for spring barley, carrots, lettuce, fodder maize, parsnips, parsley, combining peas, potatoes, onions and leeks; before transplanting for brassicas; before leaf sheaths erect for winter cereals; before bud burst for blackcurrants, gooseberries, cane fruit, hops; before flower trusses emerge for strawberries; 14 d after transplanting for leaf herbs
- Do not use on spring barley after end Mar (mid-Apr in Scotland on some labels) because dry conditions likely. Do not apply to dry seedbeds in spring unless rain imminent
- Apply to combining peas as soon as possible after sowing. Do not spray if plumule less than 13 mm below soil surface
- Apply to potatoes up to 7 d before first shoot emerges
- Apply to drilled crops as soon as possible after drilling but before crop and weed emergence
- Apply in top fruit, bush fruit and hops from autumn to early spring when crop dormant
- In cane fruit apply to weed free soil from autumn to early spring, immediately after planting new crops and after cutting out canes in established crops
- Apply in strawberries from autumn to early spring (not before Oct on newly planted bed). Do not apply pre-planting or during flower initiation period (post-harvest to mid-Sep)
- Apply pre-emergence in drilled onions or leeks, not on Sands, Very Light, organic or peaty soils or when heavy rain forecast
- Apply to brassicas after final plant-bed cultivation but before transplanting. Avoid unnecessary soil disturbance after application and take care not to introduce treated soil into the root zone when transplanting. Follow transplanting with specified post-planting treatments - see label
- Do not use on protected crops or in greenhouses

Following crops guidance

- Before ryegrass is drilled after a very dry season plough or cultivate to at least 15 cm. If treated spring crops are to be followed by crops other than cereals, plough or cultivate to at least 15 cm
- In the event of crop failure land must be ploughed or thoroughly cultivated to at least 15 cm. See label for minimum intervals that should elapse between treatment and sowing a range of replacement crops

Environmental safety

- Dangerous for the environment
- Very toxic to aquatic organisms

SEE SECTION 3 FOR PRODUCTS ALSO REGISTERED

- Dangerous to fish or other aquatic life. Do not contaminate surface waters or ditches with chemical or used container
- LERAP Category B

Hazard classification and safety precautions

Hazard Harmful [8, 9]; Irritant [4, 5]; Dangerous for the environment [1-9]; Harmful if swallowed, Very toxic to aquatic organisms [9]

Transport code 9

Packaging group III

UN Number 3082

Risk phrases H304, H315, H319 [9]; H317 [4]; R22a [8, 9]; R22b, R51 [5]; R36, R38 [8]; R50 [2, 6, 8]; R53a [2, 5, 6, 8]

Operator protection A [1-9]; C [5, 8, 9]; H [1, 3, 4, 7]; U02a [1-3, 6, 7]; U05a, U08 [1-9]; U11 [8, 9]; U13 [1-3, 6-9]; U14 [2, 4, 6, 7]; U15 [4]; U19a [1-3, 5-9]; U20b [1-4, 6, 8, 9]; U23a [5]

Environmental protection E15a [5]; E15b [1-4, 6-9]; E16a, E38 [1-9]; E16b [1, 3]; E34 [5, 8, 9]; E39 [7]; H410 [1, 3, 7, 9]; H412 [4]

Storage and disposal D01, D02, D09a [1-9]; D05 [1, 3, 5, 7]; D08 [2, 6, 8, 9]; D10b, D12a [1-3, 5-9]; D10c [4]

Medical advice M03, M05b [5, 8, 9]; M05a [1-3, 6]

348   pendimethalin + picolinafen

A post-emergence broad-spectrum herbicide mixture for winter cereals
HRAC mode of action code: K1 + F1

See also picolinafen

Products

1	Chronicle	BASF	320:16 g/l	SC	15394
2	Orient	BASF	330:7.5 g/l	SC	12541
3	Parade	BASF	320:16 g/l	SC	17246
4	PicoMax	BASF	320:16 g/l	SC	13456
5	Picona	BASF	320:16 g/l	SC	13428
6	PicoPro	BASF	320:16 g/l	SC	13454

Uses

- Annual dicotyledons in *forest nurseries* (off-label) [4]; *game cover* (off-label) [5]; *rye* [1]; *spring barley, spring wheat, triticale* [1, 3-6]; *spring barley* (off-label), *spring wheat* (off-label) [1, 2, 6]; *winter barley, winter wheat* [1-6]; *winter rye* [3-6]
- Annual grasses in *spring barley* (off-label), *spring wheat* (off-label) [1]
- Annual meadow grass in *forest nurseries* (off-label) [4]; *game cover* (off-label) [5]; *rye* [1]; *spring barley, spring wheat, triticale* [1, 3-6]; *spring barley* (off-label), *spring wheat* (off-label) [1, 2, 6]; *winter barley, winter wheat* [1-6]; *winter rye* [3-6]
- Chickweed in *rye* [1]; *spring barley, spring wheat, triticale* [1, 3-6]; *winter barley, winter wheat* [1-6]; *winter rye* [3-6]
- Cleavers in *rye* [1]; *spring barley, spring wheat, triticale* [1, 3-6]; *winter barley, winter wheat* [1-6]; *winter rye* [3-6]
- Field speedwell in *rye* [1]; *spring barley, spring wheat, triticale* [1, 3-6]; *winter barley, winter wheat* [1-6]; *winter rye* [3-6]
- Ivy-leaved speedwell in *rye* [1]; *spring barley, spring wheat, triticale* [1, 3-6]; *winter barley, winter wheat* [1-6]; *winter rye* [3-6]
- Loose silky bent in *spring barley* [5]
- Rough-stalked meadow grass in *forest nurseries* (off-label) [4]; *game cover* (off-label) [5]; *rye* [1]; *spring barley, spring wheat, triticale* [1, 3-6]; *winter barley, winter wheat* [1-6]; *winter rye* [3-6]

Extension of Authorisation for Minor Use (EAMUs)

- *forest nurseries* 20082863 [4]
- *game cover* 20082864 [5]

FOR FULL CONDITIONS OF USE ALWAYS READ THE PRODUCT LABEL

- **spring barley** *20130724* [1], *20130386* [2], *20130725* [6]
- **spring wheat** *20130724* [1], *20130386* [2], *20130725* [6]

Approval information
- Pendimethalin and picolinafen included in Annex I under EC Regulation 1107/2009

Efficacy guidance
- Best results obtained on crops growing in a fine, firm tilth and when rain falls within 7 d of application
- Loose or cloddy seed beds must be consolidated prior to application otherwise reduced weed control may occur
- Residual weed control may be reduced on soils with more than 6% organic matter or under prolonged dry conditions
- Always follow WRAG guidelines for preventing and managing herbicide resistant weeds. See Section 5 for more information

Restrictions
- Maximum number of treatments 1 per crop
- Do not treat undersown cereals or those to be undersown
- Do not roll emerged crops before treatment nor autumn treated crops until the following spring
- Do not use on stony or gravelly soils, on soils that are waterlogged or prone to waterlogging, or on soils with more than 10% organic matter
- Do not treat crops under stress from any cause
- Do not apply pre-emergence to crops drilled after 30 Nov [2]
- Consult processor before treating crops for processing [2]

Crop-specific information
- Latest use: before pseudo-stem erect stage (GS 30)
- Transient bleaching may occur after treatment but it does not lead to yield loss
- Do not use pre-emergence on crops drilled after 30th Nov [4]
- For pre-emergence applications, seed should be covered with a minimum of 3.2 cm settled soil [5, 6]

Following crops guidance
- After normal harvest any crop except ryegrass can be sown. Cultivation or ploughing to at least 15 cm is required before sowing ryegrass.
- In the event of crop failure following autumn application spring wheat, spring barley or winter field beans can be sown after the land is ploughed and at least 8 weeks have elapsed since treatment. The following spring oilseed rape, peas or spring field beans can be sown after non-inversion tillage to at least 5 cm but maize should not be planted until at least 5 months have elapsed since application.
- In the event of crop failure following spring application spring wheat, spring barley, peas, spring field beans, maize or spring oilseed rape can be sown after the land is ploughed to at least 15 cm and at least 8 weeks have elapsed since application.

Environmental safety
- Dangerous for the environment
- Very toxic to aquatic organisms
- Product binds strongly to soil minimising likelihood of movement into groundwater
- LERAP Category B

Hazard classification and safety precautions
Hazard Dangerous for the environment [1-6]; Very toxic to aquatic organisms [1-5]
Transport code 9
Packaging group III
UN Number 3082
Risk phrases R50 [4, 6]; R53a [6]
Operator protection A, H [1-4]; U02a [1-3, 5, 6]; U05a, U20c [1-6]
Environmental protection E15a [2]; E15b [1, 3-6]; E16a, E34, E38 [1-6]; H410 [1-5]
Storage and disposal D01, D02, D05, D09a, D10c, D12a
Medical advice M03

SEE SECTION 3 FOR PRODUCTS ALSO REGISTERED

349   penthiopyrad

An SDHI fungicide for disease control in cereals, apples and pears
FRAC mode of action code: 7

See also chlorothalonil + penthiopyrad
cyproconazole + penthiopyrad

Products

1 Fontelis	DuPont	200 g/l	SC	17465
2 Intellis	DuPont	200 g/l	EC	17520
3 Vertisan	DuPont	200 g/l	EC	17485

Uses

- Brown rust in **rye**, **spring barley**, **spring wheat**, **triticale**, **winter barley**, **winter wheat** [2, 3]
- Leaf blotch in **rye** *(moderate control)*, **triticale** *(moderate control)* [2, 3]
- Net blotch in **spring barley**, **winter barley** [2, 3]
- Powdery mildew in **apples** *(suppression)*, **pears** *(suppression)* [1]
- Ramularia leaf spots in **winter barley** *(moderate control)* [2, 3]
- Rhynchosporium in **rye**, **spring barley**, **winter barley** [2, 3]
- Scab in **apples**, **pears** [1]
- Septoria leaf blotch in **spring wheat**, **winter wheat** [2, 3]
- Tan spot in **spring wheat** *(reduction)*, **winter wheat** *(reduction)* [2, 3]
- Yellow rust in **rye** *(moderate control)*, **spring barley** *(moderate control)*, **spring wheat** *(moderate control)*, **triticale** *(moderate control)*, **winter barley** *(moderate control)*, **winter wheat** *(moderate control)* [2, 3]

Approval information

- Penthiopyrad is included in Annex 1 under EC Regulation 1107/2009
- Accepted by BBPA for use on malting barley

Efficacy guidance

- Always use in mixture with another product from a different fungicide resistance group that is used for control of the same target disease.
- Best results achieved from applications made before the disease is established in the crop

Restrictions

- Do not use more than two foliar applications of SDHI fungicides on any cereal crop.
- Must be applied in at least 100 l/ha since it is a skin sensitiser.
- Avoid application in either frosty or hot, sunny weather conditions [1]
- Do not apply to any crop suffering from stress as a result of drought, waterlogging, low temperatures, nutrient or lime deficiency or other factors reducing crop growth.
- In order to prevent contamination of groundwater, the second application to all crops must be made after GS39 [2, 3]
- A maximum total dose of not more than 500g penthiopyrad/hectare may be applied in a two year period on the same field

Environmental safety

- Broadcast air-assisted LERAP 40m [1]
- LERAP Category B

Hazard classification and safety precautions

Hazard Irritant, Dangerous for the environment [2, 3]; Very toxic to aquatic organisms [1]
Transport code 9
Packaging group III
UN Number 3082
Risk phrases H317, H319 [2, 3]
Operator protection A [1-3]; C, H [2, 3]; U05a, U09a, U11, U14, U15, U20a [2, 3]
Environmental protection E15b, E16a [1-3]; E17a [1] (40 m); E34, H410 [1]; E38, E40b, E40d, H411 [2, 3]
Storage and disposal D01, D02, D10c [2, 3]; D05 [1]; D09a, D12a [1-3]

FOR FULL CONDITIONS OF USE ALWAYS READ THE PRODUCT LABEL

350   phenmedipham

A contact phenyl carbamate herbicide for beet crops and strawberries
HRAC mode of action code: C1

See also desmedipham + ethofumesate + lenacil + phenmedipham
desmedipham + ethofumesate + metamitron + phenmedipham
desmedipham + ethofumesate + phenmedipham
desmedipham + phenmedipham
ethofumesate + metamitron + phenmedipham
ethofumesate + phenmedipham

Products

1	Beetup Flo	UPL Europe	160 g/l	SC	14328
2	Betasana SC	UPL Europe	160 g/l	SC	14209
3	Corzal	UPL Europe	157 g/l	SE	16952
4	Parish	Adama	320 g/l	SC	17464

Uses

- Annual dicotyledons in **chard** *(off-label)*, **herbs (see appendix 6)** *(off-label)*, **seakale** *(off-label)*, **spinach** *(off-label)*, **spinach beet** *(off-label)* [3]; **fodder beet, mangels, red beet, sugar beet** [1-3]; **winter oilseed rape** [4]
- Black bindweed in **ornamental plant production** *(off-label)* [2]
- Chickweed in **fodder beet, mangels, red beet, sugar beet** [1, 2]; **ornamental plant production** *(off-label)* [2]
- Fat hen in **ornamental plant production** *(off-label)* [2]
- Field pansy in **ornamental plant production** *(off-label)* [2]
- Knotgrass in **ornamental plant production** *(off-label)* [2]
- Penny cress in **ornamental plant production** *(off-label)* [2]
- Red dead-nettle in **winter oilseed rape** *(moderately susceptible)* [4]
- Scentless mayweed in **ornamental plant production** *(off-label)* [2]

Extension of Authorisation for Minor Use (EAMUs)

- **chard** *20151519* [3]
- **herbs (see appendix 6)** *20151519* [3]
- **ornamental plant production** *20152050* [2]
- **seakale** *20151519* [3]
- **spinach** *20151519* [3]
- **spinach beet** *20151519* [3]

Approval information

- Phenmedipham included in Annex I under EC Regulation 1107/2009

Efficacy guidance

- Best results achieved by application to young seedling weeds, preferably cotyledon stage, under good growing conditions when low doses are effective
- If using a low-dose programme 2-3 repeat applications at 7-10 d intervals are recommended on mineral soils, 3-5 applications may be needed on organic soils
- Addition of adjuvant oil may improve effectiveness on some weeds
- Various tank-mixtures with other beet herbicides recommended. See label for details
- Use of certain pre-emergence herbicides is recommended in combination with post-emergence treatment. See label for details

Restrictions

- Maximum number of treatments and maximum total dose varies with crop and product used. See label for details
- At high temperatures (above 21°C) reduce rate and spray after 5 pm
- Do not apply immediately after frost or if frost expected
- Do not spray wet foliage or if rain imminent
- Do not spray crops stressed by wind damage, nutrient deficiency, pest or disease attack etc. Do not roll or harrow for 7 d before or after treatment

SEE SECTION 3 FOR PRODUCTS ALSO REGISTERED

- Do not use on strawberries under cloches or polythene tunnels
- Consult processor before use on crops for processing
- Do not apply to crops with wet leaves [4]

Crop-specific information
- Latest use: before crop leaves meet between rows for beet crops; before flowering for strawberries
- Apply to beet crops at any stage as low dose/low volume spray or from fully developed cotyledon stage with full rate. Apply to red beet after fully developed cotyledon stage
- Apply to strawberries at any time when weeds in susceptible stage, except in period from start of flowering to picking

Following crops guidance
- Beet crops may follow at any time after a treated crop. 3 mth must elapse from treatment before any other crop is sown and must be preceded by mould-board ploughing to 15 cm

Environmental safety
- Dangerous for the environment
- Very toxic to aquatic organisms
- Harmful to fish or other aquatic life. Do not contaminate surface waters or ditches with chemical or used container
- LERAP Category B [1-3]

Hazard classification and safety precautions
Hazard Irritant [1, 2]; Dangerous for the environment [1-3]
Transport code 9
Packaging group III
UN Number 3082
Risk phrases H317, H319 [1, 2]; H318 [4]
Operator protection A [1-4]; C, H [1, 2, 4]; U05a [3]; U08 [1-4]; U11, U19a [1, 2, 4]; U14, U20c [1, 2]; U20b [4]
Environmental protection E15b [1-4]; E16a [1-3]; E16b, E38 [3]; E34, H410 [3, 4]; H411 [1, 2]
Storage and disposal D01, D09a, D12a [1-4]; D02 [1-3]; D05 [1, 2, 4]; D10b [3, 4]; D10c [1, 2]; D12b [4]

351   d-phenothrin

A non-systemic pyrethroid insecticide available only in mixtures
IRAC mode of action code: 3A

352   d-phenothrin + tetramethrin

A pyrethroid insecticide mixture for control of flying insects
IRAC mode of action code: 3 + 3

See also tetramethrin

Products

1 Killgerm ULV 1500	Killgerm	4.8:2.4% w/w	UL	H6961
2 Killgerm ULV 500	Killgerm	4.8:2.4% w/w	UL	H4647

Uses
- Flies in **agricultural premises**
- Grain storage mite in **grain stores**
- Mosquitoes in **agricultural premises**
- Wasps in **agricultural premises**

Approval information
- Product approved for ULV application. See label for details
- D-phenothrin and tetramethrin are not included in Annex 1 under EC Regulation 1107/2009

FOR FULL CONDITIONS OF USE ALWAYS READ THE PRODUCT LABEL

Efficacy guidance
- Close doors and windows and spray in all directions for 3-5 sec. Keep room closed for at least 10 min

Restrictions
- For use only by professional operators
- Do not use space sprays containing pyrethrins or pyrethroid more than once per week in intensive or controlled environment animal houses in order to avoid development of resistance. If necessary, use a different control method or product

Crop-specific information
- May be used in the presence of poultry and livestock

Environmental safety
- Dangerous for the environment
- Toxic to aquatic organisms
- Do not apply directly to livestock/poultry
- Remove exposed milk and collect eggs before application. Protect milk machinery and containers from contamination

Hazard classification and safety precautions
> **Hazard** Harmful, Dangerous for the environment
> **Transport code** 9
> **Packaging group** III
> **UN Number** 3082
> **Risk phrases** R22b, R51, R53a
> **Operator protection** A, C, D, E, H; U02b, U09b, U14, U19a, U20a, U20b
> **Environmental protection** E05a, E15a, E38
> **Consumer protection** C06, C07, C08, C09, C11, C12
> **Storage and disposal** D01, D05, D06a, D09a, D10a, D12a
> **Medical advice** M05b

353   Phlebiopsis gigantea

A fungal protectant against root and butt rot in conifers

Products

PG Suspension	Forest Research	0.5% w/w	SC	17009

Uses
- Fomes root and butt rot in *farm forestry*, *forest*, *forest nurseries*

Approval information
- Phlebiopsis gigantea included in Annex 1 under EC Regulation 1107/2009

Efficacy guidance
- Store sachets in the original box in a cool dry place between 2°C and 15°C out of the sun.
- Use the diluted solution within 24 hours of mixing. Ensure complete coverage of stumps for optimum efficacy.

Hazard classification and safety precautions
> **UN Number** N/C
> **Operator protection** A, D, H; U20c
> **Environmental protection** E15a
> **Storage and disposal** D11a

SECTION 2

SEE SECTION 3 FOR PRODUCTS ALSO REGISTERED

354   physical pest control

Products that work by physical action only

Products

1	SB Plant Invigorator	Fargro	-	SL	-
2	Silico-Sec	Interfarm	>90% w/w	DS	-
3	Vazor DE	Killgerm	n/a	PO	-
4	Vazor Liquid Mosquito Film	Killgerm	n/a	RH	-

Uses
- Aphids in *all edible crops (outdoor and protected), ornamental plant production, protected ornamentals* [1]
- Insect control in *agricultural premises* [3]
- Mealybugs in *all edible crops (outdoor and protected), ornamental plant production, protected ornamentals* [1]
- Mites in *stored grain* [2]
- Mosquitoes in *enclosed waters* [4]
- Powdery mildew in *all edible crops (outdoor and protected), ornamental plant production, protected ornamentals* [1]
- Scale insects in *all edible crops (outdoor and protected), ornamental plant production, protected ornamentals* [1]
- Spider mites in *all edible crops (outdoor and protected), ornamental plant production, protected ornamentals* [1]
- Suckers in *all edible crops (outdoor and protected), ornamental plant production, protected ornamentals* [1]
- Whitefly in *all edible crops (outdoor and protected), ornamental plant production, protected ornamentals* [1]

Approval information
- Products included in this profile are not subject to the Control of Pesticides Regulations/Plant Protection Products Regulations because they act by physical means only

Efficacy guidance
- Products act by physical means following direct contact with spray. They may therefore be used at any time of year on all pest growth stages
- Ensure thorough spray coverage of plant, paying special attention to growing points and the underside of leaves
- Treat as soon as target pests are seen and repeat as often as necessary

Restrictions
- No limit on the number of treatments and no minimum interval between applications
- Before large scale use on a new crop, treat a few plants to check for crop safety
- Do not treat ornamental crops when in flower

Crop-specific information
- HI zero

Environmental safety
- May be used in conjunction with biological control agents. Spray 24 h before they are introduced

Hazard classification and safety precautions
   Hazard Irritant [1]; Highly flammable [3]
   UN Number N/C
   Risk phrases H319, H336, R36, R67 [3]
   Operator protection A, H [3]; F [2]; U10, U14, U15 [1]
   Storage and disposal D01, D02, D05, D06h [1]

355   pinoxaden

A phenylpyrazoline grass weed herbicide for cereals
HRAC mode of action code: A

See also clodinafop-propargyl + pinoxaden
florasulam + pinoxaden

Products

.1 Axial	Syngenta	100 g/l	EC	12521
2 Clayton Rugby	Clayton	100 g/l	EC	16941
3 Pinball	Pan Agriculture	100 g/l	EC	16070
4 PureRecovery	Pure Amenity	45 g/l	EC	15137
5 Rescue	Syngenta	45 g/l	EC	14518

Uses

- Annual grasses in **game cover** *(off-label)* [1]
- Blackgrass in **spring barley, winter barley** [1-3]
- Italian ryegrass in **spring barley, spring wheat, winter barley, winter wheat** [1-3]
- Perennial grasses in **game cover** *(off-label)* [1]
- Perennial ryegrass in **amenity grassland** *(off-label)*, **managed amenity turf** *(off-label)* [5]; **spring barley, spring wheat, winter barley, winter wheat** [1-3]
- Ryegrass in **amenity grassland** *(reduction only)*, **managed amenity turf** *(reduction only)* [4, 5]
- Wild oats in **spring barley, spring wheat, winter barley, winter wheat** [1-3]

Extension of Authorisation for Minor Use (EAMUs)

- **amenity grassland** *20111290* [5]
- **game cover** *20082815* [1]
- **managed amenity turf** *20111290* [5]

Approval information

- Pinoxaden included in Annex 1 under EC Regulation 1107/2009

Efficacy guidance

- Best results obtained from treatment when all grass weeds have emerged. There is no residual activity
- Broad-leaved weeds are not controlled
- Treat before emerged weed competition reduces yield
- Blackgrass in winter and spring barley is also controlled when used as part of an integrated control strategy. Product is not recommended for blackgrass control in winter wheat
- Grass weed control may be reduced if rain falls within 1 hr of application
- Pinoxaden is an ACCase inhibitor herbicide. To avoid the build up of resistance do not apply products containing an ACCase inhibitor herbicide more than twice to any crop. In addition do not use any product containing pinoxaden in mixture or sequence with any other product containing the same ingredient
- Use these products as part of a resistance management strategy that includes cultural methods of control and does not use ACCase inhibitors as the sole chemical method of grass weed control
- Applying a second product containing an ACCase inhibitor to a crop will increase the risk of resistance development; only use a second ACCase inhibitor to control different weeds at a different timing
- Always follow WRAG guidelines for preventing and managing herbicide resistant weeds. See Section 5 for more information

Restrictions

- Maximum number of treatments 1 per crop
- Do not spray crops under stress or suffering from waterlogging, pest attack, disease or frost damage
- Do not spray crops undersown with grass mixtures

- Avoid the use of hormone-containing herbicides in mixture or in sequence. Allow 21 d after, or 7 d before, hormone application
- Product must be used with prescribed adjuvant. See label

Crop-specific information
- Latest use: before flag leaf extending stage (GS 41)
- Spray cereals in autumn, winter or spring from the 2-leaf stage (GS 12)

Following crops guidance
- There are no restrictions on succeeding crops in a normal rotation
- In the event of failure of a treated crop ryegrass, maize, oats or any broad-leaved crop may be planted after a minimum interval of 4 wk from application

Environmental safety
- Dangerous for the environment
- Toxic to aquatic organisms
- Do not allow spray to drift onto neighbouring crops of oats, ryegrass or maize

Hazard classification and safety precautions
Hazard Harmful [4, 5]; Irritant [1-3]; Dangerous for the environment [1-5]
Transport code 9
Packaging group III
UN Number 3082
Risk phrases H304, H317 [1, 2, 4, 5]; H315, H320 [3]; R22b [5]
Operator protection A, H [1-5]; C [1-3]; U05a, U20b [1-5]; U09a [1-3]
Environmental protection E15b, E38 [1-5]; H410 [1, 2, 4, 5]; H411 [3]
Storage and disposal D01, D02, D05, D09a, D10c, D12a [1-5]; D11a [1-3]
Medical advice M05b [4, 5]

356　pirimicarb

A carbamate insecticide for aphid control
IRAC mode of action code: 1A

See also lambda-cyhalothrin + pirimicarb

Products

1 Aphox	Syngenta	50% w/w	WG	17401
2 Clayton Pirimicarb	Clayton	50% w/w	WG	17704
3 Clayton Pirimicarb	Clayton	50% w/w	WG	17759
4 Pirimate 500	Becesane	50% w/w	WG	17657

Uses
- Aphids in **broad beans, combining peas, spring field beans, vining peas, winter field beans** [1-4]; *celeriac (off-label)*, *celery (outdoor)* *(off-label)*, *dwarf beans* *(off-label)*, *edible podded peas* *(off-label)*, *florence fennel* *(off-label)*, *french beans* *(off-label)*, *horseradish* *(off-label)*, *parsley root* *(off-label)*, *protected chilli peppers* *(off-label)*, *protected courgettes* *(off-label)*, *protected gherkins* *(off-label)*, *protected peppers* *(off-label)*, *protected summer squash* *(off-label)*, *radishes* *(off-label)*, *red beet* *(off-label)*, *rhubarb* *(off-label)*, *runner beans* *(off-label)* [1]

Extension of Authorisation for Minor Use (EAMUs)
- *celeriac 20161609* [1]
- *celery (outdoor) 20161610* [1]
- *dwarf beans 20171445* [1]
- *edible podded peas 20171445* [1]
- *florence fennel 20161606* [1]
- *french beans 20171445* [1]
- *horseradish 20161605* [1]
- *parsley root 20161605* [1]
- *protected chilli peppers 20171444* [1]
- *protected courgettes 20161611* [1]

FOR FULL CONDITIONS OF USE ALWAYS READ THE PRODUCT LABEL

- ***protected gherkins*** *20161611* [1]
- ***protected peppers*** *20171444* [1]
- ***protected summer squash*** *20161611* [1]
- ***radishes*** *20161605* [1]
- ***red beet*** *20161605* [1]
- ***rhubarb*** *20161610* [1]
- ***runner beans*** *20171445* [1]

Approval information

- Pirimicarb included in Annex I under EC Regulation 1107/2009
- Accepted by BBPA for use on malting barley
- In 2006 CRD required that all products containing this active ingredient should carry the following warning in the main area of the container label: "Pirimicarb is an anticholinesterase carbamate. Handle with care"

Efficacy guidance

- Chemical has contact, fumigant and translaminar activity
- Best results achieved under warm, calm conditions when plants not wilting and spray does not dry too rapidly. Little vapour activity at temperatures below 15°C
- Apply as soon as aphids seen or warning issued and repeat as necessary
- Addition of non-ionic wetter recommended for use on brassicas
- On cucumbers and tomatoes a root drench is preferable to spraying when using predators in an integrated control programme
- Where aphids resistant to pirimicarb occur control is unlikely to be satisfactory

Restrictions

- Contains an anticholinesterase carbamate compound. Do not use if under medical advice not to work with such compounds
- Maximum number of treatments not specified in some cases but normally 2-6 depending on crop

Crop-specific information

- Latest use in accordance with harvest intervals below
- HI protected courgettes and gherkins 24 h; other edible crops 3 d;

Environmental safety

- Dangerous for the environment
- Very toxic to aquatic organisms
- Dangerous to fish or other aquatic life. Do not contaminate surface waters or ditches with chemical or used container
- Chemical has little effect on bees, ladybirds and other insects and is suitable for use in integrated control programmes on apples and pears
- Keep all livestock out of treated areas for at least 7 d. Bury or remove spillages
- LERAP Category B

Hazard classification and safety precautions

Hazard Toxic, Dangerous for the environment, Toxic if swallowed, Harmful if inhaled
Transport code 6.1
Packaging group III
UN Number 2757
Risk phrases H319
Operator protection A, C, D, H, J, M; U05a, U08, U19a, U20b
Environmental protection E06c (7 d); E15b, E16a, E34, E38, H410
Storage and disposal D01, D02, D09a, D11a, D12a
Medical advice M02, M03, M04a

357 pirimiphos-methyl

A contact, fumigant and translaminar organophosphorus insecticide
IRAC mode of action code: 1B

Products

1	Actellic 50 EC	Syngenta	500 g/l	EC	12726
2	Actellic Smoke Generator No. 20	Syngenta	22.5% w/w	FU	15739
3	Clayton Galic	Clayton	500 g/l	EC	15374
4	Flycatcher	AgChem Access	500 g/l	EC	14135
5	Pan PMT	Pan Agriculture	500 g/l	EC	13996

Uses
- Flour beetle in **stored grain** [1, 3-5]
- Flour moth in **stored grain** [1, 3-5]
- Grain beetle in **stored grain** [1, 3-5]
- Grain storage mite in **stored grain** [1, 3-5]
- Grain storage pests in **grain stores** [2]
- Grain weevil in **stored grain** [1, 3-5]
- Insect pests in **grain stores** [2]
- Mites in **grain stores** [2]
- Warehouse moth in **stored grain** [1, 3-5]

Approval information
- Pirimiphos-methyl included in Annex I under EC Regulation 1107/2009
- In 2006 PSD required that all products containing this active ingredient should carry the following warning in the main area of the container label: "Pirimiphos-methyl is an anticholinesterase organophosphate. Handle with care"
- Accepted by BBPA for use in stores for malting barley

Efficacy guidance
- Chemical acts rapidly and has short persistence in plants, but persists for long periods on inert surfaces
- Best results for protection of stored grain achieved by cleaning store thoroughly before use and employing a combination of pre-harvest and grain/seed treatments [1, 2]
- Best results for admixture treatment obtained when grain stored at 15% moisture or less. Dry and cool moist grain coming into store but then treat as soon as possible, ideally as it is loaded [1]
- Surface admixture can be highly effective on localised surface infestations but should not be relied on for long-term control unless application can be made to the full depth of the infestation [1]
- Where insect pests resistant to pirimiphos-methyl occur control is unlikely to be satisfactory

Restrictions
- Contains an organophosphorus anticholinesterase compound. Do not use if under medical advice not to work with such compounds
- Maximum number of treatments 2 per grain store [1, 2]; 1 per batch of stored grain [1]
- Do not apply surface admixture or complete admixture using hand held equipment [1]
- Surface admixture recommended only where the conditions for store preparation, treatment and storage detailed in the label can be met [1]

Crop-specific information
- Latest use: before storing grain [1, 2]
- Disinfect empty grain stores by spraying surfaces and/or fumigation and treat grain by full or surface admixture. Treat well before harvest in late spring or early summer and repeat 6 wk later or just before harvest if heavily infested. See label for details of treatment and suitable application machinery
- Treatment volumes on structural surfaces should be adjusted according to surface porosity. See label for guidelines
- Treat inaccessible areas with smoke generating product used in conjunction with spray treatment of remainder of store

FOR FULL CONDITIONS OF USE ALWAYS READ THE PRODUCT LABEL

- Predatory mites (*Cheyletus*) found in stored grain feeding on infestations of grain mites may survive treatment and can lead to rejection by buyers. They do not remain for long but grain should be inspected carefully before selling [1, 2]

Environmental safety
- Dangerous for the environment
- Very toxic to aquatic organisms [1]
- Toxic to aquatic organisms [2]
- Highly flammable [2]
- Flammable [1]
- Ventilate fumigated or fogged spaces thoroughly before re-entry
- Unprotected persons must be kept out of fumigated areas within 3 h of ignition and for 4 h after treatment
- Wildlife must be excluded from buildings during treatment [1]
- Keep away from combustible materials [2]

Hazard classification and safety precautions

Hazard Harmful, Dangerous for the environment [1-5]; Highly flammable [2]; Flammable [1, 3-5]; Flammable liquid and vapour [1]; Harmful if swallowed [1, 2, 5]; Harmful if inhaled [5]; Very toxic to aquatic organisms [1, 2]

Transport code 3 [1, 3-5]; 9 [2]

Packaging group III

UN Number 1993 [1, 3-5]; 3077 [2]

Risk phrases H304, H317, H318, H336 [1]; H320, H371 [5]; H335 [1, 5]; R20, R22a, R22b, R36, R37, R50, R53a [3, 4]

Operator protection A, C, J, M [1, 3-5]; D, H [1-5]; U04a, U09a, U16a, U24 [1, 3-5]; U05a, U19a, U20b [1-5]; U14 [2]; U23a [1, 4]

Environmental protection E02a [2] (4 h); E15a, E34, E38 [1-5]; H410 [2, 5]; H411 [1]

Consumer protection C09 [1, 3-5]; C12 [1-5]

Storage and disposal D01, D02, D09a, D12a [1-5]; D05, D10c [1, 3-5]; D07, D11a [2]

Treated seed S06a [1, 3-5]

Medical advice M01 [1-5]; M03 [2]; M05b [1, 3-5]

358   potassium bicarbonate (commodity substance)

An inorganic fungicide for use in horticulture.

Products

1	Atilla	Q-Chem	85% w/w	SP	16026
2	Karma	Certis	85.42% w/w	WP	16363
3	potassium bicarbonate	various	100% w/w	AP	00000

Uses
- Botrytis in *protected aubergines* (off-label), *protected chilli peppers* (off-label), *protected peppers* (off-label), *protected tomatoes* (off-label), *table grapes* (off-label), *wine grapes* (off-label) [2]
- Disease control in *all edible crops (outdoor and protected)*, *all non-edible crops (outdoor)*, *all protected non-edible crops* [3]
- Downy mildew in *lettuce* (off-label), *protected lettuce* (off-label), *table grapes* (off-label), *wine grapes* (off-label) [2]
- Fungus diseases in *apples*, *nursery fruit trees and bushes* [2]
- Insect pests in *pears* [1]
- Pear sucker in *pears* (reduction) [1]
- Powdery mildew in *courgettes* (off-label), *cucumbers* (off-label), *melons* (off-label), *protected aubergines* (off-label), *protected chilli peppers* (off-label), *protected courgettes* (off-label), *protected cucumbers* (off-label), *protected melons* (off-label), *protected peppers* (off-label), *protected tomatoes* (off-label), *strawberries* (off-label), *table grapes* (off-label), *wine grapes* (off-label) [2]
- Sooty moulds in *apples*, *nursery fruit trees and bushes* [2]; *pears* (reduction) [1]

SEE SECTION 3 FOR PRODUCTS ALSO REGISTERED

SECTION 2

Extension of Authorisation for Minor Use (EAMUs)
- *courgettes* *20151941* [2]
- *cucumbers* *20151941* [2]
- *lettuce* *20151940* [2]
- *melons* *20151941* [2]
- *protected aubergines* *20171299* [2]
- *protected chilli peppers* *20171299* [2]
- *protected courgettes* *20151941* [2]
- *protected cucumbers* *20151941* [2]
- *protected lettuce* *20151940* [2]
- *protected melons* *20151941* [2]
- *protected peppers* *20171299* [2]
- *protected tomatoes* *20171299* [2]
- *strawberries* *20150901* [2]
- *table grapes* *20161327* [2]
- *wine grapes* *20161327* [2]

Approval information
- Approval for the use of potassium bicarbonate as a commodity substance was granted on 26 July 2005 by Ministers under regulation 5 of the Control of Pesticides Regulations 1986
- Potassium bicarbonate is being supported for review in the fourth stage of the EC Review Programme under EC Regulation 1107/2009. It was included in Annex 1 under the alternative name of potassium hydrogen carbonate and now products containing it will need to gain approval in the normal way if they carry label claims for pesticidal activity
- Accepted by BBPA for use on hops

Efficacy guidance
- Adjuvants authorised by CRD may be used in conjunction with potassium bicarbonate
- Crop phytotoxicity can occur

Restrictions
- Food grade potassium bicarbonate must only be used

Hazard classification and safety precautions
UN Number N/C
Operator protection C [2]; U05a [1, 2]; U19a [2]
Environmental protection E15a [1, 2]
Storage and disposal D09b [1, 2]

359 prochloraz + propiconazole

A broad spectrum fungicide mixture for wheat, barley and oilseed rape
FRAC mode of action code: 3 + 3

See also propiconazole

Products

Cirkon	Adama	400:90 g/l	EC	16643

Uses
- Eyespot in **spring barley, spring wheat, winter barley, winter wheat**
- Light leaf spot in **spring oilseed rape, winter oilseed rape**
- Phoma in **spring oilseed rape, winter oilseed rape**
- Rhynchosporium in **spring barley, winter barley**
- Septoria leaf blotch in **spring wheat, winter wheat**

Approval information
- Propiconazole and prochloraz included under EC Regulation 1107/2009
- Accepted by BBPA for use on malting barley

Efficacy guidance
- Best results obtained from treatment when disease is active but not well established

FOR FULL CONDITIONS OF USE ALWAYS READ THE PRODUCT LABEL

- Treat wheat normally from flag leaf ligule just visible stage (GS 39) but earlier if there is a high risk of Septoria
- Treat barley from the first node detectable stage (GS 31)
- If disease pressure persists a second application may be necessary
- Prochloraz and propiconazole are DMI fungicides. Resistance to some DMI fungicides has been identified in Septoria leaf blotch which may seriously affect performance of some products. For further advice contact a specialist advisor and visit the Fungicide Resistance Action Group (FRAG)-UK website

Restrictions
- Maximum number of treatments 2 per crop

Crop-specific information
- Latest use: before grain watery ripe (GS 71) for cereals; before most seeds green for winter oilseed rape

Environmental safety
- Dangerous for the environment
- Very toxic to aquatic organisms

Hazard classification and safety precautions
Hazard Irritant, Dangerous for the environment
Transport code 9
Packaging group III
UN Number 3082
Risk phrases R36
Operator protection A, C; U05a, U08, U11, U15, U20b
Environmental protection E15a
Storage and disposal D01, D02, D05, D09a, D10b
Medical advice M05a

360 prochloraz + proquinazid + tebuconazole

A broad spectrum systemic fungicide mixture for cereals
FRAC mode of action code: 3 + U7 + 3

See also proquinazid
tebuconazole

Products

Vareon	DuPont	320:40:160 g/l	EC	14376

Uses
- Brown rust in **spring barley**, **spring wheat**, **winter barley**, **winter wheat**
- Crown rust in **spring oats** (qualified minor use recommendation), **winter oats** (qualified minor use recommendation)
- Ear diseases in **spring wheat** (reduction only), **winter wheat** (reduction only)
- Eyespot in **spring barley**, **spring wheat**, **winter barley**, **winter wheat**
- Glume blotch in **spring wheat**, **winter wheat**
- Net blotch in **spring barley** (moderate control), **winter barley** (moderate control)
- Powdery mildew in **spring barley**, **spring oats**, **spring wheat**, **winter barley**, **winter oats**, **winter wheat**
- Rhynchosporium in **spring barley** (moderate control), **winter barley** (moderate control)
- Septoria leaf blotch in **spring wheat** (moderate control), **winter wheat** (moderate control)
- Yellow rust in **spring barley**, **spring wheat**, **winter barley**, **winter wheat**

Approval information
- Prochloraz, proquinazid and tebuconazole included in Annex 1 under EC Regulation 1107/2009
- Accepted by BBPA for use on malting barley

SEE SECTION 3 FOR PRODUCTS ALSO REGISTERED

SECTION 2

Efficacy guidance

* Resistance to DMI fungicides has been identified in Septoria leaf blotch and may seriously affect the performance of some products. Advice on resistance management is available from the Fungicide Action Group (FRAG) web site

Restrictions

* Maximum number of applications is 2 per crop
* Newer authorisations for tebuconazole products require application to cereals only after GS 30 and applications to oilseed rape and linseed after GS20 - check label

Crop-specific information

* Latest application in wheat is before full flowering (GS65)
* Latest application in barley is before beginning of heading (GS49)

Environmental safety

* Avoid spraying within 5 m of the field boundary to reduce the effect on non-target insects or other arthropods
* LERAP Category B

Hazard classification and safety precautions

Hazard Harmful, Dangerous for the environment
Transport code 9
Packaging group III
UN Number 3082
Risk phrases H317, H351, H361, R63
Operator protection A, C, H
Environmental protection E15a, E16a, E22c, E34, E38, H410
Storage and disposal D01, D02, D09a, D12a

361   prochloraz + tebuconazole

A broad spectrum systemic fungicide mixture for cereals, oilseed rape and amenity use
FRAC mode of action code: 3 + 3

See also tebuconazole

Products

1	Monkey	Adama	267:133 g/l	EW	12906
2	Orius P	Adama	267:133 g/l	EW	12880

Uses

* Brown rust in *spring barley, spring rye, spring wheat, winter barley, winter rye, winter wheat*
* Eyespot in *spring barley, spring rye, spring wheat, winter barley, winter rye, winter wheat*
* Glume blotch in *spring wheat, winter wheat*
* Late ear diseases in *spring wheat, winter wheat*
* Net blotch in *spring barley, winter barley*
* Powdery mildew in *spring barley, spring rye, spring wheat, winter barley, winter rye, winter wheat*
* Rhynchosporium in *spring barley, spring rye, winter barley, winter rye*
* Sclerotinia stem rot in *spring oilseed rape, winter oilseed rape*
* Septoria leaf blotch in *spring wheat, winter wheat*
* Yellow rust in *spring barley, spring rye, spring wheat, winter barley, winter rye, winter wheat*

Approval information

* Prochloraz and tebuconazole included in Annex I under EC Regulation 1107/2009
* Accepted by BBPA for use on malting barley

Efficacy guidance

* Protection of flag leaf and ear from Septoria diseases requires treatment at flag leaf emergence and repeated when ear fully emerged [1]

FOR FULL CONDITIONS OF USE ALWAYS READ THE PRODUCT LABEL

- Two treatments separated by a 2 wk gap may be needed for difficult Fusarium patch attacks
- To avoid resistance, do not apply repeated applications of the single product on the same crop against the same disease
- Optimum application timing is normally when disease first seen but varies with main target disease - see labels
- Prochloraz and tebuconazole are DMI fungicides. Resistance to some DMI fungicides has been identified in Septoria leaf blotch which may seriously affect performance of some products. For further advice contact a specialist advisor and visit the Fungicide Resistance Action Group (FRAG)-UK website

Restrictions

- Maximum total dose on cereals equivalent to two full dose treatments
- Maximum number of treatments on managed amenity turf 2 per yr
- Newer authorisations for tebuconazole products require application to cereals only after GS 30 and applications to oilseed rape and linseed after GS 20 - check label
- Do not exceed 450 g prochloraz/hectare per application for outdoor uses

Crop-specific information

- Latest use: before grain milky ripe (GS 73) for cereals [1]; most seeds green for oilseed rape [1]
- HI 6 wk for cereals [1]
- Occasionally transient leaf speckling may occur after treating wheat. Yield responses should not be affected

Environmental safety

- Dangerous for the environment
- Very toxic to aquatic organisms

Hazard classification and safety precautions

Hazard Harmful, Dangerous for the environment, Harmful if swallowed, Very toxic to aquatic organisms

Transport code 9
Packaging group III
UN Number 3082
Risk phrases H319, H361
Operator protection A, C, H; U05a, U09b, U14, U20a
Environmental protection E13b, E34, E38, H410
Storage and disposal D01, D02, D05, D09a, D10c, D12a
Medical advice M03

---

362   prohexadione-calcium

A cyclohexanecarboxylate growth regulator for use in apples

See also mepiquat chloride + prohexadione-calcium

Products

1	Kudos	Fine	10% w/w	WG	17578
2	Regalis Plus	BASF	10% w/w	WG	16485

Uses

- Botrytis in **table grapes** *(off-label)*, **wine grapes** *(off-label)* [2]
- Control of shoot growth in **apples**, **pears** *(off-label)* [1, 2]; **hops in propagation** *(off-label)*, **nursery fruit trees** *(off-label)*, **ornamental plant production** *(off-label)*, **plums** *(off-label)*, **protected ornamentals** *(off-label)* [2]

Extension of Authorisation for Minor Use (EAMUs)

- **hops in propagation** *20150175* [2]
- **nursery fruit trees** *20150175* [2]
- **ornamental plant production** *20150175* [2]
- **pears** *20171068* [1], *20150182* [2]
- **plums** *20171174* [2]

SEE SECTION 3 FOR PRODUCTS ALSO REGISTERED

- ***protected ornamentals*** *20150181* [2]
- ***table grapes*** *20150180* [2]
- ***wine grapes*** *20150180* [2]

Approval information
- Prohexadione-calcium included in Annex I under EC Regulation 1107/2009

Efficacy guidance
- For a standard orchard treat at the start of active growth and again after 3-5 wk depending on growth conditions
- Alternatively follow the first application with four reduced rate applications at two wk intervals

Restrictions
- Maximum total dose equivalent to two full dose treatments
- Do not apply in conjunction with calcium based foliar fertilisers

Crop-specific information
- HI 55 d for apples

Environmental safety
- Harmful to aquatic organisms
- Avoid spray drift onto neighbouring crops and other non-target plants
- Product does not harm natural insect predators when used as directed

Hazard classification and safety precautions
UN Number N/C
Risk phrases H317
Operator protection U02a, U08, U19a, U20b [1]; U05a [1, 2]
Environmental protection E15b, H411 [1]; E38 [1, 2]
Storage and disposal D01, D02, D09a, D12a [1, 2]; D10b [1]; D10c [2]
Medical advice M05a [1]

363 prohexadione-calcium + trinexapac-ethyl

A plant growth regulator for use in cereal crops

See also trinexapac-ethyl

Products

Medax Max	BASF	0.5:0.75% w/w	SG	17263

Uses
- Growth regulation in ***durum wheat*, *spring barley*, *spring oats*, *spring wheat*, *triticale*, *winter barley*, *winter oats*, *winter rye*, *winter wheat***

Hazard classification and safety precautions
Transport code 9
Packaging group III
UN Number 3077
Operator protection A; U08, U19a, U20a
Environmental protection E15b, E34, H412
Storage and disposal D01, D02, D05, D09a, D10c
Medical advice M03

FOR FULL CONDITIONS OF USE ALWAYS READ THE PRODUCT LABEL

364   propamocarb hydrochloride

A translocated protectant carbamate fungicide
FRAC mode of action code: 28

See also chlorothalonil + propamocarb hydrochloride
fenamidone + propamocarb hydrochloride
fluopicolide + propamocarb hydrochloride
fosetyl-aluminium + propamocarb hydrochloride
mancozeb + propamocarb hydrochloride

SECTION 2

Products

Proplant	Fargro	722 g/l	SL	15422

Uses

- Damping off in **protected broccoli, protected brussels sprouts, protected calabrese, protected cauliflowers, protected tomatoes** *(off-label)*
- Downy mildew in **protected broccoli, protected brussels sprouts, protected calabrese, protected cauliflowers, protected watercress** *(off-label)*, **radishes** *(off-label)*
- Peronospora spp in **forest nurseries** *(off-label)*, **ornamental plant production** *(off-label)*, **protected broccoli, protected brussels sprouts, protected calabrese, protected cauliflowers, protected ornamentals** *(off-label)*
- Phytophthora in **forest nurseries** *(off-label)*, **ornamental plant production** *(off-label)*, **protected broccoli, protected brussels sprouts, protected calabrese, protected cauliflowers, protected ornamentals** *(off-label)*, **protected tomatoes** *(off-label)*, **protected watercress** *(off-label)*
- Pythium in **forest nurseries** *(off-label)*, **ornamental plant production** *(off-label)*, **protected broccoli, protected brussels sprouts, protected calabrese, protected cauliflowers, protected ornamentals** *(off-label)*, **protected watercress** *(off-label)*
- Root rot in **protected broccoli, protected brussels sprouts, protected calabrese, protected cauliflowers, protected tomatoes** *(off-label)*

Extension of Authorisation for Minor Use (EAMUs)

- **forest nurseries** *20160785*
- **ornamental plant production** *20160785*
- **protected ornamentals** *20160785*
- **protected tomatoes** *20160784*
- **protected watercress** *20160783*
- **radishes** *20160782*

Approval information

- Propamocarb hydrochloride included in Annex I under EC Regulation 1107/2009

Efficacy guidance

- Chemical is absorbed through roots and translocated throughout plant
- Incorporate in compost before use or drench moist compost or soil before sowing, pricking out, striking cuttings or potting up
- Concentrated solution is corrosive to all metals other than stainless steel
- May also be applied in trickle irrigation systems

Restrictions

- Maximum number of treatments 1 per crop for listed brassicas;
- When applied over established seedlings rinse off foliage with water and do not apply under hot, dry conditions
- Do not apply in a recirculating irrigation/drip system
- Store away from seeds and fertilizers
- Consult processor before use on crops grown for processing

Crop-specific information

- Latest use: before transplanting for brassicas

SEE SECTION 3 FOR PRODUCTS ALSO REGISTERED

Hazard classification and safety precautions
Hazard Irritant
UN Number N/C
Risk phrases H317
Operator protection A, H, M; U02a, U05a, U08, U19a, U20b
Environmental protection E15a, E34
Storage and disposal D01, D02, D05, D09a, D10c

365   propaquizafop

A phenoxy alkanoic acid foliar acting herbicide for grass weeds in a range of crops
HRAC mode of action code: A

Products

1 Clayton Enigma	Clayton	100 g/l	EC	17391
2 Clayton Enigma	Clayton	100 g/l	EC	17435
3 Clayton Satchmo	Clayton	100 g/l	EC	17061
4 Falcon	Adama	100 g/l	EC	16459
5 Longhorn	AgChem Access	100 g/l	EC	16951
6 Shogun	Adama	100 g/l	EC	16527

Uses
- Annual grasses in *all edible seed crops grown outdoors* (off-label), *all non-edible seed crops grown outdoors* (off-label), *game cover* (off-label), *poppies for morphine production* (off-label) [4, 6]; *broad beans, bulb onions, carrots, combining peas, dwarf beans, early potatoes, fodder beet, french beans, linseed, maincrop potatoes, spring field beans, spring oilseed rape, sugar beet, swedes, turnips, winter field beans, winter oilseed rape* [1-6]; *forest* (off-label), *forest nurseries* (off-label) [4]; *garlic* (off-label), *lupins* (off-label), *red beet* (off-label), *shallots* (off-label) [6]
- Annual meadow grass in *all edible seed crops grown outdoors* (off-label), *all non-edible seed crops grown outdoors* (off-label), *game cover* (off-label), *poppies for morphine production* (off-label), *red beet* (off-label) [6]; *garlic* (off-label), *lupins* (off-label), *shallots* (off-label) [4, 6]
- Grass weeds in *garlic* (off-label), *lupins* (off-label), *shallots* (off-label) [4]
- Perennial grasses in *all edible seed crops grown outdoors* (off-label), *all non-edible seed crops grown outdoors* (off-label), *game cover* (off-label), *poppies for morphine production* (off-label) [4, 6]; *broad beans, bulb onions, carrots, combining peas, dwarf beans, early potatoes, fodder beet, french beans, linseed, maincrop potatoes, spring field beans, spring oilseed rape, sugar beet, swedes, turnips, winter field beans, winter oilseed rape* [1-6]; *forest* (off-label), *forest nurseries* (off-label) [4]; *garlic* (off-label), *lupins* (off-label), *red beet* (off-label), *shallots* (off-label) [6]
- Volunteer cereals in *poppies for morphine production* (off-label), *red beet* (off-label) [4, 6]
- Volunteer rye in *red beet* (off-label) [4]

Extension of Authorisation for Minor Use (EAMUs)
- *all edible seed crops grown outdoors* 20140751 [4], 20141283 [4], 20141927 [6], 20142509 [6]
- *all non-edible seed crops grown outdoors* 20140751 [4], 20141283 [4], 20141927 [6], 20142509 [6]
- *forest* 20163458 [4]
- *forest nurseries* 20163458 [4]
- *game cover* 20140751 [4], 20141283 [4], 20141927 [6], 20142509 [6]
- *garlic* 20140701 [4], 20141281 [4], 20141926 [6]
- *lupins* 20140702 [4], 20141282 [4], 20141928 [6]
- *poppies for morphine production* 20152056 [4], 20160285 [6]
- *red beet* 20151827 [4], 20160286 [6]
- *shallots* 20140701 [4], 20141281 [4], 20141926 [6]

Approval information
- Propaquizafop included in Annex 1 under EC Regulation 1107/2009

FOR FULL CONDITIONS OF USE ALWAYS READ THE PRODUCT LABEL

Efficacy guidance

- Apply to emerged weeds when they are growing actively with adequate soil moisture
- Activity is slower under cool conditions
- Broad-leaved weeds and any weeds germinating after treatment are not controlled
- Annual meadow grass up to 3 leaves checked at low doses and severely checked at highest dose
- Spray barley cover crops when risk of wind blow has passed and before there is serious competition with the crop
- Various tank mixtures and sequences recommended for broader spectrum weed control in oilseed rape, peas and sugar beet. See label for details
- Severe couch infestations may require a second application at reduced dose when regrowth has 3-4 leaves unfolded
- Products contain surfactants. Tank mixing with adjuvants not required or recommended
- Propaquizafop is an ACCase inhibitor herbicide. To avoid the build up of resistance do not apply products containing an ACCase inhibitor herbicide more than twice to any crop. In addition do not use any product containing propaquizafop in mixture or sequence with any other product containing the same ingredient
- Use these products as part of a resistance management strategy that includes cultural methods of control and does not use ACCase inhibitors as the sole chemical method of grass weed control
- Applying a second product containing an ACCase inhibitor to a crop will increase the risk of resistance development; only use a second ACCase inhibitor to control different weeds at a different timing
- Always follow WRAG guidelines for preventing and managing herbicide resistant weeds. See Section 5 for more information

Restrictions

- Maximum number of treatments 1 or 2 per crop or yr. See label for details
- See label for list of tolerant tree species
- Do not treat seed potatoes

Crop-specific information

- Latest use: before crop flower buds visible for winter oilseed rape, linseed, field beans; before 8 fully expanded leaf stage for spring oilseed rape; before weeds are covered by the crop for potatoes, sugar beet, fodder beet, swedes, turnips; when flower buds visible for peas
- HI onions, carrots, early potatoes, parsnips 4 wk; combining peas, early potatoes 7 wk; sugar beet, fodder beet, maincrop potatoes, swedes, turnips 8 wk; field beans 14 wk
- Application in high temperatures and/or low soil moisture content may cause chlorotic spotting especially on combining peas and field beans
- Overlaps at the highest dose can cause damage from early applications to carrots and parsnips

Following crops guidance

- In the event of a failed treated crop an interval of 2 wk must elapse between the last application and redrilling with winter wheat or winter barley. 4 wk must elapse before sowing oilseed rape, peas or field beans, and 16 wk before sowing ryegrass or oats

Environmental safety

- Dangerous for the environment
- Toxic to aquatic organisms
- Risk to certain non-target insects or other arthropods. See directions for use
- LERAP Category B

Hazard classification and safety precautions

Hazard Harmful, Dangerous for the environment
Transport code 9
Packaging group III
UN Number 3082
Risk phrases H304, H319 [1-3]; R20, R36, R38, R40, R51, R53a [4-6]
Operator protection A, C [1-6]; H [5, 6]; U02a, U05a, U08, U11, U13, U14, U15, U19a, U20b
Environmental protection E15a, E16a, E22b, E34, E38 [1-6]; H411 [1-3]

SEE SECTION 3 FOR PRODUCTS ALSO REGISTERED

SECTION 2

Storage and disposal D01, D02, D05, D09a, D12a, D14 [1-6]; D10b [1-4]; D12b [5, 6]
Medical advice M05b

366   propiconazole

A systemic, curative and protectant triazole fungicide
FRAC mode of action code: 3

See also azoxystrobin + propiconazole
chlorothalonil + cyproconazole + propiconazole
chlorothalonil + fludioxonil + propiconazole
chlorothalonil + propiconazole
cyproconazole + propiconazole
difenoconazole + propiconazole
fenbuconazole + propiconazole
fenpropidin + propiconazole
fenpropidin + propiconazole + tebuconazole
prochloraz + propiconazole
propiconazole + tebuconazole
propiconazole + trifloxystrobin

Products

1	Banner Maxx	Syngenta	156 g/l	EC	13167
2	Banner Maxx II	Syngenta	162 g/l	EC	18038
3	Bounty	AgChem Access	250 g/l	EC	14616
4	Bumper 250 EC	Adama	250 g/l	EC	14399
5	GF Propiconazole 250	AgChem Access	250 g/l	EC	17154
6	Mascot Exocet	Rigby Taylor	156 g/l	EC	15598
7	Propi 25 EC	Sharda	250 g/l	EC	14939
8	Span	ProKlass	156 g/l	EC	17165
9	Spaniel	Pan Amenity	156 g/l	EC	15530

Uses

- Anthracnose in **amenity grassland** *(qualified minor use)* [1, 6, 8, 9]; **managed amenity turf** *(qualified minor use)* [1, 2, 6, 8, 9]
- Brown patch in **amenity grassland** *(qualified minor use)* [1, 6, 8, 9]; **managed amenity turf** *(qualified minor use)* [1, 2, 6, 8, 9]
- Brown rust in **spring barley, spring wheat, winter barley, winter rye, winter wheat** [3-5, 7]
- Cladosporium leaf blotch in **garlic** *(off-label)*, **onions** *(off-label)*, **shallots** *(off-label)* [4]
- Crown rust in **grass seed crops, permanent grassland, spring oats, winter oats** [3-5, 7]
- Disease control in **forest nurseries** *(off-label)*, **garlic** *(off-label)*, **honesty** *(off-label)*, **hops** *(off-label)*, **onions** *(off-label)*, **ornamental plant production** *(off-label)*, **protected forest nurseries** *(off-label)*, **protected ornamentals** *(off-label)*, **shallots** *(off-label)*, **soft fruit** *(off-label)* [4]
- Dollar spot in **amenity grassland** [1, 6, 8, 9]; **managed amenity turf** [1, 2, 6, 8, 9]
- Downy mildew in **garlic** *(off-label)*, **onions** *(off-label)*, **shallots** *(off-label)* [4]
- Drechslera leaf spot in **grass seed crops, permanent grassland** [3-5, 7]
- Fusarium patch in **amenity grassland** [1, 6, 8, 9]; **managed amenity turf** [1, 2, 6, 8, 9]
- Glume blotch in **spring wheat, winter rye, winter wheat** [3-5, 7]
- Light leaf spot in **spring oilseed rape** *(reduction)*, **winter oilseed rape** *(reduction)* [3-5, 7]
- Mildew in **grass seed crops, permanent grassland** [3-5, 7]
- Powdery mildew in **spring barley, spring oats, spring wheat, winter barley, winter oats, winter rye, winter wheat** [3-5, 7]
- Ramularia leaf spots in **sugar beet** *(reduction)* [3-5, 7]
- Rhynchosporium in **grass seed crops, permanent grassland, spring barley, winter barley, winter rye** [3-5, 7]
- Rust in **garlic** *(off-label)*, **onions** *(off-label)*, **ornamental plant production** *(off-label)*, **protected ornamentals** *(off-label)*, **shallots** *(off-label)* [4]; **sugar beet** [3-5, 7]
- Septoria leaf blotch in **spring wheat, winter rye, winter wheat** [3-5, 7]
- Sooty moulds in **spring wheat, winter wheat** [3-5, 7]

FOR FULL CONDITIONS OF USE ALWAYS READ THE PRODUCT LABEL

- White rust in *ornamental plant production* *(off-label)*, *protected ornamentals* *(off-label)* [4]
- Yellow rust in *spring barley*, *spring wheat*, *winter barley*, *winter wheat* [3-5, 7]

Extension of Authorisation for Minor Use (EAMUs)
- *forest nurseries* *20111133* [4], *20141275* [4]
- *garlic* *20090705* [4], *20141272* [4]
- *honesty* *20090706* [4], *20141273* [4]
- *hops* *20111133* [4], *20141275* [4]
- *onions* *20090705* [4], *20141272* [4]
- *ornamental plant production* *20090707* [4], *20141274* [4]
- *protected forest nurseries* *20111133* [4], *20141275* [4]
- *protected ornamentals* *20090707* [4], *20141274* [4]
- *shallots* *20090705* [4], *20141272* [4]
- *soft fruit* *20111133* [4], *20141275* [4]

Approval information
- Propiconazole included in Annex I under EC Regulation 1107/2009
- Accepted by BBPA for use on malting barley

Efficacy guidance
- Best results achieved by applying at early stage of disease development. Recommended spray programmes vary with crop, disease, season, soil type and product.
- For optimum turf quality and control use in conjunction with turf management practices that promote good plant health [1]
- Propiconazole is a DMI fungicide. Resistance to some DMI fungicides has been identified in Septoria leaf blotch which may seriously affect performance of some products. For further advice contact a specialist advisor and visit the Fungicide Resistance Action Group (FRAG)-UK website

Restrictions
- On oilseed rape do not apply during flowering.
- Grass seed crops must be treated in yr of harvest
- A minimum interval of 14 d must elapse between treatments on leeks and 21 d on sugar beet
- Avoid spraying crops or turf under stress, eg during cold weather or periods of frost

Crop-specific information
- Latest use: before 4 pairs of true leaves for honesty
- HI cereals, grass seed crops 35 d; oilseed rape, sugar beet, garlic, onions, shallots, grass for ensiling 28 d
- May be used on all common turf grass species [1]

Environmental safety
- Dangerous for the environment
- Toxic to aquatic organisms
- Risk to non-target insects or other arthropods [1]
- When treating turf with vehicle mounted or drawn hydraulic boom sprayers avoid spraying within 5 m of unmanaged land to reduce effects on non-target arthropods [1]

Hazard classification and safety precautions
Hazard Harmful [1, 2, 6, 8, 9]; Dangerous for the environment [1-9]; Harmful if swallowed [1, 2]; Harmful if inhaled [6, 9]; Very toxic to aquatic organisms [5]
Transport code 9
Packaging group III
UN Number 3082
Risk phrases H317 [6, 9]; R20, R43, R52 [8]; R51 [3, 4]; R53a [3, 4, 8]
Operator protection A [1-6, 8, 9]; C [3-5]; H [1, 2, 6, 8, 9]; U02a, U05a, U19a, U20b [1-9]; U08 [3-5, 7]; U09a, U14 [1, 2, 6, 8, 9]
Environmental protection E15a, E34 [3-5, 7]; E15b, E22c, E38 [1, 2, 6, 8, 9]; H410 [5]; H411 [1, 2, 7]; H412 [6, 9]
Storage and disposal D01, D02, D05, D09a [1-9]; D10b [3-5, 7]; D10c, D12a [1, 2, 6, 8, 9]
Treated seed S06a [3-5, 7]
Medical advice M05a [3-5, 7]

SEE SECTION 3 FOR PRODUCTS ALSO REGISTERED

367   propoxycarbazone-sodium

A sulfonylaminocarbonyl triazolinone residual grass weed herbicide for winter wheat
HRAC mode of action code: B

See also iodosulfuron-methyl-sodium + propoxycarbazone-sodium
mesosulfuron-methyl + propoxycarbazone-sodium

Products

| Attribut | Interfarm | 70% w/w | SG | 14749 |

Uses

- Annual grasses in **miscanthus** *(off-label)*
- Blackgrass in **winter wheat**
- Couch in **miscanthus** *(off-label)*, **winter wheat**

Extension of Authorisation for Minor Use (EAMUs)

- **miscanthus** *20093179*

Approval information

- Propoxycarbazone-sodium included in Annex I under EC Regulation 1107/2009

Efficacy guidance

- Activity by root and foliar absorption but depends on presence of sufficient soil moisture to ensure root uptake by weeds. Control is enhanced by use of an adjuvant to encourage foliar uptake
- Best results obtained from treatments applied when weed grasses are growing actively. Symptoms may not become apparent for 3-4 wk after application
- Ensure good even spray coverage
- To achieve best control of couch and reduction of infestation in subsequent crop treat between 2 true leaves and first node stage of the weed
- Blackgrass should be treated after using specialist blackgrass herbicides with a different mode of action
- Propoxycarbazone-sodium is a member of the ALS-inhibitor group of herbicides. To avoid the build up of resistance do not use any product containing an ALS-inhibitor herbicide with claims for control of grass weeds more than once on any crop
- Use these products as part of a resistance management strategy that includes cultural methods of control and does not use ALS inhibitors as the sole chemical method of grass weed control

Restrictions

- Maximum total dose equivalent to one full dose treatment
- Do not use in a programme with other aceto-lactase synthesis (ALS) inhibitors
- Do not apply when temperature near or below freezing
- Avoid treatment under dry soil conditions

Crop-specific information

- Latest use: third node detectable in wheat (GS 33)

Following crops guidance

- Only winter wheat, field beans or winter barley may be sown in the autumn following spring treatment
- Any crop may be grown in the spring on land treated during the previous calendar yr

Environmental safety

- Dangerous for the environment
- Very toxic to aquatic organisms
- Take extreme care to avoid drift onto adjacent plants or land as this could result in severe damage
- LERAP Category B

Hazard classification and safety precautions

Hazard Dangerous for the environment
Transport code 9
Packaging group III

FOR FULL CONDITIONS OF USE ALWAYS READ THE PRODUCT LABEL

UN Number 3077
Operator protection A; U05a, U20b
Environmental protection E15b, E16a, E16b, E38, H410
Storage and disposal D01, D02, D09a, D11a, D12a

368   propyzamide

A residual benzamide herbicide for use in a wide range of crops
HRAC mode of action code: K1

See also aminopyralid + propyzamide
*   clopyralid + propyzamide*

Products

1	Artax Flo	Stefes	400 g/l	SC	16317
2	Barclay Propyz	Barclay	400 g/l	SC	15083
3	Careca	UPL Europe	500 g/l	SC	16976
4	Cohort	Adama	400 g/l	SC	15035
5	Judo	Headland	400 g/l	SC	15346
6	Kerb Flo	Dow	400 g/l	SC	13716
7	Kerb Flo 500	Dow	500 g/l	SC	15586
8	Kerb Granules	Barclay	4% w/w	GR	14213
9	Levada	Certis	400 g/l	SC	15743
10	Proper Flo	Globachem	400 g/l	SC	15102
11	PureFlo	Pure Amenity	400 g/l	SC	15138
12	Relva	Belcrop	400 g/l	SC	14873
13	Relva Granules	Belcrop	4% w/w	GR	16744
14	Setanta 50 WP	Mitsui	50% w/w	WP	16181
15	Setanta Flo	Certis	400 g/l	SC	15791
16	Solitaire	Certis	400 g/l	SC	15792
17	Stroller	AgChem Access	400 g/l	SC	14603
18	Stymie	Becesane	400 g/l	SC	16328
19	Zamide 80 WG	Albaugh UK	80% w/w	WG	14723
20	Zamide Flo	Albaugh UK	400 g/l	SC	14679

Uses

- Annual and perennial weeds in *amenity vegetation* [9, 15, 16]
- Annual dicotyledons in *all edible seed crops grown outdoors* (off-label), *all non-edible seed crops grown outdoors* (off-label), *almonds* (off-label), *bilberries* (off-label), *blueberries* (off-label), *bog myrtle* (off-label), *broccoli* (off-label), *calabrese* (off-label), *cauliflowers* (off-label), *cherries* (off-label), *chestnuts* (off-label), *chicory root* (off-label), *cob nuts* (off-label), *corn gromwell* (off-label), *courgettes* (off-label), *cranberries* (off-label), *cress* (off-label), *endives* (off-label), *frise* (off-label), *game cover* (off-label), *hazel nuts* (off-label), *herbs (see appendix 6)* (off-label), *hops* (off-label), *lamb's lettuce* (off-label), *leaf brassicas* (off-label), *marrows* (off-label), *mirabelles* (off-label), *protected endives* (off-label), *protected forest nurseries* (off-label), *protected herbs (see appendix 6)* (off-label), *protected lettuce* (off-label), *pumpkins* (off-label), *quinces* (off-label), *radicchio* (off-label), *salad brassicas* (off-label), *scarole* (off-label), *squashes* (off-label), *table grapes* (off-label), *walnuts* (off-label), *wine grapes* (off-label) [6]; *amenity vegetation* [2-4, 6, 8, 11, 13, 17, 18]; *apple orchards, clover seed crops, pear orchards, plums, rhubarb* [2-4, 6, 9, 11, 15-18]; *blackberries, blackcurrants, gooseberries, loganberries, raspberries* (England only), *redcurrants, strawberries* [2, 4, 6, 9, 11, 15-18]; *chicory, endives, witloof* [3]; *farm forestry* [1-4, 6, 8, 9, 11, 13-20]; *fodder rape seed crops, kale seed crops, sugar beet seed crops, turnip seed crops* [2-4, 6, 9-11, 15-18]; *forest* [1-4, 6, 8-11, 13-20]; *forest nurseries* [1-4, 6, 9, 11, 14-20]; *hedges* [2-4, 6, 9, 11, 14-18]; *lettuce* [1, 3, 4, 6, 9, 10, 14-20]; *lucerne* [4, 6, 9, 11, 15-18]; *ornamental plant production* [8, 13]; *red clover, white clover* [10]; *winter field beans* [1, 4, 6, 7, 9-11, 14-20]; *winter oilseed rape* [1-7, 9-12, 14-20]
- Annual grasses in *all edible seed crops grown outdoors* (off-label), *all non-edible seed crops grown outdoors* (off-label), *almonds* (off-label), *bilberries* (off-label), *blueberries* (off-label),

bog myrtle (off-label), *broccoli* (off-label), *calabrese* (off-label), *cauliflowers* (off-label), *cherries* (off-label), *chestnuts* (off-label), *chicory root* (off-label), *cob nuts* (off-label), *corn gromwell* (off-label), *courgettes* (off-label), *cranberries* (off-label), *cress* (off-label), *endives* (off-label), *frise* (off-label), *game cover* (off-label), *hazel nuts* (off-label), *herbs (see appendix 6)* (off-label), *hops* (off-label), *lamb's lettuce* (off-label), *leaf brassicas* (off-label), *marrows* (off-label), *mirabelles* (off-label), *ornamental plant production* (off-label), *protected endives* (off-label), *protected forest nurseries* (off-label), *protected herbs (see appendix 6)* (off-label), *protected lettuce* (off-label), *pumpkins* (off-label), *quinces* (off-label), *radicchio* (off-label), *salad brassicas* (off-label), *scarole* (off-label), *squashes* (off-label), *table grapes* (off-label), *walnuts* (off-label), *wine grapes* (off-label) [6]; *amenity vegetation* [2-4, 6, 8, 11, 13, 17, 18]; *apple orchards, clover seed crops, pear orchards, plums, rhubarb* [2-4, 6, 9, 11, 15-18]; *blackberries, blackcurrants, gooseberries, loganberries, raspberries* (England only), *redcurrants, strawberries* [2, 4, 6, 9, 11, 15-18]; *chicory, endives, witloof* [3]; *farm forestry* [1-4, 6, 8, 9, 11, 13-20]; *fodder rape seed crops, kale seed crops, sugar beet seed crops, turnip seed crops* [2-4, 6, 9-11, 15-18]; *forest* [1-4, 6, 8-11, 13-20]; *forest nurseries* [1-4, 6, 9, 11, 14-20]; *hedges* [2-4, 6, 9, 11, 14-18]; *lettuce* [1, 3, 4, 6, 9, 10, 14-20]; *lucerne* [4, 6, 9, 11, 15-18]; *ornamental plant production* [8, 13]; *red clover, white clover* [10]; *winter field beans* [1, 4, 6, 9-11, 14-20]; *winter oilseed rape* [1-6, 9-12, 14-20]

- Annual meadow grass in *winter field beans, winter oilseed rape* [7]
- Blackgrass in *winter field beans, winter oilseed rape* [7]
- Fat hen in *ornamental plant production* (off-label) [6]
- Perennial grasses in *all edible seed crops grown outdoors* (off-label), *all non-edible seed crops grown outdoors* (off-label), *almonds* (off-label), *bilberries* (off-label), *blueberries* (off-label), *bog myrtle* (off-label), *broccoli* (off-label), *calabrese* (off-label), *cauliflowers* (off-label), *cherries* (off-label), *chestnuts* (off-label), *chicory root* (off-label), *cob nuts* (off-label), *corn gromwell* (off-label), *courgettes* (off-label), *cranberries* (off-label), *cress* (off-label), *endives* (off-label), *frise* (off-label), *game cover* (off-label), *hazel nuts* (off-label), *herbs (see appendix 6)* (off-label), *hops* (off-label), *lamb's lettuce* (off-label), *leaf brassicas* (off-label), *marrows* (off-label), *mirabelles* (off-label), *protected endives* (off-label), *protected forest nurseries* (off-label), *protected herbs (see appendix 6)* (off-label), *protected lettuce* (off-label), *pumpkins* (off-label), *quinces* (off-label), *radicchio* (off-label), *salad brassicas* (off-label), *scarole* (off-label), *squashes* (off-label), *table grapes* (off-label), *walnuts* (off-label), *wine grapes* (off-label) [6]; *amenity vegetation* [2-4, 6, 8, 11, 13, 17, 18]; *apple orchards, clover seed crops, forest nurseries, hedges, pear orchards, plums, rhubarb* [2-4, 6, 9, 11, 15-18]; *blackberries, blackcurrants, gooseberries, loganberries, raspberries* (England only), *redcurrants, strawberries* [2, 4, 6, 9, 11, 15-18]; *chicory, endives, witloof* [3]; *farm forestry* [2-4, 6, 8, 9, 11, 13, 15-18]; *fodder rape seed crops, kale seed crops, sugar beet seed crops, turnip seed crops* [2-4, 6, 9-11, 15-18]; *forest* [2-4, 6, 8-11, 13, 15-18]; *lettuce* [3, 4, 6, 9, 10, 15-18]; *lucerne* [4, 6, 9, 11, 15-18]; *ornamental plant production* [8, 13]; *red clover, white clover* [10]; *winter field beans* [4, 6, 9-11, 15-18]; *winter oilseed rape* [2-6, 9-12, 15-18]
- Volunteer cereals in *winter field beans, winter oilseed rape* [7]
- Wild oats in *winter field beans, winter oilseed rape* [7]

Extension of Authorisation for Minor Use (EAMUs)
- *all edible seed crops grown outdoors* 20082942 [6]
- *all non-edible seed crops grown outdoors* 20082942 [6]
- *almonds* 20082420 [6]
- *bilberries* 20082419 [6]
- *blueberries* 20082419 [6]
- *bog myrtle* 20120509 [6]
- *broccoli* 20091902 [6]
- *calabrese* 20091902 [6]
- *cauliflowers* 20091902 [6]
- *cherries* 20082418 [6]
- *chestnuts* 20082420 [6]
- *chicory root* 20091530 [6]
- *cob nuts* 20082420 [6]
- *corn gromwell* 20122395 [6]
- *courgettes* 20082416 [6]

- **cranberries** *20082419* [6]
- **cress** *20082410* [6]
- **endives** *20082410* [6]
- **frise** *20082411* [6]
- **game cover** *20082942* [6]
- **hazel nuts** *20082420* [6]
- **herbs (see appendix 6)** *20082412* [6]
- **hops** *20082414* [6]
- **lamb's lettuce** *20082410* [6]
- **leaf brassicas** *20082410* [6]
- **marrows** *20082416* [6]
- **mirabelles** *20082418* [6]
- **ornamental plant production** *20130207* [6]
- **protected endives** *20082415* [6]
- **protected forest nurseries** *20082942* [6]
- **protected herbs (see appendix 6)** *20082412* [6]
- **protected lettuce** *20082415* [6]
- **pumpkins** *20082416* [6]
- **quinces** *20082413* [6]
- **radicchio** *20082411* [6]
- **salad brassicas** *20082410* [6]
- **scarole** *20082411* [6]
- **squashes** *20082416* [6]
- **table grapes** *20082417* [6]
- **walnuts** *20082420* [6]
- **wine grapes** *20082417* [6]

Approval information
- Propyzamide included in Annex I under EC Regulation 1107/2009
- Some products may be applied through CDA equipment. See labels for details
- Accepted by BBPA for use on hops
- Approval expiry 30 Apr 2018 [14]

Efficacy guidance
- Active via root uptake. Weeds controlled from germination to young seedling stage, some species (including many grasses) also when established
- Best results achieved by winter application to fine, firm, moist soil. Rain is required after application if soil dry
- Uptake is slow and and may take up to 12 wk
- Excessive organic debris or ploughed-up turf may reduce efficacy
- For heavy couch infestations a repeat application may be needed in following winter
- Always follow WRAG guidelines for preventing and managing herbicide resistant weeds. See Section 5 for more information

Restrictions
- Maximum number of treatments 1 per crop or yr
- Maximum total dose equivalent to one full dose treatment for all crops
- Do not treat protected crops
- Apply to listed edible crops only between 1 Oct and the date specified as the latest time of application (except lettuce)
- Do not apply in windy weather and avoid drift onto non-target crops
- Do not use on soils with more than 10% organic matter except in forestry

Crop-specific information
- Latest use: labels vary but normally before 31 Dec in year before harvest for rhubarb, lucerne, strawberries and winter field beans; before 31 Jan for other crops
- HI: 6 wk for edible crops
- Apply as soon as possible after 3-true leaf stage of oilseed rape (GS 1,3) and seed brassicas, after 4-leaf stage of sugar beet for seed, within 7 d after sowing but before emergence for field beans, after perennial crops established for at least 1 season, strawberries after 1 yr

SEE SECTION 3 FOR PRODUCTS ALSO REGISTERED

- Only apply to strawberries on heavy soils. Do not use on matted row crops
- Only apply to field beans on medium and heavy soils
- Only apply to established lucerne not less than 7 d after last cut
- In lettuce lightly incorporate in top 25 mm pre-drilling or irrigate on dry soil
- See label for lists of ornamental and forest species which may be treated

Following crops guidance
- Following an application between 1 Apr and 31 Jul at any dose the following minimum intervals must be observed before sowing the next crop: lettuce 0 wk; broad beans, chicory, clover, field beans, lucerne, radishes, peas 5 wk; brassicas, celery, leeks, oilseed rape, onions, parsley, parsnips 10 wk
- Following an application between 1 Aug and 31 Mar at any dose the following minimum intervals must be observed before sowing the next crop: lettuce 0 wk; broad beans, chicory, clover, field beans, lucerne, radishes, peas 10 wk; brassicas, celery, leeks, oilseed rape, onions, parsley, parsnips 25 wk or after 15 Jun, whichever occurs sooner
- Cereals or grasses or other crops not listed may be sown 30 wk after treatment up to 840 g.a.i/ha between 1 Aug and 31 Mar or 40 wk after treatment at higher doses at any time and after mouldboard ploughing to at least 15 cm
- A period of at least 9 mth must elapse between applications of propyzamide to the same land

Environmental safety
- Dangerous for the environment
- Very toxic to aquatic organisms
- Some pesticides pose a greater threat of contamination of water than others and propyzamide is one of these pesticides. Take special care when applying propyzamide near water and do not apply if heavy rain is forecast

Hazard classification and safety precautions
Hazard Harmful, Dangerous for the environment [1-20]; Very toxic to aquatic organisms [9]
Transport code 9
Packaging group III
UN Number 3077 [8, 13, 14, 19]; 3082 [1-7, 9-12, 15-18, 20]
Risk phrases H351 [1, 3-7, 9-14, 16, 18]; R40, R53a [2, 8, 15, 17, 19, 20]; R50 [15, 17, 20]; R51 [2, 8, 19]
Operator protection A, H [1-20]; D [2, 4-6, 11, 14, 17-20]; M [2, 4-6, 9, 11, 12, 14-20]; U05a [1, 3, 8, 13]; U20c [1-7, 9-12, 14-20]
Environmental protection E15a [1-8, 10, 11, 13, 17-20]; E15b [9, 12, 14-16]; E34 [3, 10, 14]; E38 [1-13, 15-20]; E39 [9, 12, 15, 16]; H410 [1, 9, 10, 12, 14, 16]; H411 [3-7, 11, 13, 16, 18]
Consumer protection C02a [1-6, 10, 11, 14, 17-20] (6 wk)
Storage and disposal D01, D02 [1, 3, 8, 13]; D05, D11a [1-7, 10, 11, 14, 17-20]; D07 [3, 10]; D09a [1-8, 10, 11, 13, 14, 17-20]; D12a [1-8, 10, 11, 13, 17-20]
Medical advice M03 [8, 13]

369   proquinazid

A quinazolinone fungicide for powdery mildew control in cereals
FRAC mode of action code: U7

See also chlorothalonil + proquinazid
*   prochloraz + proquinazid + tebuconazole*

Products

1 Justice	DuPont	200 g/l	EC	12835
2 Talius	DuPont	200 g/l	EC	12752

Uses
- Powdery mildew in **apples** *(off-label)*, **durum wheat** *(off-label)*, **pears** *(off-label)*, **spring barley**, **spring oats**, **spring rye**, **spring wheat**, **triticale**, **winter barley**, **winter oats**, **winter rye**, **winter wheat** [1, 2]; **courgettes** *(off-label)*, **cucumbers** *(off-label)*, **forest nurseries** *(off-label)*, **marrows** *(off-label)*, **protected aubergines** *(off-label)*, **protected courgettes** *(off-label)*,

---

FOR FULL CONDITIONS OF USE ALWAYS READ THE PRODUCT LABEL

protected cucumbers (off-label), *protected marrows* (off-label), *protected pumpkins* (off-label), *protected squashes* (off-label), *protected strawberries* (off-label), *protected tomatoes* (off-label), *pumpkins* (off-label), *soft fruit* (off-label), *squashes* (off-label), *top fruit* (off-label) [2]; *grapevines* (off-label) [1]

Extension of Authorisation for Minor Use (EAMUs)

- *apples* 20170726 [1], 20170725 [2]
- *courgettes* 20152627 [2]
- *cucumbers* 20152627 [2]
- *durum wheat* 20090418 [1], 20090419 [2]
- *forest nurseries* 20090420 [2]
- *grapevines* 20111763 [1]
- *marrows* 20152627 [2]
- *pears* 20170726 [1], 20170725 [2]
- *protected aubergines* 20152627 [2]
- *protected courgettes* 20152627 [2]
- *protected cucumbers* 20152627 [2]
- *protected marrows* 20152627 [2]
- *protected pumpkins* 20152627 [2]
- *protected squashes* 20152627 [2]
- *protected strawberries* 20170210 [2]
- *protected tomatoes* 20152627 [2]
- *pumpkins* 20152627 [2]
- *soft fruit* 20090420 [2]
- *squashes* 20152627 [2]
- *top fruit* 20090420 [2]

Approval information

- Proquinazid included in Annex 1 under EC Regulation 1107/2009
- Accepted by BBPA on malting barley

Efficacy guidance

- Best results obtained from preventive treatment before disease is established in the crop
- Where mildew has already spread to new growth a tank mix with a curative fungicide with an alternative mode of action should be used
- Use as part of an integrated crop management (ICM) strategy incorporating other methods of control or fungicides with different modes of action

Restrictions

- Maximum number of treatments 2 per crop
- Do not apply to any crop suffering from stress from any cause
- Avoid application in either frosty or hot, sunny conditions

Crop-specific information

- Latest use: before beginning of heading (GS 49) for barley, oats, rye, triticale; before full flowering (GS 65) for wheat

Environmental safety

- Dangerous for the environment
- Toxic to aquatic organisms
- Dangerous to fish or other aquatic life. Do not contaminate surface waters or ditches with chemical or used container
- LERAP Category B

Hazard classification and safety precautions

Hazard Harmful, Dangerous for the environment
Transport code 9
Packaging group III
UN Number 3082
Risk phrases H315, H318, H351
Operator protection A, C, H; U05a, U11

SEE SECTION 3 FOR PRODUCTS ALSO REGISTERED

Environmental protection E13b, E16a, E16b, E34, E38, H410
Storage and disposal D01, D02, D05, D09a, D10b, D11a, D12a
Medical advice M03, M05a

370 prosulfocarb

A thiocarbamate herbicide for grass and broad-leaved weed control in cereals and potatoes
HRAC mode of action code: N

See also clodinafop-propargyl + prosulfocarb

Products

1	Bokken	Syngenta	800 g/l	EC	17557
2	Clayton Comply	Clayton	800 g/l	EC	17702
3	Clayton Heed	Clayton	800 g/l	EC	17701
4	Clayton Obey	Clayton	800 g/l	EC	16405
5	Clayton Obey	Clayton	800 g/l	EC	16626
6	Defy	Syngenta	800 g/l	EC	16202
7	Fade	Life Scientific	800 g/l	EC	17982
8	Fidox 800 EC	Belchim	800 g/l	EC	17904
9	Moose 800 EC	Belchim	800 g/l	EC	17968
10	Quidam	UPL Europe	800 g/l	EC	17595
11	Roxy 800 EC	Belchim	800 g/l	EC	17859

Uses

- Annual dicotyledons in *almonds (off-label)*, *apples (off-label)*, *apricots (off-label)*, *bulb onions (off-label)*, *cherries (off-label)*, *chestnuts (off-label)*, *hazel nuts (off-label)*, *herbs (see appendix 6) (off-label)*, *medlar (off-label)*, *nectarines (off-label)*, *peaches (off-label)*, *pears (off-label)*, *plums (off-label)*, *quinces (off-label)*, *walnuts (off-label)* [2]; *carrots (off-label)*, *celeriac (off-label)*, *celery (outdoor) (off-label)*, *forest nurseries (off-label)*, *game cover (off-label)*, *garlic (off-label)*, *horseradish (off-label)*, *leeks (off-label)*, *miscanthus (off-label)*, *ornamental plant production (off-label)*, *parsley root (off-label)*, *parsnips (off-label)*, *poppies for morphine production (off-label)*, *rye (off-label)*, *salsify (off-label)*, *shallots (off-label)*, *spring field beans (off-label)*, *spring onions (off-label)*, *triticale (off-label)*, *winter field beans (off-label)*, *winter linseed (off-label)* [2, 6]; *durum wheat (off-label)*, *onion sets (off-label)*, *spring barley (off-label)*, *top fruit (off-label)* [6]; *potatoes, winter barley, winter wheat* [1-11]
- Annual grasses in *almonds (off-label)*, *apples (off-label)*, *apricots (off-label)*, *bulb onions (off-label)*, *cherries (off-label)*, *chestnuts (off-label)*, *hazel nuts (off-label)*, *herbs (see appendix 6) (off-label)*, *medlar (off-label)*, *nectarines (off-label)*, *peaches (off-label)*, *pears (off-label)*, *plums (off-label)*, *poppies for morphine production (off-label)*, *quinces (off-label)*, *spring field beans (off-label)*, *walnuts (off-label)*, *winter field beans (off-label)* [2]; *carrots (off-label)*, *celeriac (off-label)*, *celery (outdoor) (off-label)*, *forest nurseries (off-label)*, *game cover (off-label)*, *garlic (off-label)*, *horseradish (off-label)*, *leeks (off-label)*, *miscanthus (off-label)*, *ornamental plant production (off-label)*, *parsley root (off-label)*, *parsnips (off-label)*, *rye (off-label)*, *salsify (off-label)*, *shallots (off-label)*, *spring onions (off-label)*, *triticale (off-label)*, *winter linseed (off-label)* [2, 6]; *durum wheat (off-label)*, *onion sets (off-label)*, *top fruit (off-label)* [6]
- Annual meadow grass in *carrots (off-label)*, *celeriac (off-label)*, *celery (outdoor) (off-label)*, *durum wheat (off-label)*, *forest nurseries (off-label)*, *game cover (off-label)*, *horseradish (off-label)*, *miscanthus (off-label)*, *ornamental plant production (off-label)*, *parsley root (off-label)*, *parsnips (off-label)*, *poppies for morphine production (off-label)*, *rye (off-label)*, *salsify (off-label)*, *spring barley (off-label)*, *top fruit (off-label)*, *triticale (off-label)*, *winter linseed (off-label)* [6]; *potatoes, winter barley, winter wheat* [1-11]
- Black nightshade in *herbs (see appendix 6) (off-label)* [2, 6]
- Chickweed in *celeriac (off-label)* [2, 6]; *potatoes, winter barley, winter wheat* [1-11]
- Cleavers in *carrots (off-label)*, *herbs (see appendix 6) (off-label)*, *horseradish (off-label)*, *parsley root (off-label)*, *parsnips (off-label)*, *salsify (off-label)*, *spring field beans (off-label)*, *winter field beans (off-label)* [2, 6]; *potatoes, winter barley, winter wheat* [1-11]
- Fat hen in *carrots (off-label)*, *celeriac (off-label)*, *herbs (see appendix 6) (off-label)*, *horseradish (off-label)*, *parsley root (off-label)*, *parsnips (off-label)*, *salsify (off-label)* [2, 6]

FOR FULL CONDITIONS OF USE ALWAYS READ THE PRODUCT LABEL

- Field speedwell in **potatoes, winter barley, winter wheat** [1-11]
- Fumitory in **herbs (see appendix 6)** *(off-label)* [2, 6]
- Ivy-leaved speedwell in **potatoes, winter barley, winter wheat** [1-11]
- Knotgrass in **carrots** *(off-label)*, **herbs (see appendix 6)** *(off-label)*, **horseradish** *(off-label)*, **parsley root** *(off-label)*, **parsnips** *(off-label)*, **salsify** *(off-label)* [2, 6]
- Loose silky bent in **potatoes, winter barley, winter wheat** [1-11]
- Mayweeds in **carrots** *(off-label)*, **celeriac** *(off-label)*, **herbs (see appendix 6)** *(off-label)*, **horseradish** *(off-label)*, **parsley root** *(off-label)*, **parsnips** *(off-label)*, **salsify** *(off-label)* [2, 6]
- Polygonums in **celeriac** *(off-label)* [2, 6]
- Rough-stalked meadow grass in **potatoes, winter barley, winter wheat** [1-11]
- Thistles in **celeriac** *(off-label)* [2, 6]

Extension of Authorisation for Minor Use (EAMUs)
- **almonds** *20162930* [2]
- **apples** *20162930* [2]
- **apricots** *20162930* [2]
- **bulb onions** *20162925* [2]
- **carrots** *20162922* [2], *20131354* [6]
- **celeriac** *20162933* [2], *20131355* [6]
- **celery (outdoor)** *20162924* [2], *20131353* [6]
- **cherries** *20162930* [2]
- **chestnuts** *20162930* [2]
- **durum wheat** *20131427* [6]
- **forest nurseries** *20162926* [2], *20131432* [6]
- **game cover** *20162931* [2], *20131430* [6]
- **garlic** *20162925* [2], *20131357* [6]
- **hazel nuts** *20162930* [2]
- **herbs (see appendix 6)** *20162923* [2], *20152809* [6]
- **horseradish** *20162922* [2], *20131354* [6]
- **leeks** *20162925* [2], *20131357* [6]
- **medlar** *20162930* [2]
- **miscanthus** *20162931* [2], *20131430* [6]
- **nectarines** *20162930* [2]
- **onion sets** *20131357* [6]
- **ornamental plant production** *20162927* [2], *20131431* [6]
- **parsley root** *20162922* [2], *20131354* [6]
- **parsnips** *20162922* [2], *20131354* [6]
- **peaches** *20162930* [2]
- **pears** *20162930* [2]
- **plums** *20162930* [2]
- **poppies for morphine production** *20162934* [2], *20131385* [6]
- **quinces** *20162930* [2]
- **rye** *20162929* [2], *20131429* [6]
- **salsify** *20162922* [2], *20131354* [6]
- **shallots** *20162925* [2], *20131357* [6]
- **spring barley** *20131373* [6]
- **spring field beans** *20162932* [2], *20131356* [6]
- **spring onions** *20162925* [2], *20131357* [6]
- **top fruit** *20131433* [6]
- **triticale** *20162929* [2], *20131429* [6]
- **walnuts** *20162930* [2]
- **winter field beans** *20162932* [2], *20131356* [6]
- **winter linseed** *20162928* [2], *20131428* [6]

Approval information
- Prosulfocarb included in Annex I under EC Regulation 1107/2009
- Accepted by BBPA for use on malting barley

Efficacy guidance
- Best results obtained in cereals from treatment of crops in a firm moist seedbed, free from clods

SEE SECTION 3 FOR PRODUCTS ALSO REGISTERED

- Pre-emergence use will reduce blackgrass populations but the product should only be used against this weed as part of a management strategy involving sequences with products of alternative modes of action
- Always follow WRAG guidelines for preventing and managing herbicide resistant weeds. See Section 5 for more information
- Can be used up to 10% emergence in potatoes but must be mixed with diquat [6]

Restrictions
- Maximum number of treatments 1 per crop for winter barley, winter wheat and potatoes
- Do not apply to crops under stress from any cause. Transient yellowing can occur from which recovery is complete
- Winter cereals must be covered by 3 cm of settled soil

Crop-specific information
- Latest use: at emergence (soil rising over emerging potato shoots) for potatoes, up to and including early tillering (GS 21) for winter barley, winter wheat
- When applied pre-emergence to cereals crop emergence may occasionally be slowed down but yield is not affected
- For potatoes, complete ridge formation before application and do not disturb treated soil afterwards
- Peas are severely damaged or killed.

Following crops guidance
- Do not sow field or broad beans within 12 mth of treatment
- In the event of failure of a treated cereal crop, winter wheat, winter barley or transplanted brassicas may be re-sown immediately. Seed brassicas can be sown 14 weeks after application provided that the land is ploughed first. In the following spring sunflowers, maize, flax, spring cereals, peas, oilseed rape or soya beans may be sown without ploughing, and carrots, lettuce, onions, sugar beet or potatoes may be sown or planted after ploughing

Environmental safety
- Dangerous for the environment
- Very toxic to aquatic organisms
- LERAP Category B

Hazard classification and safety precautions
Hazard Irritant, Dangerous for the environment
Transport code 9
Packaging group III
UN Number 3082
Risk phrases H304 [1-8, 10, 11]; H315, H317 [1-7, 10]; H319 [1-11]
Operator protection A, C, H; U02a, U05a, U08, U14, U15, U20b [1-11]; U23a [7]
Environmental protection E15b, E16a, E38, H410 [1-11]; E16b [1-6, 8-11]
Storage and disposal D01, D02, D05, D09a, D10c, D12a
Medical advice M05a

---

371   prosulfuron

A contact and residual sulfonyl urea herbicide for use in maize crops
HRAC mode of action code: B

See also bromoxynil + prosulfuron
*   dicamba + prosulfuron*

Products

1 Clayton Kibo	Clayton	75% w/w	WG	15822
2 Peak	Syngenta	75% w/w	WG	15521

Uses
- Annual dicotyledons in *forage maize, grain maize* [1, 2]; *game cover* (off-label) [2]
- Black bindweed in *forage maize, grain maize* [1, 2]

- Chickweed in **forage maize, grain maize** [1, 2]
- Fumitory in **forage maize, grain maize** [1, 2]
- Groundsel in **forage maize, grain maize** [1, 2]
- Knotgrass in **forage maize, grain maize** [1, 2]
- Mayweeds in **forage maize, grain maize** [1, 2]
- Redshank in **forage maize, grain maize** [1, 2]
- Scarlet pimpernel in **forage maize, grain maize** [1, 2]
- Shepherd's purse in **forage maize, grain maize** [1, 2]
- Sowthistle in **forage maize, grain maize** [1, 2]

Extension of Authorisation for Minor Use (EAMUs)
- **game cover** 20120906 [2]

Approval information
- Prosulfuron included in Annex I under EC Regulation 1107/2009

Efficacy guidance
- For optimum efficacy, apply when weeds are at the 2 - 4 leaf stage

Restrictions
- Do not apply to forage maize or grain maize grown for seed production
- Do not apply with organo-phosphate insecticides

Following crops guidance
- In the event of crop failure, wait 4 weeks after treatment and then re-sow
- After normal harvest wheat, barley and winter beans may be sown as a following crop in the autumn once the soil has been ploughed to 15 cms. In spring, wheat, barley, peas or beans may be sown but do not sow any other crop at this time.

Environmental safety
- LERAP Category B

Hazard classification and safety precautions
Hazard Harmful, Dangerous for the environment [1, 2]; Harmful if swallowed, Very toxic to aquatic organisms [2]
Transport code 9
Packaging group III
UN Number 3077
Risk phrases R22a, R50, R53a [1]
Operator protection A, H; U05a, U14, U20c
Environmental protection E15b, E16a, E38 [1, 2]; H410 [2]
Storage and disposal D01, D02, D05, D09b, D10c, D12a

372   prothioconazole

A systemic, protectant and curative triazole fungicide
FRAC mode of action code: 3

See also benzovindiflupyr + prothioconazole
bixafen + fluopyram + prothioconazole
bixafen + fluoxastrobin + prothioconazole
bixafen + prothioconazole
bixafen + prothioconazole + spiroxamine
bixafen + prothioconazole + tebuconazole
clothianidin + prothioconazole
clothianidin + prothioconazole + tebuconazole + triazoxide
fluopyram + prothioconazole
fluopyram + prothioconazole + tebuconazole
fluoxastrobin + prothioconazole
fluoxastrobin + prothioconazole + tebuconazole
fluoxastrobin + prothioconazole + trifloxystrobin

SEE SECTION 3 FOR PRODUCTS ALSO REGISTERED

Products

1	Proline 275	Bayer CropScience	275 g/l	EC	14790
2	Rudis	Bayer CropScience	480 g/l	SC	14122

Uses

- Alternaria in **broccoli, brussels sprouts, cabbages, calabrese, cauliflowers** [2]
- Alternaria blight in **carrots, parsnips, swedes, turnips** [2]
- Brown rust in **spring barley, spring wheat, winter barley, winter rye, winter wheat** [1]
- Crown rust in **spring oats, winter oats** [1]
- Disease control in **linseed** *(off-label)*, **mustard** *(off-label)* [1]
- Eyespot in **spring barley, spring oats, spring wheat, winter barley, winter oats, winter rye, winter wheat** [1]
- Glume blotch in **spring wheat, winter wheat** [1]
- Kabatiella lini in **linseed** *(off-label)* [1]
- Late ear diseases in **spring barley, spring wheat, winter barley, winter wheat** [1]
- Leaf blotch in **leeks** *(useful reduction)* [2]
- Light leaf spot in **broccoli, brussels sprouts, cabbages, calabrese, cauliflowers** [2]; **winter oilseed rape** [1]
- Net blotch in **spring barley, winter barley** [1]
- Phoma in **broccoli, brussels sprouts, cabbages, calabrese, cauliflowers** [2]; **corn gromwell** *(off-label)*, **winter oilseed rape** [1]
- Powdery mildew in **broccoli, brussels sprouts, cabbages, calabrese, carrots, cauliflowers, parsnips, swedes, turnips** [2]; **corn gromwell** *(off-label)*, **spring barley, spring oats, spring wheat, winter barley, winter oats, winter rye, winter wheat** [1]
- Purple blotch in **leeks** [2]
- Rhynchosporium in **spring barley, winter barley, winter rye** [1]
- Ring spot in **broccoli, brussels sprouts, cabbages, calabrese, cauliflowers** [2]
- Rust in **leeks** [2]
- Sclerotinia in **corn gromwell** *(off-label)* [1]
- Sclerotinia rot in **carrots, parsnips** [2]
- Sclerotinia stem rot in **winter oilseed rape** [1]
- Septoria leaf blotch in **spring wheat, winter wheat** [1]
- Septoria linicola in **linseed** *(off-label)* [1]
- Stem canker in **winter oilseed rape** [1]
- Stemphylium in **leeks** *(useful reduction)* [2]
- Tan spot in **spring wheat, winter wheat** [1]
- Yellow rust in **spring wheat, winter barley, winter wheat** [1]

Extension of Authorisation for Minor Use (EAMUs)

- **corn gromwell** *20152427* [1]
- **linseed** *20162863* [1]
- **mustard** *20111003* [1]

Approval information

- Accepted by BBPA for use on malting barley
- Prothioconazole included in Annex I under EC Regulation 1107/2009

Efficacy guidance

- Seed treatments must be applied by manufacturer's recommended treatment application equipment
- Treated cereal seed should preferably be drilled in the same season
- Follow-up treatments will be needed later in the season to give protection against air-borne and splash-borne diseases
- Best results on cereal foliar diseases obtained from treatment at early stages of disease development. Further treatment may be needed if disease attack is prolonged [1]
- Foliar applications to established infections of any disease are likely to be less effective [1]
- Best control of cereal ear diseases obtained by treatment during ear emergence [1]
- Treat oilseed rape at early to full flower [1]
- Prothioconazole is a DMI fungicide. Resistance to some DMI fungicides has been identified in Septoria leaf blotch which may seriously affect performance of some products. For further

FOR FULL CONDITIONS OF USE ALWAYS READ THE PRODUCT LABEL

advice contact a specialist advisor and visit the Fungicide Resistance Action Group (FRAG)-UK website

Restrictions
- Maximum number of seed treatments one per batch
- Maximum total dose equivalent to two full dose treatments on barley, oats and oilseed rape, three full dose treatments on wheat and rye [1]
- Seed treatment must be fully re-dispersed and homogeneous before use
- Do not use on seed with more than 16% moisture content, or on sprouted, cracked or skinned seed
- All seed batches should be tested to ensure they are suitable for treatment
- Treated winter barley seed must be used within the season of treatment; treated winter wheat seed should preferably be drilled in the same season
- Do not make repeated treatments of the product alone to the same crop against pathogens such as powdery mildew. Use tank mixtures or alternate with fungicides having a different mode of action

Crop-specific information
- Latest use: pre-drilling for seed treatments; before grain milky ripe for foliar sprays on winter rye, winter wheat; beginning of flowering for barley, oats [1]
- HI 56 d for winter oilseed rape [1]

Environmental safety
- Dangerous for the environment [1]
- Toxic to aquatic organisms [1]
- Harmful to aquatic organisms
- LERAP Category B

Hazard classification and safety precautions
Hazard Irritant [1]; Dangerous for the environment, Very toxic to aquatic organisms [1, 2]
Transport code 9
Packaging group III
UN Number 3082
Risk phrases H319, H335 [1]
Operator protection A, H [1, 2]; C [1]; U05a, U20b [1, 2]; U09b [1]
Environmental protection E15a [1]; E15b [2]; E16a, E34, E38, H410 [1, 2]
Storage and disposal D01, D02, D09a, D10b, D12a [1, 2]; D05 [1]
Medical advice M03

373   prothioconazole + spiroxamine

A broad spectrum fungicide mixture for cereals
FRAC mode of action code: 3 + 5

See also spiroxamine

Products
1	Hale	Pan Agriculture	160:300 g/l	EC	15924
2	Helix	Bayer CropScience	160:300 g/l	EC	12264
3	Spiral	AgChem Access	160:300 g/l	EC	15437

Uses
- Brown rust in **spring barley, winter barley, winter rye, winter wheat**
- Crown rust in **spring oats, winter oats**
- Eyespot in **spring barley, spring oats, winter barley, winter oats, winter rye, winter wheat**
- Glume blotch in **winter wheat**
- Late ear diseases in **winter rye, winter wheat**
- Net blotch in **spring barley, winter barley**
- Powdery mildew in **spring barley, spring oats, winter barley, winter oats, winter rye, winter wheat**
- Rhynchosporium in **spring barley, winter barley, winter rye**

SEE SECTION 3 FOR PRODUCTS ALSO REGISTERED

- Septoria leaf blotch in **winter wheat**
- Tan spot in **winter wheat**
- Yellow rust in **spring barley**, **winter barley**, **winter wheat**

Approval information
- Prothioconazole and spiroxamine included in Annex I under EC Regulation 1107/2009
- Accepted by BBPA for use on malting barley

Efficacy guidance
- Best results obtained from treatment at early stages of disease development. Further treatment may be needed if disease attack is prolonged
- Applications to established infections of any disease are likely to be less effective
- Best control of cereal ear diseases obtained by treatment during ear emergence
- Prothioconazole is a DMI fungicide. Resistance to some DMI fungicides has been identified in Septoria leaf blotch which may seriously affect performance of some products. For further advice contact a specialist advisor and visit the Fungicide Resistance Action Group (FRAG)-UK website

Restrictions
- Maximum total dose equivalent to two full dose treatments on barley and oats and three full dose treatments on wheat and rye

Crop-specific information
- Latest use: before grain watery ripe for rye, oats, winter wheat; up to beginning of anthesis for barley

Environmental safety
- Dangerous for the environment
- Very toxic to aquatic organisms
- LERAP Category B

Hazard classification and safety precautions
Hazard Harmful, Dangerous for the environment [1-3]; Harmful if swallowed, Harmful if inhaled [1, 2]; Very toxic to aquatic organisms [2]
Transport code 9
Packaging group III
UN Number 3082
Risk phrases H315 [1, 2]; H319, H335 [2]; H320 [1]; R20, R22a, R36, R38, R50, R53a [3]
Operator protection A, C, H; U05a, U09b, U11, U19a, U20b
Environmental protection E15a, E16a, E34, E38 [1-3]; H410 [1]
Storage and disposal D01, D02, D05, D09a, D10b, D12a
Medical advice M03

374   prothioconazole + spiroxamine + tebuconazole

A broad spectrum fungicide mixture for cereals
FRAC mode of action code: 3 + 5 + 3

See also spiroxamine
tebuconazole

Products
Cello	Bayer CropScience	100:250:100 g/l	EC	13178

Uses
- Brown rust in **spring barley**, **winter barley**, **winter rye**, **winter wheat**
- Crown rust in **spring oats**, **winter oats**
- Eyespot in **spring barley** (reduction), **spring oats**, **winter barley** (reduction), **winter oats**, **winter rye** (reduction), **winter wheat** (reduction)
- Glume blotch in **winter wheat**
- Late ear diseases in **spring barley**, **winter barley**, **winter wheat**

FOR FULL CONDITIONS OF USE ALWAYS READ THE PRODUCT LABEL

- Net blotch in **spring barley, winter barley**
- Powdery mildew in **spring barley, spring oats, winter barley, winter oats, winter rye, winter wheat**
- Rhynchosporium in **spring barley, winter barley, winter rye**
- Septoria leaf blotch in **winter wheat**
- Yellow rust in **spring barley, winter barley, winter wheat**

Approval information
- Prothioconazole, spiroxamine and tebuconazole included in Annex I under EC Regulation 1107/2009

Efficacy guidance
- Best results obtained from treatment at early stages of disease development. Further treatment may be needed if disease attack is prolonged
- Applications to established infections of any disease are likely to be less effective
- Best control of cereal ear diseases obtained by treatment during ear emergence
- Prothioconazole and tebuconazole are DMI fungicides. Resistance to some DMI fungicides has been identified in Septoria leaf blotch which may seriously affect performance of some products. For further advice contact a specialist advisor and visit the Fungicide Resistance Action Group (FRAG)-UK website

Restrictions
- Maximum total dose equivalent to two full dose treatments
- Newer authorisations for tebuconazole products require application to cereals only after GS 30 and applications to oilseed rape and linseed after GS 20 - check label

Crop-specific information
- Latest use: before grain milky ripe stage for rye, wheat; up to beginning of anthesis for barley and oats

Environmental safety
- Dangerous for the environment
- Very toxic to aquatic organisms
- LERAP Category B

Hazard classification and safety precautions
Hazard Harmful, Dangerous for the environment, Harmful if inhaled, Very toxic to aquatic organisms
Transport code 9
Packaging group III
UN Number 3082
Risk phrases H315, H317, H319, H335, H361
Operator protection A, C, H; U05a, U09b, U11, U14, U19a, U20b
Environmental protection E15a, E16a, E34, E38, H410
Storage and disposal D01, D02, D05, D09a, D10b, D12a
Medical advice M03

375   prothioconazole + tebuconazole

A triazole fungicide mixture for cereals
FRAC mode of action code: 3 + 3

See also tebuconazole

Products
1	Clayton Zorro Pro	Clayton	125:125 g/l	EC	17073
2	Corinth	Bayer CropScience	80:160 g/l	EC	16742
3	Harvest Proteb	Harvest	125:125 g/l	EC	17692
4	Kestrel	Bayer CropScience	160:80 g/l	EC	16751
5	Prosaro	Bayer CropScience	125:125 g/l	EC	16732
6	Redigo Pro	Bayer CropScience	150:20 g/l	FS	15145

SEE SECTION 3 FOR PRODUCTS ALSO REGISTERED

Uses

- Black stem rust in **grass seed crops** *(off-label)* [5]
- Blue mould in **durum wheat, rye, spring oats, spring wheat, triticale, winter oats, winter wheat** [6]
- Brown rust in **spring barley, spring wheat, winter barley, winter rye, winter wheat** [1, 3-5]
- Bunt in **durum wheat, rye, spring oats, spring wheat, triticale, winter oats, winter wheat** [6]
- Covered smut in **spring barley, spring oats, winter barley, winter oats** [6]
- Crown rust in **spring oats, winter oats** [1, 3-5]
- Ergot in **rye** *(qualified minor use)* [6]
- Eyespot in **spring barley, winter barley, winter rye, winter wheat** [4]; **spring barley** *(reduction)*, **winter barley** *(reduction)*, **winter rye** *(reduction)*, **winter wheat** *(reduction)* [1, 3, 5]; **spring oats, spring wheat, winter oats** [1, 3-5]
- Fusarium ear blight in **spring barley, spring wheat, winter barley, winter wheat** [4]
- Fusarium seedling blight in **durum wheat, rye, spring barley, spring oats, spring wheat, triticale, winter barley, winter oats, winter wheat** [6]
- Glume blotch in **spring wheat, winter wheat** [1, 3-5]
- Late ear diseases in **spring barley, spring wheat, winter barley, winter wheat** [1, 3, 5]
- Leaf spot in **grass seed crops** *(off-label)* [5]
- Leaf stripe in **spring barley, winter barley** [6]
- Light leaf spot in **oilseed rape** [4]; **oilseed rape** *(moderate control only)* [2]; **spring oilseed rape, winter oilseed rape** [1, 3, 5]
- Loose smut in **durum wheat, rye, spring barley, spring oats, spring wheat, triticale, winter barley, winter oats, winter wheat** [6]
- Net blotch in **spring barley, winter barley** [1, 3-5]
- Phoma in **spring oilseed rape, winter oilseed rape** [1, 3, 5]
- Phoma leaf spot in **oilseed rape** [2, 4]
- Powdery mildew in **grass seed crops** *(off-label)* [5]; **spring barley, spring oats, spring wheat, winter barley, winter oats, winter rye, winter wheat** [1, 3-5]
- Rhynchosporium in **grass seed crops** *(off-label)* [5]; **spring barley, winter barley, winter rye** [1, 3-5]
- Sclerotinia stem rot in **oilseed rape** [2, 4]; **spring oilseed rape, winter oilseed rape** [1, 3, 5]
- Septoria leaf blotch in **spring wheat, winter wheat** [1, 3-5]
- Sooty moulds in **spring barley, spring wheat, winter barley, winter wheat** [4]
- Stem canker in **oilseed rape** [2]
- Tan spot in **spring wheat, winter wheat** [1, 3-5]
- Yellow rust in **spring barley, spring wheat, winter barley, winter wheat** [1, 3-5]

Extension of Authorisation for Minor Use (EAMUs)

- **grass seed crops** *20160083* [5]

Approval information

- Prothioconazole and tebuconazole included in Annex I under EC Regulation 1107/2009
- Accepted by BBPA for use on malting barley

Efficacy guidance

- Best results on cereal foliar diseases obtained from treatment at early stages of disease development. Further treatment may be needed if disease attack is prolonged
- On oilseed rape apply a protective treatment in autumn/winter for Phoma followed by a further spray in early spring from the onset of stem elongation, if necessary. For control of Sclerotinia apply at early to full flower
- Applications to established infections of any disease are likely to be less effective
- Best control of cereal ear diseases obtained by treatment during ear emergence
- Prothioconazole and tebuconazole are DMI fungicides. Resistance to some DMI fungicides has been identified in Septoria leaf blotch which may seriously affect performance of some products. For further advice contact a specialist advisor and visit the Fungicide Resistance Action Group (FRAG)-UK website
- Treated seed should be drilled to a depth of 40mm. Ensure no seed is left on the soil surface . After drilling if conditions allow, field should be harrowed and then rolled to ensure good incorporation [6]

FOR FULL CONDITIONS OF USE ALWAYS READ THE PRODUCT LABEL

Restrictions
- Maximum total dose equivalent to two full dose treatments on barley, oats, oilseed rape; three full dose treatments on wheat and rye
- Newer authorisations for tebuconazole products require application to cereals only after GS 30 and applications to oilseed rape and linseed after GS20 - check label
- Treated seed should not be left on the soil surface. Bury or remove spillages [6]

Crop-specific information
- Latest use: before grain milky ripe for rye, wheat; beginning of flowering for barley, oats
- HI 56 d for oilseed rape

Environmental safety
- Dangerous for the environment
- Toxic to aquatic organisms
- LERAP Category B [1-5]

Hazard classification and safety precautions
Hazard Harmful [2, 4]; Irritant [1, 3, 5]; Dangerous for the environment [1-5]; Harmful if inhaled [2]; Very toxic to aquatic organisms [3, 5]
Transport code 9
Packaging group III
UN Number 3082
Risk phrases H315, H361 [1-5]; H317 [1, 2, 4]; H319 [2, 3, 5]; H335, R43 [3, 5]
Operator protection A, H [1-6]; C [2, 4]; D [6]; U05a [1-6]; U09b, U20b [1, 3-5]; U11, U20c [2]; U19a [1-3, 5]; U20a, U20e [6]
Environmental protection E15a, E16a [1-5]; E15b [6]; E22b [2]; E22c [1, 3-5]; E34, E38 [1-6]; H410 [3, 5, 6]; H411 [1, 2, 4]
Storage and disposal D01, D02, D10b [1-5]; D05, D09a, D12a [1-6]; D10d, D21 [6]
Treated seed S01, S02, S03, S04a, S04b, S05, S07, S08 [6]
Medical advice M03 [1-5]

376 prothioconazole + trifloxystrobin

A triazole and strobilurin fungicide mixture for cereals
FRAC mode of action code: 3 + 11

See also trifloxystrobin

Products

1	Mobius	Bayer CropScience	175:150 g/l	SC	13395
2	Zephyr	Bayer CropScience	175:88 g/l	SC	13174

Uses
- Brown rust in *durum wheat, rye, spring barley, triticale, winter barley, winter wheat* [1, 2]; *spring wheat* [2]
- Ear diseases in *durum wheat, rye, triticale, winter wheat* [1]
- Eyespot in *durum wheat, rye, triticale, winter wheat* [1, 2]; *spring barley (reduction), spring wheat, winter barley (reduction)* [2]; *spring barley (reduction in severity), winter barley (reduction in severity)* [1]
- Glume blotch in *durum wheat, rye, triticale, winter wheat* [1, 2]; *spring wheat* [2]
- Net blotch in *spring barley, winter barley* [1, 2]
- Powdery mildew in *durum wheat, rye, spring barley, triticale, winter barley, winter wheat* [1, 2]; *spring wheat* [2]
- Rhynchosporium in *spring barley, winter barley* [1, 2]
- Septoria leaf blotch in *durum wheat, rye, triticale, winter wheat* [1, 2]; *spring wheat* [2]
- Yellow rust in *durum wheat, rye, spring barley, triticale, winter barley, winter wheat* [1, 2]; *spring wheat* [2]

Approval information
- Prothioconazole and trifloxystrobin included in Annex I under EC Regulation 1107/2009
- Accepted by BBPA for use on malting barley

SEE SECTION 3 FOR PRODUCTS ALSO REGISTERED

Efficacy guidance

- Best results obtained from treatment at early stages of disease development. Further treatment may be needed if disease attack is prolonged
- Applications to established infections of any disease are likely to be less effective
- Best control of cereal ear diseases obtained by treatment during ear emergence
- Prothioconazole is a DMI fungicide. Resistance to some DMI fungicides has been identified in Septoria leaf blotch which may seriously affect performance of some products. For further advice contact a specialist advisor and visit the Fungicide Resistance Action Group (FRAG)-UK website
- Trifloxystrobin is a member of the QoI cross resistance group. Product should be used preventatively and not relied on for its curative potential
- Use product as part of an Integrated Crop Management strategy incorporating other methods of control, including where appropriate other fungicides with a different mode of action. Do not apply more than two foliar applications of QoI containing products to any cereal crop
- There is a significant risk of widespread resistance occurring in *Septoria tritici* populations in UK. Failure to follow resistance management action may result in reduced levels of disease control
- Strains of wheat and barley powdery mildew resistant to QoIs are common in the UK. Control of wheat mildew can only be relied on from the triazole component
- Where specific control of wheat mildew is required this should be achieved through a programme of measures including products recommended for the control of mildew that contain a fungicide from a different cross-resistance group and applied at a dose that will give robust control

Restrictions

- Maximum total dose equivalent to two full dose treatments

Crop-specific information

- Latest use: before grain milky ripe for wheat; beginning of flowering for barley

Environmental safety

- Dangerous for the environment
- Very toxic to aquatic organisms
- LERAP Category B

Hazard classification and safety precautions

Hazard Irritant, Dangerous for the environment, Very toxic to aquatic organisms
Transport code 9
Packaging group III
UN Number 3082
Risk phrases H317
Operator protection A, C, H; U05a, U09b, U19a [1, 2]; U20b [2]; U20c [1]
Environmental protection E15a, E16a, E38, H410 [1, 2]; E34 [2]
Storage and disposal D01, D02, D09a, D10c, D12a
Medical advice M03

377 pymetrozine

A novel azomethine insecticide
IRAC mode of action code: 9B

Products

1 Chess WG	Syngenta	50% w/w	WG	17580
2 Plenum WG	Syngenta	50% w/w	WG	10652

Uses

- Aphids in **beetroot** *(off-label)*, **bilberries** *(off-label)*, **blackberries** *(off-label)*, **blackcurrants** *(off-label)*, **blueberries** *(off-label)*, **celeriac** *(off-label)*, **celery (outdoor)** *(off-label)*, **chard** *(off-label)*, **chinese cabbage** *(off-label)*, **choi sum** *(off-label)*, **collards** *(off-label)*, **cranberries** *(off-label)*, **forest nurseries** *(off-label)*, **frise** *(off-label)*, **gooseberries** *(off-label)*, **kale** *(off-label)*, **lamb's lettuce** *(off-label)*, **lettuce** *(off-label)*, **loganberries** *(off-label)*, **pak choi** *(off-label)*, **radicchio** *(off-label)*, **raspberries** *(off-label)*, **redcurrants** *(off-label)*, **rubus hybrids** *(off-label)*, **salad brassicas** *(off-label - for baby leaf production)*, **seed potatoes**, **soft fruit** *(off-label)*, **spinach** *(off-label)*, **spinach beet** *(off-label)*, **strawberries** *(off-label)*, **swedes** *(off-label)*, **sweetcorn** *(off-label)*, **turnips** *(off-label)*, **vaccinium spp.** *(off-label)*, **ware potatoes**, **whitecurrants** *(off-label)* [2]; **brassicas** *(off-label)*, **ornamental plant production**, **protected almonds** *(off-label)*, **protected apple** *(off-label)*, **protected apricots** *(off-label)*, **protected aubergines** *(off-label)*, **protected bilberries** *(off-label)*, **protected blackberries** *(off-label)*, **protected blackcurrants** *(off-label)*, **protected blueberry** *(off-label)*, **protected celery** *(off-label)*, **protected cherries** *(off-label)*, **protected chestnuts** *(off-label)*, **protected chilli peppers** *(off-label)*, **protected choi sum** *(off-label)*, **protected courgettes** *(off-label)*, **protected cranberries** *(off-label)*, **protected cucumbers**, **protected endives** *(off-label)*, **protected forest nurseries** *(off-label)*, **protected gherkins** *(off-label)*, **protected gooseberries** *(off-label)*, **protected hazelnuts** *(off-label)*, **protected hops** *(off-label)*, **protected loganberries** *(off-label)*, **protected medlars** *(off-label)*, **protected melons** *(off-label)*, **protected nectarines** *(off-label)*, **protected oriental cabbage** *(off-label)*, **protected ornamentals** *(off-label)*, **protected peaches** *(off-label)*, **protected pear** *(off-label)*, **protected peppers** *(off-label)*, **protected plums** *(off-label)*, **protected pumpkins** *(off-label)*, **protected quince** *(off-label)*, **protected raspberries** *(off-label)*, **protected redcurrants** *(off-label)*, **protected rubus hybrids** *(off-label)*, **protected strawberries** *(off-label)*, **protected summer squash** *(off-label)*, **protected tomatoes** *(off-label)*, **protected walnuts** *(off-label)*, **protected winter squash** *(off-label)* [1]

- Cabbage aphid in **chinese cabbage** *(off-label)*, **choi sum** *(off-label)*, **collards** *(off-label)*, **kale** *(off-label)*, **pak choi** *(off-label)* [2]
- Cotton aphids in **protected aubergines** *(off-label)*, **protected chilli peppers** *(off-label)*, **protected courgettes** *(off-label)*, **protected cucumbers** *(off-label)*, **protected gherkins** *(off-label)*, **protected melons** *(off-label)*, **protected okra** *(off-label)*, **protected peppers** *(off-label)*, **protected pumpkins** *(off-label)*, **protected summer squash** *(off-label)*, **protected tomatoes** *(off-label)*, **protected winter squash** *(off-label)* [1]
- Damson-hop aphid in **hops** *(off-label)* [2]
- Glasshouse whitefly in **protected aubergines** *(off-label)*, **protected chilli peppers** *(off-label)*, **protected courgettes** *(off-label)*, **protected cucumbers** *(off-label)*, **protected gherkins** *(off-label)*, **protected melons** *(off-label)*, **protected okra** *(off-label)*, **protected peppers** *(off-label)*, **protected pumpkins** *(off-label)*, **protected summer squash** *(off-label)*, **protected tomatoes** *(off-label)*, **protected winter squash** *(off-label)* [1]
- Green peach aphid in **protected aubergines** *(off-label)*, **protected chilli peppers** *(off-label)*, **protected courgettes** *(off-label)*, **protected cucumbers** *(off-label)*, **protected gherkins** *(off-label)*, **protected melons** *(off-label)*, **protected okra** *(off-label)*, **protected peppers** *(off-label)*, **protected pumpkins** *(off-label)*, **protected summer squash** *(off-label)*, **protected tomatoes** *(off-label)*, **protected winter squash** *(off-label)* [1]
- Green potato aphids in **protected aubergines** *(off-label)*, **protected chilli peppers** *(off-label)*, **protected courgettes** *(off-label)*, **protected cucumbers** *(off-label)*, **protected gherkins** *(off-label)*, **protected melons** *(off-label)*, **protected okra** *(off-label)*, **protected peppers** *(off-label)*, **protected pumpkins** *(off-label)*, **protected summer squash** *(off-label)*, **protected tomatoes** *(off-label)*, **protected winter squash** *(off-label)* [1]
- Mealy aphid in **broccoli** *(useful levels of control)*, **brussels sprouts** *(useful levels of control)*, **cabbages** *(useful levels of control)*, **calabrese** *(useful levels of control)*, **cauliflowers** *(useful levels of control)*, **chinese cabbage** *(off-label)*, **choi sum** *(off-label)*, **collards** *(off-label)*, **kale** *(off-label)*, **pak choi** *(off-label)* [2]
- Peach-potato aphid in **broccoli**, **brussels sprouts**, **cabbages**, **calabrese**, **cauliflowers**, **chinese cabbage** *(off-label)*, **choi sum** *(off-label)*, **collards** *(off-label)*, **kale** *(off-label)*, **pak choi** *(off-label)* [2]; **protected aubergines** *(off-label)*, **protected chilli peppers** *(off-label)*, **protected herbs (see appendix 6)** *(off-label)*, **protected lamb's lettuce** *(off-label)*, **protected leaf brassicas** *(off-label)*, **protected lettuce** *(off-label)*, **protected peppers** *(off-label)*, **protected tomatoes** *(off-label)* [1]
- Pollen beetle in **oilseed rape** [2]
- Red aphids in **protected aubergines** *(off-label)*, **protected chilli peppers** *(off-label)*, **protected courgettes** *(off-label)*, **protected cucumbers** *(off-label)*, **protected gherkins** *(off-label)*, **protected melons** *(off-label)*, **protected okra** *(off-label)*, **protected peppers** *(off-label)*, **protected pumpkins** *(off-label)*, **protected summer squash** *(off-label)*, **protected tomatoes** *(off-label)*, **protected winter squash** *(off-label)* [1]

SEE SECTION 3 FOR PRODUCTS ALSO REGISTERED

- Tobacco whitefly in **protected aubergines** *(off-label)*, **protected chilli peppers** *(off-label)*, **protected peppers** *(off-label)*, **protected tomatoes** *(off-label)* [1]
- Whitefly in **protected almonds** *(off-label)*, **protected apple** *(off-label)*, **protected apricots** *(off-label)*, **protected bilberries** *(off-label)*, **protected blackberries** *(off-label)*, **protected blackcurrants** *(off-label)*, **protected blueberry** *(off-label)*, **protected cherries** *(off-label)*, **protected chestnuts** *(off-label)*, **protected cranberries** *(off-label)*, **protected forest nurseries** *(off-label)*, **protected gooseberries** *(off-label)*, **protected hazelnuts** *(off-label)*, **protected hops** *(off-label)*, **protected loganberries** *(off-label)*, **protected medlars** *(off-label)*, **protected nectarines** *(off-label)*, **protected ornamentals** *(off-label)*, **protected peaches** *(off-label)*, **protected pear** *(off-label)*, **protected plums** *(off-label)*, **protected quince** *(off-label)*, **protected raspberries** *(off-label)*, **protected redcurrants** *(off-label)*, **protected rubus hybrids** *(off-label)*, **protected strawberries** *(off-label)*, **protected walnuts** *(off-label)* [1]

Extension of Authorisation for Minor Use (EAMUs)

- **beetroot** *20131567* [2]
- **bilberries** *20061702* [2]
- **blackberries** *20061633* [2]
- **blackcurrants** *20060946* [2]
- **blueberries** *20061702* [2]
- **brassicas** *20161259* [1]
- **celeriac** *20051062* [2]
- **celery (outdoor)** *20051062* [2]
- **chard** *20111664* [2]
- **chinese cabbage** *20082246* [2]
- **choi sum** *20082246* [2]
- **collards** *20082081* [2]
- **cranberries** *20061702* [2]
- **forest nurseries** *20082920* [2]
- **frise** *20070060* [2]
- **gooseberries** *20061702* [2]
- **hops** *20031423* [2]
- **kale** *20082081* [2]
- **lamb's lettuce** *20070060* [2]
- **lettuce** *20070060* [2]
- **loganberries** *20061702* [2]
- **pak choi** *20082246* [2]
- **protected almonds** *20170862* [1]
- **protected apple** *20170862* [1]
- **protected apricots** *20170862* [1]
- **protected aubergines** *20161271* [1], *20161272* [1]
- **protected bilberries** *20170862* [1]
- **protected blackberries** *20161258* [1], *20170862* [1]
- **protected blackcurrants** *20161248* [1], *20170862* [1]
- **protected blueberry** *20170862* [1]
- **protected celery** *20161267* [1]
- **protected cherries** *20170862* [1]
- **protected chestnuts** *20170862* [1]
- **protected chilli peppers** *20161271* [1], *20161272* [1]
- **protected choi sum** *20161262* [1]
- **protected courgettes** *20161272* [1], *20170881* [1]
- **protected cranberries** *20170862* [1]
- **protected cucumbers** *20161272* [1]
- **protected endives** *20161264* [1]
- **protected forest nurseries** *20170862* [1]
- **protected gherkins** *20161272* [1], *20170881* [1]
- **protected gooseberries** *20161248* [1], *20170862* [1]
- **protected hazelnuts** *20170862* [1]
- **protected herbs (see appendix 6)** *20161268* [1]
- **protected hops** *20170862* [1]

FOR FULL CONDITIONS OF USE ALWAYS READ THE PRODUCT LABEL

- *protected lamb's lettuce* 20161268 [1]
- *protected leaf brassicas* 20161268 [1]
- *protected lettuce* 20161268 [1]
- *protected loganberries* 20161248 [1], 20170862 [1]
- *protected medlars* 20170862 [1]
- *protected melons* 20161272 [1], 20170881 [1]
- *protected nectarines* 20170862 [1]
- *protected okra* 20161272 [1]
- *protected oriental cabbage* 20161262 [1]
- *protected ornamentals* 20170862 [1]
- *protected peaches* 20170862 [1]
- *protected pear* 20170862 [1]
- *protected peppers* 20161271 [1], 20161272 [1]
- *protected plums* 20170862 [1]
- *protected pumpkins* 20161272 [1], 20170881 [1]
- *protected quince* 20170862 [1]
- *protected raspberries* 20161258 [1], 20170862 [1]
- *protected redcurrants* 20161248 [1], 20170862 [1]
- *protected rubus hybrids* 20161248 [1], 20170862 [1]
- *protected strawberries* 20161249 [1], 20170862 [1]
- *protected summer squash* 20161272 [1], 20170881 [1]
- *protected tomatoes* 20161271 [1], 20161272 [1]
- *protected walnuts* 20170862 [1]
- *protected winter squash* 20161272 [1], 20170881 [1]
- *radicchio* 20070060 [2]
- *raspberries* 20061633 [2]
- *redcurrants* 20061702 [2]
- *rubus hybrids* 20061702 [2]
- *salad brassicas* (for baby leaf production) 20070060 [2]
- *soft fruit* 20082920 [2]
- *spinach* 20111664 [2]
- *spinach beet* 20111664 [2]
- *strawberries* 20060461 [2]
- *swedes* 20131567 [2]
- *sweetcorn* 20041318 [2]
- *turnips* 20131567 [2]
- *vaccinium spp.* 20061702 [2]
- *whitecurrants* 20061702 [2]

Approval information
- Pymetrozine included in Annex I under EC Regulation 1107/2009
- Accepted by BBPA for use on hops

Efficacy guidance
- Pymetrozine moves systemically in the plant and acts by preventing feeding leading to death by starvation in 1-4 d. There is no immediate knockdown
- Aphids controlled include those resistant to organophosphorus and carbamate insecticides
- To prevent development of resistance do not use continuously or as the sole method of control
- Best results achieved by starting spraying as soon as aphids seen in crop and repeating as necessary.
- To limit spread of persistent viruses such as potato leaf roll virus, apply from 90% crop emergence

Restrictions
- Maximum total dose equivalent to two full dose treatments on ware potatoes; three full dose treatments on seed potatoes, leaf spinach [2]; four full dose treatments on cucumbers, ornamentals
- Do not apply to potatoes when crop is in flower
- Consult processors before use on potatoes for processing

SEE SECTION 3 FOR PRODUCTS ALSO REGISTERED

- Check tolerance of ornamental species before large scale use. See label for list of species known to have been treated without damage. Visible spray deposits may be seen on leaves of some species

Crop-specific information
- HI cucumbers 3 d; potatoes, leaf spinach 7 d

Environmental safety
- High risk to bees (outdoor use only). Do not apply to crops in flower or to those in which bees are actively foraging. Do not apply when flowering weeds are present
- Avoid spraying within 6 m of field boundaries to reduce effects on non-target insects or arthropods. Risk to certain non-target insects and arthropods.

Hazard classification and safety precautions
>**Hazard** Harmful
>**UN Number** N/C
>**Risk phrases** H351
>**Operator protection** A, H; U05a, U20c
>**Environmental protection** E12a, E12e, E15a, H410
>**Storage and disposal** D01, D02, D09a, D11a

378   pyraclostrobin

A protectant and curative strobilurin fungicide for cereals
FRAC mode of action code: 11

See also boscalid + epoxiconazole + pyraclostrobin
boscalid + pyraclostrobin
dimethomorph + pyraclostrobin
dithianon + pyraclostrobin
epoxiconazole + fenpropimorph + pyraclostrobin
epoxiconazole + fluxapyroxad + pyraclostrobin
epoxiconazole + kresoxim-methyl + pyraclostrobin
epoxiconazole + pyraclostrobin
fenpropimorph + pyraclostrobin
fluxapyroxad + pyraclostrobin

Products

1	Comet 200	BASF	200 g/l	EC	12639
2	Flyer 200	BASF	200 g/l	EC	17293
3	Halley	AgChem Access	200 g/l	EC	15916
4	Mascot Eland	Rigby Taylor	20% w/w	WG	14549
5	Tucana	BASF	250 g/l	EC	10899
6	Vanguard	Sherriff Amenity	20% w/w	WG	13838
7	Vivid	BASF	250 g/l	EC	10898

Uses
- Brown rust in **spring barley, spring wheat, winter barley, winter wheat** [1-3, 5, 7]
- Crown rust in **spring oats, winter oats** [1-3, 5, 7]
- Disease control in **forage maize, grain maize** [2]
- Dollar spot in **managed amenity turf** *(useful reduction)* [4, 6]
- Eyespot in **forage maize, grain maize** [1, 2]; **forage maize** *(off-label)* [7]
- Foliar disease control in **durum wheat** *(off-label)*, **grass seed crops** *(off-label)*, **spring rye** *(off-label)*, **triticale** *(off-label)*, **winter rye** *(off-label)* [5, 7]; **ornamental plant production** *(off-label)* [7]
- Fusarium patch in **managed amenity turf** *(moderate control)* [4]; **managed amenity turf** *(moderate control only)* [6]
- Glume blotch in **spring wheat, winter wheat** [1-3, 5, 7]
- Net blotch in **spring barley, winter barley** [1-3, 5, 7]

FOR FULL CONDITIONS OF USE ALWAYS READ THE PRODUCT LABEL

- Northern leaf blight in **forage maize**, **grain maize** [1]; **forage maize** *(moderate control)*, **grain maize** *(moderate control)* [2]
- Red thread in **managed amenity turf** [4, 6]
- Rhynchosporium in **spring barley** *(moderate control)*, **winter barley** *(moderate control)* [1-3, 5, 7]
- Septoria leaf blotch in **spring wheat**, **winter wheat** [1-3, 5, 7]
- Yellow rust in **spring barley**, **spring wheat**, **winter barley**, **winter wheat** [1-3, 5, 7]

Extension of Authorisation for Minor Use (EAMUs)
- **durum wheat** *20062652* [5], *20062649* [7]
- **forage maize** *20131684* [7]
- **grass seed crops** *20062652* [5], *20062649* [7]
- **ornamental plant production** *20082884* [7]
- **spring rye** *20062652* [5], *20062649* [7]
- **triticale** *20062652* [5], *20062649* [7]
- **winter rye** *20062652* [5], *20062649* [7]

Approval information
- Pyraclostrobin included in Annex I under EC Regulation 1107/2009
- Accepted by BBPA for use on malting barley (before ear emergence only) and on hops

Efficacy guidance
- For best results apply at the start of disease attack on cereals [1, 5, 7]
- For Fusarium Patch treat early as severe damage to turf can occur once the disease is established [4, 6]
- Regular turf aeration, appropriate scarification and judicious use of nitrogenous fertiliser will assist the control of Fusarium Patch [4, 6]
- Best results on Septoria glume blotch achieved when used as a protective treatment and against Septoria leaf blotch when treated in the latent phase [1, 5, 7]
- Yield response may be obtained in the absence of visual disease symptoms [1, 5, 7]
- Pyraclostrobin is a member of the QoI cross resistance group. Product should be used preventatively and not relied on for its curative potential
- Use product as part of an Integrated Crop Management strategy incorporating other methods of control, including where appropriate other fungicides with a different mode of action. Do not apply more than two foliar applications of QoI containing products to any cereal crop or to grass
- There is a significant risk of widespread resistance occurring in *Septoria tritici* populations in UK. Failure to follow resistance management action may result in reduced levels of disease control [1, 5, 7]
- On cereal crops product must always be used in mixture with another product, recommended for control of the same target disease, that contains a fungicide from a different cross resistance group and is applied at a dose that will give robust control [1, 5, 7]
- Late application to maize crops can give a worthwhile increase in biomass

Restrictions
- Maximum number of treatments 2 per crop on cereals [1, 5, 7]
- Maximum total dose on turf equivalent to two full dose treatments [4, 6]
- Do not apply during drought conditions or to frozen turf [4, 6]

Crop-specific information
- Latest use: before grain watery ripe (GS 71) for wheat; up to and including emergence of ear just complete (GS 59) for barley and oats [1, 5, 7]
- Avoid applying to turf immediately after cutting or 48 h before mowing [4, 6]

Environmental safety
- Dangerous for the environment
- Very toxic to aquatic organisms
- LERAP Category B

Hazard classification and safety precautions
Hazard Harmful, Dangerous for the environment [1-7]; Harmful if swallowed [1, 2, 5, 7]; Harmful if inhaled, Very toxic to aquatic organisms [1, 2, 4-7]
Transport code 6.1 [1-3, 5, 7]; 9 [4, 6]

SEE SECTION 3 FOR PRODUCTS ALSO REGISTERED

Packaging group III
UN Number 2902 [1-3, 5, 7]; 3077 [4, 6]
Risk phrases H304, H319 [1, 2]; H315, H317 [1, 2, 5, 7]; H335 [5, 7]; R20, R22a, R38, R50, R53a [3]
Operator protection A [1-7]; D [4, 6]; U05a, U14, U20b [1-3, 5, 7]
Environmental protection E15a [6]; E15b, E16a, E38 [1-7]; E16b [1-5, 7]; E34 [1-3, 5, 7]; H410 [1, 2, 5, 7]
Storage and disposal D01, D02 [1-3, 5, 7]; D08, D09a, D10c, D12a [1-7]
Medical advice M03, M05a [1-3, 5, 7]

379   pyraflufen-ethyl

A phenylpyrazole herbicide for potatoes
HRAC mode of action code: E

See also glyphosate + pyraflufen-ethyl

Products

1 Gozai	Belchim	26.5 g/l	EC	17381
2 Kabuki	Belchim	26.5 g/l	EC	18187

Uses
* Annual dicotyledons in **hops** *(off-label)* [1]; **potatoes** [1, 2]
* Desiccation in **hops** *(off-label)* [1]; **potatoes** [1, 2]

Extension of Authorisation for Minor Use (EAMUs)
* **hops** *20171023* [1]

Approval information
* Pyraflufen-ethyl included in Annex I under EC Regulation 1107/2009

Efficacy guidance
* Use with a methylated vegetable oil adjuvant

Following crops guidance
* After cultivation to at least 20 cms winter cereals can be sown as a following crop but the safety to following broad-leaved crops has not yet been established.

Environmental safety
* Buffer zone requirement 20 m in hops [1]
* LERAP Category B

Hazard classification and safety precautions
Hazard Harmful, Dangerous for the environment, Harmful if inhaled, Very toxic to aquatic organisms
Transport code 9
Packaging group III
UN Number 3082
Risk phrases H304, H315, H317, H318
Operator protection A, C, H; U05a, U09a, U11, U14, U15
Environmental protection E15b, E16a, E34, H410
Storage and disposal D01, D02, D05, D09a, D10a, D12a, D12b
Medical advice M03, M05b

380   pyrethrins

A non-persistent, contact acting insecticide extracted from Pyrethrum
IRAC mode of action code: 3

Products

1 Dairy Fly Spray	B H & B	0.75 g/l	AL	H5579

FOR FULL CONDITIONS OF USE ALWAYS READ THE PRODUCT LABEL

Products – continued

2 Pyblast	Agropharm	30 g/l	UL	H7485
3 Pyrethrum 5 EC	Agropharm	50 g/l	EC	12685
4 Spruzit	Certis	4.59 g/l	EC	13438

Uses
- Aphids in *all edible crops (outdoor), all non-edible crops (outdoor), protected crops* [4]; *broccoli, brussels sprouts, bush fruit, cabbages, calabrese, cane fruit, cauliflowers, lettuce, ornamental plant production, protected broccoli, protected brussels sprouts, protected bush fruit, protected cabbages, protected calabrese, protected cane fruit, protected cauliflowers, protected cayenne peppers* (off-label), *protected chilli peppers* (off-label), *protected lettuce, protected ornamentals, protected peppers* (off-label), *protected tomatoes, tomatoes (outdoor), watercress* (off-label) [3]
- Caterpillars in *all edible crops (outdoor), all non-edible crops (outdoor), protected crops* [4]; *broccoli, brussels sprouts, bush fruit, cabbages, calabrese, cane fruit, cauliflowers, lettuce, ornamental plant production, protected broccoli, protected brussels sprouts, protected bush fruit, protected cabbages, protected calabrese, protected cane fruit, protected cauliflowers, protected lettuce, protected ornamentals, protected tomatoes, tomatoes (outdoor)* [3]
- Flea beetle in *broccoli, brussels sprouts, cabbages, calabrese, cauliflowers, protected broccoli, protected brussels sprouts, protected cabbages, protected calabrese, protected cauliflowers, protected tomatoes, tomatoes (outdoor), watercress* (off-label) [3]; *protected crops* [4]
- Insect pests in *dairies, farm buildings, livestock houses, poultry houses* [1, 2]
- Macrolophus caliginosus in *protected tomatoes, protected tomatoes* (off-label) [3]
- Mealybugs in *protected crops* [4]; *protected tomatoes, protected tomatoes* (off-label) [3]
- Mustard beetle in *watercress* (off-label) [3]
- Scale insects in *protected crops* [4]
- Spider mites in *all edible crops (outdoor), all non-edible crops (outdoor), protected crops* [4]
- Thrips in *all edible crops (outdoor), all non-edible crops (outdoor), protected crops* [4]
- Whitefly in *broccoli, brussels sprouts, cabbages, calabrese, cauliflowers, protected broccoli, protected brussels sprouts, protected cabbages, protected calabrese, protected cauliflowers, protected tomatoes, tomatoes (outdoor)* [3]; *protected crops* [4]

Extension of Authorisation for Minor Use (EAMUs)
- *protected cayenne peppers 20091005* [3]
- *protected chilli peppers 20091005* [3]
- *protected peppers 20091005* [3]
- *protected tomatoes 20063026* [3]
- *watercress 20171128* [3]

Approval information
- Pyrethrins have been included in Annex 1 under EC Regulation 1107/2009
- Some products formulated for ULV application [1, 2]. May be applied through fogging machine or sprayer [1, 2]. See label for details
- Accepted by BBPA for use in empty grain stores, malting barley and hops

Efficacy guidance
- For indoor fly control close doors and windows and spray or apply fog as appropriate [1, 2]
- For best fly control outdoors spray during early morning or late afternoon and evening when conditions are still [1, 2]
- For all uses ensure good spray coverage of the target area or plants by increasing spray volume where necessary
- Best results on outdoor or protected crops achieved from treatment at first signs of pest attack in early morning or evening [4]
- To avoid possibility of development of resistance do not spray more frequently than once per week

Restrictions
- Maximum number of treatments on edible crops 3 per crop or season [3]

SEE SECTION 3 FOR PRODUCTS ALSO REGISTERED

- For use only by professional operators [1, 2]
- Do not allow spray to contact open food products or food preparing equipment or utensils [1, 2]
- Remove exposed milk and collect eggs before application [1, 2]
- Do not treat plants [1, 2]
- Avoid direct application to open flowers [4]
- Do not mix with, or apply closely before or after, products containing dithianon or tolylfluanid [4]
- Do not use space sprays containing pyrethrins or pyrethroid more than once per week in intensive or controlled environment animal houses in order to avoid development of resistance. If necessary, use a different control method or product [1, 2]
- Store away from strong sunlight [3]

Crop-specific information
- HI: 24 h for edible crops [3]
- Some plant species including *Ageratum*, ferns, *Ficus, Lantana, Poinsettia, Petroselium crispum*, and some strawberries may be sensitive especially where more than one application is made. Test before large scale treatment [4]

Environmental safety
- Dangerous for the environment [2-4]
- Very toxic to aquatic organisms [1, 3]; toxic to aquatic organisms [2, 4]
- High risk to bees. Do not apply to crops in flower or to those in which bees are actively foraging. Do not apply when flowering weeds are present [3]
- Risk to non-target insects or other arthropods [3]
- Do not apply directly to livestock and exclude all persons and animals during treatment [1, 2]
- Wash spray equipment thoroughly after use to avoid traces of pyrethrum causing damage to susceptible crops sprayed later [3]
- LERAP Category A [3]; LERAP Category B [4]

Hazard classification and safety precautions
Hazard Harmful [1]; Irritant [1-3]; Dangerous for the environment [2-4]
Transport code 9
Packaging group III
UN Number 3082
Risk phrases R22a, R36, R38 [1]; R41, R53a [2, 3]; R50 [3]; R51 [2]
Operator protection A [1-4]; B [1]; C, D [2]; E [1, 2]; H [2-4]; U05a [1, 3]; U09a [1]; U11 [2, 3]; U15 [2]; U19a [1-3]; U20b [1, 2]; U20c [3]; U20d [4]
Environmental protection E02c [2]; E05a [1, 2]; E12a, E12e, E16c, E16d, E22c [3]; E12c, E15b, E16a, H411 [4]; E13c [1]; E15a [2, 3]; E19b [2, 4]; E38 [2-4]
Consumer protection C04, C06 [1]; C07, C08, C09, C10, C11 [1, 2]; C12 [2]
Storage and disposal D01, D02, D09a [1-4]; D05 [3]; D11a [1, 3, 4]; D12a [2-4]
Medical advice M05a [2]

---

381   pyridate

A contact phenylpyridazine herbicide for bulb onions and brassicas
HRAC mode of action code: C3

Products

1 Diva	Belchim	600 g/l	EC	17774
2 Gyo	Belchim	600 g/l	EC	17875
3 Lentagran WP	Belchim	45% w/w	WP	14162

Uses
- Annual dicotyledons in **asparagus** *(off-label)*, **broccoli** *(off-label)*, **calabrese** *(off-label)*, **cauliflowers** *(off-label)*, **chives** *(off-label)*, **collards** *(off-label)*, **fodder rape** *(off-label)*, **game cover** *(off-label)*, **garlic** *(off-label)*, **kale** *(off-label)*, **leeks** *(off-label)*, **lupins** *(off-label)*, **oilseed rape** *(off-label)*, **salad onions** *(off-label)*, **shallots** *(off-label)*, **spring greens** *(off-label)* [3]; **forage maize, grain maize** [1, 2]
- Black nightshade in **asparagus** *(off-label)*, **broccoli** *(off-label)*, **brussels sprouts**, **bulb onions**, **cabbages**, **calabrese** *(off-label)*, **cauliflowers** *(off-label)*, **chives** *(off-label)*, **collards** *(off-label)*,

fodder rape *(off-label),* **game cover** *(off-label),* **garlic** *(off-label),* **kale** *(off-label),* **leeks** *(off-label),* **lupins** *(off-label),* **oilseed rape** *(off-label),* **salad onions** *(off-label),* **shallots** *(off-label),* **spring greens** *(off-label)* [3]; **forage maize, grain maize** [1, 2]

- Cleavers in **asparagus** *(off-label),* **broccoli** *(off-label),* **brussels sprouts, bulb onions, cabbages, calabrese** *(off-label),* **cauliflowers** *(off-label),* **chives** *(off-label),* **collards** *(off-label),* **fodder rape** *(off-label),* **game cover** *(off-label),* **garlic** *(off-label),* **kale** *(off-label),* **leeks** *(off-label),* **lupins** *(off-label),* **oilseed rape** *(off-label),* **salad onions** *(off-label),* **shallots** *(off-label),* **spring greens** *(off-label)* [3]; **forage maize, grain maize** [1, 2]
- Fat hen in **asparagus** *(off-label),* **broccoli** *(off-label),* **brussels sprouts, bulb onions, cabbages, calabrese** *(off-label),* **cauliflowers** *(off-label),* **chives** *(off-label),* **collards** *(off-label),* **fodder rape** *(off-label),* **game cover** *(off-label),* **garlic** *(off-label),* **kale** *(off-label),* **leeks** *(off-label),* **lupins** *(off-label),* **oilseed rape** *(off-label),* **salad onions** *(off-label),* **shallots** *(off-label),* **spring greens** *(off-label)* [3]; **forage maize, grain maize** [1, 2]
- Fumitory in **asparagus** *(off-label),* **broccoli** *(off-label),* **calabrese** *(off-label),* **cauliflowers** *(off-label),* **chives** *(off-label),* **collards** *(off-label),* **fodder rape** *(off-label),* **game cover** *(off-label),* **garlic** *(off-label),* **kale** *(off-label),* **leeks** *(off-label),* **lupins** *(off-label),* **oilseed rape** *(off-label),* **salad onions** *(off-label),* **shallots** *(off-label),* **spring greens** *(off-label)* [3]; **forage maize, grain maize** [1, 2]
- Groundsel in **asparagus** *(off-label),* **broccoli** *(off-label),* **calabrese** *(off-label),* **cauliflowers** *(off-label),* **chives** *(off-label),* **collards** *(off-label),* **fodder rape** *(off-label),* **game cover** *(off-label),* **garlic** *(off-label),* **kale** *(off-label),* **leeks** *(off-label),* **lupins** *(off-label),* **oilseed rape** *(off-label),* **salad onions** *(off-label),* **shallots** *(off-label),* **spring greens** *(off-label)* [3]; **forage maize, grain maize** [1, 2]

Extension of Authorisation for Minor Use (EAMUs)

- **asparagus** *20091039* [3]
- **broccoli** *20090786* [3]
- **calabrese** *20090786* [3]
- **cauliflowers** *20090786* [3]
- **chives** *20120267* [3]
- **collards** *20090785* [3]
- **fodder rape** *20093230* [3]
- **game cover** *20090788* [3]
- **garlic** *20092862* [3]
- **kale** *20090785* [3]
- **leeks** *20090784* [3]
- **lupins** *20090787* [3]
- **oilseed rape** *20093230* [3]
- **salad onions** *20120267* [3]
- **shallots** *20092862* [3]
- **spring greens** *20090785* [3]

Approval information

- Pyridate included in Annex I under EC Regulation 1107/2009

Efficacy guidance

- Best results achieved by application to actively growing weeds at 6-8 leaf stage when temperatures are above 8°C before crop foliage forms canopy

Restrictions

- Do not apply in mixture with or within 14 d of any other product which may result in dewaxing of crop foliage
- Do not use on crops suffering stress from frost, drought, disease or pest attack

Crop-specific information

- Latest use: before 6 true leaf stage of Brussels sprouts and cabbage; before 7 true leaves for onions; before flower buds visible for oilseed rape
- Apply to cabbages and Brussels sprouts after 4 fully expanded leaf stage.   Allow 2 wk after transplanting before treating

SEE SECTION 3 FOR PRODUCTS ALSO REGISTERED

Environmental safety
- Dangerous for the environment
- Toxic to aquatic organisms
- Buffer zone requirement 12 m [1, 2]
- LERAP Category B [1, 2]

Hazard classification and safety precautions
Hazard Irritant, Dangerous for the environment [3]; Flammable liquid and vapour, Very toxic to aquatic organisms [1, 2]
Transport code 9
Packaging group III
UN Number 3077 [3]; 3082 [1, 2]
Risk phrases H315, H317, H319, H370 [1, 2]; R43, R51, R53a [3]
Operator protection A [1-3]; C [3]; H [1, 2]; U05a, U14, U22a
Environmental protection E15a [3]; E16a, H410 [1, 2]; E38 [1-3]
Storage and disposal D01, D02, D09a, D10c, D12a

382 pyriofenone

A protectant fungicide with some curative activity against mildew in the latent phase
FRAC mode of action code: U8

Products
Property 180SC	Belchim	180 g/l	SC	17666

Uses
- Powdery mildew in **spring wheat**, **winter wheat**

Approval information
- Pyrifenone is included in Annex 1 under EC Regulation 1107/2009

Efficacy guidance
- Best results obtained from application at the start of disease attack. Acts preventatively against powdery mildew with some curative activity during the latent phase. Apply before disease spreads on to new growth with a follow-up after 2 - 4 weeks according to climatic conditions.
- Should be used as part of a mildew control programme that incorporates other methods of control including fungicides active against mildew via a different mode of action.

Restrictions
- Only two applications of benzoylpyridine fungicides are permitted per crop per season. Where a second application is required it should be applied in mixture with fungicides using a different mode of action such as fenpropimorph.

Crop-specific information
- Latest use: mid-flowering (GS65) in spring and winter wheat

Following crops guidance
- Oilseed rape, maize, potato, sugar beet, kidney bean, soybean, pea, onion, turnip, lettuce, flax, oat, barley and wheat may be sown as a following crop.

Hazard classification and safety precautions
Hazard Harmful, Dangerous for the environment
Transport code 9
Packaging group III
UN Number 3082
Risk phrases H351, R53a
Operator protection A, H; U05a, U20b
Environmental protection E15b, E34, E38, H411
Storage and disposal D01, D02, D05, D09a, D12a, D20
Medical advice M05a

383   quinmerac

A residual herbicide available only in mixtures
HRAC mode of action code: O

See also chloridazon + quinmerac
dimethenamid-p + metazachlor + quinmerac
dimethenamid-p + quinmerac
imazamox + metazachlor + quinmerac
imazamox + quinmerac
metamitron + quinmerac
metazachlor + quinmerac

384   quinoclamine

A selective moss killer for moderate control of moss in managed amenity turf
HRAC mode of action code: Not classified

Products

Mogeton	Certis	25% w/w	WP	15837

Uses

* Moss in **managed amenity turf** *(moderate control)*, **protected ornamentals** *(off-label)*

Extension of Authorisation for Minor Use (EAMUs)

* **protected ornamentals** 20152273

Approval information

* Quinoclamine included in Annex 1 under EC Regulation 1107/2009

Environmental safety

* When treating managed amenity turf, do not apply within 20 m of a static or flowing water body
* LERAP Category A

Hazard classification and safety precautions

Hazard Harmful, Dangerous for the environment, Harmful if swallowed
Transport code 9
Packaging group III
UN Number 3077
Risk phrases H317, H320, H361, H373
Operator protection A, C, G, H; U05a, U11, U19a, U19e
Environmental protection E15a, E16c, H410
Storage and disposal D01, D02, D12a
Medical advice M03, M05a

385   quizalofop-P-ethyl

An aryl phenoxypropionic acid post-emergence herbicide for grass weed control
HRAC mode of action code: A

Products

1	Leopard 5EC	Adama	50 g/l	EC	16915
2	Pilot Ultra	Nissan	50 g/l	SC	17136
3	Targa Max	Nissan	100 g/l	EC	17135
4	Targa Super	Nissan	50 g/l	EC	17134

Uses

* Annual grasses in **combining peas, linseed, spring field beans, vining peas, winter field beans** [1, 2]; **fodder beet, mangels, potatoes, red beet, spring oilseed rape, sugar beet, winter oilseed rape** [1-4]; **forest nurseries** *(off-label)* [1]

SEE SECTION 3 FOR PRODUCTS ALSO REGISTERED

- Couch in **combining peas, linseed, spring field beans, vining peas, winter field beans** [1, 2]; **fodder beet, mangels, potatoes, red beet, spring oilseed rape, sugar beet, winter oilseed rape** [1-4]
- Perennial grasses in **combining peas, linseed, spring field beans, vining peas, winter field beans** [1, 2]; **fodder beet, mangels, potatoes, red beet, spring oilseed rape, sugar beet, winter oilseed rape** [1-4]; **forest nurseries** *(off-label)* [1]
- Volunteer cereals in **combining peas, linseed, spring field beans, vining peas, winter field beans** [1, 2]; **fodder beet, mangels, potatoes, red beet, spring oilseed rape, sugar beet, winter oilseed rape** [1-4]

Extension of Authorisation for Minor Use (EAMUs)
- **forest nurseries** *20151181* [1]

Approval information
- Quizalofop-P-ethyl has been included in Annex 1 under EC Regulation 1107/2009

Efficacy guidance
- Best results achieved by application to emerged weeds growing actively in warm conditions with adequate soil moisture
- Weed control may be reduced under conditions such as drought that limit uptake and translocation
- Annual meadow-grass is not controlled
- For effective couch control do not hoe beet crops within 21 d after spraying
- At least 2 h without rain should follow application otherwise results may be reduced
- Quizalofop-P-ethyl is an ACCase inhibitor herbicide. To avoid the build up of resistance do not apply products containing an ACCase inhibitor herbicide more than twice to any crop. In addition do not use any product containing quizalofop-P-ethyl in mixture or sequence with any other product containing the same ingredient
- Use these products as part of a resistance management strategy that includes cultural methods of control and does not use ACCase inhibitors as the sole chemical method of grass weed control
- Applying a second product containing an ACCase inhibitor to a crop will increase the risk of resistance development; only use a second ACCase inhibitor to control different weeds at a different timing
- Always follow WRAG guidelines for preventing and managing herbicide resistant weeds. See Section 5 for more information

Restrictions
- Maximum number of treatments 1 on all recommended crops
- Do not spray crops under stress from any cause or in frosty weather
- Consult processor before use on crops recommended for processing
- An interval of at least 3 d must elapse between treatment and use of another herbicide on beet crops, 14 d on oilseed rape, 21 d on linseed and other recommended crops
- Avoid drift onto neighbouring crops

Crop-specific information
- HI 16 wk for beet crops; 11 wk for oilseed rape, linseed; 8 wk for field beans; 5 wk for peas
- In some situations treatment can cause yellow patches on foliage of peas, especially vining varieties. Symptoms usually rapidly and completely outgrown
- May cause taint in peas

Following crops guidance
- In the event of failure of a treated crop broad-leaved crops may be resown after a minimum interval of 2 wk, and cereals after 2-6 wk depending on dose applied
- Onions, leeks and maize are not recommended to follow a failed treated crop

Environmental safety
- Dangerous for the environment
- Toxic to aquatic organisms
- Flammable

FOR FULL CONDITIONS OF USE ALWAYS READ THE PRODUCT LABEL

Hazard classification and safety precautions

Hazard Harmful, Flammable [1]; Dangerous for the environment [1-4]; Harmful if swallowed, Very toxic to aquatic organisms [4]

Transport code 3 [1]; 9 [2-4]

Packaging group III

UN Number 1993 [1]; 3082 [2-4]

Risk phrases H304 [1, 4]; H315, H319, H336, R65 [1]; H317 [4]; H318 [3, 4]

Operator protection A, C, H; U05a, U11, U14, U15, U19a [1]; U09a [2-4]; U20b [1-4]

Environmental protection E13b [2-4]; E15a, H411 [1]; E34, E38 [1-4]; H410 [4]

Storage and disposal D01, D02, D05, D10b [1]; D09a, D12a [1-4]; D10c [2-4]

Medical advice M03, M05b [1]

386   quizalofop-P-tefuryl

An aryloxyphenoxypropionate herbicide for grass weed control
HRAC mode of action code: A

Products

Rango	Certis	40 g/l	EC	17006

Uses

- Blackgrass in *combining peas, fodder beet, potatoes, spring field beans, spring linseed, spring oilseed rape, sugar beet, winter field beans, winter linseed, winter oilseed rape*
- Couch in *combining peas, fodder beet, potatoes, spring field beans, spring linseed, spring oilseed rape, sugar beet, winter field beans, winter linseed, winter oilseed rape*
- Italian ryegrass in *combining peas, fodder beet, potatoes, spring field beans, spring linseed, spring oilseed rape, sugar beet, winter field beans, winter linseed, winter oilseed rape*
- Perennial ryegrass in *combining peas, fodder beet, potatoes, spring field beans, spring linseed, spring oilseed rape, sugar beet, winter field beans, winter linseed, winter oilseed rape*
- Volunteer cereals in *combining peas, fodder beet, potatoes, spring field beans, spring linseed, spring oilseed rape, sugar beet, winter field beans, winter linseed, winter oilseed rape*
- Wild oats in *combining peas, fodder beet, potatoes, spring field beans, spring linseed, spring oilseed rape, sugar beet, winter field beans, winter linseed, winter oilseed rape*

Approval information

- Quizalofop-P-tefuryl has been included in Annex 1 under EC Regulation 1107/2009

Efficacy guidance

- Best results on annual grass weeds when growing actively and treated from 2 leaves to the start of tillering
- Treat cover crops when they have served their purpose and the threat of wind blow has passed
- Best results on couch achieved when the weed is growing actively and commencing new rhizome growth
- Grass weeds germinating after treatment will not be controlled
- Treatment quickly stops growth and visible colour changes to the leaf tips appear after about 7 d. Complete kill takes 3-4 wk under good growing conditions
- Quizalofop-P-tefuryl is an ACCase inhibitor herbicide. To avoid the build up of resistance do not apply products containing an ACCase inhibitor herbicide more than twice to any crop. In addition do not use any product containing quizalofop-P-tefuryl in mixture or sequence with any other product containing the same ingredient
- Use these products as part of a resistance management strategy that includes cultural methods of control and does not use ACCase inhibitors as the sole chemical method of grass weed control
- Applying a second product containing an ACCase inhibitor to a crop will increase the risk of resistance development; only use a second ACCase inhibitor to control different weeds at a different timing
- Always follow WRAG guidelines for preventing and managing herbicide resistant weeds. See Section 5 for more information

SEE SECTION 3 FOR PRODUCTS ALSO REGISTERED

Restrictions

- Maximum number of treatments 1 per crop
- Consult processors before use on peas or potatoes for processing
- Do not treat crops and weeds growing under stress from any cause

Crop-specific information

- HI 60 d for all crops
- Treat oilseed rape from the fully expanded cotyledon stage
- Treat linseed, peas and field beans from 2-3 unfolded leaves
- Treat sugar beet from 2 unfolded leaves

Following crops guidance

- In the event of failure of a treated crop any broad-leaved crop may be planted at any time. Cereals may be drilled from 4 wk after treatment

Environmental safety

- Dangerous for the environment
- Very toxic to aquatic organisms
- Risk to certain non-target insects or other arthropods. For advice on risk management and use in Integrated Pest Management (IPM) see directions for use
- Avoid spraying within 6 m of the field boundary to reduce effects on non-target insects and other arthropods

Hazard classification and safety precautions

Hazard Toxic, Dangerous for the environment
Transport code 9
Packaging group III
UN Number 3082
Risk phrases H317, H318, H360, H371
Operator protection A, C, H; U02a, U04a, U05a, U10, U11, U14, U15, U19a, U20b
Environmental protection E15a, E22b, E38, H411
Storage and disposal D01, D02, D09a, D10b, D12a
Medical advice M04a, M05b

387   rimsulfuron

A selective systemic sulfonylurea herbicide
HRAC mode of action code: B

Products

1 Caesar	AgChem Access	25% w/w	SG	15272
2 Clayton Bramble	Clayton	25% w/w	SG	15281
3 Clayton Nero	Clayton	25% w/w	SG	15511
4 Clayton Rasp	Clayton	25% w/w	SG	15952
5 Titus	Adama	25% w/w	SG	15050

Uses

- Annual dicotyledons in *forage maize*, *potatoes* [1-5]; *forest nurseries* (off-label), *ornamental plant production* (off-label) [5]
- Charlock in *forage maize*, *potatoes* [4, 5]
- Chickweed in *forage maize*, *potatoes* [4, 5]
- Cleavers in *forage maize*, *potatoes* [4, 5]
- Hemp-nettle in *forage maize*, *potatoes* [4, 5]
- Red dead-nettle in *forage maize*, *potatoes* [4, 5]
- Redshank in *forage maize*, *potatoes* [4, 5]
- Scentless mayweed in *forage maize*, *potatoes* [4, 5]
- Small nettle in *forage maize*, *potatoes* [4, 5]
- Volunteer oilseed rape in *forage maize*, *potatoes* [1-5]

Extension of Authorisation for Minor Use (EAMUs)

- *forest nurseries* 20141911 expires *30 Jun 2018* [5], 20141912 [5]

FOR FULL CONDITIONS OF USE ALWAYS READ THE PRODUCT LABEL

- ***ornamental plant production*** *20141911 expires 30 Jun 2018* [5], *20141912* [5]

Approval information
- Rimsulfuron included in Annex I under EC Regulation 1107/2009

Efficacy guidance
- Product should be used with a suitable adjuvant or a suitable herbicide tank-mix partner. See label for details
- Product acts by foliar action. Best results obtained from good spray cover of small actively growing weeds. Effectiveness is reduced in very dry conditions
- Weed spectrum can be broadened by tank mixture with other herbicides. See label for details
- Susceptible weeds cease growth immediately and symptoms can be seen 10 d later
- Rimsulfuron is a member of the ALS-inhibitor group of herbicides

Restrictions
- Maximum number of treatments 1 per crop
- Do not treat maize previously treated with organophosphorus insecticides
- Do not apply to potatoes grown for certified seed
- Consult processor before use on crops grown for processing
- Avoid high light intensity (full sunlight) and high temperatures on the day of spraying
- Do not treat during periods of substantial diurnal temperature fluctuation or when frost anticipated
- Do not apply to any crop stressed by drought, water-logging, low temperatures, pest or disease attack, nutrient or lime deficiency
- Do not apply to forage maize treated with organophosphate insecticides
- Do not apply to forage maize undersown with grass or clover

Crop-specific information
- Latest use: before most advanced potato plants are 25 cm high; before 4-collar stage of fodder maize
- All varieties of ware potatoes may be treated, but variety restrictions of any tank-mix partner must be observed
- Only certain named varieties of forage maize may be treated. See label

Following crops guidance
- Only winter wheat should follow a treated crop in the same calendar yr
- Only barley, wheat or maize should be sown in the spring of the yr following treatment
- In the second autumn after treatment any crop except brassicas or oilseed rape may be drilled

Environmental safety
- Dangerous for the environment
- Toxic to aquatic organisms
- Extremely dangerous to fish or other aquatic life. Do not contaminate surface waters or ditches with chemical or used container
- Herbicide is very active. Take particular care to avoid drift onto plants outside the target area
- Spraying equipment should not be drained or flushed onto land planted, or to be planted, with trees or crops other than potatoes or forage maize and should be thoroughly cleansed after use - see label for instructions

Hazard classification and safety precautions
Hazard Dangerous for the environment
Transport code 9
Packaging group III
UN Number 3077
Risk phrases R51, R53a
Operator protection A [3-5]; U08, U19a, U20b
Environmental protection E13a, E34, E38
Storage and disposal D01, D09a, D11a, D12a

SEE SECTION 3 FOR PRODUCTS ALSO REGISTERED

388   S-metolachlor

A chloroacetamide residual herbicide for use in forage maize and grain maize
HRAC mode of action code: K3

See also mesotrione + s-metolachlor

Products

1 Clayton Smelter	Clayton	960 g/l	EC	14937
2 Dual Gold	Syngenta	960 g/l	EC	14649

Uses

- Annual dicotyledons in **baby leaf crops** *(off-label)*, **forest nurseries** *(off-label)*, **garlic** *(off-label)*, **herbs (see appendix 6)** *(off-label)*, **lettuce** *(off-label)*, **onions** *(off-label)*, **ornamental plant production** *(off-label)*, **red beet** *(off-label)*, **shallots** *(off-label)*, **swedes** *(off-label)*, **turnips** *(off-label)* [2]
- Annual grasses in **begonias** *(off-label)*, **broccoli** *(off-label)*, **brussels sprouts** *(off-label)*, **cabbages** *(off-label)*, **calabrese** *(off-label)*, **cauliflowers** *(off-label)*, **chinese cabbage** *(off-label)*, **collards** *(off-label)*, **kale** *(off-label)* [2]; **chicory** *(off-label)*, **dwarf beans** *(off-label)*, **endives** *(off-label)*, **french beans** *(off-label)*, **runner beans** *(off-label)*, **strawberries** *(off-label)* [1, 2]; **chicory root** *(off-label)* [1]
- Annual meadow grass in **baby leaf crops** *(off-label)*, **edible podded peas** *(off-label)*, **forest nurseries** *(off-label)*, **garlic** *(off-label)*, **herbs (see appendix 6)** *(off-label)*, **lettuce** *(off-label)*, **onions** *(off-label)*, **ornamental plant production** *(off-label)*, **red beet** *(off-label)*, **shallots** *(off-label)*, **spring field beans** *(off-label)*, **swedes** *(off-label)*, **turnips** *(off-label)*, **vining peas** *(off-label)* [2]; **forage maize, grain maize** [1, 2]
- Black nightshade in **chicory** *(off-label)*, **endives** *(off-label)* [1, 2]; **chicory root** *(off-label)* [1]
- Chickweed in **forage maize** *(moderately susceptible)*, **grain maize** *(moderately susceptible)* [1, 2]; **spring field beans** *(off-label)* [2]
- Fat hen in **forage maize** *(moderately susceptible)*, **grain maize** *(moderately susceptible)* [1, 2]; **spring field beans** *(off-label)* [2]
- Field speedwell in **forage maize, grain maize** [1, 2]; **spring field beans** *(off-label)* [2]
- Mayweeds in **forage maize, grain maize** [1, 2]; **spring field beans** *(off-label)* [2]
- Red dead-nettle in **forage maize, grain maize** [1, 2]
- Sowthistle in **forage maize, grain maize** [1, 2]

Extension of Authorisation for Minor Use (EAMUs)

- **baby leaf crops** *20120594* [2]
- **begonias** *20132574* [2]
- **broccoli** *20111258* [2]
- **brussels sprouts** *20111259* [2]
- **cabbages** *20111259* [2]
- **calabrese** *20111258* [2]
- **cauliflowers** *20111258* [2]
- **chicory** *20111262* [1], *20111257* [2]
- **chicory root** *20103107* [1]
- **chinese cabbage** *20111260* [2]
- **collards** *20111260* [2]
- **dwarf beans** *20103103* [1], *20132572* [2]
- **edible podded peas** *20140164* [2]
- **endives** *20103107* [1], *20111262* [1], *20111257* [2]
- **forest nurseries** *20120501* [2]
- **french beans** *20103103* [1], *20132572* [2]
- **garlic** *20110840* [2]
- **herbs (see appendix 6)** *20120594* [2]
- **kale** *20111260* [2]
- **lettuce** *20120594* [2]
- **onions** *20110840* [2]
- **ornamental plant production** *20120501* [2]
- **red beet** *20111006* [2]

- *runner beans* *20103103* [1], *20132572* [2]
- *shallots* *20110840* [2]
- *spring field beans* *20140163* [2]
- *strawberries* *20103104* [1], *20132573* [2]
- *swedes* *20111006* [2]
- *turnips* *20111006* [2]
- *vining peas* *20140164* [2]

Approval information
- S-metolachlor included in Annex I under EC Regulation 1107/2009

Environmental safety
- LERAP Category B

Hazard classification and safety precautions
    Hazard Irritant, Dangerous for the environment [1, 2]; Very toxic to aquatic organisms [2]
    Transport code 9
    Packaging group III
    UN Number 3082
    Risk phrases H317, H319 [2]; R43, R50, R53a [1]
    Operator protection A [1, 2]; C [2]; H [1]; U05a, U09a, U20b
    Environmental protection E15b, E16a, E16b, E38 [1, 2]; H410 [2]
    Storage and disposal D01, D02, D09a, D10c, D11a, D12a

389   sodium hypochlorite (commodity substance)

An inorganic horticultural bactericide for use in mushrooms

Products

sodium hypochlorite	various	100% w/v	ZZ	-

Uses
- Bacterial blotch in *mushrooms*

Approval information
- Sodium hypochlorite included in Annex 1 under EC Regulation 1107/2009
- Approval as a commodity substance valid until 31 Aug 2019

Restrictions
- Maximum concentration 315 mg/litre of water
- Mixing and loading must only take place in a ventilated area
- Must only be used by suitably trained and competent operators

Crop-specific information
- HI 1 d

Environmental safety
- Harmful to fish or other aquatic life. Do not contaminate surface waters or ditches with chemical or used container

Hazard classification and safety precautions
    Operator protection A, C, H
    Environmental protection E13c

390   spearmint oil

A hot fogging treatment for post-harvest sprout suppression in potatoes

Products

Biox-M	Juno	100% w/w	HN	16021

SEE SECTION 3 FOR PRODUCTS ALSO REGISTERED

Uses

- Sprout suppression in *potatoes*

Approval information

- Not listed in Annex 1 under EC Regulation 1107/2009 since it is a natural product.

Hazard classification and safety precautions

Hazard Harmful, Dangerous for the environment
UN Number N/C
Risk phrases R22b, R43, R50, R53a
Operator protection A, C, D, H, J, M
Storage and disposal D12a

391   spinosad

A selective insecticide derived from naturally occurring soil fungi (naturalyte)
IRAC mode of action code: 5

Products

1 Conserve	Fargro	120 g/l	SC	12058
2 Tracer	Landseer	480 g/l	SC	12438

Uses

- Asparagus beetle in *asparagus* *(off-label)* [2]
- Cabbage moth in *broccoli*, *brussels sprouts*, *cabbages*, *calabrese*, *cauliflowers*, *chinese cabbage*, *collards* *(off-label)*, *kale* *(off-label)*, *kohlrabi* *(off-label)*, *oriental cabbage* *(off-label)* [2]
- Cabbage root fly in *collards* *(off-label)*, *kale* *(off-label)*, *kohlrabi* *(off-label)*, *oriental cabbage* *(off-label)* [2]
- Cabbage white butterfly in *broccoli*, *brussels sprouts*, *cabbages*, *calabrese*, *cauliflowers*, *chinese cabbage*, *collards* *(off-label)*, *kale* *(off-label)*, *kohlrabi* *(off-label)*, *oriental cabbage* *(off-label)* [2]
- Caterpillars in *celery (outdoor)* *(off-label)*, *collards* *(off-label)*, *fennel* *(off-label)*, *kale* *(off-label)*, *kohlrabi* *(off-label)*, *oriental cabbage* *(off-label)* [2]
- Celery fly in *celery (outdoor)* *(off-label)*, *fennel* *(off-label)* [2]
- Celery leaf miner in *celery (outdoor)* *(off-label)*, *fennel* *(off-label)* [2]
- Codling moth in *apples* [2]
- Diamond-back moth in *broccoli*, *brussels sprouts*, *cabbages*, *calabrese*, *cauliflowers*, *chinese cabbage*, *collards* *(off-label)*, *kale* *(off-label)*, *kohlrabi* *(off-label)*, *oriental cabbage* *(off-label)* [2]
- Insect control in *celeriac* *(off-label)*, *chicory* *(off-label)*, *courgettes* *(off-label)*, *pattisons* *(off-label)*, *protected melons* *(off-label)*, *protected pumpkins* *(off-label)*, *swedes* *(off-label)*, *turnips* *(off-label)*, *witloof* *(off-label)* [2]
- Insect pests in *blackberries* *(off-label)*, *blackcurrants* *(off-label)*, *grapevines* *(off-label)*, *hops* *(off-label)*, *ornamental specimens* *(off-label)*, *protected blackberries* *(off-label)*, *protected raspberries* *(off-label)*, *protected strawberries*, *raspberries* *(off-label)*, *redcurrants* *(off-label)*, *soft fruit* *(off-label)*, *top fruit* *(off-label)*, *whitecurrants* *(off-label)* [2]; *protected tomatoes* *(off-label)* [1]
- Large white butterfly in *collards* *(off-label)*, *kale* *(off-label)*, *kohlrabi* *(off-label)*, *oriental cabbage* *(off-label)* [2]
- Light brown apple moth in *blueberries* *(off-label)*, *cranberries* *(off-label)*, *gooseberries* *(off-label)*, *haskap* *(off-label)*, *protected blackcurrants* *(off-label)*, *protected blueberry* *(off-label)*, *protected gooseberries* *(off-label)*, *protected haskap* *(off-label)*, *protected redcurrants* *(off-label)* [2]
- Liriomyza bryoniae in *protected tomatoes* *(off-label)* [1]
- Moths in *hops* *(off-label)*, *ornamental specimens* *(off-label)*, *soft fruit* *(off-label)*, *top fruit* *(off-label)* [2]
- Onion thrips in *celery leaves* *(off-label)*, *chard* *(off-label)*, *cress* *(off-label)*, *endives* *(off-label)*, *frise* *(off-label)*, *herbs (see appendix 6)* *(off-label)*, *lamb's lettuce* *(off-label)*, *leaf brassicas* *(off-label)*, *lettuce* *(off-label)*, *protected herbs (see appendix 6)* *(off-label)*, *radicchio* *(off-label)*, *red*

mustard *(off-label)*, **rocket** *(off-label)*, **scarole** *(off-label)*, **spinach** *(off-label)*, **spinach beet** *(off-label)* [2]
- Sawflies in **blackcurrants** *(off-label)*, **redcurrants** *(off-label)*, **whitecurrants** *(off-label)* [2]
- Silver Y moth in **celery (outdoor)** *(off-label)*, **celery leaves** *(off-label)*, **chard** *(off-label)*, **cress** *(off-label)*, **dwarf beans** *(off-label)*, **endives** *(off-label)*, **fennel** *(off-label)*, **french beans** *(off-label)*, **frise** *(off-label)*, **herbs (see appendix 6)** *(off-label)*, **lamb's lettuce** *(off-label)*, **leaf brassicas** *(off-label)*, **lettuce** *(off-label)*, **protected herbs (see appendix 6)** *(off-label)*, **radicchio** *(off-label)*, **red mustard** *(off-label)*, **rocket** *(off-label)*, **scarole** *(off-label)*, **spinach** *(off-label)*, **spinach beet** *(off-label)* [2]
- Small white butterfly in **broccoli**, **brussels sprouts**, **cabbages**, **calabrese**, **cauliflowers**, **chinese cabbage**, **collards** *(off-label)*, **kale** *(off-label)*, **kohlrabi** *(off-label)*, **oriental cabbage** *(off-label)* [2]
- Spotted wing drosophila in **blueberries** *(off-label)*, **cranberries** *(off-label)*, **gooseberries** *(off-label)*, **haskap** *(off-label)*, **protected blackcurrants** *(off-label)*, **protected blueberry** *(off-label)*, **protected gooseberries** *(off-label)*, **protected haskap** *(off-label)*, **protected redcurrants** *(off-label)*, **rubus hybrids** *(off-label)* [2]
- Thrips in **asparagus** *(off-label)*, **blackberries** *(off-label)*, **bulb onions**, **celeriac** *(off-label)*, **celery (outdoor)** *(off-label)*, **chicory** *(off-label)*, **courgettes** *(off-label)*, **fennel** *(off-label)*, **garlic**, **leeks**, **pattisons** *(off-label)*, **protected blackberries** *(off-label)*, **protected melons** *(off-label)*, **protected pumpkins** *(off-label)*, **protected raspberries** *(off-label)*, **protected strawberries**, **raspberries** *(off-label)*, **salad onions**, **shallots**, **swedes** *(off-label)*, **turnips** *(off-label)*, **witloof** *(off-label)* [2]; **protected tomatoes** *(off-label)* [1]
- Tortrix moths in **apples** [2]
- Western flower thrips in **protected aubergines**, **protected cucumbers**, **protected ornamentals**, **protected peppers**, **protected tomatoes** [1]
- Winter moth in **blackcurrants** *(off-label)*, **redcurrants** *(off-label)*, **whitecurrants** *(off-label)* [2]

Extension of Authorisation for Minor Use (EAMUs)
- **asparagus** *20140301* [2]
- **blackberries** *20092118* [2]
- **blackcurrants** *20113223* [2]
- **blueberries** *20150775* [2]
- **celeriac** *20171533* [2]
- **celery (outdoor)** *20131946* [2]
- **celery leaves** *20171632* [2]
- **chard** *20171632* [2]
- **chicory** *20171533* [2]
- **collards** *20150102* [2]
- **courgettes** *20171533* [2]
- **cranberries** *20150775* [2]
- **cress** *20171632* [2]
- **dwarf beans** *20150103* [2]
- **endives** *20171632* [2]
- **fennel** *20131946* [2]
- **french beans** *20150103* [2]
- **frise** *20171632* [2]
- **gooseberries** *20150775* [2]
- **grapevines** *20122222* [2]
- **haskap** *20150775* [2]
- **herbs (see appendix 6)** *20171632* [2]
- **hops** *20082908* [2]
- **kale** *20150102* [2]
- **kohlrabi** *20150102* [2]
- **lamb's lettuce** *20171632* [2]
- **leaf brassicas** *20171632* [2]
- **lettuce** *20171632* [2]
- **oriental cabbage** *20150102* [2]
- **ornamental specimens** *20082908* [2]
- **pattisons** *20171533* [2]

SECTION 2

- **protected blackberries** *20130245* [2]
- **protected blackcurrants** *20150775* [2]
- **protected blueberry** *20150775* [2]
- **protected gooseberries** *20150775* [2]
- **protected haskap** *20150775* [2]
- **protected herbs (see appendix 6)** *20171632* [2]
- **protected melons** *20171533* [2]
- **protected pumpkins** *20171533* [2]
- **protected raspberries** *20130245* [2]
- **protected redcurrants** *20150775* [2]
- **protected tomatoes** *20130325* [1]
- **radicchio** *20171632* [2]
- **raspberries** *20092118* [2]
- **red mustard** *20171632* [2]
- **redcurrants** *20113223* [2]
- **rocket** *20171632* [2]
- **rubus hybrids** *20142018* [2]
- **scarole** *20171632* [2]
- **soft fruit** *20082908* [2]
- **spinach** *20171632* [2]
- **spinach beet** *20171632* [2]
- **swedes** *20171533* [2]
- **top fruit** *20082908* [2]
- **turnips** *20171533* [2]
- **whitecurrants** *20113223* [2]
- **witloof** *20171533* [2]

Approval information
- Spinosad included in Annex I under EC Regulation 1107/2009

Efficacy guidance
- Product enters insects by contact from a treated surface or ingestion of treated plant material therefore good spray coverage is essential
- Some plants, for example Fuchsia flowers, can provide effective refuges from spray deposits and control of western flower thrips may be reduced [1]
- Apply to protected crops when western flower thrip nymphs or adults are first seen [1]
- Monitor western flower thrip development carefully to see whether further applications are necessary. A 2 spray programme at 5-7 d intervals may be needed when conditions favour rapid pest development [1]
- Treat top fruit and field crops when pests are first seen or at very first signs of crop damage [2]
- Ensure a rain-free period of 12 h after treatment before applying irrigation [2]
- To reduce possibility of development of resistance, adopt resistance management measures. See label and Section 5 for more information

Restrictions
- Maximum number of treatments 4 per crop for brassicas, onions, leeks [2]; 1 pre-blossom and 3 post blossom for apples, pears [2]
- Apply no more than 2 consecutive sprays. Rotate with another insecticide with a different mode of action or use no further treatment after applying the maximum number of treatments
- Establish whether any incoming plants have been treated and apply no more than 2 consecutive sprays. Maximum number of treatments 6 per structure per yr
- Avoid application in bright sunlight or into open flowers [1]

Crop-specific information
- HI 3 d for protected cucumbers, brassicas; 7 d for apples, pears, onions, leeks
- Test for tolerance on a small number of ornamentals or cucumbers before large scale treatment
- Some spotting of african violet flowers may occur

Environmental safety
- Dangerous for the environment
- Very toxic to aquatic organisms

FOR FULL CONDITIONS OF USE ALWAYS READ THE PRODUCT LABEL

- Whenever possible use an Integrated Pest Management system. Spinosad presents low risk to beneficial arthropods
- Product has low impact on many insect and mite predators but is harmful to adults of most parasitic wasps. Most beneficials may be introduced to treated plants when spray deposits are dry but an interval of 2 wk should elapse before introduction of parasitic wasps. See label for details
- Treatment may cause temporary reduction in abundance of insect and mite predators if present at application
- Exposure to direct spray is harmful to bees but dry spray deposits are harmless. Treatment of field crops should not be made in the heat of the day when bees are actively foraging [2]
- Buffer zone requirement 10 m for melon and pumpkin
- Broadcast air-assisted LERAP [2] (40 m); LERAP Category B [2]

Hazard classification and safety precautions

Hazard Dangerous for the environment
Transport code 9
Packaging group III
UN Number 3082
Risk phrases R50, R53a [2]
Operator protection A, H [2]; U05a [1]; U08, U20b [2]
Environmental protection E15a, E16a [2]; E15b, H410 [1]; E17b [2] (40 m); E34, E38 [1, 2]
Storage and disposal D01, D05 [1]; D02, D10b, D12a [1, 2]

392   spirodiclofen

A tetronic acid which inhibits lipid biosynthesis
IRAC mode of action code: 23

Products

Envidor	Bayer CropScience	240 g/l	SC	17518

Uses

- Mussel scale in *apples*, *pears*
- Pear sucker in *pears*
- Red spider mites in *apples*, *pears*
- Rust mite in *apples*, *pears*, *plums* (off-label)
- Spider mites in *apricots* (off-label), *blackcurrants* (off-label), *cherries* (off-label), *gooseberries* (off-label), *hops* (off-label), *nectarines* (off-label), *nursery fruit trees* (off-label), *ornamental plant production* (off-label), *peaches* (off-label), *plums* (off-label), *protected chilli peppers* (off-label), *protected cucumbers* (off-label), *protected peppers* (off-label), *protected tomatoes* (off-label), *redcurrants* (off-label), *whitecurrants* (off-label)
- Two-spotted spider mite in *apples*, *pears*, *protected strawberries* (off-label), *strawberries* (off-label)

Extension of Authorisation for Minor Use (EAMUs)

- *apricots* 20160999
- *blackcurrants* 20161000
- *cherries* 20160998
- *gooseberries* 20161000
- *hops* 20171601
- *nectarines* 20160999
- *nursery fruit trees* 20171601
- *ornamental plant production* 20160997
- *peaches* 20160999
- *plums* 20160995
- *protected chilli peppers* 20160996
- *protected cucumbers* 20160996
- *protected peppers* 20160996
- *protected strawberries* 20171600

SEE SECTION 3 FOR PRODUCTS ALSO REGISTERED

- **protected tomatoes** *20160996*
- **redcurrants** *20161000*
- **strawberries** *20171600*
- **whitecurrants** *20161000*

Approval information
- Spirodiclofen included in Annex 1 under EC Regulation 1107/2009

Restrictions
- Consult processor on crops grown for processing or for cider production

Environmental safety
- Broadcast air-assisted LERAP (18 m)

Hazard classification and safety precautions
Hazard Harmful
UN Number N/C
Risk phrases H317, H351
Operator protection A, H; U05a, U20c
Environmental protection E12c, E12f, E15b, E22c, E34, E38, H410; E17b (18 m)
Storage and disposal D01, D02, D09a, D10c, D12a
Medical advice M05a

393   spirotetramat

Tetramic acid derivative that acts on lipid synthesis
IRAC mode of action code: 23

Products

Movento	Bayer CropScience	150 g/l	OD	14446

Uses
- Aphids in **blackcurrants** *(off-label)*, **blueberries** *(off-label)*, **broccoli**, **brussels sprouts**, **cabbages**, **calabrese**, **cauliflowers**, **chard** *(off-label)*, **chinese cabbage** *(off-label)*, **choi sum** *(off-label)*, **cress** *(off-label)*, **endives** *(off-label)*, **forest nurseries** *(off-label)*, **frise** *(off-label)*, **gooseberries** *(off-label)*, **herbs (see appendix 6)** *(off-label)*, **hops** *(off-label)*, **lamb's lettuce** *(off-label)*, **leaf brassicas** *(off-label)*, **lettuce**, **mustard** *(off-label)*, **ornamental plant production** *(off-label)*, **pak choi** *(off-label)*, **protected chard** *(off-label)*, **protected cress** *(off-label)*, **protected endives** *(off-label)*, **protected frise** *(off-label)*, **protected lamb's lettuce** *(off-label)*, **protected leaf brassicas** *(off-label)*, **protected rocket** *(off-label)*, **protected scarole** *(off-label)*, **protected spinach** *(off-label)*, **red mustard** *(off-label)*, **redcurrants** *(off-label)*, **rocket** *(off-label)*, **salad greens** *(off-label)*, **scarole** *(off-label)*, **sorrel** *(off-label)*, **spinach** *(off-label)*, **spinach beet** *(off-label)*, **tatsoi** *(off-label)*, **whitecurrants** *(off-label)*, **wine grapes** *(off-label)*
- Carrot willow aphid in **carrots** *(off-label)*, **parsnips** *(off-label)*, **swedes** *(off-label)*, **turnips** *(off-label)*
- Damson-hop aphid in **hops** *(off-label)*
- Insect pests in **forest nurseries** *(off-label)*, **ornamental plant production** *(off-label)*
- Parsnip aphid in **carrots** *(off-label)*, **parsnips** *(off-label)*, **swedes** *(off-label)*, **turnips** *(off-label)*
- Peach-potato aphid in **carrots** *(off-label)*, **parsnips** *(off-label)*, **swedes** *(off-label)*, **turnips** *(off-label)*
- White tip in **chard** *(off-label)*
- Whitefly in **broccoli**, **brussels sprouts**, **cabbages**, **calabrese**, **cauliflowers**, **chinese cabbage** *(off-label)*, **choi sum** *(off-label)*, **cress** *(off-label)*, **endives** *(off-label)*, **frise** *(off-label)*, **herbs (see appendix 6)** *(off-label)*, **lamb's lettuce** *(off-label)*, **leaf brassicas** *(off-label)*, **mustard** *(off-label)*, **pak choi** *(off-label)*, **protected chard** *(off-label)*, **protected cress** *(off-label)*, **protected endives** *(off-label)*, **protected frise** *(off-label)*, **protected lamb's lettuce** *(off-label)*, **protected leaf brassicas** *(off-label)*, **protected rocket** *(off-label)*, **protected scarole** *(off-label)*, **protected spinach** *(off-label)*, **red mustard** *(off-label)*, **rocket** *(off-label)*, **salad greens** *(off-label)*, **scarole** *(off-label)*, **sorrel** *(off-label)*, **spinach** *(off-label)*, **spinach beet** *(off-label)*, **tatsoi** *(off-label)*

FOR FULL CONDITIONS OF USE ALWAYS READ THE PRODUCT LABEL

- Willow parsnip aphid in *carrots* *(off-label)*, *parsnips* *(off-label)*, *swedes* *(off-label)*, *turnips* *(off-label)*

Extension of Authorisation for Minor Use (EAMUs)
- *blackcurrants* *20121401*
- *blueberries* *20121401*
- *carrots* *20161305*
- *chard* *20102410*
- *chinese cabbage* *20101095*
- *choi sum* *20101095*
- *cress* *20102410, 20140394*
- *endives* *20102410, 20140394*
- *forest nurseries* *20111987, 20171300*
- *frise* *20102410*
- *gooseberries* *20121401*
- *herbs (see appendix 6)* *20140394*
- *hops* *20102117*
- *lamb's lettuce* *20102410*
- *leaf brassicas* *20102410*
- *mustard* *20102410*
- *ornamental plant production* *20111987, 20171300*
- *pak choi* *20101095*
- *parsnips* *20161305*
- *protected chard* *20102410*
- *protected cress* *20102410*
- *protected endives* *20102410*
- *protected frise* *20102410*
- *protected lamb's lettuce* *20102410*
- *protected leaf brassicas* *20102410*
- *protected rocket* *20102410*
- *protected scarole* *20102410*
- *protected spinach* *20102410*
- *red mustard* *20140394*
- *redcurrants* *20121401*
- *rocket* *20102410*
- *salad greens* *20140394*
- *scarole* *20102410*
- *sorrel* *20140394*
- *spinach* *20102410*
- *spinach beet* *20140394*
- *swedes* *20161305*
- *tatsoi* *20101095*
- *turnips* *20161305*
- *whitecurrants* *20121401*
- *wine grapes* *20113085*

Approval information
- Spirotetramat is included in Annex 1 under EC Regulation 1107/2009
- Accepted by BBPA for use on hops

Hazard classification and safety precautions
Hazard Irritant, Dangerous for the environment
Transport code 9
Packaging group III
UN Number 3082
Risk phrases H317, H319, H361
Operator protection A; U05a, U14, U20b
Environmental protection E12c, E15b, E22c, E34, E38, H411
Storage and disposal D02, D09a, D10c, D12a
Medical advice M03

SEE SECTION 3 FOR PRODUCTS ALSO REGISTERED

394   starch, protein, oil and water

Natural plant extracts for control of moss, liverworts and algae

Products

MossKade	Fargro	n/a g/l	SL	00000

Uses
- Algae in **hard surfaces**
- Liverworts in **hard surfaces**
- Moss in **hard surfaces**

Approval information
- No listing required on Annex 1

Efficacy guidance
- Treat surfaces at first sign of moss, lichen, liverworts or algae since young growth is more susceptible.
- Must dry on the treated surface before it begins to work
- Summer growth is less susceptible than growth developing in autumn, winter or spring
- Repeat applications may be necessary for full efficacy

Hazard classification and safety precautions
UN Number N/C
Operator protection U08, U09a, U19a
Storage and disposal D06c, D07

395   sulfosulfuron

A sulfonylurea herbicide for grass and broad-leaved weed control in winter wheat
HRAC mode of action code: B

See also glyphosate + sulfosulfuron

Products

Monitor	Interfarm	80% w/w	WG	17695

Uses
- Brome grasses in **winter wheat** *(moderate control of barren brome)*
- Chickweed in **winter wheat**
- Cleavers in **winter wheat**
- Couch in **winter wheat** *(moderate control only)*
- Loose silky bent in **winter wheat**
- Mayweeds in **winter wheat**
- Onion couch in **winter wheat** *(moderate control only)*

Approval information
- Sulfosulfuron included in Annex I under EC Regulation 1107/2009

Efficacy guidance
- For best results treat in early spring when annual weeds are small and growing actively. Avoid treatment when weeds are dormant for any reason
- Best control of onion couch is achieved when the weed has more than two leaves. Effects on bulbils or on growth in the following yr have not been examined
- An extended period of dry weather before or after treatment may result in reduced control
- The addition of a recommended surfactant is essential for full activity
- Specific follow-up treatments may be needed for complete control of some weeds
- Use only where competitively damaging weed populations have emerged otherwise yield may be reduced

FOR FULL CONDITIONS OF USE ALWAYS READ THE PRODUCT LABEL

- Sulfosulfuron is a member of the ALS-inhibitor group of herbicides. To avoid the build up of resistance do not use any product containing an ALS-inhibitor herbicide with claims for control of grass weeds more than once on any crop
- Use these products as part of a resistance management strategy that includes cultural methods of control and does not use ALS inhibitors as the sole chemical method of grass weed control

Restrictions
- Maximum total dose equivalent to one treatment at full dose
- Do not treat crops under stress
- Apply only after 1 Feb and from 3 expanded leaf stage
- Do not treat durum wheat or any undersown wheat crop
- Do not use in mixture or in sequence with any other sulfonyl urea herbicide on the same crop

Crop-specific information
- Latest use: flag leaf ligule just visible for winter wheat (GS 39)

Following crops guidance
- In the autumn following treatment winter wheat, winter rye, winter oats, triticale, winter oilseed rape, winter peas or winter field beans may be sown on any soil, and winter barley on soils with less than 60% sand
- In the next spring following application crops of wheat, barley, oats, maize, peas, beans, linseed, oilseed rape, potatoes or grass may be sown
- In the second autumn following application winter linseed may be sown, and winter barley on soils with more than 60% sand
- Sugar beet or any other crop not mentioned above must not be drilled until the second spring following application
- Where winter oilseed rape is to be sown in the autumn following treatment soil cultivation to a minimum of 10 cm is recommended

Environmental safety
- Dangerous for the environment
- Very toxic to aquatic organisms
- Extremely dangerous to fish or other aquatic life. Do not contaminate surface waters or ditches with chemical or used container
- Take extreme care to avoid drift onto broad leaved plants or other crops, or onto ponds, waterways or ditches.
- Follow detailed label instructions for cleaning the sprayer to avoid damage to sensitive crops during subsequent use
- LERAP Category B

Hazard classification and safety precautions
Hazard Dangerous for the environment, Very toxic to aquatic organisms
UN Number N/C
Operator protection U20b
Environmental protection E15a, E16a, E16b, E34, E38, H410
Storage and disposal D09a, D11a, D12a

396 sulphur

A broad-spectrum inorganic protectant fungicide, foliar feed and acaricide
FRAC mode of action code: M2

Products
1 Kumulus DF	BASF	80% w/w	WG	04707
2 Microthiol Special	UPL Europe	80% w/w	MG	16989

Uses
- Disease control/foliar feed in **bilberries** *(off-label)*, **blueberries** *(off-label)*, **cranberries** *(off-label)*, **redcurrants** *(off-label)*, **whitecurrants** *(off-label)* [1]
- Foliar feed in **spring barley**, **spring wheat**, **winter barley**, **winter wheat** [1, 2]; **spring oats**, **sugar beet**, **swedes**, **turnips**, **winter oats** [2]; **winter oilseed rape** [1]

SEE SECTION 3 FOR PRODUCTS ALSO REGISTERED

- Gall mite in **blackcurrants** [1]
- Powdery mildew in **apples**, **gooseberries**, **hops**, **strawberries**, **sugar beet** [1, 2]; **blackcurrants**, **borage for oilseed production** *(off-label)*, **combining peas** *(off-label)*, **fodder beet** *(off-label)*, **grapevines**, **grass seed crops** *(off-label)*, **parsnips** *(off-label)*, **protected aubergines** *(off-label)*, **protected cayenne peppers** *(off-label)*, **protected chilli peppers** *(off-label)*, **protected cucumbers** *(off-label)*, **protected herbs (see appendix 6)** *(off-label)*, **protected peppers** *(off-label)*, **protected tomatoes** *(off-label)*, **redcurrants**, **spring barley**, **spring oats**, **spring wheat**, **swedes**, **turnips**, **vining peas** *(off-label)*, **winter barley**, **winter oats**, **winter wheat** [2]; **wine grapes** [1]
- Scab in **apples** [1]

Extension of Authorisation for Minor Use (EAMUs)
- **bilberries** *20061042* [1]
- **blueberries** *20061042* [1]
- **borage for oilseed production** *20151521* [2]
- **combining peas** *20151524* [2]
- **cranberries** *20061042* [1]
- **fodder beet** *20151526* [2]
- **grass seed crops** *20151523* [2]
- **parsnips** *20151527* [2]
- **protected aubergines** *20151520* [2]
- **protected cayenne peppers** *20151520* [2]
- **protected chilli peppers** *20151520* [2]
- **protected cucumbers** *20151520* [2]
- **protected herbs (see appendix 6)** *20151522* [2]
- **protected peppers** *20151520* [2]
- **protected tomatoes** *20151520* [2]
- **redcurrants** *20061042* [1]
- **vining peas** *20151524* [2]
- **whitecurrants** *20061042* [1]

Approval information
- Sulphur is included in Annex 1 under EC Regulation 1107/2009
- Accepted by BBPA for use on malting barley and hops (before burr)

Efficacy guidance
- Apply when disease first appears and repeat 2-3 wk later. Details of application rates and timing vary with crop, disease and product. See label for information
- Sulphur acts as foliar feed as well as fungicide and with some crops product labels vary in whether treatment recommended for disease control or growth promotion
- In grassland best results obtained at least 2 wk before cutting for hay or silage, 3 wk before grazing
- Treatment unlikely to be effective if disease already established in crop

Restrictions
- Maximum number of treatments normally 2 per crop for grassland, sugar beet, parsnips, swedes, hops, protected herbs; 3 per yr for blackcurrants, gooseberries; 4 per crop on apples, pears but labels vary
- Do not use on sulphur-shy apples (Beauty of Bath, Belle de Boskoop, Cox's Orange Pippin, Lanes Prince Albert, Lord Derby, Newton Wonder, Rival, Stirling Castle) or pears (Doyenne du Comice)
- Do not use on gooseberry cultivars Careless, Early Sulphur, Golden Drop, Leveller, Lord Derby, Roaring Lion, or Yellow Rough
- Do not use on apples or gooseberries when young, under stress or if frost imminent
- Do not use on fruit for processing, on grapevines during flowering or near harvest on grapes for wine-making
- Do not use on hops at or after burr stage
- Do not spray top or soft fruit with oil or within 30 d of an oil-containing spray

Crop-specific information
- Latest use: before burr stage for hops; before end Sep for parsnips, swedes, sugar beet; fruit swell for gooseberries; milky ripe stage for cereals

FOR FULL CONDITIONS OF USE ALWAYS READ THE PRODUCT LABEL

- HI cutting grass for hay or silage 2 wk; grazing grassland 3 wk

Environmental safety
- Sulphur products are attractive to livestock and must be kept out of their reach
- Do not empty into drains

Hazard classification and safety precautions
  UN Number N/C
  Operator protection U20b [2]; U20c [1]
  Environmental protection E15a
  Storage and disposal D01 [2]; D09a, D11a [1, 2]

SECTION 2

397   tau-fluvalinate

A contact pyrethroid insecticide for cereals and oilseed rape
IRAC mode of action code: 3

Products

Mavrik	Adama	240 g/l	EW	10612

Uses
- Aphids in **borage for oilseed production** *(off-label)*, **durum wheat** *(off-label)*, **evening primrose** *(off-label)*, **grass seed crops** *(off-label)*, **honesty** *(off-label)*, **linseed** *(off-label)*, **mustard** *(off-label)*, **rye** *(off-label)*, **spring barley**, **spring linseed** *(off-label)*, **spring oilseed rape**, **spring rye** *(off-label)*, **spring wheat**, **triticale** *(off-label)*, **winter barley**, **winter oilseed rape**, **winter rye** *(off-label)*, **winter wheat**
- Barley yellow dwarf virus vectors in **winter barley**, **winter wheat**
- Cabbage stem flea beetle in **winter oilseed rape**
- Pollen beetle in **spring oilseed rape**, **winter oilseed rape**

Extension of Authorisation for Minor Use (EAMUs)
- **borage for oilseed production** *20061366, 20141329*
- **durum wheat** *20061365, 20141328*
- **evening primrose** *20061366, 20141329*
- **grass seed crops** *20061365, 20141328*
- **honesty** *20061366, 20141329*
- **linseed** *20141329*
- **mustard** *20061366, 20141329*
- **rye** *20141328*
- **spring linseed** *20061366*
- **spring rye** *20061365*
- **triticale** *20061365, 20141328*
- **winter rye** *20061365*

Approval information
- Tau-fluvalinate included in Annex 1 under EC Regulation 1107/2009
- Accepted by BBPA for use on malting barley

Efficacy guidance
- For BYDV control on winter cereals follow local warnings or spray high risk crops in mid-Oct and make repeat application in late autumn/early winter if aphid activity persists
- For summer aphid control on cereals spray once when aphids present on two thirds of ears and increasing
- On oilseed rape treat peach potato aphids in autumn in response to local warning and repeat if necessary
- Best control of pollen beetle in oilseed rape obtained from treatment at green to yellow bud stage and repeat if necessary
- Good spray cover of target essential for best results
- Do not mix with boron or products containing boron [1]

Restrictions
- Maximum total dose equivalent to two full dose treatments on oilseed rape. See label for dose rates on cereals
- A minimum of 14 d must elapse between applications to cereals

Crop-specific information
- Latest use: before caryopsis watery ripe (GS 71) for barley; before flowering for oilseed rape; before kernel medium milk (GS 75) for wheat

Environmental safety
- Dangerous for the environment
- Very toxic to aquatic organisms
- High risk to non-target insects or other arthropods. Do not spray within 6 m of the field boundary [1]
- Avoid spraying oilseed rape within 6 m of field boundary to reduce effects on certain non-target species or other arthropods
- Must not be applied to cereals if any product containing a pyrethroid insecticide or dimethoate has been sprayed after the start of ear emergence (GS 51)
- LERAP Category A

Hazard classification and safety precautions

Hazard Dangerous for the environment
Transport code 9
Packaging group III
UN Number 3082
Risk phrases R50, R53a
Operator protection A, C, H; U05a, U10, U11, U19a, U20a
Environmental protection E15a, E16c, E16d, E22a
Storage and disposal D01, D02, D05, D09a, D10c

398   tebuconazole

A systemic triazole fungicide for cereals and other field crops
FRAC mode of action code: 3

See also azoxystrobin + tebuconazole
bixafen + prothioconazole + tebuconazole
bromuconazole + tebuconazole
carbendazim + tebuconazole
chlorothalonil + tebuconazole
clothianidin + prothioconazole + tebuconazole + triazoxide
difenoconazole + fludioxonil + tebuconazole
fenpropidin + prochloraz + tebuconazole
fenpropidin + propiconazole + tebuconazole
fenpropidin + tebuconazole
fludioxonil + tebuconazole
fluopyram + prothioconazole + tebuconazole
fluoxastrobin + prothioconazole + tebuconazole
imidacloprid + tebuconazole + triazoxide
prochloraz + proquinazid + tebuconazole
prochloraz + tebuconazole
propiconazole + tebuconazole
prothioconazole + spiroxamine + tebuconazole
prothioconazole + tebuconazole
prothioconazole + tebuconazole + triazoxide
spiroxamine + tebuconazole

Products

1	Buzz Ultra DF	Clayton	75% w/w	WG	17924
2	Clayton Tebucon 250 EW	Clayton	250 g/l	EW	17823

FOR FULL CONDITIONS OF USE ALWAYS READ THE PRODUCT LABEL

Products – continued

3	Deacon	Adama	200 g/l	EW	17604
4	Erase	Headland	200 g/l	EC	17507
5	Fathom	Headland	200 g/l	EC	17452
6	Fezan	Sipcam	250 g/l	EW	17082
7	Folicur	Bayer CropScience	250 g/l	EW	16731
8	Harvest Teb 250	Harvest	250 g/l	EW	17706
9	Hecates	Ventura	430 g/l	SC	17403
10	Odin	Rotam	250 g/l	EW	18016
11	Orian	ChemSource	200 g/l	EW	14803
12	Orius	Adama	200 g/l	EW	17414
13	Rosh	Harvest	250 g/l	EW	17673
14	Savannah	Rotam	250 g/l	EW	18058
15	Sokol	Becesane	430 g/l	SC	15782
16	Spekfree	Rotam	430 g/l	SC	16004
17	Starpro	Rotam	250 g/l	EW	14823
18	Starpro	Rotam	250 g/l	EW	18059
19	Tebucur 250	Globachem	255.1 g/l	EW	13975
20	Tebusha 25 EW	Sharda	250 g/l	EW	18034
21	Tebuzol 250	UPL Europe	250 g/l	EW	17417
22	Teson	Belchim	250 g/l	EW	17128
23	Tesoro	Belchim	250 g/l	EW	17222
24	Tharsis	Generica	250 g/l	EW	17201
25	Toledo	Rotam	430 g/l	SC	14036
26	Tubosan	Belchim	250 g/l	EW	17127
27	Ulysses	Rotam	430 g/l	SC	16052

SECTION 2

Uses

- Alternaria in **brussels sprouts** [11]; **cabbages** [4, 5, 9-11, 14-18, 23, 25, 27]; **carrots, horseradish** [4, 5, 9, 10, 14-18, 23, 25, 27]; **parsnips** [10, 14, 18]; **spring oilseed rape, winter oilseed rape** [1-5, 7-23, 25-27]; **spring wheat, winter wheat** [9, 15, 16, 25, 27]
- Blight in **blackberries** *(off-label)*, **raspberries** *(off-label)*, **rubus hybrids** *(off-label)* [17]
- Botrytis in **daffodils grown for galanthamine production** *(off-label)* [17, 25]; **linseed** [9, 15, 16, 25, 27]; **linseed** *(reduction)* [4, 5, 17, 23]
- Botrytis fruit rot in **blackberries** *(off-label)*, **loganberries** *(off-label)*, **raspberries** *(off-label)*, **rubus hybrids** *(off-label)* [7]
- Brown rust in **rye** [9, 15, 16, 25, 27]; **spring barley, spring wheat, winter barley** [2-27]; **spring oats, winter oats** [10, 14, 18]; **spring rye** [3-6, 8, 10-12, 14, 17-20, 22-24, 26]; **triticale** [4-6, 9, 10, 14-16, 18, 25, 27]; **winter rye** [2-8, 10-14, 17-24, 26]; **winter wheat** [1-27]
- Cane blight in **blackberries** *(off-label)*, **raspberries** *(off-label)*, **rubus hybrids** *(off-label)* [7, 17, 25]; **loganberries** *(off-label)* [7]
- Canker in **apples** *(off-label)*, **pears** *(off-label)*, **quinces** *(off-label)* [4, 5, 7, 12, 17, 25]; **chestnuts** *(off-label)*, **walnuts** *(off-label)* [7, 17, 25]; **cob nuts** *(off-label)* [17, 25]; **crab apples** *(off-label)* [12, 17, 25]; **hazel nuts** *(off-label)* [7]
- Chocolate spot in **spring field beans, winter field beans** [4, 5, 7, 9-21, 23, 25, 27]
- Cladosporium in **spring wheat, winter wheat** [9, 15, 16, 25, 27]
- Crown rust in **spring oats** [3-6, 9, 11, 15, 16, 19, 20, 24, 25, 27]; **spring oats** *(reduction)*, **winter oats** *(reduction)* [2, 7, 8, 13, 17, 21-23, 26]; **triticale** [24]; **winter oats** [3-6, 9, 11, 12, 15, 16, 19, 20, 24, 25, 27]
- Dark leaf spot in **spring oilseed rape, winter oilseed rape** [9, 15, 16, 25, 27]
- Disease control in **borage for oilseed production** *(off-label)*, **canary flower (echium spp.)** *(off-label)*, **evening primrose** *(off-label)*, **grass seed crops** *(off-label)*, **honesty** *(off-label)*, **mallow (althaea spp.)** *(off-label)*, **mustard** *(off-label)*, **parsley root** *(off-label)*, **triticale** *(off-label)* [19]; **linseed** [2, 7, 13]; **spring field beans, winter field beans** [2]
- Ear diseases in **spring wheat, winter wheat** [9, 15, 16, 25, 27]
- Fusarium in **spring wheat, winter wheat** [9, 15, 16, 25, 27]
- Fusarium ear blight in **spring wheat, winter wheat** [2-5, 7, 8, 11-13, 17, 19-23, 26]; **triticale** [4, 5]

SEE SECTION 3 FOR PRODUCTS ALSO REGISTERED

- Glume blotch in **spring wheat**, **winter wheat** [2-9, 11-13, 15-17, 19-27]; **triticale** [4-6, 9, 15, 16, 25, 27]
- Light leaf spot in **brussels sprouts** [11]; **cabbages** [4, 5, 9-11, 14-18, 23, 25, 27]; **spring oilseed rape**, **winter oilseed rape** [1-27]
- Lodging control in **spring oilseed rape**, **winter oilseed rape** [2, 4, 5, 7, 8, 11-13, 17, 21-23, 26]
- Net blotch in **spring barley**, **winter barley** [2, 4-27]; **spring oats**, **spring rye**, **spring wheat**, **triticale**, **winter oats**, **winter rye** [10, 14, 18]; **winter wheat** [1, 10, 14, 18]
- Phoma in **spring oilseed rape**, **winter oilseed rape** [1, 2, 4-8, 10-14, 17-23, 26]; **spring oilseed rape** *(good reduction)*, **winter oilseed rape** *(good reduction)* [24]
- Phoma leaf spot in **spring oilseed rape**, **winter oilseed rape** [9, 15, 16, 25, 27]
- Powdery mildew in **apples** *(off-label)*, **pears** *(off-label)*, **quinces** *(off-label)* [4, 5, 7, 17, 25]; **blackberries** *(off-label)*, **chestnuts** *(off-label)*, **raspberries** *(off-label)*, **rubus hybrids** *(off-label)*, **walnuts** *(off-label)* [7, 17, 25]; **borage for oilseed production** *(off-label)*, **broad beans** *(off-label)*, **bulb onions** *(off-label)*, **canary flower (echium spp.)** *(off-label)*, **chives** *(off-label)*, **choi sum** *(off-label)*, **cob nuts** *(off-label)*, **crab apples** *(off-label)*, **daffodils grown for galanthamine production** *(off-label)*, **dwarf beans** *(off-label)*, **evening primrose** *(off-label)*, **french beans** *(off-label)*, **garlic** *(off-label)*, **grass seed crops** *(off-label)*, **honesty** *(off-label)*, **kohlrabi** *(off-label)*, **mallow (althaea spp.)** *(off-label)*, **mustard** *(off-label)*, **pak choi** *(off-label)*, **parsley root** *(off-label)*, **runner beans** *(off-label)*, **salad onions** *(off-label)*, **shallots** *(off-label)* [17, 25]; **brussels sprouts** [11]; **bulb onion sets** *(off-label)* [25]; **cabbages**, **swedes**, **turnips** [4, 5, 9-11, 14-18, 23, 25, 27]; **carrots**, **parsnips** [4, 5, 9, 10, 14-18, 23, 25, 27]; **hazel nuts** *(off-label)*, **loganberries** *(off-label)* [7]; **linseed** [4, 5, 9, 15-17, 19, 20, 23, 25, 27]; **rye** [9, 15, 16, 25, 27]; **spring barley**, **spring wheat**, **winter barley** [2, 4-27]; **spring barley** *(moderate control)*, **spring wheat** *(moderate control)*, **winter barley** *(moderate control)*, **winter wheat** *(moderate control)* [3]; **spring oats**, **winter oats** [2-27]; **spring rye** [3-6, 8, 10-12, 14, 17-20, 22-24, 26]; **triticale** [4-6, 10, 14, 18, 24]; **triticale** *(off-label)* [17]; **winter rye** [2-8, 10-14, 17-24, 26]; **winter wheat** [1, 2, 4-27]
- Rhynchosporium in **rye** [9, 15, 16, 25, 27]; **spring barley**, **winter barley** [2, 4-27]; **spring barley** *(moderate control)*, **spring rye** *(moderate control)*, **winter barley** *(moderate control)*, **winter rye** *(moderate control)* [3]; **spring oats**, **spring wheat**, **triticale**, **winter oats** [10, 14, 18]; **spring rye** [4-6, 8, 10-12, 14, 17-20, 22-24, 26]; **winter rye** [2, 4-8, 10-14, 17-24, 26]; **winter wheat** [1, 10, 14, 18]
- Ring spot in **broccoli** *(off-label)*, **calabrese** *(off-label)*, **cauliflowers** *(off-label)*, **choi sum** *(off-label)*, **pak choi** *(off-label)* [17, 25]; **brussels sprouts** [11]; **cabbages** [4, 5, 9-11, 14-18, 23, 25, 27]; **spring oilseed rape**, **winter oilseed rape** [9, 15, 16, 25, 27]; **spring oilseed rape** *(reduction)*, **winter oilseed rape** *(reduction)* [2, 4, 5, 7, 8, 11-13, 17, 21-23, 26]
- Rust in **borage for oilseed production** *(off-label)*, **canary flower (echium spp.)** *(off-label)*, **evening primrose** *(off-label)*, **grass seed crops** *(off-label)*, **honesty** *(off-label)*, **mallow (althaea spp.)** *(off-label)*, **parsley root** *(off-label)*, **triticale** *(off-label)* [19]; **broad beans** *(off-label)* [7, 17, 25]; **chives** *(off-label)* [17, 25]; **dwarf beans** *(off-label)*, **french beans** *(off-label)*, **runner beans** *(off-label)* [4, 5, 7, 17, 25]; **leeks** [9, 11, 15-17, 23, 25, 27]; **linseed** [2, 7, 13]; **mustard** [4, 5, 19]; **spring field beans**, **winter field beans** [2, 4, 5, 7, 9-11, 13-21, 23, 25, 27]
- Sclerotinia in **carrots** [4, 5, 9, 10, 14-18, 23, 25, 27]; **parsnips** [10, 14, 18]; **spring oilseed rape**, **winter oilseed rape** [3, 6, 9, 15, 16, 25, 27]; **spring oilseed rape** *(reduction)*, **winter oilseed rape** *(reduction)* [24]
- Sclerotinia stem rot in **spring oilseed rape**, **winter oilseed rape** [1, 2, 4, 5, 7, 8, 10-14, 17-23, 26]
- Septoria leaf blotch in **spring wheat**, **winter wheat** [2, 4-9, 11-13, 15-17, 19-27]; **triticale** [4-6, 9, 15, 16, 25, 27]
- Sooty moulds in **spring wheat**, **winter wheat** [2-5, 7, 8, 11-13, 17, 19-23, 26]; **triticale** [4, 5]
- Stem canker in **spring oilseed rape**, **winter oilseed rape** [1, 2, 4, 5, 7-23, 25-27]
- White rot in **bulb onion sets** *(off-label)* [4, 5, 17, 25]; **bulb onions** *(off-label)* [7, 17, 25]; **garlic** *(off-label)*, **salad onions** *(off-label)*, **shallots** *(off-label)* [17, 25]
- Yellow rust in **rye** [9, 15, 16, 25, 27]; **spring barley**, **spring wheat**, **winter barley** [2-27]; **spring oats**, **winter oats** [10, 14, 18]; **spring rye** [3-6, 8, 10-12, 14, 17-20, 22-24, 26]; **triticale** [4-6, 10, 14, 18]; **winter rye** [2-8, 10-14, 17-24, 26]; **winter wheat** [1-27]

Extension of Authorisation for Minor Use (EAMUs)
- **apples** *20171102* [4], *20170712* [5], *20162647* [7], *20170558* [12], *20101221* [17], *20090572* [25]

SECTION 2

- **blackberries** *20162648* [7], *20101225* [17], *20090568* [25]
- **borage for oilseed production** *20101229* [17], *20122270* [19], *20090563* [25]
- **broad beans** *20162203* [7], *20101228* [17], *20090567* [25]
- **broccoli** *20141889* [17], *20141888* [25]
- **bulb onion sets** *20171355* [4], *20170693* [5], *20101223* [17], *20090569* [25]
- **bulb onions** *20162045* [7], *20101213* [17], *20090562* [25]
- **calabrese** *20141889* [17], *20141888* [25]
- **canary flower (echium spp.)** *20101229* [17], *20122270* [19], *20090563* [25]
- **cauliflowers** *20141889* [17], *20141888* [25]
- **chestnuts** *20162647* [7], *20101219* [17], *20090573* [25]
- **chives** *20101224* [17], *20090574* [25]
- **choi sum** *20101220* [17], *20090576* [25]
- **cob nuts** *20101219* [17], *20090573* [25]
- **crab apples** *20170558* [12], *20101221* [17], *20090572* [25]
- **daffodils grown for galanthamine production** *20101217* [17], *20090575* [25]
- **dwarf beans** *20171341* [4], *20170692* [5], *20162203* [7], *20101222* [17], *20090565* [25]
- **evening primrose** *20101229* [17], *20122270* [19], *20090563* [25]
- **french beans** *20171341* [4], *20170692* [5], *20162203* [7], *20101222* [17], *20090565* [25]
- **garlic** *20101213* [17], *20090562* [25]
- **grass seed crops** *20101226* [17], *20122273* [19], *20090570* [25]
- **hazel nuts** *20162647* [7]
- **honesty** *20101229* [17], *20122270* [19], *20090563* [25]
- **kohlrabi** *20101210* [17], *20090571* [25]
- **loganberries** *20162648* [7]
- **mallow (althaea spp.)** *20101211* [17], *20122272* [19], *20090566* [25]
- **mustard** *20171103* [4], *20170691* [5], *20101229* [17], *20122270* [19], *20090563* [25]
- **pak choi** *20101220* [17], *20090576* [25]
- **parsley root** *20101211* [17], *20122272* [19], *20090566* [25]
- **pears** *20171102* [4], *20170712* [5], *20162647* [7], *20170558* [12], *20101221* [17], *20090572* [25]
- **quinces** *20171102* [4], *20170712* [5], *20162647* [7], *20170558* [12], *20101221* [17], *20090572* [25]
- **raspberries** *20162648* [7], *20101225* [17], *20090568* [25]
- **rubus hybrids** *20162648* [7], *20101225* [17], *20090568* [25]
- **runner beans** *20171341* [4], *20170692* [5], *20162203* [7], *20101222* [17], *20090565* [25]
- **salad onions** *20101212* [17], *20090577* [25]
- **shallots** *20101213* [17], *20090562* [25]
- **triticale** *20101218* [17], *20122271* [19]
- **walnuts** *20162647* [7], *20101219* [17], *20090573* [25]

Approval information
- Tebuconazole is included in Annex 1 under EC Regulation 1107/2009
- Accepted by BBPA for use on malting barley
- Approval expiry 31 Oct 2018 [17]

Efficacy guidance
- For best results apply at an early stage of disease development before infection spreads to new crop growth
- To protect flag leaf and ear from Septoria diseases apply from flag leaf emergence to ear fully emerged (GS 37-59). Earlier application may be necessary where there is a high risk of infection
- Improved control of established cereal mildew can be obtained by tank mixture with fenpropimorph
- For light leaf spot control in oilseed rape apply in autumn/winter with a follow-up spray in spring/summer if required
- For control of most other diseases spray at first signs of infection with a follow-up spray 2-4 wk later if necessary. See label for details
- For disease control in cabbages a 3-spray programme at 21-28 d intervals will give good control
- Tebuconazole is a DMI fungicide. Resistance to some DMI fungicides has been identified in Septoria leaf blotch which may seriously affect performance of some products. For further advice contact a specialist advisor and visit the Fungicide Resistance Action Group (FRAG)-UK website

SEE SECTION 3 FOR PRODUCTS ALSO REGISTERED

Restrictions

- Maximum total dose equivalent to 1 full dose treatment on linseed; 2 full dose treatments on wheat, barley, rye, oats, field beans, swedes, turnips, onions; 2.25 full dose treatments on cabbages; 2.5 full dose treatments on oilseed rape; 3 full dose treatments on leeks, parsnips, carrots
- Do not treat durum wheat
- Apply only to listed oat varieties (see label) and do not apply to oats in tank mixture
- Do not apply before swedes and turnips have a root diameter of 2.5 cm, or before heart formation in cabbages, or before button formation in Brussels sprouts
- Consult processor before use on crops for processing
- Newer authorisations for tebuconazole products require application to cereals only after GS 30 and applications to oilseed rape and linseed after GS 20 - check label

Crop-specific information

- Latest use: before grain milky-ripe for cereals (GS 71); when most seed green-brown mottled for oilseed rape (GS 6,3); before brown capsule for linseed
- HI field beans, swedes, turnips, linseed, winter oats 35 d; market brassicas, carrots, horseradish, parsnips 21 d; leeks 14 d
- Some transient leaf speckling on wheat or leaf reddening/scorch on oats may occur but this has not been shown to reduce yield response to disease control

Environmental safety

- Dangerous for the environment
- Toxic to aquatic organisms
- LERAP Category B [1, 2, 4-8, 10, 12-14, 18-24, 26]

Hazard classification and safety precautions

Hazard Harmful, Dangerous for the environment [1, 2, 7-10, 13-27]; Irritant [3-6, 11, 12]; Harmful if swallowed [2, 7, 13, 20, 21]; Harmful if inhaled [7, 13, 20, 21]; Very toxic to aquatic organisms [1, 2, 7, 13, 20]

Transport code 9

Packaging group III

UN Number 3077 [1]; 3082 [2-27]

Risk phrases H317 [4, 5]; H318 [4-8, 13, 19-23, 26]; H319 [2, 3, 12]; H335 [2]; H361 [1-10, 12-16, 18-23, 26, 27]; R20, R22a, R41 [17, 24]; R36, R38, R52 [11]; R43, R63 [25]; R51 [17, 24, 25]; R53a [11, 17, 24, 25]

Operator protection A, H [1-27]; C [2-27]; U05a, U20a [1-27]; U09a, U12 [9, 15, 16, 25, 27]; U11 [1-8, 10-14, 17-24, 26]; U14, U15 [3-6, 9, 11, 12, 15, 16, 25, 27]; U19a [1, 10, 14, 18-20]

Environmental protection E15a [1, 3-22, 25, 27]; E16a [1, 2, 4-8, 10, 12-14, 18-24, 26]; E34 [1-27]; E38 [2, 7-9, 13, 15-17, 21-27]; H410 [1-5, 7, 12, 13, 20, 21]; H411 [6, 8-10, 14-16, 18, 19, 22, 23, 26, 27]

Storage and disposal D01, D02, D09a [1-27]; D05, D10b [1, 2, 7, 8, 10, 13, 14, 17-24, 26]; D10c [9, 15, 16, 25, 27]; D12a [2, 7-9, 13, 15-17, 21-27]

Medical advice M03 [1, 2, 7-10, 13-27]; M05a [7, 9, 13, 15, 16, 21, 25, 27]

399   tebuconazole + trifloxystrobin

A conazole and stobilurin fungicide mixture for wheat and vegetable crops
FRAC mode of action code: 3 + 11

See also trifloxystrobin

Products

1 Clayton Bestow	Clayton	200:100 g/l	SC	17185
2 Dedicate	Bayer CropScience	200:100 g/l	SC	17003
3 Dualitas	ProKlass	200:100 g/l	SC	18000
4 Dualitas	ProKlass	200:100 g/l	SC	18070
5 Fortify	Rigby Taylor	200:100 g/l	SC	17718
6 Fusion	Rigby Taylor	200:100 g/l	SC	17225

FOR FULL CONDITIONS OF USE ALWAYS READ THE PRODUCT LABEL

Uses

- Anthracnose in *amenity grassland*, *managed amenity turf* [6]
- Dollar spot in *amenity grassland*, *managed amenity turf* [1-6]
- Fusarium in *amenity grassland*, *managed amenity turf* [1-5]
- Fusarium patch in *amenity grassland*, *managed amenity turf* [6]
- Melting out in *amenity grassland*, *managed amenity turf* [1-5]
- Red thread in *amenity grassland*, *managed amenity turf* [1-6]
- Rust in *amenity grassland*, *managed amenity turf* [1-5]

Approval information

- Tebuconazole and trifloxystrobin included in Annex I under EC Regulation 1107/2009

Efficacy guidance

- Best results obtained from treatment at early stages of disease development. Further treatment may be needed if disease attack is prolonged
- Applications to established infections of any disease are likely to be less effective
- Treatment may give some control of *Stemphylium botryosum* and leaf blotch on leeks
- Tebuconazole is a DMI fungicide. Resistance to some DMI fungicides has been identified in Septoria leaf blotch which may seriously affect performance of some products. For further advice contact a specialist advisor and visit the Fungicide Resistance Action Group (FRAG)-UK website
- Trifloxystrobin is a member of the QoI cross resistance group. Product should be used preventatively and not relied on for its curative potential
- Use product as part of an Integrated Crop Management strategy incorporating other methods of control, including where appropriate other fungicides with a different mode of action. Do not apply more than two foliar applications of QoI containing products to any cereal crop, broccoli, calabrese or cauliflower. Do not apply more than three applications to Brussels sprouts, cabbage, carrots or leeks
- There is a significant risk of widespread resistance occurring in *Septoria tritici* populations in UK. Failure to follow resistance management action may result in reduced levels of disease control
- Strains of wheat powdery mildew resistant to QoIs are common in the UK. Control of wheat mildew can only be relied on from the triazole component
- Where specific control of wheat mildew is required this should be achieved through a programme of measures including products recommended for the control of mildew that contain a fungicide from a different cross-resistance group and applied at a dose that will give robust control
- Do not apply to grass during drought conditions nor to frozen turf

Restrictions

- Maximum number of treatments 2 per crop for broccoli, calabrese, cauliflowers; 3 per crop for Brussels sprouts, cabbages, carrots, leeks
- In addition to the maximum number of treatments per crop a maximum of 3 applications may be applied to the same ground in one calendar year
- Consult processor before use on vegetable crops for processing
- Newer authorisations for tebuconazole products require application to cereals only after GS 30 and applications to oilseed rape and linseed after GS 20 - check label

Crop-specific information

- Latest use: before milky ripe stage on winter wheat
- HI 35 d for wheat; 21 d for all other crops
- Performance against leaf spot diseases of brassicas, *Alternaria* leaf blight of carrots, and rust in leeks may be improved by mixing with an approved sticker/wetter

Environmental safety

- Dangerous for the environment
- Very toxic to aquatic organisms
- Buffer zone requirement 12 m
- LERAP Category B

Hazard classification and safety precautions

Hazard Harmful, Dangerous for the environment, Very toxic to aquatic organisms
Transport code 9

SEE SECTION 3 FOR PRODUCTS ALSO REGISTERED

SECTION 2

Packaging group III
UN Number 3082
Risk phrases H335 [6]; H361 [1-6]; R37, R70 [1-5]
Operator protection A [1-6]; H [6]; P [1-5]; U05a, U09a [1-5]; U09b [6]; U19a, U20b [1-6]
Environmental protection E15a, E16a, E34, E38, H410
Storage and disposal D01, D02, D09a, D10b, D12a [1-6]; D05 [6]
Medical advice M03

400 tebufenpyrad

A pyrazole mitochondrial electron transport inhibitor (METI) aphicide and acaricide
IRAC mode of action code: 21

Products

1	Clayton Bonsai	Clayton	20% w/w	WB	17983
2	Masai	BASF	20% w/w	WB	13082

Uses

* Bud mite in **almonds** *(off-label)*, **chestnuts** *(off-label)*, **hazel nuts** *(off-label)*, **walnuts** *(off-label)* [2]
* Damson-hop aphid in **hops** [1, 2]; **plums** *(off-label)* [2]
* Gall mite in **bilberries** *(off-label)*, **blackberries** *(off-label)*, **blackcurrants** *(off-label)*, **blueberries** *(off-label)*, **cranberries** *(off-label)*, **gooseberries** *(off-label)*, **protected rubus hybrids** *(off-label)*, **raspberries** *(off-label)*, **redcurrants** *(off-label)*, **vaccinium spp.** *(off-label)*, **whitecurrants** *(off-label)* [2]
* Red spider mites in **apples**, **pears** [1, 2]; **blackberries** *(off-label)*, **protected rubus hybrids** *(off-label)*, **raspberries** *(off-label)* [2]
* Two-spotted spider mite in **hops, protected roses, strawberries** [1, 2]

Extension of Authorisation for Minor Use (EAMUs)

* **almonds** *20080133* [2]
* **bilberries** *20080131* [2]
* **blackberries** *20100204* [2]
* **blackcurrants** *20080131* [2]
* **blueberries** *20080131* [2]
* **chestnuts** *20080133* [2]
* **cranberries** *20080131* [2]
* **gooseberries** *20080131* [2]
* **hazel nuts** *20080133* [2]
* **plums** *20080132* [2]
* **protected rubus hybrids** *20100204* [2]
* **raspberries** *20100204* [2]
* **redcurrants** *20080131* [2]
* **vaccinium spp.** *20080131* [2]
* **walnuts** *20080133* [2]
* **whitecurrants** *20080131* [2]

Approval information

* Tebufenpyrad included in Annex 1 under EC Regulation 1107/2009
* Accepted by BBPA for use on hops

Efficacy guidance

* Acts on eggs (except winter eggs) and all motile stages of spider mites up to adults
* Treat spider mites from 80% egg hatch but before mites become established
* For effective control total spray cover of the crop is required
* Product can be used in a programme to give season-long control of damson-hop aphids coupled with mite control
* Where aphids resistant to tebufenpyrad occur in hops control is unlikely to be satisfactory and repeat treatments may result in lower levels of control. Where possible use different active ingredients in a programme

FOR FULL CONDITIONS OF USE ALWAYS READ THE PRODUCT LABEL

Restrictions
- Maximum total dose equivalent to one full dose treatment on apples, pears, strawberries; 3 full dose treatments on hops
- Other mitochondrial electron transport inhibitor (METI) acaricides should not be applied to the same crop in the same calendar yr either separately or in mixture
- Do not treat apples before 90% petal fall
- Small-scale testing of rose varieties to establish tolerance recommended before use
- Inner liner of container must not be removed

Crop-specific information
- Latest use: end of burr stage for hops
- HI strawberries 3 d; apples, bilberries, blackcurrants, blueberries, cranberries, gooseberries, pears, redcurrants, whitecurrants 7d; blackberries, plums, raspberries 21 d
- Product has no effect on fruit quality or finish

Environmental safety
- Dangerous for the environment
- Very toxic to aquatic organisms
- High risk to bees. Do not apply to crops in flower or to those in which bees are actively foraging. Do not apply when flowering weeds are present
- Broadcast air-assisted LERAP (18 m); LERAP Category B

Hazard classification and safety precautions
Hazard Harmful, Dangerous for the environment
Transport code 9
Packaging group III
UN Number 3077
Risk phrases R20, R22a, R50, R53a
Operator protection A; U02a, U05a, U09a, U13, U14, U20b
Environmental protection E12a, E12e, E15a, E16a, E16b, E38; E17b (18 m)
Storage and disposal D01, D02, D09a, D11a, D12a
Medical advice M05a

---

401   tefluthrin

A soil acting pyrethroid insecticide seed treatment
IRAC mode of action code: 3

See also fludioxonil + tefluthrin

Products
Force ST	Syngenta	200 g/l	CF	11752

Uses
- Bean seed fly in **baby leaf crops** (off-label - seed treatment), **bulb onions** (off-label), **chard** (off-label - seed treatment), **chives** (off-label - seed treatment), **green mustard** (off-label - seed treatment), **herbs (see appendix 6)** (off-label - seed treatment), **leaf spinach** (off-label - seed treatment), **leeks** (off-label), **miduna** (off-label - seed treatment), **mizuna** (off-label - seed treatment), **pak choi** (off-label - seed treatment), **parsley** (off-label - seed treatment), **red mustard** (off-label - seed treatment), **salad onions** (off-label), **shallots** (off-label), **spinach beet** (off-label - seed treatment), **tatsoi** (off-label - seed treatment)
- Carrot fly in **carrots** (off-label - seed treatment), **parsnips** (off-label - seed treatment)
- Millipedes in **fodder beet** (seed treatment), **sugar beet** (seed treatment)
- Onion fly in **bulb onions** (off-label), **leeks** (off-label), **salad onions** (off-label), **shallots** (off-label)
- Pygmy beetle in **fodder beet** (seed treatment), **sugar beet** (seed treatment)
- Springtails in **fodder beet** (seed treatment), **sugar beet** (seed treatment)
- Symphylids in **fodder beet** (seed treatment), **sugar beet** (seed treatment)
- Wireworm in **chicory** (off-label), **chicory root** (off-label), **witloof** (off-label)

Extension of Authorisation for Minor Use (EAMUs)
- **baby leaf crops** (seed treatment) 20050545

SEE SECTION 3 FOR PRODUCTS ALSO REGISTERED

- **bulb onions** *20120604*
- **carrots** *(seed treatment) 20050547*
- **chard** *(seed treatment) 20050545*
- **chicory** *20142248*
- **chicory root** *20142248*
- **chives** *(seed treatment) 20050545*
- **green mustard** *(seed treatment) 20050545*
- **herbs (see appendix 6)** *(seed treatment) 20050545*
- **leaf spinach** *(seed treatment) 20050545*
- **leeks** *20120604*
- **miduna** *(seed treatment) 20050545*
- **mizuna** *(seed treatment) 20050545*
- **pak choi** *(seed treatment) 20050545*
- **parsley** *(seed treatment) 20050545*
- **parsnips** *(seed treatment) 20050547*
- **red mustard** *(seed treatment) 20050545*
- **salad onions** *20120604*
- **shallots** *20120604*
- **spinach beet** *(seed treatment) 20050545*
- **tatsoi** *(seed treatment) 20050545*
- **witloof** *20142248*

Approval information
- Tefluthrin included in Annex 1 under EC Regulation 1107/2009
- Accepted by BBPA for use on malting barley

Efficacy guidance
- Apply during process of pelleting beet seed. Consult manufacturer for details of specialist equipment required
- Micro-capsule formulation allows slow release to provide a protection zone around treated seed during establishment

Restrictions
- Maximum number of treatments 1 per batch of seed
- Sow treated seed as soon as possible. Do not store treated seed from one drilling season to next
- If used in areas where soil erosion by wind or water likely, measures must be taken to prevent this happening
- Can cause a transient tingling or numbing sensation to exposed skin. Avoid skin contact with product, treated seed and dust throughout all operations in the seed treatment plant and at drilling

Crop-specific information
- Latest use: before drilling seed
- Treated seed must be drilled within the season of treatment

Environmental safety
- Dangerous for the environment
- Very toxic to aquatic organisms
- Keep treated seed secure from people, domestic stock/pets and wildlife at all times during storage and use
- Treated seed harmful to game and wild life. Bury spillages
- In the event of seed spillage clean up as much as possible into the related seed sack and bury the remainder completely
- Do not apply treated seed from the air
- Keep livestock out of areas drilled with treated seed for at least 80 d

Hazard classification and safety precautions
Hazard Harmful, Dangerous for the environment, Harmful if inhaled, Very toxic to aquatic organisms
Transport code 9
Packaging group III

UN Number 3082
Risk phrases H317
Operator protection A, D, E, H; U02a, U04a, U05a, U08, U20b
Environmental protection E06a (80 d); E15a, E34, E38, H410
Storage and disposal D01, D02, D05, D09a, D11a, D12a
Treated seed S02, S03, S04b, S05, S07
Medical advice M05b

402   terbuthylazine

A triazine herbicide available only in mixtures
HRAC mode of action code: C1

See also bromoxynil + terbuthylazine
*   isoxaben + terbuthylazine*
*   mesotrione + terbuthylazine*
*   pendimethalin + terbuthylazine*

403   tetradecadienyl acetate

A pheromone insecticide for use in glasshouses

Products

Isonet T	Fargro	n/a	VP	17929

Uses
* Tomato leaf miner (Tuta absoluta) in **protected aubergines**, **protected chilli peppers**, **protected peppers**, **protected tomatoes**

Approval information
* Tetradecadienyl acetate included in Annex I under EC Regulation 1107/2009

Hazard classification and safety precautions
Hazard Irritant, Dangerous for the environment
Transport code 9
Packaging group III
UN Number 3082
Risk phrases H315
Operator protection A; U05a, U09c
Environmental protection H410
Storage and disposal D01, D12a, D12b, D22

404   thiabendazole

A systemic, curative and protectant benzimidazole (MBC) fungicide
FRAC mode of action code: 1

See also imazalil + thiabendazole
*   metalaxyl + thiabendazole*

Products

Tezate 220 SL	UPL Europe	220 g/l	SL	16979

Uses
* Basal stem rot in **narcissi** *(off-label)*
* Dry rot in **potatoes**, **seed potatoes**
* Fusarium basal neck rot in **narcissi**
* Gangrene in **potatoes**, **seed potatoes**
* Neck rot in **narcissi** *(off-label)*

SEE SECTION 3 FOR PRODUCTS ALSO REGISTERED

- Silver scurf in **potatoes, seed potatoes**
- Skin spot in **potatoes, seed potatoes**

Extension of Authorisation for Minor Use (EAMUs)
- **narcissi** *20151528*

Approval information
- Thiabendazole included in Annex I under EC Regulation 1107/2009
- Use of thiabendazole products on ware potatoes requires that the discharge of thiabendazole to receiving water from washing plants is kept within emission limits set by the UK monitoring authority

Efficacy guidance
- For best results tuber treatments should be applied as soon as possible after lifting and always within 24 hr
- Dust treatments should be applied evenly over the whole tuber surface
- Thiabendazole should only be used on ware potatoes where there is a likely risk of disease during the storage period and in combination with good storage hygiene and maintenance
- Benzimidazole tolerant strains of silver scurf and skin spot are common in UK and tolerant strains of dry rot have been reported. To reduce the chance of such strains increasing benzimidazole based products should not be used more than once in the cropping cycle
- On narcissi treat as the crop goes into store and only if fungicidal treatment is essential. Adopt a resistance management strategy

Restrictions
- Maximum number of treatments 1 per batch for ware or seed potato tuber treatments; 1 per yr for narcissi bulbs
- Treated seed potatoes must not be used for food or feed
- Do not remove treated potatoes from store for sale, processing or consumption for at least 21 d after application
- Do not mix with any other product
- Off-label use as post-lifting cold water dip or pre-planting hot water dip not to be used on narcissus bulbs grown in the Isles of Scilly

Crop-specific information
- Latest use: 21 d before removal from store for sale, processing or consumption for ware potatoes
- Apply to potatoes as soon as possible after harvest using suitable equipment and always within 2 wk of lifting provided the skins are set. See label for details
- Potatoes should only be treated by systems that provide an accurate dose to tubers not carrying excessive quantities of soil
- Use as a post-lifting treatment to reduce basal and neck rot in narcissus bulbs. Ensure bulbs are clean

Environmental safety
- Dangerous for the environment
- Toxic to aquatic organisms

Hazard classification and safety precautions
Hazard Dangerous for the environment
Transport code 9
Packaging group III
UN Number 3082
Operator protection A, C, D, H; U05a, U19a, U20b
Environmental protection E15b, E38, H411
Storage and disposal D01, D02, D09a, D10c, D12a
Treated seed S03, S04a, S05

FOR FULL CONDITIONS OF USE ALWAYS READ THE PRODUCT LABEL

405   thiacloprid

A chloronicotinyl insecticide for use in agriculture and horticulture
IRAC mode of action code: 4A

Products

1	Agrovista Reggae	Agrovista	480 g/l	SC	13706
2	Biscaya	Bayer CropScience	240 g/l	OD	15014
3	Exemptor	Everris Ltd	10% w/w	GR	15615
4	Sonido	Bayer CropScience	400 g/l	SC	16368
5	Zubarone	AgChem Access	240 g/l	OD	15798

Uses

- Aphids in *all edible seed crops grown outdoors* (off-label), *all non-edible seed crops grown outdoors* (off-label), *celeriac* (off-label), *chinese cabbage* (off-label), *choi sum* (off-label), *combining peas*, *pak choi* (off-label), *swedes* (off-label), *tatsoi* (off-label), *turnips* (off-label), *vining peas*, *winter wheat* [2]; *bedding plants*, *hardy ornamental nursery stock*, *ornamental plant production*, *pot plants*, *protected ornamentals* [3]; *carrots*, *parsnips*, *potatoes*, *seed potatoes* [2, 5]; *leaf brassicas* (off-label - baby leaf production), *protected herbs (see appendix 6)* (off-label), *protected lamb's lettuce* (off-label), *protected lettuce* (off-label) [1]; *ornamental plant production* (off-label) [2, 3]; *peas*, *ware potatoes* [5]
- Beetles in *bedding plants*, *hardy ornamental nursery stock*, *ornamental plant production*, *ornamental plant production* (off-label), *pot plants*, *protected ornamentals* [3]
- Bruchid beetle in *spring field beans*, *winter field beans* [2]
- Bud mite in *almonds* (off-label), *chestnuts* (off-label), *cob nuts* (off-label), *hazel nuts* (off-label), *walnuts* (off-label) [1]
- Cabbage aphid in *broccoli*, *brussels sprouts*, *cabbages*, *calabrese*, *cauliflowers* [2, 5]; *chinese cabbage* (off-label), *choi sum* (off-label), *collards* (off-label), *kale* (off-label), *pak choi* (off-label), *spring greens* (off-label), *tatsoi* (off-label) [2]
- Capsids in *strawberries* (off-label) [1]
- Damson-hop aphid in *plums* (off-label) [1]
- Gall midge in *protected blueberry* (off-label) [1]
- Insect pests in *bilberries* (off-label), *blackberries* (off-label), *blackcurrants* (off-label), *blueberries* (off-label), *cherries* (off-label - under temporary protective rain covers), *courgettes* (off-label), *cranberries* (off-label), *gherkins* (off-label), *gooseberries* (off-label), *hops* (off-label), *marrows* (off-label), *mirabelles* (off-label - under temporary protective rain covers), *ornamental plant production* (off-label), *pears* (off-label), *protected blackberries* (off-label), *protected forest nurseries* (off-label), *protected raspberries* (off-label), *protected strawberries* (off-label), *raspberries* (off-label), *redcurrants* (off-label), *rubus hybrids* (off-label), *soft fruit* (off-label), *top fruit* (off-label), *whitecurrants* (off-label) [1]
- Leaf miner in *protected aubergines* (off-label), *protected courgettes* (off-label), *protected cucumbers* (off-label), *protected ornamentals* (off-label), *protected peppers* (off-label), *protected tomatoes* (off-label) [1]
- Mealy aphid in *broccoli*, *brussels sprouts*, *cabbages*, *calabrese*, *cauliflowers* [2, 5]; *chinese cabbage* (off-label), *choi sum* (off-label), *collards* (off-label), *kale* (off-label), *pak choi* (off-label), *spring greens* (off-label), *swedes* (off-label), *tatsoi* (off-label), *turnips* (off-label) [2]
- Moths in *protected blueberry* (off-label) [1]
- Pea midge in *combining peas*, *vining peas* [2]; *peas* [5]
- Peach-potato aphid in *leaf brassicas* (off-label - baby leaf production), *protected herbs (see appendix 6)* (off-label), *protected lamb's lettuce* (off-label), *protected lettuce* (off-label) [1]
- Pear midge in *pears* (off-label) [1]
- Pollen beetle in *mustard*, *spring oilseed rape*, *winter oilseed rape* [2, 5]
- Rosy apple aphid in *apples* [1]
- Sciarid flies in *ornamental plant production* (off-label) [3]
- Thrips in *protected aubergines* (off-label), *protected courgettes* (off-label), *protected cucumbers* (off-label), *protected ornamentals* (off-label), *protected peppers* (off-label), *protected tomatoes* (off-label) [1]
- Vine weevil in *bedding plants*, *hardy ornamental nursery stock*, *ornamental plant production*, *ornamental plant production* (off-label), *pot plants*, *protected ornamentals* [3]

SEE SECTION 3 FOR PRODUCTS ALSO REGISTERED

- Western flower thrips in **protected aubergines** *(off-label)*, **protected courgettes** *(off-label)*, **protected cucumbers** *(off-label)*, **protected ornamentals** *(off-label)*, **protected peppers** *(off-label)*, **protected tomatoes** *(off-label)* [1]
- Wheat-blossom midge in **spring wheat**, **winter wheat** [2, 5]
- Whitefly in **bedding plants**, **hardy ornamental nursery stock**, **ornamental plant production**, **ornamental plant production** *(off-label)*, **pot plants**, **protected ornamentals** [3]; **protected aubergines** *(off-label)*, **protected courgettes** *(off-label)*, **protected cucumbers** *(off-label)*, **protected ornamentals** *(off-label)*, **protected peppers** *(off-label)*, **protected tomatoes** *(off-label)* [1]
- Willow aphid in **horseradish** *(off-label)*, **parsley root** *(off-label)*, **red beet** *(off-label)*, **salsify** *(off-label)* [2]
- Wireworm in **fodder maize**, **grain maize**, **sweetcorn** [4]

Extension of Authorisation for Minor Use (EAMUs)

- **all edible seed crops grown outdoors** *20142041* [2]
- **all non-edible seed crops grown outdoors** *20142041* [2]
- **almonds** *20080471* [1]
- **bilberries** *20080466* [1]
- **blackberries** *20080475* [1]
- **blackcurrants** *20080466* [1]
- **blueberries** *20080466* [1]
- **celeriac** *20152075* [2]
- **cherries** *(under temporary protective rain covers)* *20080469* [1]
- **chestnuts** *20080471* [1]
- **chinese cabbage** *20142063* [2]
- **choi sum** *20142063* [2]
- **cob nuts** *20080471* [1]
- **collards** *20142249* [2]
- **courgettes** *20102032* [1]
- **cranberries** *20080466* [1]
- **gherkins** *20102032* [1]
- **gooseberries** *20080466* [1]
- **hazel nuts** *20080471* [1]
- **hops** *20102034* [1]
- **horseradish** *20142062* [2]
- **kale** *20142249* [2]
- **leaf brassicas** *(baby leaf production)* *20080470* [1]
- **marrows** *20102032* [1]
- **mirabelles** *(under temporary protective rain covers)* *20080469* [1]
- **ornamental plant production** *20102034* [1], *20142041* [2], *20170555* [3]
- **pak choi** *20142063* [2]
- **parsley root** *20142062* [2]
- **pears** *20080464* [1]
- **plums** *20080468* [1]
- **protected aubergines** *20080474* [1]
- **protected blackberries** *20080467* [1]
- **protected blueberry** *20102035* [1]
- **protected courgettes** *20080474* [1]
- **protected cucumbers** *20080474* [1]
- **protected forest nurseries** *20102034* [1]
- **protected herbs (see appendix 6)** *20080472* [1]
- **protected lamb's lettuce** *20080472* [1]
- **protected lettuce** *20080472* [1]
- **protected ornamentals** *20080474* [1]
- **protected peppers** *20080474* [1]
- **protected raspberries** *20080467* [1]
- **protected strawberries** *20102033* [1]
- **protected tomatoes** *20080474* [1]
- **raspberries** *20080475* [1]

FOR FULL CONDITIONS OF USE ALWAYS READ THE PRODUCT LABEL

- **red beet** *20142062* [2]
- **redcurrants** *20080466* [1]
- **rubus hybrids** *20080475* [1]
- **salsify** *20142062* [2]
- **soft fruit** *20102034* [1]
- **spring greens** *20142249* [2]
- **strawberries** *20080465* [1]
- **swedes** *20142061* [2]
- **tatsoi** *20142063* [2]
- **top fruit** *20102034* [1]
- **turnips** *20142061* [2]
- **walnuts** *20080471* [1]
- **whitecurrants** *20080466* [1]

Approval information
- Thiacloprid included in Annex I under EC Regulation 1107/2009

Efficacy guidance
- Best results in apples obtained by a programme of sprays commencing pre-blossom at the first sign of aphids
- For best pear midge control treat under warm conditions (mid-day) when the adults are flying (temperatures higher than 12°C). Applications on cool damp days are unlikely to be successful
- Best results in field crops obtained from treatments when target pests reach threshold levels. For wheat blossom midge use pheromone or sticky yellow traps to monitor adult activity in crops at risk [2]
- Treatment of ornamentals and protected ornamentals is by incorporation into peat-based growing media prior to sowing or planting using suitable automated equipment [3]
- Ensure thorough mixing in the compost to achieve maximum control. Top dressing is ineffective [3]
- Unless transplants have been treated with a suitable pesticide prior to planting in treated compost, full protection against target pests is not guaranteed [3]
- Minimise the possibility of the development of resistance by alternating insecticides with different modes of action in the programme
- In dense canopies and on larger trees increase water volume to ensure full coverage
- It can give effective contact activity against adult CSFB "present and active" in oilseed rape at the time of an application to control Myzus persicae [2]

Restrictions
- Maximum number of treatments 2 per yr on apples, seed potatoes; 1 per yr on other field crops and for compost incorporation
- Consult processor before use on crops for processing [2]
- Do not mix treated compost with any other bulky materials such as perlite or bark [3]
- Use treated compost as soon as possible, and within 4 wk of mixing [3]

Crop-specific information
- Latest use: up to and including flowering just complete (GS 69) for wheat; before sowing or planting for treated compost
- HI apples and potatoes 14 d; oilseed rape and mustard 30 d
- Test tolerance of ornamental species before large scale use [3]

Environmental safety
- Dangerous for the environment
- Very toxic or toxic to aquatic organisms
- Risk to certain non-target insects or other arthropods. See directions for use
- Broadcast air-assisted LERAP [1] (18 m)

Hazard classification and safety precautions
Hazard Harmful [1-3, 5]; Dangerous for the environment [1-5]; Harmful if swallowed [1, 2, 5]; Harmful if inhaled [1]
Transport code 6.1 [1]; 9 [3, 4]
Packaging group III [1, 3, 4]

SEE SECTION 3 FOR PRODUCTS ALSO REGISTERED

UN Number 2902 [1]; 3077 [3]; 3082 [4]; N/C [2, 5]
Risk phrases H315, H319 [2, 5]; H317 [1]; H336, H351, H360 [1, 2, 5]; R22a, R40b [4]; R40, R43 [3]; R50, R53a [3, 4]
Operator protection A [1-5]; C [2, 5]; D [3, 4]; H [1, 4]; U05a [1-5]; U11, U16a, U19a [2, 5]; U14 [1, 4]; U20b [1, 2, 4, 5]; U20c [3]
Environmental protection E10b [4]; E15a [3]; E15b [1, 2, 4, 5]; E17b [1] (18 m); E22c [1]; E34 [1-3, 5]; E38 [3, 4]; H410 [1, 2, 5]
Storage and disposal D01, D02 [1-5]; D09a [1-3, 5]; D09b, D10d, D12b, D14 [4]; D10b [1]; D10c [2, 5]; D11a [3]; D12a [1, 2, 5]
Treated seed S01, S02, S03, S04d, S05, S07, S08 [4]; S04a [3]
Medical advice M03 [1-5]; M05a [1, 4]

406   thiamethoxam

A neonicotinoid insecticide for apples, pears, potatoes and beet crops with a 2 year ban from Dec 2013 on use in many crops due to concerns over effects on bees.
IRAC mode of action code: 4A

See also fludioxonil + metalaxyl-M + thiamethoxam

Products

1	Actara	Syngenta	25% w/w	WG	13728
2	Cruiser SB	Syngenta	600 g/l	FS	15012

Uses
- Aphids in *fodder beet*, *sugar beet* [2]; *hops in propagation* (off-label), *nursery fruit trees* (off-label), *potatoes*, *potatoes grown for seed* [1]
- Beet leaf miner in *fodder beet* (seed treatment), *sugar beet* (seed treatment) [2]
- Cabbage root fly in *kale* (off-label), *swedes* (off-label) [2]
- Cabbage stem flea beetle in *kale* (off-label), *swedes* (off-label) [2]
- Flea beetle in *fodder beet* (seed treatment), *sugar beet* (seed treatment), *swedes* (off-label) [2]
- Millipedes in *fodder beet* (seed treatment), *sugar beet* (seed treatment) [2]
- Peach-potato aphid in *kale* (off-label), *swedes* (off-label) [2]
- Pygmy beetle in *fodder beet* (seed treatment), *sugar beet* (seed treatment) [2]
- Springtails in *fodder beet* (seed treatment), *sugar beet* (seed treatment) [2]
- Symphylids in *fodder beet* (seed treatment), *sugar beet* (seed treatment) [2]
- Virus yellows vectors in *kale* (off-label), *swedes* (off-label) [2]
- Western flower thrips in *hops in propagation* (off-label), *nursery fruit trees* (off-label), *protected ornamentals* (off-label) [1]
- Wireworm in *fodder beet* (seed treatment - reduction), *sugar beet* (seed treatment - reduction) [2]

Extension of Authorisation for Minor Use (EAMUs)
- *hops in propagation* 20170713 [1]
- *kale* 20131877 [2]
- *nursery fruit trees* 20170713 [1]
- *protected ornamentals* 20140186 [1]
- *swedes* 20111292 [2], 20131877 [2]

Approval information
- Thiamethoxam included in Annex I under EC Regulation 1107/2009

Efficacy guidance
- Apply to sugar beet or fodder beet seed as part of the normal commercial pelleting process using special treatment machinery [2]
- Seed drills must be suitable for use with polymer-coated seeds. Standard drill settings should not need to be changed [2]
- Drill treated seed into a firm even seedbed. Poor seedbed quality or seedbed conditions may results in delayed emergence and poor establishment [2]

FOR FULL CONDITIONS OF USE ALWAYS READ THE PRODUCT LABEL

- Where very high populations of soil pests are present protection may be inadequate to achieve an optimum plant stand [2]
- Use in line with latest IRAG guidelines

Restrictions
- Maximum number of treatments 1 per seed batch [2]
- Avoid deep or shallow drilling which may adversely affect establishment and reduce the level of pest control [2]
- Do not use herbicides containing lenacil pre-emergence on treated crops [2]

Crop-specific information
- Latest use: before drilling for fodder beet, sugar beet [2]

Environmental safety
- Dangerous for the environment
- Very toxic to aquatic organisms

Hazard classification and safety precautions
Hazard Dangerous for the environment, Very toxic to aquatic organisms
UN Number N/C
Operator protection A, D, H [2]; U05a, U07 [2]; U05b, U20c [1]
Environmental protection E03, E34, E36a [2]; E12c, E12e [1]; E15b, E38, H410 [1, 2]
Storage and disposal D01, D02, D09a, D12a [1, 2]; D05, D11a, D14 [2]; D10c [1]
Treated seed S02, S04b, S05, S06a, S06b, S08 [2]

407   thifensulfuron-methyl

A translocated sulfonylurea herbicide
HRAC mode of action code: B

See also carfentrazone-ethyl + thifensulfuron-methyl
flupyrsulfuron-methyl + thifensulfuron-methyl
fluroxypyr + metsulfuron-methyl + thifensulfuron-methyl
fluroxypyr + thifensulfuron-methyl + tribenuron-methyl
metsulfuron-methyl + thifensulfuron-methyl
nicosulfuron + thifensulfuron-methyl

Products

Pinnacle	DuPont	50% w/w	SG	12285

Uses
- Annual dicotyledons in **soya beans** *(off-label)*, **spring barley**, **spring wheat**, **winter barley**, **winter wheat**
- Docks in **grassland**
- Green cover in **land temporarily removed from production**

Extension of Authorisation for Minor Use (EAMUs)
- **soya beans** *20112063*

Approval information
- Thifensulfuron-methyl included in Annex I under EC Regulation 1107/2009
- Accepted by BBPA for use on malting barley

Efficacy guidance
- Best results achieved from application to small emerged weeds when growing actively. Broad-leaved docks are susceptible during the rosette stage up to onset of stem extension
- Ensure good spray coverage and apply to dry foliage
- Susceptible weeds stop growing almost immediately but symptoms may not be visible for about 2 wk
- Only broad-leaved docks (*Rumex obtusifolius*) are controlled; curled docks (*Rumex crispus*) are resistant
- Docks with developing or mature seed heads should be topped and the regrowth treated later

SEE SECTION 3 FOR PRODUCTS ALSO REGISTERED

SECTION 2

- Established docks with large tap roots may require follow-up treatment
- High populations of docks in grassland will require further treatment in following yr
- Thifensulfuron-methyl is a member of the ALS-inhibitor group of herbicides and products should be used in a planned Resistance Management strategy. See Section 5 for more information

Restrictions
- Maximum number of treatments 1 per year for grassland. Must only be applied from 1 Feb in year of harvest
- Do not treat new leys in year of sowing
- Do not treat where nutrient imbalances, drought, waterlogging, low temperatures, lime deficiency, pest or disease attack have reduced crop or sward vigour
- Do not roll or harrow within 7 d of spraying
- Do not graze grass crops within 7 d of spraying
- Specific restrictions apply to use in sequence or tank mixture with other sulfonylurea or ALS-inhibiting herbicides. See label for details
- Only one application of a sulfonylurea product may be applied per calendar yr to grassland and green cover on land temporarily removed from production

Crop-specific information
- Latest use: before 1 Aug on grass
- On grass apply 7-10 d before grazing and do not graze for 7 d afterwards
- Product may cause a check to both sward and clover which is usually outgrown

Following crops guidance
- Only grass or cereals may be sown within 4 wk of application to grassland or setaside, or in the event of failure of any treated crop
- No restrictions apply after normal harvest of a treated cereal crop

Environmental safety
- Dangerous for the environment
- Very toxic to aquatic organisms
- Keep livestock out of treated areas for at least 7 d following treatment
- Take extreme care to avoid drift onto broad-leaved plants outside the target area or onto surface waters or ditches, or land intended for cropping
- Spraying equipment should not be drained or flushed onto land planted, or to be planted, with trees or crops other than cereals and should be thoroughly cleansed after use - see label for instructions

Hazard classification and safety precautions
Hazard Dangerous for the environment, Very toxic to aquatic organisms
Transport code 9
Packaging group III
UN Number 3077
Operator protection U19a, U20b
Environmental protection E06a (7 d); E15b, E38, H410
Storage and disposal D09a, D11a, D12a

408   thifensulfuron-methyl + tribenuron-methyl

A mixture of two sulfonylurea herbicides for cereals
HRAC mode of action code: B + B

See also tribenuron-methyl

Products

1	Hiatus	Rotam	40:15% w/w	WG	16059
2	Inka SX	DuPont	25:25% w/w	SG	13601
3	Parana	Certis	33.3:16.7% w/w	SG	16258
4	Ratio SX	DuPont	40:10% w/w	SG	12601
5	Seduce	Certis	33.3:16.7% w/w	SG	16261

FOR FULL CONDITIONS OF USE ALWAYS READ THE PRODUCT LABEL

Uses

- Annual dicotyledons in *game cover* *(off-label)*, *spring oats*, *spring rye* [2]; *spring barley*, *spring wheat*, *winter barley*, *winter wheat* [1-5]; *triticale*, *winter rye* [1, 2]; *winter oats* [2, 4]
- Black bindweed in *spring barley*, *spring wheat*, *triticale*, *winter barley*, *winter rye*, *winter wheat* [1]
- Charlock in *spring barley*, *spring wheat*, *winter barley*, *winter wheat* [2-5]; *spring oats*, *spring rye*, *triticale*, *winter rye* [2]; *winter oats* [2, 4]
- Chickweed in *spring barley*, *spring wheat*, *winter barley*, *winter wheat* [1-5]; *spring oats*, *spring rye* [2]; *triticale*, *winter rye* [1, 2]; *winter oats* [2, 4]
- Docks in *spring barley*, *spring oats*, *spring rye*, *spring wheat*, *triticale*, *winter barley*, *winter oats*, *winter rye*, *winter wheat* [2]
- Fat hen in *spring barley*, *spring wheat*, *triticale*, *winter barley*, *winter rye*, *winter wheat* [1]
- Mayweeds in *spring barley*, *spring wheat*, *winter barley*, *winter wheat* [1-5]; *spring oats*, *spring rye* [2]; *triticale*, *winter rye* [1, 2]; *winter oats* [2, 4]
- Poppies in *spring barley*, *spring wheat*, *triticale*, *winter barley*, *winter rye*, *winter wheat* [1]

Extension of Authorisation for Minor Use (EAMUs)

- *game cover* 20101408 [2]

Approval information

- Thifensulfuron-methyl and tribenuron-methyl included in Annex I under EC Regulation 1107/2009
- Accepted by BBPA for use on malting barley

Efficacy guidance

- Apply when weeds are small and actively growing
- Apply after end of Feb in year of harvest
- Ensure good spray cover of the weeds
- Apply in a volume of 200 - 400 l/ha [1]
- Susceptible weeds cease growth almost immediately after application and symptoms become evident 2 wk later
- Effectiveness reduced by rain within 4 h of treatment and in very dry conditions
- Various tank mixtures recommended to broaden weed control spectrum
- Thifensulfuron-methyl and tribenuron-methyl are members of the ALS-inhibitor group of herbicides and products should be used in a planned Resistance Management strategy. See Section 5 for more information

Restrictions

- Maximum number of treatments 1 per crop
- Do not apply to cereals undersown with grass, clover or other legumes, or any other broad-leaved crop
- Do not apply within 7 d of rolling
- Specific restrictions apply to use in sequence or tank mixture with other sulfonylurea or ALS-inhibiting herbicides. See label for details
- Do not apply to any crop suffering from stress
- Consult contract agents before use on crops grown for seed

Crop-specific information

- Latest use: before flag leaf ligule first visible (GS 39) for all crops

Following crops guidance

- Only cereals, field beans, grass or oilseed rape may be sown in the same calendar year as harvest of a treated crop.
- In the event of failure of a treated crop sow only a cereal crop within 3 mth of product application and after ploughing and cultivating to at least 15 cm. After 3 mth field beans or oilseed rape may also be sown [2]
- Only cereals, oilseed rape or field beans may be sown in the same calendar year after harvest. In the following spring only cereals, oilseed rape or sugar beet may be sown. In case of crop failure for any reason, sow only spring small grain cereal. Before sowing, soil should be ploughed and cultivated to a depth of at least 15 cm[1]

Environmental safety

- Dangerous for the environment

SEE SECTION 3 FOR PRODUCTS ALSO REGISTERED

SECTION 2

- Very toxic to aquatic organisms
- Buffer zone requirement 6m [1]
- Spraying equipment should not be drained or flushed onto land planted, or to be planted, with trees or crops other than cereals and should be thoroughly cleansed after use - see label for instructions
- Take particular care to avoid damage by drift onto broad-leaved plants outside the target area or onto surface waters or ditches
- LERAP Category B

Hazard classification and safety precautions

Hazard Irritant [3-5]; Dangerous for the environment [1-5]; Very toxic to aquatic organisms [3, 5]
Transport code 9
Packaging group III
UN Number 3077
Operator protection A, H [3-5]; U05a, U08, U20a [3-5]; U14 [4]; U19a [1]; U20c [2]
Environmental protection E15a [3-5]; E15b [1, 2]; E16a, E38, H410 [1-5]; E16b [3, 5]
Storage and disposal D01, D02 [1, 3-5]; D09a, D11a, D12a [1-5]

409   thiophanate-methyl

A thiophanate fungicide with protectant and curative activity
FRAC mode of action code: 1

See also iprodione + thiophanate-methyl

Products

1	Cercobin WG	Certis	70% w/w	WG	13854
2	Taurus	Certis	70% w/w	WG	15508
3	Topsin WG	Certis	70% w/w	WG	13988

Uses

- Disease control in **hops** *(off-label - do not harvest for human or animal consumption (including idling) within 12 months of treatment.)*, **soft fruit** *(off-label)* [1]
- Fusarium in **durum wheat, spring oilseed rape, winter oilseed rape** [3]; **durum wheat** *(reduction)*, **spring wheat** *(reduction)*, **triticale** *(reduction)*, **winter wheat** *(reduction)* [2]; **spring wheat, triticale, winter wheat** [1, 3]
- Fusarium diseases in **protected ornamentals** *(off-label)* [1]
- Mycotoxins in **durum wheat** *(reduction)* [2, 3]; **spring oilseed rape** *(reduction)*, **winter oilseed rape** *(reduction)* [3]; **spring wheat** *(reduction)*, **triticale** *(reduction)*, **winter wheat** *(reduction)* [1-3]
- Root diseases in **container-grown ornamentals** *(off-label)*, **protected ornamentals** *(off-label)* [1]
- Sclerotinia stem rot in **spring oilseed rape, winter oilseed rape** [2]
- Verticillium wilt in **protected tomatoes** *(off-label)* [1]

Extension of Authorisation for Minor Use (EAMUs)

- **container-grown ornamentals** *20111655* [1]
- **hops** *(do not harvest for human or animal consumption (including idling) within 12 months of treatment.)* *20082833* [1]
- **protected ornamentals** *20111887* [1]
- **protected tomatoes** *20091969* [1]
- **soft fruit** *20082833* [1]

Approval information

- Thiophanate-methyl included in Annex I under EC Regulation 1107/2009

Efficacy guidance

- To avoid the development of resistance, a maximum of 2 applications of any MBC product (thiophanate-methyl or carbendazim) are allowed in any one crop. Avoid using MBC fungicides alone.

FOR FULL CONDITIONS OF USE ALWAYS READ THE PRODUCT LABEL

Environmental safety
- Dangerous for the environment
- Very toxic to aquatic organisms
- LERAP Category B

Hazard classification and safety precautions

Hazard Harmful, Dangerous for the environment, Harmful if swallowed, Harmful if inhaled [2]

Transport code 9

Packaging group III

UN Number 3077

Risk phrases H317, H371 [2]

Operator protection A; U19a [2, 3]

Environmental protection E15b, E16a, E16b [1-3]; E38 [2, 3]; H411 [2]

Storage and disposal D01, D02, D12a [2, 3]; D10b [1-3]

Medical advice M05a [2, 3]

410   thiram

A protectant dithiocarbamate fungicide
FRAC mode of action code: M3

See also carboxin + thiram
prochloraz + thiram

Products

1	Agrichem Flowable Thiram	Agrichem	600 g/l	FS	17887
2	Thyram Plus	Agrichem	600 g/l	FS	17890

Uses
- Anthracnose in *lupins (off-label)* [1]
- Damping off in *borage (off-label)*, *french beans (seed treatment)*, *frise (off-label)*, *lamb's lettuce (off-label)*, *lupins (off-label)*, *swedes (off-label)* [1]; *bulb onions (seed treatment)*, *cabbages (seed treatment)*, *cauliflowers (seed treatment)*, *combining peas (seed treatment)*, *dwarf beans (seed treatment)*, *edible podded peas (seed treatment)*, *forage maize (seed treatment)*, *grass seed (seed treatment)*, *kale (seed treatment)*, *leeks (seed treatment)*, *lettuce (seed treatment)*, *radishes (seed treatment)*, *runner beans (seed treatment)*, *salad onions (seed treatment)*, *soya beans (seed treatment - qualified minor use)*, *spring field beans (seed treatment)*, *spring oilseed rape (seed treatment)*, *turnips (seed treatment)*, *vining peas (seed treatment)*, *winter field beans (seed treatment)*, *winter oilseed rape (seed treatment)* [1, 2]
- Phoma in *swedes (off-label)* [1]
- Seed-borne diseases in *baby leaf crops (seed soak)*, *carrots (seed soak)*, *celery (outdoor) (seed soak)*, *fodder beet (seed soak)*, *mangels (seed soak)*, *parsnips (seed soak)*, *red beet (seed soak)*, *spinach (seed soak)*, *sugar beet (seed soak)* [1]

Extension of Authorisation for Minor Use (EAMUs)
- *borage* 20170453 [1]
- *frise* 20163410 [1]
- *lamb's lettuce* 20163410 [1]
- *lupins* 20163408 [1]
- *swedes* 20163409 [1]

Approval information
- Thiram included in Annex I under EC Regulation 1107/2009

Efficacy guidance
- Spray before onset of disease and repeat every 7-14 d. Spray interval varies with crop and disease. See label for details
- Seed treatments may be applied through most types of seed treatment machinery, from automated continuous flow machines to smaller batch treating apparatus
- Co-application of 175 ml water per 100 kg of seed likely to improve evenness of seed coverage

Restrictions
- Maximum number of treatments 3 per crop for protected winter lettuce (thiram based products only; 2 per crop if sequence of thiram and other EBDC fungicides use); 2 per crop for protected summer lettuce; 1 per batch of seed for seed treatments
- Do not apply to hydrangeas
- Notify processor before dusting or spraying crops for processing
- Do not dip roots of forestry transplants
- Do not treat seed of tomatoes, peppers or aubergines
- Soya bean seed treatments restricted to crops that are to be harvested as a mature pulse crop only
- Treated seed should not be stored from one season to the next

Crop-specific information
- Latest use: pre-drilling for seed treatments and seed soaks; 21 d after planting out or 21 d before harvest, whichever is earlier, for protected winter lettuce; 14 d after planting out or 21 d before harvest, whichever is earlier, for spraying protected summer lettuce
- Seed to be treated should be of satisfactory quality and moisture content
- Follow label instructions for treating small quantities of seed
- For use on tulips, chrysanthemums and carnations add non-ionic wetter

Environmental safety
- Dangerous for the environment
- Very toxic to aquatic organisms
- Do not use treated seed as food or feed
- Treated seed harmful to game and wildlife
- A red dye is available from manufacturer to colour treated seed, but not recommended as part of the seed steep. See label for details

Hazard classification and safety precautions
Hazard Harmful, Dangerous for the environment, Harmful if inhaled, Very toxic to aquatic organisms
Transport code 9
Packaging group III
UN Number 3082
Risk phrases H315, H317, H319, H373
Operator protection A [1, 2]; C, H [2]; U02a, U05a, U08, U11, U20b
Environmental protection E15a, E34, H410
Storage and disposal D01, D02, D05, D09a, D10c, D12b
Treated seed S01, S02, S03, S04a, S04b, S05, S06a, S07
Medical advice M04a

411   tri-allate

A soil-acting thiocarbamate herbicide for grass weed control
HRAC mode of action code: N

Products
1	Avadex Excel 15G	Gowan	15% w/w	GR	17872
2	Avadex Factor	Gowan	450 g/l	CS	17877

Uses
- Annual grasses in *canary seed* (off-label), *linseed* (off-label), *miscanthus* (off-label), *triticale* (off-label), *winter linseed* (off-label), *winter rye* (off-label) [1]; *spring barley, winter barley, winter wheat* [1, 2]
- Annual meadow grass in *canary seed* (off-label), *linseed* (off-label), *miscanthus* (off-label), *winter linseed* (off-label) [1]
- Blackgrass in *canary seed* (off-label), *linseed* (off-label), *miscanthus* (off-label), *triticale* (off-label), *winter linseed* (off-label), *winter rye* (off-label) [1]; *spring barley, winter barley, winter wheat* [1, 2]

FOR FULL CONDITIONS OF USE ALWAYS READ THE PRODUCT LABEL

- Meadow grasses in **spring barley**, **triticale** *(off-label)*, **winter barley**, **winter rye** *(off-label)*, **winter wheat** [1]
- Wild oats in **canary seed** *(off-label)*, **linseed** *(off-label)*, **miscanthus** *(off-label)*, **spring barley**, **triticale** *(off-label)*, **winter barley**, **winter linseed** *(off-label)*, **winter rye** *(off-label)*, **winter wheat** [1]

Extension of Authorisation for Minor Use (EAMUs)
- **canary seed** *20170467* [1]
- **linseed** *20170468* [1]
- **miscanthus** *20170466* [1]
- **triticale** *20171361* [1]
- **winter linseed** *20170468* [1]
- **winter rye** *20171361* [1]

Approval information
- Tri-allate included in Annex 1 under EC Regulation 1107/2009
- Accepted by BBPA for use on malting barley

Efficacy guidance
- Apply to soil surface pre-emergence
- For maximum activity apply to well-prepared moist seedbeds
- Do not use on soils with more than 10% organic matter
- Wild oats controlled up to 2-leaf stage
- Do not apply with spinning disc granule applicator; see label for suitable types [1]
- Do not apply to cloddy seedbeds
- Use sequential treatments to improve control of barren brome and annual dicotyledons (see label for details)

Restrictions
- Maximum number of treatments 1 per crop
- Drill wheat well below treated layer of soil (see label for safe drilling depths)
- Do not apply to shallow-drilled wheat crops
- Do not undersow grasses into treated crops
- Do not sow oats or grasses within 1 yr of treatment

Crop-specific information
- Latest use: pre-drilling for beet crops; before crop emergence for field beans, spring barley, peas, forage legumes; before first node detectable stage (GS 31) for winter wheat, winter barley, durum wheat, triticale, winter rye

Following crops guidance
- Pre-emergence use in winter wheat and winter barley
- Pre-drilling or pre-emergence use in spring barley

Environmental safety
- Irritating to eyes and skin
- May cause sensitization by skin contact
- Harmful to fish or other aquatic life. Do not contaminate surface waters or ditches with chemical or used container
- Buffer zone requirement 10 m [1] and 15 m
- Avoid drift on to non-target plants outside the treated area
- LERAP Category B

Hazard classification and safety precautions
Hazard Harmful, Dangerous for the environment, Very toxic to aquatic organisms [1]
Transport code 9
Packaging group III
UN Number 3077 [1]; 3082 [2]
Risk phrases H317 [2]; H373 [1, 2]
Operator protection A, C, H; U02a, U05a, U19a, U20a [1]
Environmental protection E16a [1, 2]; H410 [1]; H411 [2]
Storage and disposal D01, D02, D09a, D11a, D19 [1]
Medical advice M05a [1]

SEE SECTION 3 FOR PRODUCTS ALSO REGISTERED

412   tribenuron-methyl

A foliar acting sulfonylurea herbicide with some root activity for use in cereals
HRAC mode of action code: B

See also florasulam + tribenuron-methyl
fluroxypyr + thifensulfuron-methyl + tribenuron-methyl
metsulfuron-methyl + tribenuron-methyl
thifensulfuron-methyl + tribenuron-methyl

Products

1	Flame	Albaugh UK	50% w/w	WG	17842
2	Thor	Nufarm UK	50% w/w	WG	15239
3	Triad	Headland	50% w/w	TB	12751

Uses

- Annual dicotyledons in **durum wheat**, **farm forestry** (off-label), **forest nurseries** (off-label), **green cover on land temporarily removed from production**, **miscanthus** (off-label), **spring oats**, **spring wheat**, **triticale**, **winter oats**, **winter rye** [2]; **grassland**, **grassland** (off-label), **undersown spring barley** [3]; **spring barley** [1-3]; **winter barley**, **winter wheat** [1, 2]
- Chickweed in **grassland**, **grassland** (off-label), **spring barley**, **undersown spring barley** [3]

Extension of Authorisation for Minor Use (EAMUs)

- **farm forestry** *20122064* [2]
- **forest nurseries** *20122064* [2]
- **grassland** *20122493* [3]
- **miscanthus** *20122064* [2]

Approval information

- Tribenuron-methyl included in Annex I under EC Regulation 1107/2009
- Accepted by BBPA for use on malting barley

Efficacy guidance

- Best control achieved when weeds small and actively growing
- Good spray cover must be achieved since larger weeds often become less susceptible
- Susceptible weeds cease growth almost immediately after treatment and symptoms can be seen in about 2 wk
- Weed control may be reduced when conditions very dry
- Tribenuron-methyl is a member of the ALS-inhibitor group of herbicides and products should be used in a planned Resistance Management strategy. See Section 5 for more information

Restrictions

- Maximum number of treatments 1 per crop
- Specific restrictions apply to use in sequence or tank mixture with other sulfonylurea or ALS-inhibiting herbicides. See label for details
- Do not apply to crops undersown with grass, clover or other broad-leaved crops
- Do not apply to any crop suffering stress from any cause or not actively growing
- Do not apply within 7 d of rolling
- Do not apply at rates greater than 10 g product./hectare before end of February in the year of harvest [2]
- Must not be applied before end of February in the year of harvest [2, 3]

Crop-specific information

- Latest use: up to and including flag leaf ligule/collar just visible (GS 39)
- Apply in autumn or in spring from 3 leaf stage of crop

Following crops guidance

- Only cereals, field beans or oilseed rape may be sown in the same calendar yr as harvest of a treated crop
- In the event of crop failure sow only a cereal within 3 mth of application. After 3 mth field beans or oilseed rape may also be sown

FOR FULL CONDITIONS OF USE ALWAYS READ THE PRODUCT LABEL

Environmental safety
- Dangerous for the environment
- Very toxic to aquatic organisms
- Take extreme care to avoid drift onto broad-leaved plants outside the target area or onto surface waters or ditches, or land intended for cropping
- Spraying equipment should not be drained or flushed onto land planted, or to be planted, with trees or crops other than cereals and should be thoroughly cleansed after use - see label for instructions

Hazard classification and safety precautions

Hazard Irritant, Dangerous for the environment [1-3]; Very toxic to aquatic organisms [1]
Transport code 9
Packaging group III
UN Number 3077
Risk phrases H317 [1-3]; H319 [1]
Operator protection A [1-3]; C [1]; H [1, 2]; U05a, U08, U20b
Environmental protection E07b [3] (3 weeks); E07e [3]; E15a, H410 [1-3]; E38 [1, 2]
Storage and disposal D01, D02, D09a, D11a [1-3]; D12a [1, 2]

413   Trichoderma asperellum (Strain T34)

A biological control agent

Products

T34 Biocontrol	Fargro	10.832% w/w	WP	17290

Uses
- Fusarium in **protected baby leaf crops** (off-label), **protected bilberries** (off-label), **protected blackberries** (off-label), **protected blackcurrants** (off-label), **protected blueberry** (off-label), **protected broccoli** (off-label), **protected brussels sprouts** (off-label), **protected cabbages** (off-label), **protected calabrese** (off-label), **protected cauliflowers** (off-label), **protected choi sum** (off-label), **protected collards** (off-label), **protected courgettes** (off-label), **protected cranberries** (off-label), **protected cucumbers** (off-label), **protected elderberries** (off-label), **protected forest nurseries** (off-label), **protected gherkins** (off-label), **protected gooseberries** (off-label), **protected herbs (see appendix 6)** (off-label), **protected kale** (off-label), **protected kohlrabi** (off-label), **protected lettuce** (off-label), **protected loganberries** (off-label), **protected melons** (off-label), **protected mulberry** (off-label), **protected oriental cabbage** (off-label), **protected ornamentals** (off-label), **protected pumpkins** (off-label), **protected raspberries** (off-label), **protected redcurrants** (off-label), **protected rose hips** (off-label), **protected rubus hybrids** (off-label), **protected strawberries** (off-label), **protected summer squash** (off-label), **protected watercress** (off-label), **protected winter squash** (off-label)
- Fusarium foot rot and seedling blight in **aubergines, chillies, ornamental plant production, peppers, tomatoes**
- Pythium in **protected baby leaf crops** (off-label), **protected bilberries** (off-label), **protected blackberries** (off-label), **protected blackcurrants** (off-label), **protected blueberry** (off-label), **protected broccoli** (off-label), **protected brussels sprouts** (off-label), **protected cabbages** (off-label), **protected calabrese** (off-label), **protected cauliflowers** (off-label), **protected choi sum** (off-label), **protected collards** (off-label), **protected courgettes** (off-label), **protected cranberries** (off-label), **protected cucumbers** (off-label), **protected elderberries** (off-label), **protected forest nurseries** (off-label), **protected gherkins** (off-label), **protected gooseberries** (off-label), **protected herbs (see appendix 6)** (off-label), **protected kale** (off-label), **protected kohlrabi** (off-label), **protected lettuce** (off-label), **protected loganberries** (off-label), **protected melons** (off-label), **protected mulberry** (off-label), **protected oriental cabbage** (off-label), **protected ornamentals** (off-label), **protected pumpkins** (off-label), **protected raspberries** (off-label), **protected redcurrants** (off-label), **protected rose hips** (off-label), **protected rubus hybrids** (off-label), **protected strawberries** (off-label), **protected summer squash** (off-label), **protected watercress** (off-label), **protected winter squash** (off-label)

Extension of Authorisation for Minor Use (EAMUs)
- **protected baby leaf crops** 20161809

SEE SECTION 3 FOR PRODUCTS ALSO REGISTERED

SECTION 2

- *protected bilberries* *20161808*
- *protected blackberries* *20161808*
- *protected blackcurrants* *20161808*
- *protected blueberry* *20161808*
- *protected broccoli* *20161809*
- *protected brussels sprouts* *20161809*
- *protected cabbages* *20161809*
- *protected calabrese* *20161809*
- *protected cauliflowers* *20161809*
- *protected choi sum* *20161809*
- *protected collards* *20161809*
- *protected courgettes* *20161805*
- *protected cranberries* *20161808*
- *protected cucumbers* *20161805*
- *protected elderberries* *20161808*
- *protected forest nurseries* *20161810*
- *protected gherkins* *20161805*
- *protected gooseberries* *20161808*
- *protected herbs (see appendix 6)* *20161809*
- *protected kale* *20161809*
- *protected kohlrabi* *20161809*
- *protected lettuce* *20161809*
- *protected loganberries* *20161808*
- *protected melons* *20161805*
- *protected mulberry* *20161808*
- *protected oriental cabbage* *20161809*
- *protected ornamentals* *20161810*
- *protected pumpkins* *20161805*
- *protected raspberries* *20161808*
- *protected redcurrants* *20161808*
- *protected rose hips* *20161808*
- *protected rubus hybrids* *20161808*
- *protected strawberries* *20161808*
- *protected summer squash* *20161805*
- *protected watercress* *20161809*
- *protected winter squash* *20161805*

Approval information
- Trichoderma asperellum (Strain T34) included in Annex 1 under EC Regulation 1107/2009

Efficacy guidance
- May be applied by spraying, through an irrigation system or by root dipping.

Restrictions
- Efficacy has been demonstrated on peat and coir composts but should be checked on the more unusual composts before large scale use.

Crop-specific information
- Crop safety has been demonstrated on a range of carnation species. It is advisable to check the safety to other species on a sample of the population before large-scale use.

Hazard classification and safety precautions
   Hazard Harmful
   UN Number N/C
   Risk phrases H317
   Operator protection A, D, H; U05a, U09a, U19a, U20b
   Environmental protection E15a
   Storage and disposal D01, D02, D05, D09a, D09b, D10c
   Medical advice M03, M04a

FOR FULL CONDITIONS OF USE ALWAYS READ THE PRODUCT LABEL

414   trifloxystrobin

A protectant strobilurin fungicide for cereals and managed amenity turf
FRAC mode of action code: 11

See also cyproconazole + trifloxystrobin
fluopyram + trifloxystrobin
fluoxastrobin + prothioconazole + trifloxystrobin
iprodione + trifloxystrobin
propiconazole + trifloxystrobin
prothioconazole + trifloxystrobin
tebuconazole + trifloxystrobin

Products

1	Martinet	AgChem Access	500 g/l	SC	14614
2	Mascot Defender	Rigby Taylor	500 g/l	SG	14065
3	Pan Aquarius	Pan Agriculture	50% w/w	WG	13018
4	PureFloxy	Pure Amenity	50% w/w	WG	15473
5	Scorpio	Bayer CropScience	50% w/w	WG	12293
6	Swift SC	Bayer CropScience	500 g/l	SC	11227

Uses

- Brown rust in **spring barley, winter barley, winter wheat** [1, 6]
- Foliar disease control in **durum wheat** *(off-label)*, **grass seed crops** *(off-label)*, **ornamental specimens** *(off-label)*, **spring rye** *(off-label)*, **triticale** *(off-label)*, **winter rye** *(off-label)* [6]
- Fusarium patch in **amenity grassland, managed amenity turf** [2-5]
- Glume blotch in **winter wheat** [1, 6]
- Net blotch in **spring barley, winter barley** [1, 6]
- Red thread in **amenity grassland, managed amenity turf** [2-5]
- Rhynchosporium in **spring barley, winter barley** [1, 6]
- Septoria leaf blotch in **winter wheat** [1, 6]

Extension of Authorisation for Minor Use (EAMUs)

- **durum wheat** *20061287* [6]
- **grass seed crops** *20061287* [6]
- **ornamental specimens** *20082882* [6]
- **spring rye** *20061287* [6]
- **triticale** *20061287* [6]
- **winter rye** *20061287* [6]

Approval information

- Trifloxystrobin included in Annex I under EC Regulation 1107/2009
- Accepted by BBPA for use on malting barley

Efficacy guidance

- Should be used protectively before disease is established in crop. Further treatment may be necessary if disease attack prolonged
- Treat grass after cutting and do not mow for at least 48 h afterwards to allow adequate systemic movement [3, 5]
- Trifloxystrobin is a member of the QoI cross resistance group. Product should be used preventatively and not relied on for its curative potential
- Use product as part of an Integrated Crop Management strategy incorporating other methods of control, including where appropriate other fungicides with a different mode of action. Do not apply more than two foliar applications of QoI containing products to any cereal crop
- There is a significant risk of widespread resistance occurring in *Septoria tritici* populations in UK. Failure to follow resistance management action may result in reduced levels of disease control
- On cereal crops product must always be used in mixture with another product, recommended for control of the same target disease, that contains a fungicide from a different cross resistance group and is applied at a dose that will give robust control
- Strains of barley powdery mildew resistant to QoIs are common in the UK

SEE SECTION 3 FOR PRODUCTS ALSO REGISTERED

Restrictions
- Maximum number of treatments 2 per crop per yr
- Do not apply to turf during dought conditions or to frozen turf [3, 5]

Crop-specific information
- HI barley, wheat 35 d [6]

Environmental safety
- Dangerous for the environment
- Very toxic to aquatic organisms
- LERAP Category B [2-5]

Hazard classification and safety precautions
Hazard Irritant [2-5]; Dangerous for the environment [1-6]; Very toxic to aquatic organisms [1, 4-6]
Transport code 9
Packaging group III
UN Number 3077 [3-5]; 3082 [1, 2, 6]
Risk phrases H317 [2-5]
Operator protection A, H; U02a, U09a, U19a [1, 6]; U05a, U20b [1-6]; U11, U13, U14, U15 [2-5]
Environmental protection E13b [1, 6]; E15b, E16a, E16b, E34 [2-5]; E38, H410 [1-6]
Consumer protection C02a [1, 6] (35 d)
Storage and disposal D01, D02, D09a, D10c, D12a
Medical advice M03 [2-5]

---

415   triflusulfuron-methyl

A sulfonyl urea herbicide for beet crops
HRAC mode of action code: B

See also lenacil + triflusulfuron-methyl

Products

1	Debut	DuPont	50% w/w	WG	07804
2	Shiro	UPL Europe	50% w/w	WG	17439
3	Upbeet	DuPont	50% w/w	WG	17544

Uses
- Annual dicotyledons in *chicory* (off-label) [1]; *fodder beet*, *sugar beet* [1-3]
- Charlock in *red beet* (off-label) [1]
- Cleavers in *red beet* (off-label) [1]
- Flixweed in *red beet* (off-label) [1]
- Fool's parsley in *red beet* (off-label) [1]
- Nipplewort in *red beet* (off-label) [1]
- Wild chrysanthemum in *red beet* (off-label) [1]

Extension of Authorisation for Minor Use (EAMUs)
- *chicory* 20142535 [1]
- *red beet* 20142536 [1]

Approval information
- Triflusulfuron-methyl included in Annex I under EC Regulation 1107/2009

Efficacy guidance
- Product should be used with a recommended adjuvant or a suitable herbicide tank-mix partner - see label for details
- Product acts by foliar action. Best results obtained from good spray cover of small actively growing weeds
- Susceptible weeds cease growth immediately and symptoms can be seen 5-10 d later
- Best results achieved from a programme of up to 4 treatments starting when first weeds have emerged with subsequent applications every 5-14 d when new weed flushes at cotyledon stage
- Weed spectrum can be broadened by tank mixture with other herbicides. See label for details

FOR FULL CONDITIONS OF USE ALWAYS READ THE PRODUCT LABEL

- Product may be applied overall or via band sprayer
- Triflusulfuron-methyl is a member of the ALS-inhibitor group of herbicides

Restrictions
- Maximum number of treatments 4 per crop
- Do not apply to any crop stressed by drought, water-logging, low temperatures, pest or disease attack, nutrient or lime deficiency

Crop-specific information
- Latest use: before crop leaves meet between rows
- HI 4 wk for red beet
- All varieties of sugar beet and fodder beet may be treated from early cotyledon stage until the leaves begin to meet between the rows

Following crops guidance
- Only winter cereals should follow a treated crop in the same calendar yr. Any crop may be sown in the next calendar yr
- After failure of a treated crop, sow only spring barley, linseed or sugar beet within 4 mth of spraying unless prohibited by tank-mix partner

Environmental safety
- Dangerous for the environment
- Very toxic to aquatic organisms
- Extremely dangerous to fish or other aquatic life. Do not contaminate surface waters or ditches with chemical or used container
- Take extreme care to avoid drift onto broad-leaved plants outside the target area or onto surface waters or ditches, or land intended for cropping
- Spraying equipment should not be drained or flushed onto land planted, or to be planted, with trees or crops other than sugar beet and should be thoroughly cleansed after use - see label for instructions
- LERAP Category B

Hazard classification and safety precautions
Hazard Irritant, Dangerous for the environment [1-3]; Very toxic to aquatic organisms [2, 3]
Transport code 9
Packaging group III
UN Number 3077
Risk phrases H351 [2]; R43, R50, R53a [1]
Operator protection A; U05a, U08, U19a, U20a
Environmental protection E13a, E15a, E16a, E16b, E38 [1-3]; H410 [2, 3]
Storage and disposal D01, D02, D05, D09a, D11a, D12a

416 trinexapac-ethyl

A novel cyclohexanecarboxylate plant growth regulator for cereals, turf and amenity grassland

See also prohexadione-calcium + trinexapac-ethyl

Products

1	Clipless NT	Headland Amenity	120 g/l	ME	17558
2	Confine	Headland	250 g/l	EC	16109
3	Cutaway	Syngenta	121 g/l	SL	14445
4	Freeze NT	Headland	250 g/l	EC	17478
5	Gyro	UPL Europe	250 g/l	EC	17426
6	Iceni	UPL Europe	250 g/l	EC	16835
7	Limitar	Belcrop	250 g/l	EC	16301
8	Modan 250 EC	Belchim	250 g/l	EC	17060
9	Moddus	Syngenta	250 g/l	EC	15151
10	Moxa	Globachem	250 g/l	EC	16105
11	Moxa 250 EC	Belchim	250 g/l	EC	16176
12	Optimus	Adama	175 g/l	EC	15249

SEE SECTION 3 FOR PRODUCTS ALSO REGISTERED

Products – continued

13	Pan Tepee	Pan Agriculture	250 g/l	EC	15867
14	Primo Maxx II	Syngenta	121 g/l	SL	17509
15	Shorten	Becesane	250 g/l	EC	15377
16	Shrink	AgChem Access	250 g/l	EC	15660
17	Sudo	Life Scientific	250 g/l	EC	17979
18	Tempo	Syngenta	250 g/l	EC	15170
19	Tridus	Globachem	250 g/l	EC	16938

Uses

- Growth regulation in *canary seed (off-label)*, *spring wheat (off-label)* [12]; *durum wheat*, *spring barley, spring oats, triticale, winter barley, winter oats, winter wheat* [5-7, 15-17]; *forest nurseries (off-label)*, *ornamental plant production (off-label)* [3, 9]; *grass seed crops* [5, 7, 17]; *grass seed crops (off-label)*, *red clover (off-label)* [9]; *lettuce (off-label)* [3]; *rye, spring wheat* [7]; *ryegrass seed crops* [5, 6, 15, 16]; *spring red wheat (off-label)* [2]; *spring rye, winter rye* [5, 6, 15-17]
- Growth retardation in *amenity grassland* [1, 3, 14]; *managed amenity turf* [1, 14]; *spring red wheat (off-label)* [18]
- Lodging control in *durum wheat* [2, 4, 9-13, 18, 19]; *grass seed crops* [2, 4, 10, 11, 19]; *grass seed crops (off-label)*, *red clover (off-label)* [9]; *rye* [4, 10, 11, 19]; *ryegrass seed crops* [9, 12, 13, 18]; *spring barley, spring oats, triticale, winter barley, winter oats, winter wheat* [2, 4, 8-13, 18, 19]; *spring red wheat (off-label)* [2, 9, 18]; *spring rye, winter rye* [2, 8, 9, 12, 13, 18]; *spring wheat* [4, 9-12, 19]

Extension of Authorisation for Minor Use (EAMUs)

- *canary seed* *20111263* [12], *20141226* [12]
- *forest nurseries* *20162140* [3], *20122055* [9]
- *grass seed crops* *20171035* [9]
- *lettuce* *20162140* [3]
- *ornamental plant production* *20162140* [3], *20103062* [9]
- *red clover* *20171035* [9]
- *spring red wheat* *20130322* [2], *20121385* [9], *20113201* [18]
- *spring wheat* *20111264* [12], *20141227* [12]

Approval information

- Trinexapac-ethyl included in Annex I under EC Regulation 1107/2009
- Accepted by BBPA for use on malting barley

Efficacy guidance

- Best results on cereals and ryegrass seed crops obtained from treatment from the leaf sheath erect stage [9, 13]
- Best results on turf achieved from application to actively growing weed free turf grass that is adequately fertilized and watered and is not under stress. Adequate soil moisture is essential
- Turf should be dry and weed free before application
- Environmental conditions, management and cultural practices that affect turf growth and vigour will influence effectiveness of treatment
- Repeat treatments on turf up to the maximum approved dose may be made as soon as growth resumes

Restrictions

- Maximum total dose equivalent to one full dose on cereals and ryegrass seed crops [9, 13]
- Maximum number of treatments on turf equivalent to five full dose treatments
- Do not apply if rain or frost expected or if crop wet. Products are rainfast after 12 h
- Only use on crops at risk of lodging [9, 13]
- Do not apply within 12 h of mowing turf
- Do not treat newly sown turf
- Not to be used on food crops
- Do not compost or mulch grass clippings
- Some products stipulate that application should not be made before GS30

FOR FULL CONDITIONS OF USE ALWAYS READ THE PRODUCT LABEL

Crop-specific information
- Latest use: before 2nd node detectable (GS 32) for oats, ryegrass seed crops; before 3rd node detectable (GS 33) for durum wheat, spring barley, triticale, rye; before flag leaf sheath extending (GS 41) for winter barley, winter wheat
- On wheat apply as single treatment between leaf sheath erect stage (GS 30) and flag leaf fully emerged (GS 39) [9, 13]
- On barley, rye, triticale and durum wheat apply as single treatment between leaf sheath erect stage (GS 30) and second node detectable (GS 32), or on winter barley at higher dose between flag leaf just visible (GS 37) and flag leaf fully emerged (GS 39) [9, 13]
- On oats and ryegrass seed crops apply between leaf sheath erect stage (GS 30) and first node detectable stage (GS 32) [9, 13]
- Treatment may cause ears of cereals to remain erect through to harvest [9, 13]
- Turf under stress when treated may show signs of damage
- Any weed control in turf must be carried out before application of the growth regulator

Environmental safety
- Dangerous for the environment [9, 13]
- Toxic to aquatic organisms [9, 13]

Hazard classification and safety precautions
Hazard Irritant, Dangerous for the environment [2, 4-6, 8-13, 15-19]; Flammable liquid and vapour [10, 19]; Harmful if inhaled [10, 14, 19]; Very toxic to aquatic organisms [5]

Transport code 9 [1, 2, 4-6, 8-13, 15-19]

Packaging group III [1, 2, 4-6, 8-13, 15-19]

UN Number 3082 [1, 2, 4-6, 8-13, 15-19]; N/C [3, 7, 14]

Risk phrases H315 [5, 12]; H317 [4-6, 8, 9, 11-14, 17, 18]; H318 [5, 8]; H319 [4, 7, 10, 12, 19]; H335 [7, 8, 10, 19]; R36 [2]; R43, R51, R53a [2, 15, 16]

Operator protection A [1-19]; C [1-4, 7-19]; H [1, 3-6, 8, 14]; K [1, 3, 14]; U05a [1-6, 8-19]; U08, U09c, U20a [7]; U10, U14 [8]; U11, U19a [7, 8]; U15, U20c [2, 4-6, 8-13, 15-19]; U20b [1, 3, 14]

Environmental protection E15a [2, 4-6, 8-13, 15-19]; E15b, E34 [7]; E38 [2, 4-13, 15-19]; H410 [9]; H411 [6-8, 10, 11, 13, 17-19]; H412 [1, 4, 12, 14]

Consumer protection C01 [1, 3, 14]

Storage and disposal D01 [1-19]; D02, D05, D10c [1-6, 8-19]; D09a, D12a [2, 4-13, 15-19]; D11a, D12b [7]

Medical advice M03 [7]

417   urea

Commodity substance for fungicide treatment of cut tree stumps

418   warfarin

A coumarin anti-coagulant rodenticide

Products

Sakarat Warfarin Whole Wheat	Killgerm	0.05% w/w	RB	UK17-1059

Uses
- Mice in *farm buildings/yards*
- Rats in *farm buildings/yards*

Approval information
- Warfarin included in Annex I under EC Regulation 1107/2009

Efficacy guidance
- For rodent control place ready-to-use or prepared baits at many points wherever rats active. Out of doors shelter bait from weather
- Inspect baits frequently and replace or top up as long as evidence of feeding. Do not underbait

Restrictions
- For use only by local authorities, professional operators providing a pest control service and persons occupying industrial, agricultural or horticultural premises

Environmental safety
- Harmful to wildlife
- Prevent access to baits by children and animals, especially cats, dogs and pigs
- Rodent bodies must be searched for and burned or buried, not placed in refuse bins or rubbish tips. Remains of bait and containers must be removed after treatment and burned or buried
- Bait must not be used where food, feed or water could become contaminated
- Warfarin baits must not be used outdoors where pine martens are known to occur naturally

Hazard classification and safety precautions
Risk phrases H360, H373
Operator protection A; U13, U20b
Storage and disposal D05, D07, D09a, D11a
Vertebrate/rodent control products V01b, V02, V03b, V04b
Medical advice M03

419   zeta-cypermethrin

A contact and stomach acting pyrethroid insecticide
IRAC mode of action code: 3

Products

1	Angri	AgChem Access	100 g/l	EW	13730
2	Chimpanzee	Pan Agriculture	100 g/l	EW	16043
3	Fury 10 EW	Headland	100 g/l	EW	17255
4	Minuet EW	Headland	100 g/l	EW	17250

Uses
- Aphids in **broad beans** *(off-label)*, **durum wheat** *(off-label)*, **fodder beet** *(off-label)*, **grass seed crops** *(off-label)*, **lupins** *(off-label)*, **rye** *(off-label)*, **triticale** *(off-label)* [3]; **spring barley, spring oats, spring wheat, winter barley, winter oats, winter wheat** [1-4]
- Barley yellow dwarf virus vectors in **spring barley, spring wheat, winter barley, winter wheat** [1-4]
- Cabbage seed weevil in **spring oilseed rape, winter oilseed rape** [1-4]
- Cabbage stem flea beetle in **spring oilseed rape, winter oilseed rape** [1-4]
- Cutworms in **potatoes, sugar beet** [1-4]
- Flax flea beetle in **linseed** [1-4]
- Flea beetle in **spring oilseed rape, winter oilseed rape** [1-4]
- Insect pests in **broad beans** *(off-label)*, **durum wheat** *(off-label)*, **fodder beet** *(off-label)*, **grass seed crops** *(off-label)*, **lupins** *(off-label)*, **rye** *(off-label)*, **triticale** *(off-label)* [3]
- Large flax flea beetle in **linseed** [1-4]
- Pea and bean weevil in **combining peas, spring field beans, vining peas, winter field beans** [1-4]
- Pea aphid in **combining peas, vining peas** [1-4]
- Pea moth in **combining peas, vining peas** [1-4]
- Pod midge in **spring oilseed rape, winter oilseed rape** [1-4]
- Pollen beetle in **spring oilseed rape, winter oilseed rape** [1-4]
- Rape winter stem weevil in **spring oilseed rape, winter oilseed rape** [1-4]

Extension of Authorisation for Minor Use (EAMUs)
- **broad beans** *20160125* [3]
- **durum wheat** *20160124* [3]
- **fodder beet** *20160126* [3]
- **grass seed crops** *20160124* [3]
- **lupins** *20160127* [3]
- **rye** *20160124* [3]
- **triticale** *20160124* [3]

FOR FULL CONDITIONS OF USE ALWAYS READ THE PRODUCT LABEL

Approval information
- Zeta-cypermethrin included in Annex I under EC Regulation 1107/2009
- Accepted by BBPA for use on malting barley

Efficacy guidance
- On winter cereals spray when aphids first found in the autumn for BYDV control. A second spray may be required on late drilled crops or in mild conditions
- For summer aphids on cereals spray when treatment threshold reached
- For listed pests in other crops spray when feeding damage first seen or when treatment threshold reached. Under high infestation pressure a second treatment may be necessary
- Best results for pod midge and seed weevil control in oilseed rape obtained from treatment after pod set but before 80% petal fall
- Pea moth treatments should be applied according to ADAS/PGRO warnings or when economic thresholds reached as indicated by pheromone traps
- Treatments for cutworms should be made at egg hatch and repeated no sooner than 10 d later

Restrictions
- Maximum total dose equivalent to two full dose treatments on all crops
- Consult processors before use on crops for processing

Crop-specific information
- Latest use: before 4 true leaves for linseed, before end of flowering for oilseed rape, cereals
- HI potatoes, field beans, peas 14 d; sugar beet 60 d

Environmental safety
- Dangerous for the environment
- Very toxic to aquatic organisms
- High risk to non-target insects or other arthropods. Do not spray within 6 m of the field boundary
- LERAP Category A

Hazard classification and safety precautions
Hazard Harmful, Dangerous for the environment, Harmful if swallowed, Harmful if inhaled [1-4]; Very toxic to aquatic organisms [1, 3]
Transport code 6.1 [1, 2]; 9 [3, 4]
Packaging group III
UN Number 3082 [3, 4]; 3352 [1, 2]
Risk phrases H317
Operator protection A, C, H; U05a, U08, U14, U15, U19a, U20b
Environmental protection E15b, E16c, E16d, E22a, E34, E38, H410
Storage and disposal D01, D02, D09a, D10b, D12a
Medical advice M05a

420   zoxamide

A substituted benzamide fungicide available only in mixtures
FRAC mode of action code: 22

See also cymoxanil + zoxamide
*   dimethomorph + zoxamide*
*   mancozeb + zoxamide*

SECTION 3
PRODUCTS ALSO REGISTERED

Products also Registered

Products listed in the table below have not been notified for inclusion in Section 2 of this edition of the *Guide*. These products may legally be stored and used in accordance with their label until their approval expires, but they may not still be available for purchase.

Product	Approval holder	MAPP No.	Expiry Date
1 abamectin			
Abamex 18 EC	MAC	15806	31 Dec 2021
Abamite Pro	CMI	16138	31 Dec 2021
Biotrine	Russell	16367	31 Dec 2021
Ramectin	RAAT	17876	31 Dec 2021
2 acetamiprid			
Aceta 20 SG	Euro	16919	31 Oct 2020
Acetamex 20 SP	MAC	15888	31 Oct 2020
Antelope	Gemini	18041	31 Oct 2020
Clayton Vault	Clayton	18181	31 Oct 2020
Gazelle	Certis	12909	31 Oct 2020
Persist	RAAT	17633	31 Oct 2019
3 acetic acid			
New-Way Weed Spray	Punya	15319	31 Aug 2019
OWK	UK Organic	15363	28 Feb 2022
4 acibenzolar-S-methyl			
Bion	Syngenta	09803	31 Dec 2018
Inssimo	Syngenta	16870	15 Jan 2019
5 adoxophyes orana gv			
Capex	Andermatt	18258	31 Oct 2019
6 alpha-cypermethrin			
Alert	BASF	16785	31 Jan 2020
Alpha C 6 ED	Techneat	13611	31 Jan 2020
Contest	BASF	16764	31 Jan 2020
Fastac	BASF	16761	31 Jan 2020
Fastac ME	BASF	17686	31 Jan 2020
7 aluminium ammonium sulphate			
Guardsman	Chiltern	05494	31 Dec 2021
8 aluminium phosphide			
Detia Gas Ex-P	Rentokil	17036	28 Feb 2022
Detia Gas-Ex-T	Rentokil	17034	28 Feb 2022
Quickphos Pellets 56% GE	UPL Europe	16987	28 Feb 2022
11 ametoctradin + dimethomorph			
Resplend	BASF	14975	31 Jan 2021
12 ametoctradin + mancozeb			
Diablo	BASF	16084	31 Jul 2020
16 aminopyralid			
Pro-Banish	Dow	14730	31 Dec 2019
17 aminopyralid + fluroxypyr			
Halcyon	Dow	14709	31 Dec 2019

Product	Approval holder	MAPP No.	Expiry Date
18 aminopyralid + halauxifen-methyl			
Trezac	Dow	18253	04 Oct 2021
20 aminopyralid + propyzamide			
Galactic Pro	Chem-Wise	17180	31 Dec 2019
Milepost	Terrechem	17537	31 Dec 2019
Milestone	PSI	17494	31 Dec 2019
Pyzamid Universe	RealChemie	18169	31 Dec 2019
Pyzamid Universe	Euro	16956	31 Dec 2019
21 aminopyralid + triclopyr			
Speedline Pro	Bayer CropScience	16208	31 Dec 2019
22 amisulbrom			
Gachinko	Nissan	18112	30 Sep 2021
Gachinko	DuPont	18219	31 Dec 2026
Leimay	Syngenta	18243	28 Nov 2020
Leimay	Nissan	17412	28 Nov 2020
Sanblight	Nissan	17411	28 Nov 2020
25 Ampelomyces quisqualis (Strain AQ10)			
AQ 10	Fargro	15518	31 May 2019
27 azoxystrobin			
5504	Syngenta	12351	30 Jun 2024
Amistar	Syngenta	10443	31 Dec 2018
Astrobin 250	Euro	15336	31 Aug 2018
Astrobin 250	Euro	15354	31 Dec 2018
Aubrac	AgChem Access	13483	31 Dec 2018
Azaka	Headland	16422	30 Jun 2024
Azoshy	Sharda	18072	30 Jun 2024
Azzox	Goldengrass	14296	31 Dec 2018
Clayton Bastille	Clayton	17012	31 Dec 2018
Clayton Belfry	Clayton	12886	31 Dec 2018
Clayton Belfry	Clayton	18154	30 Jun 2024
Cleancrop Celeb	Agrii	18040	30 Jun 2024
Cleancrop Celeb	Agrii	16519	31 Dec 2018
EA Azoxystrobin	European Ag	15048	31 Dec 2018
Hi Strobin 25	Hockley	15468	31 Dec 2018
Hill-Star	Stefes	18150	30 Jun 2024
Legado	Industrias Afrasa	18087	22 Aug 2021
Life Scientific Azoxystrobin	Life Scientific	16782	31 Jan 2020
MS Sansa	Micromix	17147	31 Dec 2018
Ortiva	Syngenta	10542	30 Jun 2024
Pantha 250	Terrechem	17953	31 Dec 2018
Phloem	Capital CP	17754	30 Jun 2024
Priori	Syngenta	10543	30 Jun 2024
Puma	Ascot Pro-G	17678	30 Jun 2024
Reconcile	Becesane	15379	31 Dec 2018
RouteOne Roxybin 25	Albaugh UK	15289	31 Dec 2018
Toran	Becesane	18239	30 Jun 2024
Xylem	EuroChem	17677	30 Jun 2024
Zoxis	Arysta	15483	30 Jun 2019
28 azoxystrobin + chlorothalonil			
Amistar Opti	Syngenta	14582	31 Jan 2019
Curator	Syngenta	14955	31 Jan 2019
Mount	Euro	16878	31 Jan 2019
Olympus	Syngenta	13797	31 Jan 2019
Opti Azoxy	Harvest	18088	28 Feb 2019

Product	Approval holder	MAPP No.	Expiry Date
Ortiva Opti	Syngenta	17839	28 Feb 2019
Ortiva Opti	Syngenta	18208	30 Apr 2020
Quadris Opti	Syngenta	18207	30 Apr 2020
Quadris Opti	Syngenta	17752	28 Feb 2019

30 azoxystrobin + difenoconazole

Amistar Top	Syngenta	12761	31 Jan 2019

35 azoxystrobin + tebuconazole

Seraphin	Adama	16248	28 Feb 2022

37 Bacillus firmus I - 1582

Flocter	Bayer CropScience	16480	31 Mar 2026

38 Bacillus subtilis

Solani	Russell	16585	31 Oct 2020

39 Bacillus thuringiensis

Bruco	Progreen	17919	23 Mar 2020

40 Bacillus thuringiensis aizawai GC-91

Agree 50 WG	Mitsui	17502	31 Oct 2021

42 Bacillus thuringiensis israelensis, strain AM65-52

Gnatrol SC	Interfarm	17802	24 Oct 2020
Gnatrol SC	Resource Chemicals	17998	24 Oct 2020

43 Beauveria bassiana

Naturalis-L	Belchim	14655	31 Mar 2020

45 benalaxyl + mancozeb

Galben M	Headland	17247	26 Jul 2020
Intro Plus	Headland	17253	26 Jul 2020
Tairel	Headland	17254	26 Jul 2020

47 bentazone

Basagran	BASF	00188	31 Dec 2019
Benta 480 SL	Sharda	14940	31 Dec 2018
Benta 480 SL	Nufarm UK	17355	31 Dec 2019
Bentazone 480	Goldengrass	14423	31 Dec 2019
Bently	Chem-Wise	14210	31 Dec 2019
Euro Benta 480	Euro	14707	31 Dec 2019
Hockley Bentazone 48	Hockley	15543	31 Dec 2019
I T Bentazone 480	I T Agro	07458	31 Dec 2019
IT Bentazone	I T Agro	13132	31 Dec 2019
Mac-Bentazone 480 SL	AgChem Access	13598	31 Dec 2019
Master Zone	Generica	16441	31 Dec 2019
RouteOne Benta 48	Albaugh UK	15266	31 Dec 2019
RouteOne Bentazone 48	Albaugh UK	15269	31 Dec 2019
Troy 480	UPL Europe	12341	30 Jun 2018
UPL B Zone	UPL Europe	14808	31 Dec 2019

51 bentazone + pendimethalin

Impuls	BASF	13372	31 Dec 2019

53 benthiavalicarb-isopropyl + mancozeb

En-Garde	Certis	14901	31 Jul 2020

54 benzoic acid

MENNO Florades	Brinkman	15091	31 Jul 2019

SECTION 3

Product	Approval holder	MAPP No.	Expiry Date
56 benzovindiflupyr + prothioconazole			
Elate	Albaugh UK	18091	13 Dec 2020
58 benzyladenine + gibberellin			
Floralife Bulb 100	Oasis	17995	28 Feb 2023
62 bifenazate			
Floramite 240 SC	Arysta	17958	31 Jan 2020
Inter Bifenazate 240 SC	Iticon	14543	31 Jan 2020
63 bifenox			
Cleancrop Diode	Clayton	14620	31 Dec 2021
Wolf	RAAT	17378	31 Dec 2021
68 bixafen			
Bixafen EC125	Bayer CropScience	15951	31 Mar 2026
71 bixafen + prothioconazole			
Saltri	Euro	16999	31 Jan 2021
75 boscalid			
Coli	Gemini	17616	31 Jan 2021
77 boscalid + epoxiconazole			
Whistle	BASF	16093	31 Jan 2021
80 boscalid + pyraclostrobin			
Daisy	RAAT	17492	31 Jul 2020
Insignis	Euro	17116	31 Jul 2020
82 bromadiolone			
Romax B Rat & Mouse Killer	Zapi	UK14-0814	31 Aug 2020
Romax Bromablock	Zapi	UK13-0777	31 Aug 2020
Romax Bromadiolone Whole Wheat	PelGar	UK12-0640	31 Aug 2020
83 bromoxynil			
Akocynil 225 EC	Aako	15688	31 Jan 2020
Alpha Bromolin 225 EC	Adama	14864	31 Jan 2020
Flagon 400 EC	Adama	14921	31 Jan 2020
Xinca	Nufarm UK	16833	31 Jan 2021
85 bromoxynil + diflufenican			
Nessie	Nufarm UK	17815	31 Jan 2021
92 bromoxynil + prosulfuron			
Jester	Syngenta	13500	31 Dec 2019
97 buprofezin			
Applaud 25 SC	Certis	17196	30 Jun 2023
Applaud 25 WP	Certis	17197	30 Jun 2023
98 Candida oleophila Strain 0			
Nexy 1	BioNext	16961	31 Mar 2026
99 captan			
Akotan 80 WG	Aako	16307	31 Jan 2021
Malvin WG	Arysta	16308	31 Jan 2021
Orthocide WG	Arysta	16309	31 Jan 2021
Ratan 80 WG	RAAT	17252	31 Jan 2021

Product	Approval holder	MAPP No.	Expiry Date
109 carbetamide			
Kartouch 60 WG	Aako	16897	30 Nov 2023
112 carboxin + thiram			
Anchor	Arysta	08684	31 Dec 2021
Anchor	Certis	16488	31 Dec 2021
113 carfentrazone-ethyl			
Aurora	Belchim	11613	31 Jan 2020
Aurora 40 WG	Belchim	11614	31 Jan 2019
Carzone 60 ME	Euro	16469	31 Jan 2020
Harrier	Belchim	14164	31 Jan 2019
Platform	Headland	17809	31 Jan 2020
Platform	Belchim	11615	31 Jan 2020
Shark	Belchim	12762	31 Jan 2019
Spotlight Plus	Belchim	12436	31 Jan 2019
114 carfentrazone-ethyl + flupyrsulfuron-methyl			
Lexus Class	DuPont	10809	13 Dec 2018
115 carfentrazone-ethyl + mecoprop-P			
Platform S	Belchim	14380	31 Jan 2019
116 carfentrazone-ethyl + metsulfuron-methyl			
Ally Express	DuPont	08640	31 Dec 2018
Asset Express	Mitsui	14378	31 Dec 2018
119 chloridazon			
Better Flowable	Sipcam	15740	30 Jun 2021
Parador	UPL Europe	16892	15 Dec 2018
Takron	BASF	16769	15 Dec 2018
124 chlormequat			
Agrovista 3 See 750	Agrovista	15975	03 Dec 2018
Belcocel	Taminco	17773	31 May 2022
Silk 750 SL	Terrechem	17496	31 May 2020
125 chlormequat + 2-chloroethylphosphonic acid			
Socom	SFP Europe	17155	31 Jan 2021
133 chlorothalonil			
Alternil 500	Arysta	16813	30 Apr 2020
Alternil Excel 720	Arysta	17434	30 Apr 2020
Asterix	Barclay	17383	30 Apr 2020
Balear 720 SC	Arysta	15545	30 Apr 2020
Banko 720	Arysta	17728	30 Apr 2020
Barclay Chloroflash	Barclay	16651	30 Apr 2020
Chlorthalis	Arysta	16513	30 Apr 2020
Claw 500	Gemini	17332	30 Apr 2020
Cleancrop Wanderer 2	Agrii	16302	30 Apr 2020
CTL 500	Goldengrass	14812	30 Apr 2020
Daconil	Syngenta	17778	30 Apr 2020
Damocles	AgChem Access	15799	30 Jun 2019
Doolin	Barclay	17676	30 Apr 2020
Duster	JT Agro	17517	30 Apr 2020
FSA Chlorothalonil	Synergy	17182	30 Apr 2020
GF Chlorothalonil	Genfarm	17011	30 Apr 2020
IT Chlorothalonil	I T Agro	15845	30 Apr 2020
Life Scientific Chlorothalonil 500	Life Scientific	15813	30 Apr 2020
LS Chlorothalonil XL	Agrovista	15443	30 Apr 2020

SECTION 3

Product	Approval holder	MAPP No.	Expiry Date
MS Catelyn	Micromix	17148	30 Apr 2020
Optio 500	Novastar	17216	30 Apr 2020
Renew Chlorothalonil 500SC	Renew	16706	30 Apr 2020
Sinconil	Agrii	16268	30 Apr 2020
Supreme	Certis	16505	30 Apr 2020
Thalonil 500	Euro	14926	30 Apr 2020
Tymoon 500	Novastar	17336	30 Apr 2020
X-SEPT	Albaugh UK	15431	31 Oct 2018
Zambesi	Terrechem	17840	30 Apr 2020

135 chlorothalonil + cyproconazole

Baritone	AgChem Access	15910	31 May 2021
Bravo Xtra	Syngenta	11824	31 Dec 2021
Citadelle	Syngenta	17886	31 Dec 2021
Cleancrop Cyprothal	United Agri	09580	31 Dec 2021

141 chlorothalonil + fluxapyroxad

Divexo	BASF	17937	30 Apr 2020

145 chlorothalonil + penthiopyrad

Aylora	DuPont	16187	31 Aug 2018
Intellis Plus	DuPont	16419	30 Jun 2018
Treoris	DuPont	16126	31 Aug 2018

146 chlorothalonil + picoxystrobin

Credo	DuPont	14577	30 Nov 2018
Zimbrail	DuPont	14735	30 Nov 2018

148 chlorothalonil + propiconazole

Barclay Avoca Premium	Barclay	15619	30 Apr 2020
Oxana	Arysta	15177	30 Apr 2020
SIP 313	Sipcam	14755	30 Apr 2020

151 chlorothalonil + tebuconazole

Confucius	Rotam	17257	30 Apr 2020
Nectar	Nufarm UK	16872	30 Apr 2020

152 chlorothalonil + tetraconazole

Eminent Star	Isagro	10447	31 Jul 2018

155 chlorotoluron + diflufenican + pendimethalin

Tribal	Adama	17075	31 Jan 2020

157 chlorpropham

Aceto Sprout Nip	Aceto	14156	31 Jul 2019
Aliacine 400 EC	Arysta	16283	31 Jan 2020
BL 500	UPL Europe	14387	31 Jul 2019
CIPC Fog 300	AgChem Access	15904	31 Jul 2019
CIPC Gold	Dormfresh	15674	31 Jul 2019
Cleancrop Amigo	United Agri	15292	31 Jan 2020
Gro-Stop Electro	Certis	17445	26 Jul 2020
Gro-Stop HN	Certis	14146	31 Jul 2019
Gro-Stop Innovator	Certis	14147	31 Jul 2019
MSS CIPC 50 M	UPL Europe	14388	31 Jul 2019
MSS Sprout Nip	UPL Europe	15676	31 Jul 2019
Neo-Stop 120 RTU	Agriphar	17420	31 Jan 2020
Neo-Stop 300 HN	Arysta	15261	31 Jul 2019
Neo-Stop Starter	Arysta	15651	31 Jan 2020
Prolonger	UPL Europe	17453	31 Jul 2019
Tuberprop Easy	UPL Europe	15701	31 Jan 2020

Product	Approval holder	MAPP No.	Expiry Date
160 chlorpyrifos			
Akofos 480 EC	Aako	14211	31 Aug 2018
Akofos 480 EC	Aako	17986	22 Jan 2021
Ballad	Headland	17999	31 Jul 2020
Ballad	Headland	11659	31 Aug 2018
Chlobber	AgChem Access	13723	31 Jul 2018
Chlorpyrat 480 EC	RAAT	17333	31 Dec 2021
Clayton Pontoon	Clayton	13219	31 Jul 2018
Cyren	Headland	11028	31 Jul 2018
Dursban WG	Dow	09153	31 Jul 2018
Equity	Dow	12465	31 Jul 2018
Pyrifos 480	Becesane	15991	31 Jul 2018
Pyrinex 48 EC	Adama	13534	31 Jul 2018
Pyrinex 48 EC	Adama	17935	22 Jan 2021
165 clethodim			
Balistik	Interfarm	18129	09 Nov 2020
Centurion Max	Interfarm	17911	19 Nov 2020
Select Prime	Arysta	16304	09 Nov 2020
166 clodinafop-propargyl			
Life Scientific Clodinafop 240	Life Scientific	17181	31 Jan 2020
Viscount	Syngenta	17495	31 Oct 2020
Wildcat	Matrix	17317	31 Oct 2020
Wildcat	UKChem	17783	31 Oct 2020
169 clodinafop-propargyl + pinoxaden			
Traxos Pro	Syngenta	16713	30 Jun 2019
170 clodinafop-propargyl + prosulfocarb			
Auxiliary	BASF	14576	30 Apr 2021
172 clofentezine			
Acaristop 500 SC	Aako	17232	30 Jun 2021
Ariane	RAAT	17296	30 Jun 2021
173 clomazone			
Centium 360 CS	Belchim	16237	12 May 2018
Cirrus CS	FMC	16579	12 May 2018
Cleancrop Chicane	Belchim	17015	12 May 2018
Cleancrop Covert	Belchim	17014	12 May 2018
Clozone	Euro	16826	12 May 2018
Czar	Terrechem	17423	30 Apr 2021
Gambit	Harvest	17747	12 May 2019
Gamit 36 CS	Belchim	16563	12 May 2018
Hobby	Ascot Pro-G	17667	30 Apr 2021
Mazone 360	Euro	16827	12 May 2019
Mazone 360	RealChemie	18168	12 May 2019
Noble	Rovogate	17162	30 Apr 2021
Regal	UKChem	17782	30 Apr 2021
Regal	Matrix	16617	30 Apr 2021
Regal Gold	Matrix	17065	12 May 2018
Retribute	Agform	16739	12 May 2018
Standon Soulmate	Standon	16879	30 Apr 2021
Titan	Goldengrass	16471	30 Apr 2021
Zulkon	Unisem	17610	30 Apr 2021
175 clomazone + linuron			
Lingo	Belchim	16564	03 Jun 2018
Linzone	Belchim	16578	03 Jun 2018

SECTION 3

Product	Approval holder	MAPP No.	Expiry Date
176 clomazone + metazachlor			
Circuit Synctec	Headland	17118	02 Aug 2019
Nimbus CS	BASF	16573	30 Apr 2021
177 clomazone + metazachlor + napropamide			
Colzor SyncTec	Syngenta	17113	28 Jul 2019
Colzor SyncTec	Syngenta	18092	30 Apr 2021
Tribeca Sync Tec	Headland	17312	28 Jul 2019
179 clomazone + napropamide			
Altiplano DAMtec	FMC	16957	31 Jul 2019
181 clopyralid			
Bariloche	Proplan-Plant	17577	31 Oct 2020
Cliophar 400	Arysta	15008	31 Oct 2020
Cliophar 600 SL	Arysta	16735	31 Oct 2020
GF-2895	Dow	16821	31 Oct 2020
Glopyr 400	Globachem	15009	31 Oct 2020
Life Scientific Clopyralid	Life Scientific	16259	03 Oct 2018
Lontrel 72SG	Dow	15235	31 Oct 2020
Vivendi 200	UPL Europe	15017	31 Mar 2019
182 clopyralid + 2,4-D + MCPA			
Esteem	Vitax	15181	31 Dec 2018
Redeem	Headland Amenity	17096	31 Dec 2018
185 clopyralid + florasulam + fluroxypyr			
Mogul	Aremie	17289	09 Sep 2099
186 clopyralid + fluroxypyr + MCPA			
Interfix	Iticon	15762	30 Apr 2020
188 clopyralid + picloram			
Curlew	Goldengrass	16791	31 Oct 2020
Piccant	Goldengrass	16794	31 Oct 2020
Pyralid Extra	Euro	16606	31 Oct 2020
190 clopyralid + triclopyr			
Blaster Pro	Headland Amenity	15752	31 Oct 2020
Blaster Pro	Dow	18074	31 Oct 2020
Headland Flail 2	Headland	16007	31 Oct 2020
Tor	Dow	17777	31 Oct 2020
192 clothianidin			
NipsIT Inside	Interfarm	14744	31 Jul 2020
202 coumatetralyl			
Romax Rat CP	Bayer CropScience	UK16-1003	31 Aug 2020
204 cyazofamid			
Rithfir	DuPont	14816	31 Jan 2020
RouteOne Roazafod	Albaugh UK	15271	31 Jan 2020
208 Cydia pomonella GV			
Cyd-X	Certis	17019	31 Oct 2021
Cyd-X Duo	Certis	16779	31 Oct 2021
Madex Top	Andermatt	18227	31 Oct 2021
209 cyflufenamid			
Cyflu EW	Euro	16472	30 Sep 2022

Product	Approval holder	MAPP No.	Expiry Date
Diego	Star	16980	30 Sep 2022
NF-149 SC	Certis	15835	30 Sep 2022

210 cymoxanil
Cymbal Flow	Belchim	17843	28 Feb 2022
Cymostraight 45	Belchim	17437	28 Feb 2022
Drum Flow	Belchim	17906	28 Feb 2022

211 cymoxanil + famoxadone
Tanos	DuPont	17160	31 Dec 2019

212 cymoxanil + fluazinam
Kunshi	Syngenta	15631	31 Mar 2019
Shirlan Forte	Syngenta	15542	31 Aug 2021

214 cymoxanil + mancozeb
Globe	Sipcam	17175	17 Sep 2019
Manconil 725	Euro	15997	30 Apr 2018
Master Cyman	Generica	16686	31 Jul 2020
Nautile WP	UPL Europe	16468	31 Jul 2020
Rhapsody	DuPont	11958	31 Jul 2020
Solace Max	Nufarm UK	16697	31 Jul 2020
Zetanil	Sipcam	17285	17 Sep 2019

216 cymoxanil + propamocarb
Axidor	Arysta	16830	31 Jan 2021
Proxanil	Arysta	16664	31 Jan 2021

217 cymoxanil + zoxamide
Reboot	Gowan	16909	31 Jul 2020
Reboot	Gowan	18202	31 Jul 2020

218 cypermethrin
Afrisect 10	Arysta	17069	17 Mar 2019
Afrisect 500 EC	Arysta	17137	16 Mar 2019
Cyperkill 10	Arysta	17068	17 Mar 2019
Cythrin Max	Arysta	17138	16 Mar 2019
Langis 300 ES	Arysta	18178	30 Apr 2020
Permasect 500 EC	Arysta	17132	16 Mar 2019
Permasect C	Arysta	16994	17 Mar 2019
Signal 300 ES	Certis	16481	30 Apr 2020
Signal 300 ES	Arysta	15949	30 Apr 2020
Supasect	Arysta	17070	17 Mar 2019
Supasect 500 EC	Arysta	17131	16 Mar 2019
Talisma EC	Arysta	16541	30 Apr 2020
Talisma UL	Arysta	16542	30 Apr 2020
Toppel 100	UPL Europe	17071	17 Mar 2019

222 cyproconazole + penthiopyrad
Cielex	DuPont	17758	30 Nov 2023

223 cyproconazole + picoxystrobin
Furlong	DuPont	17184	30 Nov 2018

227 cyprodinil
Unix	Syngenta	14846	31 Oct 2020

228 cyprodinil + fludioxonil
Button	Euro	16842	31 Oct 2020
Chrome	Pan Agriculture	18214	31 Oct 2020
Reversal	RAAT	17359	31 Oct 2020

SECTION 3

Product	Approval holder	MAPP No.	Expiry Date
230 cyprodinil + picoxystrobin			
Acanto Prima	DuPont	14971	30 Nov 2018
231 2,4-D			
2,4-D Amine 500	Nufarm UK	14360	31 Dec 2018
Damine	Agriphar	13366	31 Dec 2018
Depitox 500	Nufarm UK	17597	09 Sep 2099
Growell 2,4-D Amine	GroWell	13146	31 Dec 2018
Herboxone 60	Headland	14080	09 Sep 2099
Maton	Headland	13234	09 Sep 2099
Zip	UPL Europe	14806	30 Jun 2018
Zip	UPL Europe	16968	31 Dec 2018
233 2,4-D + dicamba			
Compo Floranid + Herbicide	Compo	14503	31 Dec 2018
Landscaper Pro Weed Control + Fertilizer	Everris Ltd	16609	30 Jun 2018
Landscaper Pro Weed Control + Fertilizer	Everris Ltd	17239	09 Sep 2099
236 2,4-D + dicamba + MCPA + mecoprop-P			
Dicophar	Arysta	16460	30 Apr 2020
237 2,4-D + dicamba + triclopyr			
Broadshot	Arysta	16655	15 Sep 2018
Cleancrop Broadshot	Arysta	16756	15 Sep 2018
Kaskara	Arysta	16757	15 Sep 2018
240 2,4-D + glyphosate			
Rasto	Nufarm UK	16795	09 Sep 2099
241 2,4-D + MCPA			
Agroxone Combi	Nufarm UK	14907	09 Sep 2099
244 2,4-D + triclopyr			
Genoxone ZX EC	Agriphar	17016	09 Sep 2099
245 daminozide			
B-Nine SG	Arysta	14434	30 Apr 2020
246 dazomet			
Basamid	Kanesho	12895	31 Dec 2021
247 2,4-DB			
Butoxone DB	Nufarm UK	14840	30 Apr 2020
Butoxone DB	Nufarm UK	18172	30 Apr 2020
CloverMaster	Nufarm UK	18251	30 Apr 2020
DB Straight	UPL Europe	18256	30 Apr 2020
Embutone 24DB	Nufarm UK	18255	30 Apr 2020
Embutone 24DB	Nufarm UK	17596	30 Apr 2020
Headland Spruce	Headland	18259	30 Apr 2020
249 2,4-DB + MCPA			
Butoxone DB Extra	Nufarm UK	15741	30 Apr 2020
250 deltamethrin			
CMI Delta 2.5 EC	CMI	16695	30 Apr 2020
Deltason-D	DAPT	17735	30 Apr 2020
GAT Decline 2.5 EC	Headland	17703	30 Apr 2020
GAT Decline 2.5 EC	GAT Micro	15769	30 Apr 2019

Product	Approval holder	MAPP No.	Expiry Date
Grain-Tect ULV	Barrettine	18076	30 Apr 2020
Grain-Tect ULV	Limagrain	15253	30 Apr 2020

254 desmedipham + ethofumesate + phenmedipham

Betanal Expert	Bayer CropScience	14034	31 Jan 2020
Beta-Team	UPL Europe	15423	31 Jan 2019
Bison 400 SE	Aako	16213	31 Jan 2020
Conqueror	UPL Europe	16972	31 Jan 2020
Conqueror	AgriChem BV	16157	31 Jan 2019
D.E.P. 205	Goldengrass	16431	09 Sep 2099
D.E.P. 251	Goldengrass	14922	31 Jan 2020

255 desmedipham + phenmedipham

Betanal Maxxim	Bayer CropScience	14186	31 Jan 2020

259 dicamba + MCPA + mecoprop-P

Nomix Tribute Turf	Nomix Enviro	16313	30 Apr 2020
Relay Turf Elite	Headland	17192	30 Apr 2020

260 dicamba + mecoprop-P

Dimeco	UPL Europe	17190	31 Jul 2020
Headland Swift	Headland	11945	31 Jul 2020
Mircam	Nufarm UK	11707	31 Jul 2020
Optica Forte	Nufarm UK	14845	31 Jul 2020

267 dichlorprop-P + MCPA + mecoprop-P

Duplosan Super	Nufarm UK	18231	30 Apr 2020
Hymec Triple	Agrichem	15753	30 Apr 2020

270 difenacoum

Romax D Block	PelGar	UK11-0077	31 Aug 2020
Romax Difenacoum Whole Wheat Bait	PelGar	UK11-0109	31 Aug 2020
Romax DP	Rentokil	UK13-0804	31 Aug 2020
Romax DP Pasta Sachets	Zapi	UK14-0843	30 Jun 2018
Romax Mouse DP	PelGar	UK11-0089	31 Aug 2020
Romaz D Rat & Mouse Killer	Rentokil	UK14-0808	31 Aug 2020

272 difenoconazole

Bogard	Syngenta	17310	30 Jun 2021
Change	RAAT	17400	30 Jun 2021
Diconazol 250	Euro	16387	30 Jun 2021
Difenostar	Life Scientific	17598	28 Jun 2020
Septuna	Novastar	17390	30 Jun 2021
Slick	Syngenta	17331	30 Jun 2021

274 difenoconazole + fludioxonil

Instrata Elite	Syngenta	17976	22 Mar 2021

275 difenoconazole + fludioxonil + tebuconazole

Celest Trio	Syngenta	15510	30 Apr 2021

279 diflubenzuron

Diflox Flow	Hockley	15640	31 Dec 2021
Dimilin Flo	Arysta	08769	31 Dec 2021

280 diflufenican

Beeper 500 SC	Aako	15358	30 Jun 2021
Beluga	Ascot Pro-G	17806	30 Jun 2021
Diecot	Gemini	17191	30 Jun 2021
Diflufen 500	Euro	16687	30 Jun 2021

SECTION 3

Product	Approval holder	MAPP No.	Expiry Date
Diflufenican GL 500	Globachem	17084	30 Jun 2021
Dina 50	Globachem	17083	30 Jun 2021
Flash	Sharda	17943	30 Jun 2021
Flyflo	Gemini	17694	30 Jun 2021
Goshawk	Goldengrass	16514	30 Jun 2021
Inter Diflufenican	UPL Europe	17342	29 Feb 2020
Overlord	Adama	13521	30 Jun 2021
Prefect	Novastar	17713	30 Jun 2021
Sempra	AgriChem BV	16824	31 Mar 2019
Solo D500	Sipcam	16098	30 Jun 2021
Terrier	Terrechem	17761	30 Jun 2021
Tornado	UKChem	17781	30 Jun 2021
Tornado	Matrix	17425	31 Oct 2020
Twister	Unique Marketing	16374	30 Jun 2021

281 diflufenican + florasulam

Product	Approval holder	MAPP No.	Expiry Date
Bow	Headland	16845	30 Jun 2021

282 diflufenican + flufenacet

Product	Approval holder	MAPP No.	Expiry Date
Adept	Agform	17628	30 Apr 2020
Amaranth	PSI	16949	30 Apr 2020
Ambush	Agform	18136	30 Apr 2020
Cleancrop Marauder	United Agri	15933	30 Apr 2020
Deliverer	Euro	16692	30 Apr 2020
Dephend	Headland	18144	30 Apr 2020
Diflufenastar	Life Scientific	17611	27 Jun 2020
Diflufenastar 100	Life Scientific	17475	30 Apr 2020
Duke	Chem-Wise	16738	30 Apr 2020
Fenican	Euro	16790	30 Apr 2020
Firebird	Bayer CropScience	14826	30 Apr 2020
Flufenican	Euro	15386	30 Apr 2020
Life Scientific Flufenacet + DFF	Life Scientific	16805	30 Apr 2019
Navigate	Headland	17506	09 Mar 2020
Pincer Plus	Agform	17306	30 Jun 2018
Saviour	Chem-Wise	16702	30 Apr 2020

283 diflufenican + flufenacet + flurtamone

Product	Approval holder	MAPP No.	Expiry Date
Flufenamone	Euro	17260	30 Apr 2020
Voyage	Chem-Wise	17238	30 Apr 2020

284 diflufenican + flupyrsulfuron-methyl

Product	Approval holder	MAPP No.	Expiry Date
Absolute	DuPont	12558	13 Dec 2018
Excalibur	DuPont	16477	13 Dec 2018

286 diflufenican + glyphosate

Product	Approval holder	MAPP No.	Expiry Date
Pistol	Bayer CropScience	12173	28 Feb 2018
Pistol Rail	Bayer CropScience	17450	14 Feb 2020
Proshield	Everris Ltd	16118	31 Mar 2018

287 diflufenican + iodosulfuron-methyl-sodium + mesosulfuron-methyl

Product	Approval holder	MAPP No.	Expiry Date
Kalenkoa	Bayer CropScience	17733	30 Apr 2020

290 diflufenican + metribuzin

Product	Approval holder	MAPP No.	Expiry Date
Tavas	Nufarm UK	18213	31 Jan 2021

295 dimethenamid-p + metazachlor

Product	Approval holder	MAPP No.	Expiry Date
Caribou	RealChemie	18195	30 Apr 2020
Muntjac	BASF	16893	30 Apr 2020

Product	Approval holder	MAPP No.	Expiry Date
296 dimethenamid-p + metazachlor + quinmerac			
Luna	Goldengrass	17053	30 Apr 2020
Medirac	Euro	17214	30 Apr 2020
300 dimethomorph			
Morph	Adama	15121	31 Jan 2021
Murphy 500 SC	Aako	15203	31 Jan 2021
Navio	Headland	17827	31 Jan 2021
Rigel WP	Servem	17347	31 Mar 2020
302 dimethomorph + mancozeb			
Saracen	BASF	15250	31 Jul 2020
303 dimethomorph + pyraclostrobin			
Optimo Tech	BASF	16455	31 Jul 2020
304 dimethomorph + zoxamide			
Presidium	Gowan	17372	15 Dec 2019
308 diquat			
A1412A2	Syngenta	13440	31 Dec 2019
Balista	Novastar	15186	31 Dec 2019
Barclay D-Quat	Barclay	14833	31 Dec 2019
BD-200	Barclay	14918	31 Dec 2019
Belquat	Agroquimicos	18238	31 Dec 2019
Brogue	Mitsui	15638	28 Feb 2018
Brogue	Certis	16421	31 Dec 2019
CleanCrop Flail	United Agri	14870	31 Dec 2019
CropStar Diquat	CropStar	17679	31 Dec 2019
Diquanet	Belcrop	16868	31 Dec 2019
Diquash	Agrichem	14849	31 Dec 2019
Di-Quattro	Arysta	15712	31 Dec 2019
Hockley Diquat 20	Hockley	15585	31 Dec 2019
I.T. Diquat	I T Agro	13557	30 Jun 2018
I.T. Diquat	UPL Europe	16970	31 Dec 2019
Inter Diquat	UPL Europe	16986	31 Dec 2019
Inter Diquat	I T Agro	14196	30 Jun 2018
Knoxdoon	AgChem Access	13987	31 Dec 2019
Life Scientific Diquat 200	Life Scientific	15811	31 Dec 2019
Mission 200 SL	UPL Europe	16475	30 Jun 2018
Quad S	Q-Chem	13578	31 Dec 2019
Quat 200	Euro	15042	31 Dec 2019
Shrike 200	Goldengrass	16197	31 Dec 2019
Standon Googly	Standon	12995	31 Dec 2019
Synchem Diquat 200	Synchem	16817	31 Dec 2019
Tuber Mission	UPL Europe	16965	31 Dec 2019
Tuber Mission	UPL Europe	14542	30 Jun 2018
UPL Diquat	UPL Europe	14944	30 Jun 2018
309 dithianon			
Alcoban	Globachem	18151	30 Nov 2023
Rathianon 70 WG	RAAT	17471	30 Nov 2023
317 dodine			
Radspor 400	Agriphar	16001	31 Dec 2021
Syllit 400 SC	Arysta	13363	31 Dec 2021
318 epoxiconazole			
Amber	Adama	16285	31 Oct 2021
Aska	Terrechem	17461	31 Oct 2021

SECTION 3

Product	Approval holder	MAPP No.	Expiry Date
Barret 125 SL	Agrifarm	16709	31 Oct 2021
Bassoon	BASF	14402	31 Oct 2021
Bowman	Adama	16286	31 Oct 2021
Corral	Headland	16942	31 Oct 2021
Epoxi 125 SC	Goldengrass	16356	31 Oct 2021
Epzole 125	Unique Marketing	16412	31 Dec 2018
Ignite	BASF	15205	31 Oct 2021
Intercord	Iticon	16155	31 Oct 2021
Life Scientific Epoxiconazole	Life Scientific	16907	03 Mar 2019
Ollyx	Novastar	16380	31 Oct 2021
Opus	BASF	12057	31 Oct 2021
Paladin	Novastar	16397	31 Oct 2021
Propov	Syngenta	17926	31 Oct 2021
Rapier	Ascot Pro-G	17857	31 Oct 2021
Regent	Matrix	17087	31 Oct 2020
Regent	UKChem	17784	31 Oct 2021
Spike	Q-Chem	17910	31 Oct 2021
Voodoo	Sipcam	16072	31 Oct 2021
Warlock	UKChem	17775	31 Oct 2021
Warlock	Matrix	17026	31 Oct 2020

321 epoxiconazole + fenpropimorph + metrafenone

Product	Approval holder	MAPP No.	Expiry Date
Polaca	Euro	16436	31 Dec 2021
Stiletto	BASF	14729	31 Dec 2021

322 epoxiconazole + fenpropimorph + pyraclostrobin

Product	Approval holder	MAPP No.	Expiry Date
Diamant	BASF	14149	31 Dec 2021
Sapphire	Euro	16648	31 Dec 2021

323 epoxiconazole + fluxapyroxad

Product	Approval holder	MAPP No.	Expiry Date
Adexar	PSI	17441	31 Oct 2021
Apex Pro	Chem-Wise	17497	31 Oct 2021
Apex Pro	Chem-Wise	17632	31 Oct 2021
Cougar	UKChem	17785	31 Oct 2021
Cougar	Matrix	17442	31 Oct 2021
Morex	BASF	17307	31 Oct 2021
Standon Coolie	Standon	17382	31 Oct 2021
ZORO	Terrechem	17462	31 Oct 2021
Zoro	Terrechem	17621	31 Oct 2021

324 epoxiconazole + fluxapyroxad + pyraclostrobin

Product	Approval holder	MAPP No.	Expiry Date
Ceriax	PSI	17527	31 Jul 2020
Chusan	Terrechem	17528	31 Jul 2020
Smaragdin	PSI	17532	31 Jul 2020

329 epoxiconazole + metconazole

Product	Approval holder	MAPP No.	Expiry Date
Ettu	Euro	16435	31 Oct 2020
Icarus	BASF	14471	31 Oct 2020

332 epoxiconazole + pyraclostrobin

Product	Approval holder	MAPP No.	Expiry Date
Ambassador	Euro	16410	31 Jul 2020
Ibex	BASF	16457	31 Jul 2020

333 esfenvalerate

Product	Approval holder	MAPP No.	Expiry Date
Barclay Alphasect	Barclay	16053	31 Aug 2019
Greencrop Cajole Ultra	Clayton	12967	31 Dec 2018
Kingpin	Belchim	18176	09 Sep 2099
Standon Hounddog	Standon	13201	31 Dec 2018

Product	Approval holder	MAPP No.	Expiry Date
335 ethanol			
Ethy-Gen II	Ripe Rite	15839	31 Dec 2021
Restrain Fuel	Restrain	14520	31 Dec 2021
336 ethephon			
Chrysal Plus	Chrysal	17847	31 Jan 2021
Ethefon 480	Goldengrass	16102	31 Jan 2021
Floralife Tulipa	Oasis	17996	31 Jan 2021
Hi-Phone 48	Hockley	15506	31 Jan 2021
Rogan	Novastar	17186	31 Jan 2021
Telsee	SFP Europe	16325	31 Jan 2021
337 ethephon + mepiquat chloride			
Gunbar	Terrechem	17460	31 Jan 2021
Mepicame	Unisem	17541	31 Jan 2021
Riggid	Gemini	17522	31 Jan 2021
Terpal	PSI	17440	31 Jan 2021
338 ethofumesate			
Alpha Ethofumesate 50SC	Adama	13055	31 Jan 2020
Barclay Keeper 500 SC	Barclay	14016	31 Jan 2020
Barclay Keeper 500 SC	Barclay	13430	31 Jan 2020
Etho 500 SC	Goldengrass	15256	31 Jan 2020
Ethofol 500 SC	UPL Europe	15179	31 Jan 2020
Ethosat 500	Adama	13050	31 Jan 2020
Kubist Flo	Nufarm UK	12987	31 Jan 2020
Master Ethos	Generica	16442	31 Jan 2020
Nortron Flo	Bayer CropScience	12986	09 Sep 2099
Oblix 500	UPL Europe	12349	30 Sep 2018
Stelga 500 SC	Novastar	14737	31 Jan 2020
339 ethofumesate + metamitron			
Goltix Plus	Aako	16538	31 Jan 2020
Goltix Super	Adama	16537	31 Jan 2020
Metafol Super	UPL Europe	16881	09 Sep 2099
340 ethofumesate + metamitron + phenmedipham			
Phemo	UPL Europe	16352	31 Jan 2019
341 ethofumesate + phenmedipham			
Betosip Combi FL	Sipcam	14403	31 Jan 2020
Fenlander 2	Adama	14030	31 Jan 2020
Gemini	EZCrop	15670	31 Jan 2020
Magic Tandem	Bayer CropScience	17358	09 Sep 2099
Powertwin	Adama	14004	31 Jan 2020
Teamforce	UPL Europe	14923	31 Jan 2019
Teamforce SE	UPL Europe	15403	31 Jan 2019
Teamforce SE	UPL Europe	16983	31 Jan 2020
Thunder	Adama	14031	31 Jan 2020
348 fatty acids			
Finalsan	Certis	13102	31 Jul 2021
Hydro Coco Houseplant Pest & Spider Mite Killer	151 Products Ltd	16898	31 Dec 2021
NEU 1170 H	Sinclair	15754	31 Jul 2021
Safers Insecticidal Soap	Woodstream	07197	30 Jun 2018
356 fenazaquin			
Matador 200 SC	Margarita	11058	31 Dec 2018

SECTION 3

Product	Approval holder	MAPP No.	Expiry Date
Matador 200 SC	Gowan	16875	31 Dec 2020
Matador 200 SC	Gowan	17870	31 Dec 2021

360 fenhexamid

Product	Approval holder	MAPP No.	Expiry Date
Agrovista Fenamid	Agrovista	13733	30 Jun 2018
RouteOne Fenhex 50	Albaugh UK	13665	09 Sep 2099
RouteOne Fenhex 50	Albaugh UK	15282	09 Sep 2099

370 fenpropimorph

Product	Approval holder	MAPP No.	Expiry Date
Cleancrop Fenpro	United Agri	09885	31 Dec 2021
Cleancrop Fenpropimorph	United Agri	09445	31 Dec 2021
Fenprop 750	Goldengrass	15388	31 Dec 2021
Marnoch Phorm	Me2	11087	31 Dec 2021
Propimorf 750	Goldengrass	14410	31 Dec 2021
Standon Fenpropimorph 750	Standon	08965	31 Dec 2021

378 ferric phosphate

Product	Approval holder	MAPP No.	Expiry Date
Aristo IP	De Sangosse	17166	22 Jun 2019
Ferramol Max	Certis	14463	30 Jun 2018
Ferramol Slug Killer	Growing Success	12274	30 Jun 2018
Ferrox	Neudorff	14356	30 Jun 2018
NEU 1181 M	Neudorff	14355	30 Jun 2033
NEU 1185	Neudorff	14736	30 Jun 2018
Slugger	Goldengrass	16771	30 Jun 2033
Sluggo	Omex	14788	31 Dec 2018
Sluxx	Certis	14462	31 Jul 2018

379 ferrous sulphate

Product	Approval holder	MAPP No.	Expiry Date
Aitken Lawn Sand Mosskiller	Aitken	17044	08 Mar 2019
Landscaper Pro Moss Control + Fertiliser	Everris Ltd	16723	28 Feb 2023
LSTF Pro	LSTF	16992	08 Mar 2019
Maxicrop No. 2 Moss Killer and Conditioner	Maxicrop	17300	02 Nov 2019
Moss-Kill Pro	Bioservices	17029	08 Mar 2019
Rigby Taylor Turf Moss Killer	Rigby Taylor	17219	28 Feb 2022
Sinclair Lawn Sand	Westland Horticulture	17562	08 Mar 2019
Sinclair Lawn Sand	Sinclair	17028	08 Mar 2019

382 flazasulfuron

Product	Approval holder	MAPP No.	Expiry Date
Chikara	Nomix Enviro	13775	31 Jul 2020
Flazasulf 25	Euro	15723	31 Jul 2020
Flazasulf 25	RAAT	17912	31 Jul 2020
Paradise	Pan Agriculture	14504	31 Jul 2020
Railtrax	PSI	15139	31 Jul 2020

384 flonicamid

Product	Approval holder	MAPP No.	Expiry Date
Primeman	RAAT	17656	28 Feb 2023

385 florasulam

Product	Approval holder	MAPP No.	Expiry Date
Barton WG	Dow	13284	30 Jun 2033
Boxer	Dow	09819	09 Sep 2099
Flora 50	Euro	15700	09 Sep 2099
Flora 50 II	Euro	17716	30 Jun 2033
Life Scientific Florasulam	Life Scientific	16581	30 Jun 2019
RouteOne Florasul 50	Albaugh UK	15285	09 Sep 2099
Solstice	Headland	17164	30 Jun 2033
Suprime	Nufarm UK	18110	09 Sep 2099
Troller	Nufarm UK	16627	14 Aug 2018

Product	Approval holder	MAPP No.	Expiry Date
386 florasulam + fluroxypyr			
Flatline	Aremie	17422	09 Sep 2099
Flurosulam XL	Euro	15236	09 Sep 2099
GF 184	Dow	10878	09 Sep 2099
Hunter	Dow	12836	09 Sep 2099
Nevada	Dow	17349	09 Sep 2099
Sickle	Dow	17923	30 Jun 2024
388 florasulam + pinoxaden			
Axial One	Syngenta	17262	30 Jun 2019
390 florasulam + tribenuron-methyl			
Bolt	Headland	17363	30 Apr 2020
Paramount Max	Headland	17271	30 Apr 2020
392 fluazifop-P-butyl			
A12791B	Syngenta	16178	31 Dec 2021
Greencrop Bantry	Greencrop	12737	31 Dec 2021
RouteOne Fluazifop +	Albaugh UK	13641	31 Dec 2021
393 fluazinam			
Boyano	Belchim	16621	31 Aug 2021
Deltic Pro	Headland	17855	31 Aug 2021
Float	Headland	16718	31 Aug 2021
FOLY 500 SC	Aako	16486	31 Aug 2021
Frowncide	ISK Biosciences	16619	31 Aug 2021
Ibiza 500	Belchim	16620	31 Aug 2021
Legacy	Syngenta	16622	31 Aug 2021
Ohayo	Syngenta	16588	31 Aug 2021
Shirlan Programme	Syngenta	16623	31 Aug 2021
Smash	RAAT	17488	31 Aug 2021
Winby	ISK Biosciences	16618	31 Aug 2021
Zinam 500	Euro	16831	31 Aug 2021
Zinam II	Euro	16886	31 Aug 2021
395 fludioxonil			
Maxim 480FS	Syngenta	16725	11 Sep 2018
399 fludioxonil + sedaxane			
A20078F	Syngenta	17945	19 Jan 2021
402 flufenacet			
Firecloud	Certis	18175	30 Apr 2020
Ranger	Headland	17123	07 Jul 2019
Realm	Headland	17198	07 Jul 2019
Starfire	Certis	18179	30 Apr 2020
Steeple	Albaugh UK	18082	14 Dec 2019
404 flufenacet + isoxaflutole			
Cadou Star	Bayer CropScience	13242	31 Jan 2020
RouteOne Oxanet 481	Albaugh UK	15309	31 Jan 2020
406 flufenacet + pendimethalin			
Cleancrop Hector	Agrii	15932	31 Jan 2020
Frozen	Harvest	17691	31 Jan 2020
Ice	BASF	13930	31 Jan 2020
Kruos	Chem-Wise	15880	31 Jan 2020
Latice	Euro	16693	31 Jan 2020
Rock	Goldengrass	14928	31 Jan 2020

SECTION 3

Product	Approval holder	MAPP No.	Expiry Date
Standon Diwana	Standon	17221	31 Jan 2020
Standon Sparkle	Standon	17114	31 Jan 2020

407 flufenacet + picolinafen

Equs	BASF	17736	09 Sep 2099
Kudu	BASF	17804	09 Sep 2099

408 flumioxazin

Guillotine	Interfarm	13562	31 Dec 2019

410 fluopicolide + propamocarb hydrochloride

Itofin	Euro	16843	31 Jan 2021

412 fluopyram + prothioconazole

Propulse	Bayer CropScience	15735	31 May 2018
Recital	Bayer CropScience	16086	30 Jun 2018

413 fluopyram + prothioconazole + tebuconazole

Raxil Star	Bayer CropScience	15197	31 May 2018

415 fluoxastrobin

Bayer UK 831	Bayer CropScience	12091	31 Jul 2018

416 fluoxastrobin + prothioconazole

Curuni	Euro	16860	31 Jan 2020
Exploit	Euro	16851	31 Jan 2021
Fandango	Bayer CropScience	12276	31 Dec 2019
Firefly	Bayer CropScience	13692	31 Jan 2021
Prostrob 20	Chem-Wise	15698	31 Jan 2021
Unicur	Bayer CropScience	14776	31 Jan 2020

419 flupyrsulfuron-methyl

Bullion	DuPont	14058	13 Dec 2018
Ductis SX	DuPont	15426	13 Dec 2018
Exceed SX	DuPont	15427	13 Dec 2018
Lexus SX	DuPont	12979	13 Dec 2018
Oklar SX	Nufarm UK	15037	13 Dec 2018
Oriel 50SX	Nufarm UK	14640	13 Dec 2018
Spelio SX	DuPont	15424	13 Dec 2018
Staka SX	DuPont	15428	13 Dec 2018

422 flupyrsulfuron-methyl + pyroxsulam

GF-2070	Dow	17472	13 Dec 2018
Unite	Dow	17385	13 Dec 2018

423 flupyrsulfuron-methyl + thifensulfuron-methyl

Lancer	Headland	13031	13 Dec 2018
Lexus Millenium	DuPont	09206	13 Dec 2018

426 fluroxypyr

Awac	Globachem	17501	02 Feb 2020
Casino	Certis	17592	30 Jun 2024
Clean Crop Gallifrey 200	Globachem	17503	02 Feb 2020
Flurox 180	Stockton	13959	09 May 2020
Gf-1784	Dow	16850	30 Jun 2024
Hudson 200	Barclay	16391	31 Aug 2018
Hurler	Barclay	13458	31 Aug 2018
Klever	Globachem	17504	02 Feb 2020
Taipan	Agro Trade	17063	30 Jun 2024
Tomahawk 2	Adama	17468	30 Jun 2024
Toska EC	Novastar	17709	30 Jun 2024

Product	Approval holder	MAPP No.	Expiry Date
430 fluroxypyr + metsulfuron-methyl			
Croupier OD	Certis	17630	09 Sep 2099
431 fluroxypyr + metsulfuron-methyl + thifensulfuron-methyl			
Omnera LQM	DuPont	17769	09 Sep 2099
433 fluroxypyr + triclopyr			
Pas	Dow	17772	31 Oct 2020
436 flutolanil			
NNF-136	Nichino	14302	31 Aug 2021
437 flutriafol			
Impact	Headland	12776	31 Dec 2021
438 fluxapyroxad			
BAS 700	BASF	17106	30 Nov 2020
Flux	Chem-Wise	17602	30 Jun 2025
443 foramsulfuron + iodosulfuron-methyl-sodium			
Logo	Bayer CropScience	17624	31 Jan 2020
444 fosetyl-aluminium			
Plant Trust	Everris Ltd	15779	31 Oct 2020
445 fosetyl-aluminium + propamocarb hydrochloride			
Avatar	Everris Ltd	16608	31 Oct 2020
450 garlic extract			
Eagle Green Care	ECOspray	14989	31 Mar 2021
EGC Liquid	Rigby Taylor	17852	28 Feb 2023
EGCA Granules	Rigby Taylor	17233	28 Feb 2022
Grenadier	Certis	17617	26 Nov 2018
NEMguard A PCN Granules	ECOspray	17377	12 Nov 2018
NEMguard A PCN Granules	Certis	17922	28 Feb 2022
NEMguard granules	ECOspray	15254	28 Feb 2022
Pitcher GR	ECOspray	18126	28 Feb 2023
Pitcher SC	ECOspray	18125	28 Feb 2023
453 Gliocladium catenulatum			
Prestop	Fargro	15103	30 Sep 2019
Prestop Mix	Fargro	15104	31 Jan 2020
454 glufosinate-ammonium			
Dwindle	Ascot Pro-G	17820	31 Jan 2021
Genghis	Terrechem	18011	31 Jan 2021
Harvest	Bayer CropScience	17236	31 Jan 2021
Kurtail Gold	Progreen	17292	31 Mar 2020
Kurtail Gold	Progreen	17489	31 Jan 2021
Macro	Novastar	17864	31 Jan 2021
Pearl	Agrigem	17531	31 Jan 2021
Pearl Elite	Agrigem	17828	31 Jan 2021
Vanish	Capital CP	17768	31 Jan 2021
Weedex	Novastar	17764	31 Jan 2021
455 glyphosate			
Accelerate	Headland	13390	30 Jun 2018
Acrion	Bayer CropScience	12677	30 Jun 2018
Alekto Plus TF	Belchim	17143	30 Jun 2020
Ardee	Barclay	17167	30 Jun 2018
Barbarian XL	Barclay	16911	31 Dec 2018

SECTION 3

Product	Approval holder	MAPP No.	Expiry Date
Barclay Barbarian	Barclay	12714	30 Jun 2018
Barclay Gallup 360	Barclay	14988	30 Jun 2018
Barclay Gallup Amenity	Barclay	13250	30 Jun 2018
Barclay Gallup Biograde 360	Barclay	15188	31 Dec 2018
Barclay Gallup Hi-Aktiv	Barclay	14987	31 Dec 2018
Boom Efekt	Albaugh UK	17588	30 Jun 2020
Boom efekt	Pinus	15606	31 Dec 2018
Buggy SL	Sipcam	17962	30 Jun 2020
Charger C	AgChem Access	14216	30 Jun 2018
Cleancrop Bronco	Belchim	17056	30 Jun 2020
Cleancrop Corral	United Agri	15220	31 Dec 2018
Cleancrop Corral	Agrii	17554	30 Jun 2020
Cleancrop Corral 2	Agrii	18023	30 Jun 2020
Cleancrop Hoedown	United Agri	17563	30 Jun 2018
Cleancrop Hoedown	United Agri	12913	30 Jun 2018
CleanCrop Hoedown	United Agri	15404	30 Jun 2018
Cleancrop Tungsten	United Agri	17553	30 Jun 2020
Cleancrop Tungsten	United Agri	13049	31 Dec 2018
Clinic	Nufarm UK	12678	30 Jun 2018
Clinic Ace	Nufarm UK	12980	30 Jun 2018
Clinic Ace	Nufarm UK	14040	30 Jun 2018
Clinic TF	Nufarm UK	16716	12 Nov 2019
Cosmic NG	Arysta	15646	30 Jun 2020
Credit DST	Nufarm UK	13822	30 Jun 2018
Credit DST	Nufarm UK	14066	30 Jun 2020
Ecoplug Max	Monsanto	14741	31 Dec 2018
Ecoplug Max	Monsanto	17581	31 Dec 2018
Egret	Monsanto	17510	30 Jun 2020
Etna	UPL Europe	14674	30 Jun 2018
Etna	UPL Europe	16962	30 Jun 2018
Euro Glyfo 360	Euro	14691	30 Jun 2018
Euro Glyfo 450	Euro	14936	30 Jun 2020
Excel DF Gold	Excel	17413	28 Feb 2020
Figaro	Belchim	16947	30 Jun 2018
Fozat 360 SL	Agro-Chemie	16171	30 Jun 2018
Frontsweep	Nomix Enviro	14700	30 Jun 2018
Gallup Hi- Aktiv Amenity	Barclay	12898	30 Jun 2020
Gladiator	Barclay	16641	30 Jun 2018
Glister	Sinon EU	12990	30 Jun 2018
Glister Ultra	Sinon EU	15209	30 Jun 2020
Glister Ultramax	Sinon EU	18021	30 Jun 2020
Glycel	Excel	15068	30 Jun 2018
Glydate	Nufarm UK	12679	30 Jun 2018
Glyder	Agform	15442	30 Jun 2018
Glyder 450 TF	Belchim	16520	30 Jun 2020
Glyfer	Nouvelle Tec	17895	30 Jun 2020
Glyfo Star	Q-Chem	17457	30 Jun 2020
Glyfo_TDI	Q-Chem	14743	30 Jun 2020
Glyfos	Headland	10995	30 Jun 2018
Glyfos Gold	Nomix Enviro	10570	30 Jun 2020
Glyfos Gold ECO	Headland Amenity	16489	30 Jun 2020
Glyfos Monte	Headland	15069	30 Jun 2020
Glyfos Proactive	Nomix Enviro	11976	30 Jun 2018
Glyfos Supreme	Headland	12371	30 Jun 2018
Glyfos Supreme XL	Headland	17705	30 Jun 2020
Glyfosat 36	Goldengrass	14297	30 Jun 2018
Glyfo-TDI	Q-Chem	13940	30 Jun 2018
Glymark	Nomix Enviro	13870	30 Jun 2018
Glyper	SBM	14383	30 Jun 2018

Product	Approval holder	MAPP No.	Expiry Date
Glyphogan	Adama	12668	30 Jun 2018
Glypho-Rapid 450	Barclay	17647	30 Jun 2020
Glypho-Rapid 450	Barclay	13882	31 Dec 2018
Glyphosate 360	Monsanto	12669	30 Jun 2018
Glystar 360	Albaugh UK	16179	30 Jun 2018
Glyweed	Sabero	14444	30 Jun 2018
Habitat	Barclay	15199	31 Dec 2018
Habitat	Barclay	17662	30 Jun 2020
Hauberk	Synergy	16163	30 Jun 2018
Helosate 450 TF	Helm	16326	30 Jun 2020
Hi-Fosate	Hockley	15062	30 Jun 2018
HY-GLO 360	Agrichem	15194	30 Jun 2018
Jali 450	DAPT	15930	30 Jun 2020
Kernel	Headland	10993	30 Jun 2018
KN 540	Monsanto	12009	30 Jun 2018
Kraken	Bayer CropScience	15175	30 Jun 2020
Landmaster	Albaugh UK	15759	30 Jun 2018
Manifest	Headland	11041	30 Jun 2018
Mascot Hi-Aktiv Amenity	Rigby Taylor	17696	30 Jun 2020
Master Gly 36T	Generica	15731	30 Jun 2018
Mentor	Monsanto	16508	30 Jun 2020
MON 76473	Monsanto	15348	30 Jun 2020
MON 79351	Monsanto	12860	30 Jun 2020
MON 79376	Monsanto	12664	30 Jun 2020
MON 79545	Monsanto	12663	30 Jun 2020
MON 79632	Monsanto	14841	30 Jun 2018
MON 79991	Monsanto	16300	31 Dec 2018
Monosate G	Monsanto	17449	30 Jun 2020
Monsanto Amenity Glyphosate	Monsanto	15227	30 Jun 2018
Montana	Sapec	14843	30 Jun 2018
MS Theon	Micromix	17150	30 Jun 2018
NASA	Agria SA	17237	30 Jun 2018
NASA	Agria SA	18170	11 Jul 2021
Nomix Frontclear	Nomix Enviro	15180	30 Jun 2020
Nomix G	Nomix Enviro	13872	30 Jun 2020
Nomix Nova	Nomix Enviro	13873	30 Jun 2020
Nomix Prolite	Frontier	16132	30 Jun 2020
Nomix Revenge	Nomix Enviro	13874	30 Jun 2020
Nufosate	Nufarm UK	12699	30 Jun 2018
Nufosate Ace	Nufarm UK	13794	30 Jun 2018
Nufosate Ace	Nufarm UK	14959	30 Jun 2018
Onslaught	Procam	14564	30 Jun 2018
Onslaught	Agrii	17607	30 Jun 2018
Ormond	Barclay	17712	30 Jun 2020
Ovation	Monsanto	17476	30 Jun 2020
Oxalis NG	Arysta	15612	30 Jun 2020
Pitch	Nufarm UK	15485	30 Jun 2020
Pitch	Nufarm UK	15693	30 Jun 2020
Pitch	Nufarm UK	15485	30 Jun 2020
Pontil 360	Barclay	14322	30 Jun 2018
Preline	Linemark	12756	30 Jun 2020
Pultare	Maxwell	18242	30 Jun 2020
Pure Glyphosate 360	Pure Amenity	14834	30 Jun 2018
Reaper	AgChem Access	14924	30 Jun 2018
Rodeo	Monsanto	16242	30 Jun 2020
Romany	Greencrop	12681	30 Jun 2018
Rosate 36	Albaugh UK	14459	30 Jun 2018
Rosate 36 SL	Albaugh UK	15263	30 Jun 2018
Roundup	Monsanto	12645	30 Jun 2018

SECTION 3

Product	Approval holder	MAPP No.	Expiry Date
Roundup Ace	Monsanto	12772	30 Jun 2020
Roundup Advance	Monsanto	15540	30 Jun 2020
Roundup Amenity	Monsanto	12672	30 Jun 2018
Roundup Assure	Monsanto	15537	30 Jun 2020
Roundup Biactive	Monsanto	10320	30 Jun 2018
Roundup Biactive 3G	Monsanto	13409	30 Jun 2020
Roundup Biactive Dry	Monsanto	12646	30 Jun 2020
Roundup Bio	Monsanto	15538	30 Jun 2020
Roundup Express	Monsanto	12526	30 Jun 2020
Roundup Gold	Monsanto	10975	30 Jun 2020
Roundup Klik	Monsanto	12866	30 Jun 2020
Roundup Max	Monsanto	12952	30 Jun 2020
Roundup Metro	Monsanto	14842	30 Jun 2018
Roundup POWERMAX	Monsanto	16373	31 Dec 2018
Roundup Pro Biactive	Monsanto	10330	30 Jun 2020
Roundup Proactive	Monsanto	17380	30 Jun 2020
Roundup ProBiactive 450	Monsanto	12778	30 Jun 2020
Roundup ProBio	Monsanto	15539	31 Dec 2018
Roundup Pro-Green	Monsanto	11907	30 Jun 2020
Roundup Provide	Monsanto	12953	30 Jun 2020
Roundup Rail	Monsanto	12671	30 Jun 2018
Roundup Ultimate	Monsanto	12774	30 Jun 2020
Roundup Ultimate	Monsanto	13081	30 Jun 2018
Roundup Vista	Monsanto	17156	30 Jun 2018
RouteOne Rosate 36	Albaugh UK	15280	30 Jun 2018
RouteOne Rosate 360	Albaugh UK	15283	30 Jun 2018
RouteOne Rosate 360	Albaugh UK	15284	30 Jun 2018
RVG Glyphosate 360	Rovogate	17072	30 Jun 2018
Scorpion	Monsanto	17516	30 Jun 2020
Sherlock	Synergy	16185	30 Jun 2018
Shyfo	Sharda	15040	30 Jun 2018
Silvio	Rotam	14626	30 Jun 2018
Snapper	Nufarm UK	15489	30 Jun 2020
Spear & Jackson Professional Super Strong Concentrated Weed Killer	Assured Products	17305	30 Jun 2018
Stacato	Sipcam	12675	30 Jun 2018
Stirrup	Nomix Enviro	13875	30 Jun 2018
Surrender	Agrovista	16169	30 Jun 2018
Symbol	UPL Europe	14769	30 Jun 2018
Tamba Bio	Maxwell	17653	30 Jun 2020
Tambora	Maxwell	17591	30 Jun 2018
Tambora	Generica	16861	30 Jun 2018
Tangent	Headland Amenity	11872	30 Jun 2020
Task 360	Albaugh UK	14570	30 Jun 2018
Toro 360	Sipcam	15652	30 Jun 2018
Total 360 WDC	Dynamite	16267	30 Jun 2018
Typhoon 360	Adama	14817	30 Jun 2020
Ultramax	PSI	15931	30 Jun 2020
Vesuvius	Generica	16232	30 Jun 2018
Vival	Belchim	17048	30 Jun 2018
Vival	Belchim	14550	30 Jun 2018
Vival	Belchim	17224	30 Jun 2018
Vival	Headland	17048	30 Jun 2018
Wither 36	Albaugh UK	15768	30 Jun 2018
Wopro Glyphosate	B.V. Industrie	15462	30 Jun 2018

Product	Approval holder	MAPP No.	Expiry Date
456 glyphosate + pyraflufen-ethyl			
Hammer	Everris Ltd	15060	31 Mar 2018
Thunderbolt	Nichino	14102	31 Dec 2018
457 glyphosate + sulfosulfuron			
Nomix Blade	Nomix Enviro	13907	09 Sep 2099
Nomix Duplex	Nomix Enviro	14953	09 Sep 2099
461 halauxifen-methyl			
GF-2573	Dow	17540	05 Feb 2028
462 hymexazol			
Tachigaren 70 WP	Sumi Agro	12568	31 Aug 2018
463 imazalil			
Fungazil 50LS	Certis	14069	30 Jun 2024
464 imazalil + ipconazole			
Rancona i-MIX	Arysta	15574	30 Jun 2024
Rancona i-MIX	Certis	16492	30 Jun 2024
470 imazamox + metazachlor + quinmerac			
Clesima	BASF	16275	31 Jan 2020
471 imazamox + pendimethalin			
Nirvana	BASF	13220	31 Jan 2020
472 imazamox + quinmerac			
Clentiga	BASF	18121	31 Jan 2021
475 imidacloprid			
Couraze	Solufeed	17183	31 Jan 2022
Mido 70% WDG	Sharda	16812	23 Aug 2019
480 indoxacarb			
Picard 300 WG	Agrifarm	16714	30 Apr 2020
481 iodosulfuron-methyl-sodium			
Hussar	Bayer CropScience	12364	30 Apr 2020
482 iodosulfuron-methyl-sodium + mesosulfuron-methyl			
Diplomat	Euro	16691	30 Apr 2020
Ficap	Life Scientific	18135	30 Apr 2020
Idosem 36	Chem-Wise	15913	30 Apr 2020
Iodomeso OD	Euro	16501	30 Apr 2020
Jerico	Terrechem	17477	30 Apr 2020
Lusitania	Life Scientific	17989	30 Apr 2020
Mesiodo 35	Euro	15305	30 Apr 2020
Mesoiodostar WG	Life Scientific	17760	31 Jul 2019
Niantic	Life Scientific	18217	30 Apr 2020
Ocean	Goldengrass	14422	30 Apr 2020
Ocelot	UKChem	17790	30 Apr 2020
Ocelot	Matrix	17261	31 Jul 2019
RouteOne Seafarer	Albaugh UK	15317	30 Apr 2020
Standon Mimas WG	Standon	17170	30 Apr 2020
Teliton Ocean	Teliton	13163	30 Apr 2020
483 iodosulfuron-methyl-sodium + propoxycarbazone-sodium			
Caliban Duo	Headland	14283	31 Jul 2019

SECTION 3

Product	Approval holder	MAPP No.	Expiry Date
485 ipconazole			
Rancona 15 ME	Arysta	16136	28 Feb 2027
486 iprodione			
Acamar WG	Servem	17329	30 Apr 2020
Advance Green	UKChem	17791	30 Apr 2020
Advance Green	Matrix	17088	30 Apr 2020
Cavron	Belcrop	16168	30 Apr 2020
Emerald Turf	Agrigem	17590	30 Apr 2020
Green Turf King	Standon	14348	30 Apr 2020
Hi-Prodione 50	BASF	15498	30 Apr 2020
MAC-Iprodione 255 SC	MAC	14781	30 Apr 2020
Mascot Rayzor	Rigby Taylor	14329	30 Apr 2020
Progress Green	Matrix	16930	30 Apr 2019
Recover	Capital CP	17766	30 Apr 2020
Rovral AquaFlo	BASF	14206	30 Apr 2020
Sinpro	Sinon EU	16236	30 Apr 2020
491 isopyrazam			
A15149W	Syngenta	15046	30 Sep 2025
492 isoxaben			
Flexidor	Landseer	05121	30 Sep 2018
Pan Isoxaben 500	Pan Amenity	17980	31 Dec 2021
495 kresoxim-methyl			
Strong WG	RAAT	17470	30 Jun 2024
496 lambda-cyhalothrin			
Astec Lambda CY	Agrii	16319	31 Dec 2018
Block	Belchim	15697	31 Dec 2018
Block	Agrii	16319	31 Dec 2018
Clayton Lambada	Clayton	13231	09 Sep 2099
CleanCrop Corsair	Clayton	14124	09 Sep 2099
CM Lambaz 50 EC	CMI	16241	09 Sep 2099
Colt 10 CS	Headland	17430	09 Sep 2099
Eminentos 10 CS	Headland	17690	09 Sep 2099
Eminentos 10 CS	Headland	15854	31 Dec 2018
Euro Lambda 100 CS	Euro	14794	09 Sep 2099
Hockley Lambda 5EC	Hockley	15516	09 Sep 2099
IT Lambda	Inter-Trade	15119	31 Dec 2018
Karis 10 CS	Headland	15312	30 Jun 2018
Kendo	Syngenta	15562	09 Sep 2099
Kusti	Syngenta	16656	09 Sep 2099
Laidir 10 CS	Headland	17693	09 Sep 2099
Laidir 10 CS	Headland	15461	31 Dec 2018
Lambda 100 CS	Goldengrass	16874	31 Dec 2018
Lambd'Africa	Astec	15697	31 Dec 2018
Life Scientific Lambda-Cyhalothrin	Life Scientific	15342	31 Dec 2018
Life Scientific Lambda-Cyhalothrin 50	Life Scientific	16889	24 Feb 2019
MAC-Lambda-Cyhalothrin 50 EC	MAC	14941	09 Sep 2099
Martlet	Goldengrass	17052	09 Sep 2099
Ninja 5CS	Syngenta	16417	09 Sep 2099
Osmoze 50 CS	Aako	16416	31 Dec 2018
Reparto	Sipcam	15452	09 Sep 2099
RouteOne Lambda C	Albaugh UK	13663	09 Sep 2099
RVG Lambda-cyhalothrin	Rovogate	16806	09 Sep 2099
Sergeant Major	Adama	16175	31 Dec 2018
Stealth	Syngenta	14551	09 Sep 2099

Product	Approval holder	MAPP No.	Expiry Date
Triumph	Matrix	16929	09 Sep 2099
Triumph CS	Capital CP	18235	09 Sep 2099
Warrior	Syngenta	13857	09 Sep 2099

498 laminarin

Vacciplant	Goemar	13260	31 Jan 2020

500 lenacil

Lenazar Flo	Hermoo	14791	28 Feb 2019
Pirapama	Agroquimicos	15777	28 Feb 2019
Venzar 80 WP	DuPont	17742	30 Jun 2021
Venzar 80 WP	DuPont	09981	31 Aug 2018
Venzar Flowable	DuPont	06907	28 Feb 2019

501 lenacil + triflusulfuron-methyl

Debut Plus	DuPont	17991	30 Jun 2021
Safari Lite WSB	DuPont	12169	28 Feb 2019
Upbeet Lite	DuPont	17990	30 Jun 2021

502 linuron

Afalon	Adama	14187	03 Jun 2018
Aredios 45 SC	Novastar	14586	31 Mar 2018
Cleancrop Affable	Agrii	16321	03 Jun 2018
Datura	UPL Europe	16955	03 Jun 2018
Datura	UPL Europe	14915	03 Jun 2018
Kaedyn 500	DAPT	14739	03 Jun 2018
Linurex 50 SC	Adama	14652	03 Jun 2018
Messidor 45 SC	Novastar	14710	31 Mar 2018
Nightjar	AgChem Access	14656	03 Jun 2018

503 magnesium phosphide

Magtoxin	Detia Degesch	16996	31 Jan 2020
Magtoxin Pellets	Rentokil	17376	28 Feb 2022

504 maleic hydrazide

Cleancrop Malahide	United Agri	13629	30 Apr 2020
Crown MH	Certis	18018	30 Apr 2020
Gro-Slo	Arysta	15851	30 Apr 2020
Himalaya	Arysta	15214	30 Apr 2020
Himalaya XL	Arysta	17511	30 Apr 2020
Itcan SL 270	Gemini	17957	30 Apr 2020
Source II	Drexel	17858	30 Apr 2020
Source II	Chiltern	13618	30 Apr 2020

505 maleic hydrazide + pelargonic acid

Finalsan Plus	Certis	15147	31 Aug 2018

506 maltodextrin

Terminus	Certis	17319	03 Aug 2019

507 mancozeb

Agria Mancozeb 75WDG	Agria SA	16807	31 Jan 2019
Agria Mancozeb 80WP	Agria SA	16636	31 Jan 2019
AUK 75 WG	Goldengrass	16398	31 Jul 2020
Cleancrop Feudal	Agrii	17894	31 Jul 2020
Cleancrop Mandrake	United Agri	14792	31 Jul 2020
Dithane 945	Indofil	17269	31 Jul 2020
Dithane 945	Interfarm	15715	30 Sep 2019
Dithane Dry Flowable Neotec	Indofil	15720	30 Sep 2019
Dithane Dry Flowable Neotec	Indofil	17268	31 Jul 2020

SECTION 3

Product	Approval holder	MAPP No.	Expiry Date
Emzeb 80 WP	Sabero	17714	02 Aug 2020
Manfil 75 WG	Interfarm	15958	31 Jul 2020
Manfil 80 WP	Indofil	15956	31 Jul 2020
Manfil WP Plus	Agrii	17939	31 Jul 2020
Penncozeb 80 WP	UPL Europe	14718	28 Feb 2019
Penncozeb WDG	UPL Europe	14719	31 Jan 2019
Tridex	UPL Europe	18196	31 Jul 2020
Trimanzone	UPL Europe	14726	31 Oct 2018
Trimanzone	UPL Europe	16759	31 Jul 2020
Unizeb Gold	UPL Europe	18186	31 Jul 2020
Zebra WDG	Headland	14875	31 Jul 2020

508 mancozeb + metalaxyl-M

Clayton Mohawk	Clayton	18173	31 Dec 2019

510 mancozeb + zoxamide

Electis 75WG	Gowan	14195	31 Jul 2019
Roxam 75WG	Gowan	14191	31 Jul 2019
Unikat 75WG	Gowan	14200	31 Jul 2019

511 mandipropamid

Mandimid	Euro	17651	31 Jan 2026
Pergardo Uni	Fargro	17513	31 Jan 2026
Standon Mandor	Standon	17536	31 Jan 2026
Standon Mandor	Standon	17533	31 Jan 2026

513 MCPA

Agrichem MCPA 500	UPL Europe	16964	30 Apr 2020
Agritox 50	Nufarm UK	14814	30 Apr 2020
Dow MCPA Amine 50	Dow	14911	30 Apr 2020
Go-Low Power	DuPont	13812	30 Apr 2020
Lagoon	Gemini	17856	30 Apr 2020
MCPA 25%	Nufarm UK	14893	30 Apr 2020
MCPA 50	UPL Europe	14908	30 Apr 2020
Nufarm MCPA 750	Nufarm UK	14892	30 Apr 2020
POL-MCPA 500 SL	Zaklady	14912	30 Apr 2020
POL-MCPA 500 SL	Ciech	17808	30 Apr 2020
Tasker 75	Headland	14913	30 Apr 2020

515 MCPA + mecoprop-P

Cleanrun Pro	Everris Ltd	15073	30 Apr 2020
Greenmaster Extra	Everris Ltd	15817	30 Apr 2020

517 mecoprop-P

Clenecorn Super	Nufarm UK	14628	31 Jul 2020
Isomec	Nufarm UK	14385	31 Jul 2020

522 mepiquat chloride + metconazole

Mepicon	Euro	16844	31 Oct 2020

523 mepiquat chloride + prohexadione-calcium

Medax Top	BASF	16574	31 Aug 2021

527 mesotrione

A12739A	Syngenta	16158	31 Jan 2020
Callisto	PSI	17512	31 Jan 2020
Evolya	Syngenta	17490	31 Jan 2020
Hockley Mesotrione 10	Hockley	15519	31 Jan 2020
Kalypstowe	AgChem Access	13844	31 Jan 2020
Life Scientific Mesotrione	Life Scientific	17092	31 Jan 2019

Product	Approval holder	MAPP No.	Expiry Date
Meristo	Syngenta	16865	31 Jan 2020
Meruba	Helm	17961	31 Jan 2020
Mesotrion 100SC	Euro	16362	31 Jan 2020
Minerva	Terrechem	17514	31 Jan 2020
Osorno	Belchim	18250	31 Jan 2020
Osorno	Globachem	17321	31 Jan 2020
RouteOne Trione 10	Albaugh UK	14765	31 Jan 2020
Standon Cobmajor	Standon	16721	31 Jan 2020
Temsa SC	Belchim	18261	31 Jan 2020

528 mesotrione + nicosulfuron

Choriste	Syngenta	16899	31 Jan 2020

529 mesotrione + s-metolachlor

Camix	Syngenta	17722	10 Aug 2020
Clarido	Syngenta	17891	10 Aug 2020

530 mesotrione + terbuthylazine

Callistar	Syngenta	16864	31 Dec 2021
Orion	EZCrop	16049	31 Dec 2021
Riscala	Euro	16839	31 Dec 2021
RouteOne Mesot	Albaugh UK	15302	31 Dec 2021
Standon Cobmaster	Standon	16772	31 Dec 2021

532 metalaxyl-M

Fongarid Gold	Syngenta	12547	31 Dec 2019
Shams 465	DAPT	16107	31 Dec 2019

533 metaldehyde

Appeal	Chiltern	12022	31 Dec 2021
Attract	Chiltern	12023	31 Dec 2021
Cargo 3 GB	Aako	15784	31 Dec 2021
Doff Horticultural Slug Killer Blue Mini Pellets	Certis	11463	31 Dec 2021
Tremolo	Chiltern	16582	31 Dec 2021

534 metamitron

Atlas Metamitron	Whyte Agrochemicals	18102	29 Feb 2024
Atlas Too-Da Loo	Nufarm UK	17394	28 Feb 2022
Betra SC	Novastar	17244	28 Feb 2022
Bettix Flo SC	UPL Europe	18245	31 Dec 2021
Bettix WG	UPL Europe	16496	28 Feb 2022
Celmitron 70% WDG	UPL Europe	16990	31 Dec 2021
Celmitron 70% WDG	UPL Europe	14408	31 Mar 2019
Clayton Devoid	Clayton	18252	28 Feb 2025
Defiant SC	UPL Europe	16531	28 Feb 2022
Defiant WG	UPL Europe	16544	28 Feb 2022
Devoid	JT Agro	18067	28 Feb 2025
Goltix 90	Adama	16497	28 Feb 2022
Goltix Compact	Aako	16545	28 Feb 2022
Goltix WG	Adama	16853	28 Feb 2022
Inter Metron 700 SC	UPL Europe	17389	28 Feb 2022
Inter Metron 700 SC	Inter-Phyto	17101	31 Oct 2019
Metani SC	Phyto Services	17916	28 Feb 2022
Siskin	Goldengrass	16796	28 Feb 2022
Target SC	UPL Europe	13307	31 Mar 2019

535 metamitron + quinmerac

Goltix Titan	Adama	17301	28 Feb 2022

SECTION 3

Product	Approval holder	MAPP No.	Expiry Date
538 metazachlor			
Butisan S	BASF	16569	31 Jan 2022
Fuego	Adama	16679	31 Jan 2022
Inter Metazachlor	UPL Europe	17323	28 Jan 2019
Makila 500 SC	Novastar	16752	31 Jan 2022
Metaza	Euro	17168	31 Jan 2022
Mikado	UKChem	17788	31 Jan 2022
Mikado	Matrix	17004	31 Oct 2020
Standon New Metazachlor	Standon	17473	31 Jan 2022
539 metazachlor + quinmerac			
Katamaran	BASF	16766	31 Jan 2022
Metamerac	Euro	17139	31 Jan 2022
Metamerac-II	Euro	16810	31 Jan 2019
Metamerac-II	RealChemie	18171	31 Jan 2022
Rocket	Adama	17093	31 Jan 2022
Vesta	Goldengrass	17051	31 Jan 2022
540 metconazole			
Caramba	BASF	15337	31 Oct 2020
Life Scientific Metconazole	Life Scientific	16022	31 Oct 2020
Life Scientific Metconazole 60	Life Scientific	17037	29 Feb 2020
Metal	Arysta	17265	31 Oct 2020
Metal 60	Arysta	17074	31 Oct 2020
Rambaca 60	Euro	16841	31 Oct 2020
546 1-methylcyclopropene			
EthylBloc Tabs	Smither	16252	29 Feb 2020
EthylBloc Tabs	Oasis	17487	30 Apr 2020
Ethylene Buster	Chrysal	16222	30 Apr 2020
RipeLock Tabs	Rohm & Haas	16339	29 Feb 2020
Ripelock Tabs	AgroFresh	17480	30 Apr 2020
Ripelock Tabs 2.0	AgroFresh	17486	30 Apr 2020
RipeLock Tabs 2.0	Rohm & Haas	16547	07 Jul 2018
Ripelock VP	AgroFresh	17484	30 Apr 2020
RipeLock VP	Rohm & Haas	16338	29 Feb 2020
SmartFresh ProTabs	Landseer	16546	30 Apr 2020
SmartFresh SmartTabs	Landseer	12684	30 Apr 2020
550 metrafenone			
Attenzo	BASF	11917	31 Oct 2020
551 metribuzin			
Discus	RAAT	17552	31 Jan 2021
Python	Adama	16010	31 Jan 2021
552 metsulfuron-methyl			
Accurate	Headland	13224	09 Sep 2099
Alias	DuPont	13397	31 Dec 2018
Ally SX	DuPont	12059	09 Sep 2099
Aztec	EZCrop	16073	31 Dec 2018
Cimarron	Headland	13408	31 Dec 2018
Clayton Peak	Clayton	13547	31 Dec 2018
Cleancrop Mondial	United Agri	13353	09 Sep 2099
Finy	UPL Europe	12855	30 Jun 2018
Forge	Interfarm	12848	31 Dec 2018
Goldron-M	Goldengrass	13466	09 Sep 2099
Landgold Metsulfuron	Teliton	13421	09 Sep 2099
Life Scientific Metsulfuron-Methyl	Life Scientific	16654	29 Sep 2021
Metro 20	Mitsui	15633	30 Apr 2018

Product	Approval holder	MAPP No.	Expiry Date
Metro 20	Certis	16503	31 Dec 2018
Metro Clear	Certis	16234	31 Dec 2018
Minx PX	Nufarm UK	13337	31 Dec 2018
Pike	Nufarm UK	12746	31 Dec 2018
Pike PX	Nufarm UK	13249	31 Dec 2018
Quest	EZCrop	15659	09 Sep 2099
Revenge 20	Agform	14449	30 Jun 2018
Ricorso Premium	Rotam	16389	09 Sep 2099
RouteOne Romet 20	Albaugh UK	15277	31 Dec 2018
Standon Gusto Xtra	Standon	13886	09 Sep 2099
Standon Metso XT	Standon	13880	31 Dec 2018
Standon Mexxon Xtra	Standon	13931	09 Sep 2099
Tattler	Goldengrass	16360	31 Dec 2018
Tuko	Terrechem	15611	09 Sep 2099

553 metsulfuron-methyl + thifensulfuron-methyl

Accurate Extra	Headland	14167	09 Sep 2099
Choir	Nufarm UK	14081	31 Dec 2018

554 metsulfuron-methyl + tribenuron-methyl

Met-Tribe 286	Euro	16572	09 Sep 2099

558 napropamide

AC 650	UPL Europe	11102	31 Aug 2018
Colzamid	UPL Europe	16324	31 Aug 2018
Devrinol	UPL Europe	09374	31 Jul 2018

559 nicosulfuron

Bandera	Rotam	16576	11 May 2018
Biretta	Ventura	17408	11 May 2018
Fornet 4 SC	Belchim	16082	30 Jun 2021
Hill-Agro 40	Stefes	17654	18 Jul 2020
Nico Pro 4SC	Syngenta	15645	28 Feb 2018
Nicoron	Hillfield	18025	18 Jul 2020
Nicoron 40	Stefes	18025	18 Jul 2020
Samson	Syngenta	12141	28 Feb 2018
Samson	Belchim	16425	30 Jun 2021
Standon Frontrunner Super 6	Standon	16890	30 Jun 2021

560 nicosulfuron + thifensulfuron-methyl

Collage	DuPont	16820	09 Sep 2099

563 paclobutrazol

A10784A	Syngenta	17172	30 Nov 2023
Bonzi	Syngenta Bioline	17095	30 Apr 2020

566 penconazole

Rapas	RAAT	17740	30 Jun 2022

568 pendimethalin

Akolin 330 EC	Aako	14282	31 Jan 2020
Alpha Pendimethalin 330 EC	Adama	13815	31 Jan 2020
Aquarius	Adama	14712	31 Jan 2020
Atlas Pendimethalin 400	Nufarm UK	16815	31 Jan 2020
Blazer M	Headland	15084	31 Jan 2020
Bromo	Novastar	17936	31 Jan 2020
Bunker	Adama	13816	31 Jan 2020
Campus 400 CS	Aako	15366	31 Jan 2020
Cinder	Adama	14526	31 Jan 2020
CleanCrop Stomp	United Agri	13443	31 Jan 2020

Product	Approval holder	MAPP No.	Expiry Date
Eximus	Gemini	17174	31 Jan 2020
Fastnet	Sipcam	14068	31 Jan 2020
Hockley Pendi 330	Hockley	15513	31 Jan 2020
PDM 330 EC	BASF	13406	31 Jan 2020
Pendi 330	Euro	15807	31 Jan 2020
Pendifin	Finchimica	18164	31 Jan 2021
Pendifin 400 SC	Finchimica	18132	31 Jan 2021
Pendimet 400	Goldengrass	14715	31 Jan 2020
Pendragon	Certis	16262	31 Jan 2020
Penta	Sharda	17680	31 Jan 2020
Pre-Empt	Euro	16649	31 Jan 2020
Quarry	Gemini	17789	31 Jan 2020
RouteOne Penthal 330	Albaugh UK	15314	31 Jan 2020
Sherman	Adama	13859	31 Jan 2020
Sovereign	BASF	13442	31 Jan 2020
Stamp 330 EC	AgChem Access	13995	31 Jan 2020
Yellow Hammer	Goldengrass	16533	31 Jan 2020

569 pendimethalin + picolinafen
Product	Approval holder	MAPP No.	Expiry Date
Flight	BASF	12534	09 Sep 2099
Galivor	BASF	14986	31 Dec 2019
PicoStomp	BASF	13455	09 Sep 2099
Sienna	BASF	14991	31 Dec 2019
Sienna Pro	BASF	17047	09 Sep 2099

570 pendimethalin + pyroxsulam
Product	Approval holder	MAPP No.	Expiry Date
Broadway Sunrise	Dow	14960	31 Jan 2020

572 penflufen
Product	Approval holder	MAPP No.	Expiry Date
Emesto Prime DS	Bayer CropScience	17280	31 Jul 2026
Emesto Prime FS	Bayer CropScience	15446	31 Jul 2026

573 penthiopyrad
Product	Approval holder	MAPP No.	Expiry Date
DP 747	DuPont	17530	31 Oct 2026
Sentenza	Terrechem	17529	31 Oct 2026

574 penthiopyrad + picoxystrobin
Product	Approval holder	MAPP No.	Expiry Date
Cypher	DuPont	16639	30 Nov 2018
Frelizon	DuPont	16218	30 Nov 2018
Refinzar	DuPont	16590	30 Nov 2018
Sebrant	DuPont	17171	30 Nov 2018

575 pepino mosaic virus strain CH2 isolate 1906
Product	Approval holder	MAPP No.	Expiry Date
PMV-01	De Ceuster	17587	07 Feb 2033

578 pethoxamid
Product	Approval holder	MAPP No.	Expiry Date
Successor	Headland	17287	18 Oct 2019

580 phenmedipham
Product	Approval holder	MAPP No.	Expiry Date
Agrichem Phenmedipham EC	UPL Europe	16224	31 Jan 2020
Agrichem PMP EC	UPL Europe	16963	31 Jan 2020
Agrichem PMP EC	AgriChem BV	15230	31 Mar 2019
Agrichem PMP SE	AgriChem BV	16226	31 Mar 2019
Agrichem PMP SE	UPL Europe	16971	31 Jan 2020
Alpha Phenmedipham 320 SC	Adama	14070	31 Jan 2020
Betanal Flow	Bayer CropScience	13893	31 Jan 2020
Corzal	UPL Europe	14192	28 Feb 2019
Corzal SC	UPL Europe	17751	31 Jan 2020
Dancer Flow	Sipcam	14395	31 Jan 2020
Galcon SC	Novastar	14607	31 Jan 2020

Product	Approval holder	MAPP No.	Expiry Date
Herbasan Flow	Nufarm UK	13894	31 Jan 2020
Mandolin Flow	Nufarm UK	13895	31 Jan 2020
Pump	Adama	14087	31 Jan 2020
Rubie	EZCrop	15655	31 Jan 2020
Shrapnel	Adama	18234	31 Jan 2021
Shrapnel	Adama	14088	31 Jan 2021

586 picolinafen

AC 900001	BASF	10714	09 Sep 2099
Vixen	BASF	13621	09 Sep 2099

587 picoxystrobin

Acanto	DuPont	13043	30 Nov 2018
Flanker	DuPont	14760	30 Nov 2018
Galileo	DuPont	13252	30 Nov 2018
Oranis	DuPont	15185	30 Nov 2018

588 pinoxaden

A13814D	Syngenta	17386	30 Jun 2019
Greencrop Helvick	Greencrop	13052	30 Jun 2019
Pivot	Goldengrass	16560	30 Jun 2019
Roaxe 100	Chem-Wise	16111	30 Jun 2019
Standon Pinoxy	Standon	16355	30 Jun 2019
Wren	Matrix	16700	30 Jun 2019

590 pirimiphos-methyl

PHOBI SMOKE PRO90	Lodi UK	17117	31 Jan 2021

593 Potassium salts of fatty acids

Jaboland	Invest A. B.	17324	28 Feb 2022
Jabolim	Quimicas	17482	28 Feb 2022
Nakar	Seipasa	17483	28 Feb 2022
Tec-bom	Iberfol	17455	28 Feb 2022

594 prochloraz

Sporgon 50 WP	Sylvan	17700	26 Jul 2020

595 prochloraz + propiconazole

Bumper P	Adama	08548	31 Dec 2021
Greencrop Twinstar	Greencrop	09516	31 Dec 2021

597 prochloraz + tebuconazole

Agate EW	Adama	16515	31 Dec 2021

598 prochloraz + thiram

Agrichem Hy-Pro Duet	Agrichem	14018	31 Jan 2019

602 propamocarb hydrochloride

Edipro	Arysta	15564	31 Jan 2021
Promess	Arysta	16008	31 Jan 2021
Propamex-I 604 SL	MAC	15901	31 Jan 2021
Rival	Agria SA	18177	31 Jan 2021

603 propaquizafop

Chitral	Terrechem	17655	31 May 2022
Flanoc	RealChemie	18155	31 May 2022
Flanoc	Euro	16816	31 Dec 2018
Snipe	Goldengrass	16601	31 May 2022

604 propiconazole

Anode	Adama	14447	31 Jul 2020

Product	Approval holder	MAPP No.	Expiry Date
Apache 250 EC	Aako	17454	31 Jul 2020
Atlas Propiconazole 250	Gemini	16377	31 Jul 2020
Atlas Propiconazole 250	Whyte Agrochemicals	17416	31 Jul 2020
Barclay Propizole	Barclay	15113	09 Feb 2020
Hockley Propicon 25	Hockley	15517	31 Jul 2020
Matsuri	Sumi Agro	15544	31 Jul 2020
Nimble Pro	Capital CP	18003	31 Jul 2020
Profiol 250	PSI	15393	31 Jul 2020
Propicon 250	Goldengrass	15255	31 Jul 2020
Zolex	ITACA	15713	31 Jul 2020

609 propyzamide

Product	Approval holder	MAPP No.	Expiry Date
Careca	UPL Europe	14948	31 Mar 2019
Cleancrop Forward	Dow	18145	31 Jul 2020
Cleancrop Rumble	Agrii	17600	31 Jul 2020
Conform	AgChem Access	14602	30 Jun 2019
Dennis	Interfarm	15081	31 Jul 2020
Edge 400	Albaugh UK	14683	31 Jul 2020
Engage	Interfarm	14233	31 Jul 2020
Flomide	Interfarm	14223	31 Jul 2020
Gemstone Granules	Agrigem	17898	31 Jul 2020
Hockley Propyzamide 40	Hockley	15449	31 Jul 2020
KeMiChem - Propyzamide 400 SC	KeMiChem	14706	31 Jul 2020
Kerb 50 W	Dow	13715	31 Jul 2020
MAC-Propyzamide 400 SC	MAC	14786	31 Jul 2020
Master Lean 500 SC	Generica	16231	31 Jul 2020
Master Prop 40	Generica	15829	31 Jul 2020
Menace 80 EDF	Dow	13714	31 Jul 2020
MS Eddard	Micromix	17149	31 Jul 2020
Pizza 400 SC	Goldengrass	14430	31 Jul 2020
Pizza Flo	Goldengrass	16600	31 Jul 2020
Prova	Frontier	15641	31 Jul 2020
Pyzamid 400 SC	Euro	15257	31 Jul 2020
RouteOne Zamide Flo	Albaugh UK	14559	31 Jul 2020
Setanta 50 WP	Certis	16518	31 Jul 2020
Shamal	Rotam	15686	31 Jul 2020
Solitaire 50 WP	Mitsui	16177	30 Apr 2018
Solitaire 50 WP	Certis	16521	31 Jul 2020
Standon Santa Fe 50 WP	Standon	14966	31 Jul 2020
Verdah 400	DAPT	14680	31 Jul 2020
Verge 400	Albaugh UK	14682	31 Jul 2020
Zammo	Headland	15313	31 Jul 2020

610 proquinazid

Product	Approval holder	MAPP No.	Expiry Date
Proquin 200	Euro	16019	31 Jan 2023
Talius	PSI	16912	31 Jan 2023
Zorkem	Unisem	16906	31 Jan 2023

611 prosulfocarb

Product	Approval holder	MAPP No.	Expiry Date
A8545G	Syngenta	16204	30 Apr 2021
Atlas Pro	Whyte Agrochemicals	17734	30 Apr 2021
Elude	Chem-Wise	17195	30 Apr 2021
Fidox	Syngenta	16209	30 Jun 2021
IC1574	Syngenta	16205	30 Apr 2021
Jade	Syngenta	16203	30 Apr 2021
Nuron	Gemini	17757	30 Apr 2021
NYX	Terrechem	17737	30 Apr 2021
Prosulfix	Novastar	17925	30 Apr 2021
Prosulfocarbstar	Life Scientific	17431	31 Mar 2020

Product	Approval holder	MAPP No.	Expiry Date
Prosulfostar	Life Scientific	17542	28 Feb 2021
Wicket	Syngenta	17555	30 Apr 2021

613 prothioconazole
Banguy	PSI	15140	31 Jul 2018
Proline	Bayer CropScience	12084	31 Jul 2018
Redigo	Bayer CropScience	12085	31 Jan 2021

614 prothioconazole + spiroxamine
Pikazole 46	Chem-Wise	15691	31 Jan 2021
Pro-Spirox	Euro	16437	31 Jan 2021

616 prothioconazole + tebuconazole
Pro Fit 25	Chem-Wise	17040	31 Jan 2021
Pro-Tebu 250	Euro	17142	31 Jan 2021
Proteus	Matrix	17210	31 Jan 2021
Proteus	UKChem	17787	31 Jan 2021
Romtil	RealChemie	18117	31 Jan 2021
Sequana	Ascot Pro-G	17818	31 Jan 2021
Standon Cumer	Standon	17848	31 Jan 2021
Standon Mastana	Standon	17039	30 Nov 2020

618 prothioconazole + trifloxystrobin
Biomus	Euro	16408	31 Jan 2020

619 Pseudomonas chlororaphis MA 342
Cerall	Chemtura	14546	31 Oct 2020

620 pymetrozine
Chess WG	Syngenta Bioline	13310	31 Dec 2019
Clayton Rook	Clayton	18139	31 Dec 2019
Quorum	Chem-Wise	16707	31 Dec 2019
Rozine	Direct	16946	31 Dec 2018
Standon LBW	Standon	17213	31 Dec 2019

621 pyraclostrobin
BAS 500 06	BASF	12338	31 Jul 2020
BASF Insignia	BASF	11900	31 Jul 2020
Comet	BASF	10875	31 Jul 2020
Flyer	BASF	12654	31 Jul 2020
Insignia	Vitax	11865	31 Jul 2020
Inter Pyrastrobin 200	Inter-Phyto	18188	31 Jul 2020
LEY	BASF	14774	31 Jul 2020
Platoon	BASF	12325	31 Jul 2020
Platoon 250	BASF	12640	31 Jul 2020
Tucana 200	BASF	17294	31 Jul 2020
Vivid 200	BASF	17295	31 Jul 2020

622 pyraflufen-ethyl
OS159	Ceres	12723	31 Dec 2018
Quickdown	Certis	15396	31 Dec 2018

623 pyrethrins
Pyrethrum 5 EC	PelGar	18210	31 Dec 2021

625 pyrimethanil
Penbotec 400 SC	Janssen	17384	31 Oct 2020
Precious	Emerald	18134	31 Oct 2020
Pyrimala	Servem	17325	31 Oct 2020
Pyrus 400 SC	Arysta	16254	31 Oct 2020

SECTION 3

Product	Approval holder	MAPP No.	Expiry Date
Scala	BASF	15222	31 Oct 2020
Spectrum	RAAT	17469	31 Oct 2020
626 pyriofenone			
Property 180 SC	Belchim	15661	31 Mar 2018
627 pyroxsulam			
Avocet	Dow	14829	31 Oct 2026
628 Pythium oligandrum M1			
Polyversum	De Sangosse	17456	31 Oct 2021
631 quinoxyfen			
Apres	Dow	08881	31 Oct 2020
Fortress	Dow	08279	31 Oct 2020
633 quizalofop-P-tefuryl			
Panarex	Certis	16945	18 Feb 2018
Panarex	Arysta	17960	18 Feb 2019
Rango	Arysta	17959	18 Feb 2019
634 rimsulfuron			
Rimsulf 250	Euro	15802	31 Oct 2020
637 Sheep fat			
Trico	Kwizda	18149	28 Feb 2023
638 silthiofam			
Latitude	Certis	10695	30 Apr 2019
Latitude	Certis	18036	30 Apr 2020
Latitude XL	Certis	18030	30 Apr 2020
Latitude XL	Certis	16467	30 Apr 2020
Meridian	AgChem Access	14676	30 Apr 2019
642 spinosad			
Exocet	Matrix	17146	31 Oct 2020
Exocet	UKChem	17792	31 Oct 2020
643 spirodiclofen			
Envidor	Bayer CropScience	13947	31 Mar 2018
645 spirotetramat			
Spiro OD	Euro	16840	30 Apr 2019
649 streptomyces griseoviridis strain K61			
Mycostop	Verdera	16637	31 Oct 2021
650 sulfosulfuron			
Fosulfuron	Euro	15164	30 Jun 2018
Monitor	Interfarm	12236	30 Jun 2018
651 sulfuryl fluoride			
ProFume	Douglas	16396	30 Nov 2019
ProFume	Douglas	17309	30 Apr 2023
652 sulphur			
Microthiol Special	UPL Europe	06268	31 Mar 2019
POL-Sulphur 80 WG	Ciech	17613	30 Jun 2022
POL-Sulphur 800 SC	Ciech	17614	30 Jun 2022
Solfa WG	Nufarm UK	11602	31 Dec 2021

Product	Approval holder	MAPP No.	Expiry Date
653 tau-fluvalinate			
Greencrop Malin	Greencrop	11787	31 Dec 2021
Klartan	Adama	11074	31 Dec 2021
Revolt	Adama	13383	31 Dec 2021
654 tebuconazole			
Cezix	Rotam	18056	28 Feb 2022
Clayton Tebucon EW	Clayton	14824	30 Sep 2018
Cleancrop Teboo	Agrii	18111	28 Feb 2022
Deacon	Adama	14270	30 Apr 2018
Erasmus	Rotam	18057	28 Feb 2022
Gizmo	Nufarm UK	16502	28 Feb 2021
Hockley Tebucon 25	Hockley	15500	30 Sep 2018
Inter-Tebu	UPL Europe	17322	31 Jan 2020
Legend	JT Agro	18116	28 Feb 2022
Mitre	Adama	17606	28 Feb 2022
Odin	Rotam	13468	30 Sep 2018
Orius 20 EW	Adama	12311	31 Mar 2018
Quarta	Terrechem	17844	28 Feb 2022
Santal	Novastar	17955	28 Feb 2022
Savannah	Rotam	14821	31 Oct 2018
Standon Beamer	Standon	17350	28 Feb 2022
Tamok	Matrix	17264	28 Feb 2022
Tamok	UKChem	17786	28 Feb 2022
TEB250	Goldengrass	15755	30 Sep 2018
657 tebuconazole + trifloxystrobin			
Defusa	Rigby Taylor	18232	31 Jan 2020
Inter Tebloxy	Iticon	17311	31 Jan 2020
Nativo 75 WG	Bayer CropScience	16867	18 Dec 2018
658 tebufenpyrad			
Clan	RAAT	17491	31 Dec 2021
660 tefluthrin			
A13219F	Syngenta	16161	31 Dec 2021
661 tembotrione			
Laudis	Bayer CropScience	17302	31 Oct 2026
667 thiabendazole			
Hykeep	Agrichem	12704	31 Dec 2019
Storite Clear Liquid	Frontier	12706	31 Dec 2019
Storite Excel	Frontier	12705	09 Sep 2099
Tezate 220 SL	UPL Europe	14007	30 Jun 2018
668 thiacloprid			
Calypso	Bayer CropScience	11257	31 Oct 2020
Pintail	Matrix	16729	17 Aug 2018
Pintail	UKChem	17817	31 Oct 2020
Rana SC	Servem	17211	31 Oct 2020
Scabiya	Euro	16787	31 Oct 2020
Standon Zero Tolerance	Standon	13546	31 Oct 2019
Standon Zero Tolerance	Standon	17551	31 Oct 2020
Thiaclomex 480 SC	MAC	15900	31 Oct 2020
669 thiamethoxam			
Actus	Progreen	18211	31 Oct 2020
Centric	Syngenta	13954	31 Oct 2020
Cruiser 70 WS	Syngenta	17338	31 Oct 2020

SECTION 3

Product	Approval holder	MAPP No.	Expiry Date
Leptom 250 WDG	Agrifarm	16717	31 Oct 2020
Moxy	Harvest	18212	31 Oct 2020

670 thifensulfuron-methyl

Harmony SX	DuPont	12181	09 Sep 2099
Prospect SX	DuPont	12212	09 Sep 2099

671 thifensulfuron-methyl + tribenuron-methyl

Calibre SX	Certis	15032	09 Sep 2099

672 thiophanate-methyl

Thiofin WG	Q-Chem	15384	30 Apr 2020

673 thiram

Agrichem Flowable Thiram	Agrichem	10784	30 Jun 2018
Thiraflo	Arysta	13338	30 Jun 2018
Thiraflo	Certis	16499	31 Jul 2018
Thiraflo	Arysta	17888	31 Oct 2020
Thiraflo	Arysta	17944	31 Oct 2020
Thyram Plus	Agrichem	10785	30 Jun 2018

674 tolclofos-methyl

Basilex	Everris Ltd	16243	31 Dec 2021
Rizolex 10D	Interfarm	14204	31 Dec 2021
Rizolex 50 WP	Interfarm	14217	31 Dec 2021
Rizolex Flowable	Interfarm	14207	31 Dec 2021

678 tri-allate

Avadex Excel 15G	Gowan	16998	31 Dec 2020
Avadex Factor	Gowan	17748	17 Oct 2020

681 tribenuron-methyl

Corida	Zenith	18247	30 Apr 2020
Helmstar A 75 WG	Rotam	15070	30 Apr 2019
Helmstar A 75 WG	Belchim	17077	30 Apr 2020
Nuance	Headland	14813	30 Apr 2020
Quantum	DuPont	15190	30 Apr 2020
Quantum SX	DuPont	15189	30 Apr 2020
Taxi	Nufarm UK	15733	30 Apr 2020
TBM 75 WG	Sharda	17326	10 Nov 2019
Toscana	Proplan-Plant	17726	30 Apr 2020
Trailer	Industrias Afrasa	17903	30 Apr 2020
Tribe 500	Euro	16888	30 Apr 2020
Tribenuron-methyl 750 g/kg WDG	Zenith	18160	30 Apr 2020
Tribun 75 WG	Belchim	17090	30 Apr 2020
Tribun 75 WG	Helm	16995	31 May 2019
Trimeo 75 WG	Belchim	17076	30 Apr 2020
Trimeo 75 WG	Helm	16719	30 Apr 2019

683 Trichoderma harzianum (Strain T22)

Trianum G	Koppert	16740	31 Oct 2021
Trianum P	Koppert	16741	31 Oct 2021

684 triclopyr

Topper	Arysta	15719	31 Oct 2020

685 trifloxystrobin

Flint	Bayer CropScience	11259	31 Jan 2020
Pan Tees	Pan Agriculture	12894	31 Jan 2020

Product	Approval holder	MAPP No.	Expiry Date
687 triflusulfuron-methyl			
Safari	DuPont	17547	30 Jun 2022
Tricle	UPL Europe	17448	30 Jun 2022
Zareba	Terrechem	17987	30 Jun 2022
688 trinexapac-ethyl			
A17600C	Syngenta	17548	31 Oct 2020
Cleancrop Alatrin	Agrii	15196	31 Oct 2020
Cleancrop Alatrin Evo	Agrii	17763	31 Oct 2020
Cleancrop Cutlass	Agrii	16046	31 Oct 2020
Clipless	Headland Amenity	15435	31 Jul 2018
Confine NT	Zantra	17770	09 May 2020
Freeze	Headland	15871	31 May 2018
Inter Trinex	UPL Europe	17334	31 Jan 2020
Life Scientific Trinexapac	Life Scientific	17125	31 Oct 2020
Life Scientific Trinexapac 250	Life Scientific	16047	31 Oct 2020
Maintain	Headland	15626	31 Jul 2018
Maintain NT	Headland	18081	27 Jun 2020
Modan	Helm	16628	15 Jul 2018
Moddus ME	Syngenta	17179	31 Oct 2020
Mowless	ProKlass	17086	30 Apr 2018
Moxa New	Globachem	17830	18 Jan 2021
Next	Sharda	17797	31 Oct 2020
Paket 250 EC	Arysta	17436	31 Oct 2020
Palisade	Syngenta	17860	31 Oct 2020
Plaza	Goldengrass	16855	15 Jul 2018
Primo Maxx	Syngenta	14780	30 Apr 2018
Pure Max	Pure Amenity	14952	30 Apr 2018
Scitec	Syngenta	15588	31 Oct 2020
Seize	Headland	16780	31 Oct 2020
Sieze NT	Headland	17794	09 May 2020
Sonis	Syngenta	16891	31 Oct 2020
Staylow	Unique Marketing	16295	31 Oct 2020
Tacet	Goldengrass	15258	31 Oct 2020
Tempest	Ascot Pro-G	18027	31 Oct 2020
TP 100	Certis	17605	31 Oct 2020
Trexstar	Belchim	17062	30 Apr 2020
Trexstar	Helm	16677	15 Jul 2018
Trexxus	Helm	16781	15 Jul 2018
Trexxus	Belchim	17059	15 Jul 2018
Trinex 222	Euro	15322	31 Oct 2020
Trinexis	Arysta	16429	31 Oct 2020
Upkeep	Adama	15225	31 Oct 2020
UPL Trinexapac	UPL Europe	16645	31 Oct 2020
Zira	Terrechem	17467	31 Oct 2020
691 Verticillium alobo-atrum			
Dutch Trig	BTL Bomendienst	17481	30 Oct 2021
693 zeta-cypermethrin			
Fury 10 EW	Belchim	12248	30 Sep 2019
RouteOne Zeta 10	Albaugh UK	15308	30 Sep 2019

SECTION 3

SECTION 4
ADJUVANTS

Adjuvants

Adjuvants are not themselves classed as pesticides and there is considerable misunderstanding over the extent to which they are legally controlled under the Food and Environment Protection Act. An adjuvant is a substance other than water which enhances the effectiveness of a pesticide with which it is mixed. Consent C(i)5 under the Control of Pesticides Regulations allows that an adjuvant can be used with a pesticide only if that adjuvant is authorised and on a list published on the HSE adjuvant database: https//secure.pesticides. gov.uk/adjuvants/search.asp. An authorised adjuvant has an *adjuvant number* and may have specific requirements about the circumstances in which it may be used.

Adjuvant product labels must be consulted for full details of authorised use, but the table below provides a summary of the label information to indicate the area of use of the adjuvant. Label precautions refer to the keys given in Appendix 4, and may include warnings about products harmful or dangerous to fish. The table includes all adjuvants notified by suppliers as available in 2018.

Product	Supplier	Adj. No.	Type
Abacus	De Sangosse	A0543	vegetable oil
Contains	20% w/w alkoxylated alcohols -EAC 1, 9% w/w oil (tall oil fatty acids - EAC 2), 53.43% w/w oil (rapeseed fatty acid esters - EAC 3)		
Use with	All approved pesticides on all edible crops when used at half their recommended dose or less, and on all non-edible crops up to their full recommended dose. Also at a maximum concentration of 0.1% with approved pesticides on listed crops up to specified growth stages		
Protective clothing	A, C, H		
Precautions	R36, U05a, U11, U14, U19a, U20b, E15a, E19b, E34, D01, D02, D05, D10a, D12a, H04		
Activator 90	De Sangosse	A0547	non-ionic surfactant/wetter
Contains	375 g/kg alkoxylated alcohols (EAC 1), 375 g/kg alkoxylated alcohols (EAC 2), 150 g/kg oil (tall oil fatty acids - EAC 3)		
Use with	All approved pesticides on all edible crops when used at half their recommended dose or less, and on all non-edible crops up to their full recommended dose. Also at a maximum concentration of 0.1% with approved pesticides on listed crops up to specified growth stages		
Protective clothing	A, C, H		
Precautions	R36, R38, R53a, U02a, U04a, U05a, U10, U11, U20b, E15a, E19b, D01, D02, D05, D10a, H04		
Addit	Koppert	A0693	spreader/sticker/wetter
Contains	780.2 g/l oil (rapeseed triglycerides)		
Use with	Mycotal at 0.25% spray solution and all approved pesticides at half or less than half the approved pesticide rate		
Protective clothing	A, F, H		
Precautions	R20, R21, R36, R51, R53a		
Adigor	Syngenta	A0522	wetter
Contains	47% w/w methylated rapeseed oil		
Use with	Topik, Axial, Trazos and Amazon on cereals in accordance with recommendations on the respective herbicide labels		
Protective clothing	A, C, H		
Precautions	R43, R51, R53a, U02a, U05a, U09a, U20b, E15b, E38, D01, D02, D05, D09a, D10c, D12a, H04, H11		

SECTION 4

Product	Supplier	Adj. No.	Type
AdjiFe	Amega	A0797	wetter

Contains	345 g/l ammonium iron (III) citrate (EAC 1), 300 g/l alkyl polyglycosides (EAC 2) and 104.5 g/l ethylene oxide-propylene oxide copolymers (EAC 3)
Use with	All authorised plant protection herbicides and fungicides
Protective clothing	A
Precautions	U02a, U04a, U05a, U14, U15, U19a, E13b, E34, D01, D02, D09a, D10a, M03

Product	Supplier	Adj. No.	Type
AdjiMin	Amega	A0812	adjuvant

Contains	920 g/kg oil (petroleum oils) (EAC 1)
Use with	All approved pesticides at half the approved maximum dose on edible crops and at the full approved dose on non-edible crops and non-crop production at 1% spray volume.
Protective clothing	A
Precautions	U02a, U04a, U05a, U14, U15, U19a, E13b, E34, D01, D02, D09a, D10a, M03

Product	Supplier	Adj. No.	Type
AdjiSil	Amega	A0763	spreader/wetter

Contains	83% w/w trisiloxane organosilicone copolymers (EAC 1)
Use with	All approved pesticides at half the approved maximum dose on edible crops and at the full approved dose on non-edible crops and non-crop production at 0.15% spray volume.
Protective clothing	A, C, H
Precautions	U02a, U04a, U05a, U14, U15, U19a, E13b, E34, D01, D02, D09a, D10a, M03, H11

Product	Supplier	Adj. No.	Type
AdjiVeg	Amega	A0756	adjuvant

Contains	90% w/w fatty acid esters (EAC 1)
Use with	All approved pesticides at half the approved maximum dose on edible crops and at the full approved dose on non-edible crops and non-crop production at 1% spray volume.
Protective clothing	A, C, H
Precautions	U02a, U04a, U05a, U11, U14, U15, E15a, E34, D01, D02, D09a, D10a, M03, H04

Product	Supplier	Adj. No.	Type
Admix-P	De Sangosse	A0301	wetter

Contains	80% w/w trisiloxane organosilicone copolymers (EAC 1)
Use with	A wide range of pesticides applied as corm, tuber, onion and other bulb treatments in seed production and in seed potato treatment
Protective clothing	A, C, H
Precautions	R20, R21, R22a, R36, R43, R48, R51, R58, U11, U15, U19a, E15a, E19b, D01, D02, D05, D10a, D12a, H03, H11

Product	Supplier	Adj. No.	Type
Amber	Interagro	A0367	vegetable oil

Contains	95% w/w methylated rapeseed oil
Use with	Sugar beet herbicides, oilseed rape herbicides, cereal graminicides and a wide range of other pesticides that have a label recommendation for use with authorised adjuvant oils on specified crops. See label for details
Protective clothing	A, C
Precautions	U05a, U20b, E15a, D01, D02, D05, D09a, D10a

Product	Supplier	Adj. No.	Type
Arma	Interagro	A0306	penetrant
Contains	500 g/l alkoxylated fatty amine + 500 g/l polyoxyethylene monolaurate		
Use with	Cereal growth regulators, cereal herbicides, cereal fungicides, oilseed rape fungicides and a wide range of other pesticides on specified crops		
Protective clothing	A, C		
Precautions	R51, R58, U05a, E15a, E34, E37, D01, D02, D05, D09a, D10a, H11		
Asu-Flex	Greenaway	A0677	spreader/sticker/wetter
Contains	10.0% w/w rapeseed oil		
Use with	Asulox, Greencrop Found, I T Asulam, Inter Asulam and Spitfire		
Protective clothing	A, C		
Precautions	U11, U12, U15, U20b, E15a, E34, D02		
BackRow	Interagro	A0472	mineral oil
Contains	60% w/w refined paraffinic petroleum oil		
Use with	Pre-emergence herbicides. Refer to label or contact supplier for further details		
Protective clothing	A		
Precautions	R22b, R38, U02a, U05a, U08, U20b, E15a, D01, D02, D05, D09a, D10a, M05b, H03		
Ballista	Interagro	A0524	spreader/wetter
Contains	10% w/w alkoxylated triglycerides		
Use with	All approved pesticides and plant growth regulators for use in winter and spring cereals applied at full rate up to and including GS 52, and at half the approved rate thereafter. May be applied with approved pesticides for other specified crops at full rate up to specified growth stages and at half rate thereafter		
Protective clothing	A		
Precautions	U05a, U20b, E15a, E34, D01, D02, D09a		
Banka	Interagro	A0245	spreader/wetter
Contains	14.6 % w/w alkyl pyrrolidone copolymers and 14.6 % w/w alkyl pyrrolidones		
Use with	Potato fungicides and a wide range of other pesticides on specified crops		
Protective clothing	A, C		
Precautions	R38, R41, R52, R58, U02a, U05a, U11, U19a, U20b, E15a, E34, E37, D01, D02, D05, D09a, H04		
Barramundi	Interagro	A0376	mineral oil
Contains	95% w/w mineral oil		
Use with	Sugar beet herbicides, oilseed rape herbicides, cereal graminicides and a wide range of other pesticides that have a label recommendation for use with authorised adjuvant oils on specified crops. Refer to label or contact supplier for further details		
Protective clothing	A		
Precautions	R22b, R38, U02a, U05a, U08, U20b, E15a, D01, D02, D05, D09a, D10a, M05b, H03		
Binder	Amega	A0598	spreader/wetter
Contains	30.0 % w/w alkoxylated alcohols (EAC 1)		
Use with	All approved pesticides at half or less than half the approved pesticide rate and all approved formulations of glyphosate		
Protective clothing	A		
Precautions	U02a, U04a, U05a, U14, U15, U19a, E13b, E34, D01, D02, D09a, D10a, M03		

SECTION 4

Product	Supplier	Adj. No.	Type
Bio Syl	Intracrop	A0773	spreader/sticker/wetter
Contains	32.67% w/w alkoxylated alcohols (EAC 1) and 1.0% w/w trisiloxane organosilicone copolymers (EAC 2)		
Use with	Recommended rates of approved pesticides on non-crop and non-edible crops; when used on edible crops the dose of the approved pesticide must be half or less than half the approved dose rate		
Protective clothing	A, C		
Precautions	R36, R38, U05a, U08, U20c, E15a, D01, D02, D10a, H04		
Bioduo	Intracrop	A0606	wetter
Contains	700 g/l alkoxylated alcohols (EAC 1) and 150 g/l rapeseed fatty acids (EAC 2)		
Use with	A wide range of pesticides used in grassland, agriculture and horticulture and with pesticides used in non-crop situations		
Protective clothing	A, C		
Precautions	R22a, R36, R38, U05a, U08, U20c, E13c, E34, D01, D02, D09a, D10a, M03, H03, H04, H08		
Biofilm	Intracrop	A0634	anti-drift agent/anti-transpirant/sticker/ UV screen/wetter
Contains	96.0% w/w pinene oligomers (EAC 1)		
Use with	All approved fungicides and insecticides on edible crops and all approved formulations of glyphosate used pre-harvest on wheat, barley, oilseed rape, stubble, and in non-crop situations and grassland destruction		
Protective clothing	A, C		
Precautions	U19a, U20c, E13c, D09a, D11a		
BioPower	Bayer CropScience	A0617	wetter
Contains	6.7% w/w 3,6-dioxaeicosylsulphate sodium salt and 20.2% w/w 3,6-dioxaoctadecylsulphate sodium salt		
Use with	Atlantis and all other approved cereal herbicides		
Protective clothing	A, C		
Precautions	R36, R38, U02a, U05a, U08, U13, U19a, U20b, E13c, E34, D01, D02, D05, D09a, D10a, H04		
Biothene	Intracrop	A0633	anti-drift agent/anti-transpirant/sticker/ UV screen/wetter
Contains	96.0% w/w pinene oligomers (EAC 1);		
Use with	All approved fungicides and insecticides on edible crops up to 30 d before harvest, and all approved formulations of glyphosate used pre-harvest on wheat, barley, oilseed rape, stubble, and in non-crop situations and grassland destruction. Must not be used in mixture with adjuvant oils or surfactants		
Precautions	U19a, U20c, E13c, D09a, D11a		

Product	Supplier	Adj. No.	Type
Bond	De Sangosse	A0556	extender/sticker/wetter
Contains	10% w/w alkoxylated alcohols (EAC 1), 45% w/w styrene-butadiene copolymers (EAC 2)		
Use with	All approved potato blight fungicides. Also with all approved pesticides on all edible crops when used at half their recommended dose or less, and on all non-edible crops up their full recommended dose. Also at a maximum concentration of 0.14% with approved pesticides on listed crops up to specified growth stages		
Protective clothing	A, C, H		
Precautions	R36, R38, U11, U14, U16b, U19a, E15a, E19b, D01, D02, D05, D10a, D12a, H04		
Broad-Flex	Greenaway	A0680	spreader/sticker/wetter
Contains	10.0% w/w rapeseed oil		
Use with	Broad Sword or Green Guard		
Protective clothing	A, C		
Precautions	U11, U12, U15, U20b, E15a, E34, D02		
Byo-Flex	Greenaway	A0545	sticker/wetter
Contains	10.0% w/w rapeseed oil		
Use with	GLY 490 (MAPP 12718)		
Protective clothing	A, C		
Precautions	U11, U12, U15, U20b, E34		
C-Cure	Interagro	A0467	mineral oil
Contains	60% w/w refined mineral oil		
Use with	Pre-emergence herbicides		
Protective clothing	A		
Precautions	R22b, R38, U02a, U05a, U08, U20b, E15a, D01, D02, D05, D09a, D10a, M05b, H03		
Ceres Platinum	Interagro	A0445	penetrant/spreader
Contains	500 g/l alkoxylated fatty amine + 500 g/l polyoxyethylene monolaurate		
Use with	Cereal growth regulators, cereal herbicides, cereal fungicides, oilseed rape fungicides and a wide range of other pesticides on specified crops		
Protective clothing	A		
Precautions	R51, R58, U05a, E15a, E34, D01, D02, D05, D09a, D10a, H11		
Clayton Astra	Clayton	A0724	penetrant/spreader
Contains	500 g/l alkoxylated sorbitan esters and 500 g/l alkoxylated tallow amines		
Use with	All approved pesticides up to a maximum conc of 0.15% of the total spray volume		
Protective clothing	A, H		
Precautions	R51, R53a, U05a, E15a, E34, D01, D02, D05, D09a, D10b, D12a, H11		
Clayton Dolmen	Clayton	A0725	adjuvant/spreader
Contains	64% w/w trisiloxane organosilicone copolymers		
Use with	All approved pesticides up to a maximum conc of 0.2% of the total spray volume		
Protective clothing	A, C, H		
Precautions	R20, R21, R36, R38, R51, R53a, U02a, U05a, U09a, U11, U16b, U19a, U20b, D01, D02, D05, D09a, D10a, H03, H11		

SECTION 4

Product	Supplier	Adj. No.	Type
Clayton NFP	Clayton	A0834	sticker
Contains	96.0 % w/w pinene oligomers		
Use with	All authorised fungicides and insecticides on edible crops and ornamental plant production and all authorised formulations of glyphosate		
Protective clothing	A, H		
Precautions	H315, H317, U04b, U19a, E38, H410, D12a, H400		
Clayton Union	Clayton	A0686	adjuvant
Contains	47% w/w methylated rapeseed oil		
Use with	Clayton Tonto as 0.5% of total spray volume		
Protective clothing	A, H		
Precautions	R43, R50, R53a, U02a, U05a, U09a, U20b, E15b, D01, D02, D05, D10c, D12a, H04, H11		
Codacide Oil	Microcide	A0629	vegetable oil
Contains	95% w/w oil (rapeseed triglycerides) (EAC 1)		
Use with	All approved pesticides and tank mixes. See label for details		
Protective clothing	A, C		
Precautions	U20b, D09a, D10b		
Companion Gold	Agrovista	A0723	acidifier/buffering agent/drift retardant/extender
Contains	16% w/w ammonium sulphate + 0.95% w/w polyacrylamide		
Use with	Diquat or glyphosate on oilseed rape at 0.5% solution; glyphosate on wheat, rye or triticale at 0.5% solution and with all approved pesticides on non-edible crops at 1% solution or all approved pesticides at 50% dose rate or less on edible crops		
Protective clothing	A, C, H		
Precautions	U05a, U14, U15, U19a, U19c, U20b, E15a, D02, D05, D09a, D10a		
Compliment	United Agri	A0705	spreader/vegetable oil/wetter
Contains	75% w/w mixed fatty acid esters of rapeseed oil		
Use with	All approved pesticides in non-edible crops, all approved pesticides on edible crops when used at half or less their recommended rate, morpholine or triazine fungicides on cereals, and with all pesticides on listed crops up to specified growth stages		
Protective clothing	A, C		
Precautions	R43, R52, R53a, U14, E38, D01, D05, D07, D08, H04		
Contact Plus	Interagro	A0418	mineral oil
Contains	95% w/w mineral oil		
Use with	Sugar beet herbicides, oilseed rape herbicides, cereal graminicides and a wide range of other pesticides that have a label recommendation for use with authorised adjuvant oils on specified crops. Refer to label or contact supplier for further details		
Protective clothing	A, C		
Precautions	R22b, R38, U02a, U05a, U08, U20b, E15a, D01, D02, D09a, D10a, M05b, H03		

Product	Supplier	Adj. No.	Type
County Mark	Greenaway	A0689	sticker/wetter
Contains	10% w/w rapeseed oil		
Use with	'Greenaway Gly-490' (MAPP 12718) and all approved 490 g/l glyphosate products		
Protective clothing	A, C		
Precautions	U11, U12, U15, U20b, E34		
Cropspray 11-E	Petro-Lube	A0537	adjuvant/mineral oil
Contains	99% w/w oil (petroleum oils) (EAC 1)		
Use with	All approved pesticides on all edible crops when used at half their recommended dose or less, and on all non-edible crops up to their full recommended dose. Also with listed herbicides on a range of specified crops and with all pesticides on a range of specified crops up to specified growth stages.		
Protective clothing	A, C		
Precautions	R22a, U10, U16b, U19a, E15a, E19b, E34, E37, D01, D02, D05, D10a, D12a, M05b, H03		
Dash HC	BASF	A0729	non-ionic surfactant/wetter
Contains	348.75 g/l oil (fatty acid esters) and 209.25 g/l alkoxylated alcohols-phosphate esters		
Use with	For use with Cleranda at 1.0 l/ha in a water volume of 100 - 400 l/ha on Clearfield oilseed rape		
Protective clothing	A, C, H		
Precautions	U05a, U12, U14, D01, D02, D10c, M05b, H03		
Designer	De Sangosse	A0660	drift retardant/extender/sticker/wetter
Contains	25% w/w styrene-butadiene copolymers (EAC 1), 7.1% w/w trisiloxane organosilicone copolymers (EAC 2)		
Use with	A wide range of fungicides, insecticides and trace elements for cereals and specified agricultural and horticultural crops		
Protective clothing	A, C, H		
Precautions	R36, R38, R52, U02a, U11, U14, U15, U19a, E15a, E37, D01, D02, D05, D09a, D10a, H04		
Desikote Max	Taminco	A0666	anti-transpirant/extender/sticker/wetter
Contains	400 g/l di-1-p-menthene		
Use with	Approved pesticides on all edible crops (except herbicides on peas) and as an anti-transpirant on vegetables before transplanting, evergreens, deciduous trees, shrubs, bushes and on turf		
Protective clothing	A		
Precautions	R38, R50, R53a, U05a, U14, U20b, E15a, E34, E37, D01, D09a, D10b, D12b, M03, H04, H11		
Diagor	AgChem Access	A0671	wetter
Contains	47% w/w methylated rapeseed oil		
Use with	Topik, Axial, Trazos, Amazon and Viscount on cereals in accordance with recommendations on the respective herbicide labels		
Protective clothing	A, C, H		
Precautions	R43, R51, R53a, U02a, U05a, U09a, U20b, E15b, E38, D01, D02, D05, D09a, D10c, D12a, H04, H11		

SECTION 4

Product	Supplier	Adj. No.	Type
Drill	De Sangosse	A0544	adjuvant
Contains	15% w/w alkoxylated alcohols (EAC 1), 7.5% w/w oil (tall oil fatty acids - EAC 2), 63.34% w/w oil (rapeseed fatty acid esters) (EAC 3)		
Use with	All approved pesticides on all edible crops when used at half their recommended dose or less, and on all non-edible crops up to their full recommended dose. Also at a maximum concentration of 0.1% with approved pesticides on listed crops up to specified growth stages		
Protective clothing	A, C, H		
Precautions	R36, U05a, U11, U14, U19a, U20b, E15a, E19b, E34, D01, D02, D05, D10a, D12a, H04		
Eco-flex	Greenaway	A0696	sticker/wetter
Contains	10% w/w refined rapeseed oil		
Use with	Approved formulations of glyphosate, 2,4-D		
Protective clothing	A, C		
Precautions	U11, U12, U15, U20b, E34		
Elan Xtra	Intracrop	A0735	spreader/wetter
Contains	530 g/l ethylene oxide-propylene oxide copolymers (EAC 1) and 415 g/l trisiloxane organosilicone copolymers (EAC 2)		
Use with	All approved pesticides on non-edible crops and non-crop production and with half dose of all approved pesticides on edible crops		
Protective clothing	A		
Emerald	Intracrop	A0636	anti-drift agent/anti-transpirant/extender/ UV screen
Contains	96.0% w/w pinene oligomers (EAC 1)		
Use with	Recommended rates of approved pesticides up to the growth stage indicated for specified crops, and with half or less than the recommended rate on these crops after the stated growth stages. Also for use alone on transplants, turf, fruit crops, glasshouse crops and Christmas trees. Must not be used in mixture with adjuvant oils or surfactants		
Protective clothing	A, H		
Precautions	U19a, U20c, E13c, E37, D09a, D11a		
Euroagkem Pen-e-trate	EuroAgkem	A0564	spreader/wetter
Contains	350 g/l propionic acid (EAC 1)		
Use with	All approved pesticides which have a recommendation for use with a wetting agent; pesticides must be used at half approved dose rate or less on edible crops, up to full dose rate on non-edible crops		
Protective clothing	A, C		
Precautions	U05a, U08, U11, U14, U15, U19a, U20c, E13e, E15a, D01, D02, D09a, D10b, M03, H05		
Felix	Intracrop	A0178	spreader/wetter
Contains	600 g/l alkoxylated alcohols (EAC 1)		
Use with	Mecoprop, 2,4-D in cereals and amenity turf, and a range of grass weedkillers in agriculture. See label for details		
Protective clothing	A, C		
Precautions	R36, R38, U05a, U08, U20a, E15a, D01, D02, D09a, D10a, H04, H08		

Product	Supplier	Adj. No.	Type
Firebrand	Barclay	-	fertiliser/water conditioner
Contains	500 g/l ammonium sulphate		
Use with	Use at 0.5% v/v in the spray solution with glyphosate		
Precautions	U08, U20a, E15a, D09a, D10b		
Gateway	United Agri	A0651	extender/sticker/wetter
Contains	73% w/v synthetic latex solution and 8.5% w/v polyether modified trisiloxane		
Use with	All approved pesticides in crops not destined for human or animal consumption, all approved pesticides on edible crops when used at half or less their recommended rate, and with all pesticides on listed crops up to specified growth stages		
Protective clothing	A, C, H		
Precautions	R41, R52, R53a, U11, U15, E37, E38, D01, D05, D07, D08, H04		
Gly-Flex	Greenaway	A0588	spreader/sticker/wetter
Contains	95% w/w refined rapeseed oil		
Use with	GLY-490 (MAPP 12718) or any approved formulations of 490 g/l glyphosate		
Protective clothing	A, C		
Precautions	R36, U11, U12, U15, U20b, E13c, E34, E37, H04		
Gly-Plus A	Greenaway	A0736	spreader/sticker/wetter
Contains	10.0% w/w oil (rapeseed triglycerides) (EAC 1)		
Use with	Any approved 360 g/l glyphosate		
Protective clothing	A, C		
Precautions	R36, U05a, U08, U11, U12, U15, U20b, E13c, E34, E37, E38, H04		
Green Gold	Intracrop	A0250	spreader/wetter
Contains	950 g/l oil (rapeseed triglycerides) (EAC 1)		
Use with	All pesticides which have a recommendation for the addition of a wetter/spreader		
Precautions	U08, U20b, E13c, E34, D09a, D10b		
Grounded	Helena	A0456	mineral oil
Contains	732 g/l petroleum oils		
Use with	All approved pesticides on edible and non-edible crops when used at half their approved dose or less. Also with approved pesticides on specified crops, up to specified growth stages, at up to their full approved dose		
Protective clothing	A, C		
Precautions	R53a, U02a, U05a, U08, U20b, E15a, E34, E37, D01, D02, D05, D09a, D10c, H11		
Headland Fortune	Headland	A0703	penetrant/spreader/vegetable oil/wetter
Contains	75% w/w mixed methylated fatty acid esters of seed oil and N-butanol		
Use with	Herbicides and fungicides in a wide range of crops. See label for details		
Protective clothing	A, C		
Precautions	R43, U02a, U05a, U14, U20a, E15a, E34, E38, D01, D02, D05, D09a, D10b, H04		

SECTION 4

Product	Supplier	Adj. No.	Type
Headland Guard 2000	Headland	A0652	extender/sticker
Contains	10% w/w styrene/butadiene co-polymers		
Use with	All approved pesticides on all edible crops when used at half their recommended dose or less, and on all non-edible crops up to their full recommended dose. Also at a maximum concentration of 0.1% with approved pesticides on listed crops up to specified growth stages		
Protective clothing	A, C		
Precautions	E15a, D01, D02		
Headland Intake	Headland	A0074	penetrant
Contains	450 g/l propionic acid		
Use with	All approved pesticides on any crop not intended for human or animal consumption, and with all approved pesticides on beans, peas, edible podded peas, oilseed rape, linseed, sugar beet, cereals (except triazole fungicides), maize, Brussels sprouts, potatoes, cauliflowers. See label for detailed advice on timing on these crops		
Protective clothing	A, C		
Precautions	R34, U02a, U05a, U10, U11, U14, U15, U19a, E34, D01, D02, D05, D09b, D10b, M04a, H05		
Headland Rheus	Headland	A0328	wetter
Contains	85% w/w polyalkylene oxide modified heptamethyl siloxane		
Use with	Any herbicide, systemic fungicide, systemic insecticide or plant growth regulator (except any product applied in or near water) where the use of a wetting, spreading and penetrating surfactant is recommended to improve foliar coverage		
Protective clothing	A, C		
Precautions	R21, R22a, R38, R41, R43, R51, R58, U02a, U05a, U08, U11, U14, U19a, U20b, E13c, E34, D01, D02, D05, D09a, D10b, M03, H03, H11		
Herbi-AKtiv	Global Adjuvants	A0821	spreader/wetter
Contains	57.0 % w/w ethylene oxide-propylene oxide copolymers (EAC 1)		
Use with	All authorised plant protection products at half or less than half the authorised plant protection product rate		
Protective clothing	A, C, H, M		
Precautions	U05a, U20c, E34, D01, D02, D10a, M04a		
Intracrop Agwet GTX	Intracrop	A0646	spreader/wetter
Contains	500 g/l alkoxylated alcohols		
Use with	All approved pesticides on non-edible crops, all approved pesticides on edible crops at half or less than half the pesticide rate and all approved pesticides with a recommendation for use with a non-ionic or wetting agent.		
Protective clothing	A, C, H		
Precautions	H318, U05a, U11, U14, U15, U19a, U20a, M04c, H226, H302		

Product	Supplier	Adj. No.	Type
Intracrop BLA	Intracrop	A0655	anti-drift agent/anti-transpirant/extender/ sticker/UV screen
Contains	22.0% w/w styrene-butadiene copolymers (EAC 1)		
Use with	All potato blight fungicides. Also with recommended rates of approved pesticides in certain non-crop situations and up to the growth stage indicated for specified crops, and with half or less than the recommended rate on these crops after the stated growth stages. Also with pesticides in grassland at half or less their recommended rate		
Protective clothing	A, C		
Precautions	U08, U20b, E13c, E34, E37, D01, D05, D09a, D10a		
Intracrop Bla-Tex	Intracrop	A0656	extender/sticker
Contains	22.0% w/w styrene-butadiene copolymers (EAC 1)		
Use with	All approved pesticides on non-edible crops and non crop production, with half dose of all approved pesticides on edible crops and with blight fungicides in potatoes		
Protective clothing	A, C		
Precautions	U08, U20b, E13c, E34, E37, D01, D05, D09a, D10a		
Intracrop Boost	Intracrop	A0774	spreader/sticker/wetter
Contains	32.67% w/w alkoxylated alcohols (EAC 1) and 1.0% w/w trisiloxane organosilicone copolymers (EAC 2)		
Use with	All approved pesticides at half or less than half the approved pesticide rate on edible crops and all approved pesticides on non-edible crops		
Protective clothing	A, C		
Intracrop Cogent	Intracrop	A0775	spreader/sticker/wetter
Contains	32.67% w/w alkoxylated alcohols (EAC 1) and 1.0% w/w trisiloxane organosilicone copolymers (EAC 2)		
Use with	All approved pesticides at half or less than half the approved pesticide rate on edible crops and all approved pesticides on non-edible crops		
Protective clothing	A, C		
Intracrop Dictate	Intracrop	A0673	adjuvant
Contains	91.0% w/w oil (rapeseed fatty acid esters) (EAC 1)		
Use with	All approved pesticides on non-edible crops; all approved pesticides applied at half or less than half dose on edible crops		
Protective clothing	A, C		
Precautions	U19a, U20b, D05, D09a, D10b		
Intracrop Evoque	Intracrop	A0754	spreader/wetter
Contains	652 g/l oil (rapeseed fatty acid esters) (EAC 1), 112 g/l trisiloxane organosilicone copolymers (EAC 2), 22.5 g/l alkoxylated alcohols (EAC 3) and 19.1 g/l alkoxylated alcohols (EAC 4)		
Use with	All authorised plant protection products on non-edible crops and all authorised plant protection products at half or less than half the authorised plant protection product rate on edible crops. Max conc is 0.2% of spray volume.		
Protective clothing	A, C		

SECTION 4

Product	Supplier	Adj. No.	Type
Intracrop F16	Intracrop	A0752	spreader/wetter
Contains	652 g/l oil (rapeseed fatty acid esters) (EAC 1), 112 g/l trisiloxane organosilicone copolymers (EAC 2), 22.5 g/l alkoxylated alcohols (EAC 3) and 19.1 g/l alkoxylated alcohols (EAC 4)		
Use with	All authorised plant protection products on non-edible crops and all authorised plant protection products at half or less than half the authorised plant protection product rate on edible crops. Max conc is 0.2% of spray volume.		
Protective clothing	A, C		
Intracrop Impetus	Intracrop	A0647	spreader/wetter
Contains	50.0 % w/w alkoxylated alcohols		
Use with	All approved pesticides on non-edible crops, all approved pesticides on edible crops at half or less than half the pesticide rate and all approved pesticides with a recommendation for use with a non-ionic or wetting agent.		
Protective clothing	A, C, H		
Precautions	H318, U05a, U11, U14, U15, U19a, U20a, M04c, H226, H302		
Intracrop Inca	Intracrop	A0784	adjuvant
Contains	840 g/l oil (rapeseed fatty acid esters) (EAC 1)		
Use with	All approved pesticides at half or less than half the approved pesticide rate on edible crops and all approved pesticides on non-edible crops		
Protective clothing	A, C		
Intracrop Incite	Intracrop	A0785	adjuvant
Contains	840 g/l oil (rapeseed fatty acid esters) (EAC 1)		
Use with	All approved pesticides at half or less than half the approved pesticide rate on edible crops and all approved pesticides on non-edible crops		
Protective clothing	A, C		
Intracrop Mica AF	Intracrop	A0808	spreader/sticker/wetter
Contains	250 g/l ethylene oxide-propylene oxide copolymers (EAC 1), 70.2 g/l styrene-butadiene copolymers (EAC 2) and 41.5 g/l trisiloxane organosilicone copolymers (EAC 3)		
Use with	All approved pesticides at half or less than half the approved rate		
Protective clothing	A, C		
Precautions	U05a, U08, U20c, E13e, D01, D02, D09a, D10a		
Intracrop Neotex	Intracrop	A0657	anti-drift agent/anti-transpirant/extender/sticker/UV screen
Contains	22.0 % w/w styrene-butadiene copolymers		
Use with	Recommended rates of approved pesticides up to specified growth stages of a wide range of agricultural arable and horticultural crops, and for non-crop uses. Use on edible crops beyond specified growth stages, and in grass, should only be with half recommended rates of the pesticide or less. See label for details of growth stage restrictions. In addition may be used with all potato blight fungicides at their recommended rates of use up to the latest recommended timing of the fungicide		
Protective clothing	A, C		
Precautions	U08, U20b, E13c, E34, E37, D01, D05, D09a, D10a		

Product	Supplier	Adj. No.	Type
Intracrop Novatex	Intracrop	A0658	anti-drift agent/anti-transpirant/extender/ sticker/UV screen
Contains	22.0 % w/w styrene-butadiene copolymers		
Use with	Recommended rates of approved pesticides up to the growth stage indicated for specified crops, and with half or less than the recommended rate on these crops after the stated growth stages. Also for use with recommended rates of pesticides on grassland and specified non-crop situations		
Protective clothing	A, C		
Precautions	U08, U20b, E13c, E34, E37, D01, D09a, D10a		
Intracrop Perm-E8	Intracrop	A0565	spreader/wetter
Contains	42.0% w/w propionic acid (EAC 1)		
Use with	All approved formulations of chlormequat, all approved formulations of glyphosate, diquat, fenoxaprop-P-ethyl, tralkoxydim, clodinafop-propargyl, fluazifop-P-butyl, cycloxydim and propaquizafop at half or less than half the approved pesticide rate in edible crops, at full rate in non-edible crops.		
Protective clothing	A, C		
Precautions	U02a, U04a, U05a, U08, U10, U11, U13, U14, U15, U19a, U20b, D01, D02, D09a, D10a, H05		
Intracrop Predict	Intracrop	A0503	adjuvant/vegetable oil
Contains	91.0% w/w oil (rapeseed fatty acid esters) (EAC 1)		
Use with	All approved pesticides on non-edible crops, and all pesticides approved for use on growing edible crops when used at half recommended dose or less. On specified crops product may be used at a maximum spray concentration of 1% with approved pesticides at their full approved rate up to the growth stages shown in the label		
Protective clothing	A, C		
Precautions	U19a, U20b, D05, D09a, D10b		
Intracrop Quad	Intracrop	A0753	spreader/wetter
Contains	652 g/l oil (rapeseed fatty acid esters) (EAC 1), 112 g/l trisiloxane organosilicone copolymers (EAC 2), 22.5 g/l alkoxylated alcohols (EAC 3) and 19.1 g/l alkoxylated alcohols (EAC 4)		
Use with	All approved pesticides on non-edible crops, and all pesticides approved for use on growing edible crops when used at half recommended dose or less.		
Protective clothing	A, C		
Intracrop Quartz	Intracrop	A0776	spreader/sticker/wetter
Contains	32.67% w/w alkoxylated alcohols (EAC 1) and 1.0% w/w trisiloxane organosilicone copolymers (EAC 2)		
Use with	All approved pesticides on non-edible crops, and all pesticides approved for use on growing edible crops when used at half recommended dose or less.		
Protective clothing	A, C		

SECTION 4

Product	Supplier	Adj. No.	Type
Intracrop Questor	Intracrop	A0495	activator/non-ionic surfactant/spreader
Contains	750 g/l ethylene oxide-propylene oxide copolymers (EAC1)		
Use with	All approved pesticides on non-edible crops and pesticides used in non-crop production, and all pesticides approved for use on growing edible crops when used at half recommended dose or less. On specified crops product may be used at a maximum spray concentration of 0.3% with approved pesticides at their full approved rate up to the growth stages shown in the label		
Protective clothing	A, C		
Precautions	R36, R38, U04a, U05a, U08, U19a, E13b, D01, D02, D09a, D10b, M03, M05a, H04		
Intracrop Rapide Beta	Intracrop	A0672	wetter
Contains	350 g/l propionic acid (EAC 1) and 100 g/l alkoxylated alcohols (EAC 2)		
Use with	All approved pesticides which have a recommendation for use with a wetting agent; pesticides must be used at half approved dose rate or less on edible crops, up to full dose rate on non-edible crops		
Protective clothing	A, C		
Precautions	U02a, U04a, U05a, U08, U10, U11, U13, U14, U15, U19a, U20b, E15a, D01, D02, D09a, D10a, M05a, H05		
Intracrop Retainer NF	Intracrop	A0711	adjuvant
Contains	91.0 % w/w oil (rapeseed fatty acid esters)		
Use with	All approved pesticides at half or less than half the approved pesticide rate on edible crops and all approved pesticides on non-edible crops		
Protective clothing	A, C		
Intracrop Rigger	Intracrop	A0783	adjuvant/vegetable oil
Contains	840 g/l oil (rapeseed fatty acid esters)		
Use with	All approved pesticides on non-edible crops, and all pesticides approved for use on growing edible crops when used at half recommended dose or less. On specified crops product may be used at a maximum spray concentration of 1.78% with approved pesticides at their full approved rate up to the growth stages shown in the label		
Protective clothing	A, C		
Precautions	U19a, U20b, D05, D09a, D10b		
Intracrop Rustler	Intracrop	A0777	spreader/sticker/wetter
Contains	32.67% w/w alkoxylated alcohols (EAC 1) and 1.0% w/w trisiloxane organosilicone copolymers (EAC 2)		
Use with	All approved pesticides on non-edible crops, and all pesticides approved for use on growing edible crops when used at half recommended dose or less.		
Protective clothing	A, C		
Intracrop Salute AF	Intracrop	A0809	spreader/sticker/wetter
Contains	250 g/l ethylene oxide-propylene oxide copolymers (EAC 1), 70.2 g/l styrene-butadiene copolymers (EAC 2) and 41.5 g/l trisiloxane organosilicone copolymers (EAC 3)		
Use with	All approved pesticides on non-edible crops, and all pesticides approved for use on growing edible crops when used at half recommended dose or less.		
Protective clothing	A, C		
Precautions	U05a, U08, U20c, E13e, E40c, D01, D02, D09a, D10a		

Product	Supplier	Adj. No.	Type
Intracrop Sapper AF	Intracrop	A0810	spreader/sticker/wetter
Contains	250 g/l ethylene oxide-propylene oxide copolymers (EAC 1), 70.2 g/l styrene-butadiene copolymers (EAC 2) and 41.5 g/l trisiloxane organosilicone copolymers (EAC 3)		
Use with	All approved pesticides on non-edible crops, and all pesticides approved for use on growing edible crops when used at half recommended dose or less.		
Protective clothing	A, C		
Precautions	U05a, U08, U20c, E13e, E40c, D01, D02, D09a, D10a		
Intracrop Saturn	Intracrop	A0494	activator/non-ionic surfactant/spreader
Contains	750 g/l ethylene oxide-propylene oxide copolymers		
Use with	All approved pesticides on non-edible crops and pesticides used in non-crop production, and all pesticides approved for use on growing edible crops when used at half recommended dose or less. On specified crops product may be used at a maximum spray concentration of 0.3% with approved pesticides at their full approved rate up to the growth stages shown in the label		
Protective clothing	A, C		
Precautions	R36, R38, U04a, U05a, U08, U19a, E13b, E40c, D01, D02, D09a, D10b, M03, M05a, H04		
Intracrop Signal XL	Intracrop	A0659	anti-drift agent/anti-transpirant/extender/sticker/UV screen
Contains	22.0% w/w styrene-butadiene copolymers (EAC 1)		
Use with	All approved pesticides on non-edible crops and non crop production uses; All approved pesticides at half or less than half the approved pesticide rate on edible crops		
Protective clothing	A, C		
Precautions	U05a, U19a, U20b, E13c, E34, D01, D02, D09a, D10b, M04a, M05a		
Intracrop Sprinter	Intracrop	A0513	spreader/wetter
Contains	19.0 % w/v alkoxylated alcohols		
Use with	Recommended rates of approved pesticides up to the growth stage indicated for specified crops, and with half or less than the recommended rate on these crops after the stated growth stages. Also with herbicides on managed amenity turf at recommended rates, and on grassland at half or less than recommended rates		
Protective clothing	A, C		
Precautions	R36, R38, R41, U05a, U08, U11, U14, U15, U20c, E15a, D01, D02, D09a, D10b, M03, H04		
Intracrop Status	Intracrop	A0506	adjuvant/vegetable oil
Contains	91.0 % w/w oil (rapeseed fatty acid esters)		
Use with	All approved pesticides on non-edible crops, and all pesticides approved for use on growing edible crops when used at half recommended dose or less. On specified crops product may be used at a maximum spray concentration of 1.0% with approved pesticides at their full approved rate up to the growth stages shown in the label		
Protective clothing	A, C		
Precautions	U19a, U20b, D05, D09a, D10b		

SECTION 4

Product	Supplier	Adj. No.	Type
Intracrop Stay-Put	Intracrop	A0507	vegetable oil
Contains	91.0 % w/w oil (rapeseed fatty acid esters)		
Use with	Recommended rates of approved pesticides for non-crop uses and with recommended rates of approved pesticides up to the growth stage indicated for specified crops, and with half or less than the recommended rate on these crops after the stated growth stages		
Protective clothing	A, C		
Precautions	U19a, U20b, D05, D09a, D10b		
Intracrop Super Rapeze MSO	Intracrop	A0782	adjuvant/vegetable oil
Contains	840 g/l oil (rapeseed fatty acid esters) (EAC 1)		
Use with	All approved pesticides on non-edible crops, and all pesticides approved for use on growing edible crops when used at half recommended dose or less. On specified crops product may be used at a maximum spray concentration of 1.78% with approved pesticides at their full approved rate up to the growth stages shown in the label		
Protective clothing	A, C		
Precautions	U19a, U20b, D05, D09a, D10b		
Intracrop Tonto	Intracrop	A0778	spreader/sticker/wetter
Contains	32.67% w/w alkoxylated alcohols (EAC 1) and 1.0% w/w trisiloxane organosilicone copolymers (EAC 2)		
Use with	All approved pesticides at half or less than half the approved pesticide rate on edible crops and all approved pesticides on non-edible crops		
Protective clothing	A, C		
Precautions	U05a, U08, U20c, E13e, E40c, D01, D02, D09a, D10a		
Intracrop Warrior	Intracrop	A0514	spreader/wetter
Contains	19.0 % w/v alkoxylated alcohols		
Use with	Recommended rates of approved pesticides up to the growth stage indicated for specified crops, and with half or less than the recommended rate on these crops after the stated growth stages. Also with herbicides on managed amenity turf at recommended rates, and on grassland at half or less than recommended rates		
Protective clothing	A, C		
Precautions	R36, R38, R41, U05a, U08, U11, U14, U15, U20c, E15a, D01, D02, D09a, D10b, M03, H04		
Intracrop Zenith AF	Intracrop	A0811	spreader/sticker/wetter
Contains	250 g/l ethylene oxide-propylene oxide copolymers, 70.2 g/l styrene-butadiene copolymers and 41.5 g/l trisiloxane organosilicone copolymers.		
Use with	All approved pesticides on non-edible crops, all approved pesticides on edible crops at half or less than half the pesticide rate and all approved pesticides with a recommendation for use on non crop production.		
Protective clothing	A, C, H		
Precautions	H318, U05a, U11, U14, U15, U19a, U20a, M04c, H226, H302		

Product	Supplier	Adj. No.	Type
Intracrop Zodiac	Intracrop	A0755	spreader/wetter
Contains	652 g/l oil (rapeseed fatty acid esters), 112 g/l trisiloxane organosilicone copolymers), 22.5 g/l alkoxylated alcohols and 19.1 g/l alkoxylated alcohols		
Use with	All authorised plant protection products on all non edible crops and all authorised plant protection products on edible crops at half or less than half the authorised plant protection product rate		
Protective clothing	A, C		
Precautions	U05a, U08, U20c, E13e, E40c, D01, D02, D09a, D10a		
Kantor	Interagro	A0623	spreader/wetter
Contains	790 g/l alkoxylated triglycerides		
Use with	All approved pesticides		
Protective clothing	A, C		
Precautions	U05a, U20b, E15b, E19b, E34, D01, D02, D09a, D10a, D12a		
Katalyst	Interagro	A0450	penetrant/water conditioner
Contains	90% w/w alkoxylated fatty amine		
Use with	Glyphosate and a wide range of other pesticides on specified crops. Refer to label or contact supplier for further details		
Protective clothing	A, C		
Precautions	R22a, R36, R38, R50, R58, U02a, U05a, U08, U19a, U20b, E15a, E34, E37, D01, D02, D05, D09a, D10a, M03, H03, H11		
Kinetic	Helena	A0252	spreader/wetter
Contains	80% w/w ethylene oxide and propylene oxide copolymers, 16% w/w trisiloxane copolymers		
Use with	Approved pesticides on non-edible crops and cereals and stubbles of all edible crops when used at full recommended dose, and with pesticides approved for use on growing edible crops when used at half recommended dose or less. On specified crops product may be used at a maximum spray concentration of 0.2% with approved pesticides at their full approved rate up to the growth stages shown in the label		
Protective clothing	A, H		
Precautions	R38, R41, U02a, U05a, U08, U19a, U20b, E13c, E34, E37, D01, D02, D05, D09a, D10c, M03, H04		
Klipper	Amega	A0260	spreader/wetter
Contains	600 g/l alkoxylated alcohols (EAC 1) and isobutanol		
Use with	Listed herbicides, fungicides and chlorpyrifos on managed amenity turf or amenity grassland (see label for details).		
Protective clothing	A, C, H		
Precautions	R22a, R41, R67, U02a, U04a, U05a, U11, U14, U15, E15a, E34, D01, D02, D09a, D10a, M03, H03, H08		
Kotek	Agrovista	A0746	spreader/wetter
Contains	83.0% w/w trisiloxane organosilicone copolymers (EAC 1)		
Use with	All approved pesticides at half or less than half the approved rate on edible crops; all approved pesticides on non-edible crops or non-production targets. Max conc 0.15% of spray volume		
Protective clothing	A, H		
Precautions	R20, R41, R51, R53a, U11		

SECTION 4

Product	Supplier	Adj. No.	Type
Leaf-Koat	Helena	A0511	spreader/wetter
Contains	80% w/w ethylene oxide and propylene oxide copolymers, 16% w/w trisiloxane copolymers		
Use with	Approved pesticides on non-edible crops and cereals and stubbles of all edible crops when used at full recommended dose, and with pesticides approved for use on growing edible crops when used at half recommended dose or less. On specified crops product may be used at a maximum spray concentration of 0.2% with approved pesticides at their full approved rate up to the growth stages shown in the label		
Protective clothing	A, H		
Precautions	R38, R41, U02a, U05a, U08, U19a, U20b, E13c, E34, E37, D01, D02, D05, D09a, D10c, M03, H04		
Level	United Agri	A0654	extender/sticker
Contains	10% w/w styrene-butadiene copolymers (EAC 1)		
Use with	All approved pesticides in non-edible crops, all approved pesticides on edible crops when used at half or less of their recommended rate, potato blight fungicides and with all pesticides on listed crops up to specified growth stages		
Protective clothing	A, C		
Precautions	E37, D01, D02, D05		
Li-700	De Sangosse	A0529	acidifier/drift retardant/penetrant
Contains	9.39% w/w alkoxylated alcohols (EAC 1), 35.0% w/w propionic acid (EAC 2), 35.0% w/w soybean phospholipids (EAC 3)		
Use with	All approved pesticides on all edible crops when used at half their recommended dose or less, and on all non-edible crops up to their full recommended dose. Also with morpholine fungicides on cereals and with pirimicarb on legumes when used at full dose; with iprodione on brassicas when used at 75% dose, and at a maximum concentration of 0.5% with all pesticides on listed crops at full dose up to specified growth stages		
Protective clothing	A, C, H		
Precautions	R36, R38, U11, U14, U15, U19a, E15a, E19b, D01, D02, D05, D10a, D12a, H04		
Logic	Microcide	A0288	vegetable oil
Contains	95% w/w oil (rapeseed triglycerides) (EAC 1)		
Use with	All approved pesticides on edible and non-edible crops for ground or aerial application		
Protective clothing	A, C		
Precautions	U19a, U20b, D09a, D10b		
Logic Oil	Microcide	A0630	vegetable oil
Contains	95% w/w oil (rapeseed triglycerides) (EAC 1)		
Use with	All approved pesticides on edible and non-edible crops for ground or aerial application		
Protective clothing	A, C		
Precautions	U19a, U20b, D09a, D10b		

Product	Supplier	Adj. No.	Type
Low Down	Helena	A0459	mineral oil
Contains	732 g/l petroleum oils		
Use with	All approved pesticides on edible and non-edible crops when used at half their approved dose or less. Also with approved pesticides on specified crops, up to specified growth stages, at up to their full approved dose		
Protective clothing	A, C		
Precautions	R53a, U02a, U05a, U08, U20b, E15a, E34, E37, D01, D02, D05, D09a, D10c, H11		

Master Sil	Global Adjuvants	A0822	spreader/wetter
Contains	83% trisiloxane organosilicone copolymers		
Use with	All approved pesticides on all edible crops when used at half their recommended dose or less, and on all non-edible crops up to their full recommended dose.		
Protective clothing	A, C, H		
Precautions	H319, E34, H411, D01, D02, D09a, D10a		

Master Wett	Global Adjuvants	A0826	spreader/wetter
Contains	84% trisiloxane organosilicone copolymers		
Use with	All approved pesticide products at a dose of 0.25% spray solution.		
Protective clothing	A, C, H		
Precautions	H319, E34, H411, D01, D02, D09a, D10a		

Meco-Flex	Greenaway	A0678	spreader/sticker/wetter
Contains	10.0% w/w oil (rapeseed triglycerides) (EAC 1)		
Use with	Re-Act, Headland Relay Depitox		
Protective clothing	A, C		
Precautions	U11, U12, U15, U20b, E15a, E34, D02		

Mero	Bayer CropScience	A0818	wetter
Contains	81.4% w/w rapeseed fatty acid esters		
Use with	All approved cereal and maize herbicides and all approved brassica insecticides		
Protective clothing	A, C, H		
Precautions	R38, U05a, U09a, U13, U20b, E34, D01, D02, D09a, D10a, M05a		

Mixture B NF	Amega	A0570	non-ionic surfactant/spreader/wetter
Contains	37% w/w alkoxylated alcohols (EAC 1) and 41% w/w alkoxylated alcohols (EAC 2) and isopropanol		
Use with	All approved pesticides on non-edible crops at their full recommended rate. Also with Timbrel (MAPP 05815) and all approved formulations of glyphosate in non-crop situations. Also with a range of specified herbicides when used at less than half their recommended rate on forest and grassland, and with all approved pesticides at full rate on a wide range of specified edible crops up to specified growth stages		
Protective clothing	A, C, H		
Precautions	U02a, U04a, U05a, U11, U14, U15, E13a, E34, D01, D02, D09a, D10a, M03, H03, H11		

SECTION 4

Product	Supplier	Adj. No.	Type
Nelson	Agrovista	A0796	spreader/wetter
Contains	300 g/l alkoxylated alcohols (EAC 1) and 300 g/l alkoxylated alcohols (EAC 2)		
Use with	All approved pesticides at half the approved maximum dose on edible crops and at the full approved dose on non-edible crops and non-crop production at 0.5% spray volume.		
Protective clothing	A, C, H		
Precautions	H318, U02a, U05a, U08, U09c, U11, U13, U14, U15, U20d, M03a, M05b, H302		
Newman Cropspray 11E	De Sangosse	A0530	adjuvant
Contains	99% w/w oil (petroleum oils) (EAC 1)		
Use with	All approved pesticides on all edible crops when used at half their recommended dose or less, and on all non-edible crops up to their full recommended dose. Also with listed herbicides on a range of specified crops and with all pesticides on a range of specified crops up to specified growth stages.		
Protective clothing	A, C		
Precautions	R22a, U10, U16b, U19a, E15a, E19b, E34, E37, D01, D02, D05, D10a, D12a, M05b, H03		
Newman's T-80	De Sangosse	A0192	spreader/wetter
Contains	800 g/l alkoxylated tallow amines (EAC 1)		
Use with	Glyphosate		
Protective clothing	A, C, H		
Precautions	R22a, R37, R38, R41, R50, R58, R67, U05a, U11, U14, U19a, U20a, E15a, E19b, E34, E37, D01, D02, D05, D10a, D12a, M03, M04a, H03, H08, H11		
Nion	Amega	A0760	spreader/wetter
Contains	90% w/w alkoxylated alcohols		
Use with	All approved pesticides at half the approved maximum dose on edible crops and at the full approved dose on non-edible crops and non-crop production at 1% spray volume.		
Protective clothing	A, C		
Precautions	U02a, U04a, U05a, U11, U14, U15, E15a, E34, D01, D02, D09a, D10a, M03, H03		
Nu Film P	Intracrop	A0635	anti-drift agent/anti-transpirant/sticker/ UV screen/wetter
Contains	96.0 % w/w pinene oligomers		
Use with	Glyphosate and many other pesticides and growth regulators for which a protectant is recommended. Do not use in mixture with adjuvant oils or surfactants		
Protective clothing	A, C		
Precautions	U19a, U20b, E13c, D09a, D10a		
Pan Oasis	Pan Agriculture	A0411	vegetable oil
Contains	95% w/w methylated rapeseed oil		
Use with	A wide range of pesticides that have a label recommendation for use with authorised adjuvant oils. Contact distributor for further details		
Protective clothing	A		
Precautions	R36, U02a, U05a, U08, U20b, E15a, D01, D02, D05, D09a, D10a, H04		

Product	Supplier	Adj. No.	Type
Pan Panorama	Pan Agriculture	A0412	mineral oil
Contains	95% w/w mineral oil		
Use with	A wide range of pesticides that have a label recommendation for use with adjuvant oils. Contact distributor for details		
Protective clothing	A		
Precautions	R22b, R38, U02a, U05a, U08, U20b, E13c, D01, D02, D05, D09a, D10a, M05b, H03		
Phase II	De Sangosse	A0622	vegetable oil
Contains	95.2% w/w oil (rapeseed fatty acid esters) (EAC 1)		
Use with	Pesticides approved for use in sugar beet, oilseed rape, cereals (for grass weed control), and other specified agricultural and horticultural crops		
Protective clothing	A, C, H		
Precautions	U19a, U20b, E15a, E19b, E34, D01, D02, D05, D10a, D12a		
Pin-o-Film	Intracrop	A0637	anti-drift agent/anti-transpirant/sticker/ UV screen/wetter
Contains	96.0 % w/w pinene oligomers		
Use with	Recommended rates of approved pesticides up to the growth stage indicated for specified crops, and with half or less than the recommended rate on these crops after the stated growth stages. Also for use with pesticides on grassland at half or less than their recommended rates. Must not be used in mixture with adjuvant oils or surfactants		
Protective clothing	A, C		
Precautions	U19a, U20c, E13c, C02a, D09a, D11a		
Planet	Intracrop	A0605	non-ionic surfactant/spreader/wetter
Contains	700 g/l alkoxylated alcohols and 150 g/l rapeseed fatty acids		
Use with	Any spray for which additional wetter is recommended		
Protective clothing	A, C		
Precautions	R22a, R36, U05a, U08, U19a, U20b, E13c, D01, D09a, D10a, D11a, H03, H04, H08		
Prima	De Sangosse	A0531	mineral oil
Contains	99% w/w oil (petroleum oils) (EAC 1)		
Use with	All pesticides on all crops when used up to 50% of maximum approved dose for that use (mixtures with Roundup must only be used for treatment of stubbles). Also with all approved pesticides at full dose on listed crops (see label). Also with listed herbicides on specified crops		
Protective clothing	A, C		
Precautions	R22a, U10, U16b, U19a, E15a, E19b, E34, E37, D01, D02, D05, D10a, D12a, M05b, H03		
Profit Oil	Microcide	A0631	extender/sticker/wetter
Contains	95% w/w oil (rapeseed triglycerides) (EAC 1)		
Use with	All approved pesticides		
Protective clothing	A, C		
Precautions	U19a, U20b, D09a, D10b		

SECTION 4

Product	Supplier	Adj. No.	Type
Pryz-Flex	Greenaway	A0679	spreader/sticker/wetter
Contains	10.0% w/w oil (rapeseed triglycerides) (EAC 1)		
Use with	Kerb Flo (MAPP 13716) and all approved 490 g/l glyphosate products on permeable surfaces overlying soil		
Protective clothing	A, C		
Precautions	U11, U12, U15, U20b, E34, E37		
Remix	Agrovista	A0765	wetter
Contains	732 g/l petroleum (paraffin) oil		
Use with	All approved pesticides at half or less than half the approved rate on edible crops; all approved pesticides on non-edible crops or non-production targets. Can also be used with all approved pesticides pre-emergence on bulb and stem vegetables, legumes, carrots, parsnips, swedes and turnips. Max conc for all uses is 1% of spray solution.		
Protective clothing	A, C, H		
Precautions	U02a, U05a, U19a, E15a, D01, D02, D05, D09a, D10a, D12a		
Respond	Amega	A0486	spreader/wetter
Contains	57.0 % w/w ethylene oxide-propylene oxide copolymers (EAC 1)		
Use with	All approved pesticides on all edible crops when used at half their recommended dose or less, and on all non-edible crops up to their full recommended dose. Also at a maximum concentration of 0.5%		
Protective clothing	A		
Precautions	U02a, U04a, U05a, U14, U15, U19a, E13b, E34, D01, D02, D09a, D10a, M03		
Reward Oil	Microcide	A0632	extender/sticker/wetter
Contains	95% w/w oil (rapeseed triglycerides) (EAC 1)		
Use with	All approved pesticides		
Protective clothing	A, C		
Precautions	U19a, U20b, D09a, D10b		
Roller	Agrovista	A0748	spreader/wetter
Contains	832 g/l ethylene oxide-propylene oxide copolymers (EAC 1) and 169.3 g/l trisiloxane organosilicone copolymers (EAC 2)		
Use with	All approved pesticides up to their maximum dose at a max conc of 0.2% v/v.		
Protective clothing	A, C, H		
Precautions	R36, R38, R41, R52, R53a, U02a, U05a, U08, U11, U20b, E15a, E34, E40c, D01, D02, D05, D09a, D10a, D12b, H04		
Saracen	Interagro	A0368	vegetable oil
Contains	95% w/w methylated vegetable oil		
Use with	Cereal graminicides and a wide range of other pesticides that have a recommendation for use with authorised adjuvant oils on specified crops. Refer to label or contact supplier for further details		
Protective clothing	A, C		
Precautions	U05a, U20b, E15a, D01, D02, D09a, D10a		

Product	Supplier	Adj. No.	Type
SAS 90	Intracrop	A0740	spreader/wetter
Contains	83% w/w trisiloxane organosilicone copolymers		
Use with	All approved herbicides at a maximum concentration of 0.025% spray volume and all approved fungicides and insecticides at a maximum concentration of 0.05% spray volume		
Protective clothing	A, C		
Precautions	U02a, U08, U19a, U20b, E13c, E34, D09a, D10a		
Saturn Plus	Intracrop	A0813	spreader/wetter
Contains	70.0% w/w ethylene oxide-propylene oxide copolymers (EAC 1)		
Use with	All approved pesticides on non-edible crops, and all pesticides approved for use on growing edible crops when used at half recommended dose or less.		
Protective clothing	A, C		
Precautions	U19a, U20c, E13c, C02a, D09a, D11a		
Siltex AF	Intracrop	A0803	spreader/sticker/wetter
Contains	250 g/l ethylene oxide-propylene oxide copolymers (EAC 1), 70.2 g/l styrene-butadiene copolymers (EAC 2) and 41.5 g/l trisiloxane organosilicone copolymers (EAC 3)		
Use with	All approved pesticides on non-edible crops, and all pesticides approved for use on growing edible crops when used at half recommended dose or less.		
Protective clothing	A, C		
Precautions	U05a, U08, U20c, E13c, E37, D01, D02, D09a, D10a		
Silwet L-77	De Sangosse	A0640	drift retardant/spreader/wetter
Contains	80% w/w trisiloxane organosilicone copolymers (EAC 1).		
Use with	All approved fungicides on winter and spring sown cereals; all approved pesticides applied at 50% or less of their full approved dose. A wide range of other uses. See label or contact supplier for details		
Protective clothing	A, C, H		
Precautions	R20, R21, R22a, R36, R43, R48, R51, R58, U11, U15, U19a, E15a, E19b, D01, D02, D05, D10a, D12a, H03, H11		
Slippa	Interagro	A0206	spreader/wetter
Contains	64.0 % w/w trisiloxane organosilicone copolymers		
Use with	Cereal fungicides and a wide range of other pesticides and trace elements on specified crops		
Protective clothing	A, C, H		
Precautions	R20, R38, R41, R43, R48, R51, R58, U02a, U05a, U08, U11, U19a, U20a, E15a, E34, E37, D01, D02, D05, D09a, D10a, M03, H03, H11		
Solar Plus	Intracrop	A0802	spreader/wetter
Contains	70.0% w/w ethylene oxide-propylene oxide copolymers (EAC 1)		
Use with	All approved pesticides on non-edible crops, and all pesticides approved for use on growing edible crops when used at half recommended dose or less.		
Protective clothing	A, C		
Precautions	R36, R38, U04a, U05a, U08, U20a, E13c, E34, D01, D02, D09a, D10b, D11a, M03, H04		

SECTION 4

Product	Supplier	Adj. No.	Type
Spartan	Interagro	A0375	penetrant/water conditioner
Contains	500 g/l alkoxylated tallow amines and 400 g/l alkoxylated sorbitan esters		
Use with	Cereal growth regulators, cereal herbicides, oilseed rape fungicides and a wide range of other pesticides on specified crops. Refer to label or contact supplier for further details		
Protective clothing	A, C		
Precautions	R22a, R36, R38, R51, R58, U02a, U05a, U19a, U20b, E15a, E34, E37, D01, D02, D05, D09a, D10a, M03, H03, H11		
Speedway Total	Everris Ltd	A0820	extender/spreader/wetter
Contains	41.0 % w/w alkoxylated alcohols (EAC 1) and 37.0 % w/w alkoxylated alcohols (EAC 2)		
Use with	All authorised plant protection products on non-edible crops and non-crop production uses; all authorised plant protection products at half or less than half the authorised plant protection product rate on edible crops.		
Protective clothing	A		
Precautions	U02a, U04a, U05a, U14, U15, U19a, E13b, E34, D01, D02, D09a, D10a, M03		
Sprayfast	Taminco	A0668	extender/sticker/wetter
Contains	334 g/l di-1-p-menthene		
Use with	Glyphosate and other pesticides, growth regulators or nutrients for which a coating agent is approved and recommended		
Protective clothing	A, C		
Precautions	U20b, E13c, D09a, D10a		
Spray-fix	De Sangosse	A0559	extender/sticker/wetter
Contains	10% w/w alkoxylated alcohols (EAC 1), 45% w/w styrene-butadiene copolymers (EAC 2)		
Use with	All approved potato blight fungicides. Also with all approved pesticides on all edible crops when used at half their recommended dose or less, and on all non-edible crops up their full recommended dose. Also at a maximum concentration of 0.14% with approved pesticides on listed crops up to specified growth stages		
Protective clothing	A, C, H		
Precautions	R36, R38, U11, U14, U16b, U19a, E15a, E19b, D01, D02, D05, D10a, D12a, H04		
Spraygard	Taminco	A0732	extender/sticker/wetter
Contains	400 g/l pinene oligomers (EAC 1)		
Use with	Approved pesticides on all edible crops (except herbicides on peas) and as an anti-transpirant on vegetables before transplanting, evergreens, deciduous trees, shrubs, bushes and on turf		
Protective clothing	A		
Precautions	R38, R50, R53a, U05a, U14, U20b, E15a, E34, E37, D01, D09a, D10b, D12b, M03, H04, H11		

Product	Supplier	Adj. No.	Type
Spraymac	De Sangosse	A0549	acidifier/non-ionic surfactant

Contains 100g/l alkoxylated alcohols (EAC 1), 350g/l propionic acid (EAC 2)

Use with All approved pesticides on all edible crops when used at half their recommended dose or less, and on all non-edible crops up to their full recommended dose. Also at a maximum concentration of 0.5% with approved pesticides on listed crops up to specified growth stages

Protective clothing A, C, H

Precautions R34, U04a, U10, U11, U14, U19a, U20b, E15a, E19b, E34, D01, D02, D05, D10a, D12a, M04a, H05

Spur	Amega	A0387	spreader/wetter

Contains 300 g/l alkoxylated alcohols (EAC 1)

Use with Use with all approved herbicides and all approved pesticides at half or less than half the approved dose rate at a maximum concentration of 0.5 % spray solution.

Protective clothing A, H

Precautions R22a, R41, R67, U02a, U04a, U05a, U11, U14, U15, E15a, E34, D01, D02, D09a, D10a, M03, H03, H08

Stamina	Interagro	A0202	penetrant

Contains 100% w/w alkoxylated fatty amine

Use with Glyphosate and a wide range of other pesticides on specified crops

Protective clothing A, C

Precautions R22a, R38, R50, R58, U02a, U05a, U20a, E15a, E34, D01, D02, D05, D09a, D10a, M03, H03, H11

Standon Shiva	Standon	A0816	wetter

Contains 6.7% w/w 3,6-dioxaeicosylsulphate sodium salt and 20.2% w/w 3,6-dioxaoctadecylsulphate sodium salt

Use with Standon Mimas WG and all other approved cereal herbicides

Protective clothing A, C

Precautions R36, R38, U02a, U05a, U08, U13, U19a, U20b, E13c, E34, D01, D02, D05, D09a, D10a, H04

Standon Wicket	Standon	A0781	wetter

Contains 47% w/w rapeseed fatty acid esters

Use with Clodinafop-propargyl, pinoxaden and mixtures of the two actives in cereal crops

Protective clothing A, H

Precautions R43, R50, R53a, U02a, U05a, U09b, U20b, E15b, E38, D01, D02, D05, D09a, D10c, D12a, H04, H11

Stika	De Sangosse	A0557	extender/sticker/wetter

Contains 10% w/w alkoxylated alcohols (EAC 1), 22.5% w/w styrene-butadione copolymers (EAC 2)

Use with All approved potato blight fungicides. Also with all approved pesticides on all edible crops when used at half their recommended dose or less, and on all non-edible crops up to their full recommended dose. Also at a maximum concentration of 0.14% with approved pesticides on listed crops up to specified growth stages

Protective clothing A, C, H

Precautions R36, R38, U11, U14, U16b, U19a, E15a, E19b, D01, D02, D05, D10a, D12a, H04

SECTION 4

Product	Supplier	Adj. No.	Type
SU Wett	Global Adjuvants	A0807	spreader/wetter
Contains	57.0 % w/w ethylene oxide-propylene oxide copolymers (EAC 1)		
Use with	All authorised plant protection products at half or less than half the authorised plant protection product rate		
Protective clothing	A, C, H, M		
Precautions	U05a, U20c, E34, D01, D02, D10a, M04a		
Super Nova	Interagro	A0364	penetrant
Contains	500 g/l alkoxylated fatty amine + 500 g/l polyoxyethylene monolaurate		
Use with	Cereal growth regulators, cereal herbicides, cereal fungicides, oilseed rape fungicides and a wide range of other pesticides on specified crops		
Protective clothing	A, C		
Precautions	R51, R58, U05a, E15a, E34, D01, D02, D05, D09a, D10a, H11		
Surfer	Dow	A0800	wetter
Contains	72.0 % w/w alkoxylated alcohols (EAC 1)		
Use with	All approved pesticides on all edible crops when used at half their recommended dose or less, and on all non-edible crops up to their full recommended dose.		
Protective clothing	A, C, H		
Precautions	H318, U05a, U11, U14, U15, U20a, E15a, E34, D01, D02, D09a, D10a		
Sward	Amega	A0747	spreader/wetter
Contains	400 g/l ethylene oxide-propylene oxide copolymers (EAC 1) and 58.1 g/l trisiloxane organosilicone copolymers (EAC 2)		
Use with	For use with all approved pesticides on managed amenity turf and amenity grassland at 0.2% v/v		
Protective clothing	A, C		
Precautions	U02a, U04a, U05a, U11, U14, U15, E15a, E34, D01, D02, D09a, D10a, D12b, M03, H04		
TM 1008	Intracrop	A0675	wetter
Contains	750 g/l ethylene oxide-propylene oxide copolymers		
Use with	All approved pesticides on non-edible crops and non crop production uses; All approved pesticides at half or less than half the approved pesticide rate on edible crops		
Protective clothing	A, C		
Precautions	R36, R38, U04a, U05a, U08, U19a, U20b, E13b, E34, D01, D02, D09a, D10a, M03, M05a, H04		
Toil	Interagro	A0248	vegetable oil
Contains	95% w/w methylated rapeseed oil		
Use with	Sugar beet herbicides, oilseed rape herbicides, cereal graminicides and a wide range of other pesticides that have a label recommendation for use with authorised adjuvant oils on specified crops. Refer to label or contact supplier for further details		
Protective clothing	A		
Precautions	U05a, U20b, E15a, D01, D02, D05, D09a, D10a		

Product	Supplier	Adj. No.	Type
Torpedo-II	De Sangosse	A0541	penetrant/wetter

Contains	190g/kg alkoxylated alcohols (EAC 1), 190g/kg alkoxylated alcohols (EAC 2), 75g/kg oil (tall oil fatty acids) (EAC 3), 210 g/kg alkoxylated tallow amines (EAC 4)
Use with	All approved pesticides on all edible crops when used at half their recommended dose or less, and on all non-edible crops up their full recommended dose. Also at a maximum concentration of 0.1% with approved pesticides on listed crops up to specified growth stages
Protective clothing	A, C, H
Precautions	R36, R38, R41, R52, R58, U05a, U10, U11, U19a, E15a, E19b, E34, E37, D01, D02, D05, D10a, D12a, M03, H04

Product	Supplier	Adj. No.	Type
Transact	United Agri	A0584	penetrant

Contains	40% w/w propionic acid
Use with	Plant growth regulators in cereals, all approved pesticides in non-edible crops, all approved pesticides on edible crops when used at half or less their recommended rate, and with all pesticides on listed crops up to specified growth stages
Protective clothing	A, C
Precautions	R34, U10, U11, U14, U15, U19a, D01, D05, D09b, M04a, H05

Product	Supplier	Adj. No.	Type
Transcend	Helena	A0333	adjuvant/vegetable oil

Contains	80% w/w oil (soybean fatty acid esters), 12% w/w trisiloxane organosilicone copolymers.
Use with	Approved pesticides on non-edible crops, cereals and stubbles of all edible crops, grassland (destruction), and pesticides approved for use on growing edible crops when used at half recommended dose or less. On specified crops product may be used at a maximum spray concentration of 0.5% with approved pesticides at their full approved rate up to the growth stages shown in the label
Protective clothing	A, H
Precautions	R38, R41, U02a, U05a, U08, U11, U19a, U20b, E13c, E34, E37, D01, D02, D05, D09a, M03, H04

Product	Supplier	Adj. No.	Type
Validate	De Sangosse	A0500	adjuvant/wetter

Contains	25% w/w alkoxylated alcohols (EAC 1), 25% w/w oil (soybean fatty acid esters) (EAC 2), 50% w/w soybean phospholipids (EAC 3)
Use with	Approved pesticides on non-edible crops where the addition of a wetter/ spreader or adjuvant oil is recommended on the pesticide label, and pesticides approved for use on growing edible crops when used at half recommended dose or less. On specified crops product may be used at a maximum spray concentration of 0.5% with approved pesticides at their full approved rate up to the growth stages shown in the label
Protective clothing	A, C, H
Precautions	R51, R53a, U05a, U11, U14, U19a, U20b, E15a, E19b, E34, D01, D02, D10a, D12a, H11

Product	Supplier	Adj. No.	Type
Velocity	Agrovista	A0697	adjuvant

Contains	745g/l oil (rapeseed fatty acid esters), 103g/l trisiloxane organosilicone copolymers.
Use with	All approved pesticides up to their full approved rate at a maximum conc of 0.5% vol/vol.
Protective clothing	A, C
Precautions	R41, R52, R53a, U02a, U05a, U08, U11, U14, U15, U20b, E13c, E15a, E34, E40c, D01, D02, D09a, D10a, H04

SECTION 4

Product	Supplier	Adj. No.	Type
Velvet	Global Adjuvants	A0814	spreader/sticker/wetter
Contains	83% trisiloxane organosilicone copolymers		
Use with	All approved pesticides on all edible crops when used at half their recommended dose or less, and on all non-edible crops up to their full recommended dose.		
Protective clothing	A, C, H		
Precautions	H319, E34, H411, D01, D02, D09a, D10a		
Verdant	Agrovista	A0714	wetter
Contains	8% w/w alkoxylated alcohol, 63% w/w alkoxylated alkyl polyglucoside and 10% w/w alkyl polyglucoside		
Use with	All approved pesticides at 50% or less on edible crops and with all approved pesticides on non-edible crops, non-crop production and stubbles of edible crops at 0.25% spray sloution.		
Protective clothing	A, C, H		
Precautions	R38, R41, U02a, U05a, U11, U14, U15, U20b, E15a, E34, D02, D05, D09a, D10a, H04		
Wetcit	Plant Solutions	A0586	penetrant/wetter
Contains	8.15% w/w alcohol ethoxylate		
Use with	All approved pesticides on all edible crops when used at half their recommended dose or less, and on all non-edible crops up to their full recommended dose. Also at a maximum concentration of 0.25% with approved pesticides on listed crops up to specified growth stages at full rate, and at half rate thereafter		
Protective clothing	A, C, H		
Precautions	R22a, R38, R41, U08, U11, U19a, U20b, E13c, E15a, E34, E37, D05, D08, D09a, D10b, H03		
X-Wet	De Sangosse	A0548	non-ionic surfactant/wetter
Contains	375g/kg alkoxylated alcohols (EAC 1), 375g/kg alkoxylated alcohols (EAC 2), 150g/kg oil (tall oil fatty acids) (EAC 3)		
Use with	All approved pesticides on all edible crops when used at half their recommended dose or less, and on all non-edible crops up their full recommended dose. Also at a maximum concentration of 0.1% with approved pesticides on listed crops up to specified growth stages		
Protective clothing	A, C, H		
Precautions	R36, R38, R53a, U02a, U04a, U05a, U10, U11, U20b, E15a, E19b, D01, D02, D05, D10a, H04		
Zarado	De Sangosse	A0516	vegetable oil
Contains	70% w/w oil (rapeseed oil fatty acid esters) as emulsifiable concentrate (EAC 1).		
Use with	All approved pesticides on edible and non-edible crops when used at half their approved dose or less. Also with approved pesticides on specified crops, up to specified growth stages, at up to their full approved dose		
Protective clothing	A, C, H		
Precautions	R43, R52, U05a, U13, U19a, U20b, E15a, E19b, E34, D01, D02, D05, D10a, M05a, H04		

Product	Supplier	Adj. No.	Type
Zeal	Interagro	A0685	water conditioner/wetter
Contains	60.0 % w/w alkoxylated tallow amines and 37.0 % w/w alkoxylated sorbitan esters		
Use with	Trace elements such as calcium, copper, manganese and sulphur and with macronutrients such as phosphates		
Protective clothing	A, C, H		
Precautions	R22a, R36, R38, R51, R53a, U05a, U09a, U11, U14, U15, U19a, E15b, E34, D01, D02, D09a, D10a, M03, H03, H11		
Zigzag	Headland	A0692	extender/sticker/wetter
Contains	36.9% w/v styrene/butadiene co-polymer and 6.375% w/v polyether-modified trisiloxane		
Use with	Approved glyphosate formulations on cereal crops; All approved pesticides applied at half-rate or less; All approved pesticides for use on crops not destined for human or animal consumption; With all approved pesticides on edible crops up to the latest growth stage specified on the label.		
Protective clothing	C, H		
Precautions	U11, U15, E38, D01		
Zinzan	Certis	A0600	extender/spreader/wetter
Contains	70% w/w 1,2 bis (2-ethylhexyloxycarbonyl) ethanesulphonate		
Use with	All approved pesticides on non-edible crops, and all pesticides approved for use on growing edible crops when used at half recommended dose or less. On specified edible crops product may be used at a maximum spray concentration of 0.075% with approved pesticides at their full approved rate up to the growth stages shown in the label		
Protective clothing	A, C, P		
Precautions	R38, R41, U11, U14, U20c, D01, D09a, D10a, H04		

SECTION 4

SECTION 5
USEFUL INFORMATION

Pesticide Legislation

Anyone who advertises, sells, supplies, stores or uses a pesticide is bound by legislation, including those who use pesticides in their own homes, gardens or allotments. There are numerous UK statutory controls, but the major legal instruments are outlined below.

Regulation 1107/2009 — The Replacement for EU 91/414

Regulation 1107/2009 entered into force within the EU on 14 Dec 2009 and was applied to new approval applications from 14th June 2011. Since that date all pesticides that held a current approval under 91/414 have been deemed to be approved under 1107/2009 and the new criteria for approval will only be applied when the active substance comes up for review. The new regulation largely mirrors the previous regulation but requires additional approval criteria to be met to weed out the more hazardous chemicals. These include assessing the effect on vulnerable groups, taking into account known synergistic and cumulative effects, considering the effect on coastal and estuarine waters, effects on the behaviour of non-target organisms, biodiversity and the ecosystem. However, these effects will only be evaluated when scientific methods approved by EFSA (European Food Safety Authority) have been developed.

The additional criteria will require that no CMR's (substances which are Carcinogens, Mutagens or toxic to Reproduction), no endocrine disruptors and no POP (Persistent Organic Pollutants), PBT (chemicals that are Persistent, Bioaccumulative or Toxic) or vPVB (very Persistent and very Bioaccumulative) will gain approval unless the exposure is negligible. However, note that for EFSA the term 'negligible' has yet to receive a precise definition. A list of 66 chemicals that pose an endocrine disruptor risk has been produced. It contained maneb and vinclozolin and both actives have now been withdrawn.

If an active substance passes these hurdles, the first approval for basic substances will be granted for 10 years (15 years for 'low risk' substances, 7 years for candidates for substitution – see below) and then at each subsequent renewal a further 15 years will be granted. Actives will be divided between 5 categories – low risk substances (e.g. pheromones, semiochemicals, micro-organisms and natural plant extracts), basic substances, candidates for substitution and then safeners/synergists and co-formulants. Candidates for substitution will be those chemicals which have just scraped past the approval thresholds but are deemed to still carry some degree of hazard. They will be withdrawn when significantly safer alternatives are available. This includes physical methods of control but any alternative must not have significant economic or practical disadvantages, must minimise the risk of the development of resistance and the consequences for any 'minor uses' of the original product must also be considered.

The Water Framework Directive

This became law in the UK in 2003 but is being introduced over a number of years to allow systems to acclimatise to the new rules and required member states of the EU to achieve 'good status' in ground and surface water bodies by 22 December 2015 with reviews every six years thereafter. To achieve 'good status' as far as pesticides are concerned the concentration of any individual pesticide must not exceed 0.1 micrograms per litre and the total quantity of all pesticides must not exceed 0.5 micrograms per litre. These levels are not based on the toxicity of the pesticides but on the level of detection that can be reliably achieved. Failure to meet these limits could jeopardise the approval of the problem active ingredients and so every effort must be made to avoid contamination of surface and ground water bodies with pesticides. The pesticides that are currently being found in water samples above the 0.1 ppb limit are metaldehyde, carbetamide, propyzamide, MCPA, mecoprop, glyphosate, bentazone, terbutryne, trietazine, clopyralid, metazachlor, chlorotoluron, chloridazon and ethofumesate. These actives are identified as a risk to water in the pesticide profiles in section 2 and great care should be taken when applying them, particularly metaldehyde which is almost impossible to remove from water at treatment plants.

Classification, Labelling and Packaging (CLP) Regulation

The CLP Regulation (EC No 1272/2008) governs the classification of substances and mixtures and is intended to introduce the United Nations Globally Harmonised System (GHS) of classification to Europe. This new system and the hazard icons that appear on pesticide labels is very similar to the old system with the most obvious difference being that the hazard icons have a white background rather than an orange one. The new system of classification became mandatory for new supplies from 1st June 2015 in the UK but for products already in the supply chain the compliance date was deferred to 1st June 2017. Since that date all products placed in the market MUST comply with the CLP Regulation. The biggest change was the replacement of the old HARMFUL icon **'X'** with a new one **'!'** to notify that caution is required when handling or using a product. This brings Europe and the UK in line with the rest of the world. Details of the new phrases are given in Appendix 4.

The Food and Environment Protection Act 1986 (FEPA)

FEPA introduced statutory powers to control pesticides with the aims of protecting human beings, creatures and plants, safeguarding the environment, ensuring safe, effective and humane methods of controlling pests and making pesticide information available to the public. This was supplemented by Control of Pesticides Regulations 1986 (COPR) which has now been replaced by EC Regulation 1107/2009 – see above. Details are given on the websites of the Chemicals Regulation Directorate (CRD) and the Health and Safety Executive (HSE). Together these regulations mean that:

- Only approved products may be sold, supplied, stored, advertised or used.
- No advertisement may contain any claim for safety beyond that which is permitted in the approved label text.
- Only products specifically approved for the purpose may be applied from the air.
- A recognised Storeman's Certificate of Competence is required by anyone who stores for sale or supply pesticides approved for agricultural use.
- A recognised Certificate of Competence is required by anyone who gives advice when selling or supplying pesticides approved for agricultural use.
- Users of pesticides must comply with the Conditions of Approval relating to use.
- A recognised Certificate of Competence is required for all contractors and persons applying pesticides approved for agricultural use (unless working under direct supervision of a certificate holder).
- Only those adjuvants authorised by CRD may be used.
- For tank-mixes of convenience, when the purpose is to reduce the number of passes in the crop, no efficacy data is required but physical and chemical compatibility data are required in line with EU guidance. For mixtures that claim positive benefits efficacy data remains necessary in addition to the physical and chemical compatibility data. Note however that ALS herbicides (HRAC mode of action code 'B') may only be mixed or applied in sequence with other ALS herbicides when a label statement permits this. Mixtures of anticholinesterase products are not permitted unless CRD has fully assessed the toxicological effects.

Dangerous Preparations Directive (1999/45/EC)

The Dangerous Preparations Directive (DPD) came into force in the UK for pesticide and biocidal products on 30 July 2004. Its aim is to achieve a uniform approach across all Member States to the classification, packaging and labelling of most dangerous preparations, including crop protection products. The Directive is implemented in the UK under the Chemicals (Hazard Information and Packaging for Supply) Regulations 2002, often referred to under the acronym CHIP3. In most cases the Regulations have led to additional hazard symbols, and associated risk and safety phrases relating to environmental and health hazards, appearing on the label. All products affected by the DPD entering the supply chain from the implementation date above must be so labelled. The new environmental hazard classifications and risk phrases are included.

Under the DPD, the labels of all plant protection products now state 'To avoid risks to man and the environment, comply with the instructions for use'. As the instruction applies to all products, it has not been repeated in each fact sheet.

The Review Programme

The review programme of existing active substances will continue under EC Directive 1107/2009. The programme is designed to ensure that all available plant protection products are supported by up-to-date information on safety and efficacy. As this will now be under 1107/2009, the new standards for approval will be applied at the due date and any that are deemed to be hazardous chemicals will be withdrawn. Any that are just within the safety standards yet to be defined will be called 'candidates for substitution' and granted only a seven-year renewal to allow time for significantly safer alternatives to be developed. If these are available at the time of renewal, the approval will be withdrawn. Chemicals deemed to be low risk, such as natural plant extracts, microorganisms, semiochemicals and pheromones, will be renewed for 15 years, while most active substances will be renewed for a further 10 years.

Control of Substances Hazardous to Health Regulations 1988 (COSHH)

The COSHH regulations, which came into force on 1 October 1989, were made under the Health and Safety at Work Act 1974, and are also important as a means of regulating the use of pesticides. The regulations cover virtually all substances hazardous to health, including those pesticides classed as Very Toxic, Toxic, Harmful, Irritant or Corrosive, other chemicals used in farming or industry, and substances with occupational exposure limits. They also cover harmful micro-organisms, dusts and any other material, mixture, or compound used at work which can harm people's health.

The original Regulations, together with all subsequent amendments, have been consolidated into a single set of regulations: The Control of Substances Hazardous to Health Regulations 1994 (COSHH 1994).

The basic principle underlying the COSHH regulations is that the risks associated with the use of any substance hazardous to health must be assessed before it is used, and the appropriate measures taken to control the risk. The emphasis is changed from that pertaining under the Poisonous Substances in Agriculture Regulations 1984 (now repealed) – whereby the principal method of ensuring safety was the use of protective clothing – to the prevention or control of exposure to hazardous substances by a combination of measures. In order of preference the measures should be:

(a) substitution with a less hazardous chemical or product
(b) technical or engineering controls (e.g. the use of closed handling systems, etc.)
(c) operational controls (e.g. operators located in cabs fitted with air-filtration systems, etc.)
(d) use of personal protective equipment (PPE), which includes protective clothing.

Consideration must be given as to whether it is necessary to use a pesticide at all in a given situation and, if so, the product posing the least risk to humans, animals and the environment must be selected. Where other measures do not provide adequate control of exposure and the use of PPE is necessary, the items stipulated on the product label must be used as a minimum. It is essential that equipment is properly maintained and the correct procedures adopted. Where necessary, the exposure of workers must be monitored, health checks carried out, and employees instructed and trained in precautionary techniques. Adequate records of all operations involving pesticide application must be made and retained for at least 3 years.

Biocidal Products Regulations

These regulations concern disinfectants, preservatives and pest control products but only the latter are listed in this book. The regulation aims to harmonise the European market for biocidal products, to provide a high level of safety for humans, animals and the environment and to ensure that the products are effective against the target organisms. Products that have been reviewed and included under these regulations are issued with a new number of the format UK-2012-xxxx and these are summarised in this book as UK??-xxxx.

SECTION 5

Certificates of Competence – the roles of BASIS and NPTC

COPR, COSHH and other legislation places certain obligations on those who handle and use pesticides. Minimum standards are laid down for the transport, storage and use of pesticides, and the law requires those who act as storekeepers, sellers and advisors to hold recognised Certificates of Competence.

BASIS

BASIS is an independent Registration Scheme for the pesticide industry. It is responsible for organising training courses and examinations to enable such staff to obtain a Certificate of Competence.

In addition, BASIS undertakes an annual assessment of pesticide supply stores, enabling distributors, contractors and seedsmen to meet their obligations under the Code of Practice for Suppliers of Pesticides. Further information can be obtained from BASIS. Since 26 November 2015 all businesses must have sufficient number of staff with BASIS certificates to advise customers at **the time of sale** where previously it was just at the point of sale

Certificates of Competence

Storage

- BASIS Certificate of Competence in the Storage and Handling of Crop Protection Products

Sale and supply

- BASIS Certificate in Crop Protection (Agriculture)
- BASIS Certificate in Crop Protection (Commercial Horticulture)
- BASIS Certificate in Crop Protection (Amenity Horticulture)
- BASIS Certificate in Crop Protection (Forestry)
- BASIS Certificate in Crop Protection (Seed Treatment)
- BASIS Certificate in Crop Protection (Seed Sellers)
- BASIS Certificate in Crop Protection (Field Vegetables)
- BASIS Certificate in Crop Protection (Potatoes)
- BASIS Certificate in Crop Protection (Aquatic)
- BASIS Certificate in Crop Protection (Indoor Landscaping)
- BASIS Certificate in Crop Protection (Grassland and Forage Crops)
- BASIS/LEAF ICM Certificate
- BASIS Advanced Certificate
- FACTS (Fertiliser Advisers Certification and Training Scheme) Certificate

NPTC

Certain spray operators also require certificates of competence under the Control of Pesticides Regulations. NPTC's Pesticides Award is a recognised Certificate of Competence under COPR and it is aimed principally at those people who use pesticide products approved for use in agriculture, horticulture (including amenity horticulture) and forestry. All contractors must possess a certificate if they spray such products unless they are working under the direct supervision of a certificate holder.

Because the required spraying skills vary widely among the uses listed above, candidates are assessed under one (or more) modules that are most appropriate for their professional work. Assessments are carried out by an approved NPTC or Scottish Skills Testing Service Assessor. All candidates must first complete a foundation module (PA1), for which a certificate is not issued, before taking one of the specialist modules. Certificate holders who change their work so that a different specialist module becomes more appropriate may need to obtain a new certificate under the new module.

Holders are required to produce on demand their Certificate of Competence for inspection to any authorised person. Further information can be obtained from NPTC.

Certificates of Competence for spray operators

- PA1 Foundation Module
- PA2 Ground Crop Sprayers – mounted or trailed
- PA3 Broadcast Air Blast Sprayer. Variable Geometry Boom Air Assisted Sprayer
- PA4 Granule Applicator – Mounted or Trailed
- PA5 Boat Mounted Applicators
- PA6 Hand Held Applicators
- PA7 Aerial Application
- PA8 Mixer/Loader
- PA9 Fogging, Misting and Smokes
- PA10 Dipping Bulbs, Corms, Plant Material or Containers
- PA11 Seed Treating Equipment
- PA12 Application of Pesticides to Material as a Continuous or Batch Process

Maximum Residue Levels

A small number of pesticides are liable to leave residues in foodstuffs, even when used correctly. Where residues can occur, statutory limits, known as maximum residue levels (MRLs), have been established. MRLs provide a check that products have been used as directed; **they are not safety limits.** However, they do take account of consumer safety because they are set at levels that ensure normal dietary intake of residues presents no risk to health. Wide safety margins are built in, and eating food containing residues above the MRL does not automatically imply a risk to health. Nevertheless, it is an offence to put into circulation any produce where the MRL is exceeded.

The surrounding legislation is complex. MRLs may be specified by several different bodies. The UK has set statutory MRLs since 1988. The European Union intends eventually to introduce MRLs for all pesticide/commodity combinations. These are being introduced initially by a series of priority lists, but will subsequently be covered by the review programme under EC Regulation 1107/2009. However, in cases where no information is available, EU Regulation 2000/42/EC requires many MRLs to be set at the limit of determination (LOD). As a result, certain approvals are being withdrawn where such use would leave residues above the MRL set in the Directive.

MRLs apply to imported as well as home-produced foodstuffs. Details of those that have been set have been published in *The Pesticides (Maximum Residue Levels in Crops, Food and Feeding Stuffs) Regulations 1994*, and successive amendments to these Regulations. These Statutory Instruments are available from The Stationery Office (www.tso.co.uk).

MRLs are set for many chemicals not currently marketed in Britain. Because of this and the ever-changing information on MRLs, direct access is provided to the comprehensive MRL databases on the Chemicals Regulation Division website (www.hse.gov.uk/crd/index.htm). This online database sets out in table form the levels specified by UK Regulations, EC Directives, and the Codex Alimentarius for each commodity.

SECTION 5

Approval (On-label and Off-label)

Only officially approved pesticides can be marketed and used in the UK. Approvals are granted by UK Government Ministers in response to applications that are supported by satisfactory data on safety, efficacy and, where relevant, humaneness. The Chemicals Regulation Division (CRD), (www.hse.gov.uk/crd/index.htm.uk) comes under the Health and Safety Executive (HSE), www.hse. gov.uk), and is the UK Government Agency for regulating agricultural pesticides and plant protection products. The HSE currently fulfils the same role for other pesticides, with the two organisations now merged. The main focus of the regulatory process in both bodies is the protection of human health and the environment.

Statutory Conditions of Use

Approvals are normally granted only in relation to individual products and for specified uses. It is an offence to use non-approved products or to use approved products in a manner that does not comply with the statutory conditions of use, except where the crop or situation is the subject of an off-label extension of use (see below).

Statutory conditions have been laid down for the use of individual products and may include:

- field of use (e.g. agriculture, horticulture etc.)
- crop or situations for which treatment is permitted
- maximum individual dose
- maximum number of treatments or maximum total dose
- maximum area or quantity which may be treated
- latest time of application or harvest interval
- operator protection or training requirements
- environmental protection
- any other specific restrictions relating to particular pesticides.

Products must display these statutory conditions in a boxed area on the label entitled 'Important Information', or words to that effect. At the bottom of the boxed area must be shown a bold text statement: **'Read the label before use. Using this product in a manner that is inconsistent with the label may be an offence. Follow the Code of Practice for Using Plant Protection Products'.** This requirement came into effect in October 2006, and replaced the previous 'Statutory Box'.

Types of Approval

Where there were once three levels of approval (full; provisional; experimental permit), there will now be just two levels (authorisation for use; limited approval for research and development). Provisional approval used to be granted while further data were generated to justify label claims, but under the new regulations, decisions on authorisation will be reached much more quickly, and this 'half-way stage' is deemed to be unnecessary. The official list of approved products, excluding those approved for research and development only, are shown on the websites of CRD and the Health and Safety Executive (HSE).

Withdrawal of Approval

Product approvals may be reviewed, amended, suspended or revoked at any time. Revocation may occur for various reasons, such as commercial withdrawal, or failure by the approval holder to meet data requirements.

From September 2007, where an approval is being revoked for purely administrative, 'housekeeping' reasons, the existing approval will be revoked and in its place will be issued:

- an approval for advertisement, sale and supply by any person for 24 months, and
- an approval for storage and use by any person for 48 months.

Where an approval is replaced by a newer approval, such that there are no safety concerns with the previous approval, but the newer updated approval is more appropriate in terms of reflecting the latest regulatory standard, the existing approval will be revoked and in its place will be issued:

- an approval for advertisement, sale and supply by any person for 12 months, and
- an approval for storage and use by any person for 24 months.

Where there is a need for tighter control of the withdrawal of the product from the supply chain, for example failure to meet data submission deadlines, the current timelines will be retained:

- immediate revocation for advertisement, sale or supply for the approval holder
- approval for 6 months for advertisement, sale and supply by 'others', and
- approval for storage and use by any person for 18 months.

Immediate revocation and product withdrawal remain an option where serious concerns are identified, with immediate revocation of all approvals and approval for storage only by anyone for 3 months, to allow for disposal of product.

The expiry date shown in the product fact sheet is the final date of legal use of the product.

Off-label Extension of Use

Products may legally be used in a manner not covered by the printed label in several ways:

- In accordance with an Extension of Authorisation for Minor Use (EAMU). EAMUs are uses for which individuals or organisations other than the manufacturers have sought approval. The Notices of Approval are published by CRD and are widely available from ADAS or NFU offices. Users of EAMUs must first obtain a copy of the relevant Notice of Approval and comply strictly with the conditions laid down therein. Users of this Guide will find details of extant EAMUs and direct links to the CRD website, where EAMU notices can also be accessed.
- In tank mixture with other approved pesticides in accordance with Consent C(i) made under FEPA. Full details of Consent C(i) are given in Annex A of Guide to Pesticides on the CRD website, but there are two essential requirements for tank mixes. First, all the conditions of approval of all the components of a mixture must be complied with. Second, no person may mix or combine pesticides that are cholinesterase compounds unless allowed by the label of at least one of the pesticides in the mixture.
- In conjunction with authorised adjuvants.
- In reduced spray volume under certain conditions.
- In the use of certain herbicides on specified set-aside areas subject to restrictions, which differ between Scotland and the rest of the UK.
- By mutual recognition of a use fully approved in another Member State of the European Union and authorised by CRD.

Although approved, off-label uses are not endorsed by manufacturers and such treatments are made entirely at the risk of the user.

SECTION 5

Using Crop Protection Chemicals

Use of Herbicides In or Near Water

Products in this Guide approved for use in or near water are listed in Table 5.1. Before use of any product in or near water, the appropriate water regulatory body (Environment Agency/Local Rivers Purification Authority; or in Scotland the Scottish Environmental Protection Agency) must be consulted. Guidance and definitions of the situation covered by approved labels are given in the Defra publication Guidelines for the Use of Herbicides on Weeds in or near Watercourses and Lakes. Always read the label before use.

Table 5.1 Products approved for use in or near water

Chemical	Product
glyphosate	Asteroid, Asteroid Pro, Asteroid Pro 450, Barclay Gallup Biograde 360, Barclay Gallup Biograde Amenity, Barclay Gallup Hi-Aktiv, Barclay Glyde 144, Buggy XTG, Discman Biograde, Envision, Gallup Hi-Aktiv Amenity, Glyfos Dakar, Glyfos Dakar Pro, Klean G, Mascot Hi-Aktiv Amenity, Palomo Green, Reaper Green, Roundup Biactive GL, Surrender T/F, Trustee Amenity, Vesuvius Green

Use of Pesticides in Forestry

Table 5.2 Products in this *Guide* approved for use in forestry

Chemical	Product	Use
aluminium ammonium sulphate	Curb Crop Spray Powder, Liquid Curb Crop Spray, Sphere ASBO	Animal deterrent/repellent
chlorpyrifos	Dursban WG, Equity	Acaricide, Insecticide
cycloxydim	Laser	Herbicide
cypermethrin	Forester	Insecticide
diflubenzuron	Dimilin Flo	Insecticide
ferric phosphate	Derrex, Iroxx, Sluxx HP	Molluscicide
fluazifop-P-butyl	Clayton Crowe, Clayton Maximus, Fusilade Max, Salvo	Herbicide
glufosinate-ammonium	Whippet	Herbicide
glyphosate	Amega Duo, Azural, Barclay Gallup Biograde 360, Barclay Gallup Biograde Amenity, Barclay Gallup Hi-Aktiv, Barclay Glyde 144, Buggy XTG, Clinic UP, Credit, Crestler, Discman Biograde, Envision, Gallup Hi-Aktiv Amenity, Glyfos Dakar Pro, Hilite, Klean G, Liaison, Mascot Hi-Aktiv Amenity, Mentor, Monsanto Amenity Glyphosate, Monsanto Amenity Glyphosate XL, Motif, Nomix Conqueror Amenity, Palomo Green, Rattler, Reaper Green, Rodeo, Rosate Green, Samurai, Snapper, Surrender T/F, Tanker, Trustee Amenity, Vesuvius Green	Herbicide
imidacloprid	Merit Forest	Insecticide, Seed treatment
metazachlor	Rapsan 500 SC, Stalwart, Sultan 50 SC, Taza 500	Herbicide
Phlebiopsis gigantea	PG Suspension	Biological agent
propaquizafop	Clayton Enigma, Clayton Satchmo, Falcon, Longhorn, Shogun	Herbicide
propyzamide	Barclay Propyz, Careca, Cohort, Kerb Granules, Proper Flo, PureFlo, Relva Granules, Setanta 50 WP, Zamide 80 WG, Zamide Flo	Herbicide
pyrethrins	Spruzit	Insecticide

SECTION 5

Pesticides Used as Seed Treatments

Information on the target for these products can be found in the relevant pesticide profile in Section 2.

Table 5.3 Products used as seed treatments (including treatments on seed potatoes)

Chemical	Product	Formulation	Crop(s)
beta-cyfluthrin + clothianidin	Poncho Beta	FS	Fodder beet, Sugar beet
clothianidin	Deter	FS	Durum wheat, Triticale, Winter barley, Winter oats, Winter rye, Winter wheat
clothianidin + prothioconazole	Redigo Deter	FS	Durum wheat, Triticale, Winter barley, Winter oats, Winter rye, Winter wheat
cymoxanil + fludioxonil + metalaxyl-M	Wakil XL	WS	Carrots, Combining peas, Parsnips, Vining peas
difenoconazole	Difend	FS	Winter wheat
difenoconazole + fludioxonil	Celest Extra	FS	Winter oats, Winter rye, Winter wheat
	Difend Extra	FS	Winter wheat
fludioxonil	Beret Gold	FS	Rye, Spring barley, Spring oats, Spring wheat, Triticale, Winter barley, Winter oats, Winter wheat
	Maxim 100FS	FS	Potatoes, Seed potatoes
fludioxonil + metalaxyl-M	Maxim XL	FS	Forage maize
fludioxonil + sedaxane	Vibrance Duo	FS	Rye, Spring oats, Triticale, Winter wheat
fludioxonil + tebuconazole	Fountain	FS	Triticale, Winter barley, Winter oats, Winter rye, Winter wheat
fludioxonil + tefluthrin	Austral Plus	FS	Spring barley, Spring oats, Spring wheat, Triticale seed crop, Winter barley, Winter oats, Winter wheat
fluopyram + prothioconazole + tebuconazole	Raxil Star	FS	Winter barley
flutolanil	Rhino DS	DS	Potatoes

Chemical	Product	Formulation	Crop(s)
metalaxyl-M	Apron XL	ES	Beetroot, Brassica leaves and sprouts, Broccoli, Brussels sprouts, Bulb onions, Cabbages, Calabrese, Cauliflowers, Chard, Chinese cabbage, Herbs (see appendix 6), Kohlrabi, Ornamental plant production, Radishes, Shallots, Spinach
methiocarb	Mesurol	FS	Fodder maize, Grain maize, Sweetcorn
pencycuron	Monceren DS	DS	Potatoes
	Solaren	DS	Potatoes
physical pest control	Silico-Sec	DS	Stored grain
prothioconazole + tebuconazole	Redigo Pro	FS	Durum wheat, Rye, Spring barley, Spring oats, Spring wheat, Triticale, Winter barley, Winter oats, Winter wheat
thiamethoxam	Cruiser SB	FS	Fodder beet, Kale, Sugar beet, Swedes
thiram	Agrichem Flowable Thiram	FS	Baby leaf crops, Borage, Bulb onions, Cabbages, Carrots, Cauliflowers, Celery (outdoor), Combining peas, Dwarf beans, Edible podded peas, Fodder beet, Forage maize, French beans, Frise, Grass seed, Kale, Lamb's lettuce, Leeks, Lettuce, Lupins, Mangels, Parsnips, Radishes, Red beet, Runner beans, Salad onions, Soya beans, Spinach, Spring field beans, Spring oilseed rape, Sugar beet, Swedes, Turnips, Vining peas, Winter field beans, Winter oilseed rape
	Thyram Plus	FS	Bulb onions, Cabbages, Cauliflowers, Combining peas, Dwarf beans, Edible podded peas, Forage maize, Grass seed, Kale, Leeks, Lettuce, Radishes, Runner beans, Salad onions, Soya beans, Spring field beans, Spring oilseed rape, Turnips, Vining peas, Winter field beans, Winter oilseed rape

SECTION 5

Aerial Application of Pesticides

New aerial spraying permit arrangements came into force in June 2012 and those undertaking aerial applications must ensure that the spraying is done in line with an Approved Application Plan that is subject to approval by CRD. Plans will only be approved where there is no viable alternative method of application or where aerial application results in reduced impact on human health and/or the environment compared to land-based application. Thus applications to bracken, forestry and possibly blight sprays to potatoes may meet these requirements but few others are likely to do so. Template application plans are available from the CRD web site and once all the necessary details have been received by CRD they undertake to approve or reject the plan within 10 working days.

Table 5.4 Products approved for aerial application

Chemical	Product	Crop(s)
diflubenzuron	Dimilin Flo	Forest

Resistance Management

Pest species are, by definition, adaptable organisms. The development of resistance to some crop protection chemicals is just one example of this adaptability. Repeated use of products with the same mode of action will clearly favour those individuals in the pest population able to tolerate the treatment. This leads to a situation where the tolerant (or resistant) individuals can dominate the population and the product becomes ineffective. In general, the more rapidly the pest species reproduces and the more mobile it is, the faster is the emergence of resistant populations, although some weeds seem able to evolve resistance more quickly than would be expected. In the UK, key independent research organisations, chemical manufacturers and other organisations have collaborated to share knowledge and expertise on resistance issues through three action groups. Participants include **ADAS, the Chemicals Regulation Division (CRD),** universities, colleges and the **Home Grown Cereals Authority (HGCA).** The groups have a common aim of monitoring resistance in the UK and devising and publishing management strategies designed to combat it where it occurs. The groups are:

- **The Weed Resistance Action Group (WRAG),** formed in 1989 (Secretary: Richard Hull, Rothamsted Research, Harpenden, Herts AL5 2JQ *Tel: 01954 268219*)
  Web: https://cereals.ahdb.org.uk/WRAG.
- **The Fungicide Resistance Action Group (FRAG),** formed in 1995 (Secretary: Mr Paul Ashby, HSE, Email: paul.ashby@hse.gsi.gov.uk). Web: https://cereals.ahdb.org.uk/FRAG
- **The Insecticide Resistance Action Group (IRAG),** formed in 1997 (Secretary: Dr Sacha White ADAS Boxworth, Battlegate Road, Cambridge, Cambs, CB3 8NN. *Tel: 01954 267666*, Email: sacha.white@adas.co.uk). Web: https://cereals.ahdb.org.uk/IRAG

The above groups publish detailed advice on resistance management relevant for each sector and, in some cases, specific to a pest problem. This information, together with further details about the function of each group, can be obtained from the **Chemicals Regulation Division** website www.hse.gov.uk/crd/index.htm.

The speed of appearance of resistance depends on the mode of action of the crop protection chemicals, as well as the manner in which they are used. Resistance among insects and fungal diseases has been evident for much longer than weed resistance to herbicides, but examples in all three categories are now widespread and increasing. This has created a need for agreement on the advice given for the use of crop protection chemicals in order to reduce the likelihood of the development of resistance and to avoid the loss of potentially valuable products in the chemical armoury. Mixing or alternating modes of action is one of the guiding principles of resistance management. To assist appropriate product choices, mode of action codes, published by the international Resistance Action Committees (see below), are shown in the respective active ingredient fact sheets. The product label and/or a professional advisor should always be consulted before making decisions. The general guidelines for resistance management are similar for all three problem areas:

Preparation in advance

- Be aware of the factors that favour the development of resistance, such as repeated annual use of the same product, and assess the risk.
- Plan ahead and aim to integrate all possible means of control.
- Use cultural measures such as rotations, stubble hygiene, variety selection and, for fungicides, removal of primary inoculum sources, to reduce reliance on chemical control.
- Monitor crops regularly.
- Keep aware of local resistance problems.
- Monitor effectiveness of actions taken and take professional advice, especially in cases of unexplained poor control.

Using crop protection products

- Optimise product efficacy by using it as directed, at the right time, in good conditions.
- Treat pest problems early.
- Mix or alternate chemicals with different modes of action.
- Avoid repeated applications of very low doses.
- Keep accurate field records.

SECTION 5

Label guidance depends on the appropriate strategy for the product. Most frequently it consists of a warning of the possibility of poor performance due to resistance, and a restriction on the number of treatments that should be applied in order to minimise the development of resistance. This information is summarised in the profiles in the active ingredient fact sheets, but detailed guidance must always be obtained by reading the label itself before use.

International Action Committees

Resistance to crop protection products is an international problem. Agrochemical industry collaboration on a global scale is via three action committees whose aims are to support a coordinated industry approach to the management of resistance worldwide. In particular they produce lists of crop protection chemicals classified according to their mode of action. These lists and other information can be obtained from the respective websites:

Herbicide Resistance Action Committee (HRAC) – www.hracglobal.com

Fungicide Resistance Action Committee (FRAC) – www.frac.infosections

Insecticide Resistance Action Committee (IRAC) – www.irac-online.org

Poisons and Poisoning

Chemicals Subject to the Poison Law

Certain products are subject to the provisions of the Poisons Act 1972. The Poisons Rules 1982 have been revoked under the Deregulation Act 2015 (Poisons and Explosive Precursors) and replaced by The Control of Poisons and Explosive Precursors Regulations 2015 with appropriate amendments to The Poisons Act 1972. These Rules include provisions for the storage and sale and supply of listed non-medicine poisons. Details can be accessed on the HSE website. The nature of the formulation and the concentration of the active ingredient allow some products to be exempted from the Rules, while others with the same active ingredient are included (see below). The chemicals approved for use in the UK are specified under Parts I and II of the Poisons List as follows.

Part I Poisons (sale restricted to registered retail pharmacists and to registered non-pharmacy businesses provided sales do not take place on retail premises):

- aluminium phosphide
- chloropicrin
- magnesium phosphide.

Part II Poisons (sale restricted to registered retail pharmacists and listed sellers registered with a local authority):

- formaldehyde
- oxamyl (a).

Notes

(a) Granular formulations that do not contain more than 12% w/w of this or a combination of similarly flagged poisons are exempt.

Occupational Exposure Limits

A fundamental requirement of the COSHH Regulations is that exposure of employees to substances hazardous to health should be prevented or adequately controlled. Exposure by inhalation is usually the main hazard, and in order to measure the adequacy of control of exposure by this route various substances have been assigned occupational exposure limits.

There are two types of occupational exposure limits defined under COSHH: Occupational Exposure Standards (OES) and Maximum Exposure Limits (MEL). The key difference is that an OES is set at a level at which there is no indication of risk to health; for a MEL a residual risk may exist and the level takes socio-economic factors into account. In practice, MELs have been most often allocated to carcinogens and to other substances for which no threshold of effect can be identified and for which there is no doubt about the seriousness of the effects of exposure.

OESs and MELs are set on the recommendation of the Advisory Committee on Toxic Substances (ACTS). Full details are published by HSE in EH 40/2005 available at: http://www. hse.gov.uk/pubns/books/eh40.htm

As far as pesticides are concerned, OESs and MELs have been set for relatively few active ingredients. This is because pesticide products usually contain other substances in their formulation, including solvents, which may have their own OES/MEL. In practice inhalation of solvent may be at least, or more, important than that of the active ingredient. These factors are taken into account by the regulators when approving a pesticide product under the Control of Pesticides Regulations. This indicates one of the reasons why a change of pesticide formulation usually necessitates a new approval assessment under COPR.

First Aid Measures

If pesticides are handled in accordance with the required safety precautions, as given on the container label, poisoning should not occur. It is difficult, however, to guard completely against the occasional accidental exposure. Thus, if a person handling, or exposed to, pesticides becomes ill, it is a wise precaution to apply first aid measures appropriate to pesticide poisoning even though the cause of illness may eventually prove to have been quite different. An employer has a legal duty to make adequate first aid provision for employees. Regular pesticide users should consider appointing a trained first aider even if numbers of employees are not large, since there is a specific hazard.

The first essential in a case of suspected poisoning is for the person involved to stop work, to be moved away from any area of possible contamination and for a doctor to be called at once. If no doctor is available the patient should be taken to hospital as quickly as possible. In either event it is most important that the name of the chemical being used should be recorded and preferably the whole product label or leaflet should be shown to the doctor or hospital concerned.

Some pesticides, which are unlikely to cause poisoning in normal use, are extremely toxic if swallowed accidentally or deliberately. In such cases get the patient to hospital as quickly as possible, with all the information you have. Some labels now include Material Safety Data Sheets and these contain valuable information for both the first aider and medical staff. If not included on the label, MSDS are available from company websites.

General Measures

Measures appropriate in all cases of suspected poisoning include the following:

- Remove any protective or other contaminated clothing (taking care to avoid personal contamination).
- Wash any contaminated areas carefully with water or with soap and water if available.
- In cases of eye contamination, flush with plenty of clean water for at least 15 minutes.
- Lay the patient down, keep at rest and under shelter. Cover with one clean blanket or coat, etc. Avoid overheating.
- Monitor level of consciousness, breathing and pulse rate.
- If consciousness is lost, place the casualty in the recovery position (on his/her side with head down and tongue forward to prevent inhalation of vomit).

Reporting of Pesticide Poisoning

Any cases of poisoning by pesticides must be reported without delay to an HM Agricultural Inspector of the Health and Safety Executive. In addition any cases of poisoning by substances named in schedule 2 of The Reporting of Injuries, Diseases and Dangerous Occurrences Regulations 1985, must also be reported to HM Agricultural Inspectorate (this includes organophosphorus chemicals, mercury and some fumigants).

Cases of pesticide poisoning should also be reported to the manufacturer concerned.

Additional Information

General advice on the safe use of pesticides is given in a range of Health and Safety Executive leaflets available from HSE Books. The major agrochemical companies are able to provide authoritative medical advice about their own pesticide products. Useful information is now available from the Material Safety Data Sheet available from the manufacturer or the manufacturer's website.

New arrangements for the provision of information to doctors about poisons and the management of poisonings have been introduced as part of the modernisation of the National Poisons Information Service (NPIS). The Service provides a year-round, 24-hour-a-day service for healthcare staff on the diagnosis, treatment and management of patients who may have been poisoned. The new arrangements are aimed at moving away from the telephone as the first point of contact for poisons information, to the use by doctors of an online database, supported by a second-tier, consultant-led information service for more complex clinical advice. NPIS no longer provides direct information on poisoning to members of the public. **Anyone suspecting poisoning by pesticides should seek professional medical help immediately via their GP or NHS Direct (www.nhsdirect.nhs.uk) or NHS 24 (www.nhs24.com).**

Environmental Protection

Environmental Land Management: 8 simple steps for arable farmers

Choosing the right measures, putting them in the right place, and managing them in the right way will make all the difference to your farm environment. The general principles given here should be considered in conjunction with local priorities for soil and water protection and wildlife conservation. This approach complements best practice in soil, crop, fertiliser and pesticide management.

An improved farm environment can be achieved by good management of around 4% of the arable area where high quality habitats are maintained or created. However, the actual area required on your farm will depend on factors such as the area of vulnerable soils and length of watercourses.

If you need further advice then consult a competent environmental adviser.

What you can do

It is important to have a balance of environmental measures that contribute to each of the relevant points below to achieve improved environmental benefits.

1. **Look after established wildlife habitats**
   Start by assessing what you already have on the farm! Maintaining, or where necessary, restoring any existing wildlife habitats, such as woodland, ponds, flower-rich grassland or field margins, is critical to the survival of much of the wildlife on the farm, and may count towards some of the following measures without the need to create new habitats. Unproductive land can be used to create new habitats to complement what you already have.

2. **Maximise the environmental value of field boundaries**
   Hedgerow management and ditch management on a 2–3 years rotation boosts flowers, fruit and refuges for wildlife. This is most suited to hedges dominated by hawthorn and blackthorn, and ditches where rotational management will not compromise the drainage function. Establish new hedgerow trees to maintain or restore former numbers within the landscape.

3. **Create a network of grass margins**
   The highest priority is to buffer watercourses, ideally with a minimum of 5 m buffer strips. Grass margins can also be used to boost beneficial insects and small mammals, and buffer hedges, ponds and other environmental features. Beetle banks can be used to reduce soil erosion and run-off on slopes greater than 1:20 and boost beneficial insects in fields greater than 20 ha.

4. **Establish flower rich habitats**
   Evidence suggests that a network of flower-rich margins on 1% of arable land will support beneficial insects and a wealth of wildlife that feeds on insects. Assess whether this is best done by allowing arable plants in the seed bank to germinate, establishing perennial margins with a grass and wildflower mix, or using nectar flower mixtures. Improving the linkages between these features on the farm will also help wildlife move across the landscape.

5. **Provide winter food for birds with weedy over-wintered stubbles or wild bird cover**
   Provision of seed for wildlife is best achieved by leaving over-wintered stubbles unsprayed and uncultivated until at least mid-February on at least 5% of arable land, or growing seed-rich crops as wild bird cover on 2% of arable land.

6. **Use of spring cropping or in-field measures to help ground-nesting birds**
   Spring crops provide better habitat for a range of plants and insects, and birds such as lapwings and skylarks. Use rotational fallows, skylark plots in winter cereals or (if breeding lapwings occur) fallow plots to support ground-nesting birds where spring cropping forms less than 25% of the arable area. Fallow plots should not be created on land liable to runoff or erosion. Evidence suggests that at least 20 skylark plots or a 1 ha fallow plot per 100 ha would support ground-nesting birds.

7. **Use winter cover crops to protect water**
    You should consider whether a winter cover crop (e.g. mustard) is necessary to capture residual nitrogen on cultivated land left fallow through the winter. This is not necessary if the stubble is retained until at least mid-February and forms a green cover.
8. **Establish in-field grass areas to reduce soil erosion and run-off**
    Land liable to act as channels for soil erosion or run-off (e.g. steep slopes or field corners) should be converted to in-field grass areas.

Remember

- Right measures
- Right place
- Right management

This guidance has been produced by The Voluntary Initiative and the Campaign for the Farmed Environment in conjunction with RSPB, Natural England, GWCT, Plantlife, Butterfly Conservation, BCPC and Buglife

Protection of Bees

Honey bees

Honey bees are a source of income for their owners and important to farmers and growers as pollinators of their crops. It is irresponsible and unnecessary to use pesticides in such a way that may endanger them. Pesticides vary in their toxicity to bees, but those that present a special hazard carry a specific warning in the precautions section of the label. They are indicated in this *Guide* in the hazard classification and safety precautions section of the pesticide profile.

Product labels indicate the necessary environmental precautions to take, but where use of an insecticide on a flowering crop is contemplated the British Beekeepers Association have produced the following guidelines for growers:

- Target insect pests with the most appropriate product.
- Choose a product that will cause minimal harm to beneficial species.
- Follow the manufacturer's instructions carefully.
- Inspect and monitor crops regularly.
- Avoid spraying crops in flower or where bees are actively foraging.
- Keep down flowering weeds.
- Spray late in the day, in still conditions.
- Avoid excessive spray volume and run-off.
- Adjust sprayer pressure to reduce production of fine droplets and drift.
- Give local beekeepers as much warning of your intention as possible.

Wild bees

Wild bees also play an important role. Bumblebees are useful pollinators of spring flowering crops and fruit trees because they forage in cool, dull weather when honey bees are inactive. They play a particularly important part in pollinating field beans, red and white clover, lucerne and borage. Bumblebees nest and overwinter in field margins and woodland edges. Avoidance of direct or indirect spray contamination of these areas, in addition to the creation of hedgerows and field margins, and late cutting or grazing of meadows, all help the survival of these valuable insects.

BeeConnected

It is best practice to notify local beekeepers before using certain crop protection products where there is a risk to bees. The online BeeConnected tool (www.beeconnected.org.uk) can help you do this.

Accidental and Illegal Poisoning of Wildlife

The Campaign Against Accidental or Illegal Poisoning (CAIP) of Wildlife, was launched back in March 1991 and has now been replaced by **The Wildlife Incident Investigation Scheme.**

This has two objectives are:

SECTION 5

- To provide information to HSE on hazards to wildlife, pets and beneficial invertebrates:

- To enforce the correct use of pesticides, identifying and penalizing those who deliberately or recklessly misuse and abuse pesticides.

Full details can be found on the HSE website at www.hse.gov.uk/pesticides/topics/reducing-environmental-impact/wildlife.htm

Water Quality

Even when diluted, some pesticides are potentially dangerous to fish and other aquatic life. Not only this, but many watercourses and groundwaters are sources of drinking water, and it requires only a tiny amount of contamination to breach the stringent European Union water quality standards. The EEC Drinking Water Directive sets a maximum admissible level in drinking water for any pesticide, regardless of its toxicity, at 1 part in 10,000 million. As little as 250 grams could be enough to cause the daily supply to a city the size of London to exceed the permitted levels (although it would be very unlikely to present a health hazard to consumers).

The protection of groundwater quality is therefore vital. The Food and Environment Protection Act 1985 (FEPA) places a special obligation on users of pesticides to "safeguard the environment and in particular avoid the pollution of water". Under the Water Resources Act 1991 it is an offence to pollute any controlled waters (watercourses or groundwater), either deliberately or accidentally. Protection of controlled waters from pollution is the responsibility of the Environment Agency (in England and Wales) and the Scottish Environmental Protection Agency. Addresses for both organisations can be found under Useful Contacts.

Users of pesticides therefore have a duty to adopt responsible working practices and, unless they are applying herbicides in or near water, to prevent them getting into water. Guidance on how to achieve this is given in the Defra Code of Good Agricultural Practice for the Protection of Water.

The duty of care covers not only the way in which a pesticide is sprayed, but also its storage, preparation and disposal of surplus, sprayer washings and the container. Products that are a major hazard to fish, other aquatic life or aquatic higher plants carry one of several specific label precautions in their fact sheet, depending on the assessed hazard level.

Where to get information

For general enquiries telephone 03708 506 506 (Environment Agency) or 0131 449 7296 (SEPA). Both Agencies operate a 24-hour emergency hotline for reporting all environmental incidents relating to air, land and water:

0800 80 70 60 (for incidents in England and Wales) 0345 73 72 71 (for incidents in Scotland)

In addition, printed literature concerning the protection of water is available from the Environment Agency and the Crop Protection Association (see Useful Contacts).

Protecting surface waters

Surface waters are particularly vulnerable to contamination. One of the best ways of preventing those pesticides that carry the greatest risk to aquatic wildlife from reaching surface waters is to prohibit their application within a boundary adjacent to the water. Such areas are known as no-spray, or buffer, zones. Certain products are restricted in this way and have a legally binding label precaution to make sure the potential exposure of aquatic organisms to pesticides that might harm them is minimised.

Before 1999 the protected zones were measured from the edge of the water. The distances were 2 metres for hand-held or knapsack sprayers, 6 metres for ground crop sprayers, and a variable distance (but often 18 metres) for broadcast air-assisted applications, such as in orchards. The introduction of LERAPs (see below) has changed the method of measuring buffer zones.

'Surface water' includes lakes, ponds, reservoirs, streams, rivers and watercourses (natural or artificial). It also includes temporarily or seasonally dry ditches, which have the potential to carry water at different times of the year. Buffer zone restrictions do not necessarily apply to all products containing the same active ingredient. Those in formulations that are not likely to contaminate surface water through spray drift do not pose the same risk to aquatic life and are not subject to the restrictions.

Local Environmental Risk Assessment for Pesticides (LERAPs)

Local Environment Risk Assessments for Pesticides (LERAPs) were introduced in March 1999, and revised guidelines were issued in January 2002. They give users of most products currently subject to a buffer zone restriction the option of continuing to comply with the existing buffer zone restriction (using the new method of measurement), or carrying out a LERAP and possibly reducing the size of the buffer zone as a result. In either case, there is a new legal obligation for the user to record his decision, including the results of the LERAP.

The scheme has changed the method of measuring the buffer zone. Previously the zone was measured from the edge of the water, but it is now the distance from the top of the bank of the watercourse to the edge of the spray area. In 2013 new rules were introduced which allowed LERAP buffer zones greater than the normal 5 metres for LERAP B pesticides to avoid the loss of some important actives. The products concerned are identified in the Environmental safety section of the affected pesticide profiles.

The LERAP provides a mechanism for taking into account other factors that may reduce the risk, such as dose reduction, the use of low drift nozzles, and whether the watercourse is dry or flowing. The previous arrangements applied the same restriction regardless of whether there was actually water present. Now there is a standard zone of 1 m from the top of a dry ditch bank.

Other factors to include in a LERAP that may allow a reduction in the buffer zone are:

- The size of the watercourse, because the wider it is, the greater the dilution factor, and the lower the risk of serious pollution.

- The dose applied. The lower the dose, the less is the risk.

- The application equipment. Sprayers and nozzles are star-rated according to their ability to reduce spray drift fallout. Equipment offering the greatest reductions achieves the highest rating of three stars. The scheme was originally restricted to ground crop sprayers; new, more flexible rules introduced in February 2002 included broadcast air-assisted orchard and hop sprayers.

- Other changes introduced in 2002 allow the reduction of a buffer zone if there is an appropriate living windbreak between the sprayed area and a watercourse.

Not all products that had a label buffer zone restriction are included in the LERAP scheme. The option to reduce the buffer zone does not apply to organophosphorus or synthetic pyrethroid insecticides. This group are classified as Category A products. All other products that had a label buffer zone restriction are classified as Category B. In addition some products have a 'Broadcast air-assisted LERAP' classification. The wording of the buffer zone label precautions has been amended for all products to take account of the new method of measurement, and whether or not the particular product qualifies for inclusion in the LERAP scheme.

Products in this *Guide* that are in Category A or B, or have a broadcast air-assisted LERAP, are identified with an appropriate icon on the product fact sheet. Updates to the list are published regularly by CRD and details can be obtained from its website at www.pesticides.gov.uk.

The introduction of LERAPs was an important step forward because it demonstrated a willingness to reduce the impact of regulation on users of pesticides by allowing flexibility where local conditions make it safe to do so. This places a legal responsibility on the user to ensure the risk assessment is done either by himself or by the spray operator or by a professional consultant or advisor. It is compulsory to record the LERAP and make it available for inspection by enforcement authorities. In 2013 new rules were introduced which allowed LERAP buffer zones greater than 5 m to avoid the loss of some important actives. These products are identified in the section of Environmental safety for the pesticide profiles affected.

More details of the LERAP arrangements and guidance on how to carry out assessments are published in the Ministry booklet PB5621 *Local Environmental Risk Assessment for Pesticides*

SECTION 5

– *Horizontal Boom Sprayers,* and booklet PB6533 *Local Environment Risk Assessment for Pesticides – Broadcast Air-Assisted Sprayers.* Additional booklets, PB2088 *Keeping Pesticides Out of Water,* and PB3160 *Is Your Sprayer Fit for Work* give general practical guidance.

Groundwater Regulations

Groundwater Regulations were introduced in 1999 to complete the implementation in UK of the EU Groundwater Directive (Protection of Groundwater Against Pollution Caused by Certain Dangerous Substances – 80/68/EEC). These Regulations help prevent the pollution of groundwater by controlling discharges or disposal of certain substances, including all pesticides.

Groundwater is defined under the Regulations as any water contained in the ground below the water table. Pesticides must not enter groundwater unless it is deemed by the appropriate Agency to be permanently unsuitable for other uses. The Agricultural Waste Regulations were introduced in May 2006 and apply to the disposal of all farm wastes, including pesticides. With certain exemptions farmers are required to obtain a waste management licence for most waste disposal activities.

As far as pesticides are concerned the new Regulations made little difference to existing controls, except for the disposal of empty containers and the disposal of sprayer washings and rinsings. It remains an offence to dispose of pesticides onto land without official authorisation. Normal use of a pesticide in accordance with product approval does not require authorisation. This includes spraying the washings and rinsings back on the crop provided that, in so doing, the maximum approved dose for that product on that crop is not exceeded. However, those wishing to use a lined biobed for this purpose need to obtain an exemption from the Agency.

In practice the best advice to farmers and growers is to plan to use all diluted spray within the crop and to dispose of all washings via the same route making sure that they stay within the conditions of approval of the product. The enforcing agencies for these Regulations are the Environment Agency (in England and Wales) and the Scottish Environment Protection Agency. The Agencies can serve notice at any time to modify the conditions of an authorisation where necessary to prevent pollution of groundwater.

Integrated Farm Management (IFM)

Integrated farm management is a method of farming that balances the requirements of running a profitable farming business with the adoption of responsible and sensitive environmental management. It is a whole-farm, long-term strategy that combines the best of modern technology with some basic principles of good farming practice. It is a realistic way forward that addresses the justifiable concerns of the environmental impact of modern farming practices, at the same time as ensuring that the industry remains viable and continues to provide wholesome, affordable food. IFM embraces arable and livestock management. The phrase 'integrated crop management' is sometimes used where a farm is wholly arable.

Pest control is essential in any management system. Much can be done to minimise the incidence and impact of pests, but their presence is almost inevitable. IFM ensures that, where a pest problem needs to be contained, the action taken is the best combination of all available options. Pesticides are one of these options, and form an essential, but by no means exclusive, part of pest control strategy.

Where chemicals are to be used, the choice of product should be made not only with the pest problem in mind, but with an awareness of the environmental and social risks that might accompany its use. The aim should be to use as much as necessary, but as little as possible. A major part of the approval process aims to safeguard the environment, so that any approved product, when used as directed, will not cause long-term harm to wildlife or the environment. This is achieved by specifying on the label detailed rules for the way in which a product may be used.

The skill in implementing an IFM pest control strategy is the decision on how easily these rules may be complied with in the particular situation where use is contemplated. Although many chemical options may be available, some are likely to be more suitable than others. This *Guide* lists, under the heading **Special precautions/Environmental safety,** the key label precautions that need to be considered in this context, although the actual product label must always be read before a product is used.

Campaign for the Farmed Environment

With the demise of set-aside, it was realised that the environmental benefits achieved by this scheme could be lost. The Campaign for the Farmed Environment (CFE) scheme seeks to retain or even exceed the environmental benefits achieved by set-aside by voluntary measures.

It has three main themes:

- farming for cleaner water and a healthier soil
- helping farmland birds thrive
- improving the environment for farm wildlife.

The scheme encourages farmers to sign up for Entry Level Stewardship (ELS) schemes and to target their already extensive knowledge of habitat management to greater effect. If successful, it will help to fend off the proposal that farmers of cultivated land in England will have to adopt environmental management options on up to 6% of their land.

SECTION 6
APPENDICES

Appendix 1
Suppliers of Pesticides and Adjuvants

Aceto:
Aceto Agricultural Chemicals Corporation
(UK) Ltd
    2nd Floor, Refuge House
    33-37 Watergate Row
    Chester
    Cheshire
    CH1 2LE
    Tel: (516) 627-6000 ext.
    Email: ftrudwig@aceto.com
    Web: www.aceto.com

Adama:
Adama Agricultural Solutions UK Ltd
    Unit 15
    Thatcham Business Village
    Colthrop Way
    Thatcham
    Berks.
    RG19 4LW
    Tel: 01635 860555
    Fax: 01635 861555
    Email: ukenquiries@adama.com
    Web: www.adama.com

AgChem Access: AgChemAccess Limited
    Cedar House,
    41 Thorpe Road
    Norwich
    Norfolk
    NR1 1ES
    Tel: 0845 459 9413
    Fax: 0207 149 9815
    Email: thomas@agchemaccess.com
    Web: www.agchemaccess.com

Agform: Agform Ltd
    Hilldale Farm Research Centre
    Titchfield Lane
    Wickham
    Hampshire
    PO17 5NZ
    Tel: 01329 836933
    Fax: 023 8040 7198
    Email: info@agform.com
    Web: www.agform.com

Agrichem: Agrichem (International) Ltd
    Industrial Estate
    Station Road
    Whittlesey
    Cambs.
    PE7 2EY
    Tel: 01733 204019
    Fax: 01733 204162
    Email: admin@agrichem.co.uk
    Web: www.agrichem.co.uk

Agrii: United Agri Products Ltd
    Throws Farm
    Stebbing
    Great Dunmow
    Essex
    DM6 3AQ
    Web: www.agrii.co.uk

Agropharm: Agropharm Limited
    Buckingham Place
    Church Road
    Penn
    High Wycombe
    Bucks.
    HP10 8LN
    Tel: 01494 816575
    Fax: 01494 816578
    Email: sales@agropharm.co.uk
    Web: www.agropharm.co.uk

Agroquimicos : Agroquimicos Genericos
    Calle San Vicente Martir
    no 85, 8, 46007
    Valencia,
    Spain
    Email: thomas@agchemaccess.com

Agrovista: Agrovista UK Ltd
    Rutherford House
    Nottingham Science and Technology
    Park
    University Bulevard
    Nottingham
    NG7 2PZ
    Tel: (0115) 939 0202
    Fax: (0115) 921 8498
    Email: enquiries@agrovista.co.uk
    Web: www.agrovista.co.uk

Albaugh UK: Albaugh UK Ltd
    1 Northumberland Street
    Trafalgar Square
    London
    WC2N 5BW

Amega: Amega Sciences
Lanchester Way
Royal Oak Industrial Estate
Daventry
Northants.
NN11 5PH
Tel: (01327) 704444
Fax: (01327) 71154
Email: admin@amega-sciences.com
Web: www.amega-sciences.com

Arysta: Arysta Life Sciences
Route d'Artix, BP80
64150
Nogueres
France
Email: Don.pendergrast@arysta.com

B H & B: Battle Hayward & Bower Ltd
Victoria Chemical Works
Crofton Drive
Allenby Road Industrial Estate
Lincoln
LN3 4NP
Tel: (01522) 529206
Fax: (01522) 538960
Email: orders@battles.co.uk
Web: www.battles.co.uk

Barclay:
Barclay Chemicals Manufacturing Ltd
Damastown Way
Damastown Industrial Park
Mulhuddart
Dublin 15
Ireland
Tel: (+353) 1 811 2900
Fax: (+353) 1 822 4678
Email: info@barclay.ie
Web: www.barclay.ie

Barrier: Barrier BioTech Ltd
36/37 Haverscroft Industrial Estate
New Road
Attleborough
Norfolk
NR17 1YE
Tel: (01953) 456363
Fax: (01953) 455594
Email: sales@barrier-biotech.com
Web: www.barrier-biotech.com

BASF: BASF plc
Agricultural Divison
PO Box 4, Earl Road
Cheadle Hulme
Cheshire
SK8 6QG
Tel: (0845) 602 2553
Fax: (0161) 485 2229
Web: www.agricentre.co.uk

Bayer CropScience:
Bayer CropScience Limited
230 Cambridge Science Park
Milton Road
Cambridge
CB4 0WB
Tel: (01223) 226500
Fax: (01223) 426240
Web: www.bayercropscience.co.uk

Becesane : Becesane s.r.o.
Rohacova, 188/377
130 00
PRAHA 3
Czech Republic
Email: thomas@agchemaccess.com

Belchim: Belchim Crop Protection Ltd
Unit 1b, Fenice Court
Phoenix Park
Eaton Socon
St Neots
Cambs.
PE19 8EP
Tel: (01480) 403333
Fax: (01480) 403444
Email: info@belchim.com
Web: www.belchim.co.uk

Belcrop: Belcrop NV
Tiensestraat 300
3400 Landen
Belgium
Tel: +32 (0) 11 59 83 60
Fax: +32 (0) 11 59 83 61
Email: info@belcrop.com
Web: www.BELCROP.com

Biofresh: Biofresh Ltd
INEX Business Centre
Herschell Building
Newcastle University Campus
Newcastle on Tyne
Tyne & Wear
NE1 7RU
Tel: 0191 243 0879
Fax: 0191 243 0244
Email: info@bio-fresh.co.uk

Capital CP: Capital CP Europe Limited
Gardeners Cottage
Treemans Road
Lewes Road
Haywards Heath
West Sussex
RH17 7EA

Certis: Certis
Suite 5, 1 Riverside
Granta Park
Great Abington
Cambs.
CB21 6AD
Tel: 0845 373 0305
Fax: 01223 891210
Email: certis@certiseurope.co.uk
Web: www.certiseurope.co.uk

Chemtura: Chemtura Europe Ltd
Kennet House
4 Langley Quay
Slough
Berks.
SL3 6EH
Tel: (01753) 603000
Fax: (01753) 603077
Web: www.chemtura.com

Chem-Wise: Chem-Wise Ltd
Westwood
Rattar Mains
Thurso
KW14 8XW

Chiltern: Chiltern Farm Chemicals Ltd
East Mellwaters
Stainmore
Bowes
Barnard Castle
Co. Durham
DL12 9RH
Tel: (01833) 628282
Email: chilternfarm@aol.com
Web: www.chilternfarm.com

Chrysal : Chrysal UK Limited
Ardsley Mills
Common Lane
East Ardsley
Wakefield
West Yorkshire
WF3 2DW

Clayton: Clayton Plant Protection Ltd
Bracetown Business Park
Clonee
Dublin 15
Ireland
Tel: (+353) 1 821 0127
Fax: (+353) 81 841 1084
Email: info@cpp.ag
Web: www.cpp.ag

DAPT: DAPT Agrochemicals Ltd
14 Monks Walk
Southfleet
Gravesend
Kent
DA13 9NZ
Tel: (01474) 834448
Fax: (01474) 834449
Email: rkjltd@supanet.com

De Sangosse: De Sangosse Ltd
Hillside Mill
Quarry Lane
Swaffham Bulbeck
Cambridge
CB25 0LU
Tel: (01223) 811215
Fax: (01223) 810020
Email: info@desangosse.co.uk
Web: www.desangosse.co.uk

Dow: Dow AgroSciences Ltd
CPC2,
Capital Park
Fulbourn
Cambridge
Cambridgeshire
CB21 5XE
Tel: (01462) 457272
Fax: (01462) 426605
Email: fhihotl@dow.com
Web: www.uk.dowagro.com

DuPont: DuPont (UK) Ltd
Crop Protection Products Department
4th Floor, Kings Court
London Road
Stevenage
Herts.
SG1 2NG
Tel: (01438) 734450
Fax: (01438) 734452
Email: enquiry.agproducts@dupont.com
Web:
www.cropprotection.dupont.co.uk

EuroAgkem: EuroAgkem Ltd
Park Chambers
10 Hereford Road
Abergavenney
NP7 5PR
Tel: (01926) 634801
Fax: (01926) 634798

SECTION 6

Everris Ltd: Everris Limited
Epsilon House
West Road
Ipswich
Suffolk
IP3 9FJ
Tel: 01473 237100
Fax: 01473 830386
Email: prof.sales@everris.com

Exosect: Exosect Ltd
Leylands Business Park
Colden Common
Winchester
SO21 1TH
Tel: 02380 60 395
Email: sue.harris@exosect.com
Web: www.exosect.com

Fargro: Fargro Ltd
Vinery Fields
Arundel Road
Poling
Arundel
West Sussex
BN18 9PY
Tel: (01903) 721591
Fax: (01903) 883303
Email: info@fargro.co.uk
Web: www.fargro.co.uk

Fine: Fine Agrochemicals Ltd
Hill End House
Whittington
Worcester
WR5 2RQ
Tel: (01905) 361800
Fax: (01905) 361810
Email: enquire@fine.eu
Web: www.fine.eu

Forest Research: Forest Research
Alice Holt Lodge
Farnham
Surrey
GU10 4LH
Tel: 01420 22255
Fax: 01420 23653
Email:
Katherine.tubby@forestry.gsi.gov.uk
Web: www.forestry.gov.uk

Generica : Generica Europa Ltd
The Granary
27C Silver Street
Buckden
St Neots
Cambridgeshire
PE19 5TS
Email: office@genericaeuropa.com

Globachem: Globachem NV
Brustem Industriepark
Lichtenberglaan 2019
3800 Sint-Truiden
Belgium
Tel: (+32) 11 69 01 73
Fax: (+32) 11 68 15 65
Email: globachem@globachem.com
Web: www.globachem.com

Global Adjuvants :
Global Adjuvants Company Ltd
20-22 Wenlock Road
London
N1 7GU
Email: office@global-adjuvants.com
Web: www.global-adjuvants.com

Goldengrass: Goldengrass Ltd
PO Box 280
Ware
SG12 8XY
Tel: 01920 486253
Fax: 01920 485123
Email: contact@goldengrass.uk.com

Gowan:
Gowan Comercio e Servicos Limitada
2 Leazes Cresent
Hexham
Northumberland
NE46 3JX
Tel: 07584 052323
Email: DLamb@GOWANCO.com

Greenaway: Greenaway Amenity Ltd
7 Browntoft Lane
Donington
Spalding
Lincs.
PE11 4TQ
Tel: (01775) 821031
Fax: (01775) 821034
Email: greenawayamenity@aol.com
Web: www.greenawaycda.com

Greencrop: Greencrop Technology Ltd
c/o Clayton Plant Protection Ltd
Bracetown Business Park
Clonee
Dublin 15
Ireland
Tel: (+353) 1 821 0127
Fax: (+353) 81 841 1084
Email: info@cpp.ag
Web: www.cpp.ag

Harvest: Harvest Agrochemicals Ltd
  Carpenter Court,
  1 Maple Road,
  Bramhall,
  Stockport,
  Cheshire,
  SK7 2DH
  United Kingdom

Headland:
Headland Agrochemicals trading as
Cheminova (UK) Ltd
  Rectors Lane
  Pentre
  Deeside
  Flintshire
  CH5 2DH
  Tel: (01244) 537370
  Fax: (01244) 532097
  Email: enquiry@headlandgroup.com
  Web: www.headland-ag.co.uk

Headland Amenity:
Headland Amenity Limited
  1 Burr Elm Court
  Main Street
  Caldecote
  Cambs.
  CB23 7NU
  Tel: (01223) 481090
  Fax: (01223) 491091
  Email: info@headlandamenity.com
  Web: www.headlandamenity.com

Helena: Helena Chemical Company
  Cambridge House
  Nottingham Road
  Stapleford
  Nottingham
  NG9 8AB
  Tel: (0115) 939 0202
  Fax: (0115) 939 8031

Hermoo: Hermoo Belgium NV
  Brustem Industriepark
  Lichtenberglaan 2045
  B-3800 Sint-Truiden
  Belgium
  Tel: +32 (0) 11 68 68 66
  Fax: +32 (0) 11 67 12 05
  Email: hermoo@hermoo.be
  Web: www.hermoo.com

Industrias Afrasa: Industrias Afrasa, S.A
  Ciudad de Sevilla, 53
  Pol. Industrial Fuenta del Jarro
  46988 - Paterna
  Valencia
  Spain

Interagro: Interagro (UK) Ltd
  Thorley Wash Barn
  London Road
  Thorley
  Bishop's Stortford
  Hertfordshire
  CM23 4AT
  Tel: (01279) 714970
  Fax: (01279) 758227
  Email: info@interagro.co.uk
  Web: www.interagro.co.uk

Interfarm: Interfarm (UK) Ltd
  Kinghams's Place
  36 Newgate Street
  Doddington
  Cambs.
  PE15 0SR
  Tel: (01354) 741414
  Fax: (01354) 741004
  Email: technical@interfarm.co.uk
  Web: www.interfarm.co.uk

Intracrop: Intracrop
  Park Chambers,
  10 Hereford Road,
  Abergavenny,
  NP7 5PR
  Tel: (01926) 634801
  Fax: (01926) 634798
  Email: admin@intracrop.co.uk
  Web: www.intracrop.co.uk

Juno : Juno (Plant Protection) Ltd
  Little Mill Farm
  Underlyn Lane
  Marden
  Kent
  TN12 9AT
  Tel: 01580 765079
  Email: nick@junopp.com

Killgerm: Killgerm Chemicals Ltd
  115 Wakefield Road
  Flushdyke
  Ossett
  W. Yorks.
  WF5 9AR
  Tel: (01924) 268400
  Fax: (01924) 264757
  Email: info@killgerm.com
  Web: www.killgerm.com

SECTION 6

687

Koppert: Koppert (UK) Ltd
Unit 8, Tudor Rose Court
53 Hollands Road
Haverhill
Suffolk
CB9 8PJ
Tel: (01440) 704488
Fax: (01440) 704487
Email: info@koppert.co.uk
Web: www.koppert.com

Landseer: Landseer Ltd
Lodge Farm
Goat Hall Lane
Galleywood
Chelmsford
Essex
CM2 8PH
Tel: (01245) 357109
Fax: (01245) 494165
Web: www.lanfruit.co.uk

Life Scientific: Life Scientific Limited
Unit 12, NovaUCD
Belfield Innovation Park
University College Dublin
Dublin 4
Ireland
Tel: 00353 1 283 2024
Web: www.lifescientific.com

Maxwell: Maxwell Amenity Limited
T/A Amenity Land Solutions
Units 1-3 Allscott Park
Allscott
Telford
Shropshire
TF6 5DY

Microcide: Microcide Ltd
Shepherds Grove
Stanton
Bury St. Edmunds
Suffolk
IP31 2AR
Tel: (01359) 251077
Fax: (01359) 251545
Email: microcide@microcide co.uk
Web: www.microcide.co.uk

Mitsui:
Mitsui AgriScience International S.A
Unit 26, Block B
Northwood House,
Northwood Business Campus,
Santry
Dublin 9
Ireland
Tel: 81-3-3285-7540
Fax: 81-3-3285-9819

Monsanto: Monsanto (UK) Ltd
PO Box 663
Cambourne
Cambridge
CB1 0LD
Tel: (01954) 717575
Fax: (01954) 717579
Email:
technical.helpline.uk@monsanto.com
Web: www.monsanto-ag.co.uk

Nissan: Nissan Chemical Europe Sarl
Parc d'Affaires de Crecy
2 Rue Claude Chappe
69371 St Didier au Mont d'Or
France
Tel: (+33) 4 376 44020
Fax: (+33) 4 376 46874

Nomix Enviro:
Nomix Enviro, A Division of Frontier
Agriculture Ltd
The Grain Silos
Weyhill Road
Andover
Hampshire
SP10 3NT
Tel: 01264 388050
Fax: 01522 866176
Email: nomixenviro@frontierag.co.uk
Web: www.nomix.co.uk

Nufarm UK: Nufarm UK Ltd
Wyke Lane
Wyke
West Yorkshire
BD12 9EJ
Tel: 01274 69 1234
Fax: 01274 69 1176
Email: infouk@uk.nufarm.com
Web: www.nufarm.co.uk

Omex: Omex Agriculture Ltd
Bardney Airfield
Tupholme
Lincoln
LN3 5TP
Tel: (01526) 396000
Fax: (01526) 396001
Email: enquire@omex.com
Web: www.omex.co.uk

Pan Agriculture: Pan Agriculture Ltd
8 Cromwell Mews
Station Road
St Ives
Huntingdon
Cambs.
PE27 5HJ
Tel: (01480) 467790
Fax: (01480) 467041
Email: info@panagriculture.co.uk

Pan Amenity: Pan Amenity Ltd
8 Cromwell Mews
Station Road
St Ives
Cambs.
PE27 5HJ
Tel: 01480 467790
Fax: 01480 467041

Petro-Lube: Petro-Lube (UK) Limited
Lonsto House, Suit No: 10
276 Chase Road
Southgate
London
N14 6HA
Tel: 0208 886 8002
Fax: 0208 886 7716
Email: info@petrolube.co.uk
Web: www.petrolube.co.uk

Phyto Services: Phyto Services Limited
13 Mizpah Grove
Elton
Bury
BL8 2SD

Plant Solutions: Plant Solutions
Pyports
Downside Bridge Road
Cobham
Surrey
KT11 3EH
Tel: (01932) 576699
Fax: (01932) 868973
Email: sales@plantsolutionsltd.com
Web: www.plantsolutionsltd.com

Progreen:
Progreen Weed Control Solutions Limited
Kellington House
South Fen Business Park
South Fen Road
Bourne
Lincolnshire
PE10 0DN
Email: info@progreen.co.uk
Web: www.progreen.co.uk

ProKlass: ProKlass Products Limited
145-157 St John Street
London
EC1V 4PW
UK
Tel: 01480 810137
Email: office@proklass-products.com

Pure Amenity : Pure Amenity Limited
Pure House
64-66 Westwick Str.
Norwich
Norfolk
NR2 4SZ
Tel: 0845 257 4710
Email: martin@pureamenity.co.uk
Web: www.pureamenity.co.uk

Q-Chem: Q-CHEM nv
Leeuwerweg 138
BE-3803
Belgium
Tel: (+32) 11 785 717
Fax: (+32) 11 681 565
Email: q-chem@skynet.be

Rentokil: Rentokil Initial plc
7-8 Foundry Court,
Foundry Lane
Horsham
W. Sussex
RH13 5PY
Tel: (01403) 214122
Fax: (01403) 214101
Email: dawn.kirby@rentokil-initial.com
Web: www.rentokil.com

Resource Chemicals:
Resource Chemicals Ltd
Resource House
76 High Street
Brackley
Northants
NN13 7DS
Tel: 01280 843800

Rigby Taylor: Rigby Taylor Ltd
Rigby Taylor House
Crown Lane
Horwich
Bolton
Lancs.
BL6 5HP
Tel: (01204) 677777
Fax: (01204) 677715
Email: info@rigbytaylor.com
Web: www.rigbytaylor.com

Rotam : Rotam Europe Ltd
Hamilton House
Mabledon Place
London
WC1H 9BB
Tel: 0207 953 0447
Web: www.rotam.com/uk

SECTION 6

SFP Europe: SFP Europe SA
  11 Boulevard de la Grand Thumine
  Parc d'Ariane - Bat B
  13090 Aix en Provence
  France

Sharda:
Sharda Worldwide Exports PVT Ltd
  Domnic Holm
  29th Road
  Bandra (West)
  Mumbai
  400050
  India
  Tel: 0032 02 466 4444

Sherriff Amenity:
Sherriff Amenity Services
  The Pines
  Fordham Road
  Newmarket
  Cambs
  CB8 7LG
  Tel: (01638) 721888
  Fax: (01638) 721815
  Web: www.sherriffamenity.com

Sinclair: William Sinclair Horticulture Ltd
  Firth Road
  Lincoln
  LN6 7AH
  Tel: (01522) 537561
  Fax: (01522) 513609
  Web: www.william-sinclair.co.uk

Sipcam: Sipcam UK Ltd
  3 The Barn
  27 Kneesworth Street
  Royston
  Herts.
  SG8 5AB
  Tel: (01763) 212100
  Fax: (01763) 212101
  Email: paul@sipcamuk.co.uk
  Web: sipcamuk.co.uk

Sphere: Sphere Laboratories (London) Ltd
  c/o Mainswood
  Putley Common
  Ledbury
  Herefordshire
  HR8 2RF
  Tel: 01684 899306
  Fax: 01684 893322
  Email:
  Fiona.Homes@Sphere-London.co.uk

Standon: Standon Chemicals Ltd
  48 Grosvenor Square
  London
  W1K 2HT
  Tel: (020) 7493 8648
  Fax: (020) 7493 4219

Stefes: Stefes GMBH
  Wendenstrasse 21b
  20097 Hamburg
  Germany
  Tel: +49 (0)40 533083-0
  Fax: +49 (0)40 5330833-29
  Email: info@stefes.eu
  Web: www.stefes.eu

Sumi Agro: Sumi Agro Europe Limited
  Vintners' Place
  68 Upper Thames Street
  London
  EC4V 3BJ
  Tel: (020) 7246 3697
  Fax: (020) 7246 3799
  Email: summit@summit-agro.com

Synergy : Synergy Generics Limited
  Market House
  Church Street
  Harlston
  Norfolk
  IP20 9BB
  Tel: 07884 435 282
  Email: thomas@agchemaccess.com

Syngenta: Syngenta UK Limited
  CPC4
  Capital Park
  Fulbourn
  Cambridge
  CB21 5XE
  Tel: 01223 883400
  Fax: (01223) 493700
  Web: www.syngenta-crop.co.uk

Taminco: Taminco UK Ltd
  15 Rampton Drift
  Longstanton
  Cambridge
  CB24 3EH
  Tel: 01954 789941
  Email: philip.forster@taminco.com
  Web: www.taminco.com

Unique Marketing : Unique Marketing SL
  Calle Marques Del Duero 67 Planta 1
  Puerta A
  San Pedro Alcantara,
  29670 Marbella,
  Malaga
  Spain
  Tel: +34 674 514 601

United Agri: United Agri Products Ltd
The Crossways
Alconbury Hill
Huntingdon
PE28 4JH
Tel: (01480) 418000
Fax: (01480) 418010
Web: www.agrii.co.uk

UPL Europe: UPL Europe Ltd
The Centre
Birchwood Park
Warrington
Cheshire
WA3 6YN
Tel: (01925) 819999
Fax: (01925) 856075
Email: pam.chambers@uniphos.com
Web: www.upleurope.com

Ventura: Ventura Agroscience Ltd
20-22 Wenlock Road
London
N1 7GU
Tel: 01480 810137
Email: office@ventura-agroscience.com

Vitax: Vitax Ltd
Owen Street
Coalville
Leicester
LE67 3DE
Tel: (01530) 510060
Fax: (01530) 510299
Email: info@vitax.co.uk
Web: www.vitax.co.uk

Zantra: Zantra Limited
Westwood
Rattar mains
Scarfskerry
Thurso
Caithness
KW14 8XW
Tel: 01480 861066
Fax: 01480 861099

Appendix 2
Useful Contacts

Agriculture Industries Confederation Ltd
Confederation House
East of England Showground
Peterborough PE2 6XE
Tel: (01733) 385230
Web: www.agindustries.org.uk

AHDB Horticulture
Stoneleigh Park
Kenilworth
Warwickshire CV8 2TL
Tel: (024) 7669 2051
Web: www.horticulture.ahdb.org.uk

BASIS Ltd
St Monica's House Business Centre
39 Windmill Lane
Ashbourne
Derbyshire DE6 1EY
Tel: (01335) 343945
Fax: (01335) 301205
Web: www.basis-reg.co.uk

British Beekeepers' Association
National Beekeeping Centre
Stoneleigh
Kenilworth
Warwickshire CV8 2LG
Tel: (0871) 811 2282
Web: www.bbka.org.uk

British Beer & Pub Association
Ground Floor
Brewers' Hall
Aldermanbury Square
EC2V 7HR
Tel: (0207) 627 9191
Web: www.beerandpub.com

British Crop Production Council (BCPC)
Garden Studio
4 Hillside
Aldershot
Surrey GU11 3NB
Tel: (01252) 285223
Web: www.bcpc.org

British Pest Control Association (BPCA)
4A Mallard Way
Pride Park
Derby DE24 8GX
Tel: (01332) 294288
Fax: (01332) 225101
Web: www.bpca.org.uk

Chemicals Regulation Division
Room 1A, Mallard House
King's Pool
3 Peasholme Green
York YO1 7PX
Tel: (01904) 640500 or (03459) 335577
Fax: (01904) 455733
Web: www.hse.gov.uk/crd/

Crop Protection Association Ltd
2 Swan Court
Cygnet Park
Hampton
Peterborough PE7 8GX
Tel: (01733) 355370
Web: www.cropprotection.org.uk

CropLife International
326 Avenue Louise Box 35
B-1050 Brussels
Belgium
Tel: (+32) 2 542 0410
Fax: (+32) 2 542 0419
Web: www.croplife.org

**Department of Agriculture and Rural
Development (Northern Ireland)**
Pesticides Section
Dundonald House
Upper Newtownards Road
Belfast BT4 3SB
Tel: (028) 9052 4704 or (0300) 200 7850
Fax: (028) 9052 4059
Web: www.daera-ni.gov.uk

**Department of Environment, Food and
Rural Affairs (Defra)**
Nobel House
17 Smith Square
London SW1P 3JR
Tel: (020) 7238 6000 / 03459 335577
Web: www.defra.gov.uk

Environment Agency
National Customer Contact Centre
PO Box 544
Rotherham S60 1BY
Tel: (03708) 506 506
*Email: enquiries@environment-agency.
gov.uk*
Web: www.environment-agency.gov.uk

European Crop Protection Association (ECPA)
Avenue E van Nieuwenhuyse 6
B-1160 Brussels
Belgium
Tel: (+32) 2 663 1550
Fax: (+32) 2 663 1560
Web: www.ecpa.eu

Farmers' Union of Wales
Llys Amaeth
Plas Gogerddan
Aberystwyth
Ceredigion SY23 3BT
Tel: (01970) 820820
Fax:(01970) 820821
Web: www.fuw.org.uk

Forestry Commission
620 Bristol Business Pk
Coldharbour Lane
Bristol BS16 1EJ
Tel: (0300) 067 4000
Web: www.forestry.gov.uk

Game and Wildlife Conservation Trust (GWCT)
Head Office
Burgate Manor
Fordingbridge
Hampshire
SP6 1EF
Tel: 01425 652381
Email: info@gwct.org.uk
Web: www.gwct.org.uk/

Health and Safety Executive
Product Approvals
Chemical Regulation
Redgrave Court
Merton Road
Bootle
Merseyside L20 7HS
Tel: (0151) 951 4000
Web: www.hse.gov.uk

Health and Safety Executive – Books
PO Box 5050
Sherwood Park
Annesley,
Nottinghamshire NG15 0JD

Lantra
National Agricultural Centre
Lantra House
Stoneleigh
Kenilworth
Warwickshire CV8 2LG
Tel: (024) 7669 6996
Fax: (024) 7669 6732
Web: www.lantra.co.uk

National Association of Agricultural Contractors (NAAC)
The Old Cart Shed
Easton Lodge Farm
Old Oundle Road, Wansford
Peterborough PE8 6NP
Tel: (01780) 784631
Fax: (01780) 784933
Web: www.naac.co.uk

National Chemicals Emergency Centre
The Gemini Building
Fermi Avenue
Harwell
Didcot
Oxfordshire OX11 0RG
Tel: (01235) 753654
Fax: (01235) 753656
Email: ncec@ricardo-aea.com
Web: www.the-ncec.com

National Farmers' Union
Agriculture House Stoneleigh Park
Stoneleigh
Warwickshire CV8 2TZ
Tel: (024) 7685 8500
Web: www.nfu.org.uk

National Poisons Information Service
(Birmingham Unit)
City Hospital, Dudley Road
Birmingham B18 7QH
Tel: (0121) 507 4123
Fax: (0121) 507 5580
Web: www.npis.org

Natural England (enquiries)
County Hall
Spetchley Road
Worcester
WR5 2NP
Tel: 0300 060 3900
Suspected wildlife/pesticide poisoning
Tel: 0800 321 600
Email: enquiries@naturalengland.org.uk
Web: www.gov.uk/government/
organisations/natural-england

Processors and Growers Research Organisation
The Research Station
Great North Road, Thornhaugh
Peterborough, Cambs. PE8 6HJ
Tel: (01780) 782585
Web: www.pgro.org

SECTION 6

Scottish Beekeepers' Association
Cowiemuir
Fochabers
Moray IV32 7PS
*Email: secretary@scottishbeekeepers.
org.uk*
Web: www.scottishbeekeepers.org.uk

Scottish Environment Protection Agency (SEPA)
Clearwater House
Heriot Watt Research Park
Avenue North
Riccarton
Edinburgh EH14 4AP
Tel: (0131) 449 7296
Fax: (0131) 449 7277
Web: www.sepa.org.uk

TSO (The Stationery Office)
85 Buckingham Gate
London SW1E 6PD
Tel: (0333) 202 5070
Email: customer.services@tso.co.uk
Web: www.tsoshop.co.uk

Appendix 3
Keys to Crop and Weed Growth Stages

Decimal Code for the Growth Stages of Cereals
Illustrations of these growth stages can be found in the reference indicated below and in some company product manuals.

0 Germination

00 Dryseed
03 Imbibition complete
05 Radicle emerged from caryopsis
07 Coleoptile emerged from caryopsis
09 Leaf at coleoptile tip

1 Seedling growth

10 First leaf through coleoptile
11 First leaf unfolded
12 2 leaves unfolded
13 3 leaves unfolded
14 4 leaves unfolded
15 5 leaves unfolded
16 6 leaves unfolded
17 7 leaves unfolded
18 8 leaves unfolded
19 9 or more leaves unfolded

2 Tillering

20 Main shoot only
21 Main shoot and 1 tiller
22 Main shoot and 2 tillers
23 Main shoot and 3 tillers
24 Main shoot and 4 tillers
25 Main shoot and 5 tillers
26 Main shoot and 6 tillers
27 Main shoot and 7 tillers
28 Main shoot and 8 tillers
29 Main shoot and 9 or more tillers

3 Stem elongation

30 Ear at 1 cm
31 1st node detectable
32 2nd node detectable
33 3rd node detectable
34 4th node detectable
35 5th node detectable
36 6th node detectable
37 Flag leaf just visible
39 Flag leaf ligule/collar just visible

4 Booting

41 Flag leaf sheath extending
43 Boots just visibly swollen
45 Boots swollen
47 Flag leaf sheath opening
49 First awns visible

5 Inflorescence

51 First spikelet of inflorescence just visible
52 1/4 of inflorescence emerged
55 1/2 of inflorescence emerged
57 3/4 of inflorescence emerged
59 Emergence of inflorescence completed

6 Anthesis

60
61 Beginning of anthesis
64
65 Anthesis half way
68
69 Anthesis complete

7 Milk development

71 Caryopsis watery ripe
73 Early milk
75 Medium milk
77 Late milk

8 Dough development

83 Early dough
85 Soft dough
87 Hard dough

9 Ripening

91 Caryopsis hard (difficult to divide by thumb-nail)
92 Caryopsis hard (can no longer be dented by thumb-nail)
93 Caryopsis loosening in daytime

(From Tottman, 1987. *Annals of Applied Biology,* **110**, 441–454)

SECTION 6

Stages in Development of Oilseed Rape

Illustrations of these growth stages can be found in the reference indicated below and in some company product manuals.

0   Germination and emergence

1   Leaf production

   1,0   Both cotyledons unfolded and green
   1,1   First true leaf
   1,2   Second true leaf
   1,3   Third true leaf
   1,4   Fourth true leaf
   1,5   Fifth true leaf
   1,10 About tenth true leaf
   1,15 About fifteenth true leaf

2   Stem extension

   2,0   No internodes ('rosette')
   2,5   About five internodes

3   Flower bud development

   3,0   Only leaf buds present
   3,1   Flower buds present but enclosed by leaves
   3,3   Flower buds visible from above ('green bud')
   3,5   Flower buds raised above leaves
   3,6   First flower stalks extending
   3,7   First flower buds yellow ('yellow bud')

4   Flowering

   4,0   First flower opened
   4,1   10% all buds opened
   4,3   30% all buds opened
   4,5   50% all buds opened

5   Pod development

   5,3   30% potential pods
   5,5   50% potential pods
   5,7   70% potential pods
   5,9   All potential pods

6   Seed development

   6,1   Seeds expanding
   6,2   Most seeds translucent but full size
   6,3   Most seeds green
   6,4   Most seeds green-brown mottled
   6,5   Most seeds brown
   6,6   Most seeds dark brown
   6,7   Most seeds black but soft
   6,8   Most seeds black and hard
   6,9   All seeds black and hard

7   Leaf senescence

8   Stem senescence

   8,1   Most stem green
   8,5   Half stem green
   8,9   Little stem green

9   Pod senescence

   9,1   Most pods green
   9,5   Half pods green
   9.9   Few pods green

(From Sylvester-Bradley, 1985. *Aspects of Applied Biology*, **10**, 395–400)

Stages in Development of Peas

Illustrations of these growth stages can be found in the reference indicated below and in some company product manuals.

0 Germination and emergence

000 Dry seed
001 Imbibed seed
002 Radicle apparent
003 Plumule and radicle apparent
004 Emergence

1 Vegetative stage

101 First node (leaf with one pair leaflets, no tendril)
102 Second node (leaf with one pair leaflets, simple tendril)
103 Third node (leaf with one pair leaflets, complex tendril)

•

l0x X nodes (leaf with more than one pair leaflets, complex tendril)

•

•

10n Last recorded node

2 Reproductive stage (main stem)

201 Enclosed buds
202 Visible buds
203 First open flower
204 Pod set (small immature pod)
205 Flat pod
206 Pod swell (seeds small, immature)
207 Podfill
208 Pod green, wrinkled
209 Pod yellow, wrinkled (seeds rubbery)
210 Dry seed

3 Senescence stage

301 Desiccant application stage. Lower pods dry and brown, middle yellow, upper green. Overall moisture content of seed less than 45%
302 Pre-harvest stage. Lower and middle pods dry and brown, upper yellow. Overall moisture content of seed less than 30%
303 Dry harvest stage. All pods dry and brown, seed dry

(From Knott, 1987. *Annals of Applied Biology*, **111**, 233–244)

SECTION 6

Stages in Development of Faba Beans

Illustrations of these growth stages can be found in the reference indicated below and in some company product manuals.

0 Germination and emergence

   000 Dry seed
   001 Imbibed seed
   002 Radicle apparent
   003 Plumule and radicle apparent
   004 Emergence
   005 First leaf unfolding
   006 First leaf unfolded

1 Vegetative stage

   101 First node
   102 Second node
   103 Third node
   •
   •
   l0x X nodes
   •
   •
   10n N, last recorded node

2 Reproductive stage (main stem)

   201 Flower buds visible
   203 First open flowers
   204 First pod set
   205 Pods fully formed, green
   207 Pod fill, pods green
   209 Seed rubbery, pods pliable, turning black
   210 Seed dry and hard, pods dry and black

3 Pod senescence

   301 10% pods dry and black
   •
   •
   305 50% pods dry and black
   •
   •
   308 80% pods dry and black, some upper pods green
   309 90% pods dry and black, most seed dry. Desiccation stage.
   310 All pods dry and black, seed hard. Pre-harvest (glyphosate application stage)

4 Stem senescence

   401 10% stem brown/black
   •
   •
   405 50% stem brown/black
   •
   •
   409 90% stem brown/black
   410 All stems brown/black. All pods dry and black, seed hard.

(From Knott, 1990. *Annals of Applied Biology*, **116**, 391–404)

Stages in Development of Potato

Illustrations of these growth stages can be found in the reference indicated below and in some company product manuals.

0   Seed germination and seedling emergence

000   Dry seed
001   Imbibed seed
002   Radicle apparent
003   Elongation of hypocotyl
004   Seedling emergence
005   Cotyledons unfolded

1   Tuber dormancy

100   Innate dormancy (no sprout development under favourable conditions)
150   Enforced dormancy (sprout development inhibited by environmental conditions)

2   Tuber sprouting

200   Dormancy break, sprout development visible
21x   Sprout with 1 node
22x   Sprout with 2 nodes
•
•
29x   Sprout with 9 nodes
21x(2) Second generation sprout with 1 node
22x(2) Second generation sprout with 2 nodes
•
•
29x(2) Second generation sprout with 9 nodes

Where x = 1, sprout >2 mm;
2, 2-5 mm; 3, 5-20 mm;
4, 20-30 mm; 5, 50-100 mm;
6, 100-150 mm long

3   Emergence and shoot expansion

300   Main stem emergence
301   Node 1
302   Node 2
•
•
319   Node 19
Second order branch
321   Node 1
•
•
Nth   order branch
3N1   Node 1
•
•
3N9   Node 9

4   Flowering
Primary flower
400   No flowers
410   Appearance of flower bud
420   Flower unopen
430   Flower open
440   Flower closed
450   Berry swelling
460   Mature berry
Second order flowers
410(2) Appearance of flower bud
420(2) Flower unopen
430(2) Flower open
440(2) Flower closed
450(2) Berry swelling
460(2) Mature berry

5   Tuber development

500   No stolons
510   Stolon initials
520   Stolon elongation
530   Tuber initiation
540   Tuber bulking (>10 mm diam)
550   Skin set
560   Stolon development

6   Senescence

600   Onset of yellowing
650   Half leaves yellow
670   Yellowing of stems
690   Completely dead

(From Jefferies & Lawson, 1991. *Annals of Applied Biology*, **119**, 387–389)

SECTION 6

Stages in Development of Linseed

Illustrations of these growth stages can be found in the reference indicated below and in some company product manuals.

0 Germination and emergence

00	Dry seed
01	Imbibed seed
02	Radicle apparent
04	Hypocotyl extending
05	Emergence
07	Cotyledon unfolding from seed case
09	Cotyledons unfolded and fully expanded

1 Vegetative stage (of main stem)

10	True leaves visible
12	First pair of true leaves fully expanded
13	Third pair of true leaves fully expanded
1n	n leaf fully expanded

2 Basal branching

21	One branch
22	Two branches
23	Three branches
2n	n branches

3 Flower bud development (on main stem)

31	Enclosed bud visible in leaf axils
33	Bud extending from axil
35	Corymb formed
37	Buds enclosed but petals visible
39	First flower open

4 Flowering (whole plant)

41	10% of flowers open
43	30% of flowers open
45	50% of flowers open
49	End of flowering

5 Capsule formation (whole plant)

51	10% of capsules formed
53	30% 0f capsules formed
55	50% of capsules formed
59	End of capsule formation

6 Capsule senescence (on most advanced plant)

61	Capsules expanding
63	Capsules green and full size
65	Capsules turning yellow
67	Capsules all yellow brown but soft
69	Capsules brown, dry and senesced

7 Stem senescence (whole plant)

71	Stems mostly green below panicle
73	Most stems 30% brown
75	Most stems 50% brown
77	Stems 75% brown
79	Stems completely brown

8 Stems rotting (retting)

81	Outer tissue rotting
85	Vascular tissue easily removed
89	Stems completely collapsed

9 Seed development (whole plant)

91	Seeds expanding
92	Seeds white but full size
93	Most seeds turning ivory yellow
94	Most seeds turning brown
95	All seeds brown and hard
98	Some seeds shed from capsule
99	Most seeds shed from capsule

(From Jefferies & Lawson, 1991. *(From Freer, 1991. Aspects of Applied Biology, **28**, 33–40)*

Stages in Development of Annual Grass Weeds

Illustrations of these growth stages can be found in the reference indicated below and in some company product manuals.

0 Germination and emergence

00	Dry seed
01	Start of imbibition
03	Imbibition complete
05	Radicle emerged from caryopsis
07	Coleoptile emerged from caryopsis
09	Leaf just at coleoptile tip

1 Seedling growth

10	First leaf through coleoptile
11	First leaf unfolded
12	2 leaves unfolded
13	3 leaves unfolded
14	4 leaves unfolded
15	5 leaves unfolded
16	6 leaves unfolded
17	7 leaves unfolded
18	8 leaves unfolded
19	9 or more leaves unfolded

2 Tillering

20	Main shoot only
21	Main shoot and 1 tiller
22	Main shoot and 2 tillers
23	Main shoot and 3 tillers
24	Main shoot and 4 tillers
25	Main shoot and 5 tillers
26	Main shoot and 6 tillers
27	Main shoot and 7 tillers
28	Main shoot and 8 tillers
29	Main shoot and 9 or more tillers

3 Stem elongation

31	First node detectable
32	2nd node detectable
33	3rd node detectable
34	4th node detectable
35	5th node detectable
36	6th node detectable
37	Flag leaf just visible
39	Flag leaf ligule just visible

4 Booting

41	Flag leaf sheath extending
43	Boots just visibly swollen
45	Boots swollen
47	Flag leaf sheath opening
49	First awns visible

5 Inflorescence emergence

51	First spikelet of inflorescence just visible
53	1/4 of inflorescence emerged
55	1/2 of inflorescence emerged
57	3/4 of inflorescence emerged
59	Emergence of inflorescence completed

6 Anthesis

61	Beginning of anthesis
65	Anthesis half-way
69	Anthesis complete

(From Lawson & Read, 1992. *Annals of Applied Biology*, **12**, 211–214)

SECTION 6

Growth Stages of Annual Broad-leaved Weeds

Preferred Descriptive Phrases
Illustrations of these growth stages can be found in the reference indicated below and in some company product manuals.

Pre-emergence
Early cotyledons
Expanded cotyledons
One expanded true leaf
Two expanded true leaves
Four expanded true leaves
Six expanded true leaves
Plants up to 25 mm across/high

Plants up to 50 mm across/high
Plants up to 100 mm across/high
Plants up to 150 mm across/high
Plants up to 250 mm across/high
Flower buds visible
Plant flowering
Plant senescent

(From Lutman & Tucker, 1987. *Annals of Applied Biology*, **110**, 683–687)

Appendix 4
Key to Hazard Classifications and
Safety Precautions

Every product label contains information to warn users of the risks from using the product, together with precautions that must be followed in order to minimise the risks. A hazard classification (if any) and the symbol are shown with associated risk phrases, followed by a series of safety numbers under the heading **Hazard classification and safety precautions**.

The codes are defined below, under the same sub-headings as they appear in the pesticide profiles.

Where a product label specifies the use of personal protective equipment (PPE), the requirements are listed under the sub-heading **Operator protection**, using letter codes to denote the protective items, e.g. handling the concentrate, cleaning equipment etc., but it is not possible to list them separately. The lists of PPE are therefore an indication of what the user may need to have available to use the product in different ways. **When making a COSHH assessment it is therefore essential that the product label is consulted for information on the particular use that is being assessed**.

Where the generalised wording includes a phrase such as '... for *xx* days', the specific requirement for each pesticide is shown in brackets after the code.

Hazard
H01	Very toxic
H02	Toxic
H03	Harmful
H04	Irritant
H05	Corrosive
H06	Extremely flammable
H07	Highly flammable
H08	Flammable
H09	Oxidising agent
H10	Explosive
H11	Dangerous for the environment
H225	Highly flammable liquid and vapour
H226	Flammable liquid and vapour
H228	Flammable solid
H260	In contact with water releases flammable gases which may ignite spontaneously
H300	Fatal if swallowed
H301	Toxic if swallowed
H302	Harmful if swallowed
H311	Toxic in contact with skin
H330	Fatal if inhaled
H331	Toxic if inhaled
H332	Harmful if inhaled
H400	Very toxic to aquatic organisms
H401	Toxic to aquatic organisms

Risk phrases
H280	Contains gas under pressure; may explode if heated
H290	May be corrosive to metals
H304	May be fatal if swallowed and enters airways
H310	Fatal in contact with skin
H312	Harmful in contact with skin
H314	Causes severe skin burns and eye damage

H315	Causes skin irritation
H317	May cause an allergic skin reaction
H318	Causes serious eye damage
H319	Causes serious eye irritation
H320	Causes eye irritation
H334	May cause allergy or asthma symptoms or breathing difficulties if inhaled
H335	May cause respiratory irritation
H336	May cause drowsiness or dizziness
H340	May cause genetic defects
H341	Suspected of causing genetic defects
H351	Suspected of causing cancer
H360	May damage fertility or the unborn child
H361	Suspected of damaging fertility or the unborn child
H370	Causes damage to organs
H371	May cause damage to organs
H372	Causes damage to organs through prolonged or repeated exposure
H373	May cause damage to organs through prolonged or repeated exposure
R08	Contact with combustible material may cause fire
R09	Explosive when mixed with combustible material
R15	Contact with water liberates extremely flammable gases
R16	Explosive when mixed with oxidising substances
R19	Fatal if swallowed
R20	Harmful by inhalation
R21	Harmful in contact with skin
R22a	Harmful if swallowed
R22b	May cause lung damage if swallowed
R22c	May be fatal if swallowed and enters airways
R23a	Toxic by inhalation
R23b	Fatal if inhaled
R24	Toxic in contact with skin
R25	Toxic if swallowed
R25b	Toxic; Danger of serious damage to health by prolonged exposure if swallowed
R26	Very toxic by inhalation
R27	Very toxic in contact with skin
R28	Very toxic if swallowed
R31	Contact with acids liberates toxic gas
R34	Causes burns
R35	Causes severe burns
R36	Irritating to eyes
R37	Irritating to respiratory system
R38	Irritating to skin
R39	Danger of very serious irreversible effects
R40	Limited evidence of a carcinogenic effect
R40b	Suspected of causing cancer
R41	Risk of serious damage to eyes
R42	May cause sensitization by inhalation
R43	May cause sensitization by skin contact
R45	May cause cancer
R46	May cause heritable genetic damage
R48	Danger of serious damage to health by prolonged exposure
R50	Very toxic to aquatic organisms
R51	Toxic to aquatic organisms
R52	Harmful to aquatic organisms
R53a	May cause long-term adverse effects in the aquatic environment
R53b	Dangerous to aquatic organisms
R54	Toxic to flora
R55	Toxic to fauna
R56	Toxic to soil organisms
R57	Toxic to bees
R58	May cause long-term adverse effects in the environment
R60	May impair fertility

R60b   May cause heritable genetic damage
R61   May cause harm to the unborn child
R62   Possible risk of impaired fertility
R63   Possible risk of harm to the unborn child
R64   May cause harm to breast-fed babies
R65   Harmful: May cause lung damage if swallowed
R66   Repeated exposure may cause skin dryness or cracking
R66b   May cause damage to organs through prolonged or repeated exposure
R67   Vapours may cause drowsiness and dizziness
R68   Possible risk of irreversible effects
R69   Danger of serious damage to health by prolonged oral exposure.
R70   May produce an allergic reaction

Operator protection

A   Suitable protective gloves (the product label should be consulted for any specific requirements about the material of which the gloves should be made)
B   Rubber gauntlet gloves
C   Face-shield
D   Approved respiratory protective equipment
E   Goggles
F   Dust mask
G   Full face-piece respirator
H   Coverall
J   Hood
K   Apron/Rubber apron
L   Waterproof coat
M   Rubber boots
N   Waterproof jacket and trousers
P   Suitable protective clothing
U01   To be used only by operators instructed or trained in the use of chemical/product/type of produce and familiar with the precautionary measures to be observed
U02a   Wash all protective clothing thoroughly after use, especially the inside of gloves
U02b   Avoid excessive contamination of coveralls and launder regularly
U02c   Remove and wash contaminated gloves immediately
U03   Wash splashes off gloves immediately
U04a   Take off immediately all contaminated clothing
U04b   Take off immediately all contaminated clothing and wash underlying skin. Wash clothes before re-use
U04c   Wash clothes before re-use
U05a   When using do not eat, drink or smoke
U05b   When using do not eat, drink, smoke or use naked lights
U06   Handle with care and mix only in a closed container
U07   Open the container only as directed (returnable containers only)
U08   Wash concentrate/dust from skin or eyes immediately
U09a   Wash any contamination/splashes/dust/powder/concentrate from skin or eyes immediately
U09b   Wash any contamination/splashes/dust/powder/concentrate from eyes immediately
U09c   If on skin, wash with plenty of soap and water
U10   After contact with skin or eyes wash immediately with plenty of water
U11   In case of contact with eyes rinse immediately with plenty of water and seek medical advice
U12   In case of contact with skin rinse immediately with plenty of water and seek medical advice
U12b   After contact with skin, take off immediately all contaminated clothing and wash immediately with plenty of water
U13   Avoid contact by mouth
U14   Avoid contact with skin
U15   Avoid contact with eyes
U16a   Ensure adequate ventilation in confined spaces
U16b   Use in a well ventilated area

U18	Extinguish all naked flames, including pilot lights, when applying the fumigant/dust/liquid/product
U19a	Do not breathe dust/fog/fumes/gas/smoke/spray mist/vapour. Avoid working in spray mist
U19b	Do not work in confined spaces or enter spaces in which high concentrations of vapour are present. Where this precaution cannot be observed distance breathing or self-contained breathing apparatus must be worn, and the work should be done by trained operators
U19c	In case of insufficient ventilation, wear suitable respiratory equipment
U19d	Wear suitable respiratory equipment during bagging and stacking of treated seed
U19e	During fumigation/spraying wear suitable respiratory equipment
U19f	In case of accident by inhalation: remove casualty to fresh air and keep at rest
U20a	Wash hands and exposed skin before eating, drinking or smoking and after work
U20b	Wash hands and exposed skin before eating and drinking and after work
U20c	Wash hands before eating and drinking and after work
U20d	Wash hands after use
U20e	Wash hands and exposed skin after cleaning and re-calibrating equipment
U21	Before entering treated crops, cover exposed skin areas, particularly arms and legs
U22a	Do not touch sachet with wet hands or gloves/Do not touch water soluble bag directly
U22b	Protect sachets from rain or water
U23a	Do not apply by knapsack sprayer/hand-held equipment
U23b	Do not apply through hand held rotary atomisers
U23c	Do not apply via tractor mounted horizontal boom sprayers
U24	Do not handle grain unnecessarily
U25	Open the container only as directed
U26	Keep unprotected workers out of treated areas for at least 5 days after treatment.

Environmental protection

E02a	Keep unprotected persons/animals out of treated/fumigation areas for at least xx hours/days
E02b	Prevent access by livestock, pets and other non-target mammals and birds to buildings under fumigation and ventilation
E02c	Vacate treatment areas before application
E02d	Exclude all persons and animals during treatment
E03	Label treated seed with the appropriate precautions, using the printed sacks, labels or bag tags supplied
E05a	Do not apply directly to livestock/poultry
E05b	Keep poultry out of treated areas for at least xx days/weeks
E05c	Do not apply directly to animals
E06a	Keep livestock out of treated areas for at least xx days/weeks after treatment
E06b	Dangerous to livestock. Keep all livestock out of treated areas/away from treated water for at least xx days/weeks. Bury or remove spillages
E06c	Harmful to livestock. Keep all livestock out of treated areas/away from treated water for at least xx days/weeks. Bury or remove spillages
E06d	Exclude livestock from treated fields. Livestock may not graze or be fed treated forage nor may it be used for hay silage or bedding
E07a	Keep livestock out of treated areas for up to two weeks following treatment and until poisonous weeds, such as ragwort, have died down and become unpalatable
E07b	Dangerous to livestock. Keep livestock out of treated areas/away from treated water for at least xx weeks and until foliage of any poisonous weeds, such as ragwort, has died and become unpalatable
E07c	Harmful to livestock. Keep livestock out of treated areas/away from treated water for at least xx days/weeks and until foliage of any poisonous weeds such as ragwort has died and become unpalatable
E07d	Keep livestock out of treated areas for up to 4-6 weeks following treatment and until poisonous weeds, such as ragwort, have died down and become unpalatable
E07e	Do not take grass crops for hay or silage for at least 21 days after application
E07f	Keep livestock out of treated areas for up to 3 days following treatment and until poisonous weeds, such as ragwort, have died
E07g	Keep livestock out of treated areas for at least 50 days following treatment
E08a	Do not feed treated straw or haulm to livestock within xx days/weeks of spraying

E08b	Do not use on grassland if the crop is to be used as animal feed or bedding
E09	Do not use straw or haulm from treated crops as animal feed or bedding for at least xx days after last application
E10a	Dangerous to game, wild birds and animals
E10b	Harmful to game, wild birds and animals
E10c	Dangerous to game, wild birds and animals. All spillages must be buried or removed
E11	Paraquat can be harmful to hares; spray stubbles early in the day
E12a	High risk to bees
E12b	Extremely dangerous to bees
E12c	Dangerous to bees
E12d	Harmful to bees
E12e	Do not apply to crops in flower or to those in which bees are actively foraging. Do not apply when flowering weeds are present
E12f	Do not apply to crops in flower, or to those in which bees are actively foraging, except as directed on [crop]. Do not apply when flowering weeds are present
E12g	Apply away from bees
E13a	Extremely dangerous to fish or other aquatic life. Do not contaminate surface waters or ditches with chemical or used container
E13b	Dangerous to fish or other aquatic life. Do not contaminate surface waters or ditches with chemical or used container
E13c	Harmful to fish or other aquatic life. Do not contaminate surface waters or ditches with chemical or used container
E13d	Apply away from fish
E13e	Harmful to fish or other aquatic life. The maximum concentration of active ingredient in treated water must not exceed XX ppm or such lower concentration as the appropriate water regulatory body may require.
E14a	Extremely dangerous to aquatic higher plants. Do not contaminate surface waters or ditches with chemical or used container
E14b	Dangerous to aquatic higher plants. Do not contaminate surface waters or ditches with chemical or used container
E15a	Do not contaminate surface waters or ditches with chemical or used container
E15b	Do not contaminate water with product or its container. Do not clean application equipment near surface water. Avoid contamination via drains from farmyards or roads
E15c	To protect groundwater, do not apply to grass leys less than 1 year old
E16a	Do not allow direct spray from horizontal boom sprayers to fall within 5 m of the top of the bank of a static or flowing waterbody, unless a Local Environment Risk Assessment for Pesticides (LERAP) permits a narrower buffer zone, or within 1 m of the top of a ditch which is dry at the time of application. Aim spray away from water
E16b	Do not allow direct spray from hand-held sprayers to fall within 1 m of the top of the bank of a static or flowing waterbody. Aim spray away from water
E16c	Do not allow direct spray from horizontal boom sprayers to fall within 5 m of the top of the bank of a static or flowing waterbody, or within 1m of the top of a ditch which is dry at the time of application. Aim spray away from water. This product is not eligible for buffer zone reduction under the LERAP horizontal boom sprayers scheme.
E16d	Do not allow direct spray from hand-held sprayers to fall within 1 m of the top of the bank of a static or flowing waterbody. Aim spray away from water. This product is not eligible for buffer zone reduction under the LERAP horizontal boom sprayers scheme.
E16e	Do not allow direct spray from horizontal boom sprayers to fall within 5 m of the top of the bank of a static or flowing water body or within 1 m from the top of any ditch which is dry at the time of application. Spray from hand held sprayers must not in any case be allowed to fall within 1 m of the top of the bank of a static or flowing water body. Always direct spray away from water. The LERAP scheme does not extend to adjuvants. This product is therefore not eligible for a reduced buffer zone under the LERAP scheme.
E16f	Do not allow direct spray/granule applications from vehicle mounted/drawn hydraulic sprayers/applicators to fall within 6 m of surface waters or ditches/Do not allow direct spray/granule applications from hand-held sprayers/applicators to fall within 2 m of surface waters or ditches. Direct spray/applications away from water

SECTION 6

E16g	Do not allow direct spray from train sprayers to fall within 5 m of the top of the bank of any static or flowing waterbody.
E16h	Do not spray cereals after 31st March in the year of harvest within 6 metres of the edge of the growing crop
E16i	Do not allow direct spray from horizontal boom sprayers to fall within 12 metres of the top of the bank of a static or flowing water body.
E16j	Do not allow direct spray from horizontal boom sprayers to fall within the specified distance of the top of the bank of a static or flowing waterbody,
E17a	Do not allow direct spray from broadcast air-assisted sprayers to fall within xx m of surface waters or ditches. Direct spray away from water
E17b	Do not allow direct spray from broadcast air-assisted sprayers to fall within xx m of the top of the bank of a static or flowing waterbody, unless a Local Environmental Risk Assessment for Pesticides (LERAP) permits a narrower buffer zone, or within 5 m of the top of a ditch which is dry at the time of application. Aim spray away from water
E18	Do not spray from the air within 250 m horizontal distance of surface waters or ditches
E19a	Do not dump surplus herbicide in water or ditch bottoms
E19b	Do not empty into drains
E20	Prevent any surface run-off from entering storm drains
E21	Do not use treated water for irrigation purposes within xx days/weeks of treatment
E22a	High risk to non-target insects or other arthropods. Do not spray within 6 m of the field boundary
E22b	Risk to certain non-target insects or other arthropods. For advice on risk management and use in Integrated Pest Management (IPM) see directions for use
E22c	Risk to non-target insects or other arthropods
E23	Avoid damage by drift onto susceptible crops or water courses
E34	Do not re-use container for any purpose/Do not re-use container for any other purpose
E35	Do not burn this container
E36a	Do not rinse out container (returnable containers only)
E36b	Do not open or rinse out container (returnable containers only)
E37	Do not use with any pesticide which is to be applied in or near water
E38	Use appropriate containment to avoid environmental contamination
E39	Extreme care must be taken to avoid spray drift onto non-crop plants outside the target area
E40	To protect groundwater/soil organisms the maximum total dose of this or other products containing ethofumesate MUST NOT exceed 1.0 kg ethofumesate per hectare in any three year period
E40b	To protect aquatic organisms respect an unsprayed buffer zone to surface waters in line with LERAP requirements
E40c	This product must not be used with any pesticide to be applied in or near water
E40d	To protect non-target arthropods respect an untreated buffer zone of 5m to non crop land
E40e	To protect aquatic organisms respect an unsprayed buffer zone of 20 metres to surface water bodies
E41	Hay for silage must not be cut from treated crops for at least 21 days after treatment
H410	Very toxic to aquatic life with long-lasting effects
H411	Toxic to aquatic life with long lasting effects
H412	Harmful to aquatic life with long lasting effects
H413	May cause long lasting harmful effects to aquatic life

Consumer protection

C01	Do not use on food crops
C02a	Do not harvest for human or animal consumption for at least xx days/weeks after last application
C02b	Do not remove from store for sale or processing for at least 21 days after application
C02c	Do not remove from store for sale or processing for at least 2 days after application
C04	Do not apply to surfaces on which food/feed is stored, prepared or eaten
C05	Remove/cover all foodstuffs before application
C06	Remove exposed milk before application

C07	Collect eggs before application
C08	Protect food preparing equipment and eating utensils from contamination during application
C09	Cover water storage tanks before application
C10	Protect exposed water/feed/milk machinery/milk containers from contamination
C11	Remove all pets/livestock/fish tanks before treatment/spraying
C12	Ventilate treated areas thoroughly when smoke has cleared/Ventilate treated rooms thoroughly before occupying

Storage and disposal

D01	Keep out of reach of children
D02	Keep away from food, drink and animal feeding-stuffs
D03	Store away from seeds, fertilizers, fungicides and insecticides
D04	Store well away from corms, bulbs, tubers and seeds
D05	Protect from frost
D06a	Store away from heat
D06b	Do not store near heat or open flame
D06c	Do not store in direct sunlight
D06d	Do not store above 30/35 °C
D06e	Store in a well-ventilated place. Keep container tightly closed
D06f	Keep away from sources of ignition - No smoking
D06g	Take precautionary measures against static discharges
D06h	Store at 10 - 25 °C
D07	Store under cool, dry conditions
D08	Store in a safe, dry, frost-free place designated as an agrochemical store
D09a	Keep in original container, tightly closed, in a safe place
D09b	Keep in original container, tightly closed, in a safe place, under lock and key
D09c	Store unused sachets in a safe place. Do not store half-used sachets
D10a	Wash out container thoroughly and dispose of safely
D10b	Wash out container thoroughly, empty washings into spray tank and dispose of safely
D10c	Rinse container thoroughly by using an integrated pressure rinsing device or manually rinsing three times. Add washings to sprayer at time of filling and dispose of container safely
D10d	Do not rinse out container
D11a	Empty container completely and dispose of safely/Dispose of used generator safely
D11b	Empty container completely and dispose of it in the specified manner
D11c	Ventilate empty containers until no phosphine is detected and then dispose of as hazardous waste via an authorised waste-disposal contractor
D12a	This material (and its container) must be disposed of in a safe way
D12b	This material and its container must be disposed of as hazardous waste
D13	Treat used container as if it contained pesticide
D14	Return empty container as instructed by supplier (returnable containers only)
D15	Store container in purpose built chemical store until returned to supplier for refilling (returnable containers only)
D16	Do not store below 4 degrees Centigrade
D17	Use immediately on removal of foil
D18	Place the tablets whole into the spray tank - do not break or crumble the tablets
D19	Do not empty into drains
D20	Clean all equipment after use
D21	Open the container only as directed
D22	Collect spillage

Treated Seed

S01	Do not handle treated seed unnecessarily
S02	Do not use treated seed as food or feed
S03	Keep treated seed secure from people, domestic stock/pets and wildlife at all times during storage and use
S04a	Bury or remove spillages
S04b	Harmful to birds/game and wildlife. Treated seed should not be left on the soil surface. Bury or remove spillages

SECTION 6

S04c	Dangerous to birds/game and wildlife. Treated seed should not be left on the soil surface. Bury or remove spillages
S04d	To protect birds/wild animals, treated seed should not be left on the soil surface. Bury or remove spillages
S05	Do not reuse sacks or containers that have been used for treated seed for food or feed
S06a	Wash hands and exposed skin before meals and after work
S06b	Wash hands and exposed skin after cleaning and re-calibrating equipment
S07	Do not apply treated seed from the air
S08	Treated seed should not be broadcast
S09	Label treated seed with the appropriate precautions using printed sacks, labels or bag tags supplied
S10	Only use automated equipment for planting treated seed potatoes

Vertebrate/Rodent control products

V01a	Prevent access to baits/powder by children, birds and other animals, particularly cats, dogs, pigs and poultry
V01b	Prevent access to bait/gel/dust by children, birds and non-target animals, particularly dogs, cats, pigs, poultry
V02	Do not prepare/use/lay baits/dust/spray where food/feed/water could become contaminated
V03a	Remove all remains of bait, tracking powder or bait containers after use and burn or bury
V03b	Remove all remains of bait and bait containers/exposed dust/after treatment (except where used in sewers) and dispose of safely (e.g. burn/bury). Do not dispose of in refuse sacks or on open rubbish tips.
V04a	Search for and burn or bury all rodent bodies. Do not place in refuse bins or on rubbish tips
V04b	Search for rodent bodies (except where used in sewers) and dispose of safely (e.g. burn/bury). Do not dispose of in refuse sacks or on open rubbish tips
V04c	Dispose of safely any rodent bodies and remains of bait and bait containers that are recovered after treatment (e.g. burn/bury). Do not dispose of in refuse sacks or on open rubbish tips
V05	Use bait containers clearly marked POISON at all surface baiting points

Medical advice

M01	This product contains an anticholinesterase organophosphorus compound. DO NOT USE if under medical advice NOT to work with such compounds
M02	This product contains an anticholinesterase carbamate compound. DO NOT USE if under medical advice NOT to work with such compounds
M03	If you feel unwell, seek medical advice immediately (show the label where possible)
M03a	If exposed or concerned, get medical advice or attention
M04a	In case of accident or if you feel unwell, seek medical advice immediately (show the label where possible)
M04b	In case of accident by inhalation, remove casualty to fresh air and keep at rest
M04c	If inhaled, remove victim to fresh air and keep at rest in a position comfortable for breathing. IMMEDIATELY call Poison Centre or doctor/physician
M05a	If swallowed, seek medical advice immediately and show this container or label
M05b	If swallowed, do not induce vomiting: seek medical advice immediately and show this container or label
M05c	If swallowed induce vomiting if not already occurring and take patient to hospital immediately
M05d	If swallowed, rinse the mouth with water but only if the person is conscious
M06	This product contains an anticholinesterase carbamoyl triazole compound. DO NOT USE if under medical advice NOT to work with such compounds

Appendix 5
Key to Abbreviations and Acronyms

The abbreviations of formulation types in the following list are used in Section 2 (Pesticide Profiles) and are derived from the *Catalogue of Pesticide Formulation Types and International Coding System* (CropLife International Technical Monograph 2, 5th edn, March 2002).

1 Formulation Types

AB	Grain bait
AE	Aerosol generator
AL	Other liquids to be applied undiluted
AP	Any other powder
BB	Block bait
BR	Briquette
CB	Bait concentrate
CC	Capsule suspension in a suspension concentrate
CF	Capsule suspension for seed treatment
CG	Encapsulated granule (controlled release)
CL	Contact liquid or gel (for direct application)
CP	Contact powder (for direct application)
CR	Crystals
CS	Capsule suspension
DC	Dispersible concentrate
DP	Dustable powder
DS	Powder for dry seed treatment
EC	Emulsifiable concentrate
EG	Emulsifiable granule
EO	Water in oil emulsion
ES	Emulsion for seed treatment
EW	Oil in water emulsion
FG	Fine granules
FP	Smoke cartridge
FS	Flowable concentrate for seed treatment
FT	Smoke tablet
FU	Smoke generator
FW	Smoke pellets
GA	Gas
GB	Granular bait
GE	Gas-generating product
GG	Macrogranules
GL	Emulsifiable gel
GP	Flo-dust (for pneumatic application)
GR	Granules
GS	Grease
GW	Water soluble gel
HN	Hot fogging concentrate
KK	Combi-pack (solid/liquid)
KL	Combi-pack (liquid/liquid)
KN	Cold-fogging concentrate
KP	Combi-pack (solid/solid)
LA	Lacquer
LI	Liquid, unspecified
LS	Solution for seed treatment
ME	Microemulsion
MG	Microgranules
MS	Microemulsion for seed treatment
OD	Oil dispersion

OL	Oil miscible liquid
PA	Paste
PC	Gel or paste concentrate
PO	Powder
PS	Seed coated with a pesticide
PT	Pellet
RB	Ready-to-use bait
RC	Ready-to-use low volume CDA sprayer
RH	Ready-to-use spray in hand-operated sprayer
SA	Sand
SC	Suspension concentrate (= flowable)
SE	Suspo-emulsion
SG	Water soluble granules
SL	Soluble concentrate
SP	Water soluble powder
SS	Water soluble powder for seed treatment
ST	Water soluble tablet
SU	Ultra low-volume suspension
TB	Tablets
TC	Technical material
TP	Tracking powder
UL	Ultra low-volume liquid
VP	Vapour releasing product
WB	Water soluble bags
WG	Water dispersible granules
WP	Wettable powder
WS	Water dispersible powder for slurry treatment of seed
WT	Water dispersible tablet
XX	Other formulations
ZZ	Not Applicable

2 Other Abbreviations and Acronyms

ACP	Advisory Committee on Pesticides
ACTS	Advisory Committee on Toxic Substances
ADAS	Agricultural Development and Advisory Service
a.i.	active ingredient
AIC	Agriculture Industries Confederation
BBPA	British Beer and Pub Association
CDA	Controlled droplet application
CPA	Crop Protection Association
cm	centimetre(s)
COPR	Control of Pesticides Regulations 1986
COSHH	Control of Substances Hazardous to Health Regulations
CRD	Chemicals Regulation Directorate
d	day(s)
Defra	Department for Environment, Food and Rural Affairs
EA	Environment Agency
EBDC	ethylene-bis-dithiocarbamate fungicide
EAMU	Extension of Authorisation for Minor Use
FEPA	Food and Environment Protection Act 1985
g	gram(s)
GS	growth stage (unless in formulation column)
h	hour(s)
ha	hectare(s)
HBN	hydroxybenzonitrile herbicide
HI	harvest interval
HSE	Health and Safety Executive
ICM	integrated crop management
IPM	integrated pest management
kg	kilogram(s)
l	litre(s)

LERAP   Local Environmental Risk Assessments for Pesticides
m   metre(s)
MBC   methyl benzimidazole carbamate fungicide
MEL   maximum exposure limit
min   minute(s)
mm   millimetre(s)
MRL   maximum residue level
mth   month(s)
NA   Notice of Approval
NFU   National Farmers' Union
OES   Occupational Exposure Standard
OLA   off-label approval
PPE   personal protective equipment
PPPR   Plant Protection Products Regulations
SOLA   specific off-label approval
ULV   ultra-low volume
VI   Voluntary Initiative
w/v   weight/volume
w/w   weight/weight
wk   week(s)
yr   year(s)

3 UN Number details

UN No.	Substance	EAC	APP	Hazards Class	Sub risks	HIN
1692	Strychnine or Strychnine salts	2X		6.1		66
1760	Corrosive Liquid, N.O.S., Packing groups II & III	2X	B	8		88
1830	Sulphuric acid, with more than 51% acid	2P		8		80
1993	Flammable Liquid, N.O.S., Packing group III	•3Y		3		30
2011	Magnesium Phosphide	4W[1]		4.3	6.1	
2588	Pesticide, Solid, TOXIC, N.O.S.	2X		6.1		66/60
2757	Carbamate Pesticide, Solid, Toxic	2X		6.1		66/60
2783	Organophosphorus Pesticide, Solid, TOXIC	2X		6.1		66/60
2902	Pesticide, Liquid, TOXIC, N.O.S., Packing Groups I, II & III	2X	B	6.1		66/60
3077	Environmentally hazardous substance, Solid, N.O.S.	2Z		9		90
3082	Environmentally hazardous substance, Solid, N.O.S.	•3Z		9		90
3265	Corrosive Liquid, Acidic, Organic, N.O.S., Packing group I, II or III	2X	B	8		80/88
3351	Pyrethroid Pesticide, Liquid, Toxic, Flammable, Flash Point 23 ?C or more	•3W	A(fl)	6.1	3	663/63
3352	Pyrethroid Pesticide, Liquid, TOXIC, Packing groups I, II & III	2X	B	6.1		66/60

Ref: Dangerous Goods Emergency Action Code List 2009 (TSO).
[1] Not applicable to the carriage of dangerous goods under RID or ADR.

SECTION 6

Appendix 6
Definitions

The descriptions used in this *Guide* for the crops or situations in which products are approved for use are those used on the approved product labels. These are now standardised in a Crop Hierarchy published by the Chemicals Regulation Directorate in which definitions are given. To assist users of this *Guide* the definitions of some of the terminology where misunderstandings can occur are reproduced below.

Rotational grass: Short-term grass crops grown on land that is likely to be growing different crops in future years (*e.g. short-term intensively managed leys for one to three years that may include clover*)

Permanent grassland: Grazed areas that are intended to be permanent in nature (e.g. permanent pasture and moorland that can be grazed).

Ornamental Plant Production: All ornamental plants that are grown for sale or are produced for replanting into their final growing position (*e.g. flowers, house plants, nursery stock, bulbs grown in containers or in the ground*).

Managed Amenity Turf: Areas of frequently mown, intensively managed, turf that is not intended to flower and set seed. It includes areas that may be for intensive public use (*e.g. all types of sports turf*).

Amenity Grassland: Areas of semi-natural or planted grassland subject to minimal management. It includes areas that may be accessed by the public (*e.g. railway and motorway embankments, airfields, and grassland nature reserves*). These areas may be managed for their botanical interest, and the relevant authority should be contacted before using pesticides in such locations.

Amenity Vegetation: Areas of semi-natural or ornamental vegetation, including trees, or bare soil around ornamental plants, or soil intended for ornamental planting. It includes areas to which the public have access. It does NOT include hedgerows around arable fields.

Natural surfaces not intended to bear vegetation: Areas of soil or natural outcroppings of rock that are not intended to bear vegetation, including areas such as sterile strips around fields. It may include areas to which the public have access. It does not include the land between rows of crops.

Hard surfaces: Man-made impermeable surfaces that are not intended to bear vegetation (*e.g. pavements, tennis courts, industrial areas, railway ballast*).

Permeable surfaces overlying soil: Any man-made permeable surface (excluding railway ballast) such as gravel that overlies soil and is not intended to bear vegetation

Green Cover on Land Temporarily Removed from Production: Includes fields covered by natural regeneration or by a planted green cover crop that will not be harvested (*e.g. green cover on setaside*). It does NOT include industrial crops.

Forest Nursery: Areas where young trees are raised outside for subsequent forest planting.

Forest: Groups of trees being grown in their final positions. Covers all woodland grown for whatever objective, including commercial timber production, amenity and recreation, conservation and landscaping, ancient traditional coppice and farm forestry, and trees from natural regeneration, colonisation or coppicing. Also includes restocking of established woodlands and new planting on both improved and unimproved land.

Farm forestry: Groups of trees established on arable land or improved grassland including those planted for short rotation coppicing. It includes mature hedgerows around arable fields.

Indoors (for rodenticide use): Situations where the bait is placed within a building or other enclosed structure, and where the target is living or feeding predominantly within that building or structure.

Herbs: Reference to Herbs or Protected Herbs when used in Section 2 may include any or all of the following. The particular label or EAMU Notice will indicate which species are included in the approval.

Agastache spp.
Angelica
Applemint
Balm
Basil
Bay
Borage (except when grown for oilseed)
Camomile
Caraway
Catnip
Chervil
Clary
Clary sage
Coriander
Curry plant
Dill
Dragonhead
English chamomile
Fennel
Fenugreek
Feverfew
French lavender
Gingermint
Hyssop
Korean mint
Land cress
Lavandin
Lavender
Lemon balm
Lemon peppermint

Lemon thyme
Lemon verbena
Lovage
Marigold
Marjoram
Mint
Mother of thyme
Nasturtium
Nettle
Oregano
Origanum heracleoticum
Parsley root
Peppermint
Pineapplemint
Rocket
Rosemary
Rue
Sage
Salad burnet
Savory
Sorrel
Spearmint
Spike lavender
Tarragon
Thyme
Thymus camphoratus
Violet
Winter savory
Woodruff

Herbs for Medicinal Uses: Reference to Herbs for Medicinal Uses when used in Section 2 may include any or all of the following. The particular label or EAMU Notice will indicate which species are included in the approval

Black cohosh
Burdock
Dandelion
Echinacea
Ginseng

Goldenseal
Liquorice
Nettle
Valerian

SECTION 6

Appendix 7
References

The information given in *The UK Pesticide Guide* provides some of the answers needed to assess health risks, including the hazard classification and the level of operator protection required. However, the Guide cannot provide all the details needed for a complete hazard assessment, which must be based on the product label itself and, where necessary, the Health and Safety Data Sheet and other official literature.

Detailed guidance on how to comply with the Regulations is available from several sources.

Pesticides: Code of Practice

The sale, supply, storage and use of pesticides are strictly controlled by EU law. The key acts, Regulations and Codes of Practice setting out the duties of employers, supervisors and operators are:

- The *Plant Protection Products Regulations 2011*;
- The *Plant Protection Products (Sustainable Use) Regulations 2012*;
- The *Control of Substances Hazardous to Health Regulations 2002* (COSHH) (as amended);
- Pesticides: *Code of Practice for Using Plant Protection Products 2006*, (incorporating the former 'Green and Orange Codes') with specific Codes of Practice for Scotland and Northern Ireland;
- Classification, Labelling and Packaging (CLP) Regulations.

The Code is currently being updated (as of May 2016) and may be replaced by a series of guidance notes as opposed to a Code in future. The current Code should be read in conjunction with guidance published on the HSE website on how new pesticide legislation has affected the responsibilities of the regulated community.

Further information and details of legislation relating to water, environment protection, waste management and transportation can be found in the UK National action Plan for the Sustainable Use of Pesticides (www.gov.uk/government/publications/pesticides-uk-national-action-plan).

Other Codes of Practice

The Food and Environment Protection Act and its regulations aim to protect the health of human beings, creatures and plants, safeguard the environment and secure safe, efficient and humane methods of controlling pests.

Code of Practice for Suppliers of Pesticides to Agriculture, Horticulture and Forestry (the 'Yellow Code') (Defra Booklet PB 0091)

Code of Good Agricultural Practice for the Protection of Soil (Defra Booklet PB 0617)

Code of Good Agricultural Practice for the Protection of Water (Defra Booklet PB 0587)

Code of Good Agricultural Practice for the Protection of Air (Defra Booklet PB 0618)

Approved Code of Practice for the Control of Substances Hazardous to Health in Fumigation Operations. Health and Safety Commission (ISBN 0-717611-95-7)

Code of Best Practice: Safe use of Sulphuric Acid as an Agricultural Desiccant. National Association of Agricultural Contractors, 2002. (Also available at www.naac.co.uk/?Codes/acidcode.asp)

Safe Use of Pesticides for Non-agricultural Purposes. HSE Approved Code of Practice. HSE L21 (ISBN 0-717624-88-9)

Other Guidance and Practical Advice

HSE (by mail order from HSE Books – See Appendix 2)

COSHH – A brief guide to the Regulations 2003 (INDG136)

A Step by Step Guide to COSHH Assessment, 2004 (HSG97) (ISBN 0-717627-85-3)

Defra (from The Stationery Office – see Appendix 2)

Local Environment Risk Assessments for Pesticides (LERAP): Horizontal Boom Sprayers.

Local Environment Risk Assessments for Pesticides (LERAP): Broadcast Air-assisted Sprayers.

Crop Protection Association and the Voluntary Initiative (see Appendix 2)

Every Drop Counts: Keeping Water Clean

Best Practice Guides. A range of leaflets giving guidance on best practice when dealing with pesticides before, during and after application.

H2OK? Best Practice Advice and Decision Trees - July 2009/10

BCPC (British Crop Production Council – see Appendix 2)

ukpesticideguide.co.uk (*The UK Pesticide Guide* online) – subscription resource

The Pesticide Manual (16th edition) (ISBN 978-1-901396-86-7)

The Pesticide Manual online (www.bcpcdata.com) – subscription resource

The GM Crop Manual (ISBN 978-1-901396-20-1)

GM Crop Manual online (www.bcpcdata.com/gm) - subscription resource

The Manual of Biocontrol Agents (5th edition) (ISBN 978-1-901396-87-4)

The Manual of Biocontrol Agents online (www.bcpcdata.com/mba) – subscription resource.

IdentiPest www.identipest.co.uk

Small Scale Spraying (ISBN 1-901396-07-X)

Field Scale Spraying – 2016 (ISBN 978-1-901396-89-8)

Using Pesticides (ISBN 978-1-901396-10-2)

Safety Equipment Handbook (ISBN 1-901396-06-1)

Spreading Fertilisers and Applying Slug Pellets (ISBN 978-1-901396-15-7)

The Environment Agency (see Appendix 2)

Best Farming Practices: Profiting from a Good Environment

Use of Herbicides in or Near Water

SECTION 6

SECTION 7
INDEX

Index of Proprietary Names of Products

The references are to entry numbers, not to pages. Adjuvant names are referred to as 'Adj' and are listed separately in Section 4. Products also registered, i.e. not actively marketed, are entered as *PAR* and are listed by their active ingredient in Section 3.

REFERENCES ARE TO ENTRY NUMBERS NOT PAGES

REFERENCES ARE TO ENTRY NUMBERS NOT PAGES

REFERENCES ARE TO ENTRY NUMBERS NOT PAGES

REFERENCES ARE TO ENTRY NUMBERS NOT PAGES

REFERENCES ARE TO ENTRY NUMBERS NOT PAGES

REFERENCES ARE TO ENTRY NUMBERS NOT PAGES

NOTES

THE UK PESTICIDE GUIDE 2018

RE-ORDERS

☐ Please send me _____ more copies of *The UK Pesticide Guide 2018* at £54.00 each

Postage: single copy £6.95; orders up to £200 £9.95… orders over £200 £12.50

Outside the UK please add £10 per order for delivery.

Name _____ Position_____

Institution _____ Department_____

Address _____

City _____ Region _____ Postcode_____Country_____

Tel_____ Fax_____ E-mail_____

EU countries except UK – VAT No:_____

Payment (pre-payment is required):

☐ I enclose a cheque/draft for £_____ payable to BCPC. Please send me a receipt.

☐ I wish to pay by credit card: ☐ Visa   ☐ Mastercard   ☐ Amex   ☐ Switch

Please charge to my card £_____ and send me a receipt. Name of issuing bank_____

Card no. ☐☐☐☐ ☐☐☐☐ ☐☐☐☐ ☐☐☐☐

Expiry date ☐☐/☐☐ Security code ☐☐☐   Switch cards only: Start date ☐☐/☐☐ Issue no. ☐

Signature_____ Date_____

Name and address of cardholder if different from above:_____

_____

 BCPC **Please photocopy and return to:**
BCPC Publications Sales, Garden Studio, 4 Hillside, Aldershot, GU11 3NB, UK.
Tel: 01252 285 223, Email: publications@bcpc.org, Web: www.bcpc.org

FUTURE EDITIONS

☐ I wish to take out an annual order for _____ copies of each new edition of *The UK Pesticide Guide*

☐ Please send me advance price details for the 2019 edition of *The UK Pesticide Guide* when available

Order by phone: 01252 285 223 or online: www.bcpc.org/shop

Bookshop orders to:

Cabi
www.cabi.org

Marston Book Services Ltd
160 Milton Park
Abingdon, Oxfordshire, OX14 4SD, UK
Tel +44(0) 1235 465577 Fax: +44(0) 1235 465555
Email: direct.orders@marston.co.uk

Bulk discount:

100+ copies	Contact BCPC
50–99	15%
10–49	10%
List price £54.00	

Thank you for your order